Recent Advances and Applications of Thermoset Resins

Recent Advances and Applications of Thermoset Resins

Second Edition

Debdatta Ratna

Elsevier
Radarweg 29, PO Box 211, 1000 AE Amsterdam, Netherlands
The Boulevard, Langford Lane, Kidlington, Oxford OX5 1GB, United Kingdom
50 Hampshire Street, 5th Floor, Cambridge, MA 02139, United States

Copyright © 2022 Elsevier Inc. All rights reserved.

No part of this publication may be reproduced or transmitted in any form or by any means, electronic or mechanical, including photocopying, recording, or any information storage and retrieval system, without permission in writing from the publisher. Details on how to seek permission, further information about the Publisher's permissions policies and our arrangements with organizations such as the Copyright Clearance Center and the Copyright Licensing Agency, can be found at our website: www.elsevier.com/permissions.

This book and the individual contributions contained in it are protected under copyright by the Publisher (other than as may be noted herein).

Notices
Knowledge and best practice in this field are constantly changing. As new research and experience broaden our understanding, changes in research methods, professional practices, or medical treatment may become necessary.

Practitioners and researchers must always rely on their own experience and knowledge in evaluating and using any information, methods, compounds, or experiments described herein. In using such information or methods they should be mindful of their own safety and the safety of others, including parties for whom they have a professional responsibility.

To the fullest extent of the law, neither the Publisher nor the authors, contributors, or editors, assume any liability for any injury and/or damage to persons or property as a matter of products liability, negligence or otherwise, or from any use or operation of any methods, products, instructions, or ideas contained in the material herein.

British Library Cataloguing-in-Publication Data
A catalogue record for this book is available from the British Library

Library of Congress Cataloging-in-Publication Data
A catalog record for this book is available from the Library of Congress

ISBN: 978-0-323-85664-5

For Information on all Elsevier publications
visit our website at https://www.elsevier.com/books-and-journals

Publisher: Matthew Deans
Acquisitions Editor: Edward Payne
Editorial Project Manager: Joshua Mearns
Production Project Manager: Sojan P. Pazhayattil
Cover Designer: Greg Harris

Typeset by MPS Limited, Chennai, India

Contents

Preface xiii

Chapter 1: Chemistry and general applications of thermoset resins 1
  Abbreviations 1
  1.1 Introduction 2
  1.2 Unsaturated polyester resin 3
      1.2.1 Unsaturated polyester 3
      1.2.2 Polyester structure 6
      1.2.3 Polyesterification kinetics 8
      1.2.4 Types of polyester 9
      1.2.5 Reactive diluents or monomer 11
      1.2.6 Inhibitor 13
      1.2.7 Curing of unsaturated polyester resin 14
      1.2.8 Application of unsaturated polyester resin 17
  1.3 Bio-based unsaturated polyester 19
      1.3.1 Alkyd resin 19
  1.4 Vinyl ester resins 23
      1.4.1 Curing of vinyl ester resins 29
      1.4.2 Properties of vinyl ester resins 36
      1.4.3 Applications of vinyl ester resin 36
  1.5 Epoxy resins 37
      1.5.1 Epoxy formulation 39
      1.5.2 Curing agents 39
      1.5.3 Modified epoxy systems 46
      1.5.4 Application of epoxy resin 58
  1.6 Polyurethanes 63
      1.6.1 Polyol 63

	1.6.2	Isocyanates	66
	1.6.3	Prepolymers	67
	1.6.4	Chain extender and crosslinker	67
	1.6.5	Bio-based polyurethanes	72
	1.6.6	Applications of polyurethanes	76
1.7	Phenolic resins	82
	1.7.1	Novolac	83
	1.7.2	Resole	85
	1.7.3	Cross-linking of phenolic resin	86
	1.7.4	Applications of phenolic resin	89
	1.7.5	Phenolic resin as additives	91
	1.7.6	Addition cure phenolic resin	97
1.8	Amino resins	98
1.9	Furan resins	100
1.10	Benzoxazine resins	101
	1.10.1	Main-chain polybenzoxazines	104
	1.10.2	Polybenzoxazine-based blends	111
	1.10.3	Application of polybenzoxazine	113
1.11	Polyimides	115
	1.11.1	Addition polyimide	116
	1.11.2	In situ polymerization of monomeric reactants	117
	1.11.3	Cross-linking of polyimide	122
	1.11.4	Curing of polyimide resin	122
	1.11.5	Application of polyimide resin	124
1.12	Bismaleimide resin	126
	1.12.1	Curing of bismaleimides	129
	1.12.2	Properties of bismaleimide resins	129
	1.12.3	Structure−property relationship	129
	1.12.4	Applications of bismaleimide resin	137
1.13	Cyanate ester resin	139
	1.13.1	Curing of cyanate ester resin	140
	1.13.2	Properties of cyanate ester resins	142
	1.13.3	Modified difunctional cyanate ester resins	143
	1.13.4	Modified multifunctional cyanate ester resins	148
	1.13.5	Application of cyanate ester resins	152
1.14	Phthalonitrile resin	154
	1.14.1	Synthesis of phthalonitrile resins	155
	1.14.2	Properties of phthalonitrile resins	156
References	157

Chapter 2: Properties and processing of thermoset resin 173

Abbreviations 173
- 2.1 Introduction 175
- 2.2 Gelation and vitrification 175
- 2.3 Cure characteristics 179
  - 2.3.1 Isothermal cure kinetics 180
  - 2.3.2 Dynamic cure kinetics 184
  - 2.3.3 Effect of vitrification on polymerization rate 189
  - 2.3.4 Modeling from pressure volume temperature approach 191
- 2.4 Cure cycle 193
- 2.5 Thermal property 195
  - 2.5.1 T_g versus service temperature (T_s) 196
  - 2.5.2 Effect of cure conversion on T_g 197
- 2.6 Thermomechanical properties 199
- 2.7 Rheological properties 204
- 2.8 Liquid crystalline thermoset 210
- 2.9 Elastomeric thermoset 213
  - 2.9.1 Application of rubbery thermoset 214
- 2.10 Interpenetrating polymer networks 219
- 2.11 Smart thermoset 226
  - 2.11.1 Self-healing thermoset 226
- 2.12 Vitrimers 246
  - 2.12.1 Synthesis of vitrimers 247
  - 2.12.2 Application of vitrimers 249
- 2.13 Metal−organic frameworks 252
  - 2.13.1 Synthesis of metal−organic framework 253
  - 2.13.2 Applications of metal−organic frameworks 255
- 2.14 Flame-retardant thermoset 255
  - 2.14.1 Inorganic flame retardants 256
  - 2.14.2 Halogen-containing flame retardants 257
  - 2.14.3 Phosphorous-containing flame retardant 258
- 2.15 Additives for thermoset resin 269
  - 2.15.1 Antioxidants 269
  - 2.15.2 Filler 271
  - 2.15.3 Blowing agent 271
  - 2.15.4 Coupling agent 272
  - 2.15.5 Surfactants 273
  - 2.15.6 Colorant 273
- 2.16 Processing of thermoset resin 273

	2.16.1	Die casting 273
	2.16.2	Rotational casting 274
	2.16.3	Compression molding 274
	2.16.4	RIM process.. 275
2.17	3D printing 275
	2.17.1	Types of 3D printing.. 277
	2.17.2	3D printing of thermoplastic 280
	2.17.3	3D printing of thermoset... 282
References 284

Chapter 3: Toughened thermoset resins 293

Abbreviations 293
3.1 Introduction 294
    3.1.1 Toughening of thermoplastics 296
    3.1.2 Mechanism of toughening of brittle polymer 297
    3.1.3 Morphological aspects 299
3.2 Toughening of thermosets... 300
3.3 Liquid rubber toughening.. 300
    3.3.1 Reaction-induced phase separation.. 301
    3.3.2 Mechanism of toughening of thermoset 304
    3.3.3 Microstructural features 309
3.4 Toughening of vinyl ester resin... 313
    3.4.1 Liquid rubber toughening... 313
3.5 Modification of UPE resin 316
3.6 Toughening of phenolic resin... 320
3.7 Toughening of PI, BMI, and CE resins 323
3.8 Toughened epoxy resins.. 325
3.9 Chemical modification 326
3.10 Rubber toughening... 328
    3.10.1 Commercial toughening agents 328
    3.10.2 Rubber-based toughening agents 330
    3.10.3 Acrylate-based toughening agents 332
    3.10.4 Hyperbranched polymer-based toughening agents.................... 342
3.11 Core–shell particle toughening.. 346
3.12 Thermoplastic toughening.. 347
    3.12.1 Engineering thermoplastics 348
    3.12.2 Amorphous thermoplastics 352
    3.12.3 Crystalline thermoplastics 353
    3.12.4 Morphology and microstructural aspects 354

	3.12.5 Mechanism of toughening 355
	3.12.6 Effect of matrix crosslink density 356
3.13	Rigid particle toughening of epoxy 356
3.14	Summary and conclusion 359
	References 360

Chapter 4: Thermoset composites 371

Abbreviations 371
4.1 Introduction 372
    4.1.1 Constituents of FRP composites 374
4.2 Composite interface 375
    4.2.1 Surface tension and contact angle 376
    4.2.2 Surface treatment of fibers 377
4.3 Processing of composites 380
    4.3.1 Contact molding 381
    4.3.2 Wet lay-up molding 381
    4.3.3 Resin transfer molding 382
    4.3.4 Vacuum-assisted resin transfer molding 383
    4.3.5 Reaction injection molding 385
    4.3.6 Pultrusion 385
    4.3.7 Filament winding 387
    4.3.8 Resin film infusion 387
    4.3.9 Prepreg molding 389
4.4 Analysis and testing of composites 390
    4.4.1 Determination of glass content 390
    4.4.2 Mechanical testing of composites 391
    4.4.3 Interlaminar shear stress 391
4.5 Prediction of composite strength and rigidity 393
4.6 Thermomechanical properties of thermoset composites 396
    4.6.1 Thermal properties 396
    4.6.2 Mechanical properties 397
4.7 Toughened composites 401
    4.7.1 Resin toughening 409
References 415

Chapter 5: Thermoset nanocomposites 419

Abbreviations 419
5.1 Introduction 420
5.2 Thermoset nanocomposites 421

5.2.1	Thermoset/clay nanocomposites (PCN)	423
5.2.2	POSS and silica-based nanocomposites	443
5.2.3	Block copolymer-based nanocomposite	449
5.2.4	CNT-based nanocomposites	455
5.2.5	Graphene-based nanocomposite	468
5.2.6	Silver nanoparticle-based nanocomposites	473
5.2.7	Nanoreinforcement and toughening	475
5.2.8	Nanotechnology and flammability	481
5.2.9	Nanoclay-based flame retardant	483
5.2.10	Combination organoclay and other flame retardants	486
5.2.11	Nanocarbon-based flame retardants	488
5.2.12	Application of nanocomposites	491
5.2.13	Summary and outlook	494
References		496

Chapter 6: Characterization, performance evaluation and lifetime analysis of thermoset resin ... 503

Abbreviations ... 503

6.1 Analysis and Characterization ... 504
  6.1.1 Titration ... 504
  6.1.2 Infrared spectroscopy ... 506
  6.1.3 Nuclear magnetic resonance spectroscopy ... 507
  6.1.4 Positron annihilation lifetime spectroscopy ... 508
  6.1.5 Crosslinked density ... 508
  6.1.6 Molecular weight distribution ... 510
  6.1.7 Morphological characterization ... 514
  6.1.8 Thermal analysis ... 517
  6.1.9 Thermal stability ... 523
  6.1.10 Rheological characterization ... 524

6.2 Testing of thermoset resins ... 526
  6.2.1 Mechanical properties ... 526
  6.2.2 Fracture toughness ... 533
  6.2.3 Impact test ... 536
  6.2.4 Hardness test ... 538
  6.2.5 Thermal properties ... 540
  6.2.6 Electrical properties ... 543
  6.2.7 Flammability and smoke tests ... 545

6.3 Performance evaluation ... 547
  6.3.1 Evaluation as a surface coating ... 547

	6.3.2	Evaluation of thermoset as adhesive	549

	6.3.2	Evaluation of thermoset as adhesive 549
	6.3.3	Evaluation as a vibration damping material 556
6.4	Lifetime prediction 560
	6.4.1	Thermal degradation kinetics 562
	6.4.2	Differential form of isoconversion kinetics 564
	6.4.3	Integral form of isoconversion kinetics 566
	6.4.4	Advanced isoconversion methods 569
	6.4.5	Estimation of a lifetime from TGA kinetics 570
	6.4.6	Example of lifetime estimation from TGA kinetics 572
6.5	Conclusion..575
	6.5.1	Ageing study under stress 576
References 579

Index **583**

Preface

This book is dedicated to thermoset resins and published as a second edition of my earlier book on "*Handbook of Thermoset Resins*" from Smither Rapra Technology, UK. As a student of polymer science and a researcher in the field of thermoset resins, I always felt the lack of a self-sufficient book dedicated to thermoset resins. That was the driving force for writing the first edition published in 2009, which was the compilation of my fundamental understanding and long research experience in this specialized field. In the last two decades, considerable research works have been carried out and a lot of developments have taken place. So I realized that a revision of this book is absolutely necessary incorporating the recent development. While doing so, I have totally reorganized the book mostly focusing on the latest developments, which will be published in the new name as "*Recent Advances and Applications of Thermoset Resins*".

This book has been divided into six chapters. It is started with the chemistry and general applications of thermoset resins. The properties and processing of thermoset resins which includes network concept and additives are dealt in Chapter 2. The toughening of thermoset resins to improve their fracture toughness is elaborated in Chapter 3. The thermoset-based composites and nanocomposites have been discussed in Chapters 4 and 5, respectively. The techniques/instrumentations (their principle) used to characterize a thermoset resin, performance evaluation, and lifetime estimations are covered in Chapter 6. With such broad technical contents covering the basic concepts and recent advances, I am sure this book will serve as a useful textbook-cum-handbook for the students, researchers, engineers, and R&D scientists from academia, research laboratories, and industries. It will be extremely useful for the scientists and researchers to make a knowledge-base in the subject as well as to plan their future works because I have not only presented the review of the recent advances in this book but also highlighted the future directions of research in the various areas of thermoset resins. The bounty of information garnered in this book will serve as a fountainhead for further development in the field of thermoset resins in general and thermoset nanocomposites in particular.

I would like to place on record for the assistance extended by my senior and retired colleagues namely Dr. B.C. Chakraborty and Shri A.K. Pandey, and junior colleagues namely Dr. S.P. Mishra and Shri Ramakant Khuswaha of DRDO during the preparation of this book. I wish to express my gratitude to my parents Late Snehalata Ratna and late Lakshmikanta Ratna for their blessings and moral supports for starting my carrier. I would like to thank my in-laws Shri Nirmalendu Satpathi and Subhadra Satpathi for their moral support. I would like to acknowledge the editorial assistance offered by my elder son Saptarshi, a graduate in metallurgical engineering and materials science from IIT Bombay. For writing this book, I had to utilize much of the quality time, which I give generally, to my family. So I am sincerely thankful to my wife (Sujata) and sons (Saptarshi and Debarshi) for their patience and for always being the source of inspiration, without which this book would have not been in reality.

Debdatta Ratna

CHAPTER 1

Chemistry and general applications of thermoset resins

Abbreviations

UPE	unsaturated polyester
FTIR	Fourier transformed infrared
BPA	bisphenol-A-fumarate
MEKP	methyl ethyl ketone peroxide
BPO	benzoyl peroxide
TBO	t-butyl peroxide
BMPA	poly [(tetramethylene maleate)-*co*-(tetramethylene phthalate)]
BFPA	poly [(tetramethylene fumarate)-*co*-(tetramethylene phthalate)]
BPSA	poly [(tetramethylene phthalate)-*co*-(tetramethylene succinate)]
FRP	fiber-reinforced plastic
VOC	volatile organic compound
PHEMA	poly(2-hydroxyethyl methacrylate)
DPE	dipentaerythritol
HBP	hyperbranched polymer
UV	ultraviolet
HBA	hyperbranched alkyd
HMMM	hexamethoxymethyl melamine
MEK	methylethylketone
VE	vinyl ester
DMESS	dimethacrylated-epoxidized-sucrose-soyate
DGEBA	diglycidyl ether of bisphenol-A
ECD	epichlorohydrin
TETA	triethylenetetramine
DDM	diamino diphenyl methane
DDS	diamino diphenyl sulfone
MDEA	4,4'-Methylenebis[2,6 diethyl aniline]
MCDEA	4'-Methylene *bis* [3-chloro 2,6-diethylaniline]
DICY	dycynamide
PMMA	poly (methyl methacrylate)
MDI	diphenylmethane diisocyanate
BMIM	1-Alkyl-3-methyl imidazolium
TGA	thermogravimetric analysis
CMPA	4-Methyl pentanoic acid
ESO	epoxidized Soyabean oil
PCL	polycaprolactone
TEU-EP	trifunctional eugenol-based epoxy monomer

PI	polyimides
DPA	diphenic anhydride
GA	glutaric anhydride
PA	phthalic anhydride
CA	camphoric anhydride
SA	succinic anhydride
IA	itaconic anhydride
TMP	trimethylol phenol
TDI	2,6-Toluene diisocyanate
HDI	1,6. Hexamethylene diisocyanate
MDI	4,4′-Diphenylmethane diisocyanate
IPDI	isophorone diisocyanate
APES	3-Aminopropyltriethoxysilane
HBAU	hyperbranched alkyd urethane
PF	phenol formaldehyde
HTMA	hexamethylene tetramine
PBMA	poly (butyl methacrylate)
SMP	shape memory polymers
PEO	poly(ethylene oxide)
CPOM	cross polarized optical microscopy
UF	urea-formaldehyde
MF	melamine-formaldehyde
FFA	furfuryl alcohol
ICB	integrated circuit board
PBZ	polybenzoxazine
NMP	*N*-methyl pyrrolidone
PMR	polymerization of monomeric reactants
BTDA	3,3′,4,4′-Benzophenone tetracarboxylic acid anhydride
MDA	4,4′-Methylene dianiline
VEM	variable frequency microwave
NLO	nonlinear optical
BDM	*bis* (4-maleimidophenyl) methane
DABA	diallylbisphenol-A
BMPP	2,2′-*bis*[4-(4-maleimidephen-oxy) phenyl] propane
HSQ	hydrogen silsesquioxane
CE	cyanate ester
BAFCY	2,2-*bis*(4-cyanatophenyl)-1,1,1,3,3,3,3-hexafluoro propane
Bz-allyl	*bis*-allyl benzoxazine
DMAc	dimethylacetamide
HFDE	4,4′-(Hexafluoroisopropylidene)-dipthalic acid
3FDA	4,4′-(2,2,2-Trifluoro-1-phenylethyle dine) dipthalic anhydride

1.1 Introduction

Today, a large number of synthetic polymers are available. The synthetic polymers can be classified into two categories, thermoplastic polymers and thermoset polymers. Thermoplastic polymers soften on heating and stiffen on cooling repeatedly several times. On the other hand, thermosetting polymers undergo some chemical reaction (known as curing) on heating and get converted into an infusible and insoluble material. The infusibility and insolubility of the cured polymer arise due to the formation of three-dimensional (3D) network structure as discussed below. On the bases of molecular weight, thermoset polymers can be grouped into two types namely thermoset resins (low-molecular-weight) and rubber or elastomer (high-molecular-weight). The low-molecular-weight thermoset polymer, that is, thermoset resins are the subject of the present book.

Thermoset resins offer several technological advantages over thermoplastics. First thermoset resins are processed from a low-molecular-weight compound (resin) unlike the high-molecular-weight macromolecules in case of thermoplastics. That is why thermoset-based products can be molded, in general, at a much lower temperature and pressure compared to thermoplastics. Second, the characteristic of a thermosetting resin is that their properties are not only dependent on the chemistry and molecular weight of the resin (as in the case of thermoplastic) but also largely dependent on the cross-link density of the resin network. A wide range of properties can be achieved by simply adjusting the cross-link density of a network, without changing the chemical structure. Third, because of the presence of chemical cross-links, the cured thermoset offers better creep property and environment.

In this chapter, chemistry and general applications of individual thermosetting resins will be discussed. For the convenience of the readers, a list of various thermosetting resins to be discussed and their precursors and field of applications are presented in Table 1.1.

1.2 Unsaturated polyester resin

Unsaturated polyester (UPE) resin consists of a polyester with unsaturation in its structure, a monomer, and an inhibitor. UPEs gained wide industrial applications due to their low viscosity, which offers easy processability, low cost, and rapid cure schedules.

1.2.1 Unsaturated polyester

Polyesters are macromolecules, which are made by reacting a diacid or dianhydride with a dihydroxy compound (diols). In order to make unsaturated polyester, maleic anhydride or fumeric acid is used in addition to a saturated acid, which provides unsaturation in the structure. The most commonly used anhydrides are maleic anhydride (unsaturated) and phthalic anhydride (saturated), and the most common diols are ethylene glycol or propylene glycol. The use of an unsaturated anhydride is very critical to provide the unsaturation in the structure, which is utilized to cure the resin by a free radical polymerization. The chemical reaction for the synthesis of UPE is shown in Fig. 1.1.

Fumarate double bonds (planar trans configuration) react faster with the reactive monomer compared to the double bonds of maleate ester, because maleate esters are slightly distorted

Table 1.1: List of important thermosetting resins and their field of application.

Resin	Precursor	Applications
Unsaturated polyester	Dicarboxylic acids, diols and reactive diluents, for example, styrene	Surface coating, fiber composites for mechanical equipment and building construction, electrical and lighting industries
Alkyd resin	Dicarboxylic acids, diols, fatty acids	Paint and surface coating industries
Vinyl ester	Epoxy resin and acrylic or methacrylic acid	Fiber composites for mechanical equipment and building construction, marine construction, electrical industries
Epoxy	Epichlorohydrin and bisphenols	Engineering adhesives, paints, and surface coating, electrical laminates, fiber composites for automotive, marine construction, and aerospace applications
Polyuretahane (PU)	Diisocyanate and hydroxyl-functionalized oligomer (polyol)	Coating, adhesive, encapsulating materials, acid-resistant cements, foams, adhesives
Phenolic resin	Phenol and formaldehyde	Wood adhesive, molding compound, foundry binder, laminate moldings, electrical laminates, ablative coating casting, and fiber composites for household appliances, automotive, aircraft construction, and accessories, electrical and lighting industries
Melamine	Melamine and formaldehyde	Coatings, molding compounds, wood materials processing, friction linings, textile auxiliaries
Urea	Urea and formaldehyde	Molding compounds, textile auxiliaries, wood materials, foundry binder, foams
Furan	Furfuryl alcohol	Refractory materials processing, fiber composites, molding compounds, grinding wheels, acid-resistant cements
Polybenzoxazine	Phenol or bisohenol-A, amine and formaldehyde	Laminate for copper cladding in Printed circuit board, heat-resistant coatings, adhesives, prepregs, and encapsulants, reactive diluents for vinyl ester resin
Polyimide	Diamines and dianhydride	Coating and composite for high-temperature applications
Cyanate ester	Cyanogen halide and alcohol or phenol	Microelectronic industries, radome, and nose cone, which house radar antennas of military and weather reconnaissance plane, Fiber composites for aircraft, rocket, missiles and reentry vehicle
Bismaleimide	Bismaleic acid and amine	Fiber-reinforced composites for aircraft, rocket, missiles, and reentry vehicle
Pthalonitrile		Rocket, missiles, and reentry vehicle

from the planar configuration, which suppresses their ability to copolymerize with styrene. Hence, fumaric acid is preferable to maleic anhydride as an unsaturated anhydride precursor for the formation of a uniform network with better properties. However, maleic anhydride is mostly used because of two reasons; first maleic acid offers lower cost and easier handling. Second, a major portion of maleate double bonds isomerize to a more stable fumarate form during the synthesis of resin at high temperatures [1,2]. It was reported that the extent of isomerization depended on the structure of glycol [3,4]. For instance, isomerization is reported to be 95% with propylene glycol, 39% with 1,4 butylene glycol, and 35% with 1,6

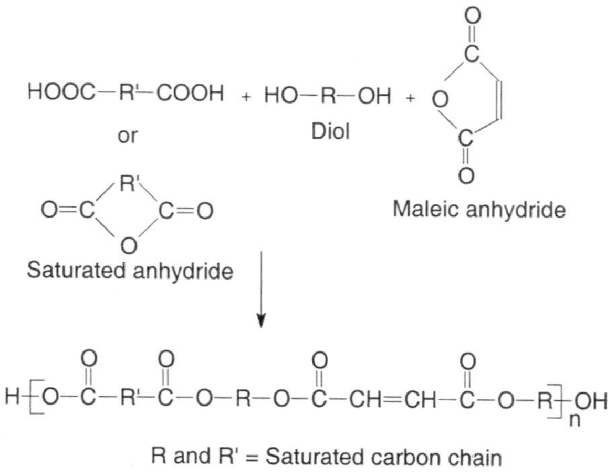

Figure 1.1
A reaction scheme for the synthesis of UPE resin. UPE, unsaturated polyester.

hexamethylene glycol at 180°C. ^{13}C NMR analysis is used to study the isomerization. The nature of acids and catalysts also play important roles in isomerization [5,6].

The rate of polyester synthesis by polycondensation depends on the chemical structure of the reactants (diacid and diol) and the stoichiometry of the reactants. Generally, glycol is used in slight excess to compensate for the potential loss of glycol via evaporation. The reaction takes place in two stages: formation of monoester followed by polycondensation at a higher temperature. The reaction is reversible and water is produced as a byproduct. Hence, it is necessary to remove the water from the reaction mixture to push the reaction toward the forward direction. The water is removed continuously from the reaction mixture either by application of vacuum or by using a solvent like xylene, which forms an azeotrope with water in the vapor state. A Dean and Stark apparatus is used for the synthesis. The azeotrope vapor is allowed to condense in a receiver tank where they are separated from each other due to the difference in density. The xylene forms the upper layer, which is fed back continuously to the reactor and the water forms the bottom layer, which is removed through an opening at the bottom of the tank. The progress of the reaction is monitored through the amount of water produced and the acid value of the reaction mixture. The acid value is checked by withdrawing a small amount of sample from the reactor and analyzing by using a standard titration method. The typical molecular weight of UPE varies from 3000 to 5000 g mole^{-1} with an acid value of the product less than 20. The polymerization being the step-growth one, the molecular weight of the polymer is highly sensitive to the purity of the reactants. At a high temperature, an alcoholysis or acidolysis of the polyester chains by hydroxyl or carboxyl groups of the monomers and/or oligomers takes place. This process is called transesterification. Transesterification reaction allows the redistribution of molecular weight and functional groups in UPE. A reaction scheme for transesterification is shown in Fig. 1.2.

1.2.2 Polyester structure

A linear structure of polyester is expected from the reaction of acid and alcohol as shown in Fig. 1.1. However, branching in polyester structure takes place as a result of side reactions. Such side reactions are first investigated by Ordelt et al. [7–9]. The electron-deficient double bond (due to the presence electron-withdrawing carbonyl groups) of maleic anhydride can react with the hydroxyl groups of glycol or oligomer via Michel addition. The reaction of glycol with the double bonds produces short branches whereas the involvement of hydroxyl groups of oligomer or macromolecules leads to the formation of long branches as represented in Fig. 1.3. Since the hydroxyl groups of glycol are more

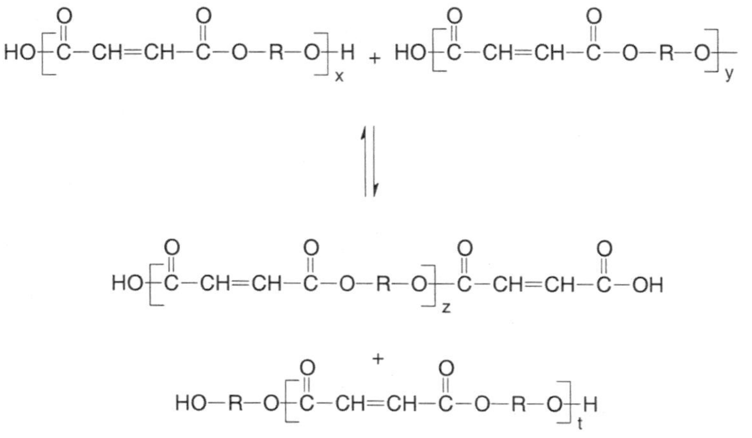

Figure 1.2
A reaction scheme for transesterification reaction.

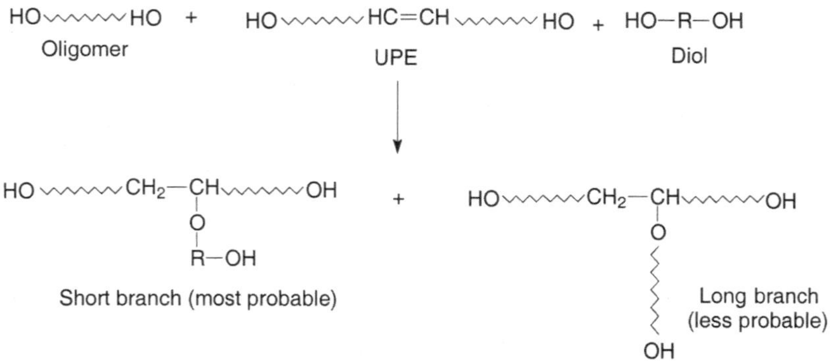

Figure 1.3
Mechanism for the formation of long and short branches in UPE resin. *UPE*, unsaturated polyester.

reactive than those of the monoesters, mostly short branches are formed. Thus the side reactions produce branching and reduce the double bond functionality of the unsaturated polyester. That is why the double bond functionality of a maleic anhydride-based polyester is found to be less than the theoretical value calculated based on the maleic acid concentration in the polyester. Hence, the polyester structure is characterized by main-chain end groups (carboxyl or hydroxyl), double bond functionality (maleate or fumarate) in the backbone, and side-chain end groups (mostly hydroxyl). It is necessary to analyze and characterize the structure of UPE with respect to the presence of functional groups and branching for proper quality control. Functional groups are identified by FTIR analysis from the characteristics peaks; carboxylic acid (1710 cm^{-1}), ester (1730 cm^{-1}), double bond (1640 cm^{-1}), ether (1100–1250 cm^{-1}), and hydroxyl (3350–3500 cm^{-1}). Maleate and fumarate double bonds are identified by ^{1}H NMR analysis. The vinylic proton of maleic acid absorbs at 6.55 ppm, whereas the same for fumeric acid absorbs at a higher chemical shift (6.98 ppm) [10].

The carboxyl and hydroxyl groups can be estimated by the standard titration methods. For the estimation of carboxyl groups, the sample is directly titrated with a standard alcoholic potassium hydroxide solution. To determine hydroxyl number, the sample is refluxed for 1 h with acetylating mixture (pyridine/acetic anhydride) and titrated back the excess acetic anhydride. The overall average functionality (f_n) (carboxyl and hydroxyl) is the sum of average carboxyl functionality (c_n), main-chain hydroxyl functionality (h_n), and branching chain hydroxyl functionality (b_n).

$$f_n = c_n + h_n + b_n \tag{1.1}$$

$$c_n + h_n = 2 \tag{1.2}$$

b_n can be determined by considering Ordelt reaction as given below [11,12]:

$$b_n = \left[\frac{2M_{n,index}X_{Ordelt}}{2M_{c=c} + 2mX_{Ordelt} - M_{n,index}X_{Ordelt}}\right] \tag{1.3}$$

$$M_{n,index} = \frac{112,200}{I_H + I_c} \tag{1.4}$$

Hence, combining Eqs. (1.1), (1.2), and (1.3), we get

$$f_n = 2 + \left[\frac{2M_{n,index}X_{Ordelt}}{2M_{c=c} + 2mX_{Ordelt} - M_{n,index}X_{Ordelt}}\right] \tag{1.5}$$

where X_{Ordelt} = Ordelt saturation degree of polyester or ratio of moles of saturated double bonds to moles of initially introduced double bonds, $M_{C=C}$ = molar mass per mole of the double bond on the polyester chain, m = molar mass of branches, I_C = carboxyl value, and I_H = hydroxyl value.

1.2.3 Polyesterification kinetics

The first studies on kinetics and mechanisms of polyesterification reactions were carried out by Flory [13]. The studies indicated a third-order self-catalyzed and a second-order acid-catalyzed reaction. For a self-catalyzed reaction, the rate of the reaction is associated with second-order dependence on carboxyl group concentration and first-order dependence on hydroxyl group concentration. For acid-catalyzed reaction, the order is one with respect to both the carboxyl group concentration and hydroxyl group concentration. A model proposed by Lin and Hsieh [14] also suggested an overall order of 3 for a self-catalyzed reaction. However, they suggested first-order dependence of the rate of reaction on carboxyl group concentration and second-order dependence of the rate of reaction on hydroxyl group concentration. For an acid-catalyzed reaction, the overall order is 2, only depending on carboxyl group concentration.

The catalytic activity of the acid catalyst is due to the presence of hydrogen ions [15]. In presence of a strong acid catalyst (e.g., p-toluene sulfonic acid), the hydrogen ions are produced mainly from the added acid. Thus the polyesterification is second-order reaction. In absence of an acid catalyst, the hydrogen ions are formed from the ionization of dicarboxylic acid and the order of the reaction is 2.5 [16]. More complicated rate equations are proposed by considering the reverse reaction, effect of the dielectric constant of the medium on ionization of diacid, etc. [17,18].

The difference in kinetics may originate from the fact that some studies were carried out under a constant oil bath and others have been carried at the constant reaction temperature. The polyesterification process is endothermic though the polyesterification reaction is exothermic because the condensates were evaporated continuously. Hence, there is a large and rapid temperature drop especially in the early stages of the reaction [17]. Thus the data generated from the experiments at a constant oil bath temperature are expected to be different from the data generated from the experiment at a constant reaction temperature.

Chen and Wu [18] proposed a rate equation considering the following factors: (1) the hydroxyl group is a stronger proton acceptor than carboxylic acid group, (2) variation in the dielectric constant of the reaction mixture affects the dissociation constant of the dicarboxylic acid, and (3) there is a possibility of reverse reaction due to presence of the unremoved water. The rate equation can be expressed as follows:

1. For self-catalyzed reaction

$$\frac{d[\text{RCOOR}^1]}{dt} = K_a \exp(\alpha p)[\text{RCOOH}]^2 [\text{R}^1\text{OH}] - K_h[\text{H}_2\text{O}][\text{RCOOR}^1] \quad (1.6)$$

where [RCOOR], [RCOOH], and [R'OH] are the concentration of polyester, carboxylic acid groups and hydroxyl groups, respectively.

Considering that α is the conversion of the acid group ($[COOH]_0-[COOH]/[COOH]_0$) and r is the ratio of diol to diacid at the beginning of the reaction ($r = [OH]_0/[COOH]_0$), we obtain:

$$\frac{d\alpha}{dt} = K_a C_0^2 \exp(x\alpha)(1-\alpha)^2(r-\alpha) - K_h[H_2O]\alpha \qquad (1.7)$$

Here K_a and K_h are forward and reverse rate constants, x is a constant related to dielectric constant, and $[H_2O]$ is the concentration of unremoved water.

For acid-catalyzed reaction

$$\frac{d[RCOOR^1]}{dt} = K_{ac}[AH]\exp(x\alpha)[RCOOH][R^1OH] - K_{hc}[H_2O][RCOOR^1] \qquad (1.8)$$

Eq. (1.9) can be expressed in terms of r and α as follows

$$\frac{d\alpha}{dt} = K_{ac} C_0 [AH]\exp(x\alpha)(1-\alpha)(r-\alpha) - K_{hc}[H_2O]\alpha \qquad (1.9)$$

Here K_{ac} and K_{hc} are forward and reverse rate constants for acid-catalyzed reaction, x is a constant related to dielectric constant, and $[H_2O]$ is the concentration of the unremoved water.

Kuo and Chen [19] applied the above equation to the reactions of adipic acid and three different diols (ethylene glycol, 1,4 butanediol, and 1, 6 hexanediol) for the self-catalyzed and acid-catalyzed reaction. They reported that the experimental data fit well with the above equation and activation energy decreased with an increase in the chain length of diols. Beigzadeh et al. [20] used the equation of Chen and Wu [18] for a two-stage production system. In the first stage, an excess amount of diol is allowed to react the saturated acid until the acid value reaches 10 and then the unsaturated acid is added. It was assumed that at the end of the first stage the oligomers formed are new diols and can react with the diacid in a similar way. They also reported good fitting of experimental data with the Chen and Wu model [18] but did not find such good fittings with other models.

1.2.4 Types of polyester

The polyester resins can be grouped broadly into two major categories based on the compositions and applications namely general-purpose polyester resin and specialty polyester resin.

1.2.4.1 General-purpose resin

The general-purpose resin is made from low cost low raw materials and used without any modification. The main criteria are that the resin can be offered at a reasonable (competitive) cost and should be able to provide a network structure with reasonable physical properties in a reasonable time. For the general-purpose UPE resins, the cost of the materials is a major consideration rather than their performance (how well they perform). Generally, propylene glycol, diethylene glycol, phthalic anhydride, and minimum amount of maleic anhydride are used for the synthesis of general-purpose polyester resin. A typical formulation of general-purpose polyester resins is given in Table 1.2. Diethylene glycol is used not only to reduce the cost but also to improve the cure rate. For further reduction in cost hydrolyzed poly (ethylene terephthalate) residue and other mixed acids or glycol by-products are used. In order to reduce the amount of maleic anhydride dicyclopentadiene is typically coupled with maleic anhydride in the initial steps. Dicyclopentadiene improves the solubility of polyester resin in styrene, which is used as reactive diluents. These resins are generally cooked to a lower molecular weight, which allows the use of fewer amounts of reactive diluents to achieve the desired viscosity.

1.2.4.2 Specialty polyester resin

In specialty polyester, the raw materials are selected judiciously to improve the properties and performance of the resin with the compromise of the cost. The specialty resins are used where high mechanical strength, chemical and corrosion resistance are required. Use of monopolar glycol like polypropylene glycol, poly (butylene glycol) contributes toward water and corrosion resistance. Three types of specialty resins are generally used namely isophthalic resin, chlorendic resin, and bisphenol-A (BPA) fumarate resin. Typical formulations of specialty UPE resins are given in Table 1.3.

Table 1.2: Typical formulation of general-purpose unsaturated polyester.

Ingredients	Molar concentration	
	Type 1	Type 2
Propylene glycol	2.1	1.75
Diethylene glycol	0	0.35
Maleic anhydride	0.6–0.8	1.75
Pthalic anhydride	1.4–1.2	0.25

Table 1.3: Typical formulations for speciality unsaturated polyester resins.

Precursor	Isopthalic	Bisphenol-A-fumarate	Chlorendic
Maleic anhydride	1.0	—	1.0
Fumaric acid	—	2.0	—
Isopthalic acid	1.0	—	—
Chlorendic anhydride	—	—	1.0
Propylene oxide	2.15	—	2.15
Propoxylated bisphenol-A	—	2.05	—

Isophthalic resins are based on isophthalic acid and maleic anhydride. The incorporation of isopthalic acid causes an increase in molecular weight of the resulting polyester, which exhibits good mechanical properties, chemical and thermal resistance. The correlation between corrosion resistance and polyester structure is not well established. In general, when the ester is sterically crowded and there are fewer number of ester groups in the chain, good chemical and corrosion resistance are achieved [21].

Bisphenol-A (BPA) fumarate resins are prepared by the reaction of propoxylated BPA with fumeric acid. The use of BPA results in significant reduction in number of ester linkages and makes the resin comparatively nonpolar. As a result, this resin shows very good corrosion and chemical resistance. Hence, bisphenol-A-fumarate resin-based composites can be suitable for replacement of metallic materials for many industrial applications such as pipe, tank, panels where materials have to serve under corrosive environment.

Chlorendic resins are prepared by reacting propylene glycol with a combination of endomethylenehexachloropthalic anhydride (chlorendic anhydride) and phthalic anhydride. This resin shows excellent corrosion resistance and fire retardancy (due to the presence of chlorine) when used as a matrix for fiber-reinforced composites. Chlorendic resins are used in applications, which require exceptional resistance to acidic and oxidizing environments. In recent years, bisphenol-A (BPA) fumarate resins have been extensively replaced by chlorendic resins for various industrial applications.

1.2.5 Reactive diluents or monomer

The unsaturated polyester as discussed above is used in combination with a
suitable comonomer, which serves as reactive diluents. The objective to add a comonomer is to reduce the viscosity, adjust the cure schedule, and improve the mechanical properties of the resulting networks. In principle, any free-radically polymerizable monomer can be used for UPE resins; however, commercial resins mostly contain styrene or in some cases divinylbenzene. The wide acceptance of styrene as a reactive diluent is due to several factors such as low cost, better solubility of polyesters in styrene compared to other monomers, and well-understood polymerization kinetics (high reactivity with 1,2-disubstituted double bonds of polyester). The good solubility of polyester and styrene is very crucial for the formation of a homogeneous network after curing. The poor binary styrene-polyester miscibility has been cited as a significant cause of heterogeneous morphology observed during the cure at low and high temperatures [22–24]. Miscibility of UPE in styrene at a constant temperature depends on physicochemical properties of the UPE such as chemical composition, molecular weight, polydispersity index, polar end groups, and concentration of ester groups in the chain [25].

Buffa and Borrajo [26] synthesized the UPEs with varying concentrations of adipic acid and phthalic anhydride and carried out the thermodynamic analysis of miscibility. They found that the miscibility of UPE in styrene increases with increasing the concentration of adipic acid. A similar composition effect was reported by Lecoinite et al. [27] using UPE prepolymers of similar molecular weights and distributions, where the polyester backbone chemical structure was changed by using different proportions of neopentyl glycol and diethylene glycol. Buffa and Borrajo [26] also investigated the effect of the molecular weight of polyester on styrene miscibility. The studies on UPE resins of a similar chemical structure and four different molecular weights (620, 1205, 1700, and 2740 g mole^{-1}) indicated that UPE miscibility in styrene changes in the following order: 1700 > 1205 > 2740 > 620 [26] as shown in Fig. 1.4. That means the miscibility increases with increasing molecular weight and decreases beyond an optimum value (1700 g mole^{-1}). This can be explained by considering the relative influence of three different contributions of chemical structure arising out of the change in molecular weight of UPE of the same composition. First, an increase in molecular weight of UPE with a fixed comonomer composition lowers the concentration of highly polar hydroxyl and carboxyl groups. This effect increases the miscibility of UPE in styrene. Second, an increase in molecular weight of UPE with a fixed comonomer composition increases the concentration of polar ester groups (in the backbone), which reduces the miscibility. Third, an increase in molecular

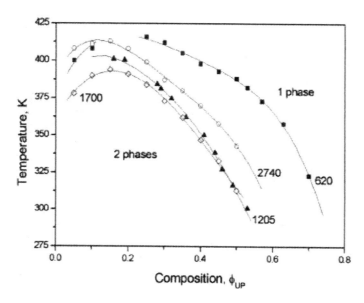

Figure 1.4
Unsaturated polyester molecular weight effect on the experimental temperature versus composition cloud point curves of a UPE resin. Study on resins of similar chemical structure and four different molecular weights (620, 1205, 1700, and 2740 g mole^{-1}) indicated that UPE miscibility in styrene changes with following order: 1700 > 1205 > 2740 > 620. UPE, unsaturated polyester. Source: *Reprinted with permission from F. Baffa, J. Borrajo, J. Appl. Polym. Sci., 102 (2006) 6064.* © *2006, John Wiley and Sons Publishers.*

weight reduces the miscibility due to the entropy factor [28]. Up to an optimum molecular weight, the first factor predominates leading to an increase in miscibility. Beyond the optimum molecular weight, the second and third factors predominate. Thus, by the judicious selection of composition and by adjusting the molecular weight, a good UPE miscibility in styrene or any other reactive diluents can be achieved.

The neat polyester is highly viscous (2000–80,000 cPs) and very difficult to process. The addition of monomer reduces the viscosity about 10 times (200–500 cPs) and makes the resin very easily processable. Polyester resin offers better processability compared to other thermoset resins like epoxy. Generally, the mole ratio of polyester to monomer is maintained in the range of 1.9 to 2.4. Further increase in styrene content reduces the cross-link density and leads to a reduction in tensile and flexural properties and an increase in curing time. The properties of unsaturated resin depend on the concentration of styrene and this can be further perfected by replacing a certain fraction of styrene with a comonomer. For instance the undiluted isophthalic resin in the cured state exhibit a T_g less than room temperature ($-20°C$ to $5°C$). By diluting with styrene, the networks with different T_g can be produced. The use of a sufficiently higher amount of styrene generated a cured network [29] with reasonably high T_g ($60°C–70°C$) and low moisture content. Sometimes, small amounts of comonomers like α-methyl styrene or vinyl toluene are used to alter the curing reaction and reduce the volatility [30,31]. Partial substitution of styrene by acrylates improves the weather resistance of the resulting UPE resin. The use of a combination of styrene and acrylonitrile instead of styrene resulted in an improvement in the mechanical properties of the cured resin. Acrylonitrile increases the miscibility of polyester in styrene by increasing the polarity and readily polymerizes with styrene.

The main problem of styrene is the high emission, which restricts the applications in closed compartments for example naval applications. Styrene emission is a health hazard (threshold value of 50 ppm). The need to reduce the volatile organic compound (VOC) is not only expressed in acute government regulation but also equally and persuasively by various social concerns for the environment. The evaporation of styrene can be reduced by the addition of waxes or pyrogenic silicic acid with hydrophilic and hydrophobic end groups. However, such additives may affect the interfacial properties of composites made out of polyester resins. Some styreneless UPE resin formulations using diacrylate monomers have also been reported [32]. However, mostly styrene is used as a reactive diluent in UPE resins.

1.2.6 Inhibitor

Another important ingredient of UPE resins is the inhibitor, which is added to prevent premature gelling or increase in molecular weight through a free radical polymerization. The type and concentration of inhibitor are very important to ensure a good shelf life at the same time the desired pot life. It must not affect the final thermomechanical properties of the cured networks, significantly. The most commonly used inhibitors are p-benzoquinone, hydroquinone, phenothiazine, etc. The addition of a very small amount of inhibitor (<500 parts per million) increases the shelf life by more than 1 year. The concentration of inhibitor should be kept to a minimum; otherwise, subsequent cross-linking with peroxide initiator will be difficult. The inhibitors basically scavenge the free radicals, which may be generated in the unsaturation sites or in the monomer. The driving force for this process is the greater stability of an inhibitor radical due to resonance. Being highly stable, the initiator radicals do not participate in the chain reactions. As a result, the possibilities of an increase in molecular weight/viscosity and gelling during storage can be avoided.

1.2.7 Curing of unsaturated polyester resin

Curing of UPE resin takes place through the free radical polymerization of unsaturated resin and monomer. Organic peroxides, azine, and azo compounds are used as to generate free radicals. UPE industry is dominated by the use of peroxides namely methyl ethyl ketone peroxide (MEKP), benzoyl peroxide (BPO) and t-butyl peroxide (TBO). MEKP decomposes at room temperature whereas BPO and TBO require higher temperatures 70°C and 140°C, respectively, to decompose. In order to ensure room temperature or low-temperature curing at a reasonable rate, it is necessary to use some accelerators, which accelerate the decomposition of peroxides. Generally, transition metal oxides are used as accelerators for curing UPE resins. The most commonly used accelerators are cobalt naphthenate and cobalt octoate. Such accelerated decomposition generates only one radical instead of two as shown below:

$$ROOH + Co^{+2} \rightarrow RO^\bullet + OH^- + CO^{+3} \tag{1.10}$$

The free radicals initiate the chain reaction, which propagates through the unsaturated sites of the polyester and monomer leading to the formation of a network structure, which is insoluble and infusible as shown in Fig. 1.5. The cobalt compound when present in excess amount reacts with the free radicals and converts them into ions.

$$RO^\bullet + Co^{+2} \rightarrow RO^- + CO^{+3} \tag{1.11}$$

Consequently, the radicals are destroyed and the curing reaction cannot proceed. That is why the accelerators are added in a very small amount (0.02 wt.%, never >0.3 wt.%).

As discussed in the earlier sections, UPE resins are a mixture of unsaturated polyester, styrene, and an inhibitor. When the resin is mixed with a peroxide initiator and activator (cobalt octoate/naphthenate), the free radicals are formed. At the initial stage, the all or most of the free radicals generated are consumed by the inhibitor. The driving force for the preferable reaction of the free radicals with the inhibitor is the higher stability of inhibitor

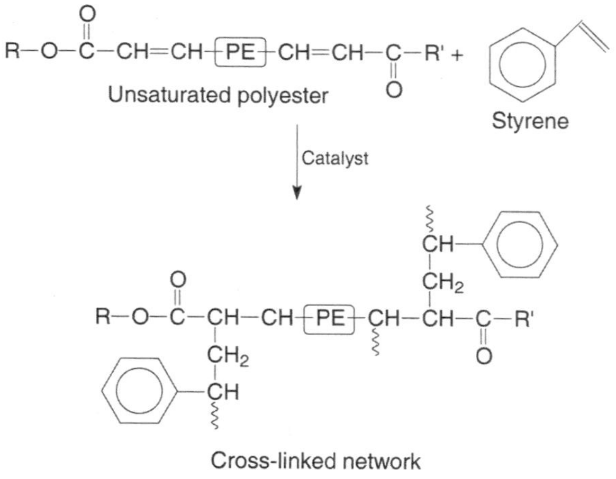

Figure 1.5
Cross-linking of UPE resin. *UPE*, unsaturated polyester.

radicals. Once the inhibitor molecules are depleted, the free radicals, produced from the initiator, initiate the polymerization of the polyesters. Styrene serves as an agent to link the adjacent polyester molecules. The curing of UPE resin (polyester + styrene) involves different types of reactions as given below:

1. intermolecular cross-linking of polyester without linking through styrene monomer
2. intermolecular cross-linking of polyester through styrene monomer
3. intermolecular cross-linking of polyester molecules without involving styrene
4. intramolecular cross-linking of polyester molecules linking through styrene monomer
5. chain branching on the polyester molecules
6. homopolymerization of styrene

The overall cure kinetics was affected by the possible reactions mentioned above. However, the chain branching and homopolymerization of styrene do not contribute to network formation. The long-chain molecules formed due to intramolecular and intermolecular reactions tend to form spherical type structures with high cyclization and cross-link density called "microgel particles" [33]. The pendant double bonds on the polyester molecules are either buried inside these primary microgels or located near the surface of the microgels. The buried double bonds participate in cross-linking reaction causing further advancement of curing reaction and increase the cross-link density of the microgels [34]. Finally, macrogelation takes place by intermolecular microgelling and microgel clusters. Phase separation may occur depending on the polyester structure and their compatibility with styrene. The factors that affect the miscibility with styrene have already been discussed in the earlier section. Muzumdar et al. [35] investigated the chemorhelogy and kinetics of cross-linking reaction. The results are shown in Fig. 1.6, where the viscosity, conversion,

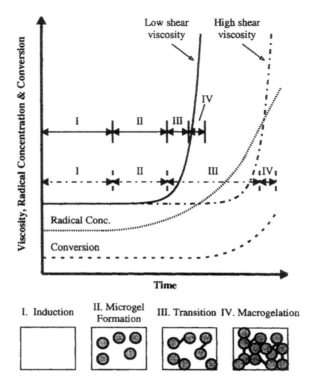

Figure 1.6
Schematic showing mechanism of viscosity build-up in UPE-styrene copolymerization. *UPE,* unsaturated polyester. Source: *Reprinted with permission from S.V. Muzumdar, L.J. Lee, Polym. Eng. Sci. 36 (1996) 943.* © *1996, John Wiley and Sons Publishers.*

and radical concentration are plotted as a function of time. In the induction stage, no cross-linking takes place due to consumption of radicals by the inhibitor and consequently, the viscosity remains almost constant. The sharp increase in viscosity takes place in macrogelation stage.

The kinetics of cross-linking reaction has been investigated. The cross-linking reaction is reported to follow the autocatalytic reaction. The detailed kinetics will be discussed in Chapter 2. The reaction rate is affected by the styrene content of the resin. Zlatanic et al. [36] carried out detailed rheological studies of acrylate-terminated UPE resins of different compositions, for example, poly [(tetramethylene maleate)-*co*-(tetramethylene phthalate)] (BMPA), poly [(tetramethylene fumarate)-*co*-(tetramethylene phthalate)] (BFPA), and poly [(tetramethylene phthalate)-*co*-(tetramethylene succinate)] (BPSA). A typical rheological plot (G' and G'' vs time) of a BFPA sample [36] at 30°C is shown in Fig. 1.7. The rheokinetic studies indicate that the cross-linking follow an autocatalytic reaction. The rate constants obtained from the autocatalytic model for BMPA and BMPA samples are 0.036 to

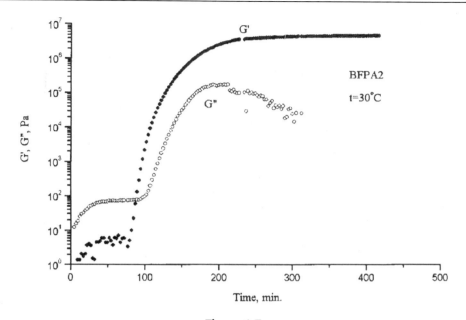

Figure 1.7
Storage and loss dynamic shear moduli (G' and G'') for the reaction mixture BFPA, plotted logarithmically as a function of total reaction time at 30°C. Source: *Reprinted with permission from A. Zlatanic, B. Dunjic, J. Djonlagic, Macromol. Chem. Phy. 200 (1999) 2048.* © 1999, Wiley-VCH Verlag GmbH & Co KgaA Publishers.

0.075 and 0.006 to 0.008 min^{-1}, respectively. A similar trend was observed using equilibrium swelling experiments of the cured samples [37]. This is expected because fumarate double bond (planar trans configuration) is more reactive than the double bond of maleate ester (distorted from the planar configuration). The effect of initiator concentration (benzoyl peroxide/*N, N*-dimethylaniline system) on the rheokinetics was also studied by Zlatanic et al. [37]. The rate of rheological conversion (β), that is, ($d\beta/dt$) versus β plot is shown in Fig. 1.8. The results indicate that the increase in BPO concentration significantly changes the rate of the reaction up to 1.5 wt.% concentration of BPO. Beyond 1.5 wt.% of BPO concentration, the influence of an increase in BPO concentration on the rate of cross-linking is less.

1.2.8 Application of unsaturated polyester resin

UPE resins can be used as clear castings or in combination with particulate fillers or fibers. The resin was developed to meet the demand for lightweight materials in military application. The first functional use of UPE was in the radome. Because of the obvious advantages of easy processability and low cost, it gained a wide range of applications in civil sectors like tanks, pipes, electronic gears, etc. Some of the important products based on cast UPE resins are the encapsulation of electronic assembly, buttons, handles for door, knife, umbrella, etc. It is also used for industrial wood and furniture finishing. Filled resin system using limestone, silica, china clay, etc., are used for floor tiles. A major use of UPE is as a matrix for fiber-reinforced composites. Such composites have received wide applications in automobile and construction industries and in a wide variety of products such as boats, water-skis, and television antennas. The examples of various applications of UPE resins are presented in Table 1.4.

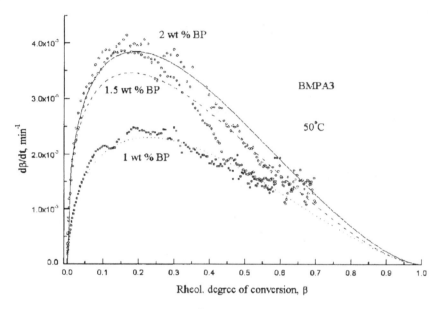

Figure 1.8
Influence of BP concentration on the rate of rheological conversion for the reaction mixture BMPA at 50°C. Symbols: experimental value, line: calculated values. BP, benzoyl peroxide. Source: *Reprinted with permission from A. Zlatanic, B. Dunjic, J. Djonlagic, Macromol. Chem. Phy. 200 (1999) 2048.* © 1999, Wiley-VCH Verlag GmbH & Co KgaA Publishers.

Table 1.4: Application of unsaturated polyester resins.

Industries/fields	Product
Automotive industry, construction industry, household appliances, electrical appliances, military application, aircraft, and aerospace industries	Automobile parts like hood, fenders, panels, bodies, etc., aircraft component, furniture, electrical encapsulation, household articles, fiber composites for boat and other marine applications, solar energy panels, building panels, fuel tank, radomes, tractor parts, transformer cover, electrical panels, helmet, grating, printed circuit board, Fenders, air conditioner panels, refrigerator parts, pressure tank, pipe and tubing, satellite dish antennae

UPE resin-based FRPs are used extensively in hull construction of various types of sail and motorboats, fishing boats, naval vessels, etc. UPEs are preferred in such applications compared to epoxy due to their better water resistance and lower cost. UPE-based FRPs are used for various applications in transportation industries namely in passenger cars, buses, trucks, construction equipment, etc. Because of lightweight, they have been replacing metallic materials for structural applications in the transportation sector like floor pans, radiator support, equipment housing, etc. The use of UPE composites in the decorative components for transportation applications is less because they cannot compete with RIM-processed PU and engineering thermoplastic in terms of quality of finish.

In construction industries, UPE-based FRP has been replacing aluminum, stainless steel, and wood because of their low cost, ease of cleaning, and working. Examples for such applications are solar panels, wall liners, automobile components, liners for truck trailers, and railroad cars. UPE-FRP products are to make sanitary ware as replacements for conventional materials like porcelainized cast iron due to their obvious advantages of lower cost and easier installation. The other related applications are bathtubs, showers, toilets, and sinks.

1.3 Bio-based unsaturated polyester

The bio-based UPE resins can be prepared by substituting different petroleum-based building blocks (e.g., polyol and polyacid) with renewable-based analog, either partially or totally [38]. For example, fatty acids or oils (chemically modified or not) were used as polyacids [39], while rigid carbohydrates such as isosorbide were used as polyols. Isosorbide is a sorbitol derivative, obtained from starch through hydrolysis and subsequent hydrogenation of glucose into sorbitol. In some cases, citric acid can be used to convert renewable resource-based polyols, for example, isosorbide into acid-functionalized polyester, yielding polyester resins with improved coating properties. The use of lignin esters as an additive in unsaturated thermosetting materials affords several advantages by acting as a toughening agent, improving the connectivity in the network, adding additional stiffening groups, and so on.

1.3.1 Alkyd resin

Alkyd resins are a class of polymers, that are introduced in 1930 as a binder for surface coating formulations [40]. Their compatibility with a wide variety of polymers, low cost, and versatility made them suitable for the production of a broad range of coating materials. The term "alkyd" was coined by Kienle and Ferguson and is derived from "al" of alcohol and "cid" of acid; "cid" was later changed to "kyd" [41]. The specific definition of alkyds are oil modified polyesters and are synthesized from three basic components: polybasic acids, polyols, and fatty acids. The oils that are mostly employed for the synthesis of alkyd resins are linseed oil, soybean oil, dehydrated castor oil, fish oil and tall oil. Linseed oil has been used for making alkyd resins since antiquity. Due to its yellowing nature, use of

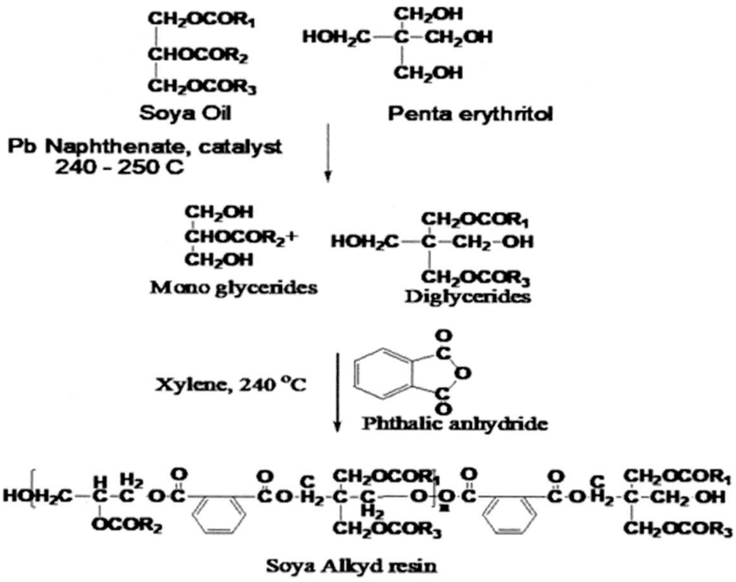

Figure 1.9
Synthesis of alkyd resin.

linseed oil is decreased, and it is replaced by dehydrated castor oil or soya oil. A typical synthesis of alkyd resin using soya oil is shown in Fig. 1.9.

Alkyd resins are mainly used as a binder in coating industries. The driving force for its industrial proliferation is its much lower cost compared to other thermosetting resins like epoxy. A conventional alkyd coating consists of a resin or a combination of resins, pigments, additives, and solvents. These coatings require a considerable amount of solvent to reduce the viscosity of the resin for an easy application due to their inherent high viscosity. The use of solvents in coating materials has adverse effects on both human health as well as on the environment. Therefore the recent trend in the coating industries is to decrease the volatile organic compound (VOC) in the coating, that is, to develop low VOC coatings. The need has been expressed not only by acute government regulations but equally and persuasively by various social concerns for the environment.

Due to the concerns over these problems, coating industries have attempted to decrease the solvent contents of coating materials for the last three decades by developing advanced technologies. Similarly, in order to address the VOC issue, a lot of research activities have also been carried out for the development of environment-friendly coatings having very low VOC content or no VOC at all [42–48]. For example, hyperbranched polyols have been used to prepare alkyd resins. Hyperbranched polymers are characterized by densely branched structures with large number of surface functional groups. Unlike dendrimers, which have perfectly branched star-like topology, hyperbranched polymers have both dendritic and linear units having irregular structures. Synthesis of dendrimer requires chromatographic separation in each step whereas hyperbranched polymers can be synthesized by a cost-effective single-step reactions resulting in their commercial proliferation. Due to the compact three-dimensional (3D) structure, hyperbranched polymers mimic the hydrodynamic volume of a sphere in solution and exhibit much lower viscosity compared to their linear analog [49–52].

For coating applications, this special relationship between molecular weight and viscosity should be highly useful in terms of the environmental issues and cost in order to formulate the low VOC coatings. Viscosity issues of conventional alkyd resins can be resolved by using hyperbranched polyols as substitutes for conventional polyols. Carlos et al. [53] have studied the rheological properties of different generations from generation two (G2) to generation five (G-5) of hyperbranched polyesters in 1-methyl 2-pyrrolidinone solvent and their blends with a linear poly(2-hydroxyethyl methacrylate) (PHEMA) counterpart. The absolute values of solution viscosity for hyperbranched polymers were much smaller than those for the linear PHEMA in the same solvent. This behavior [53] can be seen in Fig. 1.10. The difference in viscosities between PHEMA and hyperbranched polyesters can be attributed to several factors. One factor

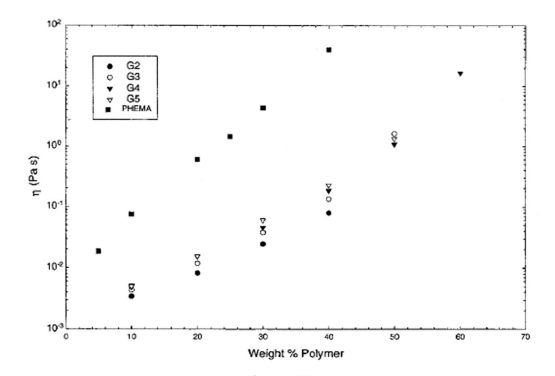

Figure 1.10
A Comparison of the zero-shear viscosities of the linear polyester with the various hyperbranched polyesters. Source: *Reprinted with permission from Carlos M. Nunez, Bor-Sen Chiou, Anthony L. Andrady, Saad A. Khan, Macromolecules 33 (2000) 1720. © 2000, American Chemical Society Publishers.*

is the absence of chain entanglements between the hyperbranched polymers. Another factor is that hyperbranched polyesters have much lower intrinsic viscosities than PHEMA, due to their branched and denser structures when compared to PHEMA.

Baloji et al. [15,54,55] have prepared hyperbranched polyols for preparing hyperbranched alkyd resins. A typical synthesis of a hyperbranched polyol is given as follows. The hyperbranched polyol was synthesized using dipentaerythritol (DPE) as a core molecule and 2, 2-*bis* (methylol) propionic acid (BMPA) as a chain extender. A mixture of DPE (38 g, 0.15 moles) and BMPA (362 g, 2.7 moles) were taken in a four-necked 1000 mL round bottom flask fitted with a mechanical stirrer, a condenser attached with a Dean & Stark apparatus, and a thermometer. 20 mL of xylene was added as an azeotropic solvent to remove the water, formed during the esterification reaction. The reaction was carried out at 140°C under a nitrogen atmosphere, using p-Toluene sulfonic acid as a catalyst. The reaction was continued till the theoretical amount of water was collected and the acid value attained below 10 mg KOH g^{-1} of the resin. The reaction scheme for the synthesis of the HBP is given in Fig. 1.11. Johansson and associates [56] synthesized hyperbranched

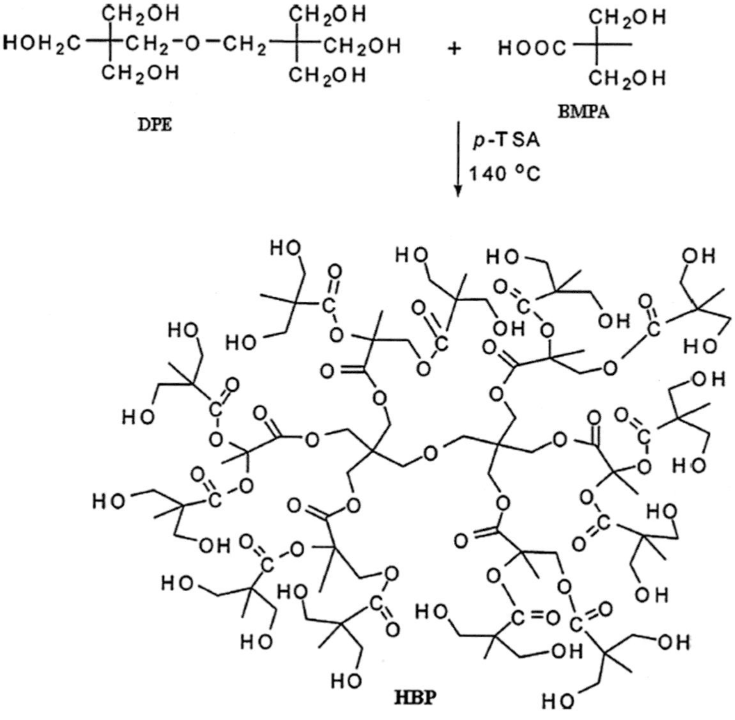

Figure 1.11
Synthesis of hyperbranched polyol. Source: *Reprinted with permission from R.B. Naik, S.B. Jagtap, D. Ratna, Prog. Org. Coat. 87 (2015) 28.* © *2015, Elsevier Publishers.*

polyesters having end hydroxyl groups that have been modified with fatty acids or acrylate, in order to obtain various resins, from liquid UV linking to powder-based, which all cures at low temperatures. The authors concluded that the use of the hyperbranched resins in coatings reduces the need for organic solvents since the products have low viscosity. Bruchmann et al. [57] synthesized hyperbranched polyurethanes with hydroxyls or isocyanates end groups, which were used as crosslinkers in two-component coatings having excellent properties, such as high hardness. Lange et al. [58] prepared hyperbranched polyester with hydroxyl functional groups that have been partially end-capped with acrylate or methacrylate units. The authors noticed that the number of hydroxyl groups in the hyperbranched resins was found to play a major role concerning the oxygen barrier, controlling both performance and water sensitivity.

Mirjana et al. [59] reported hyperbranched alkyds based on ditrimethylolpropane, dimethylolpropionic acid, and ricinoleic acid. First, hydroxy functional hyperbranched polyesters, based on ditrimethylolpropane and dimethylolpropionic acid, were synthesized. The synthesized HBPs were modified with ricinoleic acid using titanium (IV) isopropoxide as a catalyst. Terminal —OH groups of HBP were theoretically modified and prepared first-generation hyperbranched alkyd (HBA1) and second-generation hyperbranched alkyd (HBA2) resins. The reaction of second-generation hyperbranched alkyd (HBA2) formation is shown in Fig. 1.12. The synthesized hyperbranched alkyd HBA1 and HBA2 resins were mixed with two commercial melamine resins such as isobutanol (IBM), and hexamethoxymethyl melamine (HMMM) resin separately to the prepared HBA coating at room temperature in a solid ratio of 70:30. The measured properties of the cured films are shown in Table 1.5. Dry film thicknesses were in the range of 20–30 μm. The adhesion was very high for all coatings tested. All the cured coatings have satisfactory gloss values, ranging from 67% to 74%. Coating films obtained from mixtures using hexamethoxy methyl melamine resin have a higher hardness than the films obtained from the mixtures using isobutylated melamine resins (Table 1.5). The impact resistance data for all the tested films were satisfactory and ranges from 20 to 50 kg cm^{-1}. Very high values of the methylethylketone (MEK) test of coating films were obtained, which indicates that they are resistant to aggressive solvents such as MEK.

1.4 Vinyl ester resins

VE resin exhibits the desirable thermomechanical properties like epoxy and at the same, it offers processability like an UPE resin. VE resin was first commercialized by Shell Chemical Co. in 1965 under the trade name "Epocryl." Subsequently Dow Chemical Co. introduced a similar series of resins under the name "Derakane." VE resins are prepared by an addition reaction between epoxy resin (difunctional or multifunctional) with an unsaturated carboxylic acid like acrylic acid or methacrylic acid [60]. The simplest form of

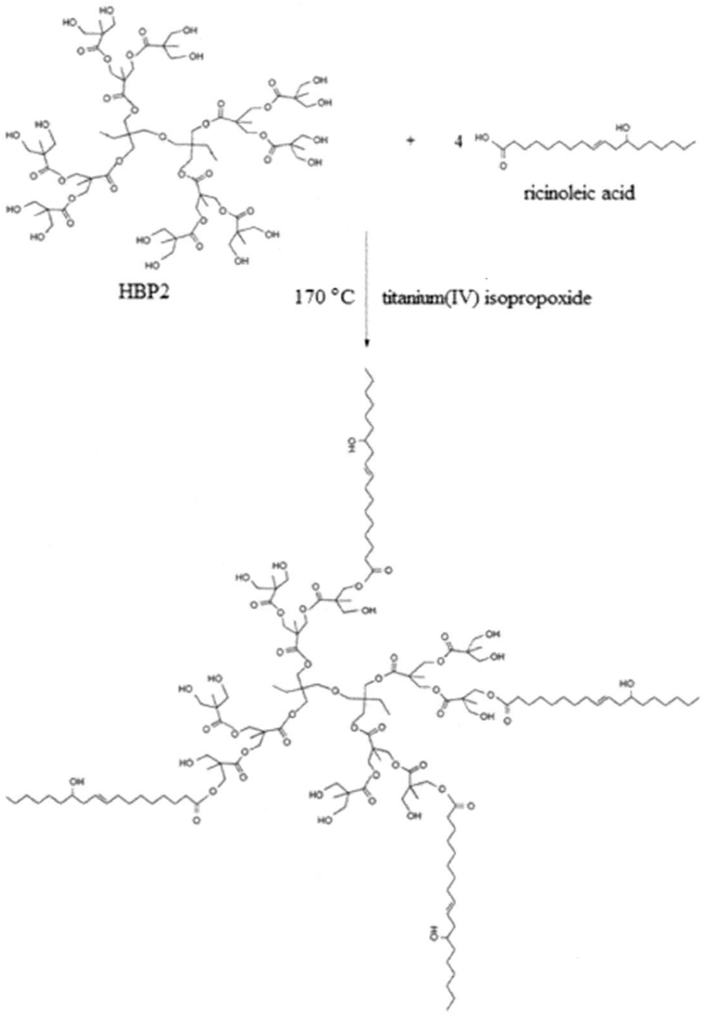

Figure 1.12
Synthesis of second-generation hyperbranched alkyd (HBA2). Source: *Reprinted with permission from Mirjana J, Radmila R, Jelena P, Oskar Bera, Dragan Govedarica, Prog. Org. Coat., 148 (2020) 105832.* © *2020, Elsevier Publishers.*

the vinyl ester is the product of the reaction between one mole of diglycidyl ether of bisphenol-A and 2 moles of methacrylic acid. The reaction scheme is shown in Fig. 1.13. Synthesis of a VE resin is usually catalyzed by tertiary amines, phosphines, and alkalies or onium salts. The reaction is carried out in a temperature range of 80°C–110°C and taking the reactants in a stoichiometric ratio. To achieve the complete esterification of the epoxy resin, in practice, is difficult because of the gelation of the product before an acid value below 10 is attained. Therefore an excess of epoxide resin is always employed in the

Table 1.5: Properties of hyperbranched alky-based cured coating films.

Mixture	Dry film thickness, ISO 2360 (um)	Cross-cut-adhesion, ISO 2409	Gloss 60° (%)	Pendulum hardness by Konig, DIN 53157 (s)	Flexibility, maximum depth (mm)	Impact test (kg cm^{-1})	Methylethylketone resistance
HBA1/IBM	20–30	0	67	74	8	20	174
HBA1/HMMM	20–30	1	74	93.8	7	35	192
HBA2/IBM	20–30	0	67	77	7	25	198
HBA2/HMMM	20–30	1	72	95.2	6	40	>200

Source: *Reprinted with permission from J. Mirjana, R. Radmila, P. Jelena, O. Bera, D. Govedarica, Prog. Org. Coat. 148 (2020) 105832. ©2020 Elsevier Publishers.*

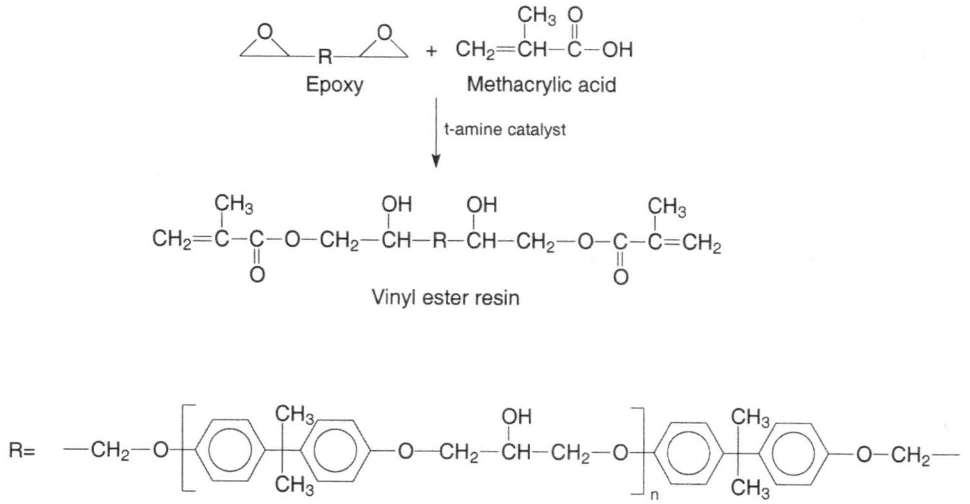

Figure 1.13
A reaction schemes for the synthesis of a VE resin. VE, vinyl ester.

esterification reactions. In order to stop the polymerization of methacrylic acid an inhibitor such as hydroquinone is used. By changing the nature of base commercial epoxy resins (as shown in Fig. 1.13), namely halogenated or phosphorous-containing resins and unsaturated acids (e.g., acrylic, methacrylic, crotonic, cinnamic, etc.), various types of vinyl ester resins can be produced.

The vinyl-ended prepolymer is a highly viscous material and cannot be used as such. It may be noted that the viscosity of a resin is a crucial factor and often a barrier in most polymer processing practices [61]. Therefore it is necessary to add reactive diluents like monofunctional vinyl compounds or acrylates to overcome the increased processing time and to reduce the high risks of inhomogeneity and consequent defect generation in the specimens [62]. The diluents are required for better control over the process and to modify the rheological and thermomechanical properties of the related VE resins. Generally, the prepolymer is dissolved in styrene to produce a polymerizable resin similar to a UPE resin. The styrene content is adjusted to provide a wide range of viscosities, typically from 100–4000 cps. Generally, 30–40 wt.% styrene is used to achieve a workable viscosity [63]. The use of styrene is associated with a potential health hazard due to the high volatility of styrene. Styrene is classified as a hazardous air pollutant and a volatile organic compound (VOC) with carcinogenic risk at long-term exposures [64]. Impacts of styrene and its derivatives on human health and the environment resulted in applying rigorous legislations to mitigate its emission by several organizations, including the US Environmental Protection Agency. Therefore attempts have been made to develop VE resins with comparatively lower or zero styrene content [65–69]. The styrene content can be reduced by developing VE resin with inherently lower viscosity compared to existing VE resins (commercially available) or using nontoxic diluents as a replacement of styrene.

With the first objective in mind, a low viscosity vinyl ester resin (viscosity = 39 Pa s) from renewable resource material, such as cardanol from cashew nut, was reported [65]. The temperature of curing the cardanol-based VE resin was found to be very close to the curing temperature for VERs from epoxidized phenolic novolac resin cured using 40% styrene. However, the developed bio-based resin shows inferior mechanical property and some issues related to color compared to VERs from epoxidized phenolic novolac resin (cured using 40% styrene). Atta et al. [66] used maleopimaric acid (MPA) and acrylopimaric acid (APA) from rosin acids to prepare VE resins. MPA is reacted with ethylene glycol to produce the hydroxymethylated derivatives. The hydroxymethylated derivative was reacted with epichlorohydrin to produce the epoxy resin and the same is finally reacted with acrylic or methacrylic acids to produce divinyl ester based on acrylate or methacrylate. The reaction scheme is shown in Fig. 1.14.

VE resins prepared from rosin-acid-based epoxy resin exhibit improved coating abilities due to faster drying and better chemical resistance. The bio-based VE resin has also been prepared [67] from myrcene, which is an oily liquid obtained by cracking on β-pinene of turpentine. These myrcene-based VE resins are capable of undergoing cure reaction rapidly (∼50 s) in the presence of UV light. These resins also show excellent thermal stability and good shore hardness. Li et al. [68] reported bio-based VE resins derived from a dimer of fatty acid polymerized with glycidyl methacrylate and maleic anhydride. The resins exhibit superior mechanical and thermal properties. Carbohydrate-derived isosorbide can be used as a core scaffold to synthesize novel and innovative VE resins with extremely low viscosity. These resins show much higher thermal stability compared to the conventional VE resin.

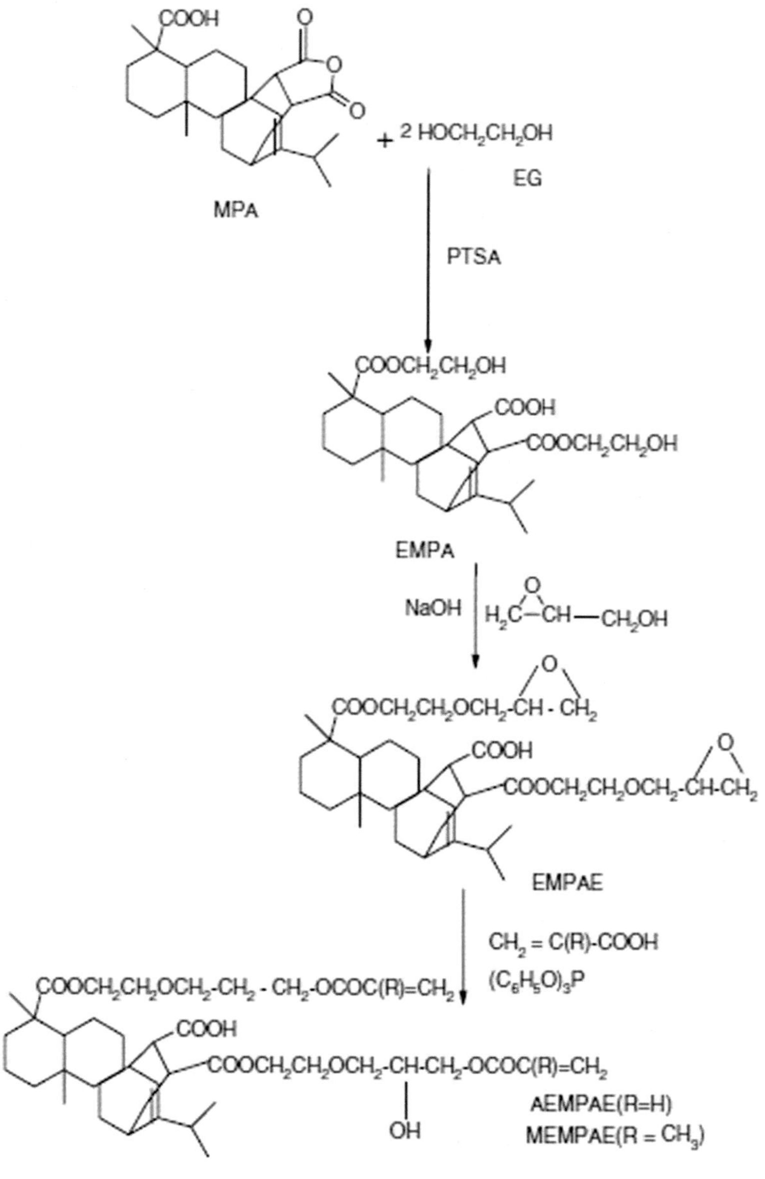

Figure 1.14
Scheme for synthesis of maleopimaric acid-based vinyl ester resin. Source: *Reprinted with permission from A.M. Atta, S.M. El-Saeed, R.K. Farag, React Funct Polym. 66 (2006)1596–1608.* © 2006, Elsevier Publishers.

The above-mentioned bio-based resins can be used as a viscosity modifier for commercial VE resin to reduce the amount of harmful styrene present in the resin. Another strategy is to develop nontoxic or less toxic bio-based reactive diluents, which can be used as a substitute for styrene. A number of reactive diluents based on plant oil, lignin, etc. have been reported [70–73]. However, the related vinyl ester resins offer inferior mechanical properties compared to the styrene-based VE resins. Some of the bio-based reactive diluents, which show considerable promise are discussed here. For example, Jahandideh et al. [74] synthesized bio-based reactive diluents from succinic acid and glycerol, which show good compatibility with VE prepolymer. The developed styrene-free VE resin shows viscosity in the range of 3.25–0.43 Pa s^{-1} at 30°C and offers a similar thermal stability and superior mechanical properties (in the cured state) in comparison to styrene-based VE resins. Yu et al. [75] reported three renewable vinyl compounds, prepared from benign bio-sourced precursors, for example, vanillin and eugenol in good overall yield and used as a reactive diluent with a renewable high-performance vinyl ester resin thermoset namely dimethacrylated-epoxidized-sucrose-soyate (DMESS). DMESS-based thermoset offer better thermomechanical properties as compared with traditional soybean oil-derived resins due to the presence of the high degree of functionality combined with rigid sucrose core [76–80]. Eugenol (4-allyl-2-methoxyphenol), an aromatic compound consisting of a rigid aromatic ring substituted by an allyl group, a hydroxyl group, and a methoxyl group extracted from plants, for example, clove oil. Vanillin (4-hydroxy-3-methoxybenzaldehyde) is obtained from lignin, which can be obtained from fibers of plants using various extraction methods. The chemical structures of lignin and vanillin are illustrated in Fig. 1.15.

It was demonstrated that out of three veratrole-diluents, the performance of 3-allyl-1, 2-dimethoxy-5-vinyl benzene is comparable to styrene in all respect. Due to its higher

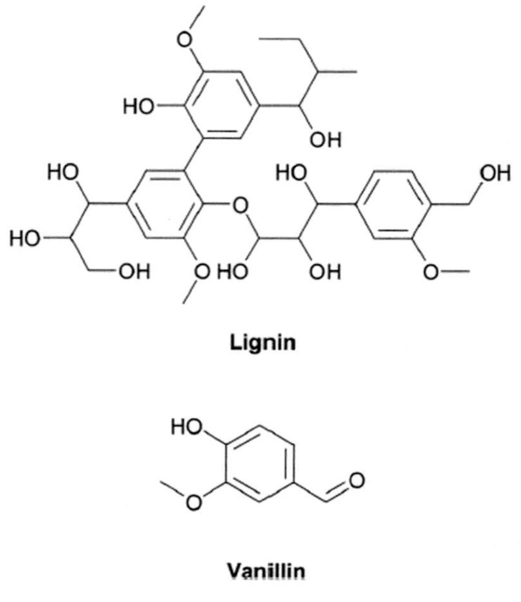

Figure 1.15
Chemical structures of lignin and vanillin.

molecular weight, exposure issues will be much less compared to styrene. It also contributes to a sustainable development of VE resin because of its natural origin. Jalliet et al. [81] attempted to prepare BPA-free VE resin using a bio-based compound namely phloroglucinol (an aromatic compound with a benzene ring and three hydroxyl groups) derived from a type of tannins known as phlorotannins. It is found in some species of brown algae such as kelps and rockweeds or sargassacean and in red algae. Phloroglucinol was reacted with epichlorohydrin to obtain the triglycidyl phloroglucinol and further reacted with the methacrylic acid to produce methacrylated phloroglucinol with a functionality of 2.8 eq mole^{-1}. The methacrylated phloroglucinol, when copolymerized with styrene in 60/40 ratio using 2% tertbutyl peroxybenzoate as an initiator, resulted in a network with higher cross-link density compared to BPA-based VE resin. This is reflected in the high T_g value (187°C) of the phloroglucinol-based VE resin.

Recently, Yadav and coworker [82] published a comprehensive review on the recent advances on plant-based vinyl ester resins and reactive diluents. Table 1.6 and Table 1.7 present an overview of bio-based vinyl ester crosslinkers and diluents along with highlighted properties and potential applications of the cured vinyl ester resins.

1.4.1 Curing of vinyl ester resins

VE resins are cured using a free radical initiator in combination with an accelerator [83–85]. Most common room temperature curing VE resin systems consist of methyl ethyl ketone peroxide (MEKP) (1–2 wt.%) and cobalt naphthenate (0.2%–0.4%). Benzoyl peroxide (BPO) and dimethylaniline system are used to achieve better corrosion resistance. BPO system requires a higher temperature to decompose and generate free radicals compared to MEKP. Thus, depending on the initiator systems used, the curing conditions and gel time will vary widely. At room temperature, promoters such as tertiary amines and salts of metals like cobalt octoate or naphthenate can be used to fasten the decomposition of initiators. The polymerization takes place through a free radical mechanism, which consists of initiation, inhibition, propagation, and termination steps. The termination step may be neglected at the later stage of polymerization due to the formation of highly crosslinked microgels [86]. In order to improve the shelf life of the resins, inhibitors like hydroquinone and tertiary butyl catechol are added in a very small amount (in ppm) to the resin formulations to inhibit premature gelation by consuming free radicals that occur naturally in the resin [65].

In the previous section, we have discussed various steps of curing UPE, elaborately. The curing of VE resin is similar to UPE resin. The curing of VE resin containing styrene basically involves three steps namely copolymerization of VE, copolymerization of VE and styrene, and homopolymerization of styrene [87–89]. Like vinyl/divinyl system, curing of VE/styrene system shows an induction period due to the presence of inhibitors, which

Table 1.6: An illustrative guide to bio-based vinyl ester crosslinkers and diluents covered along with highlighted properties and potential applications of the cured vinyl ester resins.

Vinyl ester resins	Curing procedure, initiators, and accelerators	Properties	Potential application
Vinyl ester of acrylated eposidized soybean oil	• 1 wt.% 2,5-dimethyl-2,5-di(2-Ethylhexanoyl peroxy) hexane • Thermal cure	TS: 6 MPa M: 440 MPa TS (40 wt.% styrene): 21 MPa M: 1.6 GPa $L-T_g$: 53°C	Composites coating
Vinyl ester of acrylated soybean oil	• 40 wt.% styrene • Benzoyl peroxide, dimethyl aniline • Thermal cure	$T-T_g$ (75.7% acrylation): 63.7°C G': 0.89 GPa $T-T_g$ (59.3% acrylation): 55.5°C G': 1.25 GPa	Coating
Vinyl ester of bromoacrylated soybean oil	• 1% Methyl ethyl ketone peroxide • 6% cobalt naphthenate • Thermal cure	$T-T_g$ (w/35 wt.% styrene): 55°C–65°C IRI: 5B at 200 mm	Fire retardant
Viinyl ester of cardanol-based epoxidized novolac resin (CNEVER)Vinyl ester of cardanol-based bisphenolType diepoxy (NC514)	• 40 wt.% styrene • Benzoyl peroxide • Thermal cure • 40 wt.% Several diluents • Tert-butyl peroxybenzoate • Thermal cure	Low viscosity $D-T_g$ (w/40 wt.% styrene): 104°C $T-T_g$ (w/40 wt.% styrene): 86°C	Coating Coating, friction linings, structural polymers
Vinyl ester of dimer fatty acids polymerized glycidyl methacrylate (DA-p-GMA)Vinyl ester of maleic anhydride modified dimer fatty acids polymerzed glycidyl methacrylate MA-m-Da-p-GMAVinyl ester of epoxidized myrcene	• 30–50 wt.% of styrene • Benzoyl peroxide (BPO) • Thermal cure • 30–50 wt.% of styrene • Benzoyl peroxide (BPO) • Thermal cure • 3 wt.% dimethoxybenzoin • UV cured	T_g: 67°C–71°C TS: 16–25.29 MPa $T-T_g$ (w/30 wt.% styrene): 52°C $T-T_g$ (w/40 wt.% styrene): 44.5°C $T-T_g$ (w/50 wt.% styrene): 54°C TS: 17.91–27.34 Mpa $T-T_g$: 77°C TS: 49 MPa $T-T_g$ (w/50 wt.% AESO): 58°C TS: 17.57 MPa	Coating, printing inks Coating, printing inks Coating printing inks
Vinyl ester of tung oil modified glycidyl methacrylate	• 3 wt.% dimethoxybensoin • UV cure	$T-T_g$: −34.8°C E: −0.15 GPa TS: 4.8 MPa	Coating, radiation curable inks Printed circuit board coating

(Continued)

Table 1.6: (Continued)

Vinyl ester resins	Curing procedure, initiators, and accelerators	Properties	Potential application
Vinyl ester of methacrylated of rosin [acrylopimaric acid (APA) and maleopimaric acid (MPA)]-based diepoxy	• 40 wt.% of styrene • Methyl ethyl ketone peroxide (MEKP) and cobalt octoate (CO) • Thermal cure	APA-VE $D-T_g$: 110°C Pull off resistance 4 MPa MPA-VE T_g (based on differential scanning calorimetry [DSC]): 94°C Pull off resistance: 7 MPa	Coating adhesive
Vinyl ester of acrylated rosin [acrylopimaric acid (APA) and maleopimaric acid (MPA)]-based diepoxy	• 40 wt.% of styrene • Methyl ethyl ketone peroxide (MEKP) and cobalt octoate (CO) • Thermal cure	APA-VE $D-T_g$: 105°C Pull off resistance: 7 MPa MPA-VE $D-T_g$: 85°C Pull off resistance: 9 MPa	Coating adhesive
Vinyl ester of dipentene	• 20–50 wt.% poly(ethylene glycol) di-methacrylate-200 (PEGDMA-200) • 3 wt.% dimethoxybenzoin • UV cure	$T-T_g$ (w/ 20 wt.% PEGDMA-200): 52°C TS: 6.05 MPa $T-T_g$ (w/30% PEGDMA-200): 45°C TS: 8.54 MPa $T-T_g$ (w/50% PEGDMA-200): 36°C TS: 10.05 MPa	Flame retardant
Vinyl ester of methacrylated phloroglucinol (P3M)	• 40 wt.% styrene • 3% tert-butyl peroxybenzoate • Thermal cure	$D-T_g$: 165–105°C $T-T_g$: 187–105°C T_d (N_2): 5% at 231°C T_d (N_2): 10% at 387°C	BPA replacement
Vinyl enter of methacrylated sucrose soyate (MESS)	• 2 wt.% Luperox P and 2 wt.% Luperox 10 M75 • Thermal cure	$T-T_g$ (MESS-0.8): 109°C TS: 17.8 Mpa M: 880 Mpa $T-T_g$ (MESS-0.8/20 wt.% styrene): 137°C TS: 22.7 Mpa M: 1380 MPa	Composite applications
Vinyl ester of Furan mono-2-hydroxypropyl Methacrylate (FmHPM) and di-2-hydroxypropyl methacrylate (FdHPM)	• 1% Diphenyl(2,4,6-trimethylbenzoyl) phosphine oxide • UV cure	Tensile shear strength (TSS) of FmHPM: very low, TSS of FdHPM adhered on glass: 0.6148 and on polycarnonate: 0.2067 Pencil hardness (Ph) FmHPM: H-2H and FdHPM: 2H-3H	Adhesive

(Continued)

Table 1.6: (Continued)

Vinyl ester resins	Curing procedure, initiators, and accelerators	Properties	Potential application
Vinyl ester of dimethacrylated isosorbide	• 1.5 wt.% Trigonox 239-A • 0.375 wt.% Cobalt naphthanate (CoNap) • Thermal cure	V: 157 cP at 25°C $T-T_g$: 250°C E': 2.9 GPa	Composites
Vinyl ester of glycidyl methacrylate isosorbide	• 1.5 wt.% Trigonox 239-A • 0.375 wt.% Cobalt naphthanate (CoNap) • Thermal cure	V: 620,157 cP and $T-T_g$: 90°C E': 3 GPa V (35 wt.% styrene): 172 cP $T-T_g$: 130°C E': 3.2 GPa	Composites
Vinyl ester of isosorbide methacrylate	• 2 wt.% tert-butyl peroxybenzoate • Thermal cure	Pure IM, $T-T_g$: 142°C FS: 23.5 MPa FM: 2.01 GPa IMAESO $T-T_g$: 76°C FS: 28.4 MPa FM: 1.37 GPa IMAESO $T-T_g$: 95°C FS: 344 MPa FM: 1.5 GPa	
Methacrylated vanillin-glycerol di-methacrylate (MVGDM)	• 1.5 wt.% Trigonox 239 • Thermal cure	T_d: IDT at 244°C E': 3.6 GPa at 25°C $T-T_g$: 155°C	
Methacrylated vanillin (MV) and methacrylated vanillyl alcohol (MVA)	• 2 wt.% N-tert-butyl peroxybenzoate (TBPB) • Thermal cure	T_d (MV): 10% at 280°C T_d (MVA): 10% at 341°C $T-T_g$ (MV): 90.5°C $T-T_g$ (MVA): 131.6°C E' (MV): 4.2 GPa at 25°C E' (MVA): 4.7 GPa at 25°C	
Poly (methacrylated vanillin) (PMV)	• AIBN • 2-cyano-2-propyl benzodithioate • 1,4-dioxane • Thermal cure	$D-T_g$: 120°C	
Poly (methacrylated guaiacol) (PMG) and poly(methacrylated eugenol) (PME)	• 1.5 wt.% MEKP • Thermal cure	$D-T_g$ (PMG): 92°C $D-T_g$ (PME): 103°C V (MG): 17.1 cP at 25°C V (ME): 27.9 cP at 25°C	
Methacrylated bio oil resin (MBO)	• 1.5 wt.% Trigonox 239 • Thermal cure	V: 30.3 cP at 25°C T_d: IDT at 306°C $L-T_g$: 115°C E': 3.2 GPa at 25°C	

Source: Reprinted with permission from S.K. Yadava, K.M. Schmalbachb, E. Kinacia, J.F. Stanzione IIIb, G.R. Palmesea, Eur. Polym. J. 98 (2018) 199−215. © 2018 Elsevier Publishers.

Table 1.7: An illustrative guide to bio-based diluents for vinyl ester.

Reactive diluent	Curing procedure, initiators and accelerators	Properties	Potential application
RDs of Methacrylated fatty acids (MFA)	• 65 wt.% Dimethacrylated DGBA • Trigonox 239 A • Cobalt naphthenate • Thermal cure	$L-T_g$ (styrene): 150°C E': 3 GPa $L-T_g$ (MLau): 94°C E': 1.9 GPa $L-T_g$ (MSA): 93°C E': 1.65 GPa $L-T_g$ (MOA): 90°C E': 1.7 GPa $L-T_g$ (MLin): 79°C E': 1.6 GPa	Styrene replacement
RD of methacrylated ESO (MESO)	• 48 wt.% and 30 wt.% tactix VE (TVE) • 2% tert-butyl peroxybenzoate	$T-T_g$ 52 wt.% (MESO): 128°C E': 802 MPa $T-T_g$ (70 wt.% MESO): 102°C E': 846 MPa	Binders in composite applications, Replacement of BPA and Styrene
RD of 9–10 dibrom o s teric acid glycidyl methacrylate (Di-BrSAGMA)	• 35 wt.% and 45 wt.% DGEBA VE (828) • 1.5 wt.% Trigonox 239-A • 0.375 wt.% Cobalt naphthanate (CoNap) • Thermal cure	Precursor V: 834 cP $T-T_g = 82$°C (65 wt.% DI-Br SAGMA) $T-T_g = 112$°C (55 wt.% DI-BrSAGMA)	Styrene replacement and Fire retardant
RD methacrylated cardanol (MC)	• 48 wt.% and 40 wt.% tactix VE (TVE) • 2% tert-butyl peroxybenzoate	$T-T_g$ (52 wt.% MESO): 119°C E': 967 MPa $T-T_g$ (60 wt.%): 107°C E': 960 MPa	Binders in composite applications, replacement of BPA and Styrene
RD of maleic anhydride modified dimmer (fatty) acid polymerized glycidyl methacrylate (MA-m-DA-p-GMA)	• 20–80 by weight • 3.0 wt.% benzyl dimethyl ketal (BDK) photo initiator • UV cure	$L-T_g$ (30 wt.% CDMA): 56°C $L-T_g$ (50 wt.% CDMA): 51°C $L-T_g$ (70 wt.% CDMA): 53°C	Coating and adhesives

(Continued)

Table 1.7: (Continued)

Reactive diluent	Curing procedure, initiators and accelerators	Properties	Potential application
RD of Cardanol-based his-phenol type diepoxy (NC514)	• 40 wt.% VE of mono epoxy cardanol diluents • Trigonox A & CoNap • Thermal cure	$L-T_g$: $-35°C$ E': -0.35 GPa	Coatings, composite manufacture
RDs of diglycidyl ether of bisphenol-A (DGEBAVE)	• 40 wt.% VE of mono epoxy cardanol diluent • Trigonox A & CoNap • Thermal cure	$L-T_g$: $118°C$ E': -1.0 GPa	Coatings, composite manufacture
RD of furfuryly methacrylate (FM) and furoic acid glycidyl methacrylate (FA-GM)	• 65 wt.% RDX (26936) methacrylated VE • 1.5 wt.% Trigonox 239-A • 0.375 wt.% Cobalt naphthanate (CoNap) • Thermal cure	$L-T_g$ (35 wt.% FM): $98°C$ E': 2.862 MPa $L-T_g$ (35 wt.% FA-GM): $85°C$ E': 3.193 MPa	Styrene replacement
RD of furfuryly methacrylate (FM)	• 65 wt.% glycidyl methacrylate isosorbide • 1.5 wt.% Trigonox 239-A • 0.375 wt.% Cobalt naphthanate (CoNap) • Thermal cure	$L-T_g$ (35 wt.% FA-GM): $101°C$ V: 2234 Cp	Composites

Source: Reprinted with permission from S.K. Yadava, K.M. Schmalbachb, E. Kinacia, J.F. Stanzione IIIb, G.R. Palmesea, Eur. Polym. J. 98 (2018) 199−215. © 2018 Elsevier Publishers.

abstracts the reactive free radicals as a result, polymerization practically does not take place. The analysis of curing of VE/styrene system indicates that the initial fractional rate of conversion of di-methacrylate double bonds is greater than that of the styrene double bonds. However, the fractional rate of styrene conversion increases with time and exceeds the di-methacrylate fractional conversion rate at the end of the reaction. The homopolymerization of styrene continues after the di-methacrylate reaction terminates.

Ultimately, the conversion of styrene is always greater than that of the di-methacrylates. At the initial stage, which is the period of accelerated propagation, the styrene monomer was found to react at a slower rate than the VE monomer due to higher stability of styryl free radical as a result of conjugation with the benzene ring. The kinetics of copolymerization is thought to be governed by the intrinsic reactivity of the styrene and VE double bonds and not by diffusion. In the second stage, the polymerization rate is dominated by "Trommsdorff effect" and the polymerization of VE monomer decreases and polymerization of styrene increases throughout this stage. Therefore the fractional rate of styrene conversion increases with time and exceeds the di-methacrylate fractional conversion rate at the end of the reaction. The reason for this is attributed to the higher mobility of the styrene molecule in the entire network, owing to its smaller size. The VE double bond at the end of the second becomes immobile in the network due to its larger size. Toward the end of the reaction, that is, during the third stage, the VE ceases to react while the rate of consumption of styrene continues to increase, which leads to the formation of polystyrene until vitrification takes place. The homopolymerization of styrene continues after the di-methacrylate reaction terminates. It is observed generally that a lower concentration of styrene increases the rate of cure [90]. This is due to a decrease in the termination rate constant with increased concentration of cross-linking species the di-methacrylate that raises the polymerization rate [68] or it may be caused by an increase in the propagation rate due to the replacement of the less reactive styryl radicals by the more reactive methacrylyl radicals [91].

As discussed above the polymerization of VE resin is associated with the formation of microgel, that is, domains of highly crosslinked material dispersed in the pool of unreacted monomer. Therefore a biphasic structure is developed as a result of polymerization-induced phase separation. The atomic force microscopic analysis indicates nodular morphology [92] or closely packed microgel around 100 nm in diameter during the early stages of curing reaction. Dynamic light scattering experiment conducted during the curing reaction of VE/styrene system indicates the formation aggregation of microgels of diameter between 19 and 40 nm up to the gel point, after which the macroscopic network is developed [93]. The size of the cluster was found to be greater for divinyl benzene due to the presence of a large number of pendant double bonds as compared with styrene. It was also observed that with increasing cure temperature, the size of the microgels and clusters became smaller. Scanning electron microscopic analysis of fracture surfaces shows a cluster of coral-like microgel with a maximum size of 4 microns [94].

If we compare the two-phase microstructure of cured VE resin and UPE resin then we find that the morphology is well organized in VE resin whereas in UPE it is more of a random type. Therefore the free volume of a cured VE network is lower than that a cured UPE network. As a result, VE resin shows a much lower water absorption property compared to UPE although from a polarity point of view it should be the reverse. This explains the excellent hydrolytic stability of VE resin, especially in a marine environment.

1.4.2 Properties of vinyl ester resins

Unlike UPE resins, the unsaturation occurs only at the end in vinyl ester resin resulting in lesser number of cross-link compared to UPE resin networks. Because of the lesser number of cross-links and the presence of ether linkages in the structure, VE resin networks exhibit better flexibility compared to UPE resins. During the reaction of epoxy resin with the carboxylic acid, a number of hydroxyl groups are formed along the vinyl ester chain. These hydroxyl groups allow H-bonds to form with the similar groups present in the glass fiber surfaces. That is why VE resins offer better adhesion with glass fibers, carbon fibers, nanomaterials, or other polar substrates as compared to UPE resin [95–99]. However, such adhesion is more in the case of epoxy due to the presence of more hydroxyl groups. In the case of UPE, the unreacted carbon–carbon double bonds and the ester groups, present in the backbone of UPE resins after curing, provide sites for the hydrolytic attack, oxidation, and halogenations. This makes the cured UPE resin unsuitable for aggressive environments like marine applications. In VE resin, ester linkages are present only at the terminal ends of the chain. Moreover, as discussed earlier, the nodular morphology generated due to microgel formation in the VE resin network is more ordered compared to the same in a network made from UPE resins. In the case of methacrylate ester, extra protection is provided to ester linkages by the steric hindrance caused by the pendent methyl group of methacrylic acids. VE resin incorporates the thermal stability and mechanical strength of the epoxy backbone, and the presence of double bonds at the chain ends provides the ease of cure that is comparable with that in UPE resins. Therefore VE resin offers much lower water uptake and better corrosion resistance compared to general-purpose UPE resins [100–106].

1.4.3 Applications of vinyl ester resin

VE resin can be used for all possible applications discussed for UPE resins. Specifically, VE resins have replaced metals and UPE-based FRPs for corrosion-resistant applications because of their better corrosion resistance. The applications include use in tanks, piping and ducts primarily for handling dilute acids, solvents and fuels, corrosion-resistant mixing vessels, precipitation vessels, scrubbers and process columns. Because of their inherent resistance to chloride attack and long service life, composites of VERs are adopted by the paper and pulp industries as well. VERs based on epoxy novolacs are used for chemical storage tanks, pipes, and ducting fume extraction systems and gas cleaning units, as these resins show superior chemical resistance at high temperatures. In electrical industries VE resin-based FRPs are used to make components for electrical generation stations, transmission and distribution, televisions, antennas, etc. They are also used for making electrical maintenance equipment like ladders, booms, etc. Better water resistance makes them suitable for use in air conditioners and humidifier housing and other household appliances. VE resin performs better for underwater applications than epoxy resins. Thus VE resins are extensively used for marine applications. The examples are hull construction of various types of sail and motor boats, fishing boats, naval vessels etc.

1.5 Epoxy resins

Epoxy resins are a class of thermosetting resin materials characterized by the presence of two or more oxirane rings or epoxy groups within their molecular structure [107–110]. The resins undergo curing through the reaction of epoxide groups with a suitable curing agent to produce a network structure. The most common epoxy resin is diglycidylether of bisphenol-A (DGEBA), which is prepared by the reaction of epichlorohydrin (ECD) and bisphenol-A (BPA) as shown in Fig. 1.16. The first production of epoxy resin occurred simultaneously in Europe and in the United States in the early 1940s. Today, a wide variety of epoxy resins of varying consistency are available in the market.

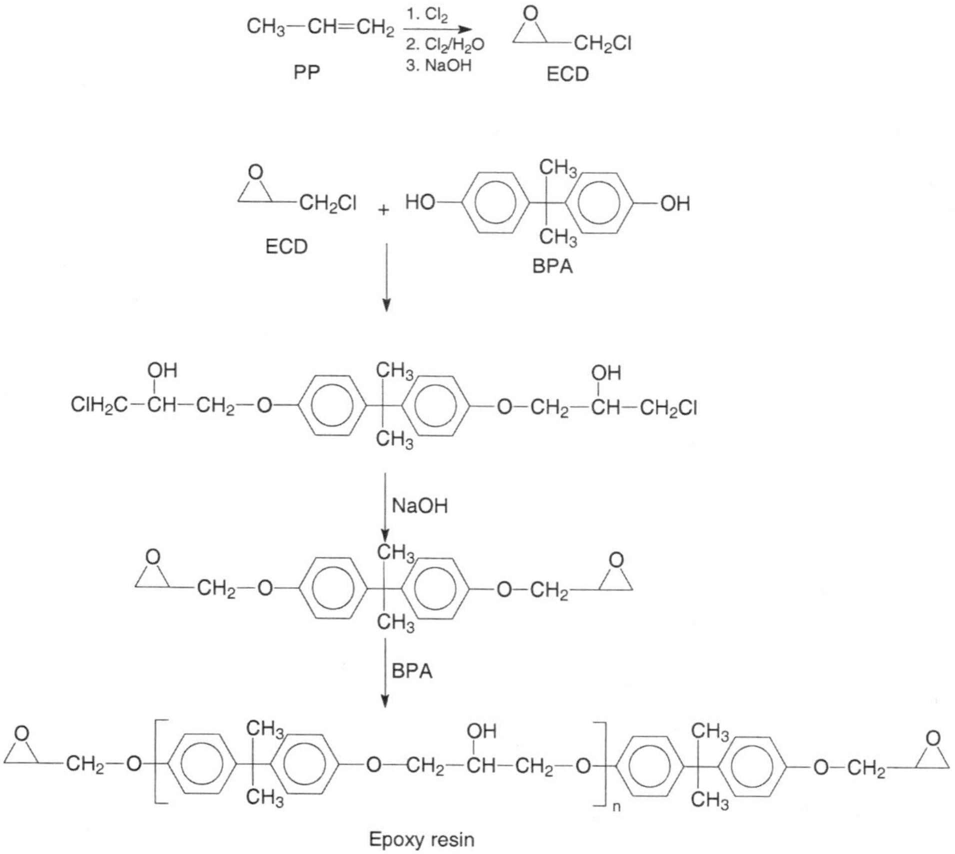

Figure 1.16
Reaction schemes for the synthesis of DGEBA-type epoxy resin. *DGEBA*, diglycidyl ether of bisphenol-A.

38 Chapter 1

Epoxy resins are unique among all the thermoset resins due to several factors [111–113]; namely, minimum pressure is needed for fabrication of products normally used for thermosetting resins, cure shrinkage is much lower and hence lower residual stress in the cured product than that encountered in the vinyl polymerization used to cure unsaturated polyester resins, use of a wide range of temperature by a judicious selection of curing agents with good control over the degree of cross-linking, availability of the resin ranging from low viscous liquid to tack-free solid, etc. Because of these unique characteristics and useful properties of the network polymers, epoxy resins are widely used [114,115] in structural adhesives, surface coatings, engineering composites, electrical laminates, etc. The majority of the composite applications utilize conventional difunctional epoxy as a matrix. However, many high-performance applications like aerospace and critical defense applications require the incorporation of epoxies of higher functionality, known as multifunctional epoxies. Tri- and tetra-functional epoxy resins are available commercially. Chemical structures of difunctional and multifunctional epoxies are shown in Fig. 1.17 and their physicochemical properties of the resins are given in Table 1.8.

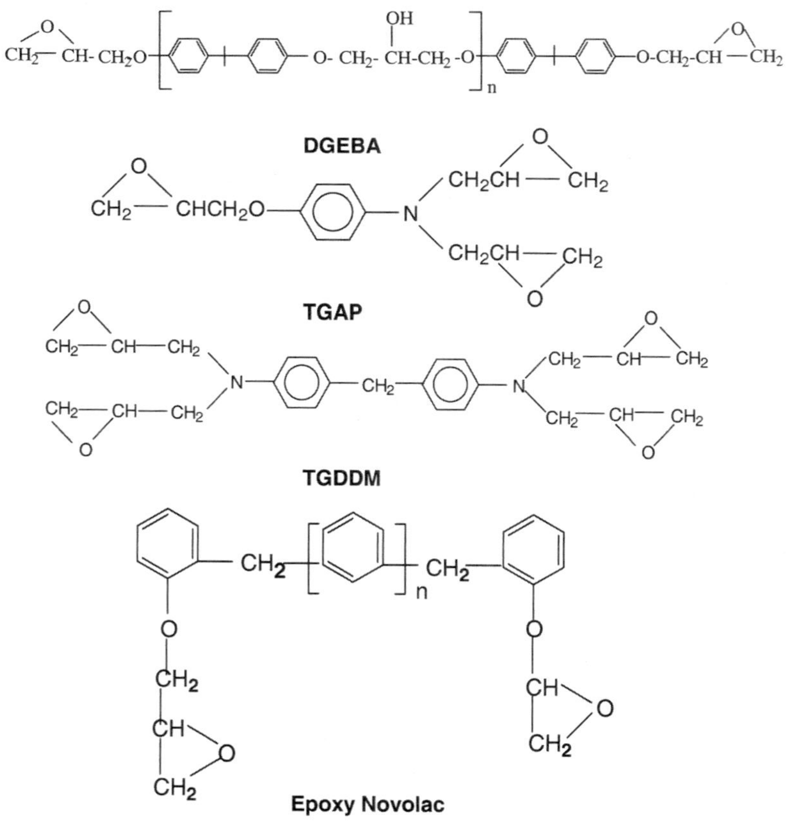

Figure 1.17
Chemical structures of difunctional and multifunctional epoxies.

Table 1.8: Physicochemical properties of difunctional and multifunctional epoxy resins.

Epoxy resin	Viscosity (Pa s)	Epoxy equivalent (g eq^{-1})	Functionality (eq mole^{-1})
DGEBA	3.5	190	2
TGAP	0.6	95	3
TGDDM	94.5	100	4

ECD is prepared from polypropylene (PP) by reacting with chlorine and sodium hydroxide. ECD is allowed to react with BPA in presence of sodium hydroxide. The first step is the cleavage of the oxirane ring of ECD by hydroxyl group of BPA. The second step is the cyclization in the base medium leading to the formation of an epoxy-ended intermediate. The intermediate then undergoes chain extension with BPA to produce an epoxy resin. A wide variety of resins can be produced by adjusting the concentration of the reactants. A liquid resin can be further chain extended with BPA to make a solid resin of higher molecular weight. Thus epoxy resins are available in various consistencies from low viscous liquid to a tack-free solid.

Undoubtedly, there exist a more number of publications and reports based on the basic and applied research on epoxy resins than that for any other commercially available thermosetting resins. The broad interest in epoxy resins originates from the versatility of the epoxy group toward a wide variety of chemical reactions and the useful properties of the network polymers [116–118] like high strength, very low creep, excellent corrosion and weather resistance, elevated temperature service capability and adequate electrical properties.

1.5.1 Epoxy formulation

An epoxy formulation must contain a suitable curing agent and some optional ingredients, which are decided considering the application in which the resin is going to be used. Examples of such ingredients are diluents, fillers or extenders, etc. Diluents are used in an epoxy formulation to reduce the viscosity or to eliminate the need for solvents. Depending on the interaction of the diluents with the epoxy, diluents are grouped into two types: reactive diluents and nonreactive diluents. Reactive diluents are monoepoxide compounds, which participate in the curing reaction and reduce the cross-link density. However, the addition of a diluent is generally associated with a reduction in mechanical strength, modulus and T_g. Therefore the amount of diluent has to be optimized by critically analyzing the thermomechanical properties keeping the technical specification for a particular application in mind. A list of common diluents and curing agents is presented in Table 1.9.

1.5.2 Curing agents

Due to the versatility of epoxy resins toward a wide variety of chemical reactions, epoxy resins can be cured using a range of materials with different types of curing conditions. The choice of curing agents (also called hardeners) depends on the curing conditions applicable and the final application of the resin. Epoxies can be cured with amines, thiols, and alcohols [119,120]. The reaction proceeds through the cleavage of the oxirane ring through nucleophilic addition reaction as shown in Fig. 1.18. Due to the involvement of addition reaction, no volatile by-products are formed during the curing of epoxy resins. As the reactions of the epoxy monomers with the respective hardener are sometimes not fast enough for a specific application, accelerators are often added to the resin formulations. Lewis acids are used for accelerating the reaction between epoxy and alcohol. The mechanism is shown in Fig. 1.19. Similarly, alcohols, phenols, or carboxylic acids are used as accelerators in amine-cured epoxies and tertiary amines (e.g., benzyl dimethylamine) are usually used in the anhydride-cured epoxy systems.

Table 1.9: List of common curing agents and diluents for epoxy resin.

Curing agent	Nonreactive diluent	Reactive diluent
Triethylene teramine (TETA) 4,4' diaminodiphenyl methane (DDM) 4,4' diaminodiphenyl sulfone (DDS) Dicynamide (Dcy) Diethyl toluene diamine (DETDA) Pthalic anhydride Hexahydropthalic acid anhydride Dicyclopentadiene dianhydride Melltic anhydride	Dioctyl phthalate Nonyl phenol Dibutyl phthalate Furfuryl alcohol Pine oil Coal tar Castor oil	Diglycidylether of 1,4 butane diol Epoxidized soybean oil epoxidized castor oil Diglycidyl ether of neopentyl glycol Butyl glycidyl ether 2-ethylhexyl glycidyl ether Phenyl glycidyl ether t-butyl glycidyl ether o-Cresyl glycidyl ether

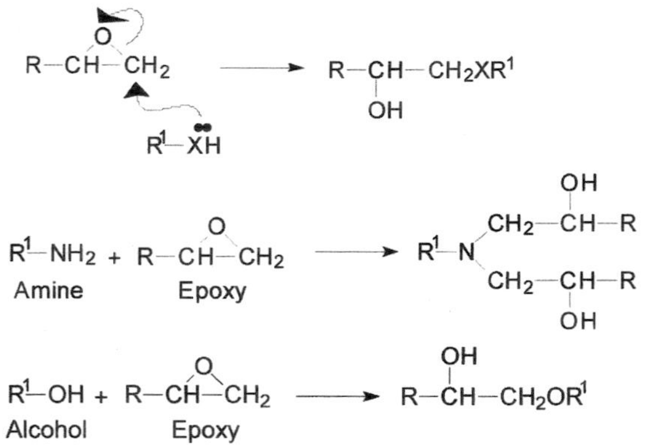

Figure 1.18
Mechanism of curing of epoxy resins.

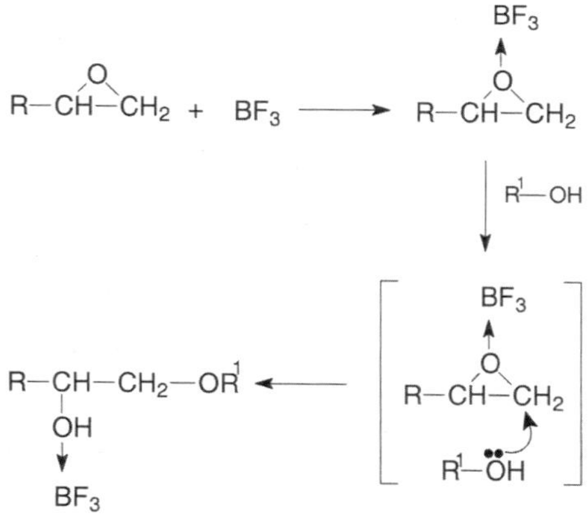

Figure 1.19
Mechanism of lewis acid-catalyzed curing of epoxy by alcohol.

Amines are widely used as hardeners for epoxy resins. Chemical structures of some commonly used amine curing agents are shown in Fig. 1.20. During curing, two epoxy rings react with a primary amine as shown in Fig. 1.18. The first step is the reaction between primary amine hydrogen with the epoxy group followed by a reaction between secondary amine hydrogen with another epoxy. Although single activation energy and heat of reaction are experimentally obtained for both the steps, reactivities of primary and secondary amino groups may be different. The hydroxyl groups generated during the cure can also react with the epoxy ring, forming ether bonds (etherification). The etherification reaction competes with the amine-epoxy cure when the reactivity of the amine is low or when there is an excess of epoxy groups. The tendency of etherification reaction to take place depends on the temperature and basicity of the diamines and increases with the concentration of epoxy in the mixture. The secondary alcohols (continuously formed during cure) also catalyze the amine-epoxy reactions, which is called autocatalysis. Thus the curing reaction occurs by two competitive mechanisms: one is autocatalytic reaction (catalyzed by the hydroxyl groups initially present in the epoxy prepolymer or those generated during the reaction) and other is a noncatalytic mechanism, which is second-order in nature. The autocatalytic mechanism involves a ternary transition complex as shown in Fig. 1.19. At high temperature, the catalytic mechanism almost vanishes due to the difficulty of forming such ternary complex. On the other hand, the second-order noncatalytic reaction takes place over the entire range of temperature. The kinetic equations related cure of thermoset in general and epoxy resin in particular will be discussed elaborately in Chapter 2.

Amines used for curing epoxy resins can be grouped into three categories namely aliphatic, cycloaliphatic, and aromatic. The advantages and disadvantages of various amines are summarized in Table 1.10. The reactivity of the amine increases with its nucleophilic

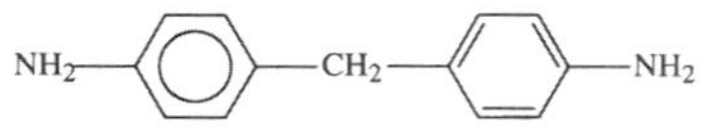

Diamino diphenyl methane (DDM)

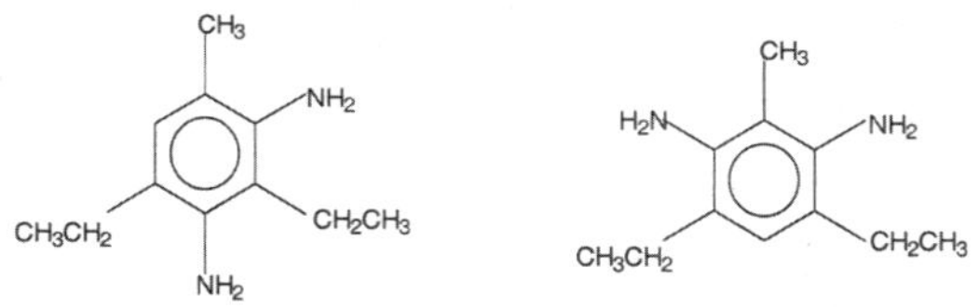

Diethyl toluene diamine (DETDA)

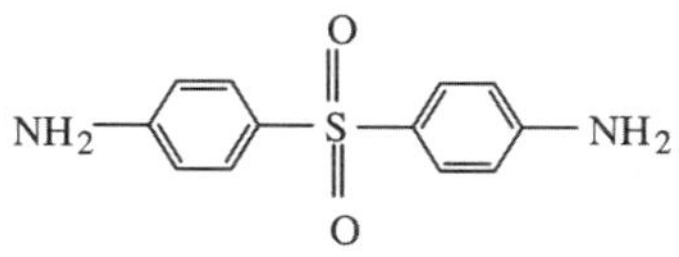

Diamino diphenyl sulfone (DDS)

$$NH_2—C_2H_4—NH—C_2H_4—NH—C_2H_4—NH_2$$

Triethylene tetramine (TETA)

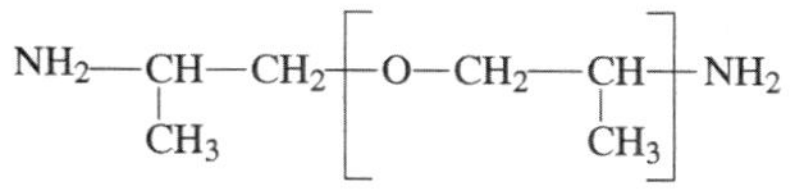

Jeffamine

Figure 1.20
Chemical structures of commonly used amine curing agents.

Table 1.10: A comparison of the properties of different types of curing agents for epoxy.

Type	Advantage	Limitation
Aliphatic amine	Low cost, low viscosity, easy to mix, room temperatue curing, fast reacting	High volatility, toxicity, short pot life, cured network can work up to 80°C not above
Cycloalipatic amine	Room temperature curing, convenient handling, long pot life, better toughness, and thermal properties of the resulting network compared to aliphatic amine-cured network	High cost, can work at a service temperature below 100°C, poor chemical resistance, poor solvent resistance
Aromatic amine	High T_g, better chemical resistance and thermal properties of the resulting network compared to aliphatic and cycloaliphatic amine-cured network	Mostly solid, difficult to mix, curing requires elevated temperature
Anhydride	High network T_g compared to amine curing agent, very good chemical and heat resistance of the resulting network	High temperature curing, long post curing, necessity of accelerator, sensitive to moisture
Dicynamide	Low volatility, improved adhesion, good flexibility, and toughness	Difficult to mix, high temperature curing, and long postcuring
Polysulfide	Flexibility of the resulting network, fast curing	Poor aging and thermal properties, odor
Polyamides	Low volatility, low toxicity, room temperature curing, good adhesion, long pot life, better flexibility, and toughness of the resulting network compared to aliphatic amine-cured	Low T_g of the resulting network, high cost and high viscosity

character: aliphatic > cycloaliphatic > aromatic. Thus appropriate curing temperatures and catalysts (called accelerators) have to be employed for curing. The network density of the cured resin can be adjusted by the careful choice of chemistry and stoichiometry of epoxy monomer and amine hardener. Increasing functionality combined with a low molecular mass of the components results in a highly crosslinked network. The advantage of aliphatic amine is that it can cure epoxy resins at an ambient temperature. Other amines mostly require heat curing. Heat curing is difficult and impractical for the fabrication of certain structures and requires a significant amount of energy. Ambient curing saves energy and is advantageous for coating or adhesive applications with intricate structures. It may be noted that though the curing takes place at room temperature, for completion of curing reaction, it is necessary to postcure at a high temperature. In this context, it is relevant to discuss the differential scanning calorimetry (DSC) analysis of a room temperature curable epoxy/ triethylenetetramine (TETA) system reported by Ratna and coworker [121]. The DSC thermogram of the epoxy/TETA system cured at room temperature for 2 days is shown in Fig. 1.21. The thermogram shows a residual exotherm indicating that the curing reaction is not complete. The same sample when postcured at 100°C for 2 h shows no residual exotherm, which indicates the completion of curing reaction.

The major limitation of aliphatic amine-cured epoxy is that they cannot provide a network system with T_g more than 120°C [122]. To make an epoxy network with high T_g, it is necessary to use aromatic amines, which require curing at high temperature [123,124] [125] Epoxies cured with aromaticdiamines can withstand high temperatures and are often used in structural composites. Reydet et al. [123] investigated the curing of epoxy using diamino diphenyl methane (DDM), diamino diphenyl sulfone (DDS), 4,4'-methylenebis[2,6 diethyl aniline] (MDEA), and 4,4'-methylene *bis* [3-chloro 2,6-diethylaniline] (MCDEA). The

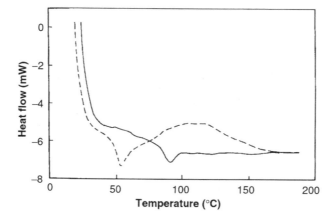

Figure 1.21
DSC thermogram of room temperature curing epoxy (-) before and (---) after post curing. *DSC*, differential scanning calorimetry.

evolution of epoxy conversion with time at 135°C for epoxy/DDS, epoxy/MDEA, and epoxy/MCDEA are shown in Fig. 1.22. The conversion versus time plots at different temperatures are shown in Fig. 1.23. The gelation (indicated by an arrow) time increases in the order MDEA > DDS > MCEDA. The reactivity of the amines can be explained in terms of basicity and steric hindrance. Diaminodiphenyl methane (DDM) is much more reactive than the amine mentioned above. That is why it is very difficult for kinetic characterization of epoxy/DDM system. The reactivity decreases comparatively in MDEA due to the steric hindrance. Due to the presence of the electron-withdrawing sulfone group, DDS is less reactive than MDEA. MCDEA is the least reactive due to the combined effects of a steric hindrance as well as the electron-withdrawing effect of chlorine present in the structure.

After the amines, acid anhydrides are the next most common hardeners for the curing of epoxy monomers. These epoxy systems are widely used for castings, especially in electronic applications. They find applications in circuit boards, as encapsulating material for integrated circuits and as insulators in power current components. In contrast to the stepwise nature of the amine-epoxy cure, the acid anhydride-epoxy reaction proceeds via a

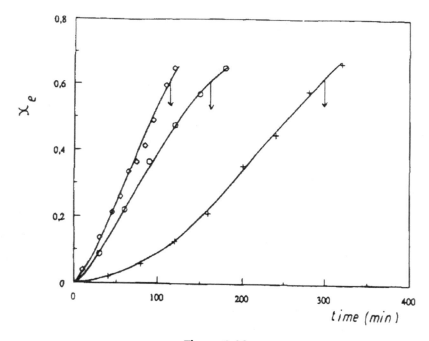

Figure 1.22
Epoxy conversion versus reaction time at 135°C for different comonomers: (◇) DGEBA-MDEA; (0) DGEBA-DDS; (+) DGEBA-MCDEA. Arrows indicate gelation times. *DDS*, diamino diphenyl sulfone; *DGEBA*, diglycidyl ether of bisphenol-A; *MCDEA*, 4'-Methylene *bis* [3-chloro 2,6-diethylaniline]; *MDEA*, 4,4'-Methylenebis[2,6 diethyl aniline]. Source: *Reprinted with permission from E.G. Reydet, C.C. Riccardi, H. Sautereau, P.J. Pascault, Macromolecule 28 (1995) 7599. © 1995, American Chemical Society Publishers.*

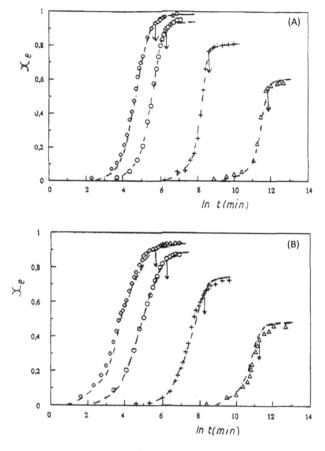

Figure 1.23
Epoxy conversion vs reaction time at different cure temperatures: (◇) 160"C; (0) 135°C; (+) 80°C; (△) 30°C; (---) diffusion model predictions. For (A) DGEBA-MCDEA; (B) DGEBA-DDS. Arrows (4) indicate vitrification times. DDS, diamino diphenyl sulfone; DGEBA, diglycidyl ether of bisphenol-A; MCDEA, 4'-Methylene bis [3-chloro 2,6-diethylaniline]. Source: *Reprinted with permission from E.G. Reydet, C.C. Riccardi, H. Sautereau, P.J. Pascault, Macromolecule 28 (1995) 7599.* © *1995, American Chemical Society Publishers.*

chainwise polymerization, comprising initiation by Lewis bases, propagation and termination, or chain transfer steps. Some of the postulated reactions are shown in Fig. 1.24. Initiation involves the ring-opening of an epoxy monomer by a tertiary amine (the most widely used Lewis base) yielding a zwitterion that contains a quaternary nitrogen cation and an active anion. The anion attacks an anhydride group at a very fast rate, leading to a species bearing a carboxylate anion as the active site. This ester is considered to be the initiator of chain wise polymerization. The amount of Lewis base initiators determines the number of active sites. The initiating species then reacts with an epoxy group leading to an alkoxide species, which again attacks a cyclic anhydride. Propagation occurs through the

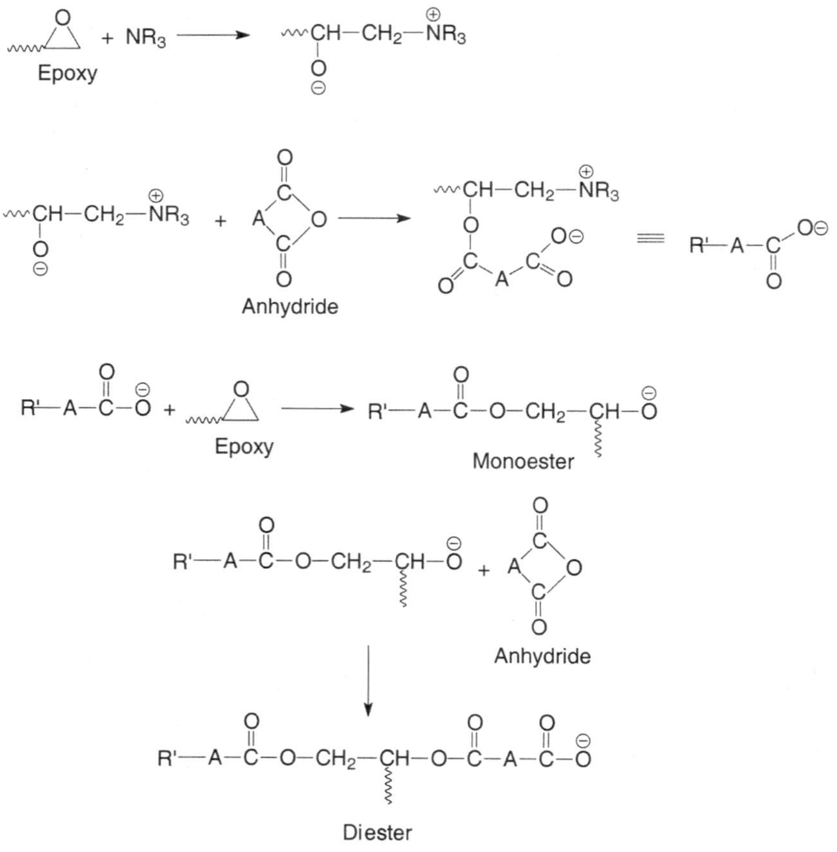

A is substituent to anhydride, A is $CH_2{=}CH_2$ for maleic anhydride

Figure 1.24
Mechanism of curing of epoxy using an anhydride.

alternating copolymerization of epoxy and acid anhydride groups. The strictly alternating copolymerization yielding polyesters was confirmed by several studies [124].

1.5.3 Modified epoxy systems

Epoxy resin offers a very wide scope of modification due to the presence of epoxide rings, which are capable of reacting with a variety of functional groups. Extensive works have been carried out by the author's group and others related to modification (chemical and physical) of epoxy resin for improvement of properties like toughness, water absorption etc., which will be discussed in Chapter 3. In this section, some of the modification strategies for structural and functional applications will be discussed.

1.5.3.1 Latent cure epoxy

Curing agents play a crucial role in determining fundamental curing reaction mechanisms, curing conditions, processability, pot life, cured network structures, end-use properties, and ultimate application of thermosets like epoxy. In the previous sections, we have discussed various curing agents of epoxy. Generally, curing epoxy at room temperature is a slow process and takes a long time for the cure reaction to complete. In many cases, heating to a higher temperature depending on the reactivity of the curing agents is needed. Aliphatic amines, thiols, and imidazoles are capable of converting epoxy into the cured network even at an ambient temperature due to their higher reactivity. Hence, such materials are to be stored separately and mix immediately before use, which not only pollute the environment but also strongly affects the processability, production efficiency, and applications of the related products. Using these materials, it is not possible to make one-component adhesives or prepregs. Single component adhesives have attracted tremendous attention in recent years, due to their obvious advantages over multicomponent adhesives. For example, one-component system does not require real-time preparation and therefore errors related to the measurement of ingredients and uneven mixing can be avoided. Another type of curing agents like anhydride and phenols requires a very high temperature for curing and obviously associated with processing difficulty [124].

This problem associated with the curing of thermoset like epoxy can be resolved by using a latent curing agent [125–128]. The use of latent curing agent is considered as an effective approach for this purpose because it not only offers good storage stability when used in a prepreg or one-component epoxy systems but undergo curing effectively under application of stimuli like heat, light, etc. The principle is that the curing agent has to be designed in such a way that it does not react with epoxy at room temperature and the reaction starts only at an elevated temperature. The various methods, generally adopted for making latent cure epoxy will be discussed below.

Dycynamide (DICY), organic hydrazines are widely used as latent curing agents. DICY is a solid powder with limited solubility in epoxides at room temperature. Since the DICY is not miscible at room temperature, it cannot react with epoxy under ambient conditions. Therefore it offers long storage stability at room temperature when used as a one-component product. At higher temperatures, it becomes soluble in epoxy and undergoes the curing reaction. Epoxy/DICY mixture exhibits excellent processability and generates a cured network with wonderful mechanical and electrical properties. Epoxy/DICY-based one-component systems are extensively used in laminates, prepregs, coatings, and adhesives [129–131]. However, a major limitation of DICY-based epoxy system is that it requires a high curing temperature, which can be as high as 150°C. The curing temperature can be reduced to 120°C by modifying the dicyandiamide with an aromatic amine in acidic conditions [132].

Imidazoles are widely used as an accelerator. It is solid at room temperature and therefore shows no catalytic activity at room temperature and converts to a liquid state and starts showing catalytic activity when the temperature reaches 90°C [133–135]. Imidazoles can also destroy the packing regularity of phenols leading to the reduction of the melting point of the phenol-based curing agents. Consequently, it can not only maintain stability at room temperature but also reduce the curing temperature of an epoxy/phenol system. Note that phenols undergo curing reaction with epoxy at a very high temperature close to 200°C, due to their high melting point (high crystallinity), which limits its commercial utilization. Imidazoles reduce the melting point of phenols by destroying their packing regularity and enhance their reactivity. The storage stability of an epoxy/DICY/Imidazole system can be further improved by introducing substituents to imidazole groups. The modified imidazole-based compounds tend to release imidazole at high temperatures to rapidly cure the epoxy, such as metal cation–imidazole salt complex [136], lanthanide–imidazole [137], and imidazolium salts [138]. The derivatives of 2-(2-hydroxyphenyl) imidazoles offer restrained nucleophilicity toward epoxy groups due to locking of the phenolic hydroxyl groups and the nitrogen atom of the imidazole ring with the help of hydrogen-bonding interaction [139,140]. This H-bonding is broken thermally around 150°C where the active imidazole initiates the curing reaction. Xu et al. [141,142] reported a feasible approach by chemically modifying imidazole with typical flame-retardant groups to develop two kinds of 1-substituted phosphorus-containing imidazole derivatives which on curing with epoxy produce a flame-retardant network. Modification of imidazole with 9,10-dihydro-9,10-oxa-10-phosphaphenanthrene-10-oxide (DOPO)-based acid suppresses their high reactivity with epoxy groups. At the same time, the efficient flame-retardant groups incorporated into epoxy systems tend to provide intrinsic flame retardance property. In addition, the curing agent offers long pot life and fast curing.

Another feasible and economic method to achieve latent curing of epoxy with a highly reactive curing agent is "microencapsulation." In this process, the curing agent is embedded in a wall material so that the curing agent is physically separated from the epoxy resin. Hence, no curing takes place during storage. Once the capsule breaks or the cell wall dissolves, the curing agent comes in contact with epoxy, and the curing reaction starts. Thus the curing reaction of epoxy resin can be controlled by the breakage or permeability of the wall material [143,144]. A lot of advancement has taken place in the last few decades on microencapsulation technology and various methods of preparation of microcapsules have been reported which include interfacial polymerization, in situ polymerization, spray drying, and solvent evaporation, etc. For example, a microcapsule-type latent curing agent was prepared by solvent evaporation method with diamino diphenyl methane (DDM) as the core material and poly (methyl methacrylate) (PMMA) as the wall material [145]. Imidazole-based curing agent was encapsulated by interfacial polymerization of triethanolamine and diphenylmethane diisocyanate (MDI). For a highly reactive system, interfacial polymerization is more effective than the solvent evaporation method as the latter requires a relatively short time for the solidification of the microcapsules. Imidazole and its derivatives like 2-methyl imidazole-based microcapsules have been prepared by spray drying and solvent evaporation as well [146–148].

Recently, Xue et al. [149] reported a novel microcapsule-type latent curing agent, which was obtained by encapsulating 1-benzyl-2-methylimidazole with double-shell material. Polydopamine (PDA) was deposited on the surface of the primary microcapsules made up of copolymers of PMMA and n-butyl acrylate to improve the shelf life of the one-component microcapsule–epoxy system. The solvent stabilities of double-shell microcapsules in toluene, acetone, and dichloromethane (DCM) were significantly improved. The double-shell microcapsules-based latent curing agent offers longer pot life at room temperature (>4 months) than the single wall microcapsule-based one. Most importantly, the mixture of epoxy and double-shell microcapsules could be cured in 1 h at 120°C.

The third approach for achieving latency in the epoxy system is to use ionic solids [150–152]. Ionic liquids (ILs) are a family of organic molten salts having melting points below 100°C. An epoxy formulation containing 3 phr (parts per 100 g of resin) of 1-alkyl-3-methyl imidazolium (BMIM) exhibited a pot life of 50–60 days at an ambient temperature where counteranions were chloride or tetrafluoroborate [153,154]. The pot life depends on the nature of anions. When the chloride or tetrafluoroborate is replaced by dicyanamide anion, the pot life of the related epoxy formulation reduces to 30 days. An epoxy composition containing 3 phr of BMIM containing thiocyanate anion [155] shows a pot life of 3 days only. Thus depending on the requirement of latency, the ionic solid can be selected. The advantage of ionic solid-based latent curing agents over others is that in addition to imparting latency, it offers low volatility, high miscibility with epoxy resin and generates a network polymer with improved mechanical properties, flame retardancy, and self-healing of abrasion damage property [156–159]. The basicity (higher pK_a) of ionic liquids determines their reactivity with epoxy resins; the higher the basicity of ionic liquids, the higher the IL reactivity with epoxy. Generally, amine-functionalized ionic liquids have high reactivity with epoxy resins. The functional design of ionic-liquid curing agents may significantly improve the overall performance of epoxy resins.

Imidazolium-type ILs are widely used due to their higher reactivity. Phosphonium-based ILs have also been used [160,161] as a crosslinker for epoxy either independently or in combination with other conventional curing agents. Phosphonium salts are preferable from availability and cost point of view. Maka et al. [162] compared the imidazolium and phosphonium-based ILs. They have used two ionic liquids, namely, 1-ethyl-3-methylimidazolium dicyanamide, [EMIM]N(CN)$_2$, and phosphonium derivative, that is, trihexyltetra-decylphosphonium dicyanamide, [THTDP]N(CN)$_2$. The chemical structures of

Table 1.11: Chemical structure of ionic liquid.

Name	Structure
1-Ethyl-3-methylimidazolium dicyanamide, [EMIM]N(CN)$_2$	(imidazolium cation with ethyl and methyl substituents; dicyanamide anion NC–N–CN)
Trihexyltetra-decylphosphonium dicyanamide, [THTDP]N(CN)$_2$	H$_3$C(H$_2$C)$_5$–P$^+$[(CH$_2$)$_5$CH$_3$]$_2$–(CH$_2$)$_{13}$CH$_3$; NC–N$^-$–CN

Source: *Reprinted with permission from H. Maka, T. Spychaj, M. Zenker, J. Ind. Eng. Chem. 216 (2015) 359–368. © 2015, Elsevier Publishers.*

[EMIM]N(CN)$_2$ and [THTDP]N(CN)$_2$ are shown in Table 1.11. It was observed that imidazolium-based IL shows higher pot life than that of phosphonium-based ILs. If we look at the thermal stability (characterized by TGA: temperature corresponding to 5% weight loss, i.e., $T_{5\%}$) of the epoxy networks cured with two different ILs, we can see that epoxy network cured with [EMIM]N(CN)$_2$ shows higher thermal stability ($T_{5\%} = 395°C$) compared to the same cured with [THTDP]N(CN)$_2$ ($T_{5\%} = 377°C$). Generally, phosphonium salt-based ionic solid, due to its inherent higher thermal stability, offers better thermal stability for the related epoxy network compared to imidazolium salt-based one. The reverse trend observed in the present case can be explained by considering the thermal decomposition of imidazolium salt (as shown in Fig. 1.25), which is associated with the formation of isocyanurate rings via condensation of cyanide groups. The isocyanurate rings present in the network contribute toward higher thermal stability of the epoxy network cured with [EMIM]N(CN)$_2$. However, the epoxy materials cured with a phosphonium-based IL showed higher thermomechanical properties and transparency (85%) in comparison to that cured with imidazolium-based IL (black opaque) [162].

It was reported that the introduction of bifunctional groups in ILs can decrease the amount of curing agents and improve the compatibility and reactivity between the epoxy resin and hardeners in case of imidazolium-based hardeners [160]. Recently, Jiang et al. [163] reported a series of dual-functionalized ionic liquids of 1-(3-amino-3-oxopropyl)-3-propynyl-imidazolium, 1-(3-amino-3-oxopropyl)-3-(2-amino-2-oxoethyl)-imidazolium, and 1-allyl-3-(2-amino-2-oxoethyl)-imidazolium combined with dicyanamide [N(CN)$_2$]$^-$ and bis-(trifluoromethylsulfonyl)imide [NTf$_2$]$^-$ anions. The melting point varies with the type

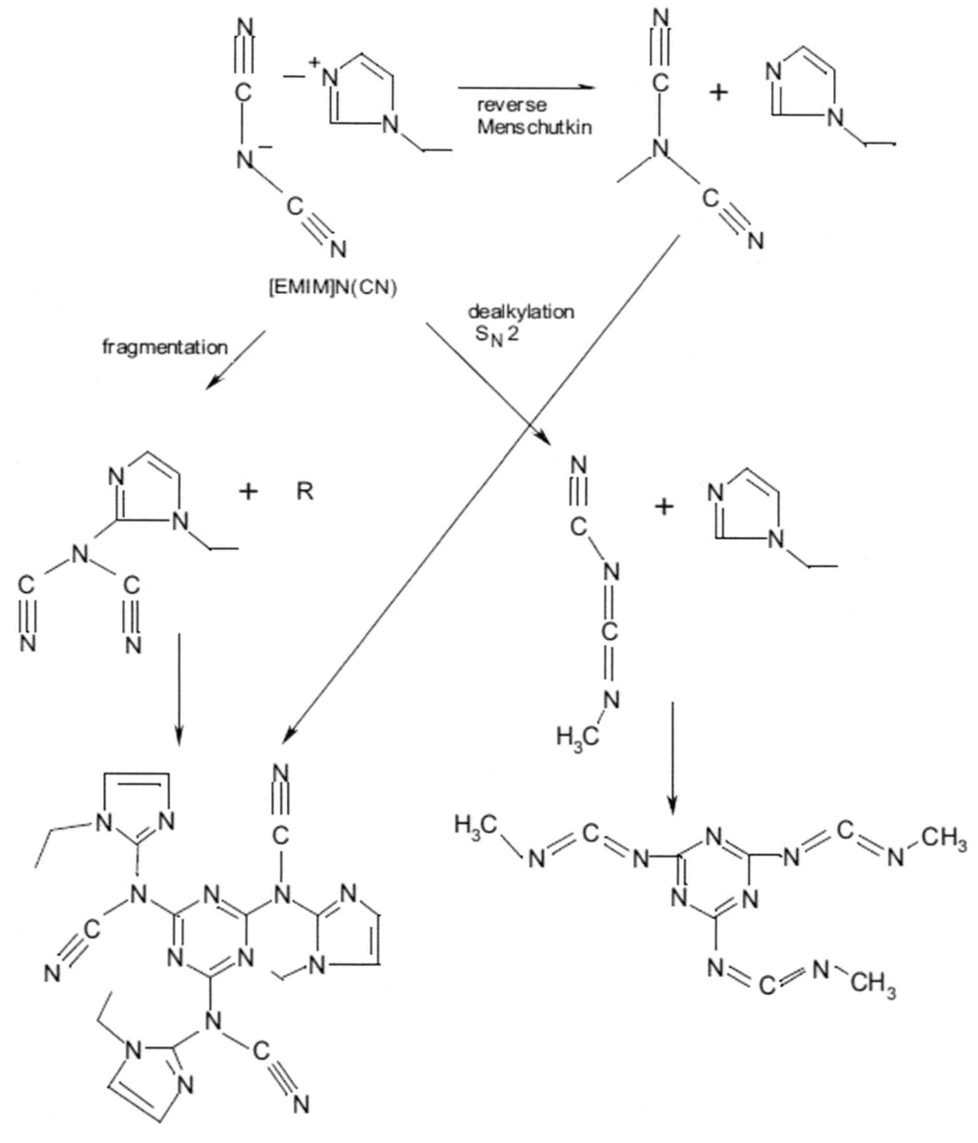

Figure 1.25
Scheme of possible mechanism of thermal decomposition of 1-ethyl-3-methylimidazolium dicyanamide with creation of isocyanurate rings. Source: *Reprinted with permission from H. Maka, T. Spychaj, M. Zenker, J. Ind. Eng. Chem. 31 (2015) 192. © 2015, Elsevier Publishers.*

of anions or cations. For example, under the same cation, the melting points of the ILs change according to the following order: $Cl^- > N(CN)_2^- > NTf_2^-$. The reaction scheme for synthesis dual-functional IL of 1-allyl-3-(2-amino-2-oxoethyl)-imidazolium is shown in Fig. 1.26. The ionic liquids were used to cure an epoxy resin, namely, diglycidyl-4,5-epoxy-

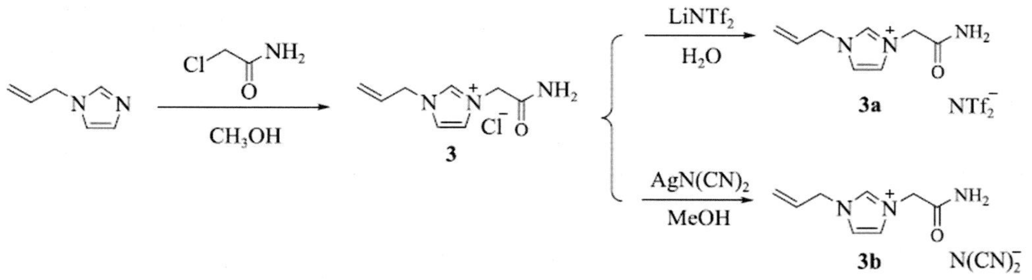

Figure 1.26
Synthesis of dual-functionalized ionic solid: 1-R-3-(2-amino-2-oxoethyl) imidazolium (R = −CH$_2$-CH=CH$_2$). Source: *Reprinted with permission from Z. Jiang, Q. Wang, L. Liu, Y. Zhang, F. Du, A. Pang, Ind. Eng. Chem. Res. 59 (2020) 3024.* © 2020, American Chemical Society Publishers.

cyclohexane-1,2-dicarboxylate (TDE-85). Detailed spectroscopic analysis indicate that [NTf$_2$]$^−$ anions do not participate in the curing reaction. The reaction occurs in three stages similar to the process of traditional organic amine curing with the epoxy resin. In the first stage (which takes place at low temperature), the primary amine groups first react with an oxirane ring to form a −OH group and secondary amine groups. In the second stage, the secondary amine group participates in the curing reaction. The third stage involves the hydroxyl termination reaction. The thermosets obtained from the bifunctional ionic liquid and TDE-85 exhibit better heat resistance with a glass transition temperature of 187°C − 212°C and a decomposition temperature at 10% weight loss of 282°C − 357°C.

1.5.3.2 Chiral epoxy

Optically active materials are those that show a positive or negative value of specific rotation. Optical rotation is measured by using a plane-polarized light with the help of a polarimeter. Epoxy resins are normally optically inactive. However, it has been demonstrated that by using chiral monomer as starting material [164] or by introducing chiral unit into a commercially available epoxy through chemical modification [165], it is possible to make optically active resin. Ratna and coworker [165] investigated the modification of epoxy with optically active L-leucine. L-leucine was modified by replacing the −NH$_2$ group with −Cl to get S-2-chloro, 4-methyl pentanoic acid (CMPA) using diazonium salt route [165]. The reaction scheme for modification is shown in Fig. 1.27. Specific rotation of CMPA is −27.32 degrees. When CMPA is attached to epoxy through a reaction between −COOH group of CMPA and the epoxide group of the epoxy resin the modified resin shows optical activity both in the uncured and cured state as presented in Table 1.12.

1.5.3.3 Bio-based epoxy

Rapid consumption of fossil-based resources (which are not renewable type) puts utilization of renewable feedstocks, especially biomasses, as sustainable fuel on top priority. If we

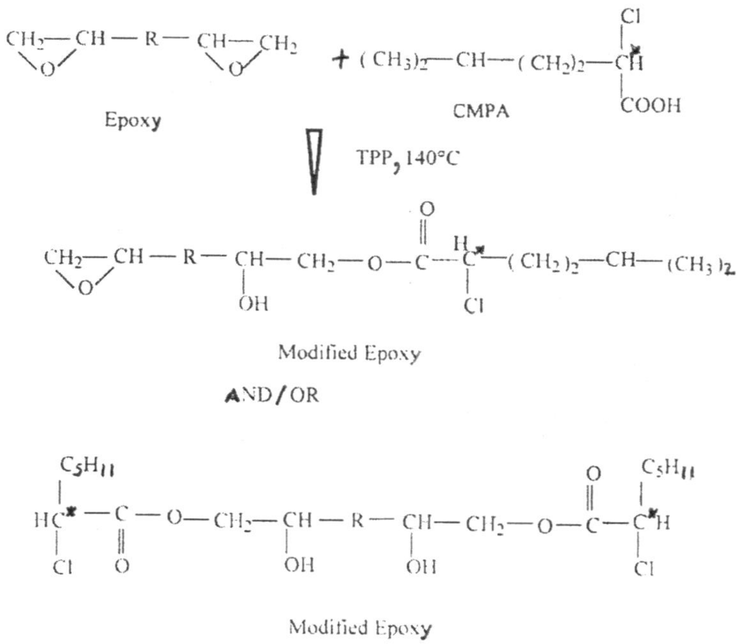

Figure 1.27
Reaction scheme for synthesis of chiral epoxy by chemical modification of DGEBA resin. Source: Reprinted with permission from A.B. Samui, D. Ratna. J. Chavan, P.C. Deb, J. Appl. Polym. Sci. 92 (2004) 1604. © 2004, John Wiley and Sons Publishers.

Table 1.12: Chiral properties of CMPA-modified epoxy.

Sr. No.	Ratio of epoxy and CMPA mole/mole	Specific rotation of solution $[\alpha]_D^a$ (°)	Optical rotation of film[a] α^b
1	0.055/0	−0.004	−0.002
2	0.055/0.006	−1.52	−0.017
3	0.055/0.009	−1.61	−0.020
4	0.055/0.012	−1.70	−0.027
5	0.055/0.015	−2.23	−0.030
6	0.055/0.018	−4.34	—
7	0.055/0.026	−5.44	—
8	0.055/0.031	−7.61	—
9	0.055/0.037	−7.71	—
10	Pure CMPA	−27.32	—

[a]Film thickness was maintained at 0.7 mm.
[b]Measured by polarimeter at 25°C.
Source: Reprinted with permission from A.B. Samui, D. Ratna. J. Chavan, P.C. Deb, J. Appl. Polym. Sci., 2004, 92, 1604. ©1604 John Wiley and Sons Publishers.

look at the production of epoxy resins worldwide, we can find approximately 90% of epoxy resins (DGEBA), are produced from petroleum-based bisphenol-A (BPA) and epichlorohydrin. BPA is synthesized from unrenewable petroleum-based phenol and acetone, and epichlorohydrin is mainly synthesized from the oxidation of allyl chloride. A major portion of allyl chloride comes from renewable glycerol. Hence about 70% of the molar mass of DGEBA is strongly dependent on fossil sources. Moreover, the use of BPA has negative impacts on human health and the environment. It has been proven that it can cause estrogen-like long-lasting effects on living organisms and acts as an endocrine disruptor. Therefore the use of BPA in the food packaging industry and food-related materials has been forbidden in some advanced countries [166,167]. Also, ECD used for the production of epoxy is known to be carcinogenic. Hence, use of ECD for the production of epoxy resin and traces of unreacted ECD present in the resin may pose an issue of health hazard. In view of the issues discussed above, a considerable amount of research and development activities (dedicated to replacing DGEBA with more environment-friendly alternatives) have been carried out in the last decade [168].

Epoxidized plant oils and fatty acids (e.g., soybean oil, canola oil, linseed oil) have been extensively utilized to make sustainable epoxy resins [169–171]. Traditionally, epoxidized plant oils are used as plasticizers and stabilizers forbrittle plastic-like poly (vinyl chloride) and as a binder in surface coating formulations [172]. On an industrial scale, the unsaturated fatty compounds are converted into epoxidized plant oils by an epoxidation method using the in situ performic acid [173]. However, the use of strong mineral acid makes the process nonselective, leading to the formation of undesirable by-products, such as ring-opened epoxides [174]. To address this issue, different catalysts have been employed, which include transition metal complexes, ion exchange resins, crown ethers and enzymes [175,176]. Epoxidation at high pressure has also been investigated [177]. However, none of these methods has shown viability for industrial production. Among epoxidized plant oils, epoxidized soybean oilsare massively produced and used for epoxy resin applications [178,179]. In practical applications, they cannot serve as an epoxy matrix alone, because of their inferior thermomechanical properties (e.g., low glass temperature, modulus) compared to the commercial DGEBA resins. The inferior thermomechanical properties of epoxidized plant oils originate from their long and highly flexible aliphatic skeleton. To improve the mechanical properties of ESO-based polymers, ESO is generally mixed with fossil-based resins to produce thermoset networks with a minimal nonrenewable components. For example, ESO was polymerized with maleate half esters of oil-soluble resoles, p-tertiary butyl phenol, and p-nonyl phenol to create tough thermosets [180]. The carboxylic acid groups of the maleic ester react with ESO through a ring-opening reaction. ESO and the biodegradable polymer polycaprolactone (PCL) were crosslinked in presence of an acid catalyst to produce a network that shows shape memory property and the properties can be tailored as a function of the concentration of PCL [181].

Natural epoxidized plant oil that contains 60%–65% vernolic acid is available in various Vernoniaspecies [182]. For example, cis-9,10-epoxy-18-hydroxy octadecanoic acid is a naturally occurring epoxidized monomer present in the outer bark of birch to about 100 g kg^{-1}. This epoxy acid can be extracted at a high yield from alkali hydrolysis of outer bark after acidification and precipitation. Cis-9,10-epoxy-18-hydroxyoctadecanoic acid can undergo Lipase-catalyzed condensation polymerization and produce epoxidized oligomers selectively [183]. Lignin and vanillin (see Fig. 1.15 for chemical structure) have been used as the raw materials to synthesize epoxy resin and the resulted epoxy resin from vanillin demonstrates good thermal and mechanical properties due to the presence of the aromatic ring in its structure [101,184]. Such bio-based epoxies suffer from the problem of slow curing. Nikafshaar and coworker reported [185] vanillin-based epoxy cured with Epicure F 205. They have reported that curing reaction can be significantly accelerated by using calcium nitrate solution as an accelerator. This is also associated with a marginal increase in the mechanical properties of the cured resin. The use of various materials produced by extraction from lignin for making bio-based epoxy has been reviewed [186,187]. Such a lignin-based epoxy resin was reported for potential application in epoxy asphalt [188]. Thus lignin-based epoxy resins are currently the subject of innovative and intense investigations [189,190].

Zhang et al. [191] reported a eugenol-based epoxy resin with a bio-based content of 62.7 wt.% via epoxidation of eugenol's double bond, followed by O-glycidylation of its hydroxyl group. The bio-based resin cured with an acid anhydride showed a glass transition temperature of 114°C. With the objective to enhance the biomass content in the bio-based epoxy resin, Wan et al. [192,193] used a strategy where two eugenol moieties were bridged by the aliphatic ether or aromatic ester bonds, to yield bifunctional eugenol-based epoxy resins having biomass content of more than 70%. The bio-based resin cured with 3,3'-diamino diphenyl sulfone (33DDS) exhibits self-extinguishing characteristics in a vertical burning test, indicatingflame retardant behavior. However, the melting temperature of the bifunctional resin was rather high (>150°C), and the glass temperature of the cured epoxy product was still lower than that of the DGEBA system. As a solution to this problem, the same research group also reported [194] a facile strategy to realize a novel trifunctional eugenol-based epoxy monomer (TEU-EP). The reaction scheme is shown in Fig. 1.28. In the first step, cyanuric chloride is reacted with the hydroxyl group of eugenol in a high efficient phase-transfer catalytic reaction system to yield an intermediate bearing three allyl groups (TEU). Then, the allyl groups of TEU were epoxidized with chloroperoxy benzoic acid in large excess to yield triazine-bridged eugenol-based epoxy monomer (TEU-EP) under a mild reaction condition (CH$_2$Cl$_2$/room temperature). The Young's modulus, hardness, and creep resistance of the cured epoxy networks are analyzed by nanoindentation technique. The results are shown in Fig. 1.29. We can see that the cured TEU-EP/33DDS exhibits significantly improved mechanical indentation resistance, increased Young's

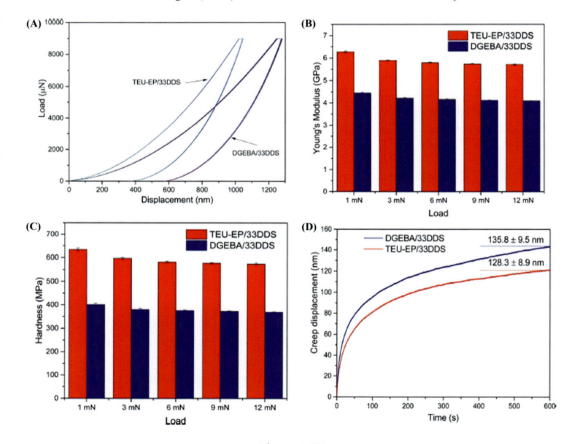

Figure 1.28
Synthesis of the triazine-bridged eugenol-based epoxy resin (TEU-EP). Source: *Reprinted with permission from J. Wan, J. Zhao, B. Gan, Cheng Li, J. Molina-Aldareguia, Y. Zhao, Y. Pan, De-Yi Wang, ACS Sustainable Chem. Eng. 4 (2016) 2869. © 2016, American Chemical Society Publishers.*

Figure 1.29
Typical results from the nanoindentation measurements. (A) Load − displacement curves (9000 μN for 5 s). The different colors represent the different intents. Comparison of E (B) and H (C) for loading of 1, 3, 6, 9, and 12 mN. (D) Creep displacement against time up to 600 s (9 mN). DGEBA/33DDS and TEU-EP/33DDS show an accumulated creep displacement of 135.8 ± 9.5 and 128.3 ± 8.9 nm, respectively. Source: *Reprinted with permission from J. Wan, J. Zhao, B. Gan, Cheng Li, J. Molina-Aldareguia, Y. Zhao, Y. Pan, De-Yi Wang, ACS Sustainable Chem. Eng. 4 (2016) 2869. © 2016, American Chemical Society Publishers.*

modulus, surface hardness, and creep resistance compared to DGEBA/33DDS system. The improved creep resistance can be attributed to a more rigid molecular structure of TEU-EP, which leads to less plastic distortion of the cured epoxy network. Francois et al. [195] reported diepoxy derivative of iso-eugenol, namely, 2-[3-methoxy-4-(2-oxiranylmethoxy) phenyl]-3-methyl oxirane (DG E-isoEu). This new bio-based monomer was prepared according to a two-step procedure, from isoeugenol (catalytically fragmented from lignin). The cured DGE-isoEu shows similar property to eugenol-based epoxy (DGE-Eu) discussed above because of their similar chemical structure.

We have discussed the above bio-based epoxy networks using bio-based resins only. In order to increase biomass content in the epoxy network, it is necessary to make bio-based hardener as well. This is especially important for anhydride-cured epoxy systems. Anhydride-based hardeners are required to be added with epoxy to a higher proportion compared to amine hardeners. In some anhydride-based epoxy formulations, the amount of anhydride is even higher than the epoxy resin. Therefore, if bio-based anhydride is not used, the biomass content of the related product cannot be more than 50%. Anhydride-based hardener offers lower viscosity of the resin system and better thermomechanical properties of cured resin compared to amine-based hardener. Anhydride-based hardeners are also preferable to amine hardeners in terms of toxicity point of view and more suitable for bio-based resin systems. Bio-based amine and anhydride hardeners have been reviewed and published in the literature [196–198]. For example, the aminated grapeseed oil (AGSO) is successfully synthesized via UV initiated thiol-ene coupling and used as a curing agent for bio-based epoxies [199,200]. Rosin-based anhydrides were made through a series of esterification and Diels–Alder reactions [201,202]. Tannic acid has been investigated for use as a bio-based curing agent with epoxy resins based upon the natural aliphatic polyols such as glycerol and sorbitol [203].

Plasseraud and coworker [204] investigated DGE-Eu and DGE-isoEU resin networks cured with six different bio-based anhydride hardeners, prepared using the procedure reported by Robert et al. [205]. The chemical structures of the bio-based anhydride hardener, namely, diphenic anhydride (DPA), glutaric anhydride (GA), phthalic anhydride (PA), camphoric anhydride (CA), succinic anhydride (SA), and itaconic anhydride (IA) are shown in Fig. 1.30. The measurement of T_g of various networks using DSC indicates that materials obtained using GA, SA, or IA hardeners exhibit lower glass transition temperatures compared to the same obtained from DPA and CA (80°C–100°C). This can be explained by considering the chemical structure of the polymer chain, the architecture of the polymer and the length and degree of cross-linking of the chains. GA, SA, or IA generates a flexible linear carbon chain when the anhydride group breaks as a result of curing. This explains the lower T_g of the related cured networks. On the other hand, when DPA, PA, and CA react with epoxy the networks with bulky and rigid side groups are formed. Therefore the epoxy networks cured with DPA, PA, or CA show T_g comparable to an amine-cured DGEBA

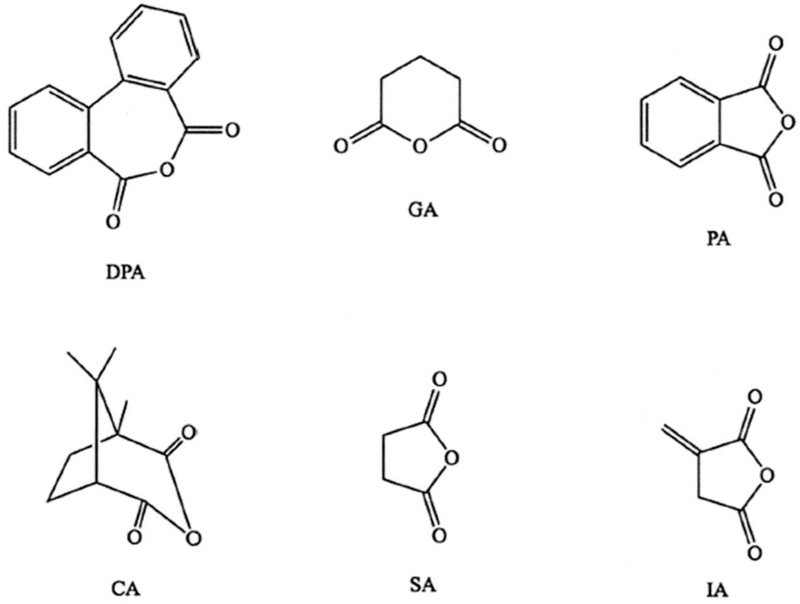

Figure 1.30
The chemical structures of the bio-based anhydride hardner, namely, DPA, GA, phthalic anhydride (PA), CA, succinic anhydride (SA), and IA. *CA*, camphoric anhydride; *DPA*, diphenic anhydride; *GA*, glutaric anhydride; *IA*, itaconic anhydride. Source: *Reprinted with permission from C. François, S. Pourchet, G. Boni, S. Rautiainen, J. Samec, L. Fournier, C. Robert, C.M. Thomas, S. Fontaine, Y. Gaillard, V. Placet, L. Plasseraud, Competus Rendus Chimie 20 (2017) 1006© 2017, Elsevier Publishers.*

(100°C–150°C). If we compare the thermal properties of the developed bio-based resin with the same for commercial bio-based resins, for example, Epobiox Sandtech (T_g = 80°C) or Greenpoxy Sicomin (T_g = 85°C), then the combination of bio-based resin and anhydride curing agents proved to be certainly very useful for biocomposite applications because of their significantly higher T_g compared to commercial bio-based resins. Barancini et al. [206] published a detailed review on bio-based epoxy resins and curing agents with an illustrative guide covering their properties and applications [207–233] as presented in Table 1.13.

1.5.4 Application of epoxy resin

Epoxy resins are widely used as adhesives, surface coatings, encapsulates, casting materials. They are used in industrial tooling applications to produce molds, master models, laminates, castings, fixtures, and other industrial production aids. This "plastic tooling" replaces metal, wood, and other traditional materials and generally improves the efficiency and either lowers the overall cost or shortens the lead-time for many industrial processes. Epoxy resins

Table 1.13: An illustrative guide to bio-based epoxy resins and epoxy curing agents covered along with highlighted properties and potential applications of the cured epoxy resin.

Epoxy resin	Curing agents, initiators, and accelerators	Properties	Potential applications	References
Epoxidized soybean oil	• Benzylsulfonium hexafluoroantimonate derivative cationic catalyst	• Max stress of 12.3 MPa • Recovery percentage at fifth cycle >90%	Shape memory polymers	[181]
Epoxidized soybean oil	• Boron frifluoride diethyl etherate catalyst	• Water vapor permeabilities of 4.5×10^{-8} g m^{-1} s^{-1} Pa^{-1}	Water-resistant paper composites	[207]
Epoxidized linseed oil	• Methylhexahydrophthalic anhydride • Benzophenone-3,3',4,4'-tetracarboxylic dianhydride	• 68°C < T_g (based on E″ peaks) < 134°C	Composites, adhesives, and laminates	[208]
Epoxidized canola oil	• Phthalic anhydride	• 13 min < Gelation time at 155°C < 28 min • 3.6°C < T_g (based on Tan & peaks) < 37.8°C	Cellulosic particle and fiberboards	[209]
Epoxidized karanja oil	• Citric acid • Tartaric acid	• 280°C < $T_{d5\%}$ (TGA in N$_2$) < 295°C • 109°C < T_g (based on differential scanning calorimetry [DSC] thermograms) < 112.7°C	Paper Coating/Lamination And mosquito deterrents	[210]
Isosorbide diepoxy (DGEDAS)	• Diaryl iodonium salt (1250)	• 4.5 MPa < TS < 10.6 MPa • 210°C < $T_{d5\%}$ (TGA in argon) < 300°C • 81°C < T_g (based on Tan & peaks) < 125°C	Coatings	[211]
Diglycidyl ethers of Isosorbide (DGEI)	• Isosorbide diamine (ISODA) hardeners	• 34°C < T_g (based on E″ peaks) < 79°C • 41 MPa < TS < 62 MPa • 1168 MPa < E bend < 4027 MPa	Epoxy resins in food industry	[212]
Furan diepoxy of 2,5-Bis(hydroxymethyl)-Furan (BHMF)	• IRGACURE 250 • Triphenylsulfonium hexafluoroantimonate salt iniators	• Joint tensile shear Strength of 3.5 MPa	Adhesives	[213]
Furan diepoxy of 2,5-Bis[(2-oxiranylme-Thoxy)- methyl]-furan (BOF)	• 4,4′—methylene bis-cyclohexanamine (PACM) • Diethyl toluene diamine (EPLKURE W)	• 71°C < T_g (based on E″ peaks) < 133°C	Polymer composites, Insulation materials, Coatings and adhesives	[214]

(Continued)

Table 1.13: (Continued)

Epoxy resin	Curing agents, initiators, and accelerators	Properties	Potential applications	References
Diglycidyl ester of 2,5-furandicarbox-Ylic acid (DGF)	• Methylhexahydrophthalic angydride (MHHPA) Poly(propylene glycol) bis(2-aminopropyl ether)	• °C < T_g (based on tan & peaks) < 152°C • 68 MPa < TS < 84 MPa • 75 MPa < FS < 96 MPa	Resins in aerospace-Coatings, adhesives And composites	[215]
Epoxidized catechin	• Methanol soluble lignin • Isophorone diamine (IPD)	• T_g (based on DSC thermograms) of 178°C • 1500 MPa < E' at 30°C < 2340 MPa • 142°C < T_g (based on tan & peaks) < 179°C	DGEBA replacements In coatings, laminates, Adhesives, Floorings, pavings, Composites, and electronics	[216,217]
Epoxidized cardanol	• IPD • Jeffamine D400 • Diamine • Jeffamine T403	• 15°C < T_g (based on E″ peaks) < 50°C • 37°C < T_g (based on DSC thermograms) < 60°C • 7 MPa < E' at 20°C < 1218 MPa • 294°C < $T_{d30\%}$ (TGA in N_2) < 366°C	Coatings	[218,219]
Terpene-maleic ester-type epoxy (TME)	• Hexamethylene diisocyanate (HDI) tripolymer	• T_gs (based on DSC thermograms) ~400°C–420°C	Waterborne polyurethane/epoxy resin	[220]
Glycidyl ether of limo-Nene alkylated Naphthol-formaldehyde	• Dicyanodiamide • Bisphenol-A-formaldehyde novo-Lac resin	• 171°C < T_g (based on DSC thermograms) <182°C • 332°C < $T_{d10\%}$ (TGA in air) < 343°C	Composite coatings DGEBA replacements In electronics	[221]
Epoxidized rosin Derivatives (tetra-glycidyl dimaleopi-maryl ketone, TGK)	• Rosin-based crosslinker • Dirosin-maleic anhydride imidodicarboxylic acid • P-phenylene diamine	• 0.71 GPa < E' at 25°C < 1.71 GPa • 122°C < T_g (based on tan & peaks) < 140°C	Replacements for BPA in coatings	[222]
Liquid epoxidized natural rubber (LENR)	• Epikure d230	• Impact strength of 7.6 kJ m^{-2} • FS of 58 MPa • E_{bend} of 3.25 GPa	Fiber-reinforced Epoxy composites in the automotive industry	[223]

Epoxy	Curing agent	Properties	Application	Ref.
Cedar, eucalyptus And bamboo-Derived epoxies	• Lignin-based curing agent • Commercial curing agent TD2131	• 259°C < $T_{d5\%}$ < 296°C (TGA in N_2)	Electronics	[224]
Epoxidized versions of de-polymerized organosolv lignin (DOL) and de-polymerized kraft lignin (DKL)	• 4,4-diaminodiphenyl methane (DDM) • Diethylenetriamine (DETA)	• 376°C < T_{max} < 416°C (TGA in N_2)	Adhesives composites Coatings and electronics	[225]
Epoxidized cardanol	• Cardanol-based amine	• 19°C < T_g < 30°C (based on DSC thermograms) • 1.4 MPa < E' at 20°C < 350 MPa	Coating	[226]
Furan-based epoxy 2,5-bis[(2-oxiranyl-methoxy) methyl-furan (BOF)	• 5,5'-methylenedifur-furylamine (DFDA) • 5,5'-ethylidenedifur-fitylamine (CH3-DFDA)	• 62°C < T_g < 142°C (based on tan δ peaks)	Coatings, adhesives, and composites	[227]
Vanillin-based diglycidyl ether of methoxyhydroqui-none (DGMHQ)	• Bis(furfurylamine)-A	• 74°C < T_g < 98°C (based on DSC thermograms)	Marine, automobile, or food industries	[228]
Epoxidized sucrose Soyate (ESS)	• Water-soluble natural acids (citric acid, malic acid, tartaric acid, malonic acid oxalic and glutaric acid)	• 17°C < T_g < 96°C (based on tan δ peaks) • 1.82 MPa < E' at 25°C < 944 MPa	Coatings and films	[229]
DGEBA	• Soybean oil-derived polyacid hardener	• T_g (based on DSC thermogram) of −3°C • E' at ∼20°C of 0.59 MPa	Coatings and bindings	[230]
Epoxidized soybean oil	• Terpene-based anhydrides (TPAn)	• 59°C < T_g < 67°C (based on DSC thermograms)	Biocomposites	[231]
Epoxidized soybean oil	• Tannic acid	• 15 MPa < TS < 26 MPa • 58°C < T_g < 61°C (based on tan δ peaks)	Biocomposites	[232]
DGEBA	• Cardanol-derived penalkamine (PKA)	• TS of 56 MPa • 8.88% elongation • Impact strength of 13.92 kJ m^{-2}	Binders, coatings and paints	[233]

Source: Reprinted with permission from E.A. Baroncini, S.K. Yadav, G.R. Palmese, J.F. Stanzione, J. Appl. Polym. Sci. (2016) 44103 © 2016 John Wiley and Sons Publishers.

are extensively used as a binder for marine paints, which is required to protect the naval structure from the corrosive marine environment.

Polydimethyl siloxane (PDMS) modified epoxy has been reported to find application as foul release coating [234]. Biofilm formation on marine vessels leads to reduced speed and carrying capacity, resulting in increased propulsive power and fuel consumption, and at the same time accelerating corrosion [235–237]. FRCs cannot inhibit the attachment of organisms, but the interfacial bond between organisms and the coating surface is weak due to their low surface free energy so that the attached organisms can be readily removed by the water shear force coming from mechanical cleaning or the ship's navigation [234] as shown in Fig. 1.31. It was reported that 15% silicone modified epoxy coating showed 25%–35% release of the fouling organisms, while 50%–60% foul release behavior was observed for the 30% silicone modified samples.

Yu and coworker [238] reported silicone modified cycloaliphatic epoxy system for use LED encapsulating materials. They have used phenyl methyl silicone, which offers good solubility with cycloalipahtic epoxy. Incorporation of 10 wt.% of silicone resulted in higher heat and ultraviolet resistance due to higher bond energy of Si—O bond than C—O and C—C bond and lower water absorption property compared to pure epoxy due to hydrophobic nature of Si—O—Si linkages. These properties are achieved as a result of silicone modification without sacrificing the glass transition temperature.

Epoxy resins can be used as a matrix for making high-strength composites with glass, carbon, and Kevlar fibers because they are chemically compatible with most substrates and tend to wet surfaces very easily. Such composites are exclusively used for aerospace applications. The aerospace industries are dominated by epoxy resins due to its excellent processability and useful property of the network polymers. This is because epoxy composites can satisfy the technical requirement of structural material for civil and military and aerospace applications namely flooring panels, ducts, wings and vertical and horizontal stabilizers. Such composites are also used to produce lightweight vehicle frames, racing cars, musical instruments, and industrial components with high strength-to-weight ratio.

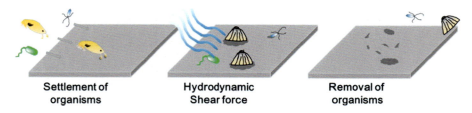

Figure 1.31
Schematic of fouling-release coatings. Source: *Reprinted with permission from P. Hu, Q. Xie, C. Ma, G. Zhang, Langmuir 36 (2020) 2170. © 2020, American Chemical Society Publishers.*

A major use of epoxy resins is in the electronics industry. They are extensively used in motors, generators, transformers, switchgear, bushings, and insulators. Epoxy resins are excellent electrical insulators and protect electrical components from short circuiting, dust and moisture. In the electronics industry, epoxy resins are the primary resin used in molding integrated circuits, transistors and hybrid circuits, and making printed circuit boards. Liquid crystalline epoxies have potential applications for electro-optics, acoustics and information technology.

1.6 Polyurethanes

Polyurethanes (PUs) are polymers containing urethane linkages and are available in both thermoplastic and thermosetting forms. In 1935, Otto Bayer and coworker [239] invented polyurethane as a product of the polyaddition reaction between a macroglycol and a diisocyanate. PUs are made in various forms such as castable elastomer, thermoplastic elastomer, and engineering thermoplastic. The precursors of standard polyurethane formulations are polyol, polyisocyanate, extender, and modifier (which are optionally used) [240,241].

1.6.1 Polyol

Polyols are hydroxyl-functionalized oligomers having a molecular weight in the range of 300–9000 g mole^{-1} and functionality in the range of 1–8 equivalent per mole. Linear and low functionality ($f = 2-3$ eq mole^{-1}) polyols such as glycerine ($f = 3$), dipropylene glycol ($f = 2$) generates flexible (low modulus) PU, whereas branched and high functionality ($f = 3-6$ eq mole^{-1}) polyols such as sorbitol ($f = 6$), Mannich bases ($f = 4$), sucrose ($f = 8$) lead to hard PU systems (high modulus). Polyols are often used as mixtures of molecules that are similar in nature but with different molecular weights. Their molecules possess different numbers of —OH groups. Therefore the hydroxyl content of a polyol is expressed in terms of average functionality. Low-molecular-weight polyols (a few hundred units) are used to make rigid Pus, whereas flexible PUs are obtained from high-molecular-weight polyols (around ten thousand and above units). Depending on the backbone structure, polyols are classified into three groups namely polyether polyol and polyester polyol and speciality polyols. Polyether polyols comprise more than 80% of the global PU. This is due to the fact that polyether polyols are much cheaper compared to polyester and speciality polyols. However, polyester polyols are used to produce PU with better abrasion, solvent resistance, and tear strength. Speciality polyols namely polysulfide polyols, polycaprolactone polyols, polycarbonate polyols, and polybutadiene polyols are utilized in the manufacturing of sealants, coating, and adhesives where the products have to withstand the harsh environment in industrial and marine applications [242,243].

Polyether polyols are prepared by an addition reaction of epoxide compound with a molecule with active hydrogen [244–246] or by ionic polymerization of alkylene oxide

[247–249]. Poly (tetramethylene ether) glycol is obtained from the polymerization of tetrahydrofuran and is mainly used for high-performance elastomers and coating applications [250,251]. Polyester polyols are produced by the esterification reaction of carboxylic acid and glycol. Unlike in chain-growth polymerization, the molecular weight of the product of a step-growth polymerization is highly sensitive to the stoichiometry of the reactants. Thus high-molecular-weight linear polyester polyols are prepared by using high purity acids and are used to produce PU with improved properties, which are not achievable by using a polyether polyol. Low-molecular-weight branched polyester polyols are prepared by glycolysis (transesterification) of recycled by-products with glycols. A list of various polyols used for the synthesis of PU resin is listed in Table 1.14. The chemical structures of some commonly used polyols are shown in Fig. 1.32.

Polyether polyols offer advantages in terms of hydrolytic stability, cost, viscosity plus flexibility, and disadvantages in terms of oxidative stability, modulus/strength, thermal instability, flammability. Aliphatic polyester polyols have advantages in terms of oxidative

Table 1.14: List of polyol and isocyanate compounds commonly used for synthesis of polyurethane resin.

Isocyanate	Polyol
4,4'-diphenylmethane diisocyanate (MDI)	Poly(ethylene oxide) (PEO)
2,4-, 2,6- toluene diisocyanate (TDI)	Poly (propylene oxide) (PPO)
1,6. hexamethylene diisocyanate (HDI)	Poly (tetramethylene oxide) (PTMO)
1,5-napthalene diisocyanate	Poly caprolactone (PCL) diol
4,4'-dicyclohexyl methane diisocyanate	1,4 polybutadiene diol
3-isocyanatomethyl-3,5,5-trimethylcyclohexyl isocyanate (isophorone diisocyanate)	Poly (ethylene adipate)
para-phenylene diisocyanate	Poly (dimethyl siloxane)
2,2,4-trimethyl-1,6-hexamethylene diisocyanate	Polyisobutylene diol
3,3'-dimethyl-diphenylmethane 4,4'-diisocyanate	

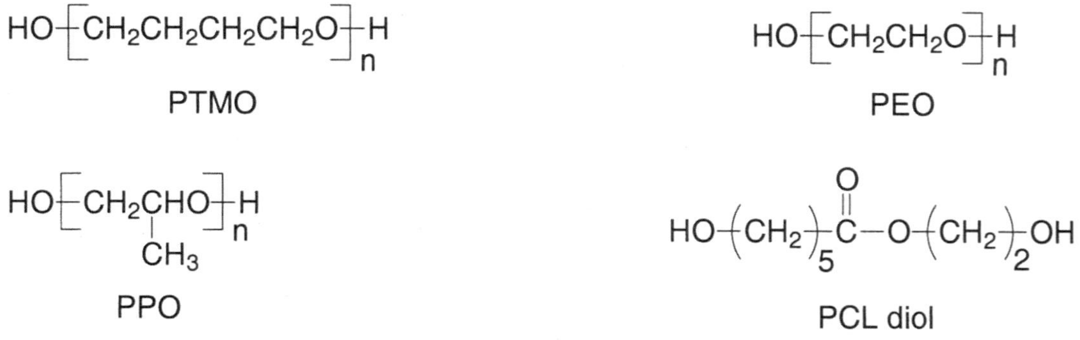

$$HO-[CH_2CH_2CH_2CH_2O]_n-H$$
PTMO

$$HO-[CH_2CH_2O]_n-H$$
PEO

$$HO-[CH_2CHO]_n-H \quad | \quad CH_3$$
PPO

$$HO-(CH_2)_5-\overset{O}{\underset{\|}{C}}-O-(CH_2)_2-OH$$
PCL diol

Figure 1.32
Chemical structures of polyols used for the synthesis of PU.

stability, mechanical strength/modulus but disadvantages with respect to viscosity and hydrolytic stability. Aromatic polyester polyols offer better mechanical properties and flame retardancy compared to their aliphatic analog. Polybutadiene-based polyols are known for their low-temperature flexibility and solvent resistance. Hydroxyl-functionalized hyperbranched polymers can also be used as polyols. They offer much lower solution viscosity compared to their linear analogs as they assume the hydrodynamic volume of a sphere in solution. The synthesis of such hyperbranched polymers was already discussed in the previous sections (see Fig. 1.11). Yi et al. [252] reported a star-shaped polyester polyol prepared by polycondensation of trimethylolpropane (TMP) and ε-caprolactone. The reaction scheme is shown in Fig. 1.33. The properties of PU made from the star-shaped polyol can be tuned by varying the ratios of ε-caprolactone/TMP and the arm number of the polyester polyols. The mechanical properties of PU films and coatings increased with decreasing the ratios of ε-caprolactone/TMP and increasing the arm number.

Polyols may be obtained from natural and renewable sources, such as vegetable oils for the sustainable development of PU materials. These renewable materials can either be fatty

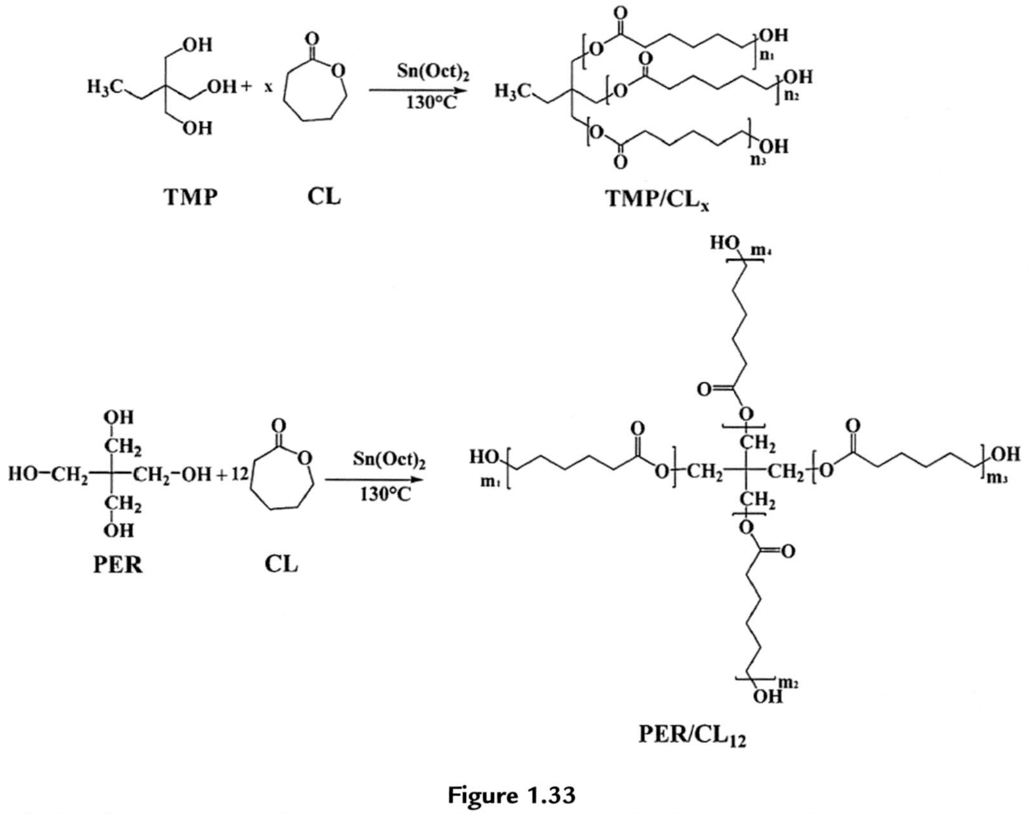

Figure 1.33
Synthesis of star-shaped polyester polyols. Source: *Reprinted with permission from J. Yi, C. Huang, H. Zhuang, H. Gong, C. Zhang, R. Ren, Y. Ma, Prog. Org. Coat. 87 (2015) 61.* © 2015, Elsevier Publishers.

acids or dimer fatty acids obtained [253] from vegetable oils, which include castor, soybean, Pongamia glabra, neem, and cotton seeds. Oils are basically triglycerides of fatty acids, which are converted to polyol by various methods [254–256] such as epoxidation and ring-opening polymerizationozonolysis, transesterification, and hydroformylation. Polyols from renewable sources can be reacted with isocyanates to produce PUs with special properties that are suitable for a wide range of applications. However, the bio-based polyols mostly contain secondary —OH groups, which are comparatively less reactive than the primary —OH groups present in petroleum-based polyols. Therefore a bio-based polyol reacts with an isocyanate at a slower rate compared to a petroleum-based polyol. For example, the addition of the presence of transesterified palm oil-based polyol reduces the reactivity of the foaming process of rigid PU [257]. Using the mixture of both bio-based polyol and petroleum-based polyols, the reaction kinetics may be adjusted and made favorable for the synthesis of speciality products such as molded foams, elastomeric coating, etc.

1.6.2 Isocyanates

The second component for PU is a monomer containing two or more isocyanate functional groups. The isocyanate compounds commonly used for the synthesis of PU are listed in Table 1.14. Generally, aromatic isocyanates especially 2,4-, 2,6-toluene diisocyanate (TDI), 4,4′-diphenylmethane diisocyanate (MDI), and 1,6. hexamethylene diisocyanate (HDI) (Fig. 1.34) are used for the synthesis of thermosetting PU. Most isocyanates are difunctional in nature (each molecule possesses two isocyanate groups), except for a few members, such as diphenylmethane diisocyanate, which comprises mixtures of molecules containing either two or more isocyanate groups. In this case, the compound usually possesses an average functionality higher than two, usually 2.7 due to the presence of a higher number of

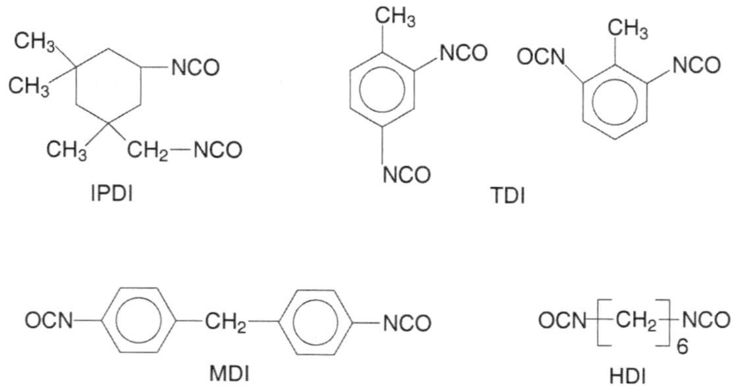

Figure 1.34

Chemical structures of some common isocyanates used for the synthesis of PU. *PU*, polyurethane.

isocyanate groups. The aliphatic and cycloaliphatic isocyanates are comparatively less used. The examples of aliphatic and cycloaliphatic isocyanates used for the synthesis of PU are 5-isocyanato-1-(isocyanatomethyl)-1,3,3-trimethylcyclohexane (Isophorone diisocyanate, IPDI), 4,4′-diisocyanato dicyclohexyl methane, (hydrogenated MDI), and 1,6-hexa methylene diisocyanate (HDI). They are particularly used in coatings and other areas where transparency and color are highly desired. It may be noted that aromatic isocyanate-based PU systems usually darken when exposed to light.

The high reactivity of isocyanates does not allow to use them as one-component system. The isocyanates are highly toxic, which is a concern for the storage of these materials. As a solution to this problem, they are reacted with a substance such as phenols, oximes, alcohols, dibutyl malonate, called blocking agents [258–262]. The blocking agent has to be selected in such a way that the blocked isocyanate will either undergo deblocking in the reaction condition and generate isocyanate, which reacts with a polyol or the blocking agent should be eliminated by the polyol during the reaction. The chemistry of reaction of blocked isocyanate and the polyol is shown in Fig. 1.35. The rate, extent, and mechanism of reaction depend on various factors such as the chemical nature of the blocking agent and polyol, catalyst, polarity of the solvent used.

1.6.3 Prepolymers

The chemistry of synthesis of polyurethane is shown in Fig. 1.36. The polyol is first reacted with excess di- or polyisocyanate to get an isocyanate-terminated intermediate, known as prepolymer. If all the hydroxyl groups are capped with the isocyanates and no free isocyanate remains in the mixture then the intermediate is called "full" prepolymer. Such prepolymers are formed if the isocyanate groups on the polyisocyanate have different reactivity as in the case of 2,4 TDI and the ratio of the equivalent of isocyanate to hydroxyl (f_{NCO}/f_{OH}) is close to 2. If the isocyanate groups are similar in reactivity or the functionality ratio (f_{NCO}/f_{OH}) is more than 2, then isocyanate groups will not be consumed fully and some amount of isocyanate will remain free. The prepolymers are to be analyzed for the isocyanate content using standard methods. This slow speed may be due to the phase incompatibility of the polar and less dense polyol phase and the relatively nonpolar and denser isocyanate phase. Therefore a suitable surfactant and suitable catalysts are required to get a faster reaction rate between them. The actions of catalysts based on the polarization of either the isocyanate or hydroxyl compound through polar interactions. The increase in rate of reaction is attributed to the enhancement of the electrophilic nature of isocyanate by the removal of electron density from the nitrogen or oxygen of the −NCO group.

1.6.4 Chain extender and crosslinker

The prepolymer is then reacted with a chain extender to get the desired polyurethane as shown in Fig. 1.36. The choice of extender plays an important role in deciding the ultimate mechanical

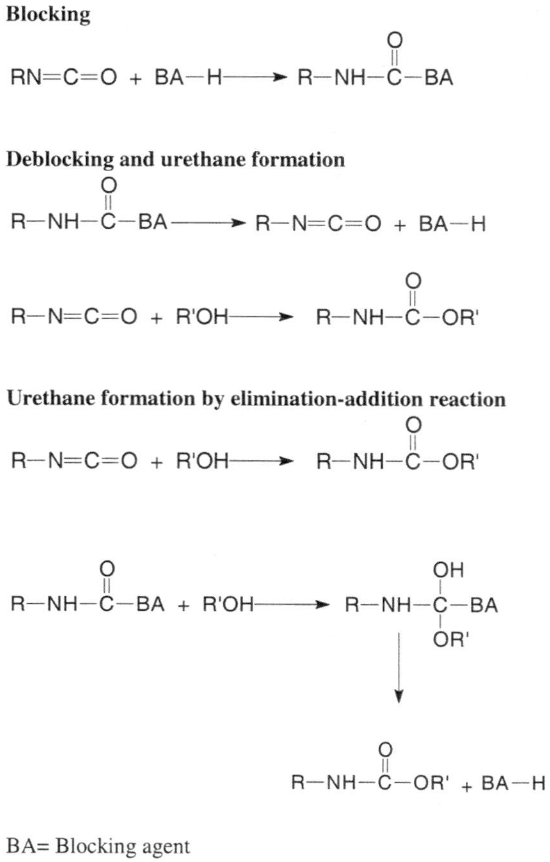

Blocking

$$RN=C=O + BA-H \longrightarrow R-NH-\overset{\overset{O}{\|}}{C}-BA$$

Deblocking and urethane formation

$$R-NH-\overset{\overset{O}{\|}}{C}-BA \longrightarrow R-N=C=O + BA-H$$

$$R-N=C=O + R'OH \longrightarrow R-NH-\overset{\overset{O}{\|}}{C}-OR'$$

Urethane formation by elimination-addition reaction

$$R-N=C=O + R'OH \longrightarrow R-NH-\overset{\overset{O}{\|}}{C}-OR'$$

$$R-NH-\overset{\overset{O}{\|}}{C}-BA + R'OH \longrightarrow R-NH-\underset{\underset{OR'}{|}}{\overset{\overset{OH}{|}}{C}}-BA$$

$$\downarrow$$

$$R-NH-\overset{\overset{O}{\|}}{C}-OR' + BA-H$$

BA= Blocking agent

Figure 1.35
The chemistry of blocking of isocyanate and reaction of blocked isocyanate with the polyol.

properties of the resulting PU. Most diols having a molecular weight less than 500 g mole^{-1} can be used as listed in Table 1.15. Extenders are of two types: diol or triol-based extender and amine-based extender. A diol extender produces urethane linkages and an amine extender produces urea linkages. The incorporation of urea linkages enhances the mechanical properties of the PU networks. The chain extender plays an important role in improving the morphology and final properties of a PU-based product. The chemicals structures of various diol and amine extenders are shown in Fig. 1.37. Water can also be used as an extender. Water reacts with excess isocyanate and produces carbon dioxide and amine as shown in Fig. 1.38. The amine further reacts with the prepolymer to form urea linkages. This reaction is utilized for making water-blown polyurethane foams, in which the rigid hard segment blocks are polydisperse and the block lengths are reported to follow the most probable distribution [263]. They exhibit morphological features, such as spheres ("islands"), cylinders, and lamellar, similar to morphologies known to exist for diblock copolymers [264,265].

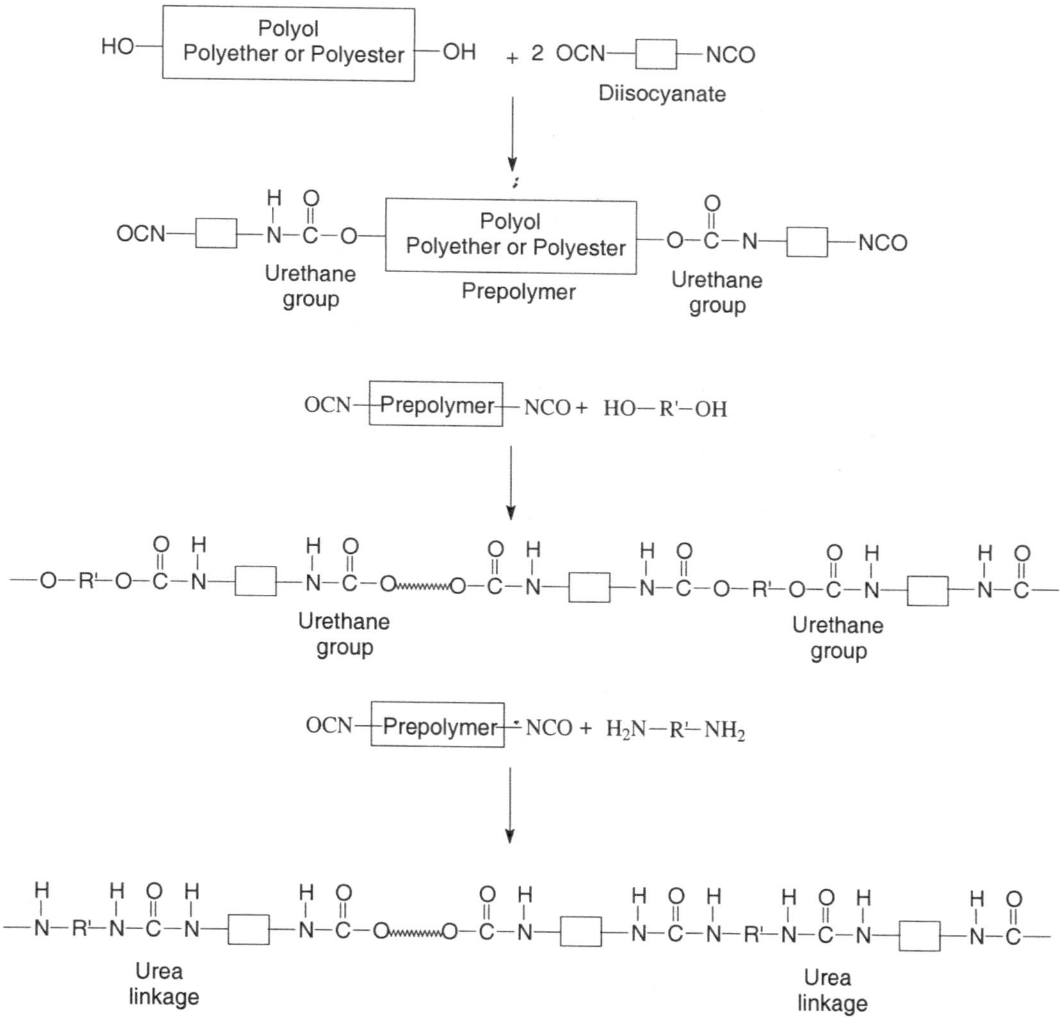

Figure 1.36
Synthesis of PU resin from polyol and diisocyanate. PU, polyurethane.

Table 1.15: List of glycol and amine extenders commonly used for synthesis of polyurethane.

Glycol extender	Amine extender
Difunctional	Diethyl toluene diamine
1,4-butane diol (1,4-BD)	t-butyl toluene diamine
1,6 hexane diol	4,4'-methylene-bis-(2-chloroaniline) (MOCA)
Ethylene glycol	4,4'-methylene-bis-(3-chloro-2,6-diethyl aniline)
Dimethylol propionic acid (DMPA)	Trimethylene glycol di-p-aminobenzoate
4, 4' bis-(2- hydroxyethoxy) biphenyl (BEBP)	4,4' diaminodiphenyl methane (DDM)
Trifunctional	4,4'-methylene-bis-(2-carbomethoxy aniline)
Glycerol	4-chloro-3,5-diamino-isobutylbenzoic acid ester
1,1,1 trimethylol propane	
1,2,6 hexane triol	

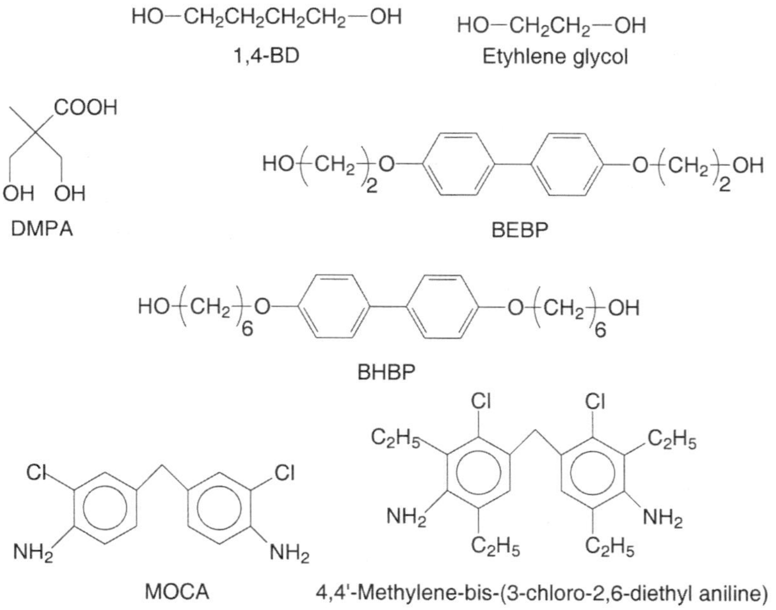

Figure 1.37
Chemical structure of some common alcohol and amine-based extenders used for synthesis of PU. PU, polyurethane.

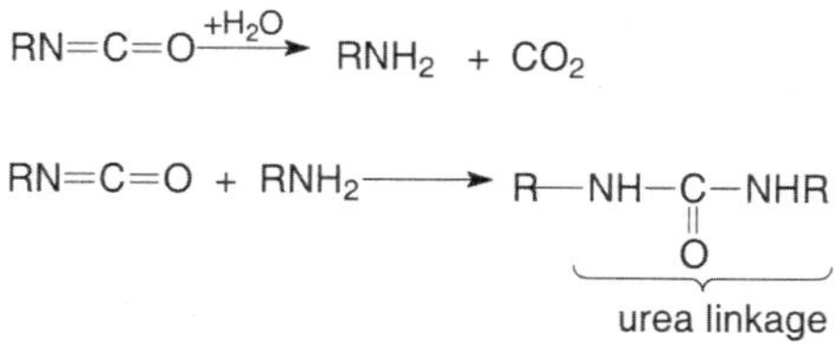

Figure 1.38
Reaction of isocyanate with water.

The extender forms the hard segment of the PU, which is immobile and stiff, while on the other hand, the soft segments, which are produced from the polyols (high-molecular-weight), can move freely. The hard segments act as physical cross-links in the PU system. An interface separation takes place between the hard and soft domains due to the incompatibility and immiscibility (while both phases are amorphous) of the soft segments (low melting and nonpolar) with the hard segments (high melting). Since the hand and soft segments of a thermosetting PU are chemically linked, it restricts the plastic flow within the polymer chains, resulting in elastomeric resiliency. Mechanical deformation of such PU leads to the uncoiling of certain portions of the stressed soft segments, making the hard segments align along the direction of the stress. The realignment of hard segments coupled with a subsequent strong hydrogen bond produces high tensile strength, tear resistance, and good elongation properties [266]. As mentioned above, judicious selection of the chain extender could also influence the chemical resistance, heat, and flexural properties of the PU. With an increase in the concentration of hard segments the hardness, modulus, tear strength, and chemical resistance of the PU system tend to increase. On the other hand, percentage elongation at break, resiliency, and low-temperature flex tend to decrease as a result of increase in the concentration of hard segment. The resiliency and low-temperature flex properties can be increased by increasing the molecular weight of the macrodiol. The extender should be selected keeping in mind the applications in which the PU materials are going to be used.

When the difunctional precursors (diisocyanate, polyol, and extender) are allowed to react in a stoichiometric amount, a thermoplastic PU is formed. The thermosetting PU are made by using excess diisocyanate (excess diisocyanate reacts with urethane structure to form allophanate bonds) or by using a trifunctional extender like glycerin, trimethylolpropane [267–269]. The unique feature of PU resin is that the change in molecular weight between cross-link offers a wide change in properties especially the strain, which reflects the flexibility. For example, a PU system with M_c 1890 shows a strain of 300% whereas when the M_c is increased to 10,000 a strain value of 750% is achieved [270].

Corcuera et al. [271] synthesized segmented PU elastomers based on castor oil as a polyol, 1,4 butanediol (BD) as a chain extender and two diisocyanates, HDI (aliphatic), and MDI (aromatic), as components of the hard segment by a two-step synthetic procedure. PUs with HS based on HDI offered a higher modulus of elasticity and higher hardness but lower elongation at break and tensile strength than those based on MDI. If we take the analogy with other thermosets like epoxy then aromatic reactant-based resin shows higher modulus compared to the same based on the aliphatic precursor. This behavior of PU can be explained by considering phase-separated morphology and a more severe strain-induced crystallization of hard segment in the case of MDI-based PUs than in the case of the more crystalline HDI analogs. For the same reason, the MDI-based PU shows better damping property [272] compared to TDI and IPDI-based PU. The flexibility or damping property decreases with the increasing concentration of isocyanate. Thus the properties of a PU are not only dependent on the chemistry of soft and hard segment but also on the method of preparation, degree of phase separation and morphology and H-bonding interaction [273–275]. This is because chemically bound hard and soft segment structures act together to affect the degree of phase separation, crystallization, morphology, and hydrogen bonding [276,277]. In most of the cases, the studies on the effect of the hard segment on properties of PU were conducted on compositions also containing soft segments, which do not give a clear idea about independent H-bonding interactions between the hard segments. Recently Jiang et al. [278] investigated the effect of the hard segments on properties and interactions of resulting PU by using monodisperse hard segments (without chain extenders). Four PU model hard segments were prepared based on different representative diisocyanates (aromatic, cycloaliphatic, and aliphatic structures) and carried out in-depth spectroscopic analysis and concluded that the crystallization behavior in the four different types of PU hard segments is closely related to their different chemical structures.

1.6.5 Bio-based polyurethanes

The main precursors for the synthesis of a PU are polyol and diisocyanate compounds. For sustainable development, research efforts have been made to produce bio-based PU. As discussed above, bio-based polyols are produced from oils (triglyceride of fatty acids) by using simple processes [251–253] such as epoxidation and ring-opening polymerization, ozonolysis, transesterification, and hydroformylation and can be used as a partial or complete replacement of petroleum-based polyols to prepare bio-based PU. Polyols can also be made from lignin. Lignin is modified into liquid polyols by oxypropylation process [279]. The polyols were reacted with a diisocyanate compound to produce PU. The PU materials were prepared from polyols derived from prepared four different types of lignin and the properties of the synthesized polyurethanes were found to be comparable with the polyurethane rigid foams prepared with commercial polyols [280]. The PU materials with It was demonstrated with high lignin contents (65–70 wt.%) were successfully developed and demonstrated that the increase in lignin contents enhances the cross-link density as well as the modulus of the PU networks. Huo et al. [281] prepared polyurethane foams by reacting lignin-amintated polyol and glycol with diphenylmethane diisocyanates in the presence of water as a blowing agent.

Another issue is related to the use of diisocyanate compounds. Isocyanate compounds are notably toxic chemicals and synthesized from an even more toxic substance phosgene, which causes environmental hazards. Moreover, exposure to isocyanates can cause health effects, such as skin irritation and long-term asthma [282]. Therefore efforts have been to develop alternative, environmentally friendly methods for the preparation of polyurethanes (nonisocyanate routes) like the reaction of cyclic carbonate with amines [283]. Javni et al. [284] prepared polyurethane by a nonisocyanate route by reacting carbonated soybean oil with diamines. Rigid PU foams have been successfully made using soybean oil-based polyol as a replacement of polypropylene-based polyol [285].

Lee and Deng [286] reported a synthetic route comprising three steps for the preparation of polyurethane using 100% bio-based materials. First, epoxidized soya epoxidized soybean oil is carbonated to synthesize carbonated soybean oil. Carbonated ESO was reacted with a coupling agent, 3-aminopropyltriethoxysilane (APES) to produce urethane monomers. Finally, prepared urethane monomers were polymerized with lignin to produce sustainable polyurethane. The schematic illustration of the reaction [286] is shown in Fig. 1.39. Lignin and soybean oil contents in the polymers were as high as 85 wt.%.

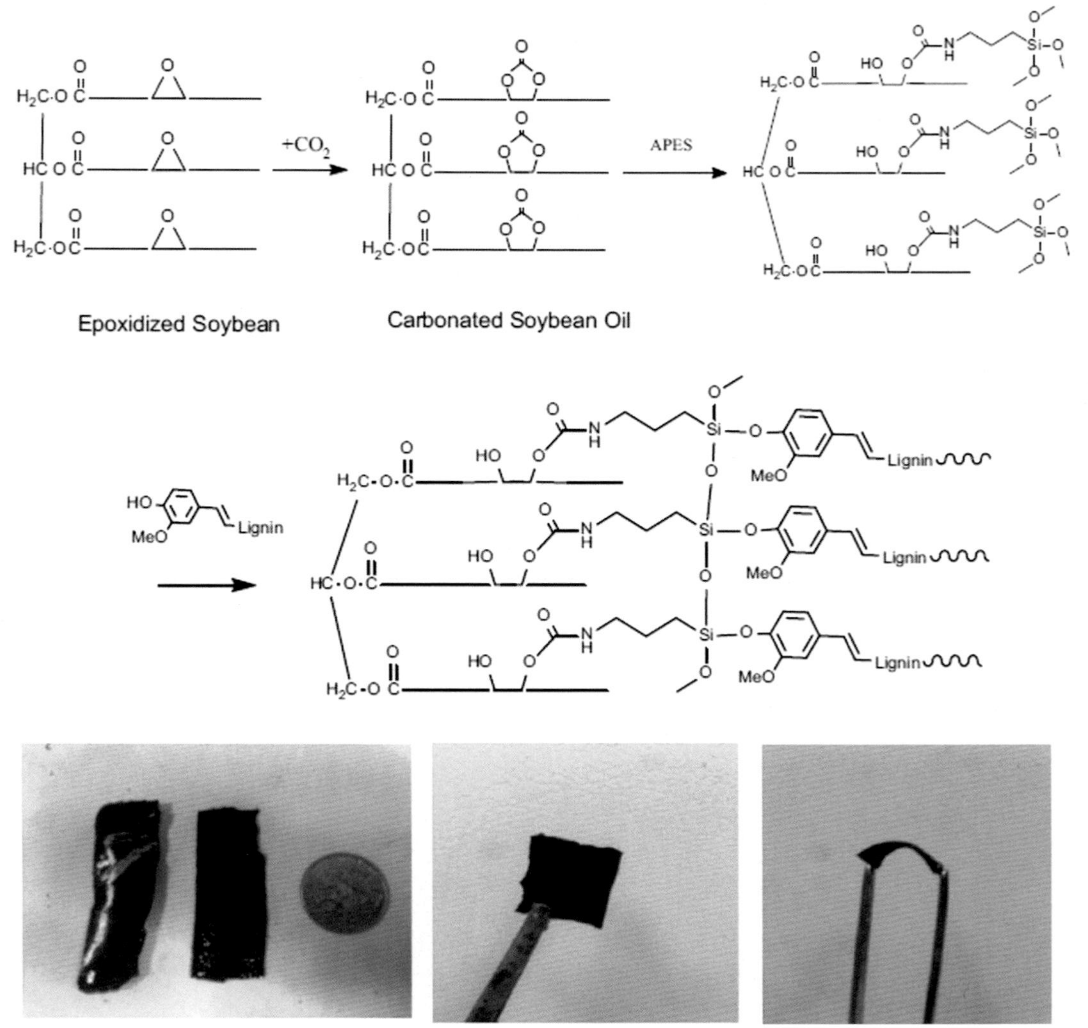

Figure 1.39
Schematic illustration of synthesis of bio-based polyurethane and corresponding photos of samples. Source: *Reprinted with permission from A. Lee, Y. Deng, Eur. Polym. J. 63 (2015) 67© 2015, Elsevier Publishers.*

The author's group has done extensive work on hyperbranched alkyd urethane (HBAU) systems for coating application especially for marine coating [287–289]. The synthesis of hyperbranched alkyd has already been discussed in section (see Fig. 1.11), in which almost all the −OH groups are converted to ester linkages. The strategy used for the synthesis of HBAU is that the −OH groups of the polyol are partially converted to ester linkages by reacting with fatty acid keeping some free reactive −OH groups. Such alkyd resin with −OH functional groups is reacted with a stoichiometric amount of an isocyanate trimers

(Desmodur N 3390) to produce a two-pack PU coating. The reaction scheme is shown in Fig. 1.40. The HBAU resin undergoes curing by both addition and condensation mechanisms. Alkyd part is cured by oxidative curing through unsaturation present in the fatty acid and condensation of —OH and —NCO resulting in the formation of urethane network. Pb-naphthenate and cobalt octoate are used as catalysts for oxidative curing of alkyd and DBTDL is used to accelerate the formation of urethane linkages. Xylene can be used as a solvent for ease of application. It may be noted that the requirement of solvent is much lower compared the commercial linear polyurethane coating. The coating with varying isocyanate contents was prepared and evaluated for surface coating applications and demonstrated that mechanical strength, abrasion, scratch, impact resistance of the HBAU increases with increasing isocyanate concentration up to 50% of concentration (Fig. 1.41).

Keeping in mind the application coating in the marine environment where two-pack PU coatings are difficult to use due to the presence of a very high humidity level, a one-pack moisture curable HBAU resin was also reported [289]. For the synthesis of moisture curable HBAU, the hydroxyl-functionalized alkyd was reacted with excess IPDI. The reaction scheme for the synthesis of —NCO-terminated HBAU resin is presented in Fig. 1.4. The resin does not require any external curing agent to be added and therefore can be used as a one-pack system. After application, it undergoes curing with the moisture present in the surface or atmosphere leading to the formation of a PU coating. Hence, it does not require a

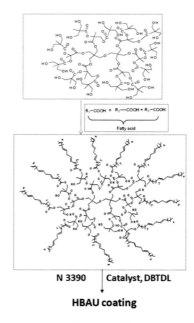

Figure 1.40
Synthesis of two-pack HBAU coating. *HBAU*, hyperbranched alkyd urethane. Source: *Reprinted with permission from R.B. Naik, D. Ratna, S.K. Singh Prog. Org. Coat. 77 (2)(2014) 369. © 2014 Elsevier.*

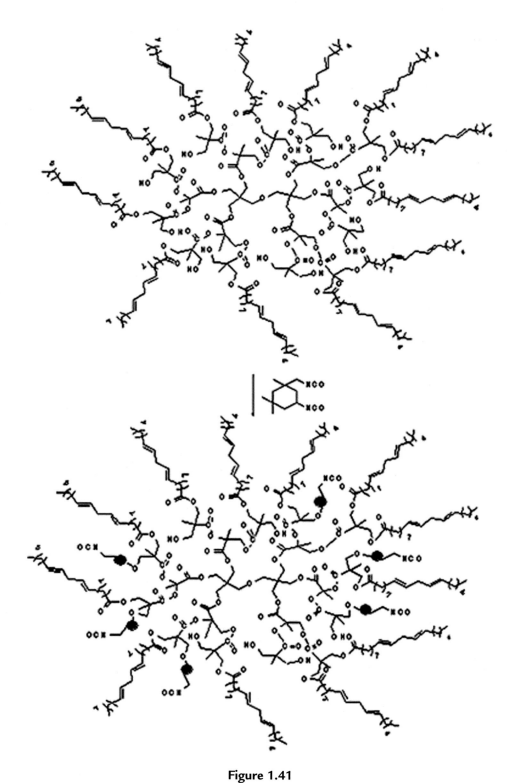

Figure 1.41
Synthesis of one-pack —NCO-terminated hyperbranched polyurethane. Source: *Reprinted with permission from R.B. Naik, N.G. Malvankar, T.K. Mahato, D. Ratna, R.S. Hastak, J. Coating Technol. Research 11 (2014) 575. © 2014, Springer Publishers.*

Table 1.16: Mechanical and weathering properties HBUA resins.

Resins	Tensile strength (MPa)	Elongation (%)	Adhesion strength (MPa)	Gloss Retention, (%) (after 300 h)
Soya alkyd	2	45	3	34
Silicone soya alkyd	9	70	6	50
Silicone acrylate soya alkyd	5	75	6	80
HBUA-30	8	55	8	76
HBUA-40	9	50	10	78
HBUA-50	10	45	11	80

Source: Reprinted with permission from R B. Naik, N.G. Malvankar, T.K. Mahato, D Ratna and R.S. Hastak J. coating Technol. Research 11 (2014) 575–586. © 2014 Springer, Publishers.

completely dry surface for an application like a two-pack PU system. These coatings should be particularly useful in the areas where surface preparation and moisture removal are difficult and the coating must be applied under humid conditions, for example, naval applications. A comparison of mechanical and gloss retention properties of HBUA coatings (with varying concentration of isocyanate) and are presented in Table 1.16. We can see from the table that the HBUA resins exhibit better gloss retention properties while showing comparable mechanical properties with soya alkyd coating.

1.6.6 Applications of polyurethanes

Polyurethanes are used in various forms namely foam (flexible or rigid), coating, adhesive and sealants, elastomeric sheets, binders, etc. Their applications in various industries are reviewed below.

1.6.6.1 Polyurethane foam

PU resins are extensively used in the form of foams, which contributes the most toward the consumption of PU for domestic and industrial applications. Two main steps namely blowing and gelling are associated with the preparation of foam. Water is used as a chemical blowing agent, which reacts with isocyanate to produce carbon dioxide and amine. The generated carbon dioxide expands and is entrapped by the reaction mixture resulting in cellular structure. Gelling takes place due to the formation of the urethane and urea linkages by the reactions of the isocyanate with the hydroxyl group of the polyol and amine group of the in situ generated amine, respectively. The morphology and microstructure of the PU foam can be controlled by changing various parameters, which include the degree of cross-linking after the reaction between the polyol and diisocyanate, the segmental movement of the urea group, the nature of the interaction between the polyol and urea, and the concentration of surfactants. A compound like polydimethylsiloxane−polyoxyalkylene, nonylphenol ethoxylates, silicone oils, etc. are generally used as a surfactant for the synthesis of a PU for the emulsification of liquid components, the regulation of cell sizes and for stabilization of cell structures to guide against collapse as well as against voids at the subsurface. The inherent chemistry of the components like polyol and isocyanate compound also plays a key role in dictating the physical properties of the foam. Therefore the components have to be selected keeping in mind the thermomechanical properties of the foam and its application area [290,291].

PU foams are characterized by two types of, namely, open-cell morphology and closed-cell morphology. The open-cell morphology corresponds to pores that are connected to each other, making the material softer and more absorbent. In the closed-cell morphology, pores are isolated from each other, which makes the foam more rigid. It was shown [292] that the type and reactivity of the petrochemical polyol had a significant influence on the foaming and cell morphology. The content of closed cells was reduced from ca. 80% in the case of the foams based on the polyether polyol to 10% for the foams obtained using the polyester polyol. It was noticed that an addition of a bio-polyol into the foam systems caused a reduction of the cell size regardless of the type of petrochemical polyol used. A significant decrease (ca. 50%) of the compressive strength of the foams based on the polyester polyol took place as a result of the modification of the polyurethane system with the bio-polyol, whereas such a modification of the polyether polyol-based system practically did not cause changes in the foam's mechanical properties [292]. The closed-cell structure with cell gases actually helps to prevent heat transfer. Radiative heat transfer happens through cell walls, so a foam with many small cells will transfer less heat than foam with a small number of large cells. The cell size has a larger impact than just changing the density. Convective heat transfer only takes place in foams with very large cell sizes and is for the most part negligible. The thermal conductivity has a minimum at a certain density. At low densities, the radiative component is more dominant since the cell size is larger, and so is the convective component. At high densities, the number of cell walls and the thickness of them are higher so the radiative component is less important, but the heat transfer through the solid phase becomes more dominant. Therefore rigid PU foams are prepared from polyether polyols using a combination of petroleum-based polyol and bio-based ones.

The major use of PU rigid foam is for refrigeration insulation like in domestic refrigerators, cold storage rooms, etc. They are also used in building and construction industries. The examples are the roof, wall, or window insulation for domestic and industrial buildings. The use of a PU foam-based insulator in a building helps to not only ensure the stable temperature but reduces the noise level for both home and commercial appliances. A major amount of energy is utilized for heating and cooling functions in domestic and industrial buildings. The rigid foam when used for insulation purpose as mentioned above, it is possible to significantly reduce energy costs. At the same time, the commercial and residential appliances can be made more comfortable and efficient. Rigid foams are also used for insulated trailers, trucks, and railway cars.

Impact absorption is a key advantage of polymer foams over the unfoamed polymer. The cells deform to absorb the energy impact and then reform when the load is removed. The PU foam should be able to resist multiple loadings. If the foam is exposed to loads that are too large, the cell walls will rupture and the amount of open-cell structure will increase and weaken the mechanical properties of the foam. Therefore PU foams are used as vibration dampers in the construction and automotive industries. Polyurethane foams with a high closed-cell structure are moisture-resistant and can be used as buoyancy in boats. The advantage is that the foam will not deflate upon a puncture, but retains its buoyancy even after many load−unload cycles.

Flexible foams are used in furniture, cushioning for transportation, and sitting and bedding applications because of their lightweight and excellent cushioning properties. The bedding applications include mattresses, topper pads, convertible sofas, mattresses of variable size and densities, packaging. PU-based packaging foams offer a wide range of product options; hence using such packaging options, most of the onsite packaging challenges can be resolved. They offer cost-effective packaging of items that demand special protection during transit, including medical equipment, electronics, large machine parts, and delicate glassware.

In automotive industries, flexible foams are used to make vehicular seats more comfortable. The comfort of the passengers can be improved by modifying the morphology and properties of such foams [293]. PU-based materials are also used in car bodies, bumpers, doors, and windows. Thus the weight of the vehicles can be reduced significantly resulting in increased fuel economy. Moreover, it provides insulation, sound absorption, and protection to the metallic parts. Because of their inherent low density of PU foams, they are suitable for the manufacture of stiff and light components, which can find applications as interior panels in aircraft, structural shapes, such as bulkhead cores, stringers, and transform cores in reinforced plastic boats [294]. They are also used as sandwich materials found in high-end sporting cars, ships, aircraft, racing cars, etc. This is due to the fact that the PU materials can help to provide heat shielding and structural stiffness as well as energy absorption [295].

1.6.6.2 Polyurethane-based coating

Polyurethane coatings are known for their excellent abrasion and weather resistance. They are used for equipment coating, textile, and leather coating. Application of PU coating on the floor resulted in improvement of durability, aesthetic appearance, and ease of cleaning and maintenance. A lifetime estimation study of a polyether-based PU [296] in a marine environment indicates that tensile property could be almost completely retained after the PU samples were kept immersed in seawater for a period of 2−5 years. Generally, the use of polymer or polymer matrix composite materials in seawater leads to different types of effects which include swelling, debonding of the fillers from the matrix, hydrolysis, plasticization, and degradation of mechanical properties. Because of better water resistance compared to epoxy, PU is preferable for surface coating and paint in marine industries. The automotive coating includes clear topcoats, plastic parts coating, and body primer. Polyurethane coatings are also used for coating military and civilian aircraft. The recent development of hyperbranched polymer-based PU [287,288] can significantly reduce the emission issues of PU-based coatings and such coatings find applications for the development of high gloss and flexible coating. Moisture curable PU coating discussed earlier [289] can find applied as anticorrosive paints for ships and submarines. The coating is one-component system and the moisture presents in the atmosphere is utilized for curing. Some speciality PUs are reported to be efficient for removing certain organic materials, for example, paraben and dye from wastewater [297].

Recent development in water-based PU dispersion has significantly enhanced the application of PU in textile industries. High solid water-based PU coating can be synthesized by a two-step emulsification process or by thiol-ene coupling [60]. Such dispersion-based coating can be easily incorporated into textile-related applications to improve various properties such as permeability, the special structure of their molecules, abrasive resistance and softness, fastness of washing and the soap fastness of reactive dyes, acid dyes, and direct dyes on dyed fabrics. The UV absorbing group can be incorporated in aqueous PU dispersion to enhance its washing fastness, UV protection, and rubbing fastness of the material. N, N-dimethyl allyl p-benzoyl ammonium bromide has been used [298] as a UV absorber to offer great UV dyeing protection to cotton fabrics and the same strategy can be extended for several other textile applications.

1.6.6.2.1 Antifouling and foul release coating

As discussed earlier, marine biofouling causes various negative impacts such as unwanted colonization, deterioration of surfaces, loss of maneuverability and seakeeping and increase in dry-docking frequency. It slows the speed and increases the fuel consumption of ships, corrodes offshore platforms, and blocks seawater pipelines. Marine antifouling paints are necessary to protect ship hulls from biofouling by releasing toxin in a controlled manner, to prevent biofouling [299–301]. The self-polishing triorganotin coatings have been the most effective solution to prevent fouling [301]. However, it was found to have adverse effects on aquatic life and more specifically on nontargeted fouling organisms such as oysters, due to their high persistence and toxicity [302,303]. Furthermore, the use of this technology has proven to be responsible for increases in the levels of organotin and other toxic compounds in or around dry docks, harbors and shipping lanes. In view of this International Maritime Organization banned the use of this type of biocide in the manufacturing of AF paints from January 1, 2003 and the presence of these paints on ship surfaces from January 1, 2008. Presently the self-polishing acrylate-based antifouling paints containing copper are the main commercial products for marine antifouling [304]. These copolymers blended with biocides

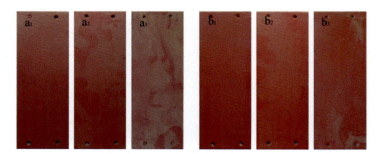

Figure 1.42

Typical images of steel panels: (A1) coated with TMP/CL9-PU before immersion, (A2) immersed in seawater for 1 month, (A3) immersed in seawater for 3 months; (B1) coated with PER/CL12-PU before immersion, (B2) immersed in sea water for 1 month, (B3) immersed in sea water for 3 months. Source: *Reprinted with permission from J. Yi, C. Huang, H. Zhuang, H. Gong, C. Zhang, R. Ren, Y. Ma, Prog. Org. Coat. 87 (2015)161.* © *2015, Elsevier Publishers.*

confer a smooth surface of the coating and the ability to control/regulate biocides leaching rate through the erosion rate of the binder. Yi et al. [252] reported a cu-containing antifouling coating based on PU using a star-shaped polyester polyol (see Fig. 1.33) where the property of the coating in terms of adhesion and toxin leaching can be tailored as a function of the composition of polyol (ratio of pentaerythritol and ε-caprolactone). The two selected formulations were evaluated by static raft testing for their antifouling performance. The photographs of steel panels coated with PU-based antifouling coating before and after exposure are shown in Fig. 1.42. We can see that the coated surfaces were almost clean (without biofouling) in three months, demonstrating the effectiveness of the degradable PU coating based on star-shaped polyester polyols.

PU-based foul release coatings were also reported. The foul release coatings do not necessarily prevent the settlement of marine organisms but permit their easy removal via application of shear to the surface due to their nonstick nature [305,306]. The ideal case is thought to be wherein the velocity of the ship creates sufficient shear to remove the fouling. Foul release coatings have attracted increasing interest because they are toxin-free and hence ecofriendly. Polydimethylsiloxane (PDMS) is mainly used for the development of foul release coating due to its low surface energy. However, because of their low mechanical strength, they are susceptible to mechanical damage such as cutting, tearing, and puncturing, which reduces their service life, limiting their applications. Although low surface energy of PDMS is beneficial foul release function, it weakens the adhesion strength between the coating and the substrates for example PDMS Sylgard 184 (0.4 MPa). A tie coat is needed to enhance the adhesion in applications, which increases the costs and operating time. This issue can be resolved by modifying PDMS with polyurethane or epoxy resins. PDMS modified epoxy systems have been discussed in the previous section and PDMS/PU systems will be discussed here.

Webster et al. [307] reported self-stratified crosslinked polysiloxane modified polyurethane-based foul release coatings. Since PDMS and PU differ in terms of solubility parameter, in solution they remain miscible; however, after application as the solvent evaporates, the stratification takes place. The PDMS segments with low surface energy migrate to the surface and the polyurethane components settle down to the sublayer. The urethane groups can form strong hydrogen bonds with the substrate resulting in good adhesion. An epoxy-modified silicone coating with isocyanate-capped PDMS was also developed [308,309], which can be cured with moisture. The bond pull strength of the coating with the primer was up to 2 MPa, which is much higher than PDMS Sylgard 184 (0.4 MPa) and enough to retain the integrity of the coating to withstand the release of fouling organisms, for example, the removal of barnacles in shear mode. Another way to modify PDMS is to incorporateurea groups into a silicone backbone, which can improve adhesion due to the formation of H-bonding interactions with the substrates. PDMS-based polyurea with PDMS soft segments and 1,6 hexane diamine hard segments was developed [310], which offer an adhesion strength of up to 2 MPa and retained its integrity in the dynamic marine field test and exhibited a nonfouled surface after three test cycles. Moreover, the removal of the fouling organism (diatoms *N. incerta*) underwater jet pressure (20 kPa) revealed that the coating had FR performance similar to that of an unmodified PDMS coating.

1.6.6.2.2 Polyurethane adhesive

Another important application of PU is in the field of adhesive. The largest consumer of PU adhesive is in the textile industry. The applications include textile lamination, integral carpet manufacture, and rebonding of foam. Rebounded foam is made using scrap PU foam bonded with a urethane prepolymer and used primarily as carpet underlay. The lifespan of carpets and their appearance can be increased through the use of PU foam underlays. The foam can also help to provide better comfort with reduced ambient noise. With the advancement of processing technology of PU, it has opened up the possibilities for producers to manufacture a wide variety of PU-based leathers, bra cups, and manmade skins, which may find applications in several sports items and a wide range of accessories. PU adhesives are also used to bond film to film, film to foil, and film to paper in a variety of packaging constructions. Other uses of PU adhesives are for laminating composite panels in truck and car applications, polycarbonate headlamp assemblies, and door panels. Polyurethane adhesives have replaced neoprene-based adhesive for footwear applications [311]. Polycaprolactone polyol and cycloaliphatic diisocyanate-based PU adhesives [312] are also used as PU—aluminum laminates for automotive applications. The adhesive properties of such PU formulations can be tailored as a function of compositions. The use of various nanofillers can further broaden the scope of applications. The structure—property relationship of such nanostructured PU materials will be discussed elaborately in Chapter 3.

1.6.6.2.3 Other applications

The polyurethane elastomer-processed by reaction injection molding (RIM) is widely used in automotive industries. The examples are bumper covers, external body panels, modular windows, and exterior and interior trim. Such elastomers are also used in equipment housing, sports equipment, and furniture. Castable PU elastomer sheets are used as vibration damping materials and acoustic window materials for various naval and civil applications. Because of higher water resistance, PU elastomers are preferable as encapsulant materials for underwater electronics. Cellular castable PU elastomers are used extensively in footwear industries. The introduction of multicolor, multidensity microcellular PU products has further broadened their applications. Noncellular products are widely used as bushings, gaskets, hoses, belts, shock damping mounts, and molded parts for automobiles. PU sealants are used for household appliances, toys, and in ships and submarines. Many electronic items, for example, transducers used for SONAR (sound navigation and ranging) are required to be deployed underwater for several years. PU because of its excellent water resistance as mentioned above can be successfully used as encapsulating material.

In the last few decades, the use of PU-based materials in building and construction application has been increased considerably due to useful properties PU such as high toughness, excellent heat insulation capacity, desirable specific strength, versatility, durability, easy availability coupled with low cost. It is possible to add flexibility and comfort to a new home by making doors based on panels with PU foam core and applying PU-based plastic roof coating. PU-based materials contribute considerably in biomedical applications due to their biocompatibility in addition to very good physicomechanical properties. Due to the low cost, good flexibility and toughness of PU materials are being used as a replacement of metal, ceramic, and metal alloys for biomedical applications. The applications include implants, surgical items (like catheters, surgical drapes) drug delivery, and cardiac tissue engineering [313]. Biodegradable and biocompatible materials were synthesized by incorporating different carbohydrates as crosslinker for propylene oxide-based PU, in which the biodegradability can be tuned as a function of the chemistry of carbohydrates [314]. The developed carbohydrate-based biodegradable and biocompatible PU systems have potential biomedical applications. The PU-based shape memory polymers can be used as a replacement of shape memory alloy (nitinol) for stent applications.

1.7 Phenolic resins

Phenolic resin or phenol formaldehyde (PF) resin is considered as the first polymer material produced commercially from a low-molecular-weight compound. In other words, phenolic resins were the first truly synthetic resins, exploited commercially. The concept of producing high-molecular-weight substances from the oligomeric resins (phenolic resin) with or without filler was first discovered by Leo H. Baekeland in 1909 [315]. Although natural macromolecules like protein, carbohydrate, natural rubber, etc. are abundant in nature, phenolic resin is the first manmade polymer. It is interesting to note that it was made without the knowledge macromolecular theory. The macromolecular hypothesis was profounded later by Staudinger in 1920. It is also interesting that modification of natural polymers using sulfur, that is, vulcanization or curing of natural rubber, was discovered much before the development of synthetic polymer, by Goodyear in 1839 [316].

This pioneering concept Baekeland has been subjected to continued technological development and perfection over the years and has reached the sophisticated phenolic resin technology, available today. Phenolic resins are prepared by a step-growth polymerization of formaldehyde and phenol or phenol derivative using an acid or a base catalyst. The product type and the quality largely depend on the ratio of the reactants used and the nature of the catalyst. Phenolic resins offer many useful properties like high mechanical strength, long-term thermal and mechanical stability, excellent fire, smoke and low toxicity characteristics, and excellent thermal insulating capability. Because of the aromatic structure, phenolic resin offers very good flame and smoke resistance. The smoke density (ASTM E662) of phenolic resin is 16 compared to 480–515 for epoxy and 530 for vinyl ester resins. The cured resins exhibit T_g more than 150°C. Phenolic resins will be discussed below in three varieties: (1) novolac, which is a thermoplastic type and can be used as it is or can be cured with hexamethylene tetramine (HTMA) to get a crosslinked structure; (2) resole, which is a multifunctional reactive compound and can be cured thermally without any catalyst or with an acid catalyst; and (3) addition cure phenolic.

1.7.1 Novolac

Novolac is produced by the reaction of formaldehyde with an excess amount of phenol or a phenol derivative in presence of an acid catalyst. The ratio of phenol to formaldehyde, generally used is in the range of 1.49–1.72. The reaction in acid medium involves initial protonation of hydrated formaldehyde followed by electrophilic substitutions in ortho and para position (electron-rich position unlike meta position). At a low temperature, only the addition of formaldehyde to phenol occurs and o or p-methylol phenols are formed. At a higher temperature, the condensation reactions of methylol phenol with phenol and/or methylol phenol occur leading to the formation of a prepolymer and the desired resin [317,318]. Water is formed as a by-product. The reaction scheme is shown in Fig. 1.43. The condensation of methylol phenol can generate three types of methylene bridges: o, o′ (condensation of two o-methylol phenol), o, p (condensation of an o-methylol phenol with a p-methylol phenol), and p, p′ (condensation of a p-methylol phenol with a p-methylol

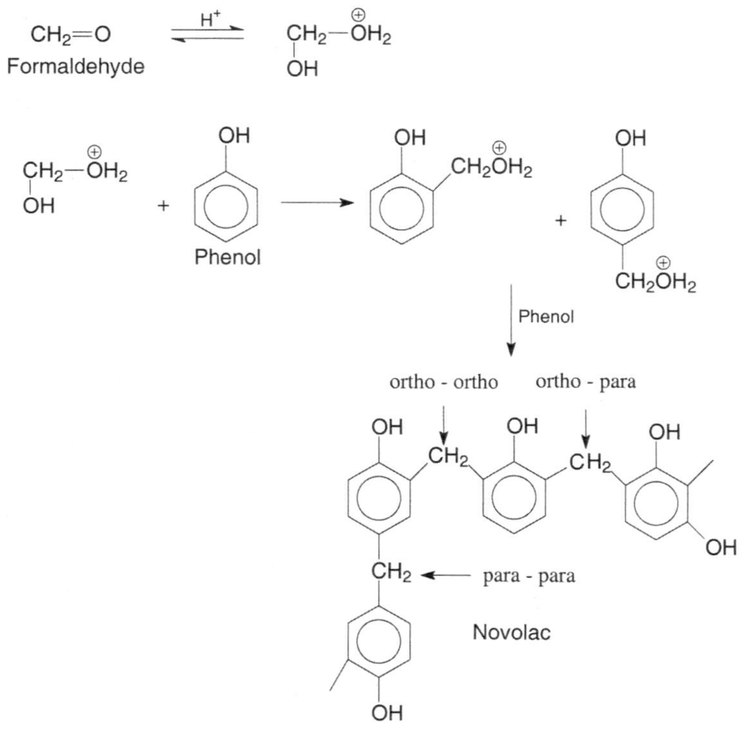

Figure 1.43
Reaction mechanism for synthesis of novolac type phenolic resin.

phenol. The reaction in presence of a strong acid like sulfuric acid, phosphoric acid produces the distribution methylene linkages (o,o′: o,p: p,p′) close to the statistical value 1:2:1. However, a weak acid catalyst like oxalic acid leads to the formation of more p,p′ linkages. A typical distribution of methylene linkages obtained for the synthesis of novolacs using sulfuric acid and oxalic acid are 25.9: 48.5: 25.9 and 23.3: 49.1: 27.6, respectively [319].

Achieving selectivity in the mixture of products with different methylene bridges is a critical issue. It is reported that the selectivity of p-isomer can be boosted by using β-cyclodextrin or crown ether [320]. The basic mechanism is that the additive form complexes with the phenoxide ions and encourages preferential attack at p-position due to the steric hindrance caused by the attack at ortho position. On the other hand, the ortho selectivity can be increased by using a metallic catalyst such as salts of zinc, magnesium, lead etc. When the reaction is carried out in presence of metallic salts, the metal ion forms a chelate intermediate as shown in Fig. 1.44. The chelate intermediate is further transformed into o-methylol phenol as confirmed by ^{1}H NMR spectra [321]. Thus high

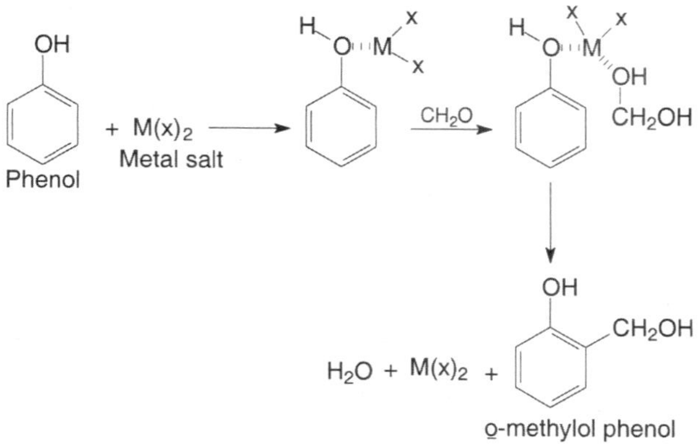

Figure 1.44
Formation chelate intermediate for phenol/formaldehyde reaction in presence of a metal salt catalyst.

ortho resins are formed with 47%–49% o,o′ and o, p methylene bridges each and less than 5% of p,p methylene bridges.

1.7.2 Resole

Resole is produced by the reaction of a phenol or a phenol derivative with an excess amount of formaldehyde in presence of a base catalyst. The reaction in basic medium proceeds through the addition of formaldehyde with phenoxide ion leading to the formation of o or p monomethylol phenol (along with some di- or trimethylol phenol) as established by complexation via cyclodextrin or crown ether [318]. The reaction scheme for the synthesis of resole is shown in Fig. 1.45. Both the novolac and resole are the products of phenol and formaldehyde and are known as phenolic resin. The differences between novolac and resole are as follows:

1. As discussed earlier, novolac is produced by using an excess phenol and an acid or metal salt catalyst whereas resole is produced by using an excess of formaldehyde and a basic catalyst.
2. Novolacs are mostly supplied in the solid form whereas resoles are mostly supplied as a solution (using water or alcohol as a solvent).
3. Novolac is not self-curing; it requires an extra curing agent whereas resole is self-curable without any additional catalyst.
4. Novolac has a long shelf life whereas resole has a limited shelf life (less than 1 year).
5. Novolac can be used as a thermoplastic and can be modified easily by reacting with the terminal hydroxyl groups. Modification of resole is comparatively difficult for gelling problem.

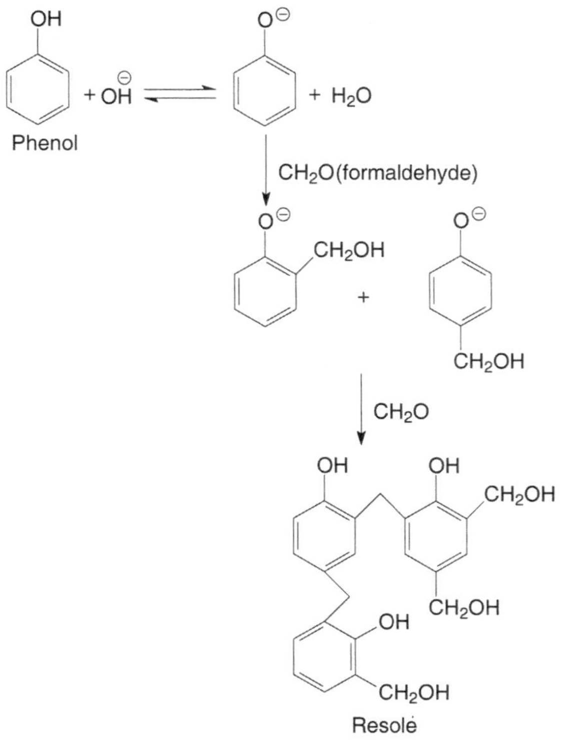

Figure 1.45
Reaction scheme for the formation of resole-type phenolic resin.

1.7.3 Cross-linking of phenolic resin

When resole solution is subjected to a thermal cure with or without an acid catalyst, the molecular weight increases leading to an intermediate gel state (B-stage), which is insoluble in the solvent (water or alcohol). As the solvent evaporates, the curing proceeds further leading to the formation of a rigid crosslinked structure.

Unlike resole, novolac is not self-curing. It requires extra formaldehyde for curing to take place. Novolac is generally cured with hexamethylene tetramine (HMTA), which is a condensation product of formaldehyde and ammonia. HMTA is added at a level between 8% and 15% [322]. Under the curing condition, HTMA decomposes to generate formaldehyde, which converts novolac to a crosslinked network structure (Fig. 1.46). The mechanism of HMTA-cured novolac has been studied extensively but numerous possible intermediates complicate the analysis. Dargavilille et al. [323], de Bruyn et al. [324], and Zang et al. [325,326] employed monomers and dimers to analyze their curing reactions with hexamine. The proposed mechanism occurs in two stages: formation of initial intermediates such as benzoxazines and benzyl amines and further reactions of these initial intermediates

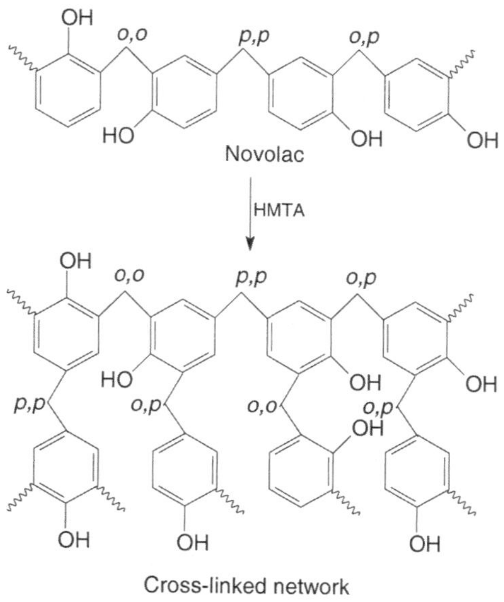

Figure 1.46
Cross-linking of novolac using HMTA. *HMTA*, hexamethylene tetramine.

into methylene bridges with residual traces of amines, imides, imines, methyl phenol, and others. When the amount of methylene is low, hexamine provides only one methylene bridge to 4, 4 site and many reactive sites remain and react to form more methylene bridges for cross-linking. Hence, larger numbers of methylene bridges are formed compared to the composition with higher hexamine content.

The decomposition of the intermediate is also affected by the pH of the medium. Novolac is acidic and hexamine is alkaline in nature. Hence, the resultant pH of the mixture depends on the composition. At a low hexamine concentration, the pH remains acidic and facilitates the decomposition of the intermediates. Another factor is the ortho para ratio of the novolac. The strategy to make preferential ortho or para substituted products has already been discussed in Section 1.7.1. Para-bridged intermediate decomposes more easily compared to the ortho-bridged one, which is more stable due to the possibility of the formation of six-membered rings. Thus, when novolac with high amount of ortho sites is reacted with a higher amount of hexamine, a stable network with high nitrogen content is formed. On the other hand, the reaction of novolac with high amount of vacant para sites with a low amount of hexamine produces a network with low nitrogen content. Since the presence of nitrogen increases the thermal stability and flame retardancy of the resin, a sufficiently high amount of hexamine should be used for curing of novolac.

Markovic et al. [327] investigated the effects of phenol to formaldehyde ratio used for the preparation of novolac and nature of methylene linkages on the curing behavior of novolac/

Table 1.17: Effect of phenol to formaldehyde ratio and content of o, o′ isomer on the reaction rate.

Phenol to formaldehyde ratio	Molecular weight[a] (g mole^{-1})	Content of o, o′ isomer (%)	Reaction rate (min^{-1})	Gel point (Min)
1:0.75	406	26.05	0.061	4.4
1:0.79	492	27.20	0.071	2.5
1:0.85	813	26.58	0.093	0.5
1:085	712	70.17	0.107	1.0

[a]Determined by ^{1}C NMR spectroscopy.
Source: Reprinted with permission from S. Markovic, B. Dunjic, A. Zlatanic, J. Djonlagic, J. Appl. Polym. Sci. 81(2001) 1902. © 2001 John Wiley and Sons Publishers.

HMTA system using rheological studies. The cure kinetics was described by a third-order phenomenological equation, which takes into account the self-acceleration effect that arises from chemical reaction and phase separation. The reaction rate was found to increase with an increase in phenol to formaldehyde ratio. For a same phenol to formaldehyde ratio, the reaction rate increased with an increase in o, o′ methylene linkages (Table 1.17). Various hydroxymethyl derivatives, for example, bismethylol cresol, bisphenol-A tetra methylol, can be used as the curing agents instead of hexamine. The networks produced by these curing agents display much higher impact strength and bending strength compared with hexamine cured networks.

The curing of novolac/HMTA is associated with an issue of less mobility of reactants, which is necessary for uniform and complete curing. HMTA is highly crystalline with a melting temperature above 280°C. Novolac resin is a rigid material due to the presence of extensive hydrogen bonding and becomes mobile only for reaction above 140°C. In order to achieve high modulus and performance at elevated temperatures, this resin must react with crosslinkers such as HMTA in order to establish a stable crosslinked structure Therefore a third component, that is, plasticizers, is used to disrupt the H-bonding and facilitate the curing reaction [328].

Two categories of plasticizers, reactive and nonreactive, can be used to facilitate cross-linking reactions. The nonreactive plasticizers (e.g., methyl benzoate) serve only as facilitators for plasticization and dissolution of the two components [329]. The reactive plasticizers (namely furfural, furfuryl alcohol, and methyl anthranilate) may react with the products of the primary reaction between the phenolic resin and HMTA leading to the formation of a crosslinked structure, which may exhibit higher modulus compared to the network made without using a plasticizer [330]. The structures of the three reactive plasticizers are shown in Fig. 1.47. Furfuryl alcohol produces a dangling end on reaction with novolac and does not contribute to the mechanical property of the final structure. The aldehyde group of furfural can react in either the ortho or para reactive sites of phenolic monomer to form two possible structures: one with a dangling end and the other with a intramethylene bridge forming the crosslinked structure. The most attractive reactive

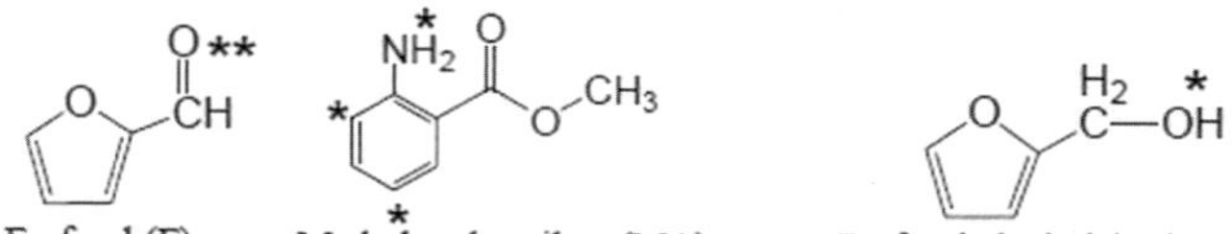

Figure 1.47
Structures of the different plasticizers for phenolic resin. Source: *Reprinted with permission from J. P. Patel, C.X. Zhao, S. Deshmukh, G.X. Zou, O. Wamuo, S. L. Hsu, A.B. Schoch, S.A. Carleen, D. Matsumoto, Polymer 107 (2016) 12© 2016, Elsevier Publishers.*

candidate is methyl anthranilate, a multifunctional compound with electron-rich rings and reactive ortho, para, ester, and amino functionalities. With this plasticizer, it is possible to connect the ortho and para sites of methyl anthranilate with the ortho or para sites of the phenolic monomer forming a stiffer crosslinked structure in comparison to the usual intramethylene bridges found between HMTA and phenolic resin.

1.7.4 Applications of phenolic resin

A major volume of phenolic resin is used in the areas of wood materials and insulating products. Such materials are mostly used in the building and automotive industries. In wood composites line panels (plywood), molded product, lumber engineered product about 10% binder resin is used which is mostly aqueous resole. The driving force for such application is that phenolic resins can offer a durable bonding, which is resistant to heat, moisture, etc. The phenolic bonded wood materials exhibit good resistance to moisture and other environmental effects and retain the properties even after a long exposure. Acoustic and thermal insulating materials should retain a constant level of functional properties over long periods of time. For example, the materials, which are used as insulating materials and in automobile constructions, should provide a load-bearing and noise damping function throughout the life of the vehicle. Phenolic resin bonded textile mats are used for such purposes. The other applications are molding compounds, foundry/refractory materials, abrasives, friction linings. Phenolic resin molding compounds continue to be in electrical as well as household appliance areas. The properties of some commercial phenolic compounds are presented in Table 1.18. One of the oldest fields of application for phenolic resin-glass fabric prepregs is the production of fishing rods [331].

The resin content in abrasive and friction line materials is about 10% whereas in foundry materials binder content is much less (approximately 2%). For friction lining applications as for example automobile braking operation, due to tremendous friction, the surface temperature of the lining may increase up to 800°C for a short time. Phenolic resins because of their high thermal stability are most suitable for such applications. The properties of some commercial molding phenolic resin being soluble

Table 1.18: Mechanical properties of phenolic molding compounds.

Molding compound	Tensile strength (MPa)	Tensile modulus (GPa)	Strain at break (%)	Flexural strength (MPa)	Compressive strength (MPa)
RX-630	82.7	13.8	0.6	193	275.8
RX-660	55.2	11.7	0.5	137.9	227.5
RX-865	68.9	17.2	0.4	186	241.3
XB-22	75.8	7.6	1.0	151.7	234.4

in low boiling point alcohol, impregnated papers, and fabric can be made easily by soaking the fabric in resin solution followed by drying and curing. Such impregnated papers are used as separators in lead-acid storage batteries and fuel cells to separate the positive and negative electrodes. Phenolic resin as an impregnation material offer adequate insulation, high resistance to acids, oxidation, and mechanical strength. Because of inherent fire-retardant properties, they are used to make heat- and fire-resistant panels.

Honeycomb structures and paper honeycombs also represent an application for phenolic resins as impregnating and dipping agents. Honeycomb structures are very effective for development of lightweight structures. Though epoxy and vinyl ester resins are used for such applications, phenolic is also preferred due to inherent fire-resistant property and low cost. Honeycomb structures were initially intended for only high-tech applications like flooring and interior paneling of aircrafts; however, nowadays such structures find applications even in passenger and freight train cars in order to reduce the weight penalty. Sliding doors for subways, shipbuildings are the example for other applications. The panels are required to be coated with suitable flame-retardant materials to reach the required fire protection standard.

Braking straps and friction pads are used in power shift transmissions, ship winch transmission, and automobile clutches. Special types of papers impregnated speciality phenolic resins are used for this application. Adequate adhesion of the paper to the metal parts is required to satisfy the parameters required for friction lining compounds. Generally, phenolic resin in combination with epoxy resins is used to develop a wide range of friction lining materials. Another application of impregnated fabric is in the foundry industry for example to filter aluminum melts. For the preparation of such impregnated fabric, woven glass fabrics are passed through an impregnating bath containing an aqueous solution of resole, dried and finally cured in stages up to 300°C. Because of the high aromatic content, phenolic resins absorb a lot of heat for their degradation. That is why it is used as an ablative material in reentry vehicles. When a space vehicle reenters into the atmosphere, due to tremendous friction with air, lot of heat is generated. Phenolic resins can absorb the heat at the cost of its degradation and save the vehicle from deterioration as a result of extremely high temperatures.

Phenolic resins are used as an alternative to conventional precursor (pitch, tar) materials for the development and production of carbon-based engineering materials for example glassy carbon, carbon fibers and carbon–carbon composites [332–334]. This application is related to the high yield of carbon (60%–70%) produced as a result of pyrolysis. Glassy carbon, which is prepared by pyrolyzing phenolic resin at about 600°C at a heating rate of 2°C–5°C per hour, exhibits distinct properties like thermal resistance up to 3000°C, very low density, alkali resistance, gas permeability, and biocompatibility. Because of high thermal stability, carbon-based engineering materials are not only used for a conventional applications like electrode materials, slip ring seals, and bearing materials but also in strategic applications like a nuclear reactor, missiles and rocket engineering. Carbon–carbon composites are made by curing the carbon fiber/phenolic prepreg followed by the carbonization at 500°C–1000°C by heat/pressure and densification by chemical vapor deposition or chemical vapor infiltration. During the process of carbonization, the precursor is converted to porous carbon fiber matrix. The unique feature of carbon–carbon composite is that both the matrix and reinforcement phases are carbon. The carbon–carbon composites are suitable where high temperature strength and high strength-to-weight ratio are of utmost importance. They can work at very high temperature (2200°C) as well as at subzero temperatures. Another important property, which carbon–carbon composites exhibit, is very low high temperature creep. On reentry of the space shuttle, the nose cone, and leading edges of the shuttle experience hot gases rushing through the nozzle/cone at high velocity, stressing, and eroding the nozzle wall. Carbon–carbon composites can resist high-temperature erosion without burning away. A major amount of carbon–carbon composites are used in commercial and military aircraft brakes and in rocket and missiles.

Phenolic resins exhibit excellent dimensional stability with a constant use temperature range of 180°C–200°C, excellent chemical, moisture, and heat resistance and favorable fire and smoke behavior. The predominant consideration in the use of phenolic resins as a matrix resin in fiber-reinforced composites is their fire behavior. Phenolic-based composites perform better under fire conditions compared to epoxy or vinyl ester-based composites. This is due to their delayed ignitability coupled with a low heat release, low smoke evolution with little, or no toxic gas emission and capability to provide significant strength retention (70%) at a service temperature of 300°C over a period of 1–2 h. Thus phenolic composites with exceptional fire resistance are suitable for use in demanding critical applications where the safety of passengers during egress in case of fire is the most important. The possible applications are composites for aircraft interiors, mass transportation, and marine/offshore areas.

1.7.5 Phenolic resin as additives

1.7.5.1 Additives for rubber

Phenolic resins are useful additives for rubbers. Rubbers also belong to the category of thermoset polymers; however, unlike thermosetting resin, they are high-molecular-weight polymers. Depending on the nature of rubber and the phenolic resin modifier, they can function as curing or vulcanizing agents, tackifiers, and reinforcing agents. The phenolic modifier can be incorporated into the rubber using the same methodology as used for incorporating carbon black and other rubber additives. Generally, rubbers are cured by sulfur for commercial purposes. Para-aminophenol resoles type of phenolic resins can be used to cure butyl rubber (containing less unsaturation) in presence of Lewis acid catalyst [335].

Unlike natural rubber, the filled synthetic rubber compounds [e.g., styrene butadiene rubber (SBR), ethylene propylene diene rubber (EPDM), etc.] exhibit poor tack property due to their inherent low tack. Tack property is very important for tire applications where the multiple layers must adhere to each other. The lack of adequate tack may lead to failure of the final product. Long-chain and branched alkyl phenol-based novolac resins have been recommended for application as a tackifier. Varghese et al. [336] investigated the effect of incorporation of different types of phenolic resins on adhesive properties of neoprene rubber for rubber (SBR)−rubber (SBR) and metal (Al)−metal (Al) bonding. It was reported that the formulation containing cardanol-based PR performs better on Al−Al bonds while the t-butyl phenol-based PR-containing formulation is better suited for SBR−SBR bonds. Nawaz et al. [337] investigated the ablative response and thermal properties of phenolic resin blended acrylonitrile-butadiene rubber (NBR) and reported a reduction of mass ablation rates from 26.1% to 21.3% increase in char content (deposition) from 0.19% to 26.8% due to incorporation of phenolic resin.

Phenolic resins in conjunction with stannous chloride have been used [338] for curing butyl-based bladder compound to achieve excellent flex and thermal aging properties. Yuan [339] used phenolic resin to develop dynamically vulcanized polylactide/natural rubber (NR) blends. The viscoelasticities of the PLA melts were largely improved by the crosslinked NR phase. Recently, blends of poly (butyl methacrylate) (PBMA) and phenolic resin were investigated [340] and observed an increasing trend of damping property with decreasing molecular weight of phenolic resins.

1.7.5.2 Modifier for poly(ethylene oxide)

Poly(ethylene oxide) has gained much importance among scientific communities due to its extensive applications in the areas like phase change materials [341,342], solid polymer electrolytes [343,344] (for battery, supercapacitors, and fuel cell), and crystallizable switching component for shape memory polymers (SMP) [345–347]. However, the inherently poor mechanical strength of PEO stands as a stumbling block for the above-mentioned applications. Phenolic resin has been used to modify the property of PEO in order to enhance its functional performance. For example, PEO is used as a solid polymer electrolyte [348] due to its ability to coordinate the alkali metal ions strongly. However, the conventional PEO-Li salt complexes show conductivities of the order of 10^{-5} S cm^{-1}, which is not sufficient for battery, capacitor, and fuel cell applications. A high crystalline phase concentration limits the conductivity of PEO-based electrolytes. Apart from high crystallinity, PEO-based electrolytes suffer from low cation transport number t_+, ion–pair formation, and inferior mechanical properties. Peter et al. [349] reported the modification of PEO with phenolic resin for improvement in mechanical property and conductivity.

In this context, let us discuss in detailed the works carried out to study the blends of PEO and phenolic resins [350–352]. The blends of novolac with both low (M_w = 20,000 g mole^{-1}) and high (M_w = 300,000 g mole^{-1}) molecular weight PEO containing 0–30 wt.% of novolac have been investigated by Ratna and coworker [350]. The blends have been proposed for their use as a vibration damping application and crystallizable switching segment for shape memory polymer system. The interaction between PEO and novolac was corroborated by Fourier transformed infrared (FTIR) spectroscopy [350]. The FTIR spectra of PEO, novolac and 20% novolac containing blend, are presented in Fig. 1.48. Novolac shows a broad band consisting of two peaks at 3500 and

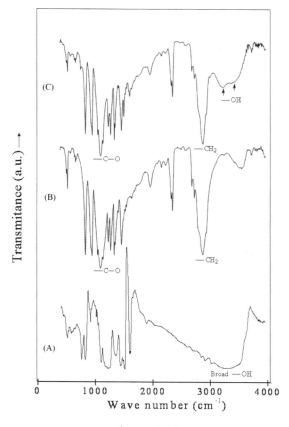

Figure 1.48
FTIR spectra of (A) novolac, (B) PEO, and (C) PEO/novolac blend (80/20 w/w). *FTIR*, Fourier transformed infrared; *PEO*, poly(ethylene oxide). Source: *Reprinted with permission from D. Ratna, T. Abraham, K. Kocsis, J. Appl. Polym. Sci. 108 (2008) 2156. © 2008, John Wiley and Sons Publishers.*

3350 cm^{-1}, which are assigned to the nonassociated free hydroxyl groups and self-associated hydroxyl groups, respectively. In the blend, the band corresponding to the self-associated −OH group remains unchanged; however, the same corresponding to nonassociated −OH group shifted to 3300 cm^{-1}. The bands corresponding to the ether groups (1200−1300 cm^{-1}) become broader compared to those in PEO. This clearly indicates that there is polar interactions between the −OH groups of PEO with ether chain of PEO via H-bonding as shown in Fig. 1.49.

The crystallization behavior of PEO/novolac blends containing various concentration of novolac is shown in Fig. 1.50. It was observed that the crystallization temperature gradually decreases with increasing concentration of novolac. This is due to the fact that amorphous novolac disturbs the crystallization of PEO as will be discussed shortly using cross polarized optical microscopy (CPOM) studies. Thus the SMPs with adjustable switching temperature can be made by using a correct blend composition as a crystallizable switching segment. A shape memory system functions only if the switching temperature matches with switching temperature. For example, medical application requires the switching temperature to be about 37°C. Hence, by using the PEO/phenolic blend composition as a switching component, the application of PEO-based shape memory polymers can be broadened.

The CPOM photographs of PEO and blends containing 20 and 30 wt.% novolac are presented in Fig. 1.51. When the PEO melt is cooled, small circular spherulites are formed, which subsequently impinge upon one another producing a typical maltese cross (a regular isotropic three-dimensional shape) as shown in Fig. 1.48. In case of blends, the spherulites are characterized by many anisotropic, nonspherulitic shapes with jagged edges. The sizes of spherulites decrease with an increase in novolac concentration in the blends. This can be attributed to the disturbance of radial

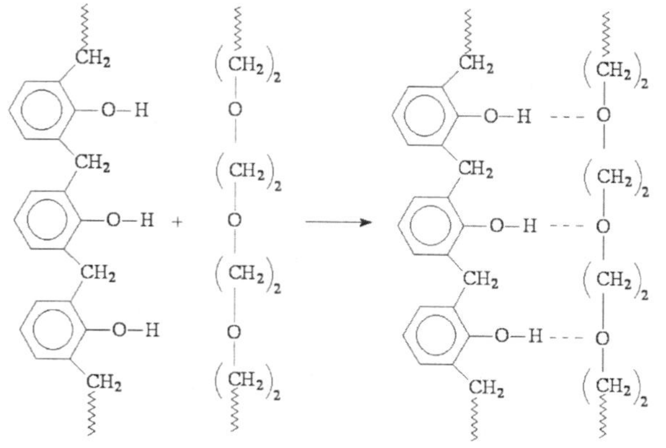

Figure 1.49
Polar interactions between the −OH groups of PEO with ether chain of PEO via H-bonding. *PEO*, poly(ethylene oxide).

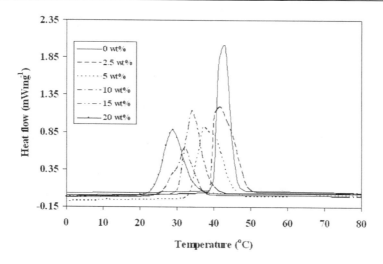

Figure 1.50
Crystallization behavior of PEO/phenolic blends with varying concentration of phenolic resin. PEO, poly(ethylene oxide). Source: *Reprinted with permission from D. Ratna, T. Abraham, K. Kocsis, J. Appl. Polym. Sci. 108 (2008) 2156.* © *2008, John Wiley and Sons Publishers.*

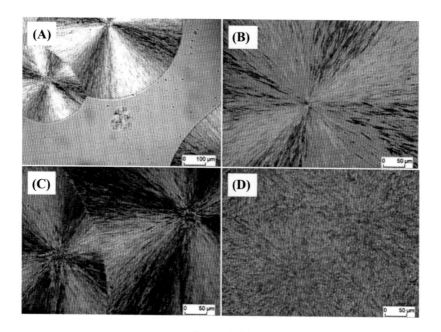

Figure 1.51
Cross polarized optical microscope images of PEO blends (A) at the beginning of crystallization, (B) spherullites with maltese cross section, (C) PEO + 20 wt.% novolac, and (D) PEO + 30 wt.% novolac. PEO, poly(ethylene oxide). Source: *Reprinted with permission from D. Ratna, T. Abraham, K. Kocsis, J. Appl. Polym. Sci. 108 (2008) 2156.* © *2008, John Wiley and Sons Publishers.*

orientation caused by amorphous novolac during the crystallization process. The blends with higher concentration (>20wt.% of novolac) show a leaf-like structure due to the anisotropy of the total crystallization process. Hence, the optimum phenolic concentration is 20 wt.% in the blend because an adequate overall crystallinity of the blend is essential for triggering the shape memory effect [353,354].

They have also investigated viscoelastic properties of PEO/novolac blends by subjecting the blend and pure PEO samples to DMA. The loss tangent versus temperature plots of PEO/novolac blends of varying composition are shown in Fig. 1.52. All the blends exhibit clear, single loss-tangent peaks (indicative of T_g) in DMA and the T_g is increased with increasing novolac content. This behavior is a typical characteristic for compatible blends. The T_g value of the blend shows a positive deviation with respect to the value calculated by the Fox equation [355]. This can be attributed to the crystallinity and interassociation H-bonding between the −OH groups of novolac with oxygen atoms of ether chain of PEO as confirmed by FTIR analysis discussed above [350]. It was also observed that the damping (loss-tangent peak value) increased dramatically with increasing novolac concentration in the blend. This can be explained by considering two factors. First, the introduction of amorphous novolac being miscible with the amorphous phase of PEO increases the overall amorphous content of the blend. Second, the interaction between PEO and novolac via the H-bonding interaction the frictional loss increases leading to the enhancement of damping. The high damping and broad loss-tangent peak of the blends with higher novolac content indicates that they may find applications as vibration damping materials for which the acoustic frequency of absorption can be adjusted by manipulating the blend composition.

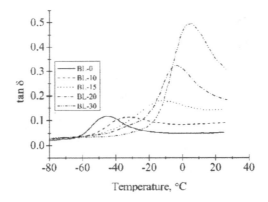

Figure 1.52
Loss factor versus temperature plots of various PEO/novolac blends. *PEO*, poly(ethylene oxide).
Source: *Reprinted with permission from D. Ratna, T. Abraham, K. Kocsis, Macromol. Chem. Phys. 209 (2008) 209.* © *2008, Wiley VCH Publishers.*

1.7.6 Addition cure phenolic resin

As discussed above, novolac and resole-based phenolic resins and their composites capture a major market in thermostructural application in the aerospace industry due to good heat and flame resistance, excellent ablative properties, and low cost. However, the conventional phenolic resin suffers from several technological issues related to the inherent chemistry of curing such resins, which limit their application in many engineering areas. These resins cure at moderately high temperature by a condensation mechanism with the evolution of volatiles, which may cause the formation of internal microvoids during processing leading to the deterioration of the final property of the product. The formation of voids is generally controlled by the application of pressure and temperature simultaneously and the process requires optimization of both pressure and temperature. The usage of strong acid or base as catalysts brings in the corrosion of processing equipment. The resole has limited shelf life at ambient temperature, which poses a storage problem required for bulk production. As discussed in the section, the chemical structure of conventional phenolic comprises rigid aromatic units tightly held by the short methylene linkages. This structural feature, though offers thermal resistance property, makes the matrix extremely brittle. This limits their use as matrix resins and adhesive/coating for any application where the structure is likely to be subjected to shock and vibration.

In order to address the above-mentioned issues of conventional phenolic resins, lots of research works have been done to develop new chemistry leading to the development of addition cure phenolic resins, which can address the shortcomings associated with the cure of phenolic resins. The major strategies in designing addition cure phenolics are: (1) incorporation of thermally stable addition-curable groups onto a novolac backbone; (2) structural modification (transformation) involving phenolic hydroxyl groups; (3) curing of novolac by suitable curatives through addition reactions of OH groups; and (4) reactive blending of structurally modified phenolic resin with a functional reactant. Novolac can also be viewed as a reactive intermediate and can be transformed into other groups and different types of structures can be generated. Various addition cure phenolic resins namely allyl-functional phenolics, bisoxazoline—phenolics, propargyl ether, ethynyl novolacs, etc. [356—360].

C.P Nair and coworker [361,362] reported novolac polymers bearing varying proportions of propargyl and phthalonitrile groups on the same backbone. The synthesis of the novolac-based polymers involves two steps. First one is the partial phthalonitrilation of novolac by reacting the novolac with 4 nitro-phthalonitrile. The reaction is basically a simple nucleophilic displacement of the nitro group as shown in Fig. 1.53. In the second step, the free phenol groups of novolac-phthalonitrile were propargylated by reaction with propargyl bromide using benzyl triethyl ammonium chloride as catalyst and acetone as the reaction medium. The synthesized novolac polymer (containing phthalonitrile and propargyl moiety)

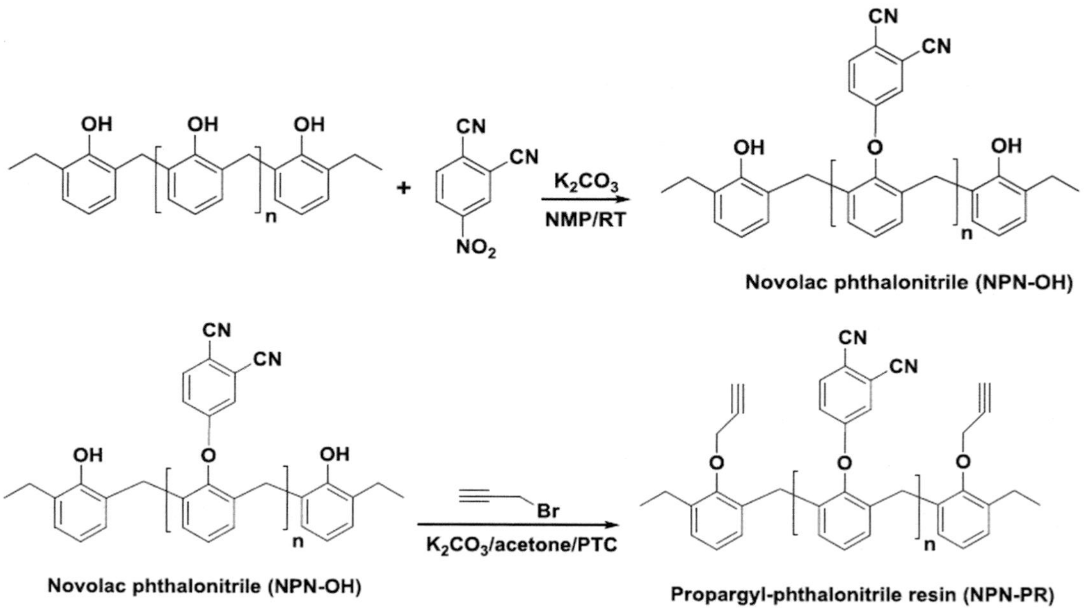

Figure 1.53
Synthetic route of phenol-phthalonitirle (NPN-OH) precursors and one-component propargyl-phthalonirile (NPN-PR) oligomers. Source: *Reprinted with permission from D. Augustine, D. Mathew, C. P. Reghunadhan Nair, Euro. Polym. J. 71 (2015) 389.* © 2015, Elsevier Publishers.

is capable of self and cocuring to produce a thermally stable material with initial decomposition temperature at around 435°C and char yield 73%–76% at 900°C. This is attributed to the formation of a network with thermally stable triazine, phthalocyanine groups. A schematic representation of the overall cocured network comprising of chromene, triazine, and phthalocyanine structures is depicted in Fig. 1.54

1.8 Amino resins

Amino resins are defined as the resins, which are produced by the reaction between amino group-containing compound and formaldehyde. The most popular amino resins are urea-formaldehyde (UF) resins and melamine-formaldehyde (MF) resins. Amino resins are considered as a supplement and complement to phenolic resins [363,364]. UF and MF resins are prepared by reacting formaldehyde with urea and melamine, respectively. The general reaction scheme for the synthesis of UF and MF resins is shown in Fig. 1.55. The liquid amino resins are prepared by reacting the corresponding methylol intermediate in n-butanol or methanol.

The amino resins are used as curing agents for the resins containing carboxyl, hydroxyl and amide groups. Amino resins can be used in all possible applications where phenolic resins

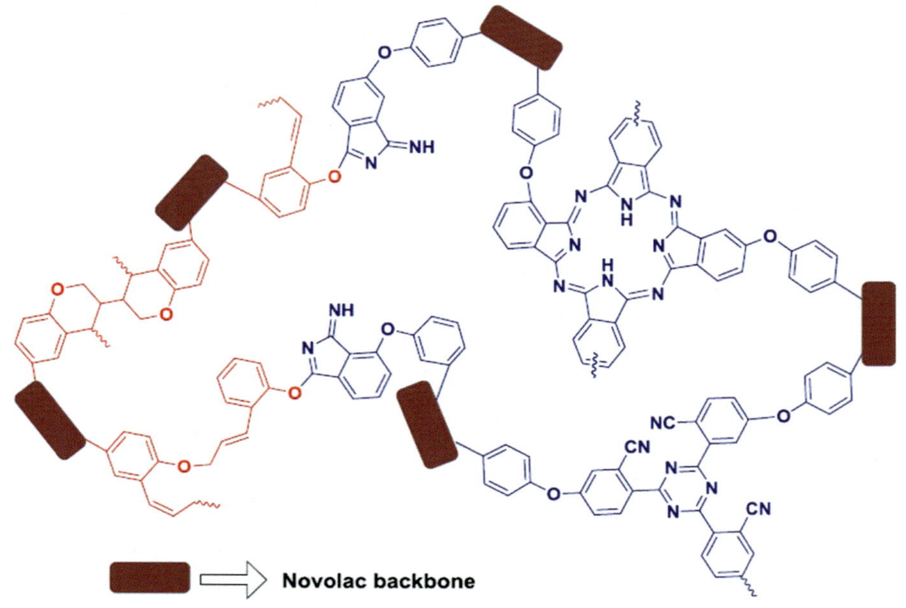

Figure 1.54
A schematic representation of overall cocured network comprising of chromene, triazine and phthalocyanine structures. Source: *Reprinted with permission from R. Nair, Euro. Polym. J. 71 (2015) 389.* © *2015, Elsevier Publishers.*

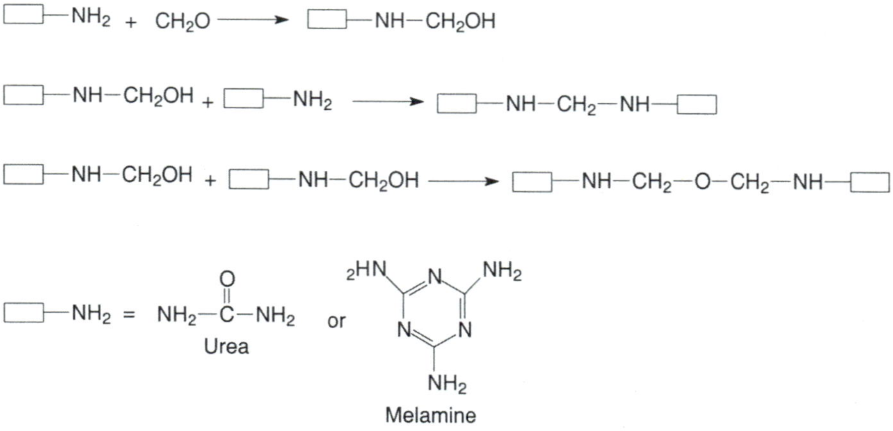

Figure 1.55
The general reaction scheme for synthesis of UF and MF resins. *MF*, melamine-formaldehyde; *UF*, urea-formaldehyde.

are used as discussed in Section 2.1.7. Unlike phenolic resin, which is dark brown in color, the amino resins are light in color. Hence, the amino resins have replaced phenolic resin in applications where bright color and aesthetic appearances are needed. MF resin shows the best performance in terms of water resistance and toughness properties. However, the use of MF resin is restricted due to its high cost. The first commercial production of MF resin was started by American Cyanamid Corporation (United State of America) in 1939. The well-known applications of phenolic resins are in domestic plugs and switches. UF resins have largely replaced phenolic resins for such purposes due to their better antitracking properties and wider color range.

1.9 Furan resins

Furan resins are also a low volume consumption resin like amino resins and used as a supplement to phenolic resins [365,366]. They are prepared by reaction between phenol and furan compounds like furfural, furfuryl alcohol, furan, etc. The furan compounds can be used in place of formaldehyde in the conventional production of phenolic resins. Most popular and viable furan resins are prepared from furfuryl alcohol (FFA). FFA undergoes homopolymerization through an addition reaction in an acid medium leading to the formation of a furan resin (Fig. 1.56). Like other thermosetting resins, furan resin undergoes cross-linking in the presence of a strong acid and forms a three-dimensional network. The cross-linking involves the reaction with methylene bridge as shown in Fig. 1.56.

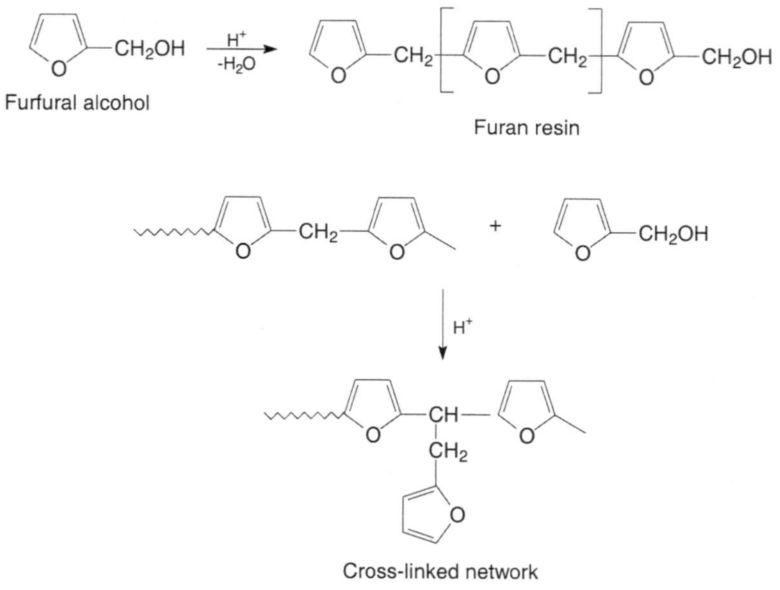

Figure 1.56
Reaction schemes for synthesis and cross-linking of furan resin.

Like amino resins, furan resins can also be used in all possible applications where phenolic resins are used as discussed above.

1.10 Benzoxazine resins

Polybenzoxazines (PBZs) are a class of addition-curable phenolic systems, which have the capability to overcome several shortcomings of conventional (novolac and resole-type) phenolic resins while retaining their beneficial properties. PBZ does not need any strong acid catalysts for curing and not associated with the release of any toxic by-products during curing as observed in the case of resole. It offers the thermal properties and flame-retardant properties comparable or better than phenolics and other advantages in terms of processing, mechanical performance and the flexibility of molecular design like an advanced epoxy system. PBZ exhibit (1) near-zero volumetric change upon curing; (2) low water absorption; (3) high char yield, excellent mechanical properties and resistance to chemicals and UV light and good dielectric property with low and stable dielectric constant and dielectric loss [367–371].

Benzoxazine monomers are characterized by the presence of oxazine rings (heterocyclic six-membered rings with oxygen and nitrogen atoms and attached to a benzene ring) and were first synthesized by Holly and Cope through Mannich condensations of phenols with formaldehyde and amines [372]. The work related to the potential of benzoxazines for the preparation of phenolic materials with improved properties was first patented by Schreiber [373]. Polybenzoxazines are prepared by reacting a phenol with formaldehyde and an aromatic amine and are cured by thermal ring-opening polymerization [374,375]. The mechanistic evidences for the formation of the benzoxazine ring structure suggest that amine reacts with formaldehyde and forms aminomethylol group as an intermediate immediately after mixing, then it further reacts to form 1,3,5-triaza-like structure. Subsequently, it reacts with phenol and formaldehyde to form the oxazine ring. At elevated temperature, benzoxazine monomer with or without initiator/catalyst undergoes cationic ring-opening polymerization and forms the polymer consisting of phenolic groups that are linked by Mannich base as repeating units. In general, the thermal curing of BZ monomers generates a PBZ structure via the attack of iminium ions on the ortho position (with respect to phenol) of the monomers. Regardless of the type of accelerator used, the mechanism of the curing process involves several steps. The various mechanisms are presented [375] in Fig. 1.57, which include (1) thermally accelerated ring-opening, (2) ring-opening by protonation, (3) ring-opening by hydrogen bond formation, (4) polymerization through the formation of an arylether structure (N,O-acetal-type linkage), and (5) Colbert reaction mechanism via two-step acid-catalyzed nucleophilic addition. The curing is always autocatalyzed as the newly produced hydroxyl groups in phenolic structure (compound III in Fig. 1.57) act as an additional initiator [375].

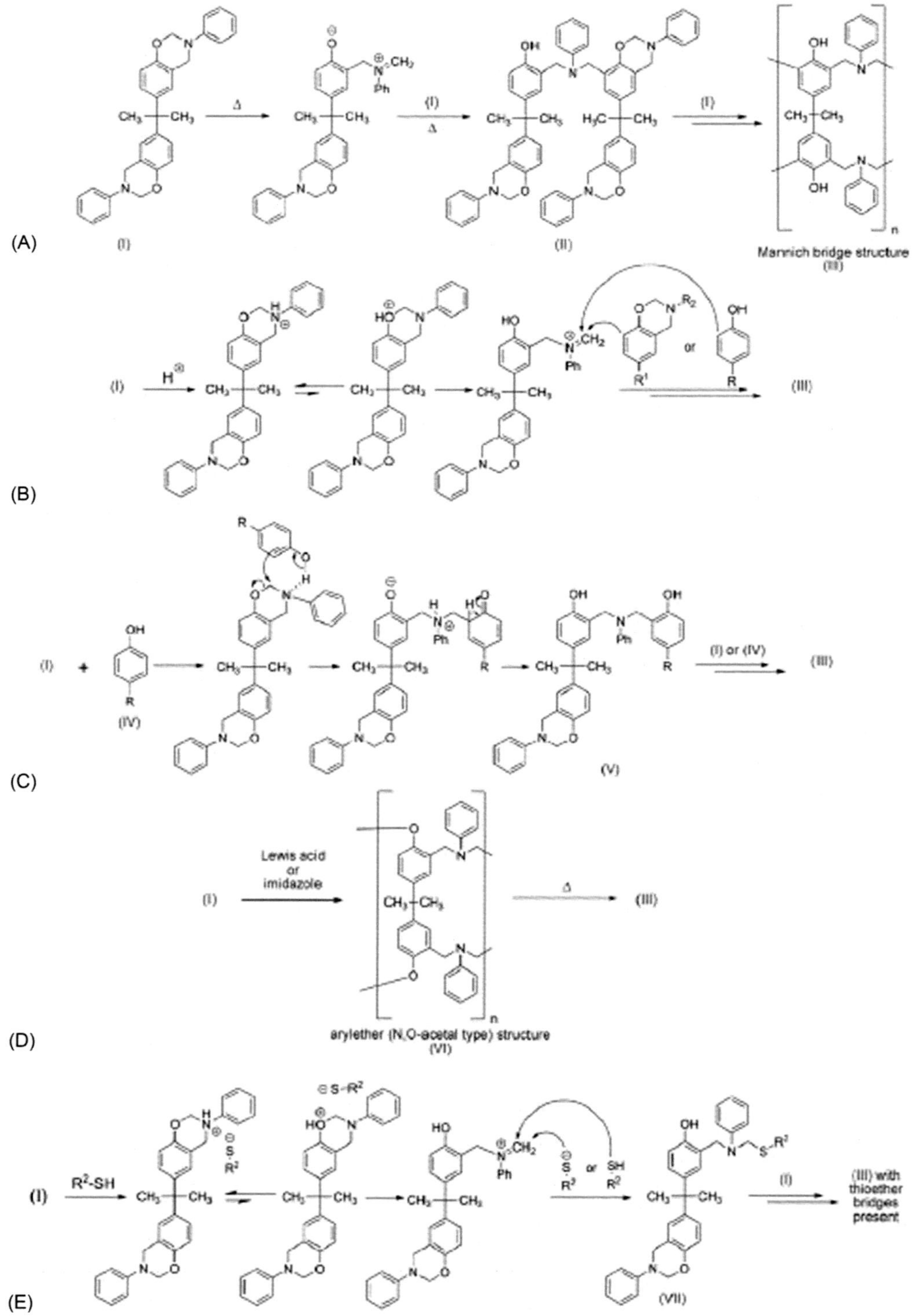
(A)
(I)
(II)
Mannich bridge structure
(III)
(B)
(C)
(IV)
(V)
(I) or (IV)
(D)
Lewis acid
or
imidazole
arylether (N,O-acetal type) structure
(VI)
(E)
R²-SH
(VII)
(III) with
thioether
bridges
present

(*Continued*)

The widely studied PBZ, that is, traditional PBZ network is synthesized from bisphenol-A (BPA) and aniline. BPA has been shown to polymerize via a thermally induced ring-opening reaction via ortho attack to form a phenolic structure characterized by a Mannich base bridge (CH_2-NR-CH_2) instead of the methylene bridge structure associated with traditional phenolic resins [376,377]. This explains why conventional phenolics are more brittle than polybezoxazines. Using different types of phenol and amine, a variety of PBZ polymer can be synthesized. In other words, the choice of phenol and amine permits design flexibility and tailoring of polymer properties. PBZ derived from monofunctional monomers (containing one oxazine ring) forms linear polymers, whereas di- or polyfunctional BZ (having two or more oxazine rings) display cross-linking networks which are very useful for commercial applications.

A combination of molecular modeling, density functional theory, chemical-shift calculations, and advanced solid-state nuclear magnetic resonance (NMR) experiments such as 1H-NMR−magic angle spinning (MAS) and double quantum MAS spectroscopy were utilized to elucidate the supramolecular and helical structure of BZ oligomers [378,379]. It was reported that PBZ systems have lower cross-link densities but they offer higher T_g values and high moduli when compared with an epoxy-based system [380]. This phenomenon is unexpected because, in general, the glass transition temperature increases with increasing the degree of cross-linking. This behavior is attributed to the hydrogen-bonding interaction in PBZ, which leads to higher stiffness, modulus and T_g. The polymers exhibit a T_g in the range of 160°C−340°C, depending upon their chemical structure. They are usually cured in the temperature window of 160°C−220°C. The ring-opening polymerization of these new materials occurs with near-zero shrinkage or even with a slight expansion upon cure. It is proposed that the volumetric expansion of the BZ resin is mostly due to the consequence of molecular packing influenced by inter- and intramolecular hydrogen bonding.

The curing reactions of BZ precursors based on bisphenol-A and aniline were extensively studied using DSC to determine the feasibility of processing them into final phenolic products. Curing was found to have an autocatalytic character with an overall reaction order of 2 [381]. The curing of BZ precursors is an autocatalyzed reaction until vitrification is reached, beyond which the reaction becomes diffusion-controlled. The activation energy is

Figure 1.57
Possible pathways of benzoxazine polymerization and cross-linking: (A) thermally accelerated ring-opening, (B) ring-opening by protonation, (C) ring-opening by hydrogen bond formation, (D) polymerization through formation of an arylether structure (N,O-acetal-type linkage), and (E) Colbert reaction mechanism via two-step acid-catalyzed nucleophilic addition. Source: *Reprinted with permission from A. Rucigaj, B. Alic, M. Krajnc, U. Sebenik, Express Polym. Lett. 9 (2015) 647−657 © 2015 Online Publishing.*

reported to be 102–116 kJ mol^{-1}. Isothermal and dynamic experiments suggest the possibility of structural rearrangements occurring at high temperatures. The isothermal heat of reaction follows a quadratic relationship with the cure temperature, whereas the onset of curing and the exotherm maximum during the isothermal tests were found to decay exponentially with temperature. The curing kinetics of polyfunctional benzoxazine resins based on 3,5-xylidine and bisphenol-A (designated as BA-35x) was also investigated [382]. The BA-a resin shows only one dominant autocatalytic curing process with an average activation energy of 81–85 kJ mol^{-1}, whereas BA-35x exhibits two dominant curing steps.

1.10.1 Main-chain polybenzoxazines

The monofunctional benzoxazine molecules do not form large-molecular-weight polymers despite the possibility of forming a linear polymer [383]. Under mild polymerization conditions, chain growth can terminate at the dimer length due to formation hydrogen bonds at the growth front [384]. Difunctional benzoxazines can be prepared by using a diamine in place of a monoamine. The difunctional benzoxazines can be cured to produce crosslinked polymers with improved thermomechanical properties. Although monofunctional and difunctional benzoxazines offer advantages in terms of processing due to their inherent low viscosity, their thermomechanical properties are not sufficient for their use as matrices for advanced high-temperature composites for aerospace and missile applications. This can be attributed to lack of molecular connectivity between the nanoaggregates present in the matrix. To improve molecular connectivity, PBZ with oxazine moiety in the main chain was developed [385–392], which is known as main-chain PBZ. During processing, the main-chain PBZ behave like a thermoplastic and can be processed as self-supporting films.

Ishida and coworker [385] reported a linear PBZ molecule with oxazine rings in the main chain by reaction of 1,4 diaminohexane, bisphenol-A, and formaldehyde with a molecular weight of ~10 kDa. The reaction scheme is shown in Fig. 1.58. The resultant polymer had a moderately broad polydispersity index of 3.0. Two distinctive exothermic peaks (161°C and 242°C) appeared in the DSC analysis. The first peak could be assigned to the crosslinking reaction due to methylol end groups. The higher temperature peak was due to conventional BZ polymerization. The polymer can be cured at elevated temperatures to form a 3D network. If we compare its properties with the same of polymer derived from a similar diamino hexane-based monomeric benzoxazinethe, crosslinked polymer derived from the main-chain oxazine rings seems to be thermally more stable. This could be caused by better decomposition resistance due to the improved network connectivity in the crosslinked material derived from the reactive polymer as shown in Fig. 1.59.

High-molecular-weight PBZ precursors have been synthesized from aromatic or aliphatic diamine and bisphenol-A with paraformaldehyde [386]. These PBZ films exhibited

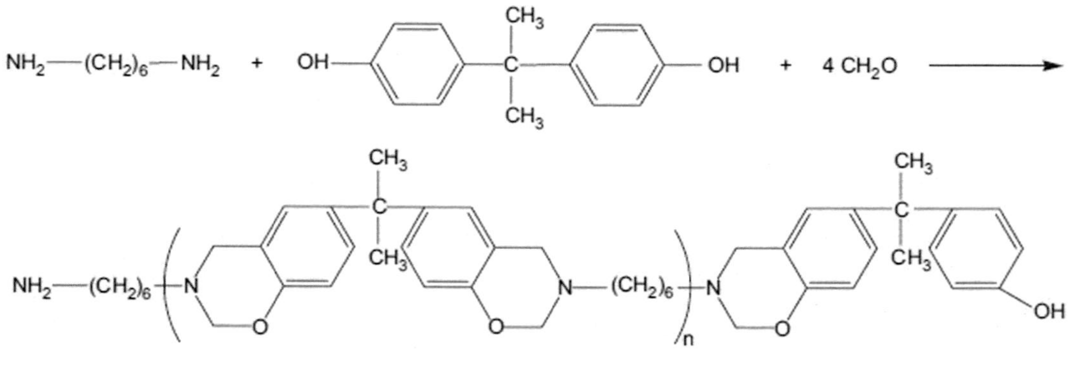

Figure 1.58
Synthesis of main-chain benzoxazine polymer. Source: *Reprinted with permission from A. Chernykh, J. Liu, H. Ishida, Polymer 47 (2006) 7664* © *2006, Elsevier Publishers.*

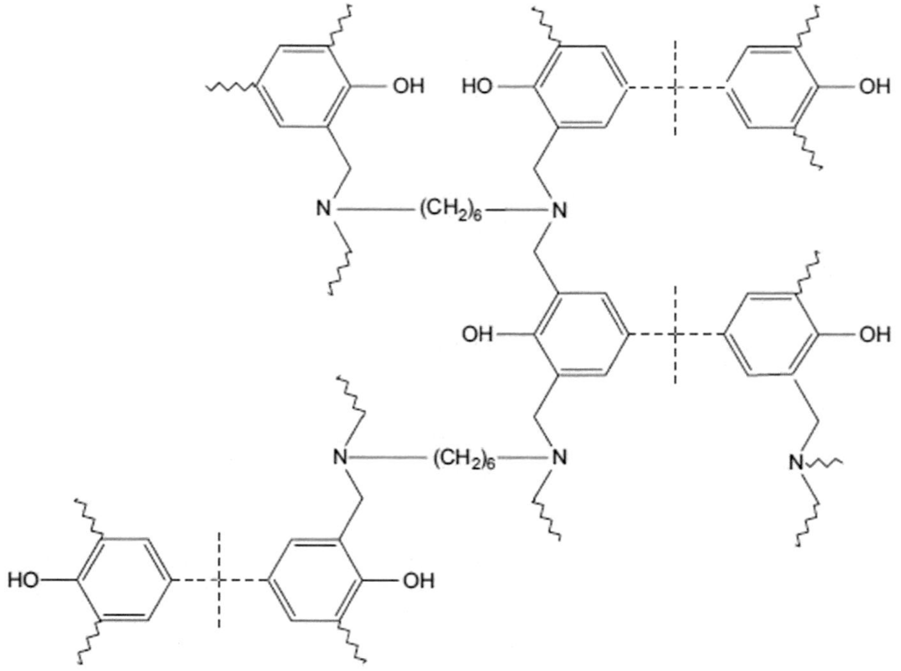

Figure 1.59
Additional connectivity (dashed) of the network structure in the crosslinked main-chain benzoxazine polymer compared to the crosslinked aliphatic diamine-based benzoxazine. Source: *Reprinted with permission from A. Chernykh, J. Liu, H. Ishida Polymer 47 (2006) 7664.* © *2006, Elsevier Publishers.*

significantly improved toughness than the traditional PBZ (PBA-a) due to the long linear backbone and high cross-link density. The bisphenol-A- and dimethyl diamine-based PBZ (PB-mda) and bisphenol-A- and hexamethyldiamine-based PBZ (PB-hda) films were very tough and showed good bending flexibility. The T_g values of PBZ from the high-molecular-weight precursors were 238°C−260°C, much higher than the T_g of the typical PBA-a (171°C). As discussed above, the thermal and mechanical performance of PBZ thermosets derived from main-chain benzoxazine polymers are superior in terms of thermomechanical properties, compared to those derived from monomers. However, the synthesis of main-chain benzoxazine polymers through Mannich base polycondensation is associated with shortcomings, such as poor solubility of the products due to molecular rigidity resulting in poor processing performance of the main-chain PBZ systems. In recent years, a lot of research works have been carried out taking advantage of the design flexibility of PBZ systems to address the processability issues. It may be noted that the precursor for a thermosetting resin should be in the prepolymer/oligomeric form rather than polymeric or monomeric form to achieve desirable mechanical properties of the cured networks especially the ductility. With this motivation in mind, a number of novel PBZ copolymer oligomers have been designed and developed which exhibit comparable thermomechanical properties like main-chain PBZ but better processability and superior dielectric properties. The precursors used for synthesis of various types of PBZ including the main-chain PBZ copolymer oligomer are presented in Table 1.19 for a better understanding of the synthetic process.

Liu et al. [393] synthesized short-chain oligomers using the isomer mixture of bisphenol-F, formaldehyde, aromatic diamine, and aniline (AN) as a reactive diluent. The reaction scheme is shown in Fig. 1.60. The pendant benzene ring in bisphenol-F/aniline-based benzoxazine monomers is relatively easy to degrade in comparison to the DDM-based benzoxazine oligomers. Hence, the polybenzoxazines derived from the oligomers show enhanced thermal stability compared with the polybenzoxazines derived from the monomers. The temperature corresponding to 5% weight loss; T_{d5} obtained from T_g A increased from 332°C to 364°C as a result of the reduced evaporation rate around 300°C. While the char yield of the monomer and oligomers do not show a significant difference, the intermediate temperature stability in the range of 300°C and 500°C is significantly higher in the oligomer-derived polybenzoxazine than the one derived from a monomer. The

Table 1.19: Precursor used for various types of polybenzoxazines.

Benzoxazine monomer	Phenol, monofunctional amine, formaldehyde
Difunctional benzoxazine monomer	Phenol or bisphenol-A, difunctional amine, formadehyde
Main-chain polybenzoxazine prepolymer	Bisphenol-A, difunctional amine, formadehyde
Main-chain polybenzoxazine Copolymer prepolymer	Bisphenol-A, difunctional amine, formaldehyde, phenol, or amine

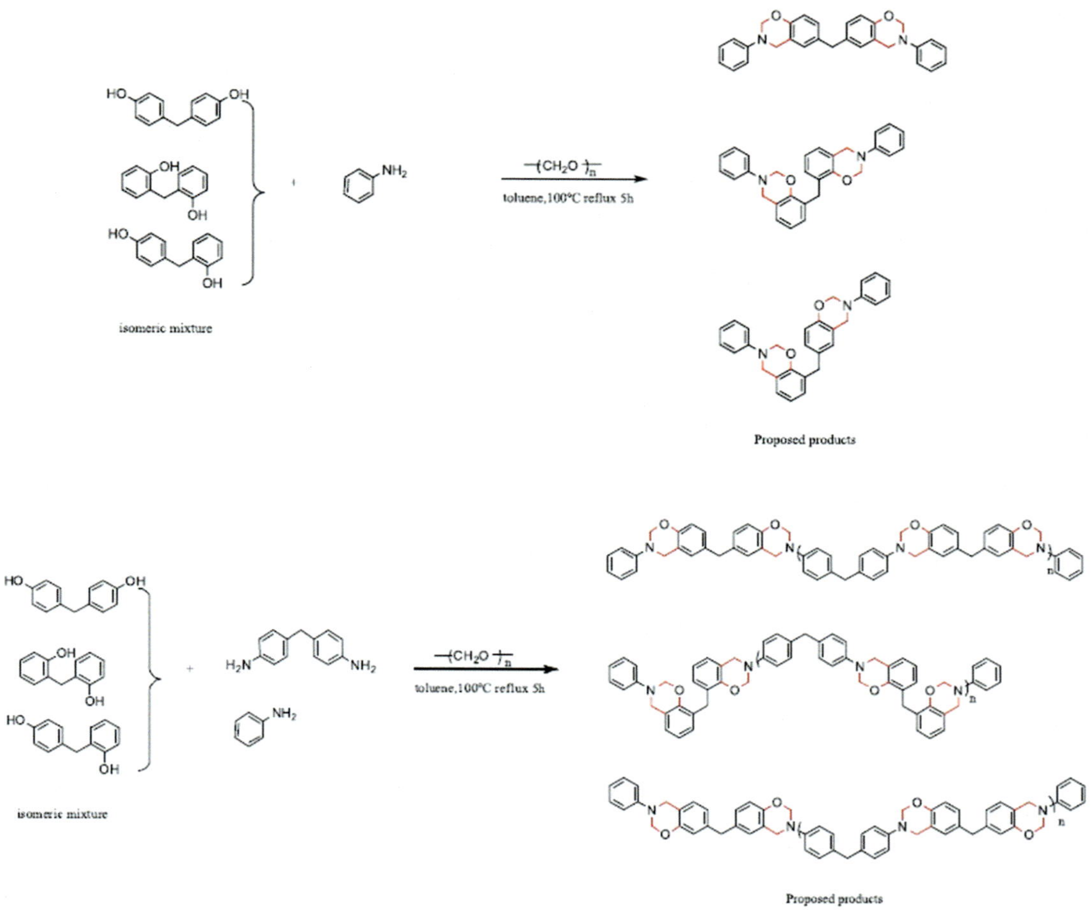

Figure 1.60
Synthesis of bisphenol-F isomer-based benzoxazine monomer and oligomers. Source: *Reprinted with permission from J. Liu, T. Agag, H. Ishida Polymer 51 (2010) 5688. © 2010, Elsevier Publishers.*

developed resins offers required low viscosity for processing of the related fiber-reinforced composites through the resin transfer molding (RTM) process. Yin and coworker [394] investigated the rheological properties of aliphatic amine-based benzoxazine and reported a stable viscosity lower than 0.8 Pa s from 85°C to 120°C and suitable for processing using RTM.

Xu et al. [395] synthesized a main-chain benzoxazine copolymer oligomers from 4,4′-diaminodiphenylmethane (DAM) from AN, and bisphenol-A with paraformaldehyde by the Mannich reaction. The oligomer is capable of undergoing ring-opening polymerization of oxazine rings to produce main-chain benzoxazine copolymer films. The aniline used in this reaction can react with the hydroxyl groups at the end of main-chain benzoxazine leading to the formation of a structure, terminated with oxazine rings, accompanied by the decrease in chain length and molecular polarity. The purity and molecular weight of aromatic diamine-based main-chain polybenzoxazine precursors depends on the polarity of the solvent used as a reaction medium. The above-mentioned main-chain benzoxazine oligomers (designated as BAM series) were prepared separately using two solvents (e.g., toluene and toluene/ethanol mixture) as described in Fig. 1.61. The properties of BAM oligomer are presented in Table 1.20. The molecular weight of the oligomers is lower than the high-molecular-weight

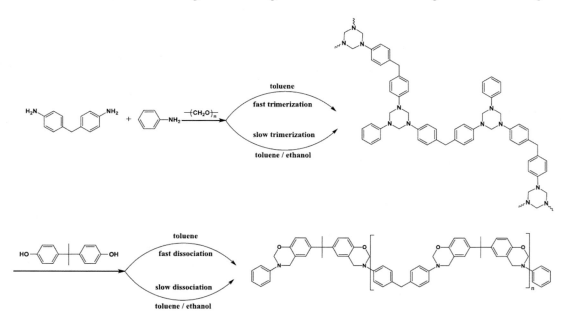

Figure 1.61
Proposed reaction pathways for the preparation of aromatic main-chain copolymer oligomers in toluene and toluene/ethanol. Source: *Reprinted with permission from Q. Xu, M. Zeng, J. Chen, S. Zeng, Y. Huang, Z. Feng, Q. Xu, C. Yan, Y. Gu, React. Funct. Polym. 122 (2018) 158.* © *2018, Elsevier Publishers.*

Table 1.20: Solvent effect on the synthesis of benzoxazine copolymer precursors.

Sample	Solvent	Reaction condition	Mn/Da	Mw/Da	Polydispersity Index (PDI)
BAM-1	Toluene	90°C/6 h	878	1775	2.02
BAM-2	Toluene	90°C/7 h	872	1852	2.12
BAM-3	Toluene	90°C/8 h	946	1903	2.01
BAM-4	Toluene	100°C/8 h	795	1517	1.91
BAM-5	Toluene	110°C/6 h	905	1744	1.93
BAM-6	Toluene	110°C/8 h	913	1684	1.84
BAM-7	Toluene/Ethanol (2:1v/v)	80°C/6 h	750	1421	1.89
BAM-8	Toluene/Ethanol (2:1v/v)	80°C/8 h	771	1434	1.86
BAM-9	Toluene/Ethanol (2:1v/v)	80°C/10 h	759	1414	1.86

Source: *Reprinted with Permission from Q. Xu, M. Zeng, J. Chen, S. Zeng, Y. Huang, Z. Feng, Q. Xu, C. Yan, Y. Gu, Funct. Polym. 122 (2018) 158 – 166.* © *2018 Elsevier Publishers.*

main-chain PBZ ($\sim$10 kDa) discussed earlier [385]. In the present case, in addition to an aromatic diamine, a monofunctional amine (aniline) is used, which not only takes part in the formation of the copolymer oligomer but also lowers the viscosity and molecular weight of the oligomers.

It is observed that a relatively lower molecular weight and higher purity of BAM copolymer prepolymers were obtained in the heterogeneous solvent (mixture of toluene and ethanol). In a similar study, Lin et al. [396] found that aromatic diamine-based main-chain PBZ copolymer oligomers, prepared in toluene/ethanol heterogeneous solvent exhibit higher purity and molecular weight leading to a better thermal property than those prepared in chloroform. The results can be explained by considering the chemistry of the synthesis of PBZ. There are two steps for the preparation of benzoxazine monomers or oligomers: (1) the formation of a hexahydro-s-triazine network and (2) the dissociation of the resulting hexahydro-s-triazine network by phenols [397]. It was found that the polar solvent could not only reduce the hexahydro-s-triazine network formation rate but also slow down the reaction rate of the intermediate product withbisphenol-A for aromatic diamine-based copolymer precursors as elaborated in Fig. 1.61 [395]. This is due to the fact that oxazine ring prefers to form in the nonpolar solvent rather than in the polar solvent [398]. Furthermore, the polar solvent could act as an effective filter to improve the purity of oligomers to some extent. Therefore it could be concluded that the reaction solvent really plays a key role in the preparation of aromatic diamine-based main-chain benzoxazine copolymer prepolymers. The polar solvent was assumed to not only dissociate the triazine networks but also reduce the rate of the formation of oxazine groups, resulting in a relatively lower molecular weight and higher purity of BAM copolymer prepolymers. Accordingly, the BAM-8 copolymer films prepared in the optimized toluene/ethanol solvent condition had the least activation energy, the minimum dielectric constant value, and a relatively high T_g. The dielectric constant value for BAM-8 (prepared in toluene/ethanol mixture at 80°C/8 h) was 2.81 whereas the same values for copolymer oligomers (prepared in toluene) were in the range of 2.91–3.32. Note that a lower value of the dielectric constant is beneficial for use in microelectronic industries, which will be elaborated in the subsequent sections.

If we compare the rates for a reaction for aromatic and aliphatic diamine-based bezoxazines in heterogeneous solvent, it is observed that the time of completing the polycondensation reaction for an aliphatic diamine-based system was shorter than that for an aromatic diamine-based system [399]. This is because unlike in the case of aromatic amine-based main-chain PBZ, a relatively fast dissociation rate of the hexahydro-s-triazine network and fast cyclization reaction rate were obtained for aliphatic diamine-based counterparts prepared in the polar/nonpolar heterogeneous solvents.

The above-mentioned PBZ copolymer oligomers have been synthesized using an aromatic amine as a chain terminator and the copolymers exhibit better processability and dielectric properties compared to high-molecular-weight main-chain PBZ. PBZ copolymer oligomers have also been synthesized using phenol as a chain terminator. For example, Wang et al. [400] reported the polybenzoxazine precursor from the Mannich condensation of 2,2′-*bis* [4-(4-aminophenoxy) phenyl] propane, paraformaldehyde, bisphenol-F, and phenol, in which phenol was used as a chain terminator. A new main-chain benzoxazine copolymer oligomer has also been synthesized from bisphenol-A, paraformaldehyde, and o-norbornene functional phenol [401]. All the oligomers mentioned above have low viscosity, while maintaining the major advantages of main-chain benzoxazines.

The PBZ copolymer oligomers are characterized by terminal oxazine ring leading to a decrease in polar functional groups. Therefore they offer very good dielectric property. The conventional PBZ offers a dielectric constant (k) of about 3.5 and a dielectric loss factor (f) of about 0.02. The dielectric property can be improved by taking advantage of the design flexibility of PBZ copolymers. In order understand the theoretical basis for dielectric property, it is necessary to understand the factors, which affect the dielectric properties, that is, k and f. The dielectric constant depends on various types of polarization and can be related using Debye equations [402]

$$\frac{k-1}{k-2} = \frac{4\pi}{3} N \left(a_e + a_d + \frac{u^2}{3k_b T} \right)$$

where k_b is the Boltzmann constant, N is the number density of dipoles, a_e is the electric polarization, a_d is the distortion polarization, u is the orientation polarization related to the dipole moment, and T and k are the temperature and the dielectric constant, respectively. From this equation, we can easily explain why aliphatic amine-based PBZ exhibit lower k compared to aromatic amine-based PBZ. An amorphous polymer shows lower k than the crystalline one with the same chemistry due to a lower value of orientation polarization (u) for the former. Similarly, the incorporation of bulky group in benzoxazine polymer will result in a decrease in molecular packing and an increase in free volume resulting in a reduction of N. Thus the k value of a PBZ polymer can be reduced by introducing nonpolar bulky moiety like $-C(CH_3)_3$ into the polymer structure. However, it is difficult to make a polymer with good mechanical properties having such a bulky group in the main chain. Using the principle as discussed above, very recently, Chen et al. [403] designed and prepared an aliphatic diamine-based PBZ copolymer oligomer using p-tert-butylphenol as a chain terminator and ethanol/toluene heterogeneous solvent. By incorporating a bulky group and selecting the difunctional amine and phenol judiciously, they have successfully developed a PBZ copolymer oligomer system with low k (<3) and ultralow f (<0.005). The presence of a bulky group in the polymer chain and amorphous aliphatic structure makes it possible to achieve such outstanding dielectric properties especially measured at higher frequency 5–10 GHz.

1.10.2 Polybenzoxazine-based blends

In the previous section, we have discussed how the property and processability of a PBZ resin can be adjusted by changing the chemistry and using a copolymerization strategy. However, the commercially available PBZ resins show some disadvantages like high brittleness, low cross-linking density, less pot life, and internal void formation in the final product. Various strategies like alloying, blending, or cocuring have been utilized to overcome the disadvantages mentioned above. Toughening of PBZ resin will be discussed elaborately in Chapter 3. In this section, few blend systems will be discussed, which offer synergistic improvement in the properties of PBZ resins. The cross-link density and mechanical properties of a PBZ network can be improved by cocuring with a DGEBA-type epoxy resin. The epoxy resin gets cured without any external curing agent like amine or anhydride. Benzoxazine itself acts as a curing agent for the epoxy resins.

Let us discuss about the curing of PBZ/epoxy blend systems. When the curing of benzoxamine and epoxy resins are considered separately, their polymerizations are relatively simple. However, when they are mixed together, the polymerizations become complex because the tertiary amine of the Mannich bridge and phenolic OH produced by the ring-opening polymerization of BZ resin catalyzes the homopolymerization or copolymerization of epoxy resin. In the present case, we can visualize three different types of curing reactions, namely, (a) the thermal-induced homopolymerization of PBZ resin; (b) the thermal-induced or tertiary amine (Mannich bridge of polybenzoxazine) catalyzed homopolymerization of DGEBA; (c) the thermal-induced or tertiary amine (Mannich bridge of polybenzoxazine) catalyzed copolymerization between the phenolic hydroxyl (−OH) of PBZ and the epoxy group of epoxy resin. The mechanism of cocuring of epoxy and PBZ is shown in Fig. 1.62.

The phenolic hydroxyl groups (OH) produced by the ring-opening polymerization of BZ can copolymerize with epoxy groups of epoxy resin. The cocured networks are reported to exhibit homogeneous morphology with no indication of phase separation [404,405]. However, Karger Kocsis and coworker [406,407] occasionally observed nanoscale (5−150 nm) phase-separated structures in a series of PBZ/epoxy systems. The possibility of phase separation occurs if epoxy resin polymerizes before BZ due to an increase in dynamic asymmetry between the benzoxamine and epoxy. Zhao et al. [408] used a series of epoxy resins having different molecular weights (M_n = 420, 492, 1220, 1854, 4370, 5116 g mol^{-1}) and investigated the morphology of PBZ/epoxy blends using various techniques such as turbidity observation, dynamic mechanical analysis and scanning electron microscope experiments. It was reported that the morphology of the blends depends on the molecular weight of the epoxy resin used for blending. If the M_n of epoxy resin exceeds 4370 g mol^{-1}, phase separation could take place during cocuring with BZ resin and different morphologies could be observed by varying composition of the blends.

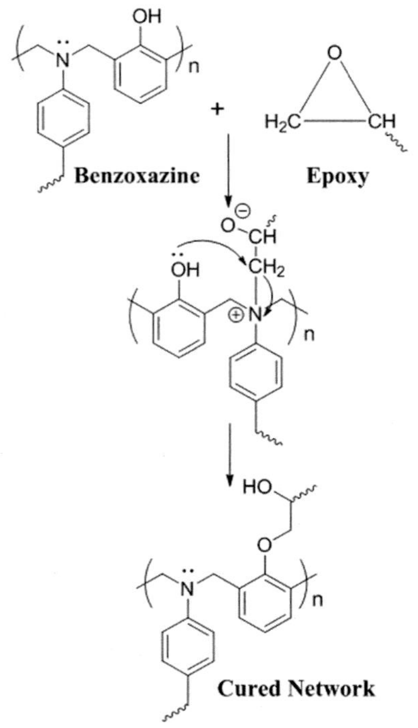

Figure 1.62
The mechanism of cocuring of epoxy and benzoxazine resins.

It is also reported that mechanical property and shape memory properties of a bisphenol-A-based polybenzoxazine can be significantly improved by cocuring with an epoxy system comprising bisphenol-A epoxy and diepoxy polyisopropylene glycol. The blending of flexible aliphatic epoxy resin to the benzoxazine resin improves fracture toughness and processing window and also helps to overcome the problem of degassing required for benzoxazines to get internal void-free products. However, the improvement in mechanical properties and shape memory properties are achieved with a significant sacrifice in thermal properties. To address this issue, bisphenol-F-based polybenzoxamine has been cocured with an epoxy system comprising bisphenol-F novolac epoxy and diepoxy polyethylene glycol [409,410]. Novolac epoxy resins of ternary mixture with its rigid aromatic structures and higher epoxy functionality (f) ($f > 2$) can enhance the cross-linking density and mechanical strength of the material which results in an increase in T_g but at the same time, aliphatic epoxy resins improve the segmental mobility, cross-link density and lower the brittleness of the material which results in lower T_g of polybenzoxazine. Hence, the developed ternary systems using a combination of rigid and flexible epoxies, with optimum composition offer improved service temperature, thermal stability, cross-linking density, processing window and fracture toughness when compared to polybenzoxazine alone. The applicability of the ternary mixture was demonstrated by preparing S-glass fabric reinforced composites using the ternary resin systems. These composites showed improved mechanical properties, hardness, and hydrolytic stability when compared to pure polybenzoxazine composites.

1.10.3 Application of polybenzoxazine

Polybenzoxazine resins can be used as an alternative to epoxy, phenolic, and bismaleimide resins. Because of their superior resistance to chemicals, low flammability, and excellent heat stability, they are an excellent choice for components that are exposed to high temperatures and corrosive media. Examples include chemical- and heat-resistant coatings, adhesives, prepregs, and encapsulants as well as halogen-free laminates for printed circuit boards. Polybenzoxazines are also used in the automotive and aerospace industries for applications where superior thermal and mechanical properties relative to conventional resins are required.

Benzoxazines are potential materials for integrated circuit board (ICB) board, which plays a critical role in performing the electrical functions of most electrical devices. ICBs are fabricated with copper-clad laminates, which comprises a polymeric matrix reinforced with glass fabrics and copper layers. The dielectric property of the resin system used for the laminate plays an important role in controlling the data transmission volume and data transmission rate of the IC and the related electronic devices [411]. Signal propagation loss is proportional to dielectric loss (f) and the square root of dielectric constant (k). The low value of k and f and other properties like high-temperature resistance, low moisture uptake, good chemical resistance good adhesion to semiconductor surfaces, near-zero cure shrinkage make them suitable for their use in copper-clad laminate of ICB. Recently developed PBZ copolymer oligomer discussed above [403] having with low k (<3) and ultralow f (<0.005) will find application for high-speed high-frequency (5–10 GHz) IC system.

Low viscosity benzoxazine monomers containing allyl groups can be used as reactive diluents to replace styrene in vinyl ester resins [412]. Styrene is used as a reactive diluent to improve the processability of VE resins. The main problem, associated with the use of styrene is the high emission, which causes health and environmental risks due to the volatile nature of styrene. It may be noted that the permissible limit of styrene in the atmosphere is 50 ppm. The emission issue restricts the application of VE resins, especially in the close compartment. The low viscosities of allyl-containing benzoxazine monomers help to maintain the good processability of traditional vinyl ester resins without using styrene. The allyl groups can copolymerize with vinyl ester resin via a free radical mechanism and benzoxazine can polymerize cationically during postcuring of the resin to form a more thermally stable network. The allyl-containing benzoxazine monomers are much safer in terms of volatile emission compared to styrene and the benzoxazine monomer-based vinyl ester resin offers better thermal stability compared to styrene-based vinyl ester resins. PBZ has been used as a latent curing agent of epoxy resin [413]. The latent cure epoxy is already discussed earlier (Section 1.5.3.1). For classical polybenzoxazines, hydrogen bonds are formed between the generated phenolic hydroxyl (−OH) and hydrogen-bonding acceptors (i.e., −OH, −N-Ar, and π). The hydrogen bond interaction can block the reactivity of −OH at a low temperature and release active free phenolic hydroxyl with increasing temperature. PBZ not only offers thermal latency, when used to cure epoxy resin but also produce copolymers of polybenzoxazines and epoxy resins, which exhibit enhanced properties. The phenomenon is described in Fig. 1.63.

PBZ systems using bio-based feedstock like cardanol, vanillin, eugenol, furfuryl amine, and stearyl amine have been synthesized [414−419]. These bio-based systems have demonstrated great potential to promote the development of "Green chemistry" due to the abundant availability and low cost of some of the feedstock like eugenol. However, the eugenol-based PBZ systems exhibit glass transition temperature in the range of

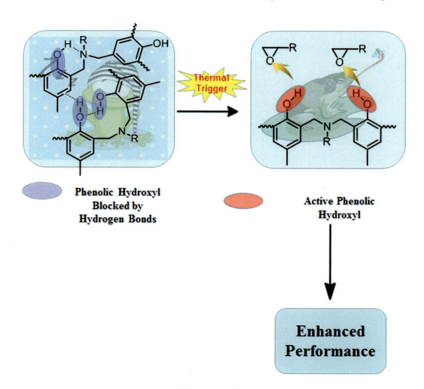

Figure 1.63
Phenomenon of thermal latency for curing of epoxy offered by bezoxazine though H-bonding.
Source: *Reprinted with permission from S. Zhang, P. Yang, Y.; Bai, T. Zhou, R. Zhu, Y. Gu, ACS Omega 2 (2017) 1529. © 2017, American Chemical Society Publishers.*

150°C–220°C, which is much lower than the same prepared from the fossil fuel-based feedstocks. The eugenol-based benzoxamines have been modified with comonomers such as furfuryl amine, bismaleimide stearyl amine to enhance their thermal stability to a certain extent [420–422]. The resins offer low viscosity (<1 Pa s at 120°C) and can be used as matrices for fiber-reinforced composites to be processed through resin transfer molding (RTM). The bio-based FRP composites may find applications where service temperature is less than 100°C. They can also find applications as diluents for traditional RTM resins.

1.11 Polyimides

Polyimides are a class of high-temperature resin with −CO-NR-CO- backbone that predominantly consists of ring structure [423–426]. Because of ring structure, polyimides offer outstanding thermal stability. The commercial polyimide materials were introduced by DuPont in the early 1960s in the trade name of Kapton, Vespel, etc. Polyimides are generally made by reacting a dianhydride with an aromatic amine. A general reaction scheme is shown in Fig. 1.64. In the first step, a polyamic acid intermediate is formed

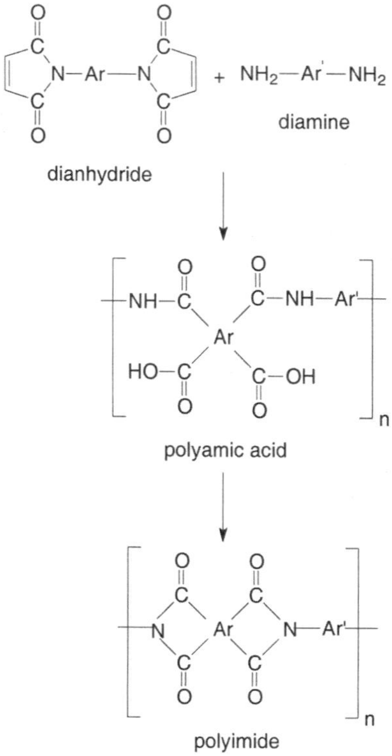

Figure 1.64
A general reaction scheme for the synthesis of polyimide.

Table 1.21: Properties of commercial thermoplastic polyimides.

Polyimide	T_g (°C)	Service temperature (°C) (max)	Tensile strength (MPa) (Mod., GPa)	Flexural strength (MPa) (Mod., GPa)	G_{IC}, (J m^{-2})	Impact strength (J m^{-1})
Torlon (BP Amoco)	267	225	186 (4.4)	211 (4.5)	3900	133.5
Kapton H (DuPont)	360	316	173 (3.0)	–	2000	23
Lark TPI (NASA Lange Centre)	265	300	166 (3.6)	–	–	21.4
Vespel SP-1 (DuPont)	360	287	72	83 (3.2)	2000	–
Pyramil PI (HD Micro-system)	400	316	113 (2.5)	–	–	–

which further polymerizes to form polyimides. They are generally prepared using N-methyl pyrrolidone (NMP) or dimethylacetamide (DMAc) as a solvent. An appropriate amount of diamine is dissolved in the solvent at room temperature under nitrogen. The dianhydride is then added as a solid or as a slurry in the solvent under stirring conditions. The reaction is allowed to continue till all the amine is consumed. Depending on the reactivity of the raw materials it may take 5–30 h to prepare the polyamic acid solution. The polyamic acid solution on further heat treatment gets converted to polyimide. The polyamic acid solution can be used to prepare polyimide powder, thin films, adhesive tape, or impregnated fibers.

The high-molecular-weight polyimides are called condensation or thermoplastic polyimide. Thermoplastic polyimides exhibit high T_g, good toughness, and excellent thermal and thermooxidative stability. Typical properties of some thermoplastic polyimides are given in Table 1.21. However, it is very difficult to process these materials because of their high softening or melting temperature. In order to solve the processing problem or to make the materials more tractable without significant sacrifice in thermal stability, extensive works have been done in last few decades. This has led to the development of polyamide-imides, polyester-imide, polybismalenimides, and polyether imides [427]. Detailed discussion on these materials is beyond the scope of this book, which is dedicated to thermosetting resins. Another class of polyimides, which have been developed to address the processability issues, is known as addition polyimides or thermosetting polyimides.

1.11.1 Addition polyimide

Addition polyimides are introduced in the late 1970s and investigated extensively in the last few decades [428–447]. These materials are low-molecular-weight imide oligomers with unsaturated functional groups (capable of undergoing an addition type of reaction) located in terminal or pendant position. Mostly, the materials contain unsaturated end caps, and in

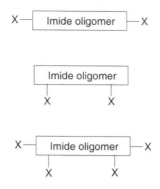

Figure 1.65
Representations of polyimide resins with terminal and pendant reactive groups.

some cases, they can also be present as pendant groups in the backbone. Schematic representations of such polyimide oligomers are shown in Fig. 1.65. The materials are synthesized in a similar way as discussed for condensation polyimide using an aromatic compound with the desired unsaturated group. A general method for the preparation of addition polyimides is shown in Fig. 1.66. A wide variety of polyimides can be synthesized by changing the unsaturated functional group (depicted as X in Fig. 1.66) for example vinyl, nadic, acetylene, phenyl ethynyl, etc. The polyimides are characterized easily by FTIR analysis by the presence of the two characteristic bands at 1790 cm^{-1} for symmetrical carbonyl stretching and at 1710–1730 cm^{-1} for asymmetrical carbonyl stretching. The disappearance of the characteristic anhydride carbonyl peak at 1860 cm^{-1} indicates the completion of the reaction.

Because the resins are oligomeric in nature, addition polyimides (PI) exhibit excellent processability during the fabrication of castings or composites. A list of thermosetting PI resin with different end-capping groups and their precursor materials is presented in Table 1.22. Upon thermal cure, the unsaturated groups undergo complex reactions, which include isomerization, chain extension, branching, addition, etc., leading to the formation of a network structure. The curing reactions of such polyimides are not completely understood. The reaction involves the formation of double bonds or polyene structures. Since the reactions are addition type no volatiles are evolved as a byproduct.

1.11.2 In situ polymerization of monomeric reactants

The polyimide oligomers are generally prepared by using high-boiling-point solvents like NMP or DMAC. Removal of these solvents from the final polymer product is very difficult. The presence of traces of such solvents causes a significant deterioration of the properties of the polyimide networks. In order to circumvent this problem, in situ polymerization of

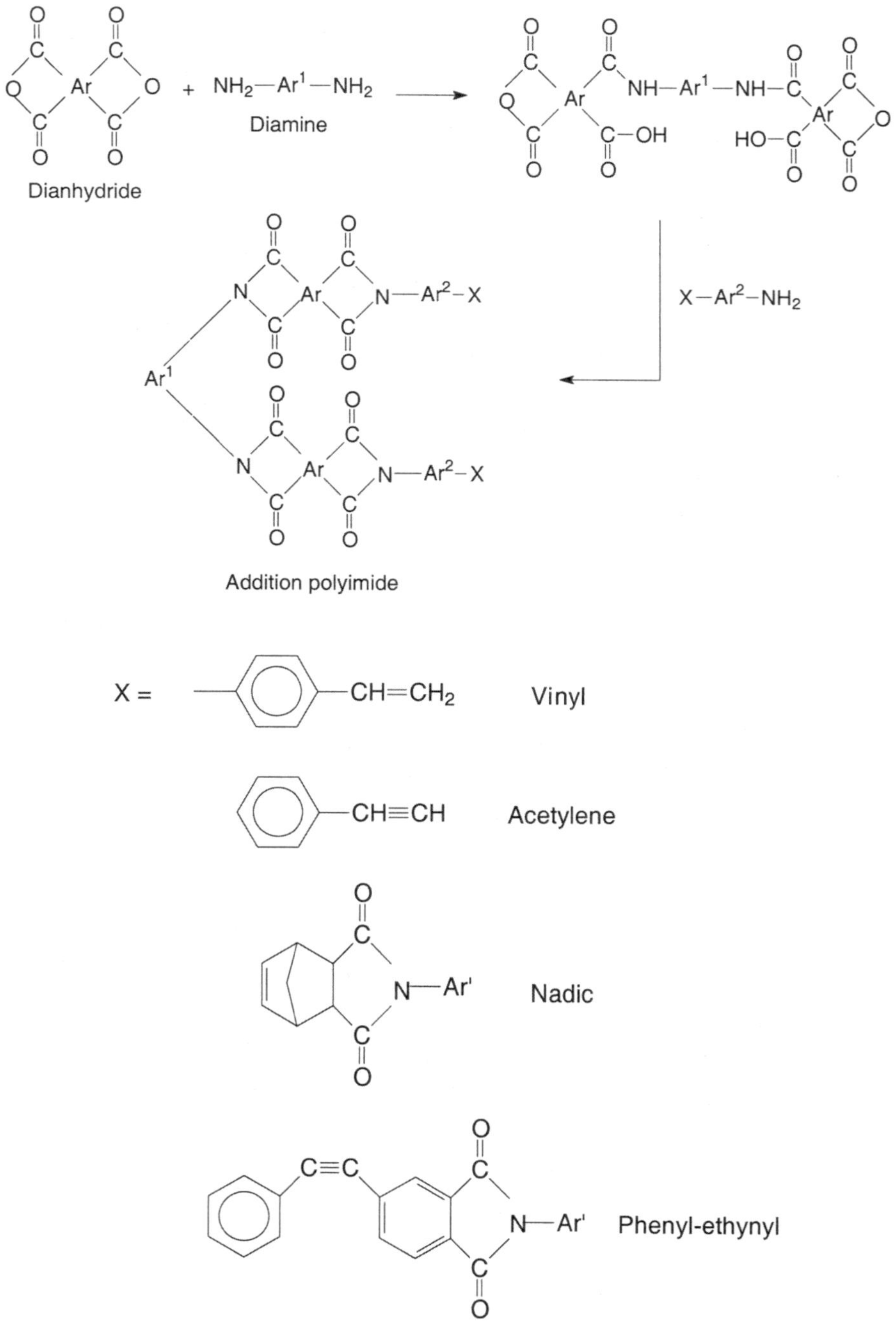

Figure 1.66
A general reaction scheme for the synthesis of addition polyimides with different end groups.

Chemistry and general applications of thermoset resins   119

Table 1.22: List of precursor, end groups, and thermal properties of addition polyimides.

Polyimide	Precursor	End cap	T_g/T_{ms} (°C)	References
PETI-5	3,4′-oxydianiline, 1,3-*bis* (3-amino phenoxy benzene) and 3,3′,4,4′-biphenyl tetracarboxylic dianhydride encapped with 4-phenylethynyl pthalic anhydride	Phenyl ethynyl	270/250	[430–432]
PMR-I (PMR-15, PMR-30, PMR-50)	Monomethyl ester of 5-norbornene-2,3 dicarboxylic acid (nadic ester), dimethyl ester of 3,3′,4,4′-benzophenone tetracarbo xylic acid dianhydride (BTDA) and 4,4′-methylene dianiline Mole ratio 2:2.087: 3.087	Nadic	340(395)/316	[433–435]
LARC-RP46	Modified PMR-I formulation: replacement of MDA by 3,4′-oxydianiline	Nadic	280(395)/300	[436,437]
PMR-II 6 F PMR	BTDA of PMR (I) is replaced by half ester of 4,4′-(hexafluoroisopropylidene)-dipthalic acid (HFDE)	Nadic	350(390)/316	[438,439]
PMR-II 3 F PMR	BTDA of PMR (I) is replaced by 4,4′-(2,2, 2-trifluoro-1-phenylethyledine) dipthalic anhydride (3FDA)	Nadic	370(380)/350	[440]
PMR-12F-71	Nadic ester, HFDE, 2,2′-*bis* (trifluoromethyl)-4,4′-diaminobiphenyl (BTDB)	Nadic	280 (318)/343	[443]
V-CAP-50	p-amino styrene, HFDE, MDA	Vinyl	311(362)/275	[445,446]
V-CAP-12F-71	p-amino styrene, HFDE, 2,2′-*bis* (trifluoromethyl)-4,4′-diaminobiphenyl (BTDB)	Vinyl	310(385)/316	[446,447]
PE Ultem 3000	2,2′,3,3′-biphenyltetracarboxylic dianhydride, 4-phenylethylnyl pthalic anhydride, m-phynelyne diamine	Phenyl ethynyl	230/250	[446,447]

T_{ms} = upper use service temperature; the T_g values reported in the bracket indicate the values obtained after post curing.

monomeric reactants (PMR) approach has been successfully adopted. The half esters of tetracarboxylic acid such as 3,3′,4,4′-benzophenone tetracarboxylic acid anhydride (BTDA) and nadic anhydride are prepared by dissolving the anhydride in methanol or ethanol. Then the amine (4, 4′-methylene dianiline) is added and the mixture is stored as a solution or in impregnated form. The formation of half ester reduces the reactivity of the anhydride and the mixture of half ester and aromatic amine can be stored for a long. When the mixture is heated to a high temperature (120°C–150°C), anhydride is reformed and reacts with the amine groups leading to the formation of a polyimide oligomer. Under this condition, the reaction between the half ester and amine does not take place. The initial resin was made using the molar ratio of the reactant 2:2.087:3.087 corresponding to a resin molecular weight of 1500 g mole^{-1} and termed as PMR-15 [433–435]. Accordingly, the higher molecular weight resins PMR-30 and PMR-50 were developed. The PMR resins made from BTDA is called first-generation PMR (PMR-I). The diamine (MDA) used for the synthesis of PMR-1 is highly toxic and reported to be carcinogenic. Necessary precautions must be adopted in regards to the handling of MDA. Hence, efforts have been made to make PMR without using MDA. For instance, LARC-RP46 resin was developed [436,437] as an alternative to PMR-1 resin by replacing MDA with 3,4′-oxydianiline. Similarly, other aromatic amines such as 2,2- dimethyl benzidine and 2,2′,6,6′-tetramethyl benzidine were also tried out as a replacement for MDA. However, such replacement often resulted in a reduction of thermooxidative stability of the polyimide product [438].

PMR-1 resins are widely used as commercial resins (marketed by SP Systems Imitec Inc., Hycomp. Inc., Eikos Inc.) for fabrication of carbon fiber-reinforced composites with an upper use temperature of 316°C [427]. However, for applications like a gas turbine engine, missiles, etc., polymer matrix having the capability to resist thermal and thermooxidative degradation at higher temperatures (350°C–400°C) is necessary. Hence, extensive research has been carried out in the last few decades to develop polymides with higher and higher temperature service capability. This has led to the development of second-generation PMR resins (PMR-II). PMR-II resins are made in a similar way as PMR-1 by replacing BTDA with fluorine-containing anhydrides such as 4,4′-(hexafluoroisopropylidene)-dipthalic acid (HFDE) and 4,4′-(2,2,2-trifluoro-1-phenylethyle dine) dipthalic anhydride (3 FDA) [439,440]. The reaction scheme for the synthesis of 3F-PMR resin using 3 FDA [440] is shown in Fig. 1.67.

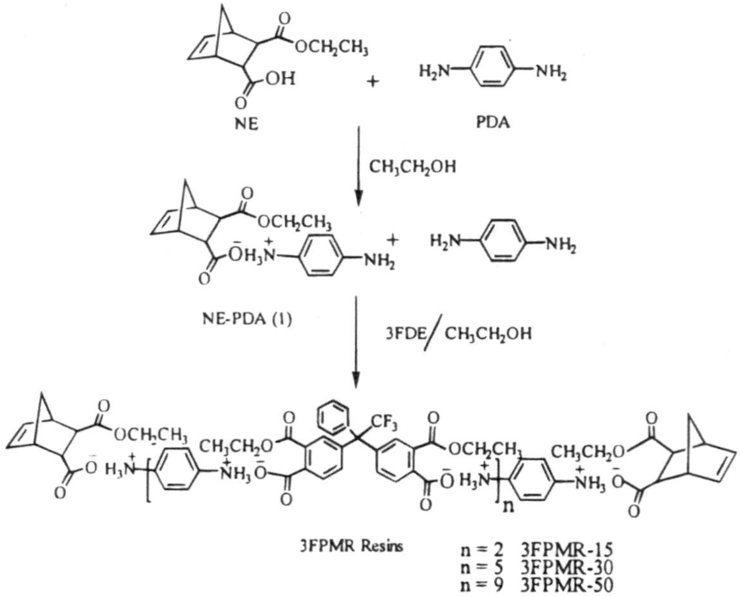

Figure 1.67
Chemistry of 3F-PMR resin. Source: *Reprinted with permission from A.J. Hu, J.Y. Hao, T. He, S.Y. Yang, Macromolecules 32 (1999) 8046.* © 1999, American Chemical Society Publishers.

Due to the presence of fluorine-containing monomer in the structure, the PMR-11 resins exhibit better thermal and thermooxidative stability compared to the PMR-1 resins of similar molecular weight and molecular weight distribution. The molecular weight of PMR prepolymer plays an important role in respect to the processability and thermooxidative stability of the cured resin [441]. Higher molecular weight formulations offer the desired thermooxidative stability and toughness. Fig. 1.68 compares the thermooxidative stability of PMR and 3F-PMR resins of different molecular weights. The figure demonstrates the order of thermooxidative stability as 3F-PMR-50 > 3F-PMR-30 > PMR-50 > PMR-30. Thus 3F-PMR shows much better thermooxidative stability compared to PMR and the stability increases with an increase in molecular weight of the oligomeric resins.

However, as the molecular weight of the prepolymer is increased, the melting temperature and melt viscosity also increase. As a result, resin flow is severely restricted during processing [442]. Thus low-molecular-weight resins are preferable from a processing point of view. However, they often cannot match the thermooxidative stability and toughness properties offered by the high-molecular-weight resin formulations. Hence, the design of a high-molecular-weight PMR prepolymer with better processibility is a great challenge. Chuang et al. [443] synthesized PMR-II (PMR-50) resin by replacing MDA with 2, 2'-*bis* (trifluoromethyl)-4, 4'-diaminobiphenyl

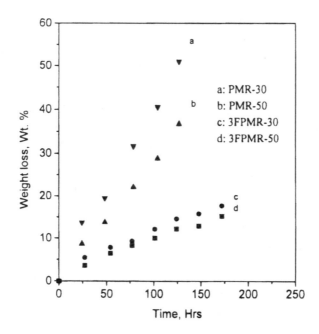

Figure 1.68
Thermooxidative stability of 3F-PMR and PMR resin powder at 371°C in air. *PMR*, polymerization of monomeric reactants. Source: *Reprinted with permission from A.J. Hu, J.Y. Hao, T. He, S.Y. Yang, Macromolecules 32 (1999) 8046.* © *1999, American Chemical Society Publishers.*

(BTDB). The substitution at the 2 and 2′ position of the biphenyl moiety forces the two phenyl rings into adopting a noncoplanar conformation, which reduces the melting point and glass transition temperature of the resin and imparts better flow characteristics [444].

1.11.3 Cross-linking of polyimide

Cross-linking mechanism of PMR and other related addition polyimides is not well understood. Several mechanisms have been proposed in the literature [448–451]. A representative curing reaction of 3F-PMR [440] resin is shown in Fig. 1.69. The reaction involves various steps such as imidization, isomerization, and double Diels–Alder adduct formation. In the case of nadic end-capped polyimides, retro Diels–Alder reactions of norborene end group take place leading to the formation of maleimide groups and cyclopentadiene [452]. The formation of cyclopentadiene and occurrence of retro Diels–Alder reaction was confirmed by a combined thermogravimetric analysis (TGA)-gas chromatography (GC)-mass spectroscopic analysis [453,454]. The maleimide and cyclopentadiene undergo thermally induced free radical polymerization reaction and produces the cured three-dimensional network structures. The cross-linking of ethynyl-terminated imide is basically a free radical propagation of the ethyne moiety to a linear conjugated polyene. The kinetic chain length of this reaction is unusually short and termination is first order. The reaction mechanism is not well understood. A major part of the ethyne groups dimerizes to form an enyne structure.

1.11.4 Curing of polyimide resin

Curing is an important process that differentiates between thermoplastic and thermosetting polyimides. As it is discussed in earlier sections that when the T_g of a thermoset resin reaches the curing temperature, the vitrification takes place and the reaction becomes diffusion controlled. Hence, one cannot expect a high T_g of a polyimide, cured at a low temperature. In order to get a high glass transition temperature, it is necessary to cure at a sufficiently high temperature for a longer time, which resulted in high cross-link density. For instances, a PMR-15 resin cured at 274°C and 316°C for a few hours exhibit T_g values 190°C and 280°C, respectively. A postcuring treatment in air at 316°C for more than 16 h is required to achieve the T_g close to 340°C as reported in Table 1.22. The different T_g values reported for cured and postcured samples indicate that further heat treatment for a longer time also causes an increase in T_g to some extent. In order to avoid a long thermal curing cycle, variable frequency microwave (VFM) processing can be used [455,456]. VFM curing offers the polyimide network with nearly identical mechanical properties to the thermally cured film. The most significant differences in properties between VFM and thermally cured polyimide involve a higher relative permittivity for the VFM film due to the slow evolution of volatile reaction by-products.

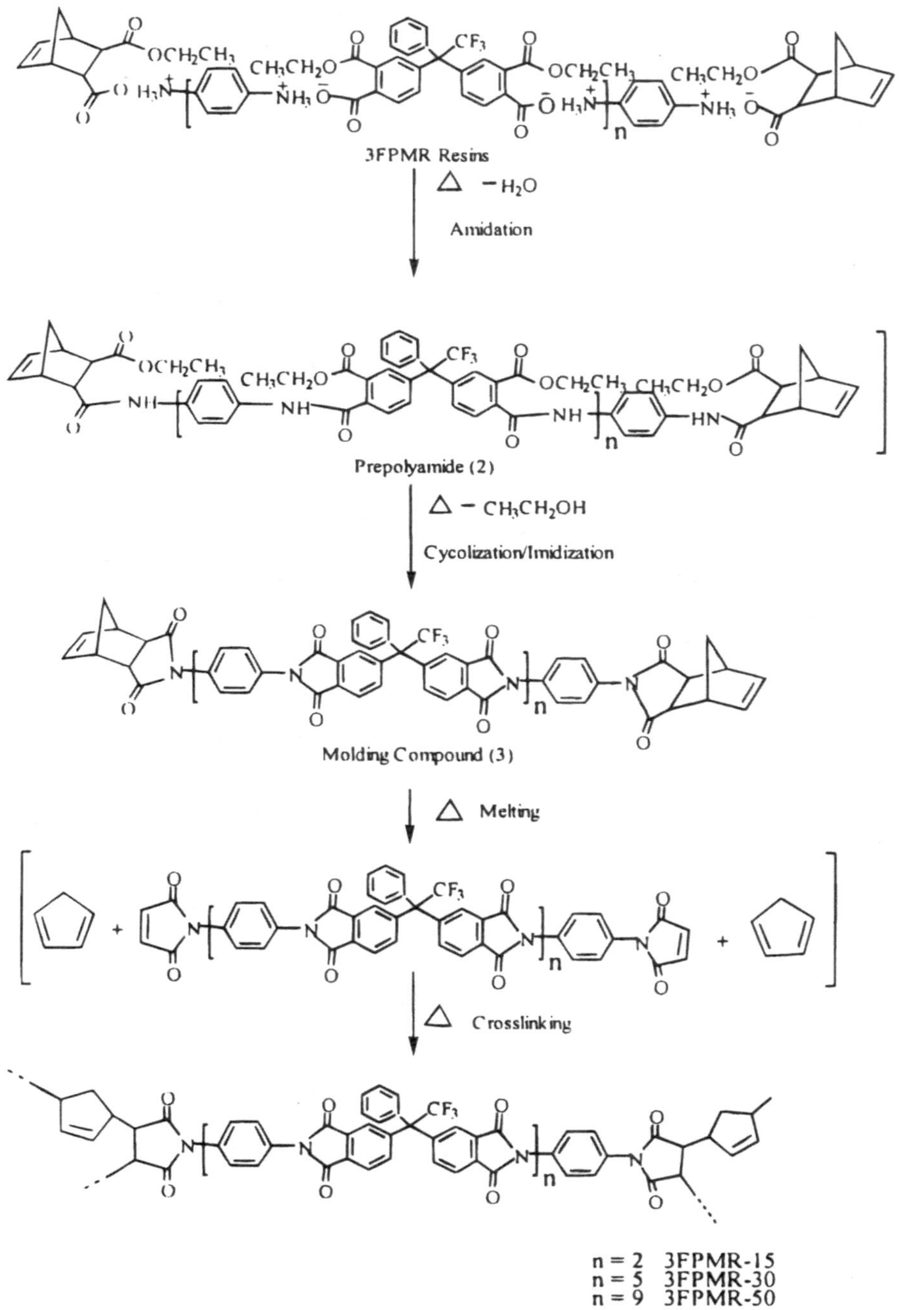

Figure 1.69
Chemical reactions of 3F-PMR resins in thermal processing. Source: *Reprinted with permission from A.J. Hu, J.Y. Hao, T. He, S.Y. Yang, Macromolecules 32 (1999) 8046.* © 1999, American Chemical Society Publishers.

1.11.5 Application of polyimide resin

The demand for polyimides originates from their outstanding thermal properties and thermooxidative resistance in combination with excellent mechanical properties. Polyimides burn but they have self-extinguishing properties. They have a very low dielectric constant and resistant to ionizing radiation. Among the low-cost thermoset resins, the phenolic resin has thermal stability comparable to polyimides. However, phenolic resin is extremely brittle and generally needs blending with other resins like epoxy/polyurethane or rubber for augmenting the toughness property. Such modifications significantly reduce their thermal stability and they cannot be used for high-temperature structural applications. Polyimides, due to their high strength and high heat resistance, often replace glass and metals, such as steel, in many demanding industrial applications. The thermosetting polyimides have aroused tremendous interest as advanced materials in both civil and defense applications. The various applications are summarized in Table 1.23.

The first application of polyimides was for wire enamel [457]. They are suitable for the parts where the materials are to be exposed to a harsh environment. The major use of PMR-based polyimides are in aircraft engines, missiles, and reentry vehicle. Because of high-temperature stability, the polyimide composites find use as compressor seal in jet engines, automobile engines, exhaust system components, self-lubricating bearing for high-temperature use. They are extensively used for bearings for appliances, seals, and gaskets for high temperature, pressure discs, sliding and guide rolls, and valve shafts in shutoff valves. They are also used for other high-temperature applications like light ductwork for high temperatures, heat-resistant panels, fire-resistant panels, etc. Because of their self-extinguishing property, polyimides are used for the development of lightweight and flame-resistant composite structures.

Table 1.23: Major applications of polyimides.

Sector	Field/industry	Application
Civil	Electronics, automobile, aircraft, mechanical engineering, aerospace, medical, fuel cell, environmental science	Wire insulation, tape, chip carrier, switches, liquid crystal display panels, foam insulation, baffles, bushings, seal rings, abrasive cutting wheel, composite structure, adhesives, gas separation membrane, high temperature adhesive for semiconductor industries, pacemakers, eye lenses implants, fuel cell membrane
Defense	Manufacturing, missile, rocket aircraft, spacecraft, mechanical engineering, energy devices	Bushings, bearing, nozzle flaps for aircraft engine, nose cones radomes, ablative coating for reentry vehicle, disk in compressor valve system, high temperature adhesive, structural composite, structural adhesives

Due to very high heat resistance and radiation resistance, polyimides are suitable for aerospace applications. The driving force to replace metallic and ceramic materials with high-temperature polymers for aerospace applications is to reduce the weight of the spacecraft. A reduction in the weight of a spacecraft significantly decreases the launching cost. It is estimated that the reduction of 1 pound from the launching payload saves expenditure of $5000 to $30,000, depending on the launch platform. Thus the use of polyimides as a replacement for metallic and ceramic materials can reduce the launching cast to a great extent. Epoxy-based composites offer excellent mechanical properties and widely used for structural applications but they cannot offer heat resistance comparable to polyimides.

Polyimides have a lot of applications in microelectronic industries. They are often used in the electronics industry for flexible cables, as friction elements in data processing equipment and as an insulating film on magnet wire. For example, in a laptop computer, the cable that connects the main logic board to the display (which must flex every time the laptop is opened or closed) is often a polyimide base with copper conductors. Polyimides are used as high-temperature adhesives in semiconductor industries and aerospace industries. They are used in the struts and chassis in cars as well as some parts under the hood because they can withstand the intense heat and corrosive lubricants, fuels, and coolants, etc. They are also used in the construction of microwave cookware and other kitchen items and food packaging because of their thermal stability, resistance to oils, greases, and transparency to microwave radiation.

Recently, polyimides have drawn considerable attention for the development of membranes for fuel cells. In polymer membrane, fuel cell nafion is widely used as a membrane. However, the nafion membrane cannot withstand high temperature (>80°C) Hence, the research on the development of polyimide-based fuel cell membranes has opened up as a challenging research field [458–462]. Polyimides also find applications in the development of gas separation membranes because of their compact ring structure and less free volume, which allows restriction toward the passage of gases.

Another potential application is for the development of nonlinear optical (NLO) polymers. NLO polymers possess dipoles, which can be aligned by application of electric field at a temperature higher than T_g of the polymer. The dipole arrangement can be fixed by cooling the polymer to room temperature, keeping the electric field on. Such poled system exhibits an interesting property of second harmonic generation. When radiation having a frequency of w is passed through the material, the radiation is converted into one with a frequency of $2w$. Thus such materials have potential use for frequency modulation and many other interesting applications. However, the diploes in a polymer with moderate T_g relax very fast leading to a loss in efficiency of the second harmonic generation. Polyimides, because of their inherently high T_g and less free volume in the structure, are the potential candidates for the development of NLO polymer system [463,464].

In recent years, colorless and transparent polyimide films have drawn great attention in the research and development of high-performance polymer optical films [465–467]. Recently, Jiang et al. [468] reported a series of semialicyclic colorless polyimides containing rigid-rod amide moieties which offer easy processability through a solution-cast method to make a film with desirable linear coefficients of thermal expansion and transparency. The colorless and transparent polyimide have potential applications in optoelectronics areas such as substrates or covering windows for flexible active-matrix organic light-emitting diodes, substrates for organic photovoltaic solar cells, and so on. Note that the wholly aromatic polyimide films discussed in earlier sections often exhibit highly colored appearance and poor optical transparency in the visible light region and not suitable for optoelectronic applications.

1.12 Bismaleimide resin

Bismaleimides are a relatively new class of polyimdes that have gained acceptance for wide industrial application. Propertywise bimaleimide resins are intermediate between PMR polyimides and epoxy resin. They offer thermal stability close to polyimides and epoxy-like autoclave processing. The unique features of bismaleimide resins include excellent physical property retention at elevated temperature and in a wet environment, nonflammabilty and processability [469]. Bismalemide resins are difunctional (or multifunctional) monomers or prepolymers with maleimide end groups. A general structure of bismaleimide and the synthetic route is presented in Fig. 1.70. Bismaleimides are prepared by chemical dehydration of corresponding bismaleamic acid (intermediate product of an amine and

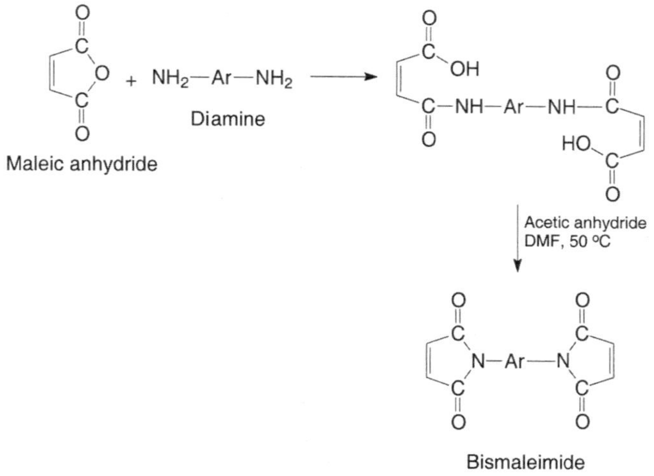

Figure 1.70
A reaction schemes for the synthesis of bismaleimide resin.

excess maleic anhydride) [470]. In the most commonly used route, the appropriate diamine is treated with two equivalents of maleic anhydride in the presence of acetic anhydride and sodium acetate to effect dehydration and cyclization of the resulting maleamic acid. DMF is generally used as a solvent for this reaction. The reaction involves several by-products such as isoimide, acetanilides, etc. Therefore the yield of pure bismaleimide (obtained by recrystallization) is generally low (<70%). By selecting the amine, it is possible to design bismaleimide having a wide variety of chemical structures. The commercial production of BMI started with employing MDA. For example, Kerimid 601 was marketed by Rhone Poulenc using a combination of MDA and bis (4-maleimidophenyl) methane (BDM).

MDA is highly toxic and reported to be a carcinogenic material. Necessary precaution must be adopted in regards to the handling of MDA during the production of 4,4-bismaleimidodiphenylmethane (BMI-MDA). At the same time, the residual free MDA present in the product may cause potential health hazards. Kerimid 601 contains about 6%–10% free MDA. The free residual MDA can be reduced by perfecting the technology to a higher level of sophistication. For example, a product named "Compimide 200" was made by Evonics Industries EG, Germany (using the chemistry as used for the synthesis of Kirimid 601) having a free MDA content of less than 0.1%. However, it is very difficult to achieve the accepted value of free MDA content. It may be noted that the Occupational Safety and Health Administration has set a permissible exposure limit at 0.01 ppm over an 8-h time-weighted average, and a short-term exposure limit at 0.10 ppm. This issue makes it increasingly difficult to market MDA-based commercial products. Hence bismaleimide based on other aromatic diamines were of increasing research interest [471,472]. It was reflected in the development of commercial products as well. A low melting amorphous resin known as "Compimide-353" was developed, which comprises bis(4-maleimidophenyl) methane (55%), 1,5-bismaleimidotoluene (30%), and 1,6-bismaleimido-2,2′-dimethyl-4-methylhexane(15%). Another product marketed originally by Ciba Geigy in the trade name of Matrimid 5292 is made by a combination of diallylbisphenol-A (DABA) and BDM. The curing reaction between DABA and BMI monomer proceeds through the generation of an intermediate through Alder–ene reaction between –C=C– groups of BMI molecule and that of DABA followed by Diels–Alder reaction at a high temperature [473,474].

Bismaleimides can be used by themselves, with other thermosetting resin as a comonomer or as a chain extended form using a suitable amine. When a bismaleimide is reacted with a stoichiometric amount of diamine, a linear polymer is formed which shows poor thermal stability [475]. The nonstoichiometric reaction of an aromatic diamine with a bismaleimide converts bismaleimide into a polyamino bismaleimide (Fig. 1.71). The chain extension of bismaleimide using an amine allows further tailoring of cross-link density and properties of the resulting crosslinked network [476–480]. Sripadaraj et al. [481] reported N, N′-(4-aminophenyl)-p-quinone-based maleimide. They have also investigated chain extension of bismaleimides containing flexible ether linkages, that is, 1,3-bis(4-maleimidophenoxy)

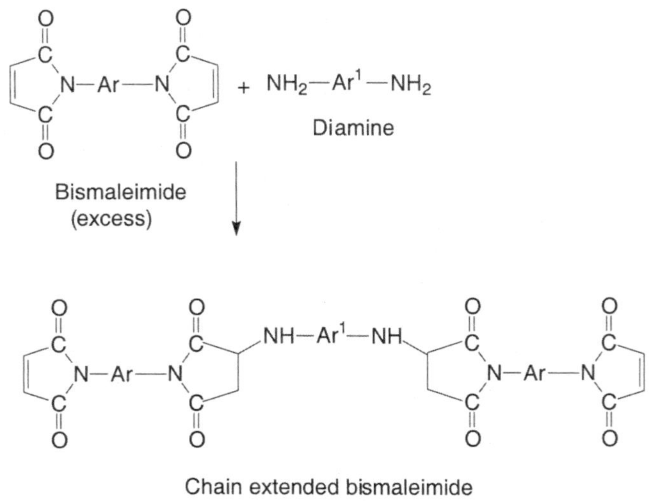

Figure 1.71
Chemistry of chain extension of bismaleimide resin.

benzene and 1,4- bus(4-maleimidophenoxy) benzene with N,N'-(4-aminophenyl)-p-quinone. The chain extended product was characterized by FTIR analysis, which indicates the appearance of a weak-intensity band at 3470 cm^{-1} due to —NH stretching (see Fig. 1.71) and at 2920 and 2860 cm^{-1} due to aliphatic stretching. The imide linkage shows the two characteristic bands at 1790 cm^{-1} for symmetrical carbonyl stretching and at 1710–1730 cm^{-1} for asymmetrical carbonyl stretching. The chain extended bismaleimide resins exhibit inferior thermal stability compared to the neat bismaleimide.

BMI and biscitraconimides with bis-allyl groups and brominated BMI also showed enhanced thermal and mechanical features [482,483]. A new type of BMI resin containing an epoxy unit and phosphorus in the main chain was synthesized [484]. The polymers were synthesized by the reaction between BMI and diamine extender and they exhibit excellent thermal properties and high char yield. Novolac resin based on 2, 2'-diallyl bisphenol-A when coreacted with bisphenol-A bismaleimide provided a resultant high-temperature resin with good adhesive strength at a higher temperature. The blend with optimum cross-link density can be used as an adhesive for aluminum substrates. The adhesive can retain almost full of its adhesive strength (measured at room temperature) at 150°C indicating its suitability for high-temperature applications [485]. The thermal stability can be further enhanced by incorporating 1, 3, 4-oxadiazole rings into the BMI skeletons without sacrificing their mechanical performance. Several BMI monomers containing a 1,3,4-oxadiazole moiety have been synthesized and characterized, which exhibit higher melting temperatures (T_m), and sharp curing exothermic peaks [486–488].

1.12.1 Curing of bismaleimides

The double bonds in bismalemide are highly electron-deficient due to the presence of two flanking carbonyl groups and are reactive toward bimolecular addition reaction. Hence, the maleimide groups of bismaleimide monomer or chain extended prepolymer can undergo homopolymerization to produce three-dimensional network structures. In order to manipulate the structure and property, different types of maleimide monomer or prepolymer can be used. The reactivity of such resins depends on their chemical structure and molecular weight. In general, the presence of electron-withdrawing groups such as CO, SO_2, and so on, in the vicinity of maleimide groups, increases the reactivity as the reaction proceeds through a nucleophilic attack. On the other hand, the presence of electron-donating groups makes the corresponding maleimide group less nucleophilic and thereby reduces the reactivity.

The bismaleimides can also be cured using a Diels–Alder comonomer [489]. When a bisdiene is reacted with an excess bismaleimide, a prepolymer carrying maleimide termination is formed as an intermediate, which can be further crosslinked to form a three-dimensional cured network. Bis-(o-propenylphenoxy) benzophenone (Compimide TM123, Technochemie, Germany) is an example of a commercial Diels–Alder comonomer. CompimideTM123 is a low melting, low viscosity material and can easily be melt blended with BMI and cured at a high temperature to get a heat-resistant network.

1.12.2 Properties of bismaleimide resins

Bismaleimide can be cured by polyaddition reactions, which do not produce any volatile side products. The cured bismeleimides can withstand the service temperature up to 180°C. By manipulating chemical structures, the thermal stability can be improved further and the service temperature can be extended up to 270°C. Thus thermal stability of bismaleimides is higher than that of epoxy but lower than that of PMR polyimides. However, bismaleimides offer better processability than PMR polyimide resins. Bismaleimide can be processed like epoxy by autoclave molding. At the same time, they exhibit mechanical properties and damage tolerance similar to epoxy resin. Another advantage of bismaleimides is that they are inherently fire resistant. Because of excellent processability, good mechanical properties, very good thermal stability and adequate electrical properties, bismaleimides gained popularity for the development of high-performance composites and microelectronic components. Mechanical characterization of bismaleimide resins is mostly studied in the composite form, which will be discussed in subsequent chapters.

1.12.3 Structure–property relationship

If we analyze the research activities in the last two decades, we can observe that a lot of efforts have been made to improve the toughness and processability of available bismaleimide resins. This is because bismaleimides based on aromatic amines are crystalline substances with a high melting point due to the strong interaction between carbonyl groups in the imide rings. For example, the melting point of BMI-MDA is about 160°C. The melting point BMI monomers containing a 1,3,4-oxadiazole moiety are 208°C. Since bismaleimides are processed in a melt state like thermoplastic, processability is directly related to the melting point of the resin. The high and sharp melting point of monomers results in only a narrow processing window and poses challenges in the processing of these materials. The melting point of a resin depends on the chemical structure and the desired processability can be achieved by judiciously selecting the precursor [490−492]. A variety of BMI resin can be made by just changing the aromatic moiety (−Ar in Fig. 1.71). The effect of the chemical structure on the melting point of BMI monomer, curing temperature (cure onset and cure peak) and of the cured network is summarized [493] in Table 1.24. We can see from the table that the introduction of ether and sulfone groups resulted in a decrease in the melting points of the related monomers. By

Table 1.24: Representative experimental monomers and selected physical properties.

-R-	M_p (°C)	Cure onset (°C)	Cure peak (°C)	Cured T_g (°C)
	155−168	174	235	342
	212	217	236	313
	158−163	203	302	312
	104	198	211	288
	239	−	252	−
	163	−	254	−

Mp, melting point.
Source: Reprinted with Permission from R.J. Iredale, C. Ward, I. Hamerton, Prog. Polym. Sci. 69 (2017) 1−21©2017 Elsevier Publishers.

introducing an aromatic moiety (containing 4 benzene rings, two ether linkages and one sulfone group) resulted in a BMI monomer having a melting point of 104°C, which is much lower compared to 155°C–168°C for BMI-MDA. The decrease in melting point a BMI monomer will certainly be beneficial from a processing point of view. Moreover, as a result of the introduction of a longer aromatic moiety containing ether and sulfone group in between the two maleimide rings, the effective cross-link density decreases. In addition, the incorporation of the ether group is known to impart more rotational freedom because the rotational energy of a −C-O− bond is lower than a −C-C− bond. Because of the combined effect of cross-link density and molecular flexibility the related BMI networks offer higher energy-absorbing ability resulting in improved impact performance. The thermal stability and char yield are improved due to the incorporation of the aromatic moiety. However, these improvements are associated with a significant decrease in glass transition temperature and modulus. The modulus can be enhanced by using a suitable reinforcement, however, reduction of glass transition temperature is certainly going to impact on the high-temperature performance of BMI. When service temperature goes beyond the glass transition temperature, the structure made of BMI loses its dimensional stability. A trade-off between the impact performance and glass transition temperature is a major challenge for the development of such materials.

For example, the toughness of a BMI network can be improved significantly by incorporating an ether containing commercially available BMI monomer 2,2′-bis[4-(4-maleimidephen-oxy) phenyl] propane (BMPP) with conventional BMI monomer [494]. Blending of BMPP (containing two linkages in the structure) optimally with BDM and DABA resulted in an increase in impact strength, critical stress intensity factor (K_{IC}), the critical strain energy release rate (G_{IC}) from 10.1 kJ m^{-2}, 0.87 MPa.m$^{0.5}$ and 160 J m^{-2} to 20.5 kJ m^{-2}, 1.21 MPa.m$^{0.5}$, 295.6 J m^{-2}, respectively. The remarkable improvement in fracture toughness is achieved while maintaining the glass transition temperature of about 300°C. Though the glass transition temperature of a BMPP-based system is lower than that of BMI-MDA (340°C), it offers very good mechanical properties up to a temperature of 270°C. In addition, the BDM/BMPP/DABA ternary system offers low viscosity (<1 Pa s) at a temperature range from 120°C to 180°C. We have discussed here how the toughness of MBI can be improved by changing the chemistry. There are some more strategies used for the improvement of the toughness of BMI resins without affecting the thermomechanical performance which will be discussed in Chapter 3.

The second issue related to the processability of BMI has also attracted the attention of researchers to a great extent. As mentioned above because of the strong interaction between carbonyl groups in the imide rings they not only exhibit a high melting point leading to a narrow processing window but also poor solubility in the low boiling point solvents. From the application point of view, poor processability greatly hinders the process of manufacturing of prepreg via wet or hot-melt impregnation method required for fabricating high-temperature components of aero engines. The research works carried out to address this issue are reviewed here briefly. Mitica and Sava [495] synthesized and characterized a series of new bismaleimides with flexible units like ether, ester, and amide units by reacting maleic anhydride with different diamines, or by reacting 4-maleimidobenzoyl chloride with bisphenol and diamine. The reactivity of the bismaleimide compounds is influenced by the chemical nature of the bridge unit between the maleimide rings. The incorporation of flexible linkages into the aromatic main chain of these monomers resulted in products with improved solubility. The products offer good thermal stability with decomposition temperatures above 400°C and 46%–58% char yield at 700°C in a nitrogen atmosphere.

A series of BMI resins modified with hydrogen silsesquioxane (HSQ), hexafluorobisphenol-A, 4, 4′-dihydroxydiphenyl ether (HSQ), dipropargyl ethers of bisphenol-A (DPBPA) were prepared [496]. The resins exhibited excellent processability, especially the BMI-HSQ-DPBPA resin, which had a low viscosity of 0.2–0.3 Pa s at temperatures between 81°C and 196°C. The cured resins showed high heat resistance, with T_g above 330°C and T_{d5} of 400°C. Xiong et al. [497] introduced phthalide cardo structure in the BMI monomer and produced crosslinked BMI network by curing thermal bulk polymerization. It was reported that such modification resulted in lowering of melting temperature and increase their solubility in a common solvent. The reported glass transition temperatures are in the range of 324°C–364°C.

Another strategy used to address the issue of processability is to use asymmetric monomer for the synthesis of BMI resin. The asymmetric monomers are used as a way to hinder the ordered packing or crystallization of the molecular chains to improve the solubility and decrease the melting point and viscosity of the uncured BMI systems. The chemical structures of asymmetric monomers used by various researchers are presented in Fig. 1.72. Chen [497,498] synthesized two asymmetric bismaleimide monomers containing 1,3,4-oxadiazole group, namely, 2-[(4-maleimidophenoxy)methyl]-5-(4-maleimidophenyl)-1,3,4-oxadiazole (Mioxd) and 2-[(4-maleimidophenoxy)phenyl]-5-(4-maleimido-phenyl)-1,3,4-oxadiazole (m-Mioxd). As mentioned above, the BMI systems are known to offer very high thermal stability; however, the symmetric BMI monomers containing 1,3,4-oxadiazole moiety exhibit high melting temperature and sharp curing exothermic peak, which indicates poor processability via melt mixing.

Moreover, the symmetric BMI monomers are not soluble in common organic solvents, therefore they cannot be processed through a solution mixing process. It may be noted that the asymmetric BMI monomers with 1,3,4-oxadiazole moiety, mentioned above are soluble in acetone and tetrahydrofuran. In addition, the cured BMI prepared using the asymmetric monomers mentioned above, exhibit excellent thermal stability with 10% weight loss above 510°C and residual weight percentage at 700°C in the range of 48%–62%. The BMI resins have also been used to make laminate with glass cloth and the fiber-reinforced composites

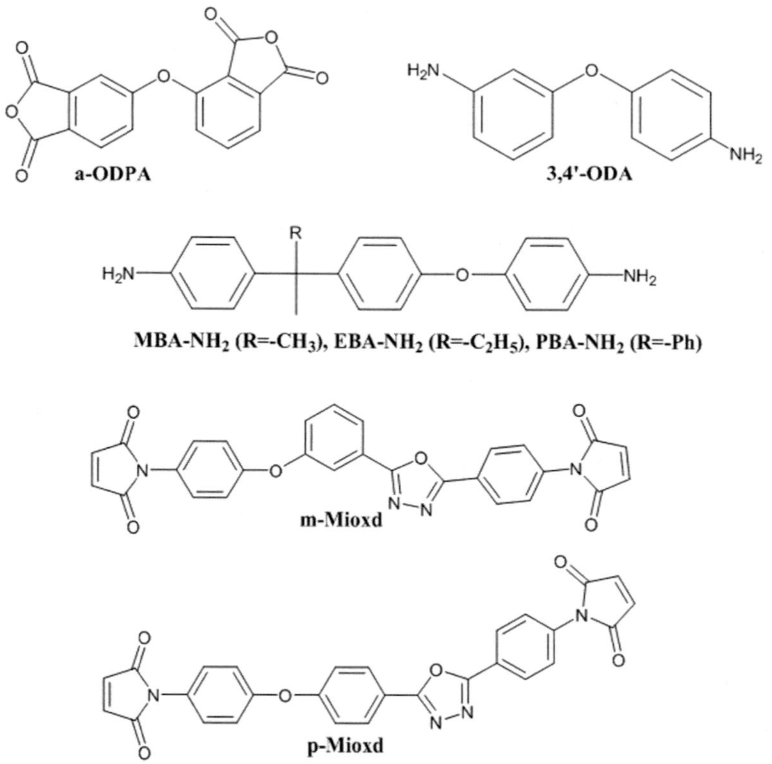

Figure 1.72
The chemical structures of asymmetric monomers for synthesis of PBI. *PBI*, polybenzimidazole.

were characterized for their thermomechanical properties by DMA. The glass fiber-reinforced composites showed a high storage modulus (>6.5 GPa) and glass transition temperature (>450°C). This indicates the excellent thermal stability of the resulting composites. However, the melting points of the three BMI monomers are higher than 200°C, which is close to the onset curing temperatures, which means the processing temperature window is very narrow for the molten resins.

Xiao and coworker [499] synthesized and characterized three different asymmetric BMI systems namely 2-(4′-maleimido) phenyl-2-[4′-(4″-maleimido) phenoxyl] phenylpropane (MBA-BMI), 2-(4′-maleimido)phenyl-2-[4′-(4″-maleimido) phenoxyl] phenylbutane (EBA-BMI), and 1-(4′-maleimido)phenyl-1-[4′-(4″-maleimido) phenoxyl] phenyl-1-phenylethane (PBA-BMI) using asymmetric amine precursors (Fig. 1.72). They have investigated the rheology, reaction kinetics of the BMI systems, and properties of the cured BMI networks and demonstrated that unsymmetrical chemical structure is helpful in improving the solubility of the monomers in common organic solvents. The asymmetric BMI systems show good solubility in various solvents, such as acetone, tetrahydrofuran, and 1,4-dioxan and their solubility characteristics are better than that of BMI system containing 1,3,4-oxadiazole group [300,301]. MBA-BMI, EBA-BMI, and PBA-BMI can be dissolved in common organic solvents up to a concentration of about 30 mg mL^{-1}, while the asymmetric bismaleimide containing 1,3,4-oxadiazole group in the chain, can be dissolved in acetone only up to a concentration of 10 mg mL^{-1}. The values $T_{5\%}$ and char yield at 800 are in the range of 466°C–471°C and 39%–42%, respectively. PBA-BMI shows the highest char yield due to the presence of an additional phenyl ring in the structure.

Yang and coworker [500] investigated the blends of MDA-BMI and EBA-BMI and with varying concentrations of EB-BMI. MDA-BMI was used as a major component for all the blends. The melt viscosity of a blend containing 20% EBA-polybenzimidazole (PBI) decreased from 474 to 51 mPa s as the temperature rises from 148°C to 180°C indicating that the blend has attractive processing performance with low molten viscosity in a wide range of temperature. Hence, the processing window is significantly broadened. As mentioned above, the BMI system with 1,3,4-oxadiazole moiety exhibits a narrow processing window. This suggests that the addition of asymmetric BMI gives rise to a lower molten viscosity and wider processing temperature window. In addition, cured blend shows remarkable thermal stability with 5% weight loss at around 470°C, outstanding storage moduli up to 3.2 GPa even at 400°C, high T_g over 400°C, and low coefficients of thermal expansion of 36–40 ppm °C^{-1} in the temperature range of 50°C–250°C.

Stenzenberger and coworker [501] reported a one-part, easy-to-process bismaleimide resin, coded as "Compimide 50LM," comprising the eutectic blend of *bis* (1,3-maleimidomethyl) benzene (Compimide-MXBI) and BMI-MDA (Compimide MDAB). The resin, when compared with Compimide-353 A, allows liquid processing at significantly reduced temperatures without affecting the other desired properties of the formulated cured products. The work was taken up based on the concept that eutectic mixtures of two or more bismaleimides, when combined with a liquid comonomer like DABA can afford low viscosity, almost liquid systems at room temperature. For example the binary eutectic blends of aromatic bismaleimides comprising BMI-MDA and 2,4-bismaleimidotoluene combined with a specific aliphatic BMI (1,6-bismaleimido-2,2,4-trimethylhexane) resulted in the formation of the ternary eutectics with extremely low melting temperatures (70°C–110°C) and very low viscosities at temperatures above 110°C [502,503].

Asymmetric BMI monomers were also synthesized using both asymmetric amine and anhydride precursors [504]. 3,4′-oxydianiline is reacted with a combination of 2,3,3′4′-oxydiphthalic dianhydride (a-ODPA) (see Fig. 1.72) and maleic anhydride to make the asymmetric BMI resin. It was observed that introducing the asymmetric groups and ether groups into the oligomer chains indeed significantly improves the solubility of the related oligomers in a range of solvents including 1,4-dioxane and chloroform with low boiling point. The BMI oligomers can be dissolved in dimethylacetamide up to a concentration of

50 wt.% and the homogeneity of the solvent remained intact up to a period of one year. This indicates excellent solution stability in the natural environment. The BMI systems offer very good toughness and ductility with elongation at break value up to 17%. This is due to the combined effect of the introduction of ether moiety and the use of both asymmetric amine and anhydride-based precursors which tend to make the BMI structure amorphous and loose-packed. The oligomer with optimum composition exhibited the lowest complex viscosity of 8.1 Pa s at 176°C. However, the use of asymmetric precursors impacted the desirable thermomechanical properties of the resins. The thermal property of the resin is not comparable to polyimide resin rather it is equivalent to a high-performance epoxy system; the glass transition temperature and the temperature of 5% weight loss in nitrogen were above 256°C and 449°C, respectively.

In earlier sections, we have seen that most of the thermoset resins have been synthesized from natural feedstock by replacing the fossil fuel-based feedstock partially or completely and BMI resin is no exception. This is a step toward a drive for ecologically sustainable methods for manufacturing composite materials for a structural and functional application covering both civil and defense applications. The naturally occurring unsaturated compound can be reacted with BDM to produce a bio-based BMI system [505,506]. The chemical structure of various naturally occurring unsaturated compounds and the properties of BMI systems made by reacting with BDM in different proportions are shown [493] in Table 1.25. We can observe from the table that the BMI systems made with a long aliphatic chain containing compound failed to retain the desired thermal properties comparable to polyimide resins. However, BMI systems made from small molecules like eugenol and turpentine variants can offer promising high-temperature properties. Shibata et al. [506] synthesized bio-based BMI by reacting eugenol with N-phenylmaleimide (PMI). Mass spectral analysis indicates that the reaction of EG and PMI involves the formation of 1/1 and 1/3 adducts in addition to copolymerized products [507,508]. A probable reaction scheme [506] of eugenol and BMI is shown in Fig. 1.73. First, the ene reaction of eugenol and PMI produces trans-rich 1/1 ene adduct, which undergoes DA reaction to generate reactive 1/2 ene–DA adducts. In this case, two regio-isomers (1/2 ene–DA adduct 1 and 1/2 ene–DA adduct 2) are possible. The 1/2 ene–DA adducts immediately rearomatize by the subsequent ene reaction with PMI to give 1/3 ene–DA–ene adducts. The presence of methane carbon evidenced by ^{13}C NMR peak at 44.1 ppm supports the formation of 1/3 ene–DA–ene adduct. Another possibility is the formation of 1/3 ene–DA–DA adducts with a-methoxy carbonyl moiety, if the 1/2 ene–DA adduct undergoes DA reaction with PMI, by the keto-enol tautomerism. However, the characteristic ^{13}C NMR peak for the carbonyl moiety generating by the keto-enol tautomerism at around 200–220 ppm was not observed [506]. Therefore the formation of 1/3 ene–DA–DA adducts is not confirmed. The developed eugenol-based BMI exhibit a glass transition temperature of 345°C, which is higher than the reported value (310°C) for petroleum-based DABA/BMI [356].

Table 1.25: Representative bio-based BMIs with selected properties when blended with BDM.

Modifier	Ratio of monomer to BDM	T_g (°C)	$T_{5\%}$ (°C)
	1/3	>400	481
	1/3	336	431
	1/2	141	441
	1/4	150	468
	1/1	365	479 ($T_{10\%}$)
	2/5	340	450 ($T_{10\%}$)
	1/3	377	478
	1/3	365	447

Source: Reprinted with Permission from R.J. Iredale, C. Ward, I. Hamerton, Prog. Polym. Sci. 69 (2017) 1–21. © 2017 Elsevier Publishers.

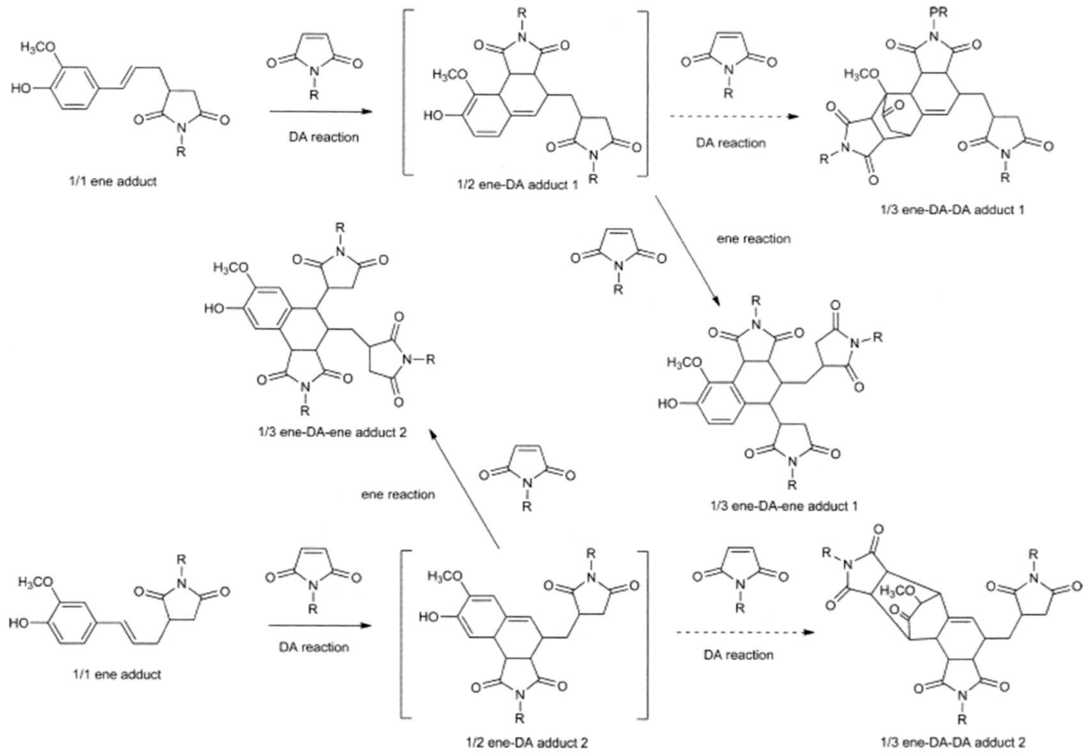

Figure 1.73
Probable reaction mechanism for euganol and BMI.

1.12.4 Applications of bismaleimide resin

Bimaleimides are often blended with other resins like epoxy, unsaturated polyester to improve their thermal stability. Aerospace composites are mostly dominated by epoxy resins. Bismaleimides, due to their higher thermal stability and comparable mechanical properties, processability, and cost with high-performance epoxies, are gaining acceptance in aerospace industries. They are used to design wings for extended supersonic aircraft, which can withstand extremely high maneuver. They are also used as preferred matrices compared to epoxy for structural hinges (attached to the trailing edges of the wings) because such structures experience engine exhaust in routine operation. Similarly, bismaleimides are used as replacements of metallic materials for applications like helicopter tail boom, where the material has to be exposed to exhaust gases from the engine as well as experience corrosive and fatigue environment.

Because of the presence of reactive maleimide groups, which are polymerizable by free radical mechanism, it is a potential blending component for free-radically cross-linkable resins like UPE and vinyl ester. Li et al. [509] have investigated the thermomechanical

Table 1.26: Thermal and mechanical properties of bismaleimide modified unsaturated polyester resin network.

Percentage of bismaleimide by weight	T_{md}/T_{50} (°C)	Heat distortion temperature (°C)	Tensile strength (MPa)	Impact strength (kJ m^{-2})
0	412/402	54	68	6.3
5	420/409	69	63	5.8
8	424/413	80	53	4.3

T_{md} = temperature corresponding to maximum rate of weight loss.
T_{50} = temperature corresponding to 50% weight loss.

properties and morphology of copolymers of 4, 4′-bismaleimido-diphenyl methane and UPE resin. The bismaleimide was readily dissolved in UPE resin and underwent free radical polymerization with styrene (reactive diluent) to form alternating copolymers [510,511]. Thus an intercross-linking network of UPE and bismaleimide is formed. The effect of the addition of bismaleimide on thermal and mechanical properties of UPE resin is presented in Table 1.26. The results clearly indicate a significant increase in thermal stability as a result of the incorporation of bismaleimide. However, the concentration of bismaleimide is to be kept to minimum (5–10 wt.%) to avoid a drastic reduction in mechanical properties.

Nair and coworker [512] carried similar studies with the aim to enhance the properties of a phenolic resin. A modified phenolic resin was synthesized by introducing an allyl group, which is capable of reacting with BMI through Alder–ene reaction to form a crosslinked networks. The reaction proceeds via an intermediate step of Wagner–Jauregg reaction. They have conducted extensive DSC experiments and reported different types of reactions at different temperature ranges during curing. It was proposed to occur Ene reaction via Claisen rearrangement at 100°C–200°C, ene, and Wagner–Jauregg reaction at 160°C–220°C, Diels–Alder reaction at 225°C–275°C, and thermal rearrangement leading to network formation at 250°C–300°C. The properties of blends can be tailored as a function of composition. As expected, the thermal stability of the blend increases with increasing BMI concentration and impact strength increases with increasing phenolic concentration. By changing the blend composition the glass transition temperature and impact strength can be varied from 153°C to 290°C and 217 to 176 kJ m^{-1}, respectively. Depending on the service temperature and structural strength requirement, an appropriate blend can be selected for a particular application.

Because of high-temperature stability, the bismaleimide composites find application in jet engines, automobile engines, exhaust system components, self-lubricating bearing for high-temperature use. They are extensively used for bearings for appliances, seals, and gaskets for high temperatures. They are also used for other high-temperature applications like light ductwork for high temperatures, heat-resistant panels, fire-resistant panels, etc. Because of their self-extinguishing property, bismaleimides are used for the development of lightweight and flame-resistant composite structures. Bismaleimide resins are extensively used in multilayer printed circuit boards. They can also be used in insulation, fibers for protective clothing. Due to their ability to adhere and retain adhesive strength at high temperatures, bismaleimides are widely used as a base material for flexible printed circuit boards in the electrical and electronics industries. Bismaleimides are often linked with rubber chains for use as flexible linking molecules to reinforcing the rubber compound for tire applications. The double bond of bismaleimides readily reacts with all hydroxy or thiol groups found on the matrix to form a stable carbon−sulfur, carbon−nitrogen, or carbon−carbon bond.

1.13 Cyanate ester resin

The fourth resin in the series of high-temperature thermosetting resins is cyanate ester resin or polycyanurates. The use of cyanate ester resin is limited to aerospace and defense applications due to its high cost. Aryl cyanate monomer was first synthesized by Grigat et al. [513] and commercial production of CE resin started in the mid-1980s. CE resins are characterized by the presence of two or more cyanate functional groups (-O-C≡N-) with an aromatic or cycloaliphatic backbone. CE resins are generally prepared by reacting cyanogen chloride with an alcohol or phenol in presence of a tertiary amine [514,515]. Diethylcyanamide is formed as a gaseous by-product. The formation of the by-product is more when cyanogen bromide is used instead of cyanogen chloride. Cyanogen bromide is preferable from a handling point of view as it is a solid substance unlike cyanogen chloride, which is a gas. In order to reduce the formation of the undesirable by-products, the reaction is carried out at low temperatures for a longer time to reduce the formation diethylcyanamide. The hydroxyl compounds used as a precursor for the reaction can also act as a catalyst for the trimerization of CE resin product. That is why for the preparation of CE resin with long shelf life, it is necessary to ensure almost complete consumption of phenol or alcohol in the reaction [516,517]. The unreacted phenol present in the resin catalyzes the curing of CE resin and reduces the shelf life.

Due to the high cost and the health hazards, associated with the handling of cyanogen halide, the source of cyanate ester resins is limited to only a few companies like Lonza, Vantico, and Mitsubishi gas chemicals. The commercial production of CE resin started in the mid-1980s. The CE resin technology was licensed to Mitsubishi Gas Chemicals and Celanese. The range of CE resin products was expanded when the CE resin activities were acquired by Vantico. CE resins are available in liquid, semisolid, or low melting solid form. CE resins contain either cyanate monomers or a prepolymer, which is prepared by a controlled cyclotrimerizaion reaction of monomer in an inert atmosphere. CE resins are also commercially available in combination with a suitable impact modifier. For example, rubber-modified CE resin is supplied by Vantico in the trade name of XU-7178. Different types of resins are available with a varying chemical structure of the backbone. The chemical structures of commercially available CE resins are presented in Fig. 1.74. 2, 2′ *bis* (4-cyanatophenyl) propane

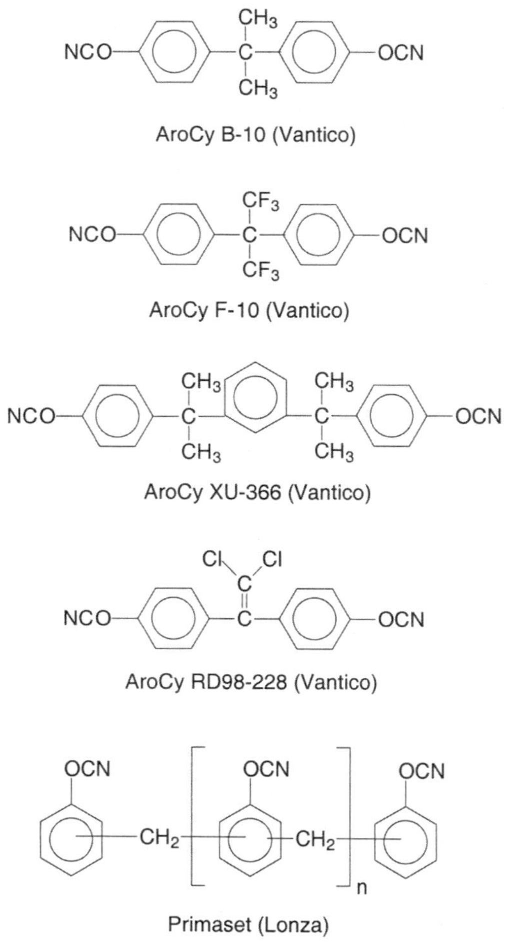

Figure 1.74
Chemical structures of commercially available cyanate ester resin.

(the dicyanate ester of bisphenol-A) known as BADCy and 1,1′-bis(4-cyanatophenyl) ethane (the dicyanate ester of bisphenol-E) known as LECy are marketed by Lonza. Allied Signal commercialized a multifunctional CE resin in the trade name of Primaset (phenolic triazine) in 1991 [518]. In 1995, the Lonza acquired the PT family of resin from Allied Signal.

1.13.1 Curing of cyanate ester resin

The curing of CE ester involves a simple cycotrimerization reaction as shown in Fig. 1.75. The curing occurs via a thermally activated addition reaction, which produces no volatiles. The process allows the preparation of void-free castings and fiber-reinforced composites with a good surface finish [519]. The cyclotrimerization addition reaction

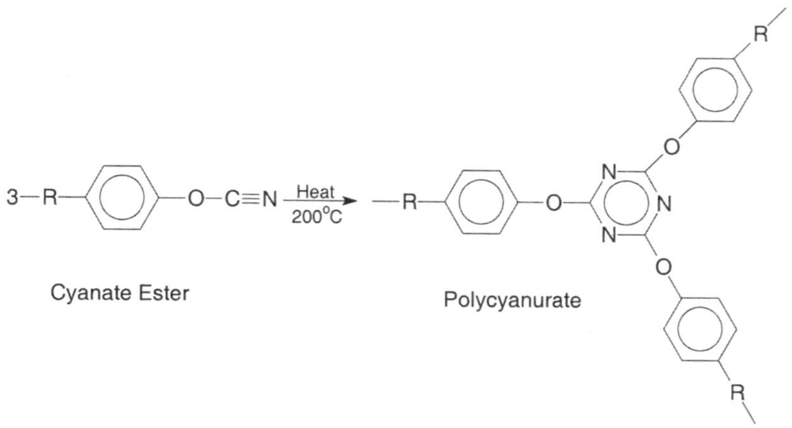

Figure 1.75
Mechanism of curing of cyanate ester resin.

leads to the formation of a three-dimensional crosslinked network consisting of triazine rings linked to the backbone structure through ether linkages. The autocatalytic reaction occurs without any catalyst at above 200°C. The use of catalyst offers the control of cross-linking reaction. The carboxylates and chelate salts of transition metals are the example of the most common catalysts. The transition metals form a complex with the cyanate groups bridging three reactive groups together and encourage trimerization. By using a catalyst, the curing temperature can be brought down to less than 100°C. The curing reaction is highly exothermic. The heat evolved for curing 1 mole of resin is about 105 kJ, which is about twice the value for epoxy (50–60 kJ mole^{-1}). Hence, care must be taken to carry out the curing of CE resin. The high heat build-up during curing may increase the temperature leading to the degradation of the resin. The CE resin degrades beyond 300°C as shown in Fig. 1.76. The CE resins thermally decompose through a common mechanism which begins with thermolytic cleavage of the resin backbone and culminates with decyclization of the cyanurate rings to produce a variety of volatile species [520] as shown in (Fig. 1.76).

This is why the curing is carried out in different stages, including two or more initial curing followed by a postcuring. The curing is generally carried out at 80°C–120°C in the first stage, 120°C–180°C in the second stage and postcuring at 250°C–300°C. It may also be noted that the curing of CE resin is sensitive to moisture and hence a dry atmosphere has to be maintained for curing the resin. The cyanate groups react with water to form a carbamate [521]. Carbamate is very unstable and decomposes into amine and carbon dioxide. The amine further reacts with the cyanate group to form an isourea (Fig. 1.77). The generated carbon dioxide produces blisters in CE resin-based castings and composites.

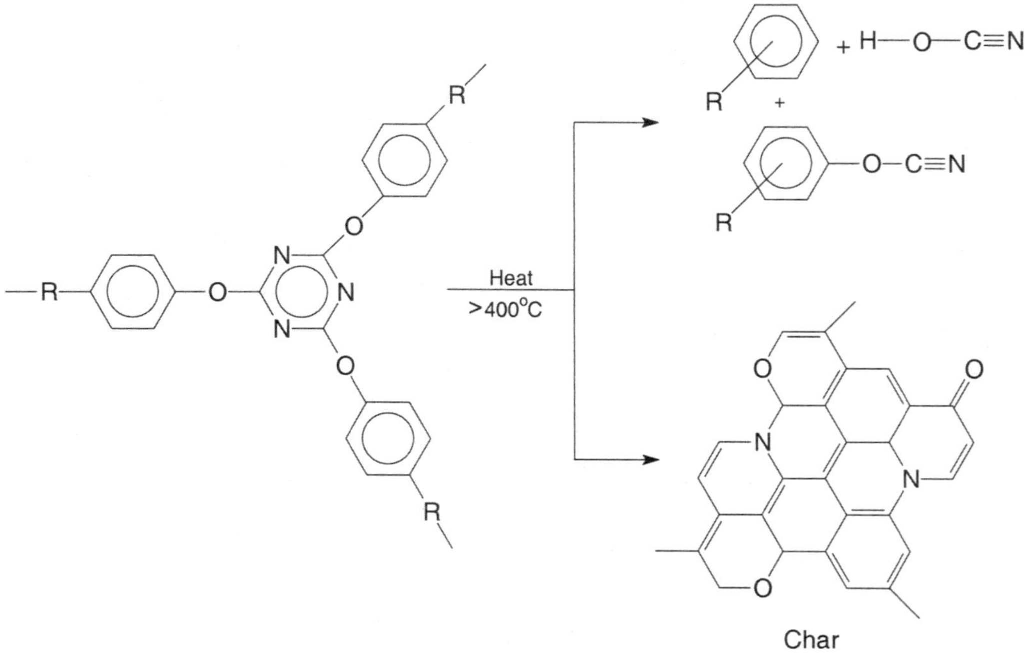
+ H—O—C≡N
+
—O—C≡N
Heat
>400°C
Char

Figure 1.76
Chemistry of degradation of cured cyanate ester resin.

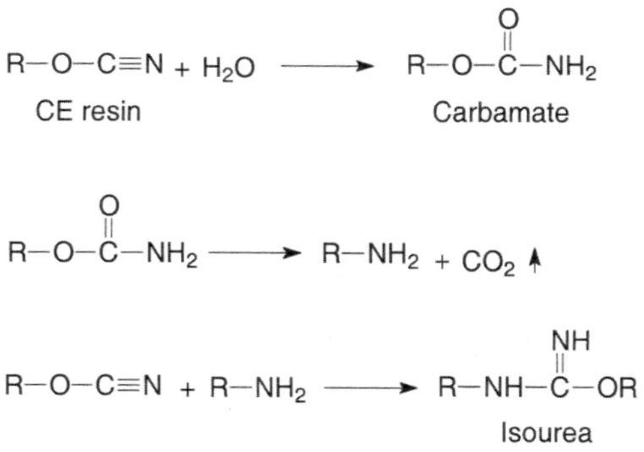
R—O—C≡N + H2O → R—O—C(=O)—NH2
CE resin
Carbamate
R—O—C(=O)—NH2 → R—NH2 + CO2 ↑
R—O—C≡N + R—NH2 → R—NH—C(=NH)—OR
Isourea

Figure 1.77
Reaction of cyanate ester resin with water to form carbamate and isourea.

1.13.2 Properties of cyanate ester resins

A cured CE resin network consists of triazine rings linked to the backbone in the structure through ether linkages. Because of the presence of flexible ether linkages cyanate esters exhibit higher flexibility compared to the unmodified epoxy and bismaleimide resins. Because of the symmetric cyclic structure and a very less number of polar groups, it shows

Table 1.27: Thermomechanical properties of cyanate ester resin.

Properties	AroCY B-10	Primaset PT-30
Tensile strength (MPa)	73	48
Tensile strain (%)	3.8	1.9
Flexural strength (MPa)	162	79
Flexural modulus (GPa)	2.9	3.59
G_{1c} (J m^{-2})	190	60
T_g (°C)	258	320
Dielectric constant at 1 GHz	2.58	2.97
Dissipation factor at 1 Hz	0.006	0.007

exceptionally low water absorption and dielectric loss. Moisture absorption for a high temperature cured epoxy when kept in 100% humidity for 100 h was more than 6% whereas under a similar condition the CE resin shows water absorption less than 2% [522]. The other interesting properties of CE resins include low cure shrinkage and low coefficient of thermal expansion. Typical thermomechanical properties of AroCy B-10 and Primaset PT-30 are presented in Table 1.27. PT resins are available as a viscous liquid, semisolid and powdered material. The availability of resin in various consistencies (liquid and solid forms) provides the broad processing capabilities of PT resins by various techniques namely hot melt (prepreg/adhesives), filament winding, resin transfer molding and powder coating. The resins require initial cure and postcure at elevated temperatures to achieve the maximum T_g of 350°C–400°C. The resins offer outstanding thermal resistance. For instance, mechanical properties remain unaltered after aging 200 h at 288°C or 100 h at 315°C.

1.13.3 Modified difunctional cyanate ester resins

As discussed above, the commercial CE resins offer useful properties like high thermooxidative stability, low water uptake, and improved mechanical toughness (in comparison to other thermosetting polymers). However, significant improvements in these properties are still needed for many advanced applications. Therefore a lot of research works have been carried out to develop a modified version of CE resins having enhanced properties compared to commercial CE resins like BADCy and LECy. CE resins containing —CF_3 groups have been shown to exhibit low dielectric constant, water resistance, and higher thermal stability compared to their hydrocarbon counterpart [523,524]. The water uptake of 2,2-bis(4-cyanatophenyl)-1,1,1,3,3,3-hexafluoro propane (BAFCY) is 1.8%, which is much lower than the value for BADCy (2.3%) [525]. The decrease in water uptake value can be explained by two factors the presence —CF_3 group and lower cyanurate ring density compared to BADCy. The water uptake is known to reduce with decreasing cyanurate ring density. For example, dicyclopentadiene-containing dicyanates and bisphenol-M analog of BADCy offer still lower water uptake values, 1.2 wt% and 0.6%, respectively, due to their considerably lower cyanurate ring density [525,526].

However, the fluorinated resins are high-melting solids. For example, the melting point of BAFCY is 87°C. Hence, the resin can be melt-processed only at high temperatures. Liquid resins have the advantage that they can be impregnated with the reinforcement and process at a lower temperature. Low-temperature processing saves energy and advantageous for application in intricate structures. The crystallinity of the CE resin can be reduced by introducing oligomeric spacer groups with a random length between the terminal reactive cyanate groups. Such modification also reduces the cross-link density of the resulting network due to an increase in molecular weight between the cross-links and expected to improve the flexibility of the network [527–529]. Laskoski et al. [530] reported the synthesis of liquid fluorinated CE resin by incorporating aryl ether spacer groups between the reactive CE resins. A modified Ullmann reaction was utilized for the synthesis. The reaction scheme is shown in Fig. 1.78. First, a hydroxyl-terminated intermediate was prepared by reacting hexafluorobisphenol-A with 1, 3-tribromo benzene. The hydroxyl-terminated intermediate is then reacted with cyanogen bromide to get the cyanate ester prepolymer. The length of the spacer can be varied by changing the composition of the reactants. The prepolymer is a liquid and offers better processability. The cured resin has a T_g of 175°C, which is lower than the T_g of cured BAFCY (270°C). The TGA results of cured resin (cured with chromium catalyst for a maximum temperature of 300°C in an inert atmosphere) are presented in Fig. 1.79. The sample showed weight retention of 95% at 430°C and the onset of thermal decomposition takes place at about 425°C. Thus the thermal stability is comparable to BAFCy. Nair et al. [531] reported polycyanate esters of an imide-modified novolac of different maleimide-content. The resins underwent a two-stage independent thermal curing through trimerization of the cyanate groups as well as the addition polymerization of the maleimide moieties. On heating, the cyanate esters were transformed into animido-phenolic-triazine network polymer. The cured resins exhibited a higher initial decomposition temperature compared to the cured maleimide novolacs. However, the thermal stability was found to be inferior to the conventional phenolic-triazine resin.

Incorporation of silicon at the molecular level has been exploited successfully for epoxy resins resulting in a considerable increase in thermooxidative stability, flame retardancy and char yield [532,533]. A similar strategy has been extended for CE resins and *bis* (4-cyanatophenyl) dimethylsilane, that is, ester of silicon-containing analog of bisphenol-A (SiMCy) was reported [534,535]. If we compare the chemical structure of SiMCy with BADCy, we can find that the only difference is that each quaternary carbon atom in the linkage between cyanate ester groups is replaced with a quaternary silicon atom. SiMCy is synthesized by a three-step reaction. In the first step, bis(4-benzyloxy phenyl) dimethylsilane is prepared by treating 4-(benzyloxy)bromobenzene with n-BuLi and dichlorodimethylsilane in tetrahydrofuran medium. Bis(4-benzyloxyphenyl) dimethyl silane is then reduced in presence of palladium catalyst (10% palladium on carbon) to prepare bis

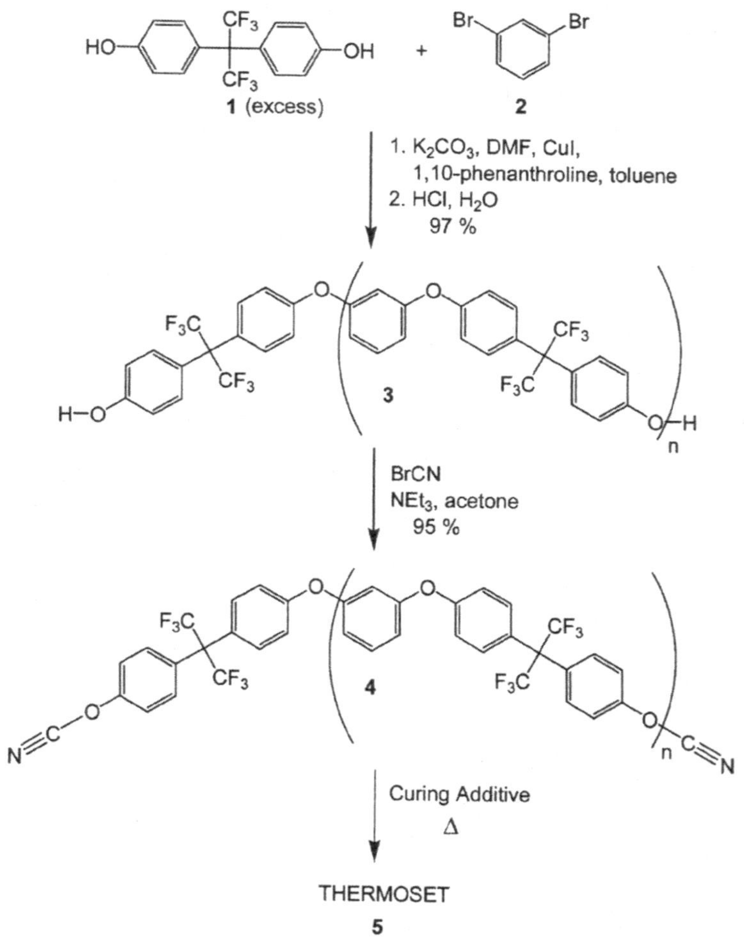

Figure 1.78
A reaction schemes for the synthesis and curing of liquid fluorinated cyanate ester resin. Source: Reproduced from M. Laskoski, D.D. Dominguez, T.M. Keller, J. Mater. Chem. 15 (2005) 1611 by the permission of The Royal Society of Chemistry Publishers.

(4-hydroxyphenyl)dimethylsilane, which is finally reacted with cyanogen bromide to produce SiMCy. The reaction scheme is shown in Fig. 1.80. The crystalline melting point of SiMCy is about 20°C lower than BADCy and the cured SiMCy exhibit T_g about 10°C lower than the cured BADCy. This can be explained by the fact that the C–Si bond lengths in the methyl side groups of the monomer, as measured by X-ray diffraction, are longer for SiMCy than for BADCy which imparts flexibility and change in molecular structure and the related interactions. In BADCy, the dipole–dipole interactions for the cyanate ester groups dominate the crystal packing forces, whereas in the case of SiMCy, the interactions responsible for crystalline packing are between the aromatic hydrogens and the cyanate

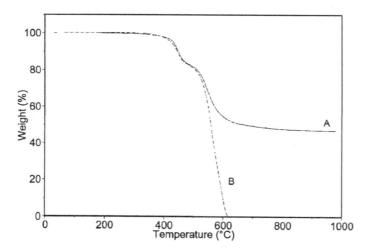

Figure 1.79
TGA thermograms of cured fluorinated cyanate ester resin under (A) nitrogen and (B) air. TGA, thermogravimetric analysis. Source: *Reproduced from M. Laskoski, D.D. Dominguez, T.M. Keller, J. Mater. Chem. 15 (2005) 1611 by the permission of The Royal Society of Chemistry Publishers.*

ester nitrogen atoms. If we study thermooxidative stability in a nitrogen atmosphere, the cured SiMCy performs substantially worse than cured BADCy, with higher weight loss rates at temperatures from 410°C to around 490°C. The char yield at 600°C is also reduced from 49% in cured BADCy to 43% in cured SiMCy. Interestingly when the same test is carried out in the air the reverse trend is observed. The cured SiMCy outperforms the cured BADCy in terms of thermooxidative stability. The char yield at 600°C is around 50% for cured SiMCy compared to 33% for cured BADCy. The decreased thermal stability of SiMCy in the nitrogen atmosphere might result from the substitution of carbon–silicon bonds for carbon–carbon in each repeating unit and higher thermooxidative stability in the air can be explained by considering the formation of SiO_2 during the thermal decomposition of cured SiMCy.

SiMCy exhibits a significantly lower moisture uptake compared to BADCy and LECy [535]. It is proposed [536] that there may be three ways for a thermoset like epoxy resins to absorb water: (1) formation of a polymer–diluents solution; (2) adsorption at hydrophilic sites; and (3) adsorption on the surface of free volume elements. In the case of cyanate ester resin, the moisture uptake phenomena appear to be more complex. Water uptake is generally explained from the free volume present in the network, which can be determined from positron annihilation lifetime spectroscopy (PALS) analysis. The readers may refer to Chapter 6, for getting details on PALS. Due to the presence of longer C—Si bonds, the total amount of free volume is higher for the cured SiMCy than the cured BADCy [535]. Moreover, the free volume of the blend of BADCy and SiMCy increases with increasing the concentration of SiMCy. If we consider the total free volume responsible for controlling

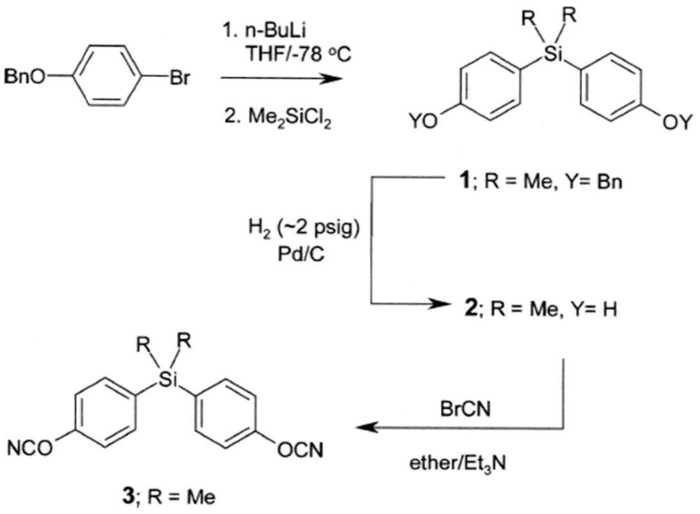

Figure 1.80
The synthetic route for production of *bis*(4-cyanatophenyl) dimethylsilane. Source: *Reprinted with permission from A. J. Guenthner, G.R. Yandek, M.E. Wright, B.J. Petteys, R. Quintana, D. Connor, R.D. Gilardi, D. Marchant, Macromolecules 39 (2006) 6046. © 2006, American Chemical Society Publishers.*

water uptake then the higher water uptake of cured SiMCy than the cured BADCy or reduction in moisture absorption of BADCy/SiMCy blends with increasing concentration of SiMCy cannot be explained. It may also be noted that PALS data suggests that a significant amount of free volume develops during the final stages of curing reaction, that is, free volume in BADCy increased with increasing conversion for the range of conversions investigated (85% − 100%) [536]. The molar volume determined from density measurement show a similar trend [537]. Regression analysis indicated a statistically significant effect ($P = .03$) of the extent of cure on density, with a decrease in density of about 0.014 g cm^{-3} for every 10% increase in the extent of cure. Therefore the degree of cure had a strong influence on moisture uptake.

From the above discussion, it appears that moisture uptake in cured cyanate esters is not simply a matter of free volume, but rather depends on the availability of a specific type of free volume created at characteristic locations near the triazine rings during the development of the cyanate ester network. These free volume sites possess a larger characteristic size, according to positron annihilation lifetime data, and cannot relax or diffuse away to a significant extent from the characteristic locations when cure temperature approaches the glass transition temperature due to vitrification (diffusion-controlled curing). These free volume sites got frozen in during the vitrifications and play an important role in dictating the moisture absorption. The free volume created due to less tight packing of the network is not going to concentrate at characteristic locations near the triazine rings and had little effect on the moisture uptake property of the network. Hence, the lower moisture uptake of SiMCy can be attributed to the reduction in cyanurate ring density. A 10% decrease in cyanurate ring density resulted in a 25% decrease in moisture uptake. Similarly, in the case of bisphenol-M analog of BADCy, a 30% decrease in cyanurate ring density leads to a 75% decrease in moisture uptake. The nonlinear relationship between moisture uptake and cyanurate ring density can be explained by considering the fact that the presence of flexible linkages (for example Si—C in SiMCy) helps to diffuse/relax the generated free volume during vitrification and lessen the amount of free volume in the characteristic location.

Guenthner and coworker [538] investigated three-component cocured networks formed from similar dicyanate ester monomers namely BADCy, LECy, and SiMCy. Since the monomers are similar in nature, they generate homogeneous network without any sign of phase separation. They have reported an interesting result, that is, an unexpected 25% reduction in moisture uptake for conetworks of equal amounts of LECy and SiMCy segments indicating a synergistic effect. The result cannot be explained in terms decrease in the number density of cyanurate rings. The deviations from linearity that appeared to be driven by interactions among components occurred in samples containing mixtures of SiMCy and LECy segments. Similar observation was reported by others [539,540]. For all other conetworks, a clear relationship between moisture uptake and the number density of cyanurate rings was observed. It was concluded that the net increase in free volume at room temperature associated with the final stages of cure, provides favorable sites for moisture uptake, whereas increased free volume due to less efficient packing of the network segments has, comparatively, much lesser influence on moisture uptake.

It is also observed [541] that inclusion of methyl groups positioned ortho to the phenyl cyanate oxygen, resulted in decreased water uptake and improved hydrolytic stability of the network analogus to BADCy or LECy. This can be explained by considering the fact that water is free to approach the cyanurate oxygen, facilitating interaction in BADCy or LECy network, whereas in the presence of methyl group in o methylated network water is effectively blocked from accessing the oxygen. The effect of chemical configuration on the properties of an isomeric series of n-propyl-bridged cyanate ester monomers has also been investigated [542]. Configurations that provide cyanurate oxygen atoms with bridge groups exhibited decreased moisture uptake by up to 50%. This is associated with about 20°C—40°C reduction in wet T_g. This can be explained by considering the fact that the bridge itself may provide some additional steric hindrance around the cyanate ester groups ortho to the bridge linkages. Various forms of steric hindrance found in the repeat structure of polycyanurate networks are illustrated in Fig. 1.81.

1.13.4 Modified multifunctional cyanate ester resins

In the previous section, we have discussed how the properties of difunctional CE resins can be enhanced by using a new synthetic strategy. As discussed earlier multifunctional

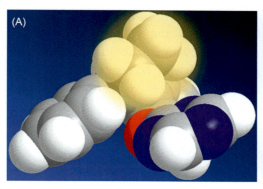 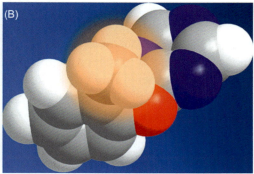

Figure 1.81
Illustration of various forms of steric hindrance found in the repeat structures of polycyanurate networks: (A) a bulky bridge group (*highlighted yellow*) may hinder access to a cyanurate oxygen atom in a manner similar to (B) a methyl group (*highlighted orange*) ortho to a cyanurate linkage. Color scheme, C: gray; H: white; N: blue; O: red. For simplicity, repeat unit structures are H-terminated. Source: *Reprinted with permission from A. J. Guenthner, B. G. Harvey, A. P. Chafin, M. C. Davis, J. J. Zavala, K. R. Lamison, J. T. Reams, K. B. Ghiassi, J. M. Mabry, Macromolecules 50 (2017) 4887–4896.* © 2017, American Chemical Society Publishers.

thermoset like epoxy offers better thermal stability compared to its difunctional counterpart. Therefore tricyanate esters have been investigated, although they received much less attention than bisphenol type dicyanate esters in terms of basic research and the development of structure–property relationships. An example of commercial multifunctional CE resin is phenyl triazine PT-30 by Lonza (Fig. 1.74). Tricyanate esters offer superior thermooxidative stability and exhibit network junction densities approximately double those of the more common bisphenol type cyanate esters, due to the fact that the central branch point in the chemical structure of the monomer forms a junction point for polymer network segments. However, the use of multifunctional CE resins is associated with two major issues. First, the high cross-link density restricts segmental motion, thereby raising the glass transition temperature of fully trimerized systems near to, and in some cases, beyond, the limit of even short-term thermal stability of the molecules. As a result, complete trimerization becomes very difficult to achieve from practical point of view. This is reflected in variation of properties of such resins, reported in the literature. Second, the cured network having very high cross-link density is extremely brittle, which hinders fabrication and evaluation of test specimens for mechanical property assessments and pose limitation in regards to use of such materials in various applications [543]. A number of attempts have been made to resolve the issues mentioned above, which will be reviewed shortly.

Yameen et al. [544] synthesized a new tricyanate ester oligomer that combined excellent flow and processing characteristics with a glass transition temperature of 280°C, equivalent to the best-performing bisphenol type cyanate esters. Similarly, Cambrea et al. [545] synthesized a tricyanate ester monomer with very good flow property at room temperature and offer a dry glass transition temperature of 305°C. The strategy used in the above-mentioned examples is to reduce the effective cross-link density compared to the cured PT-30 while the rigid phenyl ring was retained at the monomer core. Another possible strategy can be to introduce a flexible central branch point or core while maintaining a similar cross-link density comparable to that of cured PT-30.

With this concept in mind, a new tricyanate ester monomer featuring flexible chemical linkages at the central branch point, that is, [1,2,3-Tris(4-cyanatophenyl) propane] has been synthesized and characterized [546]. The reaction scheme is described in Fig. 1.82. The reaction comprises nine steps, namely, Perkin condensation of 4-methoxyphenylacetic acid on anisaldehyde, catalytic hydrogenation, standard Friedel Crafts reaction, reduction with sodium borohydride, acid-catalyzed dehydration, hydrogenation, demethylation, and reaction cyanogens bromide. The product is synthesized with an overall yield of 26%. The tricynate is capable of undergoing a cure reaction at 210°C–290°C in the same way as PT-30. The activation energy for cure is slightly lower than that PT-30. The chemical structure and network architecture of cured networks for both the developed tricyanate and PT-30 are presented in Fig. 1.83. We can clearly visualize that networks formed from PT-30 exhibit

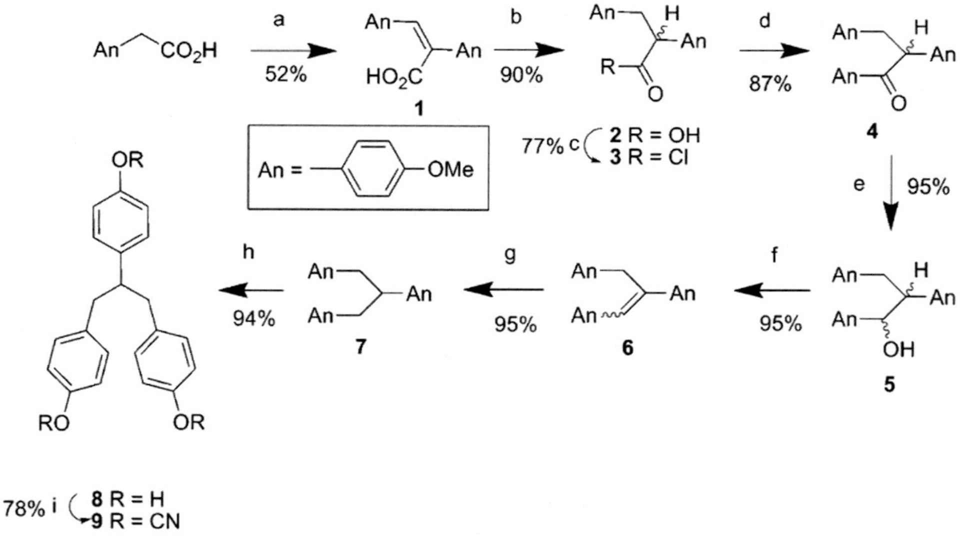

Figure 1.82
Synthesis of tricyanate monomer 1,2,3-Tris(4-cyanatophenyl) propane: (A) 4-MeOC$_6$H$_4$CHO, Ac$_2$O, TEA, reflux; (B) H$_2$, 5% Pd/C, DMF; (C) SOCl$_2$, heat; (D) anisole, AlCl$_3$; (E) NaBH$_4$, EtOH; (F) pTsOH, PhMe, reflux; (G) H$_2$, 5% Pd/C, EtOH; (H) pyridine HCl, reflux; (I) BrCN, TEA, acetone, −20°C.
Source: Reprinted with permission from A.J. Guenthner, M.C. Davis, K.R. Lamison, G.R. Yandek, L.R. Cambrea, T.J. Groshens, L.C. Baldwin, J.M. Mabry, Polymer 52 (2011) 3933. © 2011, Elsevier Publishers.

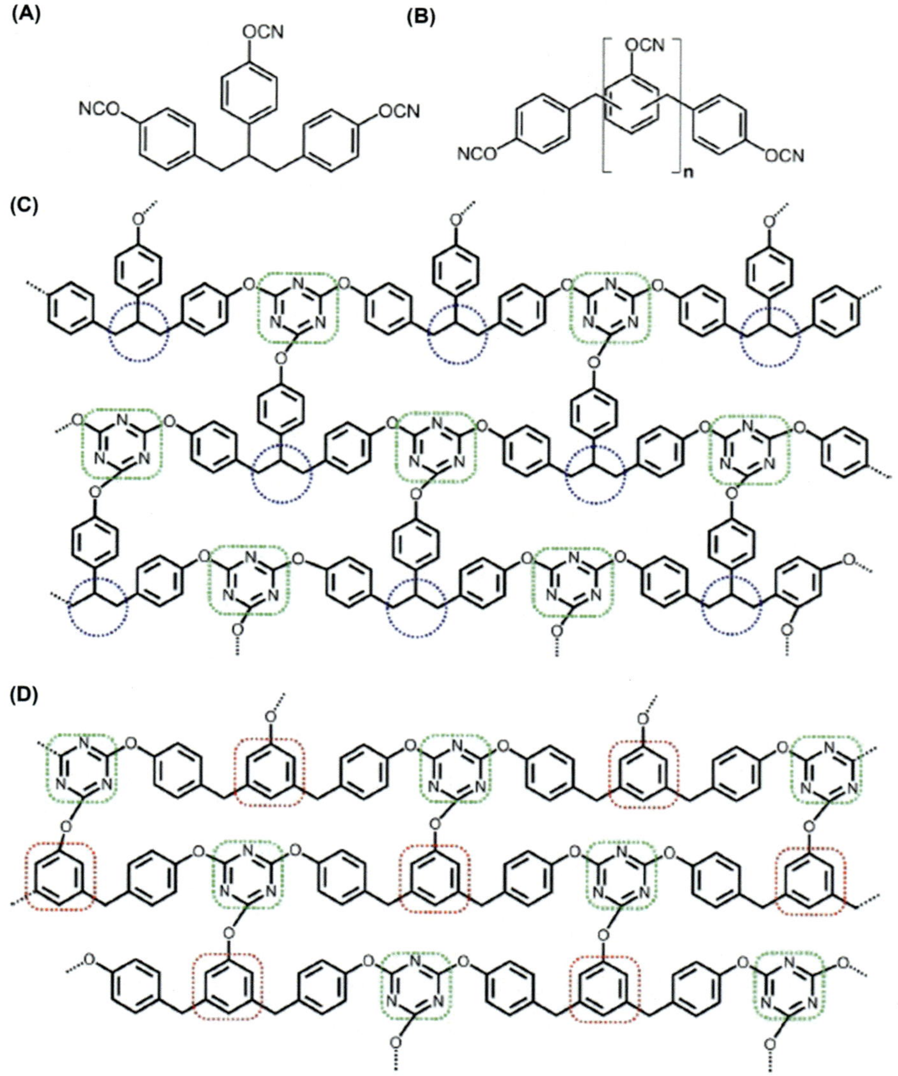

Figure 1.83

(A) Chemical structure and network architecture of cured tricyanate networks; (B) approximate chemical structure ($n = 1-2$) for Primaset_PT-30 resin (C) after cure, the idealized network architecture the tricyanate features a "checkerboard" pattern of relatively flexible tri-substituted propyl junctions (*dashed circles*) alternating with more rigid triazine ring junctions (*dashed rounded rectangles*), (D) whereas in PT-30 (shown with $n = 1$ for convenience), the network architecture features only rigid triazine ring or phenyl ring junctions. Source: *Reprinted with permission from A.J. Guenthner, M.C. Davis, K.R. Lamison, G.R. Yandek, L.R. Cambrea, T.J. Groshens, L.C. Baldwin, J.M. Mabry, Polymer 52 (2011) 3933.* © *2011, Elsevier Publishers.*

junctions from either phenyl rings or triazine rings, both of which are comparatively rigid whereas the new tricynate-based network contains a flexible core based on a 1,2,3-trisubstituted propane. The network exhibits glass transition temperature at 340°C. These results indicate that many of limitations associated with rigid thermosetting monomers having a high cross-link density may be overcome with only modest trade-offs through a carefully chosen introduction of flexible junctions in the molecular architecture of the network.

1.13.5 Application of cyanate ester resins

CE resins failed to enter into the aircraft and engineering composite industry due to their much higher cost compared to epoxy and bismaleimides, which predominate in these industries. The major use of CE resin is in aerospace applications. The driving force to replace the metallic and ceramic materials by high temperature polymer is to reduce the weight, which significantly reduces the launching cost. CE resins find better acceptance for aerospace applications compared to epoxy and bismaleimide due to their comparatively lower moisture absorption, more flexibility, and radiation resistance. CE resin-based composites can be processed in the same way as epoxy and bismaleimide and are of special interest for space applications such as antennas, signal devices, arrays, etc.

Major use of CE resins is in the electronic industry. The electronic applications include printed wiring circuit boards, thin cards, multichip module laminates, and ship encapsulants. CE resins have replaced epoxy and bismaleimide resins to a great extent for such microelectronic applications. This is due to the obvious advantages of CE resins in respect of comparatively lower moisture absorption and dielectric dissipation factor.

Another important application for CE resin is for radomes. An effective exploitation of advance radar systems, which involve communication of wide band microwave (600 MHz to 100 GHz), targeting and tracking requires the design of suitable window materials. The material should be transparent to the microwave at the same time it should withstand the high temperature generated due to passage of such wave in the service condition and also should retain the dimensional stability at high temperature. Because of tendency to micro cracking, epoxy and bismaleimides are not successful for communication satellite. The radar transparency and microcracking resistance make CE resins suitable for the conical radome nose cone, which house radar antennas of military and weather reconnaissance plane.

CE resins have the lowest dielectric constant and dissipation factor among the thermoset family. The only polymers that have a lower dielectric constant and dissipation factors are Teflon and polyethylene, which are low modulus materials. Low values of dielectric constant and dissipation factor are extremely desirable for aerospace applications, which allow an increase in signal speed and circuit density and at the same time decrease in power requirements and the amount of heat generated relative to other materials. Thus the use of CE resin resulted in increased transmission of microwaves through radome walls, antenna housings and stealth aircraft composites. Cyanate esters also find use as friction materials in brake lining, grinding wheels and as high-performance adhesive and coating [547].

PT resins are used in aircraft interior duct components, plastic ball grind arrays, encapsulation of electronic components. Similar to bismaleimides, CE resins often blended with other resins to improve their thermal stability. The high water absorption and dielectric loss of epoxies limit their applications for printed circuit boards requiring high-speed signal propagation and transport. The blending of CE resin can improve the above-mentioned properties in addition to improve thermal stability. Kim [548] studied the effect of blending of cyanate ester (AroCy L-10) on the cure behavior and thermal stability of epoxy resin. The addition of 28.6% (40 parts per 100 parts of epoxy) of CE resin resulted in an increase in T_g by 40°C and improvement of thermal stability (temperature corresponding to the onset of weight loss). Similarly, the addition of epoxy in CE resin improves the processability and tack/drape properties and reduces the cost [527]. Blends of cyanate ester and bismaleimide have been realized to derive systems bearing the good physicochemical attributes of the two components, that is, the heat resistance of bismaleimide and the mechanical properties of polycyanurates [549–551]. Commercial blend formulations of bismaleimide–cyanate known as B–T resins are available, which are extensively used as materials for engineering materials in aircraft, reinforced plastics, and injection-molding powders, as well as materials in electric motor coil windings and allied applications [552].

The addition of a small amount of allyl-functionalized aryl cyanate esters may increase the T_g of the overall network by linking the polycyanurate network to the poly-bismaleimide network. Wang et al. [553] developed a thermosetting terpolymer composition comprising bis-allyl benzoxazine (Bz-allyl), cyanate ester (BADCy), and 4,4'-bismaleimido-diphenyl methane (BMI), which can be cured via cocuring reactions. When the Bz-allyl content in the cured Bz-allyl/BMI/BADCy resins was 9 wt.%, the dielectric constant values and the dielectric loss factors of the blends reached the minimum values of 3.02 and 0.0158 at 10 MHz, respectively. In the case of cured Bz-allyl/BMI/BADCy resin containing 9 wt.% Bz-allyl, the water absorption value is only 1.77 wt.%, which is much lower than those of the BMI/BADCy resin.

However, the addition of Bz-allyl tends to increase the water-resistance property. The improvement of the water-resistant property of Bz-allyl/BMI/BADCy systems may be due to the more perfect network structure. Compared with BMI/ BADCy, Bz-allyl has phenolic hydroxyl groups, which catalyze the reaction between BMI with BADCy; Bz-allyl also has allyl groups, which react with BMI/BADCy and form a perfect network structure after curing. Therefore the Bz-allyl/BMI/ BADCy systems have fewer free volume elements, which results in the lower water absorption values of the Bz-allyl/BMI/BADCy systems. Development of the blend systems discussed above is going resolve the cost and processing issue of CE resins and broadened their scope of applications. Thus a well-developed understanding of the relationships among physical and chemical characteristics of the network, processing, and resultant solid-state properties for cured cyanate esters will certainly improve the design and performance of the resins used in many technologically important products ranging from microcapacitors [554] to interplanetary space probe [555].

The unique characteristic of CE resins is their ability to withstand extreme environments namely the radiation environment around Jupiter encountered by the Juno space probe [556], the intense neutron bombardment expected within the thermonuclear fusion reactor [557,558], and other superconducting magnets exposed to high radiation doses [559]. Keeping the various applications discussed above in mind, a lot of works have also been carried out to produce bio-based CE resins [560–562] for sustainable development through green technology. For example, Harvey et al. [563] reported the eugenol-based cyanate ester that could be well cured at a high temperate, and the cured epoxy exhibited excellent thermal stability up to 350°C with a T_g of 186°C. Therefore the use of CE resin shows every sign of showing an upward trend in the future.

1.14 Phthalonitrile resin

With the current growing demands from the key technological and industrial sectors, heat-resistant thermosetting resins are very much important for high-performance adhesives, coatings, prepregs, and composites. They are frequently being used in the defense and aerospace industries wherein metals and ceramics were the only available options for structural design. The various applications are wings, fuselage skins, engine parts, landing gears and satellite components. The performance criteria for these components in electrical and mechanical devices are above 300°C–400°C in continuous usage and to hold the temperature spikes of 500°C–600°C. Though the high-temperature-resistant polymers have lower service temperatures vis-a-vis metals and ceramics, they have better fatigue resistance than ceramics and much lighter than metals. There are many available thermosetting polymers namely polyimides, bismaleimides, cyanate esters, etc. for such applications as already discussed above.

Phthalonitriles (PN) are the one among the high-temperature-resistant thermosetting resins family well researched as compared to other resin systems. They have demonstrated excellent mechanical properties, high glass temperatures, thermooxidative stability, good moisture resistance, high char yield, and superior fire retardancy [564–568]. The key property matrix has opened up wide gate for potential applications like high-performance composites, adhesives, structural applications in defense, and aerospace industry [540,569,570]. The high-temperature resistance of PN resins is the outcome of its chemical structure having aromatic and heterocyclic rings within the backbone of a polymeric system linked with a flexible ether linkage.

1.14.1 Synthesis of phthalonitrile resins

PN resins are synthesized by the thermal polymerization reaction. Due to sluggish thermal polymerization various curing additives are employed [571]. However, the most common additives are aromatic diamines. PN monomers with several structural modifications have been synthesized to evolve a thermally stable network with good processability. Typically, PN resins are synthesized by a nucleophilic substitution reaction of an aromatic dihydroxy compound with 4 nitro-phthalonitrile in presence of a strong base like K_2CO_3 in polar aprotic solvents like DMSO as shown in Fig. 1.84. The resin so obtained is called Stage A resin. The resin in "Stage A" can be converted to a more stable prepolymer form known as Stage "B," which is dark green in color has indefinite ambient temperature stability along with ease of processability into composite components by regular prepreg consolidation, resin transfer molding, filament winding from the melt of prepolymer. Whereas other known high-temperature-resistant resin systems possess limited shelf life. More often, these resins are stored under refrigerated conditions below subzero temperatures, which is one of the major drawbacks with other thermosetting resins.

Although PN resin offers many technological advantages, it is associated with the drawbacks such of high curing temperature, slow curing rate, complex, and uncontrolled curing products [572–574]. To address these issues, various routs have been employed to overcome via using assortment of curing agents like metal salts, strong organic acid/amine salt, or active hydrogen containing hetrocyclic compounds [575–579]. It is assumed that employed methodology of curing agents effectively modulate curing process [580–582]. There are many chemical and physical factors, which cumulatively make a particular polymer heat resistant. The chemical factors include primary bond strength, resonance stabilization, mechanism of bond cleavage, molecular symmetry/structural regularity, rigid interchain structure, branching, and cross-linking. Physical factors include secondary Van der Waals forces, molecular dipolar interaction, hydrogen bonding, molecular weight, molecular weight distribution, crystallinity, molecular dipolar interaction, and purity [583]. All the components of high temperature structural features have been incorporated in one or another way by the research group to provide adequate high temperature resistance keeping the above features. The synthetic advancements considering all the aspects of structural features enumerated above some of modifications in PN monomer synthesis can be

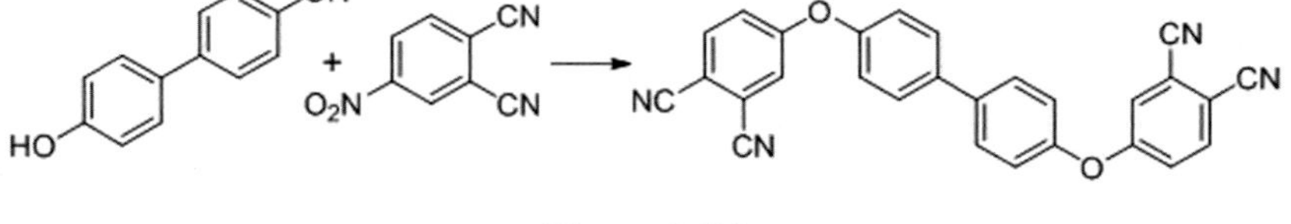

Figure 1.84
Synthesis of PN resin. *PN*, phthalonitrile.

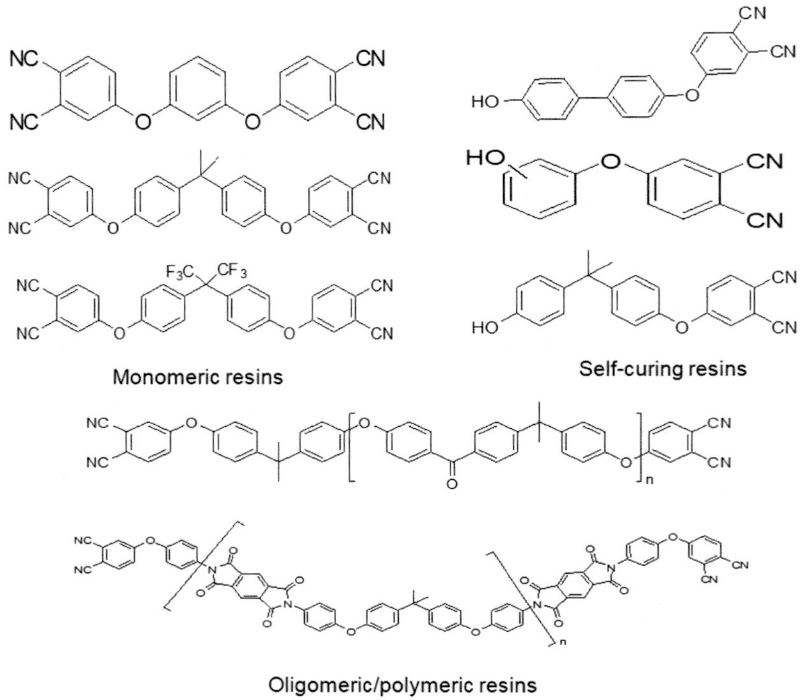

Figure 1.85
The chemical structure of different types of PN resins. *PN*, phthalonitrile.

classified into various categories namely monomeric resin (resorcinol, biphenyl-based), self-curing resins containing hydroxyl and amino groups, oligomeric/polymeric resins (poly ether or imide based) and mixed curing functionality-based resins having novolac, cyanate ester propargyl ether moieties. The chemical structure of different types of PN resins mentioned above are shown in Fig. 1.85.

1.14.2 Properties of phthalonitrile resins

If we look at commercial availability of PN resins, a number of industries are in the process of developing various types of PN resins. The international status of phthalonitrile resin bulk production reveals that mainly US-based firms are chief manufacturers of different formulations of phthalonitrile resin. PEK, biphenyl, and resorcinol-based PN resins are marketed by US-based industries namely M/s. Maverick Corporation, M/s. Cardolite Corporation, M/s. JFC Technologies. Polythalonitrile resin family has emerged as most potent matrix materials for high temperature applications over 300°C for continuous usage in thousands of hours with high char yield. A comparison of properties of PN resins with other high temperature resins is presented in Table 1.28.

Table 1.28: Comparison of properties of phthalonitrile resin with common high-performance thermosetting polymer matrices.

Property	Toughened BMI	PMR-15	Phthalonitriles
Highest continuous use temperature (°C)	350	300	400
Cure temperature (°C)	220–300	280–320	250–350
Tensile modulus (GPa)	3.4–4.1	4.4	4–5
Onset degradation temperature (°C)	360–400	~400	>450
T_g (°C)	~150–350	~327	~350 to >450
Toxic gas emission (combustion)	Yes	Minimum	Negligible
Storage life and conditions	4–5 years in solid form	~1 year in solution form at ambient conditions	> 5 years Stage B resin at ambient conditions

Phthalonitriles are in fact one of the resins that qualify the modern definition of high temperature resin system, which states that the material should retain mechanical properties for thousands of hours at 230°C, hundreds of hours at 300°C, minutes at 540°C, and seconds up to 760°C. Also, these materials retain useful properties under particular conditions like pressure or vacuum, mechanical loading, radiation, chemical, or electrical exposure, all at temperatures ranging from cryogenic to above 500°C. PN resins are known for their very long room temperature storage life (>5 years under ambient conditions) as compared to many of the resins which possess limited storage life in cryo storage (−18°C). Although PN resins offers many useful properties, there exist some technical challenges related to their high curing temperatures such as requirement of new tools and machineries to fabricate parts using PN resin systems. Efforts are needed to tailor the chemistry in a way to make these resins user friendly encompassing variety of processing methods. Wu et al. [584] reported a novel curing process using mixed curing agents to improve the curing process and thermal stability of phthalonitrile resins. Zhang et al. [585] synthesized a new type of phthalonitrile-etherified resole resin (PNR) from resole resin and 4-nitrophthalonitrile. The synergistic effect among the phenolic hydroxyl groups, hydroxymethyl and nitrile groups significantly decreased the curing time and accelerated the curing reaction of the PNR compared with those of phthalonitrile resin. Self-curing triphenol-A-based phthalonitrile resin precursor has been reported [586], which can as a flexibilizer and curing agent for phthalonitrile resin.

References

[1] C.J. Larez, G.A.P. Mendoza, Polym. Bull. 22 (1989) 513.
[2] C.J. Larez, G.A.P. Mendoza, Polym. Bull. 23 (1990) 577.
[3] V.I. Szmercsanyi, L.K. Maros, A.A. Zahran, J. Appl. Polym. Sci. 10 (1966) 513.

[4] L.G. Curtice, Ind. Eng. Chem. Prod. Res. Dev. 3 (1964) 218.
[5] S.K. Gupta, R.T. Thampy, Makromol. Chem. 139 (1970) 103.
[6] V. Zvonar, A. Sternscuss, Plast. Kautsch. 7 (1960) 228.
[7] Z. Ordelt, V. Soedin, Macromol. Chem. 4 (1) (1962) 110.
[8] Z. Ordelt, V. Soedin., Macromol. Chem. 63 (1963) 153.
[9] Z. Ordelt, F. Ciganek, Chem. Prum. 14 (1964) 141.
[10] W. Kemp, Organic Spectroscopy, 2nd Ed.), Macmillan Education Ltd, Hong Kong, 1987.
[11] Y.S. Yang, J.P. Pascault, J. Appl. Polym. Sci. 64 (1997) 133.
[12] Y.S. Yang, J.P. Pascault, J. Appl. Polym. Sci. 64 (1997) 147.
[13] P.J. Flory, J. Am. Chem. Soc. 61 (1939) 3334.
[14] C.C. Lin, K.H. Hsieh, J. Appl. Polym. Sci. 21 (1997) 2711.
[15] R.B. Naik, S.B. Jagtap, D. Ratna, Prog. Org. Coat. 87 (2015) 28–35.
[16] A.C. Tang, K.S. Yao, J. Polym. Sci. 35 (1959) 219.
[17] Y.R. Fang, C.G. Lai, J.L. Lu, M.K. Chen, Sci. Sin. 18 (1975) 72.
[18] S.A. Chen, K.C. Wu, J. Polym. Sci. Polym. Chem. (Ed.) 20 (1982) 1819.
[19] C.T. Kuo, S. Chen, J. Polym. Sci. Polym. Chem. (Ed.) 27 (1989) 2793.
[20] D. Beigzadeh, S. Sajjadi, F.A. Taromi, J. Polym. Sci. Polym. Chem. (Ed.) 33 (1995) 1505.
[21] W.F. Gum, W. Rinse, H. Ulrich (Eds.), Reaction Polymer, Hanser, 2000, pp. 153–174.
[22] C.P. Hu, L. Lee, Polymer 34 (1993) 4516.
[23] B. Mortaigne, B. Feltz, P. Laurens, J. Appl. Polym. Sci. 66 (1997) 1703.
[24] Y.J. Huang, C.C. Su, J. Appl. Polym. Sci. 55 (1995) 323.
[25] L. Suspene, D. Fourquier, Y.S. Yang, Polymer 32 (1993) 1593.
[26] F. Baffa, J. Borrajo, J. Appl. Polym. Sci. 102 (2006) 6064.
[27] J.P. Lecointe, J.P. Pascault, L. Suspene, Y.S. Yang, Polymer 33 (1990) 3226.
[28] P.J. Flory, Principles of Polymer Chemistry, Cornell University Press, Ithaca, NY, 1975.
[29] P. Eisenberg, J.C. Lucas, R.J.J. Williams, J. Appl. Polym. Sci. 65 (1997) 755.
[30] R. Bhatnagar, I.K. Verma, J. Therm. Anal. Calorim. 35 (2005) 1241.
[31] P. Siva, I.K. Verma, D.M. Patel, T.J.M. Sinha, Bull. Mater. Sci. 17 (1994) 1095.
[32] M. Patri, J.G. Chavan, V. Hande, U.G. Suryawanshi, S. Rath, P.C. Deb, Indian. Pat. 29 (2005) 0947.
[33] C.P. Hsu, L.J. Lee, Polymer 34 (1993) 4495.
[34] K. Dusek, Network formation by chain crosslinking copolymerizationCh. 4 in: R.N. Haward (Ed.), Development in Polymerization-3, Applied Science Publishers, London, New Jersey, 1982.
[35] S.V. Muzumdar, L.J. Lee, Polym. Eng. Sci. 36 (1996) 943.
[36] A. Zlatanic, B. Dunjic, J. Djonlagic, Macromol. Chem. Phys. 200 (1999) 2048.
[37] A. Zlatanic, J. Djonlagic, B. Dunjic, Macromol. Chem. Phy. 199 (1998) 2029.
[38] J. Van Haveren, E.A. Oostveen, F. Micciche, B.A.J. Noordover, C.E. Koning, R.A. van Benthlem, et al., J. Coat. Technol. Res. 4 (2007) 177–186.
[39] H. Miyagawa, Ak Mohanty, R. Burgueno, L.T. Drzal, M. Misra, J. polym. Sci. Part. B Polym. Phys. 45 (2007) 698–704.
[40] B. Bieganska, M. Zubielwiez, E. Smieszek, Prog. Org. Coat. 16 (1988) 219–229.
[41] R.H. Kienle, C.S. Ferguson, Ind. Eng. Chem. 21 (1929) 349.
[42] J. Lindeboom, Prog. Org. Coat. 34 (1998) 147–151.
[43] K. Manczyk, P. Szewczyk, Prog. Org. Coat. 44 (2002) 99–109.
[44] Zabel, Klaassen, Muizebelt, Gracey, Hallett, Brooks, Prog. Org. Coat. 35 (1999) 255–264.
[45] Johansson, Glauser, Jansson, Hult, Malmstrom, Claesson, Prog. Org. Coat. 48 (2003) 194–200.
[46] S. Asif, Eur. Poly. J. 39 (2003) 933–938.
[47] A. Staring, R. Van Bentham, Prog. Org. Coat. 45 (2002) 101–117.
[48] J.M. Desimone, Science 269 (1995) 1060–1061.
[49] B. Voit, J. Pol. Sci. Part: A: Pol. Chem. 38 (2000) 2505–2525.
[50] F.W. Billmayer, Textbook of Polymer, in: Science, third ed., Wiley, New York, 1984.

[51] K. Sasihar, R. Narayan, K.V.S.N. Raju, J. Coat. Technol. Res. 10 (2013) 609–619.
[52] B. Erhan, G. Gunduz, D. Kasakurek, I.M. Akhmedov, Prog. Org. Coat. 55 (2006) 330–336.
[53] C.M. Nunez, B.-S. Chiou, A.L. Andrady, S.A. Khan, Macromolecules 33 (2000) 1720–1726.
[54] R.B. Naik, N.G. Malvankar, T.K. Mahato, D. Ratna, R.S. Hastak, J. Coat. Technol. Res. 11 (2014) 575–586.
[55] R.B. Naik, D. Ratna, S.K. Singh, Prog. Org. Coat. 77 (2014) 369–379.
[56] M. Johansson, T. Glauser, A. Jansson, A. Hult, E. Malmström, H. Prog. Org. Coat. 48 (2003) 194–200.
[57] B. Bruchmann, R. Königer, H. Renz, Tailor Macromol. Symp. 187 (2002) 271–280.
[58] J. Lange, E. Stenroos, M. Johansson, E. Malmström, Polym. (Guildf.) 42 (2001) 7403–7410.
[59] J. Mirjana, R. Radmila, P. Jelena, O. Bera, D. Govedarica, Prog. Org. Coat. 148 (2020) 105832.
[60] M.B. Launikitis, Vinyl ester resins, in: G. Lubin (Ed.), Handbook of Composites, Section 3, Van Nostrand Reinhold Co, New York, 1982.
[61] D.R. Son, A.V. Raghu, K.R. Reddy, H.M. Jeong, J. Macromol. Sci. Part. B 55 (2016) 1099.
[62] A. Jahandideh, K. Muthukumarappan, Eur. Polym. J. 87 (2017) 360.
[63] U.R. Vaidya, V.M. Nadkarni, Ind. Eng. Chem. Res. 26 (1987) 194.
[64] S. Cousinet, A. Ghadban, E. Fleury, F. Lortie, J.P. Pascault, D. Portinha, Eur. Polym. J. 67 (2015) 539.
[65] M. Sultania, J.S.P. Rai, D. Srivastava, Eur. Polym. J. 46 (2010) 2019–2032.
[66] A.M. Atta, S.M. El-Saeed, R.K.R. Farag, Funct. Polym 66 (2006) 1596–1608.
[67] X. Yang, S. Li, X. Tang, J. Xia, Adv. Mater. Res. (2013) 2826–2830. 807–809.
[68] S. Li, J. Xia, M. Li, K. Huang, J. Am. Oil Chem. Soc. 90 (2013) 695–706.
[69] S. Jaswal, B. Gaur, Rev. Chem. Eng. 30 (6) (2014) 567–581.
[70] J.J. La Scala, J.M. Sands, A. Joshua, E. Orlicki, J. Robinette, G.R. Palmese, Polymer 45 (22) (2004) 7729.
[71] A. Campanella, J.J. La Scala, R.P. Wool, Polym. Eng. Sci. 49 12 (2009) 2384.
[72] S. Li, X. Yang, K. Huang, M. Li, J. Xia, Prog. Org. Coat. 77 (2) (2014) 388.
[73] E. Goiti, M.B. Huglin, J.M. Rego, Polymer 42 (26) (2001) 10187.
[74] A. Jahandideh, N. Moini, S. Bajgholi, M.J. Zohuriaan-Mehr, K. Kabiri, J. Appl. Polym. Sci. 49 (144) (2020) 1–16.
[75] A. Yu, E.M. Serum, A.C. Renner, J.M. Sahouani, M.P. Sibi, D.C. Webster, J. Am. Chem. Soc. 141 (12) (2019) 4928–4935.
[76] J. Yan, D.C. Webster, Green. Mater. 4 (3) (2014) 132–143.
[77] D.C. Webster, A.Z. Yu, U.S. Patent No. 9765233, 19 Sep. 2017.
[78] A.Z. Yan, A. Rahimi, D.C. Webster, Eur. Polym. J. 99 (2018) 202–211.
[79] A.Z. Yu, J.M. Sahouani, R.A. Setien, D.C. Webster, Reactive Funct. Polym. 128 (2018) 29–39.
[80] A.Z. Yu, J.M. Sahouani, D.C. Webster, Prog. Org. Coat. 122 (2018) 219–228.
[81] Jaillet F., Nouailhas H., Boutevin B. and Caillol S. 4(2) (2016) 63–71.
[82] S.K. Yadava, K.M. Schmalbachb, E. Kinacia, J.F. Stanzione IIIb, G.R. Palmesea, Eur. Polym. J. 98 (2018) 199–215.
[83] J.J. La Scala, J.M. Sands, J.A. Orlicki, E.J. Robinette, G.R. Palmese, Polymer 45 (2004) 7729–7737.
[84] S.N. Khot, J.J. Lascala, E. Can, S.S. Morye, G.I. Williams, G.R. Palmese, et al., J. Appl. Polym. Sci. 82 (2001) 703–723.
[85] P. Zhang, J. Zhang, Green. Chem. 15 (2013) 641–645.
[86] T. Eren, S.H. Küsefoğlu, J. Appl. Polym. Sci. 91 (2004) 2700–2710.
[87] F. Jaillet, E. Darroman, A. Ratsimihety, R. Auvergne, B. Boutevin, S. Caillol, Eur. J. Lipid Sci. Technol. 116 (2014) 63–73.
[88] E. Can, E. Kınacı, G.R. Palmese, Eur. Polym. J. 72 (2015) 129–147.
[89] P. Verge, V. Toniazzo, D. Ruch, J.A. Ind. Crop. Prod. 55 (2014) 180–186.
[90] F. Jaillet, H. Nouailhas, R. Auvergne, A. Ratsimihety, B. Boutevin, S. Caillol, Eur. J. Lipid Sci. Technol. 116 (2014) 928–939.
[91] X. Yang, S. Li, J. Xia, J. Song, K. Huang, M. Li, Bioresources 10 (2015) 2130–2142.
[92] X. Yang, S. Li, J. Xia, J. Song, K. Huang, M. Li, Ind. Crop. Prod. 63 (2015) 17–25.

[93] A.M. Atta, S.M. El-Saeed, R.K. Farag, React. Funct. Polym. 66 (2006) 1596–1608.
[94] W. Mao, S. Li, M. Li, X. Yang, J. Song, M. Wang, et al., J. Appl. Polym. Sci. 133 (2016) 44084.
[95] J. Yan, D.C. Webster, Green. Mater. 2 (2014) 132–143.
[96] J. Fanny, N. Hélène, B. Bernard, C. Green, Mater 4 (2016) 63–71.
[97] J. Jeong, B. Kim, S. Shin, B. Kim, J.-S. Lee, S.-H. Lee, et al., J. Appl. Polym. Sci. 127 (2013) 2483–2489.
[98] J.M. Sadler, A.-P.T. Nguyen, F.R. Toulan, J.P. Szabo, G.R. Palmese, C. Scheck, et al., J. Mater. Chem. A 1 (40) (2013) 12579–12586.
[99] G.R. Palmese, J.J.L. Scala, J.M. Sadler, A.P.T. Lam, Renewable bio-based (meth) acrylated monomers as vinyl ester cross-linkers, United States Patent 9644059, 2017.
[100] W. Liu, T. Xie, R. Qiu, ACS Sustain. Chem. Eng. 5 (2017) 774–783.
[101] J.F. Stanzione Iii, J.M. Sadler, J.J. La Scala, K.H. Reno, R.P. Wool, Green. Chem. 14 (2012) 2346–2352.
[102] C. Zhang, S.A. Madbouly, M.R. Kessler, Macromol. Chem. Phys. 216 (2015) 1816–1822.
[103] A.L. Holmberg, J.F. Stanzione, R.P. Wool, T.H. Epps, ACS Sustain. Chem. Eng. 2 (2014) 569–573.
[104] J.F. Stanzione, J.M. Sadler, J.J. La Scala, R.P. Wool, Chem. Sus. Chem 5 (2012) 1291–1297.
[105] J.F. Stanzione, P.A. Giangiulio, J.M. Sadler, J.J. La Scala, R.P. Wool, ACS Sustain. Chem. Eng. 1 (2013) 419–426.
[106] F. Jaillet, H. Nouailhas, B. Boutevin, S. Caillol, Eur. J. Lipid Sci. Technol. 118 (2016) 1336–1349.
[107] K.A. Hodd, Rapra Rev. Rep. 38 (1990) 4. 2.
[108] Kirk, Othmer (Eds.), Encycl. Chem. Technol. 9 (1994) 730.
[109] H.F. Mark (Ed.), Encyclopedia of Polymer Science and Engineering, 6 (1991) 322.
[110] R.S. Bauer, Epoxy Resin Chemistry, Advances in Chemistry Series, 114, 18, American Chemical Society, Washington, 1980, p. 395.
[111] W.G. Potter, Epoxide Resins, Springer, New York, 1970.
[112] C.A. May, G.Y. Tanka, Epoxy Resin Chemistry and Technology, Marcel Dekker, New York, 1973.
[113] Q. Guo (Ed.), Thermosets: Structure, Property and Applications, Woodhead Publishers, United Kingdom, 2012.
[114] H. Lee, K. Nevile, Handbook of Epoxy Resins, McGraw Hill, New York, 1967.
[115] A.B. Samui, B.C. Chakraborty, D. Ratna, Int. J. Plastic Technol. 8 (2004) 279.
[116] D. Ratna, Rapra Rev. Rep. 16 (5) (2005) 1.
[117] M. Murli, D. Ratna, A.B. Samuiand, B.C. Chakraborty, Effect of molecular weight, J. Appl. Polym. Sci. 103 (2007) 1723–1730.
[118] D. Ratna, R. Varley, G.P. Simon, J. Appl. Polym. Sci. 89 (2003) 2339.
[119] D. Ratnaand Ajit, K. Banthia, Polym. Int. 49 (2000) 281.
[120] D. Ratna, G.P. Simon, Polym. Eng. Sci. 41 (2001) 1815.
[121] D. Ratna, B.C. Chakraborty, P.C. Deb, J. Polym. Mater. 14 (1997) 185.
[122] P. Ortiz, R. Vendamme, W. Eevers, Molecules 25 (2020) 1158.
[123] E.G. Reydet, C.C. Riccardi, H. Sautereau, P.J. Pascault, Macromolecule 28 (1995) 7599.
[124] L. Matejka, J. Lovy, S. Pokorny, K. Bouchal, K. Dusek, J. Polym. Sci. Part. A: Polym. Chem. 21 (1983) 2873.
[125] H. Lowe, R.D. Axinte, D. Breuch, C. Hofmann, J.H. Petersen, R. Pommersheim, Chem. Eng. J. 163 (2010) 429–437.
[126] F. Wu, X. Zhou, X. Yu, RSC Adv. 8 (2018) 8248.
[127] K. Shi, Y. Shen, Y. Yang, T. Wang, J. Appl. Polym. Sci. 138 (4) (2021) 49730.
[128] X. Yang, L. Zhao, Q. Li, L. Ma, 2020 IOP Conf. Ser.: Mater. Sci. Eng. 746 (2020) 012021.
[129] C.H. Zhao, S.J. Wan, L. Wang, X.D. Liu, T. Endo, J. Polym. Sci., Part A: Polym. Chem. 52 (2014) 375–382.
[130] M. Hayaty, M.H. Beheshty, M. Esfandeh, J. Appl. Polym. Sci. 120 (2011) 1483–1489.
[131] K. Raetzke, M.Q. Shaikh, F. Faupel, P.L.M. Noeske, Int. J. Adhes. Adhes. 30 (2010) 105–110.

[132] W. Du, H. Huang, W. Pan, J. Thermochim. Acta 392 (2002) 391–394.
[133] A. Hale, C. Macosko, H. Bair, J. Appl. Polym. Sci. 38 (2010) 1253–1269.
[134] M. Pire, C. Lorthioir, E. Oikonomou, J. Polym. Chem. 3 (2012) 946–953.
[135] K. Arimitsu, S. Fuse, K. Kudo, M. Furutani, Mater. Lett. 161 (2015) 408–410.
[136] B. Yang, Y. Mao, Y. Zhang, et al., Polymer 178 (2019) 121586.
[137] M. Kaplan, A. Wayda, A. Lyons, J. Polym. Sci. Part. APolym Chem. 28 (1990) 731–740.
[138] F. Ricciardi, W.A. Romanchick, M.M. Joullié, J Polym Sci Polym Lett, 21, 1983, pp. 633–638.
[139] J.M. Barton, G.J. Buist, I. Hamerton, B.J. Howlin, J.R. Jones, S.J. Liu, Mater. Chem. 4 (1994) 379–384.
[140] J.C. Capricho, B. Fox, N. Hameed, Polym. Rev. 60 (2020) 1–41.
[141] Y.J. Xu, J. Wang, Y. Tan, M. Qi, L. Chen, Y.Z. Wang, Chem. Eng. J. 337 (2018) 30–39.
[142] Y.J. Xu, L. Chen, W.H. Rao, M. Qi, D. Guo, W. Liao, et al., Chem. Eng. J. 347 (2018) 223–232.
[143] F. Ahangaran, A.H. Navarchian, M. Hayaty, et al., Smart Mater. Struct. 25 (2016) 095035.
[144] L.I. Qi, A.K. Mishra, N.H. Kim, et al., Compos. Part. B Eng. 49 (2013) 6–15.
[145] L. Zhao, Q. X., L. Li, Ma, J. Appl. Polym. Sci. 136 (2019) 47757.
[146] J.S. Min, G.K. Jin, J.S. Shin, J. Appl. Polym. Sci. 122 (2012) 108–115.
[147] Y.R. Ham, D.H. Lee, S.H. Kim, et al., J. Ind. & Eng. Chem. 16 (2010) 728–733.
[148] J.S. Min, Y.J. Shin, S.W. Hwang, J.S. Shin, J. Appl. Polym. Sci. 129 (2013) 1036.
[149] Y. Xue, C. Y., J.C. Li, J. Tan, J. Mater. Sci. 55 (2020) 7321–7336.
[150] B. Yin, W. Xu, C. Liu, M. Kong, Y. Lv, Y. Huang, et al., RSC Adv. 10 (2020) 2085–2095.
[151] R. Ruan, Henriques, B.G. Soares, Appl. Clay Sci. 200 (2021) 105890.
[152] A.P.A. Carvalho, D.F. Santos, B.G. Soare, J. Appl. Polym. Sci. 137 (2020) 48326.
[153] A. Silva, S. Livi, D.B. Netto, B.G. Soares, J. Duchet, J.F. Gerard, Polymer 54 (2013) 2123.
[154] H. Maka, T. Spychaj, R. Pilawka, Express Polym. Lett. 8 (2014) 723.
[155] H. Maka, T. Spychaj, K. Kowalczyk, J. Appl. Polym. Sci. 131 (2014) 40401.
[156] M.Q. Kong, C.J. Liu, B. Tang, W.Q. Xu, Y.J. Huang, G.X. Li, Ind. Eng. Chem. Res. 58 (2019) 8080–8089.
[157] Y. Tan, Z.B. Shao, X.F. Chen, J.W. Long, L. Chen, Y.Z. Wang, ACS Appl. Mater. Interf. 7 (2015) 17919–17928.
[158] Y.J. Xu, J. Wang, Y. Tan, M. Qi, L. Chen, Y.Z. Wang, Chem. Eng. J. 3 (2018) 30–39.
[159] H. Gong, Y.J. Gao, S.L. Jiang, F. Sun, ACS Appl. Mater. Interf. 10 (2018) 26694–26704.
[160] F.C. Binks, G. Cavalli, M. Henningsen, B.J. Howlin, Hamerton, Polymer 150 (2018) 318–325.
[161] B.G. Soares, S. Livi, J. Duchert-Rumeau, J.F. Gerard, Preparation of epoxy/MCDEA networks modified with ionic liquids, Polymer 53 (2012) 60–66.
[162] H. Maka, T. Spychaj, M. Zenker, J. Ind. Eng. Chem. 216 (2015) 359–368.
[163] Z. Jiang, Q. Wang, L. Liu, Y. Zhang, F. Du, A. Pang, Ind. Eng. Chem, Res. 59 (2020) 3024–3034.
[164] A. Hartwig, T.K. Mahato, T. Kaese, D. Wohrle, Macromol. Chem. Phys. 206 (2005) 1718.
[165] A.B. Samui, D. Ratna, J. Chavan, P.C. Deb, J. Appl. Polym. Sci. 92 (2004) 1604.
[166] S. Flint, T. Markle, S. Thompson, E. Wallace, J. Environ. Manage. 104 (2012) 19–34.
[167] C. Aouf, H. Nouailhas, M. Fache, S. Caillol, B. Boutevin, H. Fulcrand, Eur. Polym. J. 49 (2013) 1185–1195.
[168] C. Aouf, J. Lecomte, P. Villeneuve, E. Dubreucq, H. Fulcrand, Green. Chem. 14 (2012) 2328–2336.
[169] M.A.R. Meier, J.O. Metzger, U.S. Schubert, Plant oil renewable resources as green alternatives in polymer science, Chem. Soc. Rev. 36 (2007) 1788–1802.
[170] J.S. Puglisi, M.A. Chaudhari, Epoxies (EP), in: J.N. Epel, J.M. Margolis, S. Newman, R.B. Seymour (Eds.), Engineered Materials Handbook, vol. 2, ASM International, Metals Park, OH, 1988, pp. 240–241.
[171] F.S. Guner, Y. Yagci, A.T. Erciyes, Polymers from triglyceride oils, Prog. Polym. Sci. 31 (2006) 633–670.
[172] J.V. Crivello, R. Narayan, S.S. Sternstein, J. Appl. Polym. Sci. 64 (1997) 2073–2087.
[173] U. Biermann, W. Friedt, S. Lang, W. Luhs, G. Machmuller, J.O. Metzger, et al., Angew. Chem. Int. (Ed.) 39 (2000) 2206–2224.

[174] V.V. Goud, N.C. Pradhan, A.V. Patwardhan, Epoxidation of karanja (Pongamiaglabra) oil by H_2O_2, J. Am. Oil Chem. Soc. 83 (2006) 635–640.

[175] T. Vlcek, Z. Petrovic, Optimization of the chemoenzymatic epoxidation of soybean oil, J. Am. Oil. Chem. Soc. 83 (2006) 247–252.

[176] V.V. Goud, A.V. Patwardhan, S. Dinda, N.C. Pradhan, J. Eur, Lipid Sci. Technol. 109 (2007) 575–584.

[177] D. Ratna, A.K. Banthiaand, N.C. Maity, Chlorinated rubber based nonpolluting coating, Ind.J. Chem. Tech. 2 (1995) 253–257.

[178] D. Ratna, A.K. Banthia, J. Adhes. Sci. Technol.14 (2000) 15–25.

[179] D. Ratna, Polym. Int. 50 (2001) 179–184.

[180] J. Cavusoglu, G. Cayli, J. Appl. Polym. Sci. 132 (2015) 41457.

[181] T. Tsujimoto, T. Takayama, H. Uyama, Polymers-Basel 7 (2015) 2165.

[182] L.H. Sperling, J.A. Manson, S. Qureshi, A.M. Fernandez, Ing. Eng. Chem. Prod. Res. 20 (1981) 163–166.

[183] A. Olsson, M. Lindstrom, T. Iversen, Biomacromolecules 8 (2007) 757–760.

[184] M. Fache, E. Darroman, V. Besse, R. Auvergne, S. Caillol, B. Boutevin, Green. Chem. 16 (2014) 1987–1998.

[185] S. Nikafshar, O. Zabihi, S. Hamidi, Y. Moradi, S. Barzegar, M. Ahmadi, et al., RSC Adv. 7 (2007) 8694–8701.

[186] A. Llevot, E. Grau, S. Carlotti, S. Grelier, H. Cramail, Macromol. Rapid. Commun. 37 (2016) 9–28.

[187] F. Ferdosian, Z. Yuan, M. Anderson, C. Xu, Thermochim. Acta 618 (2015) 48e55.

[188] J. Xin, M. Li, R. Li, M.P. Wolcott, J. Zhang, ACS Sustain. Chem. Eng. 4 (2016) 2754–2761.

[189] L.C. Over, E. Grau, S. Grelier, M.A.R. Meier, H. Cramail, Macromol. Chem. Phys. 218 (2017) 1600411.

[190] C. Gioia, R.,G. Lo, M. Lawoko, L. Berglund, J. Am. Chem. Soc. 140 (2018) 4054–4061.

[191] J. Qin, H. Liu, P. Zhang, M. Wolcott, Zhang, J. Polym. Int. 63 (2014) 760–765.

[192] J. Wan, B. Gan, C. Li, J.M. Molina-Aldareguia, Z. Li, X. Wang, et al., J. Mater. Chem. A 3 (2015) 21907–21921.

[193] J. Wan, B. Gan, C. Li, J. Molina-Aldareguia, E.N. Kalali, X. Wang, et al., Eng. J. 284 (2016) 1080–1093.

[194] J. Wan, J. Zhao, B. Gan, C. Li, J. Molina-Aldareguia, Y. Zhao, et al., ACS Sustain. Chem. Eng. 4 (2016) 2869–2880.

[195] C. Francois, S. Pourchet, G. Boni, S. Fontaine, Y. Gaillard, V. Placet, et al., RSC Adv. 6 (2016) 68732–68738.

[196] S. Thiyagarajan, H.C. Genuino, J.C. van der Waal, E. de Jong, B.M. Weckhuysen, J. van Haveren, et al., Angew. Chem., Int. (Ed.) 55 (2016) 1368–1371.

[197] V. Froidevaux, C. Negrell, S. Caillol, J.-P. Pascault, B. Boutevin, Chem. Rev. 116 (2016) 14181–14224.

[198] M. Fache, C. Monteremal, B. Boutevin, S. Caillol, Eur. Polym. J. 73 (2015) 344–362.

[199] M. Stemmelen, F. Pessel, V. Lapinte, S. Caillol, J.P. Habas, J.J. Robin, J. Polym. Sci. Pol. Chem. 49 (2011) 2434.

[200] M. Stemmelen, V. Lapinte, J.P. Habas, J.J. Robin, Eur. Polym. J. 68 (2015) 536.

[201] H.H. Wang, B. Liu, X.Q. Liu, J.W. Zhang, M. Xian, Green. Chem. 10 (2008) 1190.

[202] J.L. Qin, H.Z. Liu, P. Zhang, M. Wolcott, J.W. Zhang, Polym. Int. 63 (2014) 760.

[203] M. Shibata, K.J. Nakai, Polym. Sci. Pol. Phys. 48 (2010) 425.

[204] C. François, S. Pourchet, G. Boni, S. Rautiainen, J. Samec, L. Fournier, et al., Comptes. Rendus Chimie 20 (2017) 1006–1016.

[205] C. Robert, F. deMontigny, C.M. Thomas, ACS Catal. 4 (2014) 3586–3589.

[206] E.A. Baroncini, S.K. Yadav, G.R. Palmese, J.F. Stanzione, J. Appl. Polym. Sci. (1 19) (2016) 44103.

[207] S.D. Miao, K. Liu, P. Wang, Z.G. Su, S.P. Zhang, J. Appl. Polym. Sci. 132 (2015) 41575.

[208] J.M. Pin, N. Sbirrazzuoli, A. Mija, Chem. sus. chem. 8 (2015) 1232.

[209] T.S. Omonov, J.M. Curtis, J. Appl. Polym. Sci. 131 (2014) 40142.

[210] A. Kadam, M. Pawar, O. Yemul, V. Thamke, K. Kodam, Polymer 72 (2015) 82.

[211] C. Lorenzini, D.L. Versace, E. Renard, V. Langlois, React. Funct. Polym. 93 (2015) 95.
[212] J. Hong, D. Radojcic, M. Ionescu, E. Eastwood, Polym. Chem. UK 5 (2014) 5360.
[213] J.K. Cho, J.S. Lee, J. Jeong, B. Kim, B. Kim, S. Kim, et al., J. Adhes. Sci. Technol. 27 (2013) 2127.
[214] F.S. Hu, J.J. La Scala, J.M. Sadler, G.R. Palmese, Macromolecules 47 (2014) 3332.
[215] J. Deng, X.Q. Liu, C. Li, Y.H. Jiang, J. Zhu, Rsc. Adv. 5 (2015) 15930.
[216] S. Basnet, M. Otsuka, C. Sasaki, C. Asada, Y. Nakamura, Ind. Crop. Prod. 73 (2015) 63.
[217] S. Benyahya, C. Aouf, S. Caillol, B. Boutevin, J.P. Pascault, H. Fulcrand, Ind. Crop. Prod. 53 (2014) 296.
[218] E. Darroman, N. Durand, B. Boutevin, S. Caillol, Prog. Org. Coat. 83 (2015) 47.
[219] F. Jaillet, E. Darroman, A. Ratsimihety, R. Auvergne, B. Boutevin, S. Caillol, Eur. J. Lipid Sci. Tech. 116 (2014) 63.
[220] G.M. Wu, Z.W. Kong, J. Chen, S.P. Huo, G. Liu, F. Prog. Org. Coat. 77 (2014) 315.
[221] K. Xu, M.C. Chen, K. Zhang, J.W. Hu, Polymer 45 (2004) 1133.
[222] R.A. El-Ghazawy, A.M. El-Saeed, H.I. Al-Shafey, A.R.M. Abdul-Raheim, M.A. El-Sockary, Eur. Polym. J. 69 (2015) 403.
[223] Y.H. Muhammad, S. Ahmad, M.A. Abu Bakar, A.A. Mamun, H.P.J. Heim, Reinf. Plast. Comp. 34 (2015) 896.
[224] C. Asada, S. Basnet, M. Otsuka, C. Sasaki, Y. Nakamura, Int. J. Biol. Macromol. 74 (2015) 413.
[225] F. Ferdosian, Z.S. Yuan, M. Anderson, C.B. Xu, Thermochim. Acta. 618 (2015) 48.
[226] E. Darroman, L. Bonnot, R. Auvergne, B. Boutevin, S. Caillol, Eur. J. Lipid Sci. Tech. 117 (2015) 178.
[227] F.S. Hu, S.K. Yadav, J.J. La Scala, J.M. Sadler, G.R. Palmese, Macromol. Chem. Phys. 216 (2015) 1441.
[228] M. Fache, C. Monteremal, B. Boutevin, S. Caillol, Eur. Polym. J. 73 (2015) 344.
[229] S.Q. Ma, D.C. Webster, Macromolecules 48 (2015) 7127.
[230] F. Jaillet, M. Desroches, R. Auvergne, B. Boutevinu, S. Caillol, Eur. J. Lipid Sci. Technol. 115 (2013) 698.
[231] T. Takahashi, K.I. Hirayarna, N. Teramoto, M. Shibata, J. Appl. Polym. Sci. 108 (2008) 1596.
[232] M. Shibata, N. Teramoto, K. Makino, J. Appl. Polym. Sci. 120 (2011) 273.
[233] R. Panda, J. Tjong, S.K. Nayak, M.M. Sain, Bioresources 10 (2015) 4126.
[234] P. Hu, Q. Xie, C. Ma, G. Zhang, Langmuir 36 (2020) 2170−2183.
[235] R.L. Townsin, Biofouling 19 (2003) 9−15.
[236] M.P. Schultz, Biofouling 23 (2007) 331−341.
[237] G.G. Geesey, Z. Lewandowski, H.C. Flemming, Biofouling and Biocorrosion in Industrial Water Systems, CRC Press, 1994.
[238] R. Wen, J. Huo, J. Lv, Z. Liu, Y. Yu, J. Mater. Sci. Mater. Electron. 28 (2017) 14522−14535.
[239] O. Bayer, Angew. Chem. 59 (9) (1947) 257−272.
[240] R. Narayan, D.K. Chattopadhyay, B. Sreedhar, K.V. Raju, N.N. Mallikarjuna, T.M. Aminabhavi, J. Appl. Polym. Sci. 99 (2006) 368−380.
[241] A.A. Caraculacu, S. Coseri, Prog. Polym. Sci. 26 (2001) 799−851.
[242] O. John, M.D. Akindoyo, H. Beg, S. Ghazali, M.R. Islam Nitthiyah Jeyaratnam, A.R. Yuvaraj, RSC Adv. 6 (2016) 114453−114482.
[243] M. Szycher, Handbook of Polyurethanes, CRC Press, Boca Raton, FL, 1999.
[244] F.C. Schilling, A.E. Tonelli, Macromolecules 19 (1986) 1337.
[245] A. Le Borgne, N. Spassky, C.L. Jun, A. Momtaz, Macromol. Chem. 189 (1988) 637.
[246] S. Inoue, T. Aida, Ring Opening Polymerization, in: K.J. Ivin, T. Saegusa (Eds.), Vol. 1, Elsevier Applied Science Publishers, London, 1984.
[247] S.D. Gangnon, Encyclopedia of Polymer Science and Engineering, Vol 6, Wiley Interscience Publication, NY, 1986, pp. 273−307.
[248] A.M. Eastman, The Chemistry of Cationic PolymerizationChapter 10 in: P.H. Plesch (Ed.), Pergamon Press, Elmford, NY, 1963.

[249] P. Dreyfuss, M.P. Dreyfuss, 3rd ed., Kirk-Othmer, Encyclopedia of Chemical Technology, Vol 18, Wiley and Sons, Inc, NY, 1982, pp. 645–669.
[250] B.F. Richardand, B.A.R.D. Edmund, Mechanically Frothed Gel Elastomers and Methods of Making and Using Them, US 20160017084 A1, US 14/730, 867, Jan 21, 2016.
[251] G.P. Rajendran, V. Mahadevan, M. Srinivasan, Eur. Polym. J. 25 (1989) 461–463.
[252] J. Yi, C. Huang, H. Zhuang, H. Gong, C. Zhang, R. Ren, Yupu Ma Prog. Org. Coat. 87 (2015) 161–170.
[253] A. Chaudhari, A. Kuwar, P. Mahulikar, D. Hundiwale, R. Kulkarni, V. Gite, RSC Adv. 4 (2014) 17866–17872.
[254] A. Guo, Y. Choand, Z.S. Petrovíc, J. Polym. Sci., Part. A: Polym. Chem. 38 (2000) 3900–3910.
[255] Z.S. Petrovic, W. Zhang, I. Javni, Biomacromolecules 6 (2005) 713–719.
[256] U. Schuchardt, R. Sercheli, R.M. Vargas, J. Braz. Chem. Soc. 9 (2008) 199–210.
[257] M.Z. Arniza, S.S. Hoong, Z. Idris, S.K. Yeong, H.A. Hassan, K. Dinand, et al., J. Am. Oil Chem. Soc. 92 (2015) 243–255.
[258] T. Xavier, B. Didier, T. Lan, Eur. Polym. J. 36 (2000) 1745.
[259] Z.W. Wicks, F. Jones, S. Pappas, Vol 1 Organic Coating Science and Technology, John Wiley and Sons, Hoboken, New Jersey, USA, 1992p. 188.
[260] I. Ahmad, J.H. Zaidi, R. Hussain, A. Munir, Poly. Int. 56 (2007) 1521.
[261] P.T. Gnanarajan, N. Padmanava, N.A. Sultan, G. Radhakrishnan, Polym. Int. 51 (2002) 195.
[262] T. Mukaiyama, Y. Hoschino, J. Am. Chem. Soc. 78 (1956) 1946.
[263] D.J. Yontz, S.L. Hsu, Macromolecules 33 (22) (2000) 8415–8420.
[264] E.M. Christenson, J.M. Anderson, A. Hiltner, E. Baer, Polymer 46 (25) (2005) 11744–11754.
[265] E.W. Cochran, C.J. Garcia-Cervera, G.H. Fredrickson, Macromolecules 39 (7) (2006) 2449–2451.
[266] J. Blackwell, K.H. Gardner, Structure of the hard segments in polyurethane elastomers, Polymer 20 (1979) 13–17.
[267] Y. Zouohong, H. Jinlian, L. Yeqiu, Y. Lapyan, Mater. Chem. Phys. 98 (2006) 368.
[268] J. Hu, Z. Yang, L. Yeung, F. Ji, Y. Liu, Polym. Int. 54 (2005) 584.
[269] S.H. Lee, J.W. Kim, B.K. Kim, Smart Mater. Struct 12 (2004) 1345.
[270] T.L. Smith, A.B. Magnusson, J. Polym. Sci. Xiii (1960) 391.
[271] M.A. Corcuera, L. Rueda, A. Saralegui, M. Dolores, M. Borja, J. Appl. Polym. Sci. 122 (2011) 3677–3685.
[272] H.A. Hamuzana, K.H. Badria, AIP Conference Proceedings 1784, 2016 030019.
[273] X.K. Li, Y. Lu, H. Wang, E. Pöselt, Y.F. Men, Eur. Poly. J. 97 (2017) 423–436.
[274] Y.I. Odarchenko, et al., Acta Biomat 9 (2013) 6143–6149.
[275] S. Nozaki, T. Hirai, Y. Higaki, K. Yoshinaga, K. Kojio, A. Takahar, Polymer 116 (2017) 423–428.
[276] S. Nozaki, et al., Macromolecules 50 (2017) 1008–1015.
[277] K. Kojio, M. Furukawa, S. Motokucho, M. Shimada, M. Sakai, Macromolecules 42 (2009) 8322–8327.
[278] L. Jiang, Z. Ren, W. Zhao, W. Liu, H. Liu, R. Soc. Open. Sci. 5 (7) (2018) 180536.
[279] C. Cateto, M. Barreiro, A. Rodrigues, M. Belgacem, 48 (2009) 2583–2589.
[280] T. Saito, et al., RSC Adv. 3 (2013) 21832.
[281] S. Huo, M. Nie, Z. Kong, G. Wu, J. Chen, J. Appl. Polym. Sci. 125 (2012) 152–157.
[282] D.J. David, H.B. Staley, Analytical Chemistry for Polyurethanes, XVI (III), Wiley Interscience with John Wiley & Sons, New York, 1969.
[283] G. Rokicki, A. Piotrowska, Polymer 43 (2002) 2927–2935.
[284] I. Javni, D.P. Hong, Z.S. Petrovic, J. Appl. Polym. Sci. 108 (2008) 3867.
[285] S. Tan, et al., Polymer 52 (2011) 2840.
[286] A. Lee, Y. Deng, Eur. Polym. J. 63 (2015) 67–73.
[287] R. Baloji Naik, D. Ratna, S.K. Singh, Prog. Org. Coat. 77 (2) (2014) 369–379.
[288] R.B. Naik, S.B. Jagtap, R.S. Naik, N.G. Malvankar, D. Ratna, Prog. Org. Coat. 77 (11) (2014) 1883–1889.

[289] R.B. Naik, N.G. Malvankar, T.K. Mahato, D. Ratna, R.S. Hastak, J. Coat. Technol. Res. 11 (4) (2014) 575–586.
[290] G. Gaidukova, A. Ivdre, A. Fridrihsone, A. Verovkins, U. Cabulis, S. Gaidukovs, Ind. Crop. Prod. 9 (2017) 102–133.
[291] A. Agrawal, R. Kaur, R.S. Walia, Eur. Polym. J. 95 (2017) 255–274.
[292] A. Prociak, L. Szczepkowski, J. Ryszkowska, M. Kurańska, M. Auguścik, E. Malewska, et al., July J. Polym. Environ. 27 (11) (2019).
[293] R. Deng, P. Davies, A. Bajaj, J. Sound. Vib. 262 (2003) 391–417.
[294] J. Njuguna, S. Michałowski, K. Pielichowski, K. Kayvantash, A.C. Walton, Polym. Compos. 32 (2011) 6–13.
[295] A. Serrano, A.M. Borreguero, I. Garrido, J.F. Rodríguezand, M. Carmona, Appl. Therm. Eng. 103 (2016) 226–232.
[296] P. Davies, G. Evrard, Polym. Degrad. Stab. 92 (2007) 1455–1464.
[297] Y.P. Chin, S. Mohamad, M.R.B. Abas, Int. J. Mol. Sci. 11 (2010) 3459–3471.
[298] S.-J. Peng, Y. Jin, X.-F. Cheng, T.-B. Sun, R. Qiand, B.-Z. Fan, Prog. Org. Coat. 86 (2015) 1–10.
[299] S. Xinrong, W. Nanfang, S. Kunyang, D. Shaand, C. Zhen, J. Ind. Eng. Chem. 20 (2014) 3228–3233.
[300] F. Fay, I. Linossier, J. Peron, Prog. Org. Coat. 60 (2007) 194–206.
[301] D.M. Yebra, S. Kiil, K. Dam-Johansen, Prog. Org. Coat. 53 (2005) 256–275.
[302] Y. Yonehara, H. Yamashita, C. Kawamura, K. Itoh, Prog. Org. Coat. 42 (2001) 150–158.
[303] C. Hugues, C. Bressy, P. Bartolomeo, A. Margaillan, Eur. Polym. J. 39 (2003) 319–326.
[304] C. Bressy, A. Margaillan, Prog. Org. Coat. 66 (2009) 400–405.
[305] I. Omae, Chem. Rev. 103 (2003) 3431–3448.
[306] Y. Sun, S. Guo, G.C. Walker, C.J. Kavanagh, G.W. Swain, Biofouling 20 (2004) 279–289.
[307] P. Majumdar, D.C. Webster, Macromolecules 38 (2005) 5857–5859.
[308] S.K. Rath, J.G. Chavan, S. Sasane, A. Srivastava, M. Patri, A.B. Samui, et al., Prog. Org. Coat. 65 (2009) 366–374.
[309] S.K. Rath, J.G. Chavan, S. Sasane, Jagannath, M. Patri, A.B. Samui, et al., Appl. Surf. Sci. 256 (2010) 2440–2446.
[310] C. Liu, Q.Y. Xie, C.F. Ma, G.Z. Zhang, Ind. Eng. Chem. Res. 55 (2016) 6671–6676.
[311] G. Verma, A. Kaushik, A.K. Ghosh, Prog. Org. Coat. 99 (2016) 282–294.
[312] N.M. Zain, E.N. Roslin, S. Ahmad, Int. J. Adhes. Adhes. 71 (2016) 1–9.
[313] N. Baheiraei, H. Yeganeh, J. Ai, R. Gharibi, M. Azami, F. Faghihi, Mater. Sci. Eng., C. 44 (2014) 24–37.
[314] A. Solanki, S. Thakore, Int. J. Biol. Macromol. 80 (2015) 683–691.
[315] A. Gardziella, L.A. Pilato, A. Knop, Phenolic Resins, second ed., Springer-Verlag, Heidelberg, 2000.
[316] H.L. Fisher, Chemistry of Natural and Synthetic Rubbers, Reinhold Publishing Corporation, New York, 1958.
[317] J.P. Critchley, G.J. Knight, W.W. Wright, Heat Resistant Polymers—Technologically Useful Materials, Plenum Press, New York and London, 1983.
[318] A. Knop, V. Bohmer, L.A. Pilato, Phenol-formaldehyde polymersChap 35, Vol 5 in: Sir G. Allen, J.C. Bevington (Eds.), Comprehensive Polymer Science, 1989.
[319] A. Knop, L.A. Pilato, Phenolic Resins, Springer, Berlin, Heidelberg, New York, 1979.
[320] M. Komiyyama, Prog. Polym. Sci. 18 (1993) 871.
[321] Z. Hedwig, M. Szesgtay, F. Tudos, Angew. Makromol. Chemie. 57 (1996) 241.
[322] J.A. Hiltz, S.G. Kuzak, P.A. Waitkus, J. Appl. Polym. Sci. 79 (2001) 385.
[323] T.R. Dargaville, F.N. Guerzoni, M.G. Looney, D.H. Solomon, X. Zhang, J. Polym. Sci., Part. A Polym. Chem. 35 (1997) 1399.
[324] P.J. de Bruyn, L.M. Foo, A.S.C. Lim, M.G. Looney, D.H. Solomon, Tetrahedron 53 (1997) 13915.
[325] X. Zhang, A.C. Potter, D.H. Solomon, Polymer 39 (1998) 1967.
[326] X. Zhang, A.C. Potter, D.H. Solomon, Polymer 39 (1998) 399.

[327] S. Markovic, B. Dunjic, A. Zlatanic, J. Djonlagic, J. Appl. Polym. Sci. 81 (2001) 1902.
[328] J.P. Patel, Z.G. Xiang, S.L. Hsu, A.B. Schoch, S.A. Carleen, D. Matsumoto, J. Polym. Sci. Part. B Polym. Phys. 53 (21) (2015) 1519−1526.
[329] J.P. Patel, S.L. Hsu, J. Polym. Sci. Part. B Polym. Phys. (2016) (In Press).
[330] J.P. Patel, C.X. Zhao, S. Deshmukh, G.X. Zou, O. Wamuo, S.L. Hsu, et al., Polymer 107 (2016) 12−18.
[331] D.V. Rostavo, C.S. Grove Jr., Filament Winding, Interscience, New York, 1964.
[332] L. Pilato, M.J. Micano, Advanced Composite Materials, Springer-Verlag, 1994.
[333] G.M. Jenkins, K. Kawamura, Polymeric Carbons, Cambridge University Press, Cambridge, UK, 1976.
[334] A. Knop, L.A. Pilato, Phenolic Resins, Springer, Berlin, 1985, pp. 156−160.
[335] R.P. Lattimer, R.A. Kinsey, R.W. Layer, C.K. Rhee, Rubber Chem. Technol. 62 (1989) 107.
[336] B.T. Lity Alen Varghese, E. Abraham, T. Thomas, Progress in Plastic, Rubber and Recycling Technology, 22, 2006, pp. 253−268.
[337] R. Nawaz, N. Rashid, Z. Ali, N. Ul-Haq, Fibers Polym. 19 (2018) 1318−1326.
[338] S. Mahanty, Rubber Plastic N. 15 (2010).
[339] D. Yuan, Z. Chen, K. Chen, Wenjie Mou, &Y., Chen Pages (2016) 1115−1123.
[340] X. Huang, S. Chen, S. Wan, B. Niu, X. He, R. Zhang, Polymers 12 (2020) 490.
[341] C. Qi, P. Liu, Eur. Polym. J. 42 (2006) 2931.
[342] K. Pielichowska, K. Pielichowski, J. Appl. Polym. Sci. 116 (2010) 1725.
[343] B. Scrosati (Ed.), Applications of Electroactive Polymers, Chapman & Hall, New York, 1993, p. 251.
[344] W. Wieczorek, D. Raduacha, A. Zalewska, J.K. Stevens, J. Phys. Chem. B. 102 (1998) 8725.
[345] G. Liu, C. Guan, H. Xia, F. Guo, X. Ding, Y. Peng, Macromol. Rapid. Commun. 27 (2006) 1100.
[346] D. Ratna, J. Karger-Kocsis, J. Mater. Sci. 43 (2008) 254.
[347] T.N. Abraham, D. Ratna, S. Siengchin, J. Karger-Kocsis, Polym. Eng. Sci. 49 (2009) 379.
[348] D. Ratna, S. Divekar, P. Sivaraman, A.B. Samui, B.C. Chakraborty, Polym. Int. 56 (2007) 900.
[349] P.P. Chu, M.J. Ready, J. Tsai, J. Polym. Sci. Polym Phys. 42 (2004) 3866.
[350] D. Ratna, T.N. Abraham, Macromol. Chem. Phys. 209 (2008) 723−733.
[351] D. Ratna, J. Karger-Kocsis, J. Appl. Polym. Sci. 127 (2012) 4039−4043.
[352] H. Kosonen, J. Ruokolanien, M. Torkkeli, R. Serima, P. Nyholm, O. Ikkala, Macromol. Chem. Phys. 203 (2002) 388.
[353] D. Ratna, J. Karger-Kocsis, Polymer 52 (2011) 1063−1070.
[354] S.B. Jagtap, V. Dalvi, S. Kalleta, D. Ratna, Polym. Int. 68 (2019) 812−817.
[355] T.J. Fox, B. Am, Phys. Soc. 1 (1956) 123.
[356] C.P. Reghunadhan Nair, Prog. Polym. Sci. 29 (2004) 401.
[357] A. Dekar, H. Stretz, J. Koo, Polymeric Mater. Sci. Eng. 83 (2000) 105.
[358] Y. Yan, X. Shi, J. Liu, T. Zhao, Y. Yu, J. Appl. Polym. Sci. 83 (2002) 1651.
[359] G.Z. Liang, J. Fan, J. Appl. Polym. Sci. 73 (1999) 1623.
[360] B. Dao, J.H. Hodgkin, T.C. Morton, High. Perform. Polym. 9 (1997) 413.
[361] D. Augustine, D. Mathew, C.P. Reghunadhan Nair, Eur. Polym. J. 71 (2015) 389−400.
[362] D. Augustine, D. Mathew, C.P. Reghunadhan Nair, Eur. RSCAdv. 5 (2015) 91254−91261.
[363] B. Meyer, U-formaldehyde Resins, Addison-Weslay Publishing Company, Inc., Massachusetts, USA, 1979.
[364] C.P.R. Nair, R.L. Bindu, K.N. Ninan, Polym. Degrad. Stab. 73 (2001) 251−257.
[365] S.H. Solomon (Ed.), Handbook of Thermoset Plastics, 2nd ed.), Noyes Publications, New Jersey, USA, 1998.
[366] T.L. Richardson, Industrial Plastics, Theory and Applications, Delmen Publishers Inc., Albany, New York, 1989, p. 210.
[367] H. Ishida, H.Y.A. Low, Macromolecules 30 (1997) 1099−1106.
[368] X. Ning, H. Ishida, J. Polym. Sci., Part. A: Polym. Chem. 32 (1994) 1121−1129.
[369] H.J. Kim, Z. Brunovska, H. Ishida, Polymer 40 (1999) 6565−6573.
[370] I. Hamerton, B.J. Howlin, A.L. Mitchell, L.T. Mcnamara, S. Takeda, Funct. Polym. 72 (2012) 736−744.

[371] M. Zeng, T. Pang, J. Chen, Y. Huang, Q. Xu, Y.J. Gu, Mater. Sci., Mater. Electron. 29 (2018) 5391−5400.
[372] F.W. Holly, A.C. Cope, J. Am. Chem. Soc. 66 (1875) 1944.
[373] H. Schreiber, Ger. Offen 2 (1973) 225, 504.
[374] Z. Brunovska, J. Liu, Macromol. Chem. Phys. 5 (2015) 3709−3719.
[375] A. Rucigaj, B. Alic, M. Krajnc, U. Sebenik, Express Polym. Lett. 9 (2015) 647−657.
[376] R. Andreau, J.A. Reina, J.C. Ronda, J. Polym. Science, Part. A:Polymer Chem. (Ed.) 46 (2008) 3353.
[377] W.J. Burke, E.L. Glennie, C. Weatherbee, J. Org. Chem. 29 (1964) 909.
[378] I. Schnell, P.B. Steven, H.Y. Low, H. Ishida, H.W. Spiess, J. Am. Chem. Soc. 120 (1998) 11784.
[379] R.G. Gillian, S. Daniel, I. Schnell, H.W. Spiess, H.D. Kim, H. Ishida, J. Am. Chem. Soc. 125 (2003) 5792.
[380] M. Nakamura, H. Ishida, Polymer 50 (2009) 2688.
[381] H.Y. Low, H. Ishida, J. Polym. Science, Part. B: Polym. Phys. (Ed.) 36 (1998) 1935.
[382] C. Jubsilp, S. Damrongsakkul, T. Takeichi, S. Rimdusit, Thermochim. Acta 447 (2006) 131.
[383] G. Reiss, J.M. Schwob, G. Guth, M. Roche, B. Lande, in: B.M. Culbertson, J.E. McGrath (Eds.), Advances in Polymer Synthesis, Plenum, New York, 1985, p. 27.
[384] G.R. Govard, D. Sebastiani, I. Schnell, H.W. Spiess, H.D. Kim, H. Ishida, J. Am. Chem. Soc. 125 (2003) 5792.
[385] A. Chernykh, J.P. Liu, H. Ishida, Polymer 47 (2006) 766−7669.
[386] T. Takeichi, T. Kano, T. Agag, Polymer 46 (2005) 12172.
[387] T. Agag, T. Takeichi, J. Polym. Sci. Part. A Polym Chem. 45 (2007) 1878−1888.
[388] W. Gong, K. Zeng, L. Wang, S.X. Zheng, Polymer 49 (2008) 3318−3326.
[389] B. Kiskan, B. Aydogan, Y. Yagci, J. Polym. Sci. Part. A Polym Chem. 47 (2009) 804−811.
[390] L. Wang, W. Gong, S. Zheng, Polym. Int. 58 (2009) 124−132.
[391] L. Wang, S.X. Zheng, Polymer 51 (2010) 1124−1132.
[392] A. Chernykh, T. Agag, H. Ishida, Macromolecules 42 (2009) 5121−5127.
[393] J. Liu, T. Agag, H. Ishida, Polymer 51 (2010) 5688−5694.
[394] C.P. Yin, J.Y. Xiao, J.C. Zeng, J.J. Liu, Mater. Eng. Perfrom 6 (2008) 003.
[395] Q. Xu, M. Zeng, J. Chen, S. Zeng, Y. Huang, Z. Feng, et al., Funct. Polym. 122 (2018) 158−166.
[396] C. Lin, S. Chang, T. Shen, Y. Shih, H. Lin, C. Wang, Polym. Chem. 3 (2012) 935−945.
[397] H. Ishida, T. Agag, Handbook of Benzoxazine Resins, Elsevier, Amsterdam, 2011, p. 47. 47.
[398] C.I. Chou, Y.L. Liu, J. Polym. Sci., A 46 (2008) 6509−6517.
[399] M. Zeng, J. Chen, Q. Xu, Y. Huang, Z. Feng, Y. Gu, Polym. Chem. 9 (2018) 2913−2925.
[400] M. Wang, R. Jeng, C. Lin, RSC Adv. 6 (2016) 18678−18684.
[401] K. Zhang, J. Qiu, S. Li, Z. Shang, J. J. Appl. Polym. Sci. 134 (2017) 45408−45415.
[402] W. Volksen, R.D. Miller, G. Dubois, Chem. Rev. 110 (2010) 56−110.
[403] J. Chen, M. Zeng, Z. Feng, T. Pang, Y. Huang, Q. Xu, ACS Appl. Polym. Mater. 1 (2019) 625−630.
[404] N.N. Ghosh, B. Kiskan, Y. Yagci, Polym. Sci. 32 (2007) 1344−1391.
[405] R. Jain, A.K. Narula, V. Choudhary, J. Appl. Polym. Sci. 106 (2007) 3327−3334.
[406] S. Grishchuk, L. Sorochynska, O.C. Vorster, J.O. Karger- Kocsis, J. Appl. Polym. Sci. 127 (2013) 5082−5093.
[407] S. Grishchuk, Z. Mbhele, S. Schmitt, J. Karger-Kocsis, Express Polym. Lett. 5 (2011) 273−282.
[408] P. Zhao, Q. Zhou, X. Liu, R. Zhu, Q. Ran, Y. Gu, Polym. J. 45 (2013) 637−644.
[409] S. Rajesh Kumar, J. Dhanaselaran, S. Krishna, et al., RSC Adv. 5 (2015) 3709−3719.
[410] S. Rajesh Kumar, P.M. Asseref, J. Dhanasekaran, S. Krishna Mohan, RSC Adv. 4 (2014) 12526−12533.
[411] R. Gao, A. Gu, G. Liang, S. Dai, L. Yuan, J. Appl. Polym. Sci. 121 (2011) 1675−1684.
[412] L. Jin, T. Agag, H. Ishida, Polym. Int. 62 (2013) 71−78.
[413] S. Zhang, P. Yang, Y. Bai, T. Zhou, R. Zhu, Y. Gu, ACS Omega 2 (2017) 1529−1534.
[414] J. Dai, N. Teng, X. Shen, Y.L.I. Cao, J. Zou, X. Liu, Ind. & Eng. Chem. Res. 57 (2018) 3091−3102.
[415] L. Dumas, L. bonnaud, M. Olivier, M. Poorteman, P. Dubois, J. Mater. Chem., A 3 (2015) 6012−6018.

[416] P. Thirukmara, A. S. parveen, Sarojadevi, M. Polym. Compos. 36 (2015) 1973−1982.
[417] L. Dumas, L. Bonnaud, M. Oliver, M. Poorteman, P. Dubois, Eur. Polym. J. 67 (2015) 494−502.
[418] P. Thirukumaran, A. Shakila, S. Muthusamy, RSC Adv. 4 (2014) 7956−7966.
[419] Y.L. Liu, C.I. Chou, Polym. Sci. Polym. Chem. 43 (2005) 5267−5282.
[420] C. Gaina, O. Ursache, V. Gaina, V.E. Musteata, High performance thermosets based on multifunctional intermediates containing allyl, maleimide and benzoxazine groups, J. Polym. Res. 20 (2013) 263.
[421] T. Periyasamy, S.P. Asrafali, S. Muthusamy, J. Chem. 39 (2015) 1691−1702.
[422] P. Thirukumaran, A. Shakila Parveen, M. Sarojadevi, Chem. Eng. 2 (2014) 2790−2801.
[423] second ed., H.F. Mark, N.M. Biikles, C.G. Overberger, G. Menges (Eds.), Encyclopedia of Polymer Science and Engineering, Vol 10, Wiley, New York, 1989.
[424] K.L. Mittal (Ed.), Polyimides: Synthesis, Characterization and Applications, Plenum, New York, 1984.
[425] S.Z.D. Cheng, S.K. Lee, J.S. Barley, S.L.C. Hsu, F.W. Harris, Macro molecules 24 (1991) 1883.
[426] S.Z.D. Cheng, F.E. Arnold, A. Zhang, S.L.C. Hsu, F.W. Herris, Macro molecules 24 (1991) 5856.
[427] J.A. Brydson, Plastic Materials, sixth Ed., Butterworth- Heinemann, 1982, pp. 499−512.
[428] T.M. Bogert, R.R. Renshaw, in: C.A. May (Ed.), ACS Symp. Ser. 132, Vol 30, American Chemical Society, 1980.
[429] I.K. Varma, in: P. Ramakrishna (Ed.), Advances in Composite Materials, IBH, Oxford, UK, 1991, pp. 85−93.
[430] P.M. Hergenrother, J.W. Connell, J.G. Smith Jr., Polymer 41 (2000) 5073.
[431] J.G. Smith Jr., J.W. Connell, P.M. Hergenrother, J. Comp. Mater. 34 (2000) 614.
[432] B.J. Jensen, R.G. Bryant, J.G. Smith Jr., P.M. Hergenrother, J. Adhes. 54 (1995) 57.
[433] T.T. Serafini, P. Delvigs, G.R. Lightsey, J. Appl. Polym. Sci. 16 (1972) 905.
[434] P.M. Hergenrother, ACS Polym. Prepr. 25 (1984) 97.
[435] T.T. Serafini, in: C.A. May (Ed.), Resins for Aerospace, ACS, Wasington, DC, 1980, p. 15.
[436] R.D. Vannucci, SAMPE J. 19 (1987) 31. 1.
[437] W.X. Wang, Y. Takao, F.G. Yuan, J. Compos. Mater. 32 (16) (1998) 1508.
[438] J.P. Critchley, G.J. Knight, W.W. Wright, Heat resistant Polymers, Springer, 1983.
[439] S.G. Allen, J.C. Bevington, (Ed.) Comprehensive Polymer Science, 5, 1989, p. 423.
[440] A.J. Hu, J.Y. Hao, T. He, S.Y. Yang, Macromolecules 32 (24) (1999) 8046.
[441] D.L. Wilson, Brit. Polym. J. 20 (1988) 405.
[442] P.L. Dynes, T.T. Liao, C.L. Hammermesh, E.E. Witucki, in: K.L. Mittal (Ed.), Polyimides: Synthesis, Characterisation and Applications, Plenum, New York, 1984, pp. 311−326.
[443] K.C. Chuang, R.D. Vannucci, I. Ansari, L.L. Cerny, D.A. Scheiman, J. Polym. Sci., Polym. Chem. 32 (1994) 1341.
[444] H.G. Rogers, R.A. Gaudiana, W.C. Hollinsed, P.S. Kalyanaraman, J.S. Manello, C. Mcgrown, et al., Macromolecules 18 (1985) 1058.
[445] W. Hatke, H.W. Schmidt, W. Heitz, Polym. Sci., Polym. Chem. 29 (1991) 1387.
[446] C.A. Pryde, J. Polym. Sci., Polym. Chem. 27 (1989) 711.
[447] I.K. Varma, Polymers and Composites, Recent Trends, IBH, Oxford, UK, 1989.
[448] M.A. Meador, J.C. Johnston, P.J. Cavano, Macromolecules 30 (1997) 515.
[449] A.C. Wong, W.M. Ritchey, Macromolecules 14 (1981) 825.
[450] Y. Liu, X.D. Sun, X.Q. Xie, D.A. Scola, J. Polym. Sci., Polym. Chem. 36 (1998) 2653.
[451] H.R. Lubowitz, ACS Div. Org. Coat. Plast. Chem. Pap. Vol 31 (1971) 560−568. American Chemical Society.
[452] R.W. Lauver, J. Polym. Sci. 17 (1979) 2529.
[453] C. Cateto, M. Barreiro, A. Rodrigues, M. Belgacem, Ind. Eng. Chem. Res. 48 (2009) 2583−2589.
[454] D. Wilson, J.K. Wells, J.N. Hay, D. Lind, G.A. Owens, F. Johnson, SAMPE J. 18 (1987) 242.
[455] K.D. Farnsworth, R.N. Manepalli, S.B. Allen, P.A. Kohl, Int. J. Microcirc. Electron. Packag. 23 (2000) 162.
[456] K.D. Farnsworth, R.N. Manepalli, S.B. Allen, P.A. Kohl, IEEE Trans. Comp. Packaging Technol. 24 (2001) 474.

[457] M. Chanda, S.K. Roy, Plastic Tech. Handbook, (3rd Ed.), Marcel Dekker, New York, Chap. 4 (1997) 638.
[458] K. Okamoto, J. Photopolym. Sci. Technol. 16 (2003) 247.
[459] N. Asano, M. Aoki, S. Suzuki, K. Miyatake, H. Uchida, M. Watanabe, J. Am. Chem. Soc. 128 (2006) 1762.
[460] G. Meyer, G. Gebel, L. Gonon, P. Capron, D. Marscaq, C. Marestin, et al., J. Power. Source 157 (2006) 293.
[461] M. Aoki, N. Asano, K. Miyatake, H. Uchida, M. Watanabe, J. Electrochem. Soc. 153 (2006) A1154.
[462] Y. Xu, C. Chen, P. Zhang, B. Sun, J. Li, J. Chem. Eng. 51 (2006) 1841.
[463] J.Y. Do, S.K. Park, J.J. Ju, S. Park, M.H. Lee, Optical Mater. 26 (2004) 223.
[464] J.J. Ju, S.K. Park, S. Park, J. Kim, M.S. Kim, M.H. Lee, Appl. Phys. Lett. 88 (2006) 241106.
[465] X.M. Zhang, Y.Z. Song, J.G. Liu, S.Y. Yang, J. Photopolym. Sci. Technol. 29 (2016) 31–38.
[466] S.A.A. Zaidi, J.Y. Song, J.H. Lee, S.M. Kim, Kim, Y. Mater. Res. Exp. 6 (2019) 086434.
[467] H.C. Yu, J.W. Jung, J.Y. Choi, C.M. Chung, J. Polym. Sci. Part. A Polym. Chem. 54 (2016) 1593–1602.
[468] G.-L. Jiang, D.-Y. Wang, H.-P. Du, X. Wu, Y. Zhang, Y.-Y. Tan, et al., Polymers 12 (2020) 413.
[469] J.M. Margolis (Ed.), Advanced Thermoset Composites: Industrial and Commercial Applications, Van Nostrand Reinhold Co, New York, 1986.
[470] H.N. Cole, W.F. Gruber, U.S. Patent 3 (127) (1964) 414.
[471] A. Nagai, A. Takahashi, M. Suzuki, A. Mukoh, J. Appl. Polym. Sci. 44 (1992) 159.
[472] H. Stenzenberger, J. Appl. Polym. Sci. 22 (1973) 77.
[473] J.M. Barton, I. Hamerton, J.R. Jones, J.C. Stedman, Polym. Bull. 170 (1991) 163–170.
[474] J. Mijovic, S. Andjelic, Macromolecules 29 (1996) 239–246.
[475] J.V. Crivelo, J. Polym. Sci., Polym. Chem. (Ed.) 11 (1973) 313.
[476] I.K. Varma, Sangita, D. Ralli, Polym. N. 12 (1987) 294.
[477] I.K. Varma, A.K. Gupta, Sangita, J. Polym. Sci., Polym Lett. Edn 20 (1982) 621.
[478] I.K. Varma, A.K. Gupta, Sangita, D.S. Varma, J. Appl. Polym. Sci. 28 (1983) 191.
[479] I.K. Varma, S. Sharma, Eur. Polym. J. 20 (1984) 1101.
[480] I.K. Varma, D.S. Varma, J. Polym. Sci. Polym. Chem., (Ed.) 22 (1984) 1419.
[481] K. Sripadaraj, N. Gupta, I.K. Varma, Die Angew. Makromol. Chem. 260 (1998) 41.
[482] G. Viorica, High. Perform. Polym. 19 (2) (2007) 160.
[483] M. Sava, J. Appl. Polym. Sci. 101 (6) (2006) 3881.
[484] S. Minda, T. Paifeng, T. Wenyu, H. Wenhsiag, Eur. Polym. J. 42 (8) (2006) 1899.
[485] C. Gouri, C.P. Reghunadhan Nair, R. Ramaswamy, J. Appl. Polym. Sci. 50 (3) (2001) 403.
[486] X.-H. Xiong, P. Chen, Q. Yu, J.-X. Zhang, Polym. Eng. Sci. 51 (2011) 1599.
[487] H.-Y. Tang, N.-H. Song, Z.-H. Gao, Q.-F. Zhou, Polymer 48 (2007) 129.
[488] S.L. Gaonkar, K.M.L. Rai, B. Prabhuswamy, Eur. J. Med. Chem. 41 (2006) 841.
[489] H. Stenzenberger, Composite Struct. 24 (1993) 219–231.
[490] H. Stenzenberger, D. Wilson, H.D. Stenzenberger, P.M. Hergenrother (Eds.), Polyimides, Springer Science, London, 1990, pp. 79–108.
[491] I.K. Varma, D.S. Sangita Varma, J. Polym. Sci. Polym Chem. (ED.) 22 (1984) 1419–1433.
[492] S. Takeda, H. Akiyama, H. Kakiuchi, J. Appl. Polym. Sci. 35 (1988) 1341–1350.
[493] R.J. Iredale, C. Ward, I. Hamerton, Prog. Polym. Sci. 69 (2017) 1–21.
[494] C. Qu, l Zhao, D. Wang, H. Li, W. Xiao, C. Yue, J. Appl. Polym. Sci. 40395 (2014) 1–8.
[495] M. Sava, Designed Monomers Polym. 16 (2013) 14–24.
[496] F. Huang, F. Huang, Y. Zhou, Lei Du. Polym. J. 42 (2010) 261–267.
[497] X. Xiong, P. chen, Q. Yu, N. Zhu, B. wang, J. Zhang, et al., Polym. Int. 59 (2010) 1665–1672.
[498] X. Xiong, P. Chen, N. Zhu, Q. Yu, J. Zhang, B. Wang, Polym. Eng. Sci. 51 (2011) 1599.
[499] Z. Ren, Y. Cheng, Y. Li, F. Xiao, J. Appl. Polym. Sci. 43491 (2016) 1–8.
[500] Z. Ren, S. Hao, Y. Xing, C. Yang, S. Dai, High. Perform. Polym. 31 (2019) 1–8.

[501] S.E. Evsyukov, T. Pohlmann, H.D. Stenzenberger, Polym. Adv. Technol. 26 (2015) 574–580.
[502] H. Petrovicki, H.D. Stenzenberger, A. v. Harnier, Proc. 32nd SPE Ann. Techn. Conf. Society of Plastics Engineers, San Francisco, 88 (1974) 88–91.
[503] H.D. Stenzenberger, J. Appl. Polym. Sci.: Appl. Polym. Symp. 31 (1977) 91–104.
[504] P. Yu, Y.-li Zhang, X. Yang, L.-jing Pan, Z.-yao Dai, M.-zhao Xue, et al., Polym. Eng. Sci. 59 (2019) 2265–2272.
[505] M. Neda, K. Okinaga, M. Shibata, Mater. Chem. Phys. 148 (2014) 319–327.
[506] M. Shibata, T. Shimasaki, H. Satoh, M. Iwai, M. Neda, React. Funct. Polym. 97 (2015) 69–76.
[507] T. Enoki, H. Ohkubo, T. takeda, K. Ishii, J. Thermosetting Plast, Jpn. 16 (1995) 94–98.
[508] D. Reyx, I. Campistron, C. Cailaud, Macromol. Chem. Phys. 196 (1995) 775–785.
[509] Y. Li, L. Feng, L.C. Zhang, J. Appl. Polym. Sci. 100 (2006) 593.
[510] J.D. He, J. Wang, S.D. Li, Acta Polym. Sin. 5 (2000) 559.
[511] M.S. Chandran, T.M. Robert, D. Mathew, C.P.R. Nair, Polym. Adv. Technol. 25 (2014) 881–890.
[512] M. Satheesh Chandran, T.M. Robert, Sk Dona Mathew, C.P. Reghunadhan Nair, Polym. Adv. Technol. 25 (2014) 881–890.
[513] E. Grigat, R. Putter, German Patent, 1, (195) (1963) 746.
[514] I. Hamerton (Ed.), Chemistry and Technology of Cyanate Ester Resins, Blackie Academic & Professional (Chapman and Hall, London, 1994.
[515] T. Fang, D. Shimp, Prog. Polym. Sci. 20 (1995) 61.
[516] D. Shimp, Polym. Mater. Sci. Eng. 70 (1994) 561.
[517] J.C. Abed, R. Mercier, J.E. McGrath, J. Polym. Sci., Part. A, Polym. Chem. 35 (1997) 977.
[518] S. Das, C.D. Prevorsek, U.S. Patent, 4,831,086.
[519] H. Parvatareddy, P.H.T. Wilson, D.A. Dillard, Compos. Sci. Technol. 56 (1996) 1129.
[520] S.J. Ising, D.A. Shimp, Proceeding 34th International Symposium, May 8 (11), 1998, pp. 1326–1334.
[521] D.A. Shimp, J.R. Cristenson, S.J. Ising, Technical bulletin, Ciba, Ardsley, NY, 1991.
[522] S. Robitaille, ASTM Handbook, Composites, ASM, 2002, pp. 126–131.
[523] A.W. Snow, L.J. Buckley, Macromolecules 30 (1997) 394.
[524] M. Delucchi, S. Turri, A. Barbucci, M. Bassi, S. Novelli, G. Cerisola, J. Polym. Sci. Part. B Polym. Phys. 40 (1) (2002) 52.
[525] I. Hamerton (Ed.), Chemistry and Technology of Cyanate Ester Resins, Chapman & Hall, London, 1994.
[526] D.A. Shimp, S.J. Ising, J.R. Christenson Society of Plastics Engineers: Cleveland, OH, (1989) 127–140.
[527] G. Liang, M. Zhang, J. Appl. Polym. Sci. 85 (2002) 2377.
[528] T. Iijima, S. Katsurayana, W. Fukuda, M. Tomoi, J. Appl. Polym. Sci. 76 (2000) 208.
[529] J. Fan, X. Hu, C.Y. Yue, J. Polym. Sci. Part. B Polym. Phys. 41 (2003) 1123.
[530] M. Laskoski, D.D. Dominguez, T.M. Keller, J. Mater. Chem. 15 (2005) 1611.
[531] C.P.R. Nair, D. Mathew, K.N. Ninan, Eur. Polym. J. 37 (2001) 315.
[532] S.J. Park, F.L. Jin, J.R. Lee, Macromol. Res. 13 (2005) 8.
[533] E.M. Maya, A.W. Snow, L.J. Buckley, Macromolecules 35 (2002) 460.
[534] M.E. Wright, Polym. Prepr. 45 (2) (2004) 294.
[535] A.J. Guenthner, G.R. Yandek, M.E. Wright, B.J. Petteys, R. Quintana, D. Connor, et al., Macromolecules 39 (2006) 6046–6053.
[536] O. Georjon, J. Galy, Polymer 39 (1998) 339–345.
[537] D.A. Shimp, W.M. Craig, SAMPE Int. (1989) 1336–1346.
[538] A.J. Guenthner, K.R. Lamison, V. Vij, J.T. Reams, G.R. Yandek, J.M. Mabry, Macromolecules 45 (2012) 211–220.
[539] W.K. Goertzen, M.R. Kessler, Composites, Part. A 38 (2007) 779–784.
[540] D. Augustine, M.S. Chandran, D. Mathew, C.R. Nair, Polyphthalonitrile Resins and their High-End Applications, Elsevier, Amsterdam, 2018, p. 577.
[541] A.J. Guenthner, M.E. Wright, A.P. Chafin, J.T. Reams, K.R. Lamison, M.D. Ford, et al., Macromolecules 47 (2014) 7691–7700.

[542] A.J. Guenthner, B.G. Harvey, A.P. Chafin, M.C. Davis, J.J. Zavala, K.R. Lamison, et al., Macromolecules 50 (2017) 4887−4896.
[543] S. Ganguli, D. Dean, K. Jordan, G. Price, R. Vaia, Polymer 44 (4) (2003) 1315−1319.
[544] B. Yameen, H. Duran, A. Best, U. Jonas, M. Steinhart, W. Knoll, Macromol. Chem. Phys. 209 (16) (2008) 1673−1685.
[545] L.R. Cambrea, M.C. Davis, T.J. Groshens, A.J. Guenthner, K.R. Lamison, I.M. Mabry, J. Polym Sci. Polym. Chem. 48 (20) (2010) 4547−4554.
[546] A.J. Guenthner, M.C. Davis, K.R. Lamison, G.R. Yandek, Lee, R.C. Thomas, et al., Polymer 52 (2011) 3933−3942.
[547] A.S.A. Srinivasan, J.C. Abed, J.E. McGrath, Polym. Prep. 359 (1994) 537.
[548] B.S. Kim, J. Appl. Polym. Sci. 65 (1997) 85.
[549] M.J. Meyers, J. Sun, K.E. Carlson, G.A. Marriner, B.S. Katzenellenbgen, J.A. Katzenellenbogen, J. Med. Chem. 44 (24) (2001) 4230−4251.
[550] W.K. Goertzen, M.R. Kessier, Compos. Appl. Sci. Manuf. 38 (3) (2007) 779−784.
[551] K.R. Lamison, A.J. Guenthner, V. Vij, J.M. Mabry, PMSE Prepr. 104 (2011) 299−300.
[552] I. Hamerton, A.M. Emsley, B.J. Howlin, P. Klewpatinond, S. Takeda, Polymer 44 (17) (2003) 4839−4852.
[553] Y. Wang, K. Kou, G. Wu, A. Feng, L. Zhuo, RSC Adv. 5 (2015). 58821-28831.
[554] F. Chao, N. Bowler, X.L. Tan, G.Z. Liang, M.R. Kessler, Composites, Part. A 40 (2009) 1266−1271.
[555] P.D. Wienhold, D.F. Persons, SAMPE J. 39 (6) (2003) 6−17.
[556] S. Arepalli, P. Moloney, MRS Bull. 40 (2015) 804−811.
[557] K. Bittner-Rohrhofer, K. Humer, H. Fillunger, R.K. Maix, H.W. Weber, P.E. Fabian, et al., J. Nucl. Mater. 329 (2004) 1083−1087.
[558] K. Humer, R. Prokopec, H.W. Weber, Fusion. Eng. Des. 88 (2013) 350−360.
[559] A. Idesaki, T. Nakamoto, M. Yoshida, A. Shimada, M. Iio, K. Sasaki, et al., Fusion. Eng. Des. 112 (2016) 418−424.
[560] B.G. Harvey, A.J. Guenthner, W.W. Lai, H.A. Meylemans, M.C. Davis, L.R. Cambrea, et al., Macromolecules 48 (2015) 3173−3179.
[561] H.A. Meylemans, B.G. Harvey, J.T. Reams, A.J. Guenthner, L.R. Cambrea, T.J. Groshens, et al., Biomacromolecules 14 (2013) 771−780.
[562] B.G. Harvey, A.J. Guenthner, G.R. Yandek, L.R. Cambrea, H.A. Meylemans, L.C. Baldwin, et al., Polymer 55 (2014) 5073−5079.
[563] B.G. Harvey, C.M. Sahagun, A.J. Guenthner, T.J. Groshens, L.R. Cambrea, J.T. Reams, et al., Chem. Sus. Chem 7 (2014) 1964−1969.
[564] M. Derradji, N. Ramdani, T. Zhang, J. Wang, L.D. Gong, X.D. Xu, et al., Polym. Compos. 38 (2017) 1549.
[565] M. Laskoski, M.B. Schear, A. Neal, D.D. Dominguez, H.L. Ricks-Laskoski, J. Hervey, et al., Polymer 67 (2015) 185.
[566] J.-Z. Ma, K. Cheng, J.-B. Lv, C. Chen, J.-H. Hu, K. Zeng, et al., Chinese J. Polym. Sci. 36 (2018) 497.
[567] S. Shan, X. Chen, Z. Xi, X. Yu, X. Qu, Q. Zhang, High. Perform. Polym. 29 (2017) 113.
[568] S. Xu, Y. Han, Y. Guo, Z. Luo, L. Ye, Z. Li, et al., Eur. Polym. J. 95 (2017) 394.
[569] B. Bulgakov, A. Sulimov, A. Babkin, A. Kepman, A. Malakho, V. Avdeev, J. Appl. Polym. Sci. 134 (2017) 44786.
[570] Y. Hu, Z. Weng, Y. Qi, J. Wang, S. Zhang, C. Liu, et al., RSC Adv. 8 (2018) 32899.
[571] T.M. Keller, Price, J. Macromol. Sci. Chem. 18 (1982) 931.
[572] D.D. Dominguez, H.N. Jones, T.M. Keller, Polym. Compos. 25 (2004) 554.
[573] S.B. Sastri, T.M. Keller, J. Polym. Sci. Part A: Polym. Chem. 1998 (1885) 36.
[574] D. Wu, Y. Zhao, K. Zeng, G. Yang, J. Polym. Sci. Part. A: Polym. Chem. 50 (2012) 4977.
[575] P. Burchill, J.P olym. Sci. PartA: Polym. Chem. 32 (1994) 1.
[576] J. Hu, R. Sun, Y. Wu, J. Lv, Z. Wang, K. Zeng, et al., RSC Adv. 7 (2017) 43978.

[577] T.M. Keller, J.R. Griffith, J. Fluor. Chem. 13 (1979) 315.
[578] S.B. Sastri, T.M. Keller, J. Polym. Sci. Part. A: Polym. Chem 37 (1999) 2105.
[579] Z. Wu, J. Han, N. Li, Z. Weng, J. Wang, X. Jian, Polym. Int. 66 (2017) 876.
[580] D. Augustine, D. Mathew, C.R. Nair, Eur. Polym. J. 71 (2015) 389.
[581] R.M. Christie, D.D. Deans, J. Chem. Soc. Perkin Trans. 2 (1989) 193.
[582] M. Xu, D. Ren, L. Chen, K. Li, X. Liu, Polymer 143 (2018) 28.
[583] Z. Liu, Y. Liu, J. Appl. Polym. Sci. 48138 (2019) 1.
[584] Z. Wu, J. Han, N. Li, Z. Weng, J. Wang, X. Jian, Polym. Int. 66 (2017) 876−881.
[585] D. Zhang, X. Liu, X. Bai, Y. Zhang, G. Wang, Y. Zhao, et al., e-Polymers 20 (2020) 500−509.
[586] Y. Hu, Q.Y. Weng, J.Y. Wang, S.H. Zhang, C. Liu, RSC Adv. 8 (2018) 32899−32908.

CHAPTER 2

Properties and processing of thermoset resin

Abbreviations

DP	Degree of polymerization
GP	Gel point
NMR	Nuclear magnetic resonance
DSC	Differential scanning calorimetry
IR	Infrared
TCMPA	Tetrachorobisphenol-A
BPA	Bisphenols namely bisphenol-A
TBBPA	Tetra bromobisphenol-A
DMA	Dynamic mechanical analyzer
PTOH	Phenol end-capped poly(tetramethylene oxide)
DGEBA	Diglycidyl ether of bisphenol-A
TGDDM	Tetraglycidylether of 4,4′ diaminodiphenyl methane
MDI	4,4′-diphenylmethane diisocyanate
HDI	1,6-hexamethylene diisocyanate
BD	1,4 Butanediol
PU	Polyurethane
HS	Hard segment
TGAP	Triglycidyl p-amino phenol
VTES	Vinyl triethoxy silane
DETDA	Diethyltoluene diamine
DDS	4, 4′ Diaminodiphenyl sulfone
TTT	Time–temperature–transformation
LC	Liquid crystalline
LCEP	Liquid crystalline epoxy
DDM	4, 4′ Diaminodiphenyl methane
KV	Kelvin–Voigt
CLD	Constrained layer damping
FLD	Free-layer damping
IPN	Interpenetrating polymer networks
INC	Interconnected polymer networks
PEO	Poly(ethylene oxide)
p-HRR	Peak heat release rate
VE	Vinyl ester
EP	Epoxy resins
PEG	Poly(ethylene glycol)
CE	Cyanate ester

PETI	Phenyl-ethynyl-terminated imide
MMA	Methyl methacrylate
TEDGMA	Triethylene glycol methacrylate
PU	Polyurethane
TGIC	1,3,5-triglycidyl isocyanurate
PMMA	Poly(methyl methacrylate)
PALS	Positron annihilation lifetime spectroscopy
SBR	Styrene-butadiene rubber
DCPD	Dicyclopentadiene
POSS	Polyhedral oligomeric silsesquioxanes
PCL	Poly(caprolactone)
AAAB	Acrylamide azobenzene
SMA	Shape-memory alloy
SMT	Shape-memory thermoset
SMPU	Shape-memory polyurethanes
PCL	Polycaprolactone
CBMI	Chain-extended bismaleimide
DDS	Diaminodiphenyl sulfone
BMI	Bismaleimide
GA	Glutaric anhydride
HEMA	2-hydroxyethyl methacrylate
MMA	Methyl methacrylate
SMT	Shape-memory thermoset
HBPU	Hyperbranched polyurethane
PCL	Poly(ε-caprolactone)
PTMEG	Poly(tetramethylene ether) glycol
CAN	Covalent adaptable networks
CNT	Carbon nanotubes
MDI	Methylene diphenyl diisocyanate
MOF	Metal−organic frameworks
SBU	Secondary building units
ELF	Elastic layered metal−organic frameworks
DMF	Dimethylformamide
CP	Chlorinated paraffins
DB	Decabromobiphenyl oxide
DOPO	9,10 dihydro-9-oxa-10-phosphaphenanthrene-10-oxide
DPDP	10-(2,5-dihydroxyphenyl)-9,10-dihydro-9-oxa-10-phosphate phenanthrene-10-oxide
MPDA	m-phenylene diamine
CB	Chain breaking
CFCs	Chlorofluro carbon compounds
RIM	Reaction injection molding
CAD	Computer-aided design
SLA	Stereolithography
DLP	Digital light processing
FDM	Fused deposition model
SLS	Selective laser sintering
SLM	Selective laser melting
EBM	Electron beam melting
BJ	Binder jetting
FDM	Fused deposition modeling

ABS	Copolymer of acrylonitrile, butadiene, and styrene
BEBP	4, 4′-bis (2-hydroxyethoxy) biphenyl
DIW	Direct-ink write
EAA	Epoxy aliphatic acrylate
AUD	Aliphatic urethane diacrylate
CLD	Constrained layer damping
FRP	Fiber-reinforced plastic
FLD	Free-layer damping

2.1 Introduction

In the first chapter, the chemistry of thermosetting resins is discussed elaborately. A thermoset resin forms a "network" on curing and converts into solid with desirable mechanical properties. The terminology "network" is widely used to describe the structure of solid-state materials. Molecular or atomic network structure is the basis of the mechanical coherence of such materials. Lattice is an example of a network of ions in which the electrostatic ionic forces keep the cations and anions together. When a solid is melted or dissolved in any solvent, the network structure is broken and the structural integrity ceases to exist. Diamond is a classic example of covalent network, in which each carbon atom is covalently bonded to its neighbors forming a tetrahedral structure. The structure is basically a macromolecular three-dimensional network, which is responsible for its remarkable hardness in contrast to the other allotrope, graphite.

Polymer networks or crosslinked networks are molecular-based networks whose network structures depend entirely on covalent bonding or on physical intermolecular interactions between the macromolecules. Just like in diamond each pair of adjacent junction points in the network are separated by only one covalent bond, in a polymer network two junction points are separated by linear subchains of several or many covalent bonds. When the connectivity from junction point is through chemical bond then they are called chemical crosslinks. The crosslinks generated due to the entanglement of high polymer chains are known as physical crosslinks. In case of thermoset polymers, the crosslinks are chemical crosslinks.

2.2 Gelation and vitrification

When a thermoset resin cures, it encounters an interesting phenomenon, which is called gelation. The gel point is considered as a point in polymerization where network structure first occurs with a unit probability. The question is, what is the molecular basis of gelation? The basic condition of a compound to be a precursor of a polymer is that the functionality of the compound must be greater than or equal to two. When two molecules with functionality two, react with each other, the resulting product will always have a functionality of two because out of the total four functionalities two are lost due to the reaction. Thus a linear polymer is formed due to polymerization of a difunctional monomer.

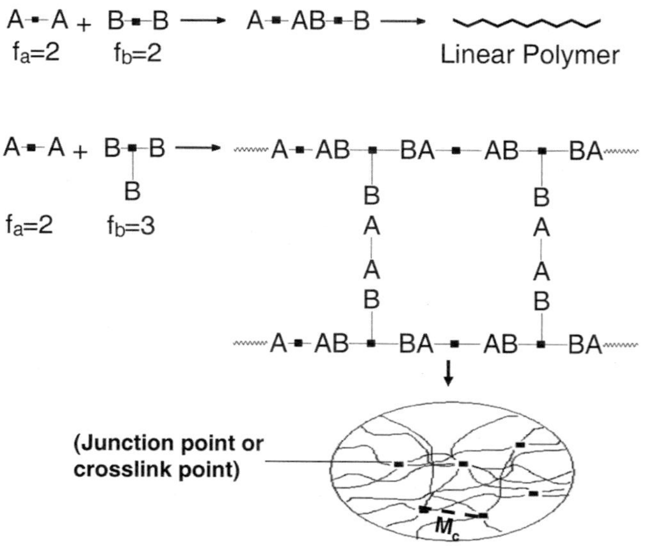

Figure 2.1
Schematic representation of network formation during reaction between two molecules with functionality more than 2.

However, if the functionality is greater than two, branching will be generated and the number of reactive functionality will increase with an increase in the branch points as shown in Fig. 2.1. As for example, if a trifunctional molecule reacts with a difunctional molecule, the number of reactive functionality of the intermediate will be 3 and when the first intermediate react with another trifunctional monomer, the number of functionality will be 4. Now if we assume that all the functional groups are equally reactive, irrespective of the attachments, then the intermediate will react preferentially and a network will be formed before completion of the reaction (Fig. 2.1). Thus if we consider a reaction between a difunctional monomer and monomer of functionality f then the number of reactive groups (N_r) in an intermediate molecule with n branch point can be expressed as:

$$N_r = n(f-1) + 2. \tag{2.1}$$

The gelation phenomena in polymerization reactions were described by Flory and Stockmayer [1–3]. In the case of polymerization reaction of A and B with functionality f_a and f_b respectively, the number of paths that can lead from a randomly chosen A group from the f_a functional groups is $(f_a - 1)$. Similarly, one B group can react in $(f_b - 1)$ a number of ways with any one of the A groups. If P_a and P_b are the extents of reaction of A and B groups, p_a and p_b are the probabilities per path for A and B groups, respectively. Hence the gel point can be statistically expressed as:

$$(f_a - 1)p_a(f_b - 1)p_b = 1. \tag{2.2}$$

The statistical approach for deriving the above equation is based on the assumption that all A and all B groups have the same probabilities of being reacted and there is no intramolecular reaction. In the case of a mixture of reactants of different functionalities, the functionality should be replaced by an average functionality. Thus expressing Eq. (1.2) in terms of average functionality we get

$$(f_{an} - 1)p_a(f_{bn} - 1)p_b = 1, \tag{2.3}$$

where f_{an} and f_{bn} are the number average functionalities of A and B, respectively.

The number average degree of polymerization (DP_n) is defined as the ratio of the number of units (number of molecules initially present) to the number of molecules present after the reaction. Thus DP_n can be expressed as below:

$$DP_n = \frac{N_a + N_b}{N_a + N_b - N_x}, \tag{2.4}$$

where N_x is the number of A or B groups reacted. N_x can be expressed in terms of the extent of reaction as follows:

$$N_x = N_a f_{an} p_a = N_b f_{bn} p_b. \tag{2.5}$$

Combining Eq. (1.4) and Eq. (1.5), we get

$$DP_n = \frac{N_a + N_b}{N_a + N_b - N_a f_{an} p_a} = \frac{1}{1 - p_a \frac{1}{\frac{1}{f_{an}} + \frac{1}{rf_{bn}}}}, \tag{2.6}$$

where r is the molar ratio of A and B, that is, N_a/N_b

Similarly,

$$DP_w = 1 + \frac{p_a p_b f_{an} f_{bn}\{p_a(f_{an} - 1) + 2\}}{\{1 - p_a p_b(f_{an} - 1)(f_{bn} - 1)\}(p_a f_{an} + p_b f_{bn})}. \tag{2.7}$$

For the resins, which self-polymerize without any curing agent namely resole type phenolic resin or cyanate ester resin ($p_a = p_b$, $r = 1$, and $f_a = f_b$), Eqs. (1.6) and (1.7) can be written as given below.

$$DP_n = \frac{1}{1 - p_a \frac{f_{an}}{2}}, \tag{2.8}$$

$$DP_w = \frac{1 + p_a}{1 - p_a(f_{an} - 1)}. \tag{2.9}$$

Carothers proposed the criterion for gelation that $DP_n \to \infty$ at gel [4]. This criterion is based on the percept that all the units in a polymerizing mixture are connected to form a single network. However, this is inconsistent with a random polymerization because it is not possible that the groups will organize in such a way that only two per original molecule react. Thus the more accepted criteria for gelation is that weight-average degree of polymerization $P_w \rightarrow \infty$, which means that the average degree of polymerization per unit is infinite. On the other hand, the condition $DP_n \rightarrow \infty$ means average DP per molecule is infinite as the number of molecules reduces to zero. Thus it can be stated that gelation is accompanied by a divergence in DP distribution in which the weight averages or higher averages become infinite.

From the application point of view, effective use of any thermosetting system requires one to be able to predict the cure kinetics of the system [5,6] to consistently obtain the maximum glass transition temperature and also to predict the flow behavior of the curing resin, in particular to precisely locate when the sol−gel transitions occur. This is because the polymer can be easily shaped or processed only before the gel point, where it can still flow and can easily be formed with stresses applied relaxed to zero thereafter. Accurate knowledge of the gel point (GP) would therefore allow estimation of the optimum temperature and time for which the sample should be heated before being allowed to set the mold. The GP is also useful in that it can be used to determine the activation energy for the cure reaction of the system.

The GP of a crosslinking system is defined unambiguously as the instant at which the weight-average molecular weight reaches infinity and as such is an irreversible reaction. Crosslinked polymer at its GP is a transition state between a liquid and a solid. The polymer reaches its GP at a critical extent of crosslinking (α_{gel}). Before the GP, that is $\alpha < \alpha_{gel}$ the polymer is called a sol, because it is typically soluble in an appropriate solvent. Beyond GP, that $\alpha > \alpha_{gel}$, at least part of the polymer is typically not soluble in any solvent and is called a gel. Kinetically, gelation does not usually inhibit the curing process so the conversion rate remains unchanged. Hence, it cannot be detected by technique sensitive only to a chemical reaction like differential scanning and thermogravimetric analysis. There are various methods to determine the gel point. Gel time for an epoxy resin is generally determined as per ASTM D 3532−76. The time when the resin ceases to form strings by contact with a pick is taken as the gel point. The most sophisticated method for the determination of gel time of an epoxy system is one by using a rheological experiment. For details of these measurements, the readers may refer to Chapter 6 where measurement techniques are elaborated. Another process, which a thermoset resin undergoes, during cure, is vitrification. It is defined as the point at which the glass transition temperature of the network has become the same as the cure temperature. At this point, the material is transformed from a rubbery gel to a gelled glass. The determination of gelation and vitrification from rheological properties will be discussed in subsequent sections.

2.3 Cure characteristics

For effective utilization of the potential of a thermoset resin, it is necessary to study the cure kinetics of the resin. The kinetic study is required to ensure proper selection of curing conditions for large-scale application. Cure kinetics of the thermosetting resins has been investigated using various thermal and spectroscopic techniques [7–10]. The principle of these techniques will be discussed in Chapter 6. When a thermoset resin cures, gradual conversion of functional groups due to chemical reaction takes place. Hence, the first step of cure kinetics is to define the conversion or extent of curing (α). The extent of curing is determined from the disappearance of functional groups using Fourier transform infrared (FTIR) or nuclear magnetic resonance (NMR) spectroscopic analysis and from the heat of reaction using differential scanning calorimetry (DSC). For example, α can be determined from the IR absorbance of the reactive groups using the following relationship [7]:

$$\alpha = 1 - \frac{A^t}{A^0}. \tag{2.10}$$

where A^t and A^0 are the normalized absorbance of the functional groups before reaction and after reaction time t. The absorbance of a group that does not participate in the reaction is used as a reference for normalization.

Although the extent of curing can be determined from any of the above-mentioned methods, in most of the cases reported in the literature, the cure kinetic of various thermoset resins has been investigated using DSC. The extent of curing can be expressed in terms of heat of reaction as follows:

$$\frac{d\alpha}{dt} = \frac{1}{H_0}\frac{dH}{dt}, \tag{2.11}$$

where H_0 is the total heat released during complete curing and H is the heat released from the onset of polymerization up to time t. If the curing reaction involves only one chemical reaction, then the crosslinking reaction will be characterized by a single heat of reaction and the extent of curing will be the same as the number of reacted groups determined by spectroscopic analysis. However, if the curing reaction involves several chemical reactions then the heat of the reaction determined by DSC will represent the average value. The details about the various techniques mentioned above will be discussed in Chapter 6.

The curing reactions of thermoset resins have been extensively studied by using isothermal and nonisothermal DSC measurements and various models have been proposed to represent the experimental data [11–14]. The phenomenological model developed by Kamal [15] is mostly used for isothermal kinetic analysis of a thermoset like epoxy resin. The differential methods using experimental data from the dynamic DSC experiments given by Friedman [16],

Flynn–Wall–Ozawa [17], and Kissinger–Akahira–Sunose [18] have been used to determine the activation energies of the curing reaction of thermoset resins. The so-called isoconversional methods were also utilized to describe the curing process of thermosets like polyurethane under isothermal conditions [19]. The differential methods [16–18] have been used and compared [19] with the model-based autocatalytic approach by Kamal–Sourour [15]. Isocoversional kinetic models studied for isothermal systems have also been extended for nonisothermal curing processes [20,21]. It may be noted that the accuracy of such types of models is strongly dependent on the quality of the experimentally determined reaction processes. It is important to ensure that the exothermic curing reactions of the considered resins are detected precisely using thermal analysis methods.

2.3.1 Isothermal cure kinetics

The general equation for an nth-order reaction can be written as [15,16]

$$\frac{d\alpha}{dt} = K_1(1-\alpha)^n, \qquad (2.12)$$

where K_1 is the rate constant and n is the order of the reaction. In many curing reactions, the new groups (produced as a result of curing) catalyze the curing reaction. For instance, the hydroxyl groups formed during the curing of epoxy resin catalyze the epoxy/amine reaction. The equation for an autocatalytic curing reaction (which considers the reaction between the functional groups of resin and curing agent plus catalytic and autocatalytic effects) can be represented as

$$\frac{d\alpha}{dt} = (K_1 + K_2\alpha^m)(1-\alpha)^n, \qquad (2.13)$$

where K_2 expresses the rate constant for an autocatalytic curing reaction, m and n are the kinetic exponents of the reaction, and $(m+n)$ gives the overall order of the curing reaction. When $K_2 = 0$, the equation reduces to a noncatalytic one. From the isothermal DSC experiment, the extent of reaction and reaction rate data have to be determined and adjusted with kinetic equations. The initial rate of reaction, that is, reaction rate at $\alpha = 0$ is used to determine K_1.

Taking the logarithm of both side of Eq. (2.13), we can write in two forms as below

$$\ln\left(\frac{d\alpha}{dt}\right) = \ln(k_1 + k_2\alpha^m) + n\ln(1-\alpha), \qquad (2.14)$$

$$\ln\left[\frac{(d\alpha/dt)}{(1-\alpha)^n}\right] = \ln k_2 + m\ln\alpha. \qquad (2.15)$$

Using Eq. (2.14), n can be determined from the slope of the plot of $\ln(d\alpha/dt)$ versus $\ln(1-\alpha)$, which will be a straight line. Similarly, substituting the value of n and k_1 and using the corresponding linear plot m and k_2 can be determined. However, fitting isothermal data is as such cumbersome due to more number of variables as mentioned above.

Kamal's model does not consider the influence of diffusion, which becomes important when the glass transition temperature of the thermoset reaches the curing temperature, that is, vitrification as discussed above. However, most of the cases of thermoset cure obey this model. Cai et al. [22] studied the curing of epoxy with two polyether amines (D-230 and T-403 from BASF, Germany) and reported validation of the autocatalytic model by measuring the heat flow rate during the isothermal scanning. The various kinetic parameters used in Eq. (2.13) for epoxy/D-230 and epoxy/T-403 systems are shown in Table 2.1. The comparison of the cure rate of both the epoxy systems at various temperatures measured experimentally and those calculated from the autocatalytic kinetic model is shown in Fig. 2.2. The result shows that the developed cure kinetic model agreed well with a small deviation from the experimental results. Zhang and coworker [23] investigated the nonisothermal cure characteristics of epoxy with three different bisphenols, namely, bisphenol-A (BPA), tetrachorobisphenol-A (TCMPA), and tetra bromobisphenol-A (TBBPA). They have used Sestak–Berggren model to fit the nonisothermal cure kinetics of epoxy/BPA, epoxy/TCMPA and epoxy/TBBPA and reported that the calculated kinetic data are in good accordance with the experimental data for all three systems (Fig. 2.3).

The kinetic constants K_1 and K_2 can be correlated to temperature according to the Arrhenius equation given below

$$K_i = A\exp\left(\frac{E_i}{RT}\right). \tag{2.16}$$

where K_i and E_i are the rate constant and the activation energy, respectively; A is a constant; R is the universal gas constant; and T is the absolute temperature. Thus the kinetic model as discussed above allows calculation of activation energy (E), using linear regression on data

Table 2.1: Parameters used in Kamal's model in DGEBA/T-403 and DGEBA/D-230 systems.

Temperature (°C)	K_1 (k^{-1})	K_2 (k^{-1})	m	n	m + n	r
T403						
70	0.011	0.058	0.68	2.40	3.08	0.9995
80	0.013	0.060	0.68	2.45	3.13	0.9934
90	0.038	0.062	0.60	2.30	2.90	0.9715
D230						
70	0.014	0.025	0.50	1.02	1.52	0.9981
80	0.025	0.040	0.48	1.09	1.57	0.9954
90	0.081	0.061	0.48	1.05	1.53	0.9988

Source: *Reprinted with permission from H. Cai, P. Li, G. Sui, Y. Yu, G. Li, X. Yang, S. Ryu, Thermochim. Acta 473 (2008) 101−105, ©2008 Elsevier Publishers.*

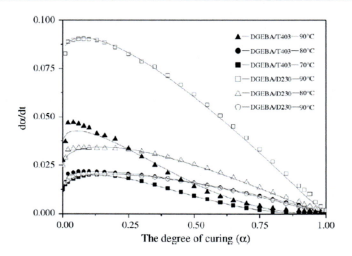

Figure 2.2
Comparisons of experimental data with the Kamal's model in DGEBA/T-403 and DGEBA/D-230 systems. *Source: Reprinted with permission from H. Cai, P. Li, G. Sui, Y. Yu, G. Li, X. Yang, S. Ryu, Thermochim. Acta 473 (2008) 101–105, ©2008 Elsevier Publishers.*

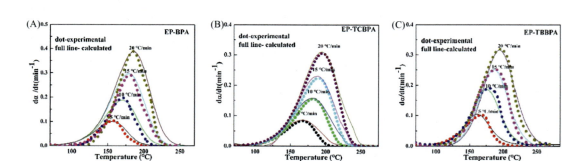

Figure 2.3
Comparison of predicted Šesták–Berggren model rates with experimental data: (A) EP-BPA; (B) EP-TCBPA; and (C) EP-TBBPA. *Source: Reprinted with permission from P. Zhang, S.A.A. Shah, F. Gao, H. Sun, Z. Cui, J. Cheng, J. Zhang, Thermochim. Acta 676 (2019) 130–138, ©2019 Elsevier Publishers.*

obtained at different temperatures. Typical plots showing the effect of temperature on reaction rate and conversion of free-radically polymerized unsaturated polyester are shown in Fig. 2.4A and B, respectively. As we can see, the rate of reaction at a constant temperature increases with time attains a maximum and decreases thereafter. The peak of reaction rate versus time shifted to lower time due to the enhanced rate of reaction at elevated temperature. It is also observed that conversion increases with increasing temperature.

Most of the studies on cure kinetics of thermoset have used DSC for the determination of α and K. Billotte et al. [24] used a novel technique to determine cure kinetic data as well as

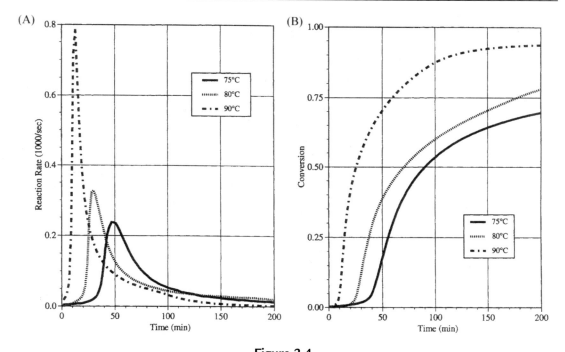

Figure 2.4
(A) Isothermal reaction rate versus time plot of a free-radically polymerized unsaturated polyester resin. (B) Conversion profile versus time plot of a free-radically polymerized unsaturated polyester resin. *Source: (A) Reprinted with permission from S.V. Muzumdar, L.J. Lee, Polym. Eng. Sci. 36 (1996) 943, ©1996 John Wiley and Sons Publishers. (B) Reprinted with permission from S.V. Muzumdar, L.J. Lee, Polym. Eng. Sci. 36 (1996) 943, ©1996 John Wiley and Sons Publishers.*

the evolution of mechanical properties and cure shrinkage of an epoxy/anhydride resin system. A heat flux cell as illustrated in Fig. 2.5 is coupled with a dynamic mechanical thermal analyzer (DMA + 450 from Metravib) and used for combined measurement. The "heat flux cell" consists of two plates (upper and lower) with a cavity between them to hold the sample. This novel heat flux cell enables direct injection of the liquid resin into a closed cavity. The temperature control is ensured by the thermal enclosure of the DMA instrument. The resin is contained in the mold where the upper and lower surfaces act as heat flux sensors. The degree of cure measured using DSC and the "heat flux cell" are presented as a function of time in Fig. 2.6. The curing reaction appeared to be faster when investigated with DSC compared to that with the above-mentioned arrangement, though a similar procedure was followed for injection of sample, in both cases. We can see differences in slopes near the plateau at the end of the cure. The difference can be attributed to the fact that the mass of the sample used in the heat flux cells is three orders of magnitude higher than the one used in DSC. This might have created the expected issue of temperature

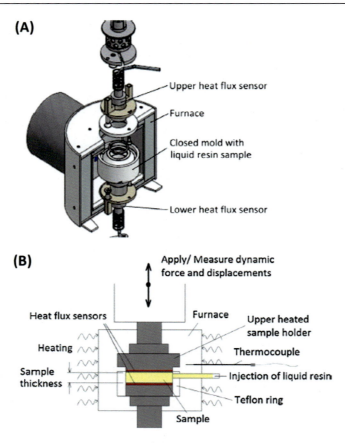

Figure 2.5
Thermal flux cell HFC200 on DMA + 450: (A) design of the fixture and (B) working principle.
Source: *Reprinted with permission from C. Billotte, F.M. Bernard, Edu Ruiz, Eur. Polym. J. 49 (2013) 3548–3560,* ©*2013 Elsevier Publishers.*

gradient in the sample. It may be noted that dissipation of the heat produced during cure is much easier in a small amount sample (about 10 mg) as used in DSC.

2.3.2 Dynamic cure kinetics

In dynamic or nonisothermal cure kinetics, the heat flow is measured as a function of temperature using different heating rates. A typical heat flow versus temperature plot of an epoxy/polyetheramine (Jeffemine ED-300) system measured at a different heating rate (5, 10, 15, and 20°C min^{-1}) are presented in Fig. 2.7. As discussed above for polymerization of polyester resin with a free radical mechanism, the maximum heat flow is observed at peak temperature. As expected, in the epoxy/Jeffemine system, the heat flow versus temperature curves show a shift of peak temperature at maximum heat flow at higher

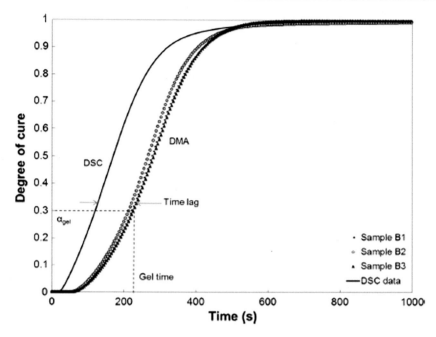

Figure 2.6
Comparison between degree of cure measured with DSC and the DMA HCF200 cell for samples of series B. Source: *Reprinted with permission from C. Billotte, F.M. Bernard, Edu Ruiz, Eur. Polym. J. 49 (2013) 3548–3560, ©2013 Elsevier.*

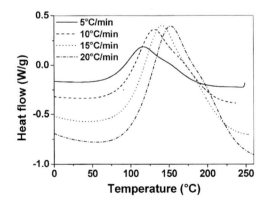

Figure 2.7
A typical heat flow versus temperature plot of an epoxy/polyetheramine Jeffemine (ED-400) system measured at a different heating rate (5, 10, 15, and 20°C min^{-1}).

temperature with increasing scanning rate. This shift of peak temperature with increasing scanning rate depends on the activation energy associated with each reaction. Based on this peak shifting phenomena, various kinetic models can be used to find the activation energy during the cure reaction. Various kinetic models which can be used for both TGA kinetics for estimation of life of a thermoset will be discussed in Chapter 6. The readers may use the same kinetic models for DSC kinetics as well. In this section two most widely used models namely Kissinger and the Flynn–Wall–Ozawa [25,26] models will be discussed in detail with experimental results. The popularity of these two methods is due to the fact that, unlike other nonisothermal models, they do not require prior knowledge of the reaction mechanism to quantify kinetic parameters.

According to Kissinger, the activation energy can be obtained from the maximum reaction rate, where $d(d\alpha/dt)/dt$ is zero at a constant-heating rate condition. The relationship between activation energy (E_a), the heating rate (β), and T_p is the peak exothermic temperature at which the rate is maximum. The resulting relationships can be expressed as

$$\frac{d[\ln(\beta/T_p^2)]}{d(1/T_p)} = -\frac{E_a}{R}, \tag{2.17}$$

where β is the heating rate and R is the universal gas constant. Therefore a plot of $\ln(\beta/T_p^2)$ versus $1/T_p$ gives the activation energy (E_a) without the need to make any assumption about the conversion-dependent function. The value of E_a can be calculated from the slope of the linear fit of experimental data. In the Kissinger method, the DSC peak exotherm is isoconversional so its value is not dependent on the heating rate. This assumption may not be valid for all systems and has to be verified. It also measures changes in the reaction mechanism via changes in the activation energy with conversion. For both nth-order and autocatalytic reactions, the values of E_a, $\ln A$, and conversion rates obtained are comparable to isothermal methods [18,27].

The Ozawa developed an alternative method to determine the relationship between the activation energy (E_a), the heating rate (β), and isoconversion temperature (T_i), under the assumption that the extent of reaction at the exothermic peak is constant and independent of heating rate. The proposed equation as:

$$E_\alpha = -\frac{R}{1.052} \frac{d(\ln \beta)}{d(1/T_i)}, \tag{2.18}$$

where β is the heating rate, T_i is the isoconversion temperature, and R is the universal gas constant. The advantage here is that the activation energy can be measured over the entire course of the reaction.

Ratna and coworker group have carried out extensive studies [28,29] on cure characteristics of epoxy/polytheramine systems and the related nanocomposites. Nonisothermal cure kinetic method was applied to study the cure kinetics of three epoxy systems (EP/J400, EP/J 600, and EP/J-900), where the same difunctional DGEBA resin and polyether amines of

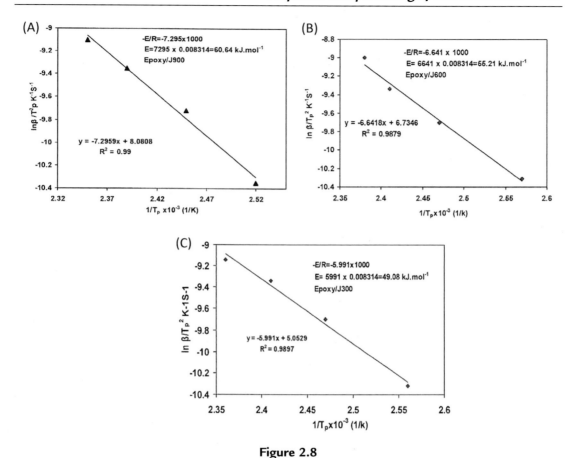

Figure 2.8
Plot of $[-\ln(\beta/T^2_p)]$ versus $1/T_p$ for different epoxy systems: (A) epoxy/J-900, (B) epoxy/J 600, and (C) epoxy/J 400 systems.

three different molecular weights (400, 600, and 900 g mole^{-1}) were used. Using the Kissinger equation $[-\ln(\beta/T^2_p)]$ and $1/T_p$ was plotted in Fig. 2.8 for the epoxy/J-900, epoxy/J600, and epoxy/J300 systems. The data is linear and the slope is equal to $-E_a/R$, indicating that this model fit the experimental data quit well. The activation energies calculated from the slope are tabulated in Table 2.2.

To perform isoconversion or Ozawa analysis, the original MDSC data were transformed into conversion (α) for each ith heating rate. The value of α can be obtained from the following expressions

$$\alpha = \Delta H_p / \Delta H_{tot}, \tag{2.19}$$

where ΔH_p is the partial heat of reaction and ΔH_{tot} is the total heat of reaction taken from the dynamic reaction. Using Ozawa equation, the logarithm of the heating rate ($\ln \beta$) versus

Table 2.2: Activation energy of curing of epoxy/Jeffemine-400, epoxy/Jeffemine 600, and epoxy/Jeffemine 900.

System	Curing enthalpy (ΔH_{total}) (J g^{-1})	Activation energy (kJ mole^{-1}) Kissinger method	Activation energy (kJ mole^{-1}) Ozawa method
DGEBA/J-900	163.8	60.6	76.0
DGEBA/J600	234.9	52.8	65.2
DGEBA/J400	306.3	49.1	52.9

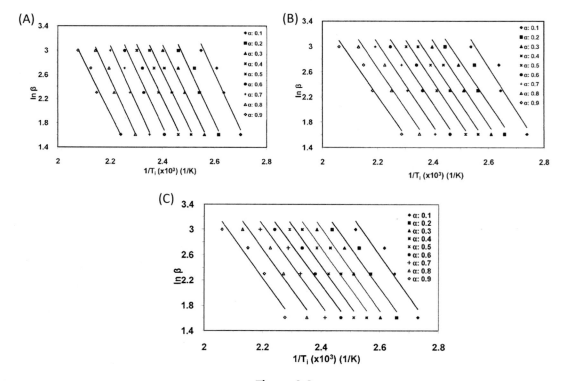

Figure 2.9
Plot of reciprocal of temperature ($1/T_p$) versus the logarithm of the heating rate ($\ln \beta$) for three different epoxy systems (A) EP/J400, (B) EP/J 60, and (C) EP/J-900 with α value varying from 0.1 to 0.9.

the reciprocal of temperature ($1/T_p$) was plotted in Fig. 2.9 for three different epoxy systems (Ep/J400, Ep/J 600, and Ep/ J-900) with α value varying from 0.1 to 0.9. There is an excellent linear fit in all the cases, indicating that this model fits the experimental data quite well. The E_a obtained from the slope of the straight line of each conversion for all the systems are averaged and tabulated in Table 2.2. We can see from the table the epoxy/J400 system shows the lowest activation energy. This is expected because with increasing the molecular weight of Jeffemine the reactivity decreases and more energy required to activate the reactions. As expected the activation energies calculated from Kissinger method are slightly lower than that of Ozawa's method. Further, the E_a values of each system obtained by Ozawa method were plotted against conversion for all three systems. It can be seen that the E_a gradually decreases with the extent of cure up to $\alpha = 0.4$ and thereafter increased. The decrease is more predominant with increasing the molecular weight of the polyether chain of the clay modifier. This could be attributed to the cooperative motion of the chain segments. On the other hand, an increase of E_a at a higher conversion (after $\alpha = 0.4$) can be attributed to the decrease of the mobility of the reactive groups of the partially cured epoxy [30,31]

2.3.3 Effect of vitrification on polymerization rate

The polymerization or curing rate is a function of conversion and temperature and can be defined as

$$r = f(x, T) = k(x, T) f_1(x), \qquad (2.20)$$

where k is the overall rate constant and $f_1(x)$ is a measure of the dependence of curing rate on the concentration of functional groups. Unlike the reaction for the synthesis of small molecules, the polymerization reaction is associated with an increase in the viscosity of the medium. As the viscosity of the medium increases, the rate of diffusion of the molecules decreases and hence there will be a competition between reactivity control and diffusion control reactions. At a critical conversion (α_c), at which the T_g of the network approaches the curing temperature, the reaction becomes totally diffusion controlled. At this stage, conversion from a rubbery gel to glassy gel takes place and the process is called vitrification. Due to a significant decrease in free volume, the rate of diffusion will be significantly reduced at this stage. A semiempirical relation can be used to express the rate constant as given in equation [32,33]

$$K_d = K_r \exp[-D(\alpha - \alpha_c)]. \qquad (2.21)$$

where K_r is the reactivity rate constant, D is a constant, and α_c is the critical extent of curing at which the glassy state is attained. Since switching from reactivity-controlled reaction takes place gradually, the overall rate constant $k(x, T)$ can be expressed using Rabinovitch model [13] in terms of Arrhenius rate constant (k) or reactivity rate constant (k_r), and diffusion rate constant (k_d) as follows [34,35]:

$$\frac{1}{k(\alpha, T)} = \frac{1}{k_r(\alpha, T)} + \frac{1}{k_d(\alpha, T)}. \qquad (2.22)$$

The ratio of $K(\alpha, T)$ to $K_r(\alpha, T)$ is defined as the diffusion factor $f(\alpha)$ for the rate of curing. Hence, by combining Eq. (2.21) and Eq. (2.22) and rearranging, we get

$$f_\alpha = \frac{K(\alpha, T)}{K_r(\alpha, T)} = \frac{1}{1 + \exp[D(\alpha - \alpha_c)]}. \tag{2.23}$$

During the initial stage of curing, α is much smaller than α_c ($\alpha << \alpha_c$), f_α is close to the unity, and the curing is reactivity controlled.

Therefore, in some cases, the catalytic or noncatalytic models discussed above cannot explain the experimental results. Because they have not considered the effect of vitrification. Recently, Stanko and Stommel [36] studied the reaction kinetics of a methylene diphenyl diisocyanate-based fast curing polyurethane resin and presented the mathematical description of the curing process using an isoconversional method (model-free approach). The crosslinking reaction of fast curing PUR resins can be modeled with high accuracy using this method for curing conditions $T_{cure} > T_{g\infty}$. Below the glass transition temperature $T_{g\infty}$, a deviation is observed between the experimentally determined data and the model prediction, which is caused by vitrification.

Hence a generalized autocatalytic and noncatalytic kinetic equation can be expressed as below in Eqs. (2.24) and (2.25), respectively

$$\frac{d\alpha}{dt} = (K_1 + K_2\alpha^m)(1-\alpha)^n \frac{1}{1 + \exp[D(\alpha - \alpha_c)]}, \tag{2.24}$$

$$\frac{d\alpha}{dt} = K_1(1-\alpha)^n \frac{1}{1 + \exp[D(\alpha - \alpha_c)]}. \tag{2.25}$$

Using diffusion cure kinetic model curing of epoxy prepreg has been studied and calibrated the model based on isothermal and dynamic DSC experiment [37]. The degree of cure values calculated from the model as well as experimentally determined from the DSC experiment are presented in Fig. 2.10. The predicted evolution of the degree of cure using the diffusion cure kinetic model at different isothermal temperatures compared closely with the corresponding values measured experimentally using isothermal and dynamic DSC experiment.

The prediction of the degree of cure and the associated heat during the curing of a large polyurethane block (size: $200 \times 200 \times 60$ mm^3) was also successfully done using a standard finite element analysis model [38]. A number of dynamic DSC experiments were conducted and kinetic data were calculated after necessary baseline correction. The data were analyzed with the model-free isoconversional method combined with the compensation effect. The same set of parameters allowed prediction of cure behavior over two orders of magnitude of time and at a curing temperature range from room temperature up to 420K. Agreement

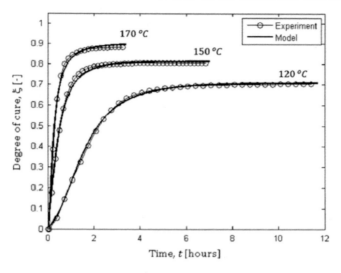

Figure 2.10
Degree of cure evolution with different isothermal temperature. Source: *Reprinted with permission from O.G. Kravchenko, S.G. Kravchenko, R.B. Pipes, Composites, Part A 80 (2016) 72–81, ©2016 Elsevier.*

between the simulations and measurements for the (DSC) was within a relative error of c. 10%. Temperature measurements in the large block were accurate within 3K. The timing of the peak temperatures was predicted within 3 min in the core of the block and 9 min at its edges, while the whole simulation spanned more than 13 h.

2.3.4 Modeling from pressure volume temperature approach

Most of the models on thermoset curing deal with the effect of temperature on the curing reaction. However, pressure is a critical parameter for molding of thermoset resins. While curing, the volume of the resin decreases and this is called a cure shrinkage. The volumetric changes of thermoset resins during the curing process can be described as a combination of thermal effects due to expansion/contraction and chemical effects associated with the chemical shrinkage of the polymer chains [39]. The chemical shrinkage takes place during the formation of the polymer network [40] and results from the movement in the molecule from Van der Waals bonding distance in monomer to covalent bonding distance when the polymer was formed [41]. Therefore the processing of thermoset resin without the application of pressure often leads to the generation of voids [42] as a result of the volumetric changes mentioned above.

Cure shrinkage for various thermosets has been analyzed by using plunger-type instrument (also called pressure volume temperature (PVT) device) or laser scanning dilatometer. Before the curing starts, the molecules of the liquid matrix move freely. At the beginning of the chemical reaction, networks of molecules are formed. After passing the gel point of the matrix, the size of these molecular networks is increased so much that their mobility is reduced. Billotte et al. [24] studied the curing of DGEBA epoxy with an anhydride-based curing agent and observed the beginning of shrinkage at 54% of cure. Thereafter, the linear shrinkage increases proportionally until nearly full cure. Maximum shrinkage was found to be 3 vol.%. Nawab et al. [43] investigated the curing of a vinyl ester resin and reported the apparent shrinkage (the difference between initial thickness and final thickness) in the range of 1.7−2.5 vol.%. It should be noted that apparent shrinkage is always lesser than chemical shrinkage. This is because the fact that thermal expansion coefficients of uncured or partially cured resin in a liquid or gelled state during heating are much higher than the thermal expansion coefficient of solid cured resin during cooling. Due to this difference, one cannot observe the whole effect of chemical shrinkage. The behavior can be modeled by the following equation:

$$l = C(\alpha - \alpha_c), \tag{2.26}$$

where l is the linear shrinkage, α the degree of cure at beginning of shrinkage, and α_c is a constant determined from the experimental results. This linear shrinkage behavior is observed until the polymerization reaches up to a conversion of 90% where the cure reaction is controlled by thermal diffusivity [44].

Kravchenko et al. [37] determined the strain during cure using a carbon-epoxy prepreg (CyCOM 5320−1 manufactured by Cytec Engineering Materials). The measured strains as a function of cure time and degree of cure are presented in Fig. 2.11. We can see a thermal shrinkage of 0.0026 strain beyond the gelation point. After reaching the hold temperature, the resin begins to shrink due to crosslinking reaction. At the end of the hold stage as per the cure cycle, the cure reaction slowed down and the strain value reaches a plateau value (−0.010). This indicates a resulting shrinkage of 0.0186 relative to the gelled state. Therefore it is interesting to study the effect of pressure on curing reactions. Unlike the huge literature dedicated to the study of the effect of temperature on curing, there are only limited studies on the effect of pressure [45−48] on the curing of a thermoset resin. For example, the curing of a vinyl ester at two different pressures was investigated [49] and it was noted that for all reinforcement types and fiber volume fractions, the effect of pressure on chemical shrinkage is significant. The value of shrinkage at higher pressure is higher than the value of shrinkage at low pressure. In this study, two different pressures were maintained for the entire curing process. However, the application of pressure over a short period of time does not affect the amount of heat released during curing and has not changed chemical shrinkage significantly.

If the curing is carried out at a constant temperature then the volume change can be attributed to the cure shrinkage and the conversion in terms of volume can be expressed as [50]

$$\alpha = \frac{v_t - v_1}{v - v_0}, \tag{2.27}$$

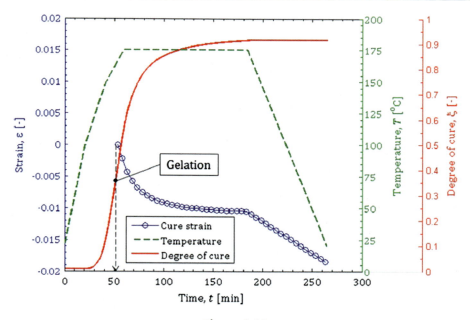

Figure 2.11
Free epoxy strain during the cure cycle. Source: *Reprinted with permission from O.G. Kravchenko, S.G. Kravchenko, R.B. Pipes, Composites, Part A 80 (2016) 72−81, ©2016 Elsevier.*

where v_t and v_1 are the specific volumes at time t and at $t = 0$, respectively and v and v_0 are the final and initial specific volume, respectively.

The dependence of the rate constant on pressure at a constant temperature can be expressed as follows [50]

$$K = K_0 \exp\left(-\frac{\Delta v}{RT} + \beta\right)(P - P_0), \qquad (2.28)$$

where K and K_0 are the rate constants at pressure P and P_0, respectively, Δv is the activation volume, and β is the compressibility factor. At the initial stage of curing when the reaction is reactivity controlled, the increase in pressure increases the probability of the mutual approach of the reactive groups and thereby enhances the rate of reaction. At a later stage of curing, when the reaction is diffusion controlled, the application of pressure further reduces the diffusion leading to the reduction in reaction rate.

2.4 Cure cycle

Thermoset resins are required to be cured at either ambient temperature or high temperature under pressure to produce a cured network with desired thermomechanical properties. In order to achieve high T_g of a cured thermoset, it is necessary to cure the same at a high temperature ($>T_g$). Curing up to an optimum level is required to obtain the desired mechanical property of casting or related composites. Therefore heat is provided for curing reactions. At the same time heat is generated in the system due to the exothermic nature of the curing reaction. Therefore the internal heat generation source no longer remains linear in nature. A nonlinear increase in internal temperature induced by the exothermic chemical reaction of a thermoset resin may result in temperature overshoot leading to the development of a complex gradient in terms of temperature and degree of cure. This affects the uniformity of the curing process. Nonuniform curing can lead to an incomplete cure or resin degradation and entrapped volatiles or voids, which may ultimately cause a reduction in the overall quality and in-service performance of the finished component.

The heat management becomes more critical when we deal with a large amount of resin for fabricating big and thick components having a large volume to surface ratio, since, due to the low heat conductivity of the matrix, the heat cannot be removed easily. When the heat cannot be removed from the inside of the component, temperature increases substantially and the properties of the polymer may degrade, even leading to a fire in the worst case. A complete cure is required to obtain good mechanical properties. Curing also causes shrinkage that can influence the surface quality of the component. Depending on the geometry of the part, thermal and curing shrinkage can also create internal stresses that can weaken the part or lead to cracking. To address the issues involved for curing reactions, thermosets are not cured at a fixed temperature rather they are cured as per a cycle (known as cure cycle) recommended by the manufacturer. The cure cycle has to be decided judiciously for producing precision parts. Good control of the production parameters is required to achieve desired properties of a molded part.

As discussed above the curing behavior changes with the mass of the resin under investigation. The cure reaction of the same resin system studied in DSC and DMA shows different results. Consequently, when we study a large amount of resin as required for bulk production of items, the heat transfer through the sample has also to be considered. In this case source of heat is the supplied heat plus the same generated due to curing reaction (always exothermic in nature). Therefore the heat conduction process has to be modeled considering a nonlinear heat generation source. The thermosetting resins are hardly used as a virgin and various fillers are used to adjust structural and functional requirements. Hence, the role of filler has to be accounted for calculating heat transfer. According to Fourier's heat conduction equation and energy conservation law, the mathematical equation for modeling can be given as below:

$$\rho_r V_r H_r \frac{d\alpha}{dt} + k_{xx}\frac{\partial^2 T}{\partial x^2} + k_{yy}\frac{\partial^2 T}{\partial y^2} + k_{zz}\frac{\partial^2 T}{\partial z^2} = \rho C_p \frac{\partial T}{\partial t}, \qquad (2.29)$$

where k_{xx}, k_{yy}, and k_{zz} are heat transfer coefficients of x, y, and z directions, respectively; ρ and C_p are the density and the specific heat of the cured material; ρ_r, V_r, α, and $d\alpha/dt$ are

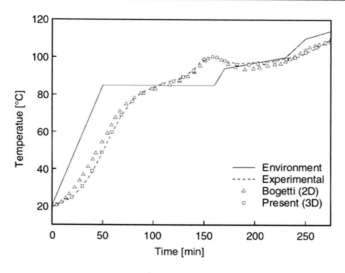

Figure 2.12
Temperature profiles at the center point of 2.54 cm thickness laminate. Source: *Reprinted with permission from J. Zhang, Y.C. Xu, P. Huang, Polym. Express Lett. 9 (2009) 534–541, ©2009 online publication.*

the density, volume fraction, degree of cure, and curing rate of resin; H_r is the exothermic enthalpy from the resin curing process.

A number of the numerical simulations were utilized to predict the temperature rise during cure and the cure simulation was applied to suggest an optimal cure cycle for specific epoxy structures [51–54]. Guo et al. [55] developed a cure model for thermosetting matrix laminates also based on ANSYS and demonstrated that the conventional cure cycles recommended by prepreg manufacturers for thin laminates should not be used as it is for fabrication of the large component to be modified to reduce out-of-plane temperature gradient. The same has to be modified based on the actual situation. Yan [56] reported a two-dimensional finite element model to simulate and analyze the mechanisms pertaining to resin flow, heat transfer, and consolidation of laminates during autoclave processing. Zhang et al. [57] investigated a composite laminate having dimension $15.24 \times 15.24 \times 2.54$ cm^3 and calculated the temperature profile at the center of the laminate using a 3D model. Fig. 2.12 gives the comparison of temperature between experimental and calculated results using 2D and 3D model. It can be concluded that the results calculated using the 3D model are closer to the experimental data than the results obtained from the 2D model [58].

2.5 Thermal property

Polymer exhibit two types of transition [5,59], namely, melting and glass transition. On cooling, crystalline polymer shows crystallization. Melting is a thermodynamically first-order change where the first derivative of free energy (enthalpy) versus temperature plot is discontinuous. Melting is a characteristic of crystalline material. Melting is characterized by a sharp peak in DSC (heat flow vs temperature plot). Thermosets as a whole cannot exhibit melting; however, some crystalline segments may undergo melting for example polyurethanes. Thermoset resins exhibit glass transition, which is thermodynamically a second-order transition at which the second derivative of free energy (volume, heat capacity) versus temperature plot is continuous. The second derivative of free energy (volume, heat capacity) versus temperature plot is discontinuous for melting. The glass transition is the consequence of segmental motion and called α-relaxation. As the temperature increases the free volume increases and at a certain temperature the free volume becomes sufficient to start the segmental motion. The temperature is called glass transition temperature (T_g). Below T_g also some relaxations take place. These are called β- and γ-relaxations.

2.5.1 T_g versus service temperature (T_s)

The maximum temperature at which a material has to work is called its service temperature. It is not necessary that service temperature will be always constant. It may so happen that a material has to work partly at a low temperature and partly at a high temperature. In that case, the higher service temperature has to be considered for assessment of the suitability of a thermoset for that application. The necessary condition for the suitability of a thermoset resin for a particular application is that the resin should be thermally stable at the service temperature. In other words, the T_i should be higher than the service temperature (T_s). This is important particularly at a higher temperature because most of the thermoset resins are stable up to 100°C. The sufficient condition for suitability of a resin is decided on the basis of its T_g value keeping the nature of application in mind. If the application is load-bearing in nature, then the T_g should be well above T_s. Because at T_g the modulus of a polymer decreases drastically and it loses its dimensional stability [5,59]. Once the resin loses its dimensional stability, it cannot be used anymore for load-bearing applications. On the other hand, if the application is for sealant where the resin has to be rubbery in nature, the T_g should be well below T_s. If T_s is less than the T_g of the resin, the material will be hard and brittle at T_s and will not be suitable for sealing applications. For adhesive and coating applications the T_g should be close to the T_s. If T_g is much higher than T_s, the film will be brittle leading to poor adhesion. At the same time, if T_g is very low, the scratch resistance and peel resistance will be very low. Hence, in order to ensure good adhesion and film properties, the T_g of the resin has to be adjusted. For vibration-damping applications, the material should have a broad glass transition covering the T_s. The strategy to design vibration-damping materials will be discussed in subsequent sections.

2.5.2 Effect of cure conversion on T_g

The state of a thermoset resin is determined by its chemical conversion as a result of curing or its extent of cure. From the process control point of view, it is important to know the relationship between the chemical conversion of a thermoset and its glass transition temperature. A unique relationship between the T_g and chemical conversion independent of cure temperature and thermal history has been reported for many thermosets, although it does not hold well for all the thermoset resins. The one-to-one relationship between T_g and the conversion implies that either the network structure does not change with cure temperature or the associated change in network structure does not have any significant effect on T_g. Such relation is found to hold good for epoxy/amine systems. However, the temperature independent relationship is reported to be not valid for epoxy/dicyanamide [60], phenolic/hexamethylene tetramine and unsaturated polyester resins [61]. For the thermosetting resin systems for which the relationship exists, measurement of conversion is equivalent to the measurement of T_g. Both the data can be generated from DSC analysis and compared. T_g can be measured from a more sophisticated instruments like dynamic mechanical thermal analysis (refer to Chapter 6 for detailed information). The T_g measurement can be used as a practical means for the determination of the degree of cure. This is especially useful where the determination of heat of reaction is erroneous due to the loss of crosslinker or for analytical errors.

With the advancement of curing reaction, it is very easy to understand that the T_g of the resin will increase. However, the goal is to quantitatively predict the T_g of a resin as a function of cure conversion. Several models have been proposed to correlate the T_g with the conversion or extent of curing (α). With the increase in conversion, the concentration of reactive functionalities decreases, crosslinks or junction points is formed leading to the departure from Gaussian behavior. The steric hindrance arises on the chain conformation at high crosslink densities. The models basically are based on the statistical description of network formation and calculation of the concentration of junction points of different functionalities as a function of conversion. However, one issue, which complicates the calculation and not fully resolved is that whether to consider all the junction point or only those which are elastically effective.

An equation (known as Dibenedetto equation), which has been successfully applied to correlate the experimental values of T_g as a function of conversion for much thermosetting resin like epoxy, phenolics are given below [62]

$$\frac{T_g - T_{g0}}{T_{g\infty} - T_{g0}} = \frac{\lambda \alpha}{1 - (1-\lambda)\alpha}, \quad (2.30)$$

where T_{g0} is the T_g of the resin mixture before cure, $T_{g\infty}$ is the T_g obtainable after maximum possible curing, and λ is an adjustable parameter. Pascault and Williams [63] derived a similar equation using Couchman's analysis [64] considering the isobaric heat capacity change as a variable as given below:

$$T_g = \frac{\alpha \Delta c_{p\infty} T_{g\infty} + (1-\alpha) \Delta c_{p0} T_{g0}}{\alpha \Delta c_{p\infty} + (1-\alpha) \Delta c_{p0}}, \quad (2.31)$$

where Δc_{p0} and $\Delta c_{p\infty}$ are the change in heat capacity corresponding to T_{g0} and $T_{g\infty}$. Comparing the two equations, we get

$$\lambda = \frac{\Delta c_{p\infty}}{\Delta c_{p0}}. \quad (2.32)$$

During the advancement of the curing reaction, it was observed that the heat capacity change decreases. Montserrat [65] proposed an equation to correlate $\Delta c_p(T_g)$ with the T_g of the network as given below:

$$\Delta c_p(T_g) = x + \frac{b}{T_g}. \quad (2.33)$$

Neglecting the constant (x), which may be applicable for a particular case, we get

$$\lambda = \frac{\Delta c_{p\infty}}{\Delta c_{p0}} = \frac{T_{g0}}{T_{g\infty}}. \quad (2.34)$$

Combining the Eqs. (2.31) and (2.34) and rearranging, we get

$$\frac{1}{T_g} = \frac{(1-\alpha)}{T_{g0}} + \frac{\alpha}{T_{g\infty}}. \quad (2.35)$$

This is an equation similar to the Fox equation, which is widely used to predict T_g of a copolymer as a function of composition. However, this simple rule-of-mixture equation cannot precisely explain the experimental results obtained in actual systems. Hence, a modified equation is proposed [66,67]

$$\frac{1}{T_g} = \frac{(1-\alpha)}{T_{g0}} + \frac{\alpha}{T_{g\infty}} + c\alpha(1-\alpha). \quad (2.36)$$

This equation predicts the increment of the glass transition temperature of cured thermoset at various stages of cure (from monomer stage to network polymer).

The predicted value of T_g from Eq. (2.36) can be validated using the values of instantaneous T_g corresponding to different levels of curing measured by a dynamic DSC experiment. Kravchenko et al. [37] compared the T_g of carbon-epoxy prepreg (CyCOM 5320–1 manufactured by Cytec Engineering Materials) at various degrees of cure predicted from the Dibenedetto equation and the same experimentally measured as shown in

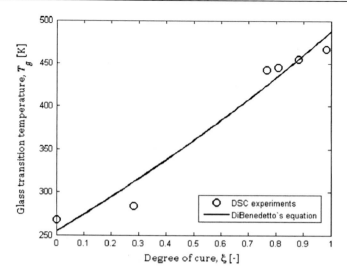

Figure 2.13

Instantaneous glass transition temperature with degree of cure. Source: *Reprinted with permission from O.G. Kravchenko, S.G. Kravchenko, R.B. Pipes, Composites, Part A 80 (2016) 72–81, ©2016 Elsevier.*

Fig. 2.13. The mechanical properties of a network also increase with an increasing degree of cure. Ruiz and Trochu [68] have shown that the evolution of mechanical properties starts just after gelation and increases, following a logarithmic function, until the sample is fully cured. The nonlinear variation of mechanical properties with a log–log function of the degree of cure can be modeled using the following equation [24].

$$\frac{\log K}{\log \alpha} \cong B \quad \log K = B\left[\frac{(1-\alpha_{gel})\log\alpha + 1}{(1-\alpha_{gel})}\right], \tag{2.37}$$

where K is the stiffness of the sample, the degree of cure, B a fitting constant obtained from the experimental results, and α_g is the degree of cure at gelation. The first equation on the left represents the general trend by a log–log function, while the equation on the right is the particular model proposed for the epoxy-anhydride resin system ($B = 11.225$, $\alpha_{gel} = 0.332$).

2.6 Thermomechanical properties

The thermomechanical properties of a cured thermoset network depend primarily on two factors namely chemistry of the resin/curing agent and crosslink density of the network. The thermoset resins can be classified into two categories namely general-purpose resin (e.g., unsaturated polyester, vinyl ester and DGEBA epoxy, phenolics, polyurethane) and high-temperature resins (e.g., phenolics, multifunctional epoxy, benzoxazines, polyimide, bismaleimide, cyanate ester, pthalonitrile, etc.). Thermomechanical properties [69–71] of

Table 2.3: Thermomechanical properties of UPE, vinyl ester, and epoxy.

Resin type	Tensile strength (MPa)	Tensile modulus (GPa)	Elongation %	Flexural strength (MPa)	Flexural modulus (GPa)	HDT (°C)
UPE (general purpose)	55	3.4	2.1	80	3.9	80
UPE (isopthalic)	75	3.2	3.3	130	3.6	90
UPE(BPA-fumarate)	40	2.8	1.4	110	3.4	130
UPE (chlorendic)	20	3.4	1.2	120	3.9	140
Vinyl ester	60	3.5	1.5	130	3.0	150
Epoxy	65	1.5	3.0	140	3.0	150

general-purpose resins namely unsaturated polyester, vinyl ester and epoxy (difunctional) are presented in Table 2.3. They offer similar thermomechanical properties although they have their own advantages and disadvantages as elaborated in Chapter 1. The resins are widely used for various applications, especially for room-temperature applications. They cannot be used for high-temperature applications. The resins which can be used for higher service temperature are termed as high-temperature resins namely phenolic, multifunctional epoxy, benzoxazine, polyimide, bismaleimide, cyanate ester, pthalonitrile, etc. Phenolic resins are also used for general-purpose applications due to their low cost. A comparison of properties of various resins used for high-temperature applications [72] are given in Table 2.4.

The difunctional epoxies are used for general-purpose applications; however, multifunctional epoxies offer high T_g and thermal stability and can compete with high-temperature thermosets like polyimide, cyanate ester, etc. The dynamic mechanical properties of difunctional and multifunctional epoxies are shown in Table 2.5. We can see from the table the T_g and dynamic modulus increase with increasing functionality of epoxy resins. Ratna and Simon [73] reported the free volume of difunctional [diglycidyl ether of bisphenol-A (DGEBA)] and tetrafunctional [Tetraglycidylether of 4,4' diaminodiphenyl methane (TGDDM)] epoxy resins using positron annihilation lifetime spectroscopy (PALS) to correlate the thermomechanical properties with molecular structure. The details of PALS are available in Chapter 6. The orthopositronium (o-Ps) lifetime (τ_3) value of DGEBA (1.76 ± 0.01 ns) is found to be greater than that of TGDDM (1.69 ± 0.01 ns). The data clearly showed a decrease in free volume as a result of the increase in crosslink density due to the increase in functionality. Because of lower free volume, multifunctional epoxy networks require higher temperature compared to the difunctional epoxy network to attain the level of the free volume necessary to start segmental motion. This explains the higher T_g of a multifunctional epoxy-based network compared to difunctional epoxy-based one.

Table 2.4: Comparative properties of various high-performance polymers.

Property	Epoxy	Phenolics	Toughened BMI	Bisox-phen (40–60)	Cyanate ester	P–T resin	PBZ
Density (g cc^{-1})	1.2–1.25	1.24–1.32	1.2–1.3	1.3	1.1–1.35	1.25	1.19
Max use temperature (°C)	180	200	~200	250	150–200	300–350	130–280
Tensile strength (MPa)	90–120	24–45	50–90	91	70–130	42	100–125
Tensile modulus (GPa)	3.1–3.8	03/05	3.5–4.5	4.6–5.1	3.1–3.4	4.1	3.8–4.5
Elongation (%)	3–4.3	0.3	3	1.8	02/04	2	2.3–2.9
Dielectric constant (1 MHz)	3.8–4.5	04/10	3.4–3.7	–	2.7–3.0	3.1	3–3.5
Cure temperature (°C)	RT-180	150–190	220–300	175–225	180–250	177–316	160–220
Cure shrinkage (%)	>3	0.002	0.007	<1	~3	~3	~0
TGA onset (°C)	260–340	300–360	360–400	370–390	400–420	410–450	380–400
T_g (°C)	150–220	170	230–380	160–295	250–270	300–400	170–340
G_{ic} (J m^{-2})	54–100	–	160–250	157–223	–	–	168
K_{ic} (MPa m$^{1/2}$)	0.6	–	0.85	–	–	–	0/94

Source: Reprinted with permission from C.P.R. Nair. Prog. Polym. Sci. 29 (2004) 401–498, ©2004 Elsevier Publishers.

Table 2.5: Viscoelastic properties of difunctional and multifunctional epoxies cured with DETDA.

Epoxy resin	E' at RT (GPa)	T_g^a (°C)	Tanδ peak temperature (°C)	Tanδ_{max}
DGEBA	1.2	195	217	0.6
TGAP	1.7	260	295	0.3
TGDDM	2.1	230	260	0.7

^{a}Temperature corresponds to the onset of modulus drop.

The T_g of an epoxy network can be varied to a great extent by changing the hardener. The T_g values for DGEBA epoxy cured with different curing agents are shown in Table 2.6, which clearly demonstrated that a tremendous effect of hardener on T_g. The adhesive property also depends on the chemistry of the hardener. The lap shear strength values of a few room-temperature curing adhesive formulations are given in Table 2.7. It is evident from the table that the adhesive strength of the same epoxy resin can be improved significantly by using polar curing agents [74]. A polyamide-based epoxy system offers better adhesive properties compared to an amine-based hardener. The adhesive strength can be further improved by various methods related to toughening, which will be discussed in the Chapter 3.

Like epoxy resins, polyurethanes also offer a wide scope for tailor-making of thermomechanical properties by changing the chemistry of the resin. Because of the

Table 2.6: Glass transition temperature (T_g) of DGEBA epoxy cured with various curing agents.

Curing agent	T_g (°C)
TETA	130
DDM	190
DDS	189
DETDA	217
Dicyanamide	139
Anhydrides	160–200
Jeffamine-230	85
Jeffamine-400	45
Jeffamine-600	25
Jeffamine-900	0
Jeffermine-2000	−30

Table 2.7: Bond strength data for epoxy resin and various hardener systems.

Adhesive system	Lap shear strength (MPa)
DGEBA/TETA	3
DGEBA/hexamethylene diamine	6
DGEBA/HY 848	7
DGEBA/HV 953u	9

HY 848 & HV 953u (Vantico) are polyamide-based hardeners.

microphase-separated nature of many polyurethanes, the volume fraction of hard domains is an important descriptor of the polymer morphology and has a large impact on polyurethane material properties [75]. However, its value is dependent not only on the hard segment content as computed from formulation but also on the degree of phase separation and therefore is difficult to determine from the formulation. The hard segment chain length can be experimentally determined by matrix-assisted laser desorption-ionization (MALDI) mass spectrometric analysis [76] and validated by Monte Carlo simulation.

Corcuera et al. [77] investigated 4,4′-diphenylmethane diisocyanate (MDI) and 1,6-hexamethylene diisocyanate (HDI)-based PU systems with varying hard segment (HS) content. The HS was varied by changing the ratio of polyol, disocyanate, and butanediol. For example, when polyol, MDI, and BD are reacted in the ratio of 1:3:2, a network with 27% HS is formed whereas % HS content is increased to 59% when the reactants are taken in the ratio of 1:11:10. The stress–strain plot of four PU systems (PU-MDI-1, PU-MDI-2, PU-MDI-3, and PU-MDI-4) with varying HS contents (27%, 39%, 48%, and 59%, respectively) are shown in Fig. 2.14. As observed in the figure, the modulus of elasticity and tensile strength increased, but the elongation at break decreased with increasing HS content of the PU network. The MDI-based PU with 59% HS content exhibits a tensile strength of 23.8 MPa, which is about 3 times higher than HDI-based one with 50% HS content (8.2 MPa). MDI-based PUs also presented a higher ductility, lower modulus than

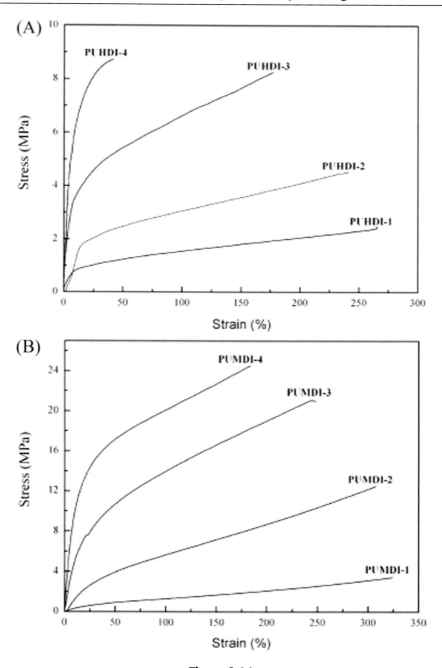

Figure 2.14
Stress–strain curves of the (A) PUHDI and (B) PUMDI systems with various HS contents. Source: Reprinted with permission from M.A. Corcuera, L. Rueda, A. Saralegui, M.D. Martín, B.F. Arlas, I. Mondragon, et al., J. Appl. Polym. Sci. 122 (2011) 3677–3685, ©2011 John Wiley and Sons Publishers.

the HDI-based PUs for the same HS content. This was related to the differences in the phase morphology. The much higher tensile strength of MDI-based PUs should have been due to the less phase-separated structure, which could have given place to an orientation of the hard domains during the strain. In the case of the HDI-based PUs, the highly crystalline structure of hard domains must have had a hindrance effect over the whole PU network and prevented the hard domains from dispersing into soft phases to give a strain-induced crystallization phenomenon [78].

2.7 Rheological properties

The processing of thermosets and thermoset matrix composites is more complicated and controlled to a lesser extent compared to thermoplastics. This is due to the chemical reactivity of thermoset resins. In these processes, shaping and network formation takes place in a single operation, which involves the conversion of liquid monomers or prepolymers into a solid crosslinked polymer. The mechanism and kinetics of cure determine the network morphology, which, in turn, dictates the physical and mechanical properties of the cured product. Thus understanding the cure kinetics of thermosets is very important for process development and quality control of the related products. An important property of polymeric materials in regard to their processing behavior is the rheological properties. Viscosity control during the processing of thermosets is particularly critical because the viscosity varies not only with changing temperature and flow conditions but also with time as a result of polymerization reactions involved for curing of the thermosets under investigation. Therefore, in order to control the curing effectively and to optimize the processing schedules as well as to achieve desired properties of the finished products, it is necessary to understand the relationship between the curing kinetics and rheological behavior. The study of the deformation properties of reactive polymer systems is known as "chemorheology." Much work has been done to determine the appropriate models that best describe the cure kinetics and chemoviscosity of thermosets [11–19]. The autocatalytic and Kamal's phenomenological models have already been discussed in Section 2.3.1. Because of the complex nature of thermosetting reactions, phenomenological models are the most popular for these systems. The chemorheology of thermoset polymers (a subset of reactive polymers) is often difficult to characterize as there are complex effects of temperature on the material. For example, at lower temperatures, the viscosity is high due to the lack of thermal mobility of chains, but as temperatures increase, the viscosity will drop due to this increase in thermal motion. However, eventually, the higher temperatures will increase the reaction and cure of the material and viscosity will increase again. The chemorheology model not only considers batch-specific viscosity versus time or temperature correlations but also includes kinetic and/or structure information [79].

It is clear that chemorheology is a very important tool for thermoset materials—both on a fundamental level in characterizing the fundamental flow behavior of the curing materials, and on a practical level, for industry via the use of chemorheological models in flow simulation tools to better design, predict, and optimize thermoset material processing. As the tolerances for products, and hence processes, tighten due to increased quality requirements, miniaturization, and use of thermoset products in extreme environments, the need for chemorheology and thermoset flow simulation as a tool for more controlled design and processing is evident.

The term chemorheology arises from the two areas of concern being studied. The first concern comes early in the cure. The processability or flow (rheology) of the polymer system is of primary importance because process changes such as heating rates hold temperatures, or pressures normally are made during this early period. The other area of concern is the chemistry of the process: the rate of reaction, the mechanisms, the kinetics, and the cessation of the chemical reaction or the end of the cure. The interactions between these two concerns are obvious because the rate of rheological change in either the liquid or solid-state and the development of the optimum glass transition temperatures cannot be separated from the chemical contribution to these effects.

Rheological testing provides a precise means for measuring intrinsic mechanical material properties that can be related to a material's processing characteristics and performance. The rheological tests for thermosetting resins are carried out in a parallel plate rheometer. The resin system is placed in between the plate and the strain is applied to it. The three rheological test modes, steady, dynamic, and transient are distinguished by the manner in which the strain is applied to the sample. A steady test uses continuous rotation to apply the strain and provide a constant shear rate. The resultant stress is then measured when the sample reaches a steady state. In a dynamic test, an oscillatory strain is applied to a sample and the resulting stress is measured. Dynamic tests can be made using free oscillations at the resonance frequency of the test material (e.g., the torsion pendulum), or with sinusoidal (or other wave forms) oscillations at a forced frequency chosen from a wide available range. In a transient test, the response of a material as a function of time is measured after subjecting the material to an instantaneous change in strain, strain rate, or stress.

Ratna et al. [80] compared the chemorheology of diglycidylether of bisphenol-A (DGEBA), triglycidyl p-amino phenol (TGAP), tetraglycidylether of 4,4′-diaminodiphenyl methane (TGDDM) cured with diethyltoluene diamine (DETDA). Rheological measurements carried out with a controlled stress rheometer using a parallel plate assembly permitted the characterization of gelation and vitrification process during curing. The evolution of loss modulus (G'') for DGEBA, TGAP, and TGDDM against the curing time at 140°C is shown in Fig. 2.15. The point at which a clear increase in G'' occurs is defined as a gel point and G'' maximum is defined as a vitrification point. It is evident from the figure that TGAP and TGDDM show both gelation and vitrification (loss modulus maximum) whereas in the case of DGEBA the vitrification is not clear. This supports the work by Gilham and

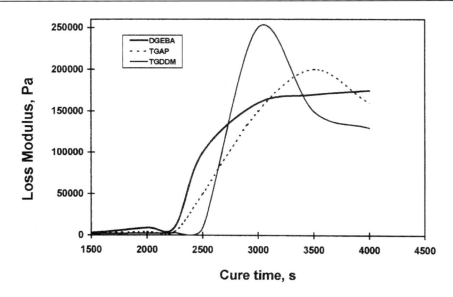

Figure 2.15
Evolution of loss modulus versus cure time 140°C (measured using parallel plate rheometer).
Source: *Reprinted with permission from D. Ratna, R. Varley, G.P. Simon, J. Appl. Polym. Sci. 92 (2004) 1604, ©2004 John Wiley and Sons Publishers.*

coworker [81] using torsional brained analysis, which showed that vitrification is not necessarily observed as it can depend upon the cure temperature and reaction kinetics. They found that at higher cure temperatures the time to vitrify increases relative to gelation while the loss modulus peak becomes less intense until it reaches a point where it is so diffuse that it is not readily observed. Conversely, Varlay et al. [82] have shown using the highly crosslinked systems such as TGAP/DDS, during flexural brand analysis, vitrification always occurs soon after gelation with no evidence of decreasing any intensity.

Gel point conversion is inversely proportional to the functionality of a monomer [83]. Hence, the gel time is expected to decrease from DGEBA to TGDDM with increasing functionality of the resin. However, a reverse trend was observed by reported by Ratna et al. [80] in their comparative study. The tetrafunctional resin (TGDDM) gels at a longer time compared to difunctional resin (DGEBA). This can be explained by considering the fact that the epoxy resins are basically prepolymer and their initial viscosity and increase in viscosity during curing plays an important role in controlling the rate of curing reaction. An increase in viscosity of the medium retards the chemical reaction attributed to the restricted molecular mobility. The increase in viscosity during cure is higher in the case of resin with higher functionality. This explains why TGDDM gels later than TGAP and DGEBA. Beyond gelation, the system becomes highly viscous and then functionality rather than viscosity controls the curing, and thus TGDDM vitrifies sooner than does TGAP (trifunctional epoxy). The maximum loss modulus value of TGDDM was found to be higher than that of TGAP. This can be explained in terms of the chemical structures of TGAP and TGDDM and their functionality. The molecular weight between crosslinks (M_c) for the TGAP network is lower than that of the TGDDM network. Therefore TGDDM generates a tighter network with a higher modulus than that of TGAP.

The gelation and vitrification are more clearly determined from a loss factor plot as shown in Fig. 2.16 for a trifunctional epoxy (TGAP/DETDA) system [84]. This experiment clearly demonstrates the time required for curing an epoxy system at a particular temperature to get a required T_g can be quantitatively estimated. This is very important for the development of prepregs. It may be noted that when the T_g of the cured system reaches the cure temperature, the curing reaction becomes very slow. That is why when a multifunctional epoxy network, without a postcuring treatment at a sufficiently high temperature, is subjected to dynamic mechanical analysis, it shows two-loss peaks; one is for the partially cured network and the other is for the fully cured network. Hence, it is very important to give a proper postcuring treatment to an epoxy system (especially multifunctional epoxy system) to get a network with desired T_g.

The different transformations, observed during the cure of epoxy with a stoichiometric amount of curing agent, are generally recorded in the form of a time–temperature–transformation (TTT) diagram [85,86]. Such a diagram is built from the experimental determinations of the time to gel (t_{gel}) and the time to vitrify (t_{vit}) under isothermal

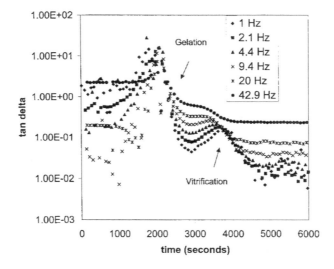

Figure 2.16
Loss factor versus temperature plot at various frequencies (measured using parallel plate rheometer) showing gelation and vitrification. Source: *Reprinted with permission from D. Ratna, R. Varley, G.P. Simon, J. Appl. Polym. Sci. 92 (2004) 1604, ©2004 John Wiley and Sons Publishers.*

conditions. As the reaction rate varies exponentially with temperature, times are represented in a logarithmic scale. It also includes a char region that is reached after degradation at high temperatures. In a general curing procedure, the prepolymer system is employed in the liquid state. The curing temperature should range between $T_{g,\text{gel}}$, the glass transition temperature of the gelled polymer, and $T_{g,\infty}$, the glass transition temperature of the fully cured thermoset. In the case of an isothermal cure, the ongoing reaction can be followed on a horizontal line in the TTT diagram. A TTT diagram for an epoxy novolac cured with a stoichiometric amount of DDS as reported by Oyanguren and Williams [87] is presented in Fig. 2.17. The temperature at which the vitrification and gelation take place simultaneously is 65°C. Since the reaction becomes diffusion controlled at vitrification, the epoxy systems must be cured at a temperature higher than 215°C, which is the T_g of the fully cured epoxy network. It is necessary to consider the fact that at a high temperature there is a possibility for degradation. In order to avoid such degradation, the postcuring temperature has to be set below the degradation temperature because degradation will cause a reduction in thermomechanical properties. The degradation temperature is chosen arbitrarily at 270°C for the present system. However, it may be noted that the degradation depends on the exposure time.

The rheological property of an anhydride cure epoxy system [24] determined by a parallel plate rheometer is shown in Fig. 2.18. We can see that before curing starts, the loss modulus is very low. With the advancement of curing, the loss modulus increases due to an

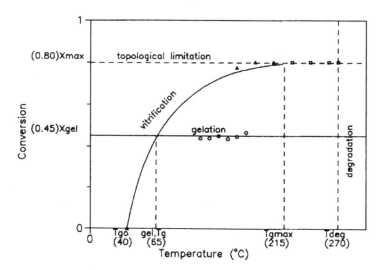

Figure 2.17
Conversion versus temperature transformation diagram showing conditions where gelation, vitrification and chemical degradation take place, as well as restriction arising from topological limitations for novolac epoxy resin cured with DDS. Source: *Reprinted with permission from P.A. Oyanguren, R.J.J. Williams, J. Appl. Polym. Sci. 47 (1993) 1361, ©1993 John Wiley and Sons Publishers.*

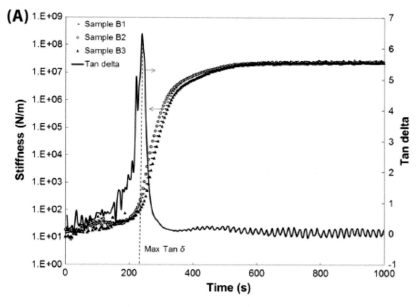
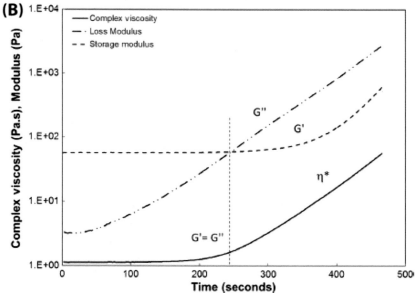

Figure 2.18
Evolution of stiffness and rheology properties with time during isothermal cure of epoxy-anhydride resin at 120°C: (A) evolution of stiffness and Tan delta for samples of series B during DMA experiment and (B) evolution of complex viscosity and storage and loss modulus during rheology SAOS experiment (1% strain amplitude, 50 rad s^{-1} frequency). Source: *Reprinted with permission from C. Billotte, F.M. Bernard, Edu Ruiz, Eur. Polym. J. 49 (2013) 3548–3560, ©2013 Elsevier.*

increase in molecular weight of the resin as a result of a chemical reaction. At one point, the loss modulus curve intersects the storage modulus versus time plot. The intersection point is considered as gel point and corresponds to 265 s. The point is also associated with a steep increase in complex viscosity. This is close to the value determined 230 s determined by DMA from loss factor versus time plot. The time that corresponds to maximum damping is considered as gel time.

2.8 Liquid crystalline thermoset

Liquid crystals exhibit a partially ordered state (anisotropic), which falls in between the completely ordered solid state and completely disordered liquid state. It is sometimes referred to as the "fourth state of matter." In recent years, the interest in liquid crystalline thermosets especially liquid crystalline epoxy has increased tremendously [88–99]. If the LC epoxy is cured in their mesophase, the liquid crystalline superstructure is fixed permanently in the polymer network even at a higher temperature. LC epoxies are prepared either by using LC monomer [88–93] or by chemical modification of epoxy resin [98], which incorporates LC unit in the epoxy structure. Liquid crystalline epoxy resins with different types of mesogen such as benzaldehyde azine [88], binaphthyl ether [89,90], phenyl ester [91,92], and azomethine ethers [93,94] were reported in the literature. Depending on the chemical nature of the mesogen, the related epoxies display a wide range of thermomechanical properties. The resins can be cured chemically with acid or amine [95,96] or by photochemical curing in presence of a photoinitiator. Broer et al. [97] demonstrated the fabrication of uniaxially oriented nematic networks from diepoxy monomer in presence of a photoinitiator.

Su [99] synthesized and studied the main chain thermotropic liquid crystalline thermosets based on biphenyl mesogen and reported that the LC epoxy networks exhibit higher T_g and dielectric strength and lower thermal expansion compared to the conventional epoxy networks. Recently, LC epoxy (monomers and polymers) containing photoresponsive moieties (e.g., azobenzene, stilbene, and chalcone) were extensively investigated [100–102] due to their interesting photochemical and photo physical properties, which may lead to a wide range of applications. Chemically or photo chemically crosslinkable epoxy groups and photo labile $-N=N-$ or $=C$ units may be utilized for the fabrication of photo-aligned liquid crystalline recordable layers without the aid of the electric or magnetic field.

LC epoxies bearing stilbene were first reported by Calundann et al. [100]. Ober et al. [101,102] reported the synthesis and characterization of liquid crystalline thermosets based on dihydroxy stilbene. Depending upon molecular weight and polydispersity of the starting material, the polymers showed either nematic or smectic mesophase. Carfagna et al. [88,89,103] reported the synthesis of epoxy-terminated stilbene monomer, which on curing with an amine curing agent generates smectic and nematic mesophases. They demonstrated the use of these materials in macroscopically oriented optical films [103]. When an electromagnetic field is applied to a LC epoxy sample in the uncured state, domains got oriented along a preferred direction. Subsequent crosslinking reaction locks the molecules in the aligned conformation achieved under the influence of external field and the structure is stabilized at the gel point. The alignment of the rigid molecular backbone remains then stable over a wide range of temperatures.

Very interestingly, Choi and Cha [104] reported the synthesis of photo-crosslinkable liquid crystalline chalcone epoxy compounds for fabrication of photo-aligned liquid crystalline layers. Photodimerization and photopolymerization of compounds under linearly polarized UV light in presence of cationic photo initiator were also reported. Koschienlny et al. [105] reported the synthesis of liquid crystalline azobenzene epoxy polymer, which has shown nematic mesophase having a narrow temperature range. Castell et al. [106] reported the synthesis of epoxy liquid crystalline monomers with azo groups in the central mesogenic core. They have developed anisotropic networks of the resulting polymers by irradiation with linearly polarized light without magnetic or electric orientation by utilizing photochemical cis-trans isomerization of azobenzene moieties.

Ratna and coworker [98] synthesized LC epoxy by chemical modification of epoxy resins. Terephthaloyl chloride was reacted with 4-hydroxy benzoic acid to get terephthaloyl bis (4-oxybenzoic) acid (PBA) which was characterized and further reacted with epoxy resin [diglycidyl ether of bisphenol-A (DGEBA)] to get a liquid crystalline epoxy resin (LCEP). The reaction scheme is shown in Fig. 2.19. The LC epoxy is characterized by an optical microcopy with hot-stage facility. The formation of smectic phase can be seen from the microphotograph, shown in Fig. 2.20. The isotropization takes place at 130°C. LCEP can be cured in the similar way as commercially available DGEBA epoxy. The LCEP has been blended in various compositions with DGEBA and cured by using a room-temperature curing hardener. The cured blends were found to show higher storage moduli and lowered glass transition temperatures (tanδ_{max}, from DMA) compared to that of the pure DGEBA network. A liquid crystalline thermoset was reported [107] based on 1,3,4-oxadiazole containing epoxy resin and diaminodiphenyl methane (DDM). The LC thermoset offers higher T_g and better fracture resistance compared to conventional thermoset indicating their potential application for advanced composites.

A family of low melting all-aromatic ester-based liquid crystal oligomers end-capped with reactive phenylethynyl end groups was reported by Iqbal and coworker [108]. The reactive end groups can be activated at high temperature and are capable of undergoing chain extension and subsequent curing at a temperature above 300°C. All the oligomers display nematic mesophases and in most cases, the nematic order is maintained after cure. It was demonstrated that the cured LC thermoset exhibits glass transition temperatures in the range of 164°C–275°C, and high storage moduli (~5 GPa at ambient temperature and ~2 GPa

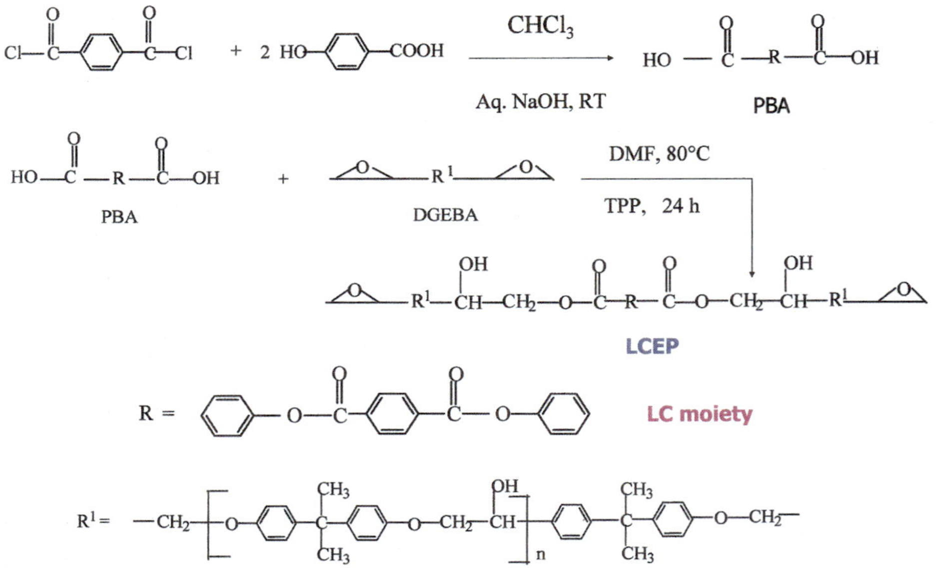

Figure 2.19
Reaction scheme for synthesis of liquid crystalline epoxy by chemical modification of DGEBA resin. Source: *Reprinted with permission from K. Sadagopan, D. Ratna, A.B. Samui, Polym. Sci. Part. A: Polym. Chem. 41 (2003) 3375, ©2003 John Wiley and Sons Publishers.*

Figure 2.20
Optical microscopy photograph showing the smectic texture of liquid crystalline epoxy. Source: *Reprinted with permission from K. Sadagopan, D. Ratna, A.B. Samui, Polym. Sci. Part. A: Polym. Chem. 41 (2003) 3375, ©2003 John Wiley and Sons Publishers.*

at 200°C). This indicates their superior performance compared to commercial engineering thermoplastic like polyetherether ketone.

2.9 Elastomeric thermoset

The commercially available thermoset resins (except polyurethanes) are glassy materials and used extensively for various applications as discussed elaborately in Chapter 1. Although epoxy resins cured with conventional amine are glassy in nature with high T_g, the T_g can be reduced by using ether containing long-chain hardener and rubbery epoxy network can be made, as demonstrated by Ratna and coworkers [109]. Loss factor versus temperature plots of epoxy networks generated by curing DGEBA epoxy with polyether amines having different chain lengths are presented in Fig. 2.21. The corresponding thermomechanical properties are presented in Table 2.8. By varying the molecular weight of the polyether segment, a broad range of properties can be obtained. As expected, the T_g of the network decreases with increase in molecular weight of the polyether amine used. The decrease in the T_g is associated with a decrease in tensile strength and an increase in elongation at break. This behavior is explained by the effect of two mechanisms: the reduction of crosslink density and the increase in the length of polyether segments. A rubbery epoxy network can therefore be made by using a polyether amine of suitable chain length. The rotational energy of O—O bond is much lower than the C—C bond; therefore

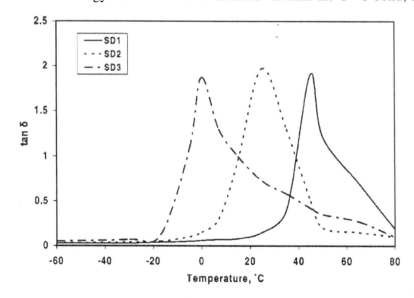

Figure 2.21
Plot of loss tangent versus temperature plots of epoxy networks cured with polyetheramine (Jeffemine) of various chain length: SD-1 (Jeff-300), SD-2 (Jeff-500), and SD-3 (Jeff-800). Source: Reprinted with permission from D. Ratna, N.R. Manoj, L. Chandrasekhar, B.C. Chakraborty, Polym. Adv. Technol. 15 (2004) 583, ©2004 John Wiley and Sons Publishers.

Table 2.8: Mechanical properties of epoxy networks cured with polyetheramine of various chain length.

Epoxy network	Mol. wt. of polyether chain of Jeffamine (g mol^{-1})	Tensile Strength (MPa)	Elongation at break (%)
SD 1	300	14.8	50
SD 2	500	2.4	75
SD 3	800	0.5	210

Source: Reprinted with permission from D. Ratna, N.R. Manoj, L. Chandrasekhar, B.C. Chakraborty, Polym. Adv. Technol. 15 (2004) 583, ©2003 John Wiley and Sons Publishers.

the presence of ether linkages enhances the flexibility. In addition, the crosslink density decreases with increasing molecular weight of the polyether amine hardener. The decrease in crosslink density also contribute toward imparting flexibilizing effect.

2.9.1 Application of rubbery thermoset

The epoxy and PU-based rubbery thermoset can be used as a substitute for conventional rubber. Unlike conventional rubber, the elastomeric thermosets do not contain unsaturation and therefore offers better thermal stability and life expectancy compared to rubbers. The evaluation methods of thermosets for various applications and lifetime estimation will be discussed in Chapter 6. The elastomeric thermosets are widely used as sealant, putty, adhesive, and coating for various civil and defense applications. Since thermosets are low molecular weight liquid, they can be used directly for such application without any solvent. On the other hand, the rubbers (because of their higher molecular weight) may require solvent for applications in above-mentioned use. One of the most important applications of elastomeric thermoset is in the field of vibration damping, which will be elaborated on here.

Vibration and its efficient control are a universal concern for product designers considering its occurrence in most machines, structures, and dynamic systems. Vibration often leads to undesirable consequences such as unpleasant motions, noise, and dynamic stresses that lead to fatigue and failure of structures, decreased reliability, and degraded performance. The vibration produced in ships and submarine, radiate as a noise in sea, which can be detected by the sound navigation system of the enemy. Hence the vibration-damping application is relevant to both the civil and defense sectors. Vibration damping is achieved by using rubber or elastomeric thermosets due to their viscoelastic nature. In order to understand the vibration-damping characteristics of such materials, it is necessary to understand the fundamental of vibration and viscoelastic behavior of polymeric materials. The dynamic viscoelasticity is briefly discussed here. The readers may refer Author's book on "polymers for vibration-damping application" [110] for fundamental understanding and applications vibration-damping polymers.

2.9.1.1 Dynamic viscoelasticity

Polymers are termed as "viscoelastic" because they are neither perfectly elastic solids nor viscous liquids, but behave as a combination of these two. Under an external force, a polymer experiences an instant strain accounting for elastic response and a time dependent strain accounting for viscous response. The elastic stress is directly proportional to the strain whereas the stress applied on viscous material is proportional to the viscous strain rate (time derivative of strain). Hence, a physical interpretation of a polymer can be either a summation of two strains or summation of two stresses as described by Maxwell's model and by Kelvin–Voigt (KV) model. In these basic models, the elastic part is represented by a spring, having a spring constant independent of time, and a dashpot which develops strain in progression with time at constant stress, so that the strain rate is directly proportional to the stress. In Maxwell's model, the spring and the dashpot is arranged in series, so that the Maxwell body behaves as a viscoelastic fluid and in KV model, the spring and dashpot is arranged in parallel, so that the KV body behaves as a viscoelastic solid. The basic Maxwell model describes the stress relaxation and defines relaxation time of a polymer while the basic KV model describes the creep phenomenon and defines retardation time of a polymer. Stress relaxation arises from the tendency of the segments to minimize the instantaneous stress when put at a constant strain as the segments can move. Retardation arises from the tendency of the molecule to deform continuously under a constant stress as the segments can move. Obviously, these phenomena can only be exhibited by a polymer above its glass transition temperature, since in the glassy state, the segmental mobility is frozen. However, neither of these models truly represents a real polymer. There are many models developed on these basic models in different ways of combinations to obtain more near feature of a real polymer. Three parameter Zenner model and four-parameter Berger model are among the mostly used ones. There are, in addition, semiempirical models too for describing the nearly accurate viscoelastic behavior of polymers. Considering dynamic loading on a polymer under shock or vibration, such dynamic models yields some useful expressions of dynamic modulus as the elastic modulus and loss modulus as the part of lost energy and the ratio of loss to elastic part as a loss factor with respect to frequency:

$$\text{Dynamic or Storage Modulus: } E' = \frac{E_0 + \omega^2 \tau^2 E_\infty}{1 + \omega^2 \tau^2}, \quad (2.38)$$

$$\text{Loss Modulus: } E'' = \frac{\omega \tau (E_\infty - E_0)}{1 + \omega^2 \tau^2}, \quad (2.39)$$

$$\text{And Loss Factor: } \tan\delta = \frac{(E_\infty - E_0)\omega\tau}{E_0 + E_\infty \omega^2 \tau^2}, \quad (2.40)$$

where E_0 is elastic modulus in the rubbery region ($\omega \to 0$), E_∞ is elastic modulus in glassy region ($\omega \to \infty$), ω is angular frequency, and τ is relaxation time. The relaxation time is defined as the time required for a molecule to reduce the stress to (1/e) of the original (instantaneous) stress at a constant strain. The dynamic properties are dependent on temperature as the relaxation time follows an Arrhenius type relationship with temperature (T) as:

$$\tau = \tau_0 \exp\left(\frac{E_a}{RT}\right), \tag{2.41}$$

where τ_0 is a constant, E_a is the activation energy of relaxation process, R is universal gas constant, and T is the temperature in K. On substituting Eq. (2.41) in Eqs. (2.38), (2.39), and (2.40), the temperature dependence of the dynamic properties can be obtained at any constant frequency.

The effect of temperature is exactly the opposite to the effect of frequency on the dynamic properties. The dynamic or storage modulus grows with increasing frequency at a constant temperature but diminishes with increasing temperature at constant frequency. The temperature and frequency (inverse of time) are related mathematically by Eq. (2.41) as well as by an expression by Williams, Landel, and Ferry (WLF) [111]. The expression defines a shift factor, which is used to relate a set of frequency and temperature to another set for same value of the dynamic viscoelastic properties. The expression assumes that the free volume of a polymer is constant below the glass transition temperature and increases linearly above this temperature so that the volume expansion coefficient is constant till any melting or flow takes place. Therefore the WLF equation is valid for temperature above the glass transition. The equation is expressed as:

$$\log(a_T) = \log\left(\frac{\tau}{\tau_{ref}}\right) = \frac{-C_1(T - T_{ref})}{C_2 + (T - T_{ref})}, \tag{2.42}$$

where τ denotes relaxation time and C_1, C_2 are constants. The values of these constants are: $C_1 = 17.44$ and $C_2 = 51.6$ when the reference temperature $T_{ref} = T_g$ and $C_1 = 8.86$ and $C_2 = 101.6$, respectively at $T_{ref} = T_g + 50$. The Arrhenius equation (Eq. 2.41) can also be used to calculate the shift factor as follows:

$$\log(a_T) = \log\left(\frac{\tau}{\tau_{ref}}\right) = \frac{E_a}{R}\left(\frac{1}{T} - \frac{1}{T_{ref}}\right). \tag{2.43}$$

The above Eqs. (2.41) and (2.43) are also valid when the species are in random motion, which is possible above the glass transition.

2.9.1.2 Vibration-damping behavior

The vibration-damping property of thermoset originated from viscoelasticity as discussed above can be explained by considering the segmental motion. The polymer segments, which consist of about 10–40 carbon atoms, undergo movement keeping end to end distance unaltered, just like snakes move keeping head to tails distance unchanged. Due to localized movement, the segments behave as a viscous liquid, whereas the molecule as a whole remains solid because mobility is forbidden. Therefore, under the influence of an applied stress, it exhibits properties of a viscous fluid as well as an elastic solid and undergoes viscoelastic deformation. Due to internal friction of the segments, energy is dissipated as heat leading to the manifestation of the vibration-damping effect. In Fig. 2.21, we can see that loss factor versus temperature plots for all the three epoxy systems passes through a maxima. At a temperature well below the peak temperature damping is very low, the segments remain in a frozen condition. When the temperature is well above the peak temperature, the damping is low as well because the molecular segments are free to move and consequently there is little resistance to flow. In the transition region, on the other hand, damping is high because of the initiation of micro-Brownian motion of the molecular chain segments and their stress relaxation. When the segments start moving, the other segments resist the same and energy is dissipated due to internal friction. Since a viscoelastic material shows a maximum damping in the glass transition region, the damping material has to be developed in such a way that the transition region covers the use temperature. As discussed in Section 2.9, the epoxy/polyetheramine offers the possibility to tailor the glass transition region as a function of molecular weight of polyetheramine. Hence they can be effectively used as vibration-damping materials. The epoxy-based vibration-damping materials offers several technological advantages compared to rubber compound-based damping materials such as high damping, easier application. Because of self-adhering in nature, they do not need any extra adhesive for application.

The time−temperature superposition as discussed above is used to asses a vibration-damping material since it assists in understanding the damping behavior at very low and very high frequencies, which are normally unavailable in machines. Fig. 2.22 shows the master curves of an epoxy-based damping material at three different reference temperatures of 0°C, 25°C, and 50°C. It is clearly seen that the peaks corresponding to the glass transition temperature shift gradually to the lower frequencies as the reference temperature decreases (from 100 Hz at a reference temperature of 50°C to 10^{-7} Hz at 0°C). Thus a material effective in damping at 50°C may not work at 0°C or below. For example, at 0°C, the present damping materials will have almost no damping at normal frequencies of interest (20−20,000 Hz). The frequency dependences of three damping material with varying polyether chains (SD-1-Jeffemine-800. SD-2-Jeffemine 500, SD-3-Jeffemine 300) are presented in Fig. 2.23. It is evident that with increase in polyether chain length, the frequency corresponding to maximum damping, increases. It is also necessary to consider the mode of application of the material as discussed below.

Polymers are commonly used in two basically different configurations to dissipate the mechanical energy of vibration as heat [112]. The first one is the free-layer damping (FLD) where energy is dissipated due to direct strains, that is, alternate extension and compression

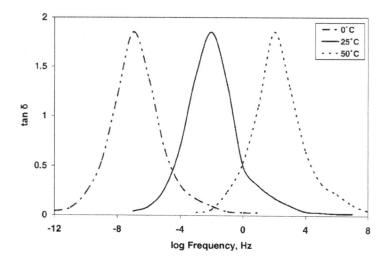

Figure 2.22
Master curves of loss tangent versus frequency a rubbery epoxy at different reference temperatures. Source: *Reprinted with permission from D. Ratna, N.R. Manoj, L. Chandrasekhar, B.C. Chakraborty, Polym. Adv. Technol. 15 (2004) 583,* ©*2004 John Wiley and Sons Publishers.*

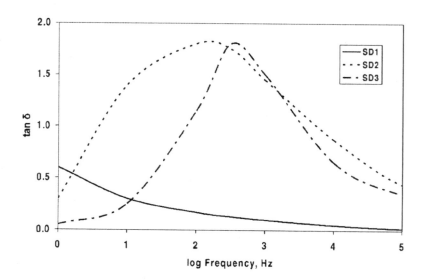

Figure 2.23
Master curve of loss tangent versus frequency plots of rubbery epoxy networks cured with polyetheramine (Jeffemine) of various chain length: SD-1 (Jeff-300), SD-2 (Jeff-500), and SD-3 (Jeff-800). Source: *Reprinted with permission from D. Ratna, N.R. Manoj, L. Chandrasekhar, B.C. Chakraborty, Polym. Adv. Technol. 15 (2004) 583,* ©*2004 John Wiley and Sons Publishers.*

of the viscoelastic layer. The second configuration is the constrained layer damping (CLD) where shear strains in the viscoelastic layer cause damping of vibrations. In a CLD system, a thin layer of a viscoelastic material is applied over the vibrating substrate and covered with a stiff constraining layer such as metal or fiber-reinforced plastic (FRP). Some of the applications of CLD are in aircraft and missile substructures, machinery supports, mounting platforms of electronic equipment, bridges, and buildings. Though CLD treatment provides higher damping compared to FLD, the latter is more convenient to apply and retain damping over a wide frequency range [113]. Hence the mode of damping has to be selected depending on the applications. For example, if the structure to be treated for vibration-damping is intricate in nature then FLD is preferable compared to CLD.

Generally, in CLD, polymeric sheets are used as viscoelastic materials, which are to adhere to the substrate as well as the constraining layer. The advantages of the compositions studied are that they can be sprayed to the substrate and no secondary adhesives are required [114]. The strength of adhesion of epoxy materials to many substrates is essentially high. This is coupled with the high loss factor over a wide range of temperature and frequencies. Hence, these materials offer less weight penalty making it ideal for aircraft applications. More compositions can be tailor-made to suit the mode and range of application by simple blending of various epoxies and amines.

When we make a vibration-damping coating using a thermoplastic resin, it normally contains either water or some organic solvent. Water initiates corrosion in metallic surfaces and solvent causes environmental pollution which poses restrictions for especially application in a closed compartment. Unlike conventional vibration-damping coating, the rubbery thermoset-based coating does not require the use of water or any organic solvent because they are liquid [115]. Therefore application of thermoset-based vibration-damping coating is not associated with any issue of corrosion or pollution. Such coatings will have potential applications for damping of thin metal panels present in modern automobiles, small unmanned aerial vehicles, drones where weight penalty and high strain rate fatigue are very important requirements to be considered for design.

2.10 Interpenetrating polymer networks

Interpenetrating polymer networks (IPNs) are polymer systems, which comprise two or more polymers that are at least partially interlaced on a polymer scale but not covalently bonded to each other [116–120]. Though the networks are not chemically connected in an IPN, the networks cannot be separated unless chemical bonds are broken due to extensive interpenetration. The description of interconnected polymer networks (ICN) is similar to that of IPN, but one important difference is that ICN consists of IPN-like systems that also include a limited amount of internetwork chemical links. When two or more networks are independently crosslinked, then the resulting IPN is called Full IPN. Semi-IPNs are formed when only one polymer system is crosslinked and other components do not participate in crosslinking. For example, in the IPNs of poly(ethylene oxide) (PEO) and methacrylate monomers [121–124] only acrylate monomers undergo curing and PEO remains in uncrosslinked form. Such IPNs are prepared by dissolving the PEO in a mixture of an acrylate monomer and a free radical initiator. The IPNs were used for shape-memory applications, which will be discussed in subsequent sections.

Very recently, an elegant account of IPNs has been given by Silverstein [125] in a review article providing a schematic representation of the most common synthesis routes for IPN, semi-IPN, and ICN (Fig. 2.24). In the scheme, Mi represents monomer i, Xi represents the

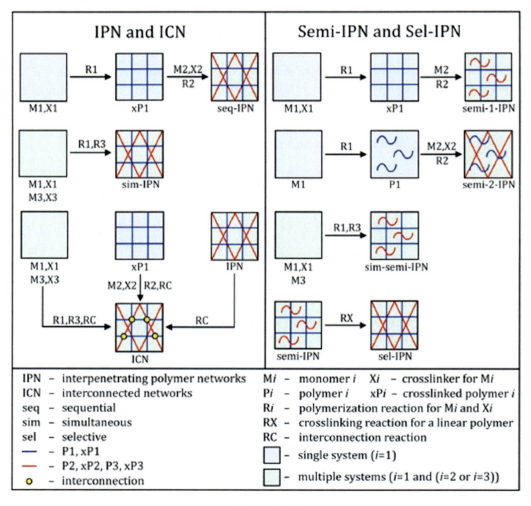

Figure 2.24

A schematic depiction of the most common synthesis routes for IPN, semi-IPN, and ICN. *Source: Reprinted with permission from M.S. Silverstein, Polymer 207 (2020) 122929, ©2020 Elsevier Publishers.*

crosslinker for monomer i, Pi represents polymer i, xPi represents crosslinked polymer i, and Ri represents the reaction used to polymerize Mi and Xi. R1 and R2 are two reactions in a sequential synthesis, and therefore they can entail the same polymerization mechanism. R1 and R3, on the other hand, are two reactions in a simultaneous synthesis, and therefore they must be mutually exclusive polymerization mechanisms. RX is a selective IPN-forming reaction that crosslinks the linear polymer in a semi-IPN. RC is a selective ICN-forming reaction that generates a limited number of internetwork bonds.

Consequent to this preparation methodology, IPN exhibits unique topological features compared to other multicomponent polymeric systems such as blends, graft or block copolymers. The constituent polymers in IPNs are held together by physical entanglements or secondary forces of interactions without any covalent bonding. The concept of making IPN is to integrate two incompatible polymers with very different behavior (through molecular level mixing) to achieve synergistic property. The different types of macromolecular architectures that can be generated depending on the precursor (M_1, M_2, and M_3) are presented in a simplistic manner in Fig. 2.24. Obviously, real systems would be much more complex than these simplistic descriptions. There will be possibilities of the existence of the copolymeric nature of the xPi and the inhomogeneities that exist within the xPi. Although thermodynamically driven phase separation between two dissimilar polymers is topologically prevented due to considerable interpenetration, it is not possible to achieve homogeneous composition or phase morphology like a random copolymer as expected for an ideal IPN. Due to the permanent interlocking of the network segments, thermodynamic incompatibility can be overcome leading to the formation of microheterogeneous morphology due to a limited nanoscale phase separation taking place during the synthesis of IPNs. The micro heterogeneous morphology observed in an IPN is quite different from that observed in an immiscible blend. This explains why IPNs often exhibit better elastic, mechanical properties and thermal stability compared to the individual thermoset components. Thus the morphology of an IPN lies in between the random copolymers and the related blends. Unlike a random copolymer or the blend, IPNs display a broad spectrum of domain compositions that can range continuously from pure P_1 to P_1-rich phase, P_2-rich phase, and pure P_2, as shown schematically in Fig. 2.25.

If we look at the literature published on IPNs, we find rubber-based IPNs and thermoset resin-based IPNs. Rubber-based IPNs are made either by the sequential method as discussed above or by latex polymerization. In sequential IPN, the second polymeric component is crosslinked after the completion of the crosslinking of the first one [117–119]. The usual procedure is to swell the crosslinked polymer with a mixture of monomer, initiator plus crosslinker and allowed to crosslinking. The latex is swelled with a mixture of monomer and initiator and allowed to polymerize to prepare the IPN in the form of latex. This process is called the "monomer flooded process." Alternatively, they can be prepared by adding the monomer (for the second polymer) into the emulsion or dispersion at a slow and

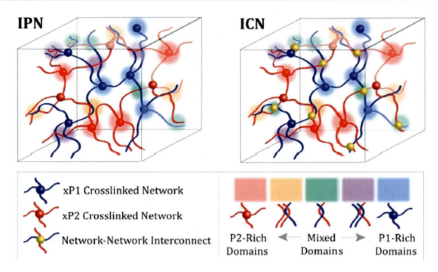

Figure 2.25
Schematic illustrations showing two physically entangled macromolecular networks, highlighting the broad spectrum of domain compositions that can range continuously from xP1-rich to xP2-rich, for IPN (left) and ICN (right). Source: *Reprinted with permission from M.S. Silverstein, Polymer 207 (2020) 122929, ©2020 Elsevier Publishers.*

controlled rate such that the rate of addition equals to rate of polymerization. This is known as the monomer starved process. Note that the monomer used in this process must be water-soluble and a water-soluble initiator like potassium or ammonium persulfate is to be used. Detailed discussion on rubber-based IPNs is beyond the scope of this book. The reader may refer author's book [110] for detailed discussion on rubber-based IPNs.

Thermoset-based IPNs can easily be made by a simultaneous method. Simultaneous IPNs are prepared by a process in which both the component networks are polymerized concurrently. They are processed as homogeneous solution containing required crosslinker for the network formation. The formation of both the network takes place simultaneously by similar or different independent polymerization mechanisms. This type of IPNs offers advantages in terms of applications and well suited for casting, brushing, spraying, and injection molding. However, main disadvantages are achieving a homogeneous solution out of a mixture containing different resins and other ingredients and interference of one curing process with the other one. In this section thermoset-based IPNs will be elaborated. Karger Kocsis and Gryshchuk [126] investigated simultaneous IPN of vinyl ester (VE) and epoxy resins (EP). Note that the VE crosslinks via free radical-induced copolymerization with styrene. On the other hand, polyaddition and homopolymerization are used for crosslinking of EP. The outcome is a simultaneous IPN in a broad VE/EP composition range. The characteristics of the IPN morphology, namely, strand width and mean roughness data (analyzed by atomic force microscopy), tend to increase with an increasing amount of the EP component in the IPN. It was observed that the fracture mechanical data pass through a maximum as a function of the VE/EP ratio, similar to that of the phase segregation term determined by dynamic mechanical analysis.

Giavery et al. [127] reported an IPN binder formed by a simultaneous polymerization of silicone and epoxide prepolymers and used for high temperature (>400°C) coating applications. When the IPN-based binder is applied on a metal surface and given a thermal treatment at 450°C, a ceramic layer comprising silica and fillers remains on the metal's surface and acts as a thermal barrier coating. The property can be improved by judiciously selecting the organic substituents of silicone or adding nanoparticles like graphene. Jin et al. [128] investigated simultaneous IPN of soybean oil-based PU and epoxy. Addition of PU turned epoxy into the rubbery elastomer resulting in a 13-fold increase of elongation at break as compared to the pure epoxy. At the beginning of the cure reaction, T_g of PU/epoxy IPN is usually lower than that of the neat epoxy. The cure rate of epoxy is usually slower than that of PU. Some of the epoxide groups remain unreacted and act as a plasticizer leading to a lowering of T_g of IPNs at the beginning of the cure reaction. When reaction proceeds, the T_g is increased due to formation of IPNs. The increase in T_g can be attributed to the miscibility/formation of graft structure through the reaction of hydroxyl groups of epoxy with the isocyanate group of PU [129]. Yu et al. [130] blended PU into EP to form the IPNs and found a significant improvement of mechanical properties (in terms of tensile strength and impact strength) and proposed for use in high-performance water-lubricated bearing materials. In the case of vegetable oil-based PU (e.g., castor oil), after curing reactions of castor oil with diisocyanate, the double bonds present along the fatty acid backbone can also participate in the radical polymerization in the presence of vinyl or acrylic monomers leading to the formation of EP/PU IPNs. The IPNs made of castor oil-based polyurethane can be copolymerized with different compounds derived from petroleum or renewable resources like nitrokonjac glucomannan, poly(ethylene glycol) (PEG), poly(hydroxyethyl methacrylate), benzyl starch, chitosan, nitrocellulose, etc. [131–136] to modulate the mechanical and chemical properties of the networks.

Meier et al. [137] investigated IPNs based on novolac-type cyanate ester (CE) resin and phenyl-ethynyl-terminated imide (PETI) with or without a compatibilizer. The thermal properties of the resultant thermosets were similar or slightly higher than those of pure polycyanurate. Qu et al. [138] studied the curing kinetics of blends of bisphenol-A dicyanate ester (BADCy) and ethynyl-terminated imide oligomers based on 2, 3, 3′, 4′-biphenyltetracarboxylic dianhydride and 3, 4′-oxydianiline. They have not observed any significant improvements in terms of thermomechanical properties and fracture toughness in comparison to the pure polycyanurate. This is attributed to the high crosslinking density, as well as the existence of macroscopic phase separation. Ballestero et al. [139] addressed the phase separation problem of IPNs by generating chemical crosslink points between two networks, thus, generating a class of IPNs called graft-IPNs. They have synthesized the graft-IPNs based on a highly stiff copolymer phase, comprising bisphenol-A bis (2-hydroxy-3-methacryloxypropyl) ether (BisGMA) resin and two acrylic monomers: methyl methacrylate (MMA) and triethylene glycol dimethacrylate (TEDGMA), and a soft, rubbery polyurethane phase (PU). The crosslinking of the two networks is accomplished by means of the BisGMA resin. This resin has terminal double bonds and secondary hydroxyl groups, which allows the resin to react with the acrylic monomers by radical polymerization, as well as undergo a polyaddition between the secondary (−OH) groups and the (NCO) groups of the isocyanate. It was demonstrated that the miscibility of the different network and fracture properties of the resulting IPN can be significantly improved by generating chemical crosslinks between the networks

Recently, Wen et al. [140] reported IPN-based highly soluble fluorinated ethynyl-terminated imide (FETI) oligomers and bisphenol-A dicyanate ester (BADCy) through a solvent-free procedure. FETI oligomers were prepared via a conventional one-step method in m-cresol, using 4, 4′-(hexafluoroisopropylidene) diphthalic anhydride and 2,2′-bis (trifluoromethyl) benzidine as the monomers, and ethynylphthalic anhydride as the end-capper. The IPNs exhibited a glass transition temperature of >300°C and improved mechanical properties (tensile strength/modulus = 87−95 MPa/3.2−4.3 GPa and impact strength = 27−37 kJ m^{-2}) compared to pure BADCY. The IPNs could be potentially utilized for high-temperature adhesive and composite applications.

As discussed above, thermoset-based IPNs reported so far are simultaneous types. The sequential IPNs are mostly made from rubbers. The reason behind this is that the molecular weight between crosslinks (M_c) for a thermoset is much lower than that of rubber. Hence, the extent swelling for a cured thermoset is also much less making it difficult to prepare the sequential thermoset-based IPNs. Recently, Ratna and coworker [141] synthesized sequential IPNs based on a polyether amine-cured rubbery epoxy and polymethyl methacrylate (PMMA). The details of the preparation of epoxy/PMMA IPNs using the sequential method are described below. The elastomeric epoxy network (EP/J-900) was first made by curing conventional difunctional epoxy with polyetheramine (Jeffamine ED 900). The cured epoxy sheet was immersed in MMA containing azobisisobutyronitrile (AIBN) (0.5 wt.%, with respect to monomer weight) as an initiator and TEGDMA (4 wt.%) as a crosslinker for swelling. The MMA was allowed to soak in the epoxy sheet till the desired weight gain is achieved. The swollen film of cured epoxy was kept 10 min for equilibrium and heated at 80°C for 4−6 h and 100°C for 2 h. Thus the polymerization and crosslinking of MMA take place in the existing cured epoxy network forming an IPN. The resultant film was hardened and kept in a vacuum for the complete removal of the unreacted MMA monomer.

In order to study the inhomogeneities of the networks at the molecular level, the IPNs with varying compositions were characterized [141] by positron annihilation lifetime spectroscopy (PALS) analysis, which provides structural information of polymers at the molecular scale [142]. The details of PALS analysis are provided in Chapter 6. It was observed that the free volume decreases with increasing MMA concentration and showed a minimum at 20 wt.%. The decrease in free volume is attributed to the interpenetration of stiffer PMMA chains with the flexible epoxy matrix. Similar studies on styrene-butadiene rubber (SBR)/PMMA IPNs [143] show a monotonous decrease in free volume up to 40 wt. % of PMMA. This difference can be explained by considering the inherently higher free volume of SBR compared to the elastomeric epoxy network. The loss tangent versus temperature plots of EP/J-900, PMMA, and the blend (80/20 w/w) and IPN (80/20 w/w) are presented in Fig. 2.26. The Loss factor versus temperature plots of EPJ 900 shows a sharp loss tangent peak at 0°C. Epoxies cured with conventionally used amines (aliphatic or aromatic) are generally glassy materials ($T_g > 50°C$). The rubbery behavior of the EP/J-900 can be explained by considering two factors: the first introduction of ether linkages having lower energy of rotation compared to C–C bond and second the reduction in crosslink density due to the higher molecular weight of Jeffamine (900 g mole^{-1}). The crosslinked PMMA also exhibits a narrow loss peak but at a higher temperature (c.a. 129°C). The epoxy/PMMA blend (composition 80/20 w/w) shows typical characteristics of an immiscible blend in which two-loss peaks appear at the respective positions of the component polymers. However, an epoxy/PMMA IPN (composition 80/20 w/w) shows a broad transition covering almost the entire range of temperature in between the T_g of epoxy and crosslinked PMMA. This clearly indicates that by utilizing IPN strategy the service temperature of a vibration-damping material effective in wide frequency range can be significantly broadened.

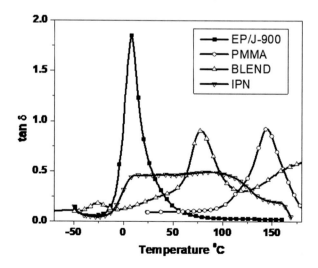

Figure 2.26
Loss tangent versus temperature plots of epoxy (EP/J-900), PMMA, the blend, and IPN.

2.11 Smart thermoset

2.11.1 Self-healing thermoset

Self-healing thermosets are a class of smart materials, which has the capability to heal the damages caused by various reason such as impact, thermal stress generated during operation or ultraviolet light exposure, and so on. Once a crack is generated in cured thermoset system, it propagates very fast leading to the catastrophic failure of the materials. Even a scratch on a thermosetting resin film reduces the mechanical strength significantly. Therefore it is necessary to heal the developed cracks to assure the reliability and safety of thermoset-based products. The concept is inspired by the phenomena, observed in biological systems; any kind of minor damage triggers an automatic healing process. For example, in the case of a bone fracture, a chemical signal released at the site of fracture initiates a systematic response that transports repair agents to the site of injury and promotes healing. Blood clotting, regrown of antlers of animals, etc. are another example of self-healing in the living kingdom.

In the last two decades, extensive works have been done on various aspects of self-healing of polymers in general and thermosetting resin in particular. If a material heals the damages without any external intervention then it is called autonomic self-healing material. When the materials require the help of any external stimulus such as light, heat, radiation, etc. then they are termed as nonautonomic or stimuli-responsive self-healing materials. If we look at the literature, we find some of them are in proof of concept level and for some, technology has been demonstrated. Some of the technologies have already been exploited for development of products. This indicates a wide range of technology readiness levels for the works carried out so far. Various methods of self-healing reported so far can be classified into two categories, namely, extrinsic and intrinsic.

2.11.1.1 Extrinsic self-healing

Extrinsic methods mainly make use of micro containers like microcapsules, microvascular networks, or nanoparticles that are embedded within the polymer matrix, which automatically release a healing agent into the crack site upon fracture. The healing agent fills the damaged area and restores mechanical integrity as a result of the curing reaction. The concept of extrinsic self-healing technology for thermosetting resin was first demonstrated in 2001 by White et al. [144]. They incorporated capsules of dicyclopentadiene (DCPD) and bis(tricyclohexylphosphene)-benzylidine ruthenium dichloride catalyst (Grubbs' catalyst) in an epoxy matrix. The idea is that when the damage event occurs, the microcapsules get ruptured and release the encapsulated healing material that reacts with the catalyst and initiates a polymerization reaction sealing the crack. They have demonstrated a 75% recovery of fracture toughness due to the incorporation of microcapsules. The same group also explored different ruthenium catalysts as alternatives to using the first generation Grubbs' catalysts for this purpose [145].

Following this pioneering work, various other DCPD-based systems have also been investigated [146,147]. Due to advancement of microencapsulation technology using various methods of preparation of microcapsules like interfacial polymerization, in situ polymerization, spray drying, solvent evaporation, etc., a lot of works have been done on microcapsule-based healing technology [148–151]. A major challenge in the development of successful microcapsule-based self-healing technology is to achieve autonomic damage restoration, which requires chemistries that are robust, cost-effective, environmentally stable, and provide high healing efficiency. Solvent-assisted self-healing technology is also reported [152]. The epoxy matrix is swollen by solvent delivered from ruptured microcapsules, which liberates residual amine groups to initiate further crosslinking. However, diffusion of hardener requires chain mobility within an undercured epoxy matrix, which is a substantial limitation of such healing mechanism especially for structural composites, which are generally cured at elevated temperatures. If the healing chemistry is the same as the curing of the resin system to be self-healed then the repair system becomes chemically and mechanically compatible with the host matrix. With this idea in mind for epoxy matrices, the use of microcapsule of epoxy and amine hardener has been investigated. Note that preparation of amine microcapsule is more difficult than that of the epoxy microcapsule.

Jin et al. [153] reported the preparation of amine microcapsules that constitute a dual-microcapsule epoxy-amine self-healing system for epoxy polymers. Epoxy resin-containing capsules are prepared by in situ polymerization of a commercial epoxy resin with a urea-formaldehyde shell wall, while amine-containing capsules are produced by vacuum infiltration of an amine hardener into polymeric hollow microcapsules. Hollow UF microcapsules that were infiltrated with EPIKURE 3274 sank in the vacuum jar and were collected by filtering. The SEM microphotographs of amine (EPIKURE 3274) microcapsules after infiltration and optical micrograph of capsules immersed in EPIKURE 3274 are as shown in Fig. 2.27. Filled microcapsules appear clear due to the matching of the refractive index while hollow microcapsules appear dark. Capsules tend to agglomerate due to the wet exterior surface (see Fig. 2.27A). Nevertheless, we can easily visualize from Fig. 2.27B that amine capsules can be evenly dispersed when manually mixed into neat EPIKURE 3274. In addition, the filled capsules appear clear and transparent in the continuous phase due to matching of refractive index, while hollow capsules appear black. About 80% healing efficiency was achieved with an epoxy system containing 4 wt.% amine capsules and 6 wt.% epoxy capsules. SEM images of a fracture surface of the unhealed epoxy specimen and the same after-healed test are shown in Fig. 2.28. We can clearly see the crack tails and the healed network failed cohesively.

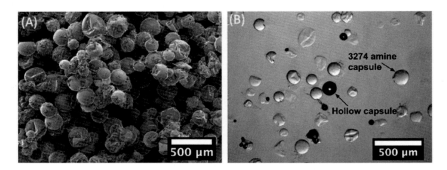

Figure 2.27
Characterization of amine microcapsules. (A) SEM image of amine microcapsules after infiltration. (B) Optical micrograph of capsules immersed in EPIKURE 3274. Filled microcapsules appear clear while hollow microcapsules appear dark. Source: *Reprinted with permission from H. Jin, C.L. Mangun, D.S. Stradley, J.S. Moore, N.R. Sottos, S.R. White, Polymer 53 (2012) 581, ©2012 Elsevier Publishers.*

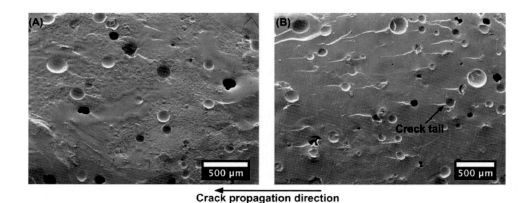

Figure 2.28
SEM images of fracture surface from a self-healing specimen containing 4 wt.% amine capsules and 6 wt.% epoxy capsules. (A) Fracture surface after-healed test. (B) Unhealed fracture surface (rinsed with ethanol immediately after fracture) revealing crack tails. Source: *Reprinted with permission from H. Jin, C.L. Mangun, D.S. Stradley, J.S. Moore, N.R. Sottos, S.R. White, Polymer 53 (2012) 581, ©2012 Elsevier Publishers.*

The advantage of this technology is that it offers the flexibility of using numerous different healing chemistries and encapsulation techniques making it possible to utilize it for various different types of thermoset matrices including highly crosslinked thermosets. Note that for effective self-healing performance in such systems, healing rate should match with the damage rate. However, in the extrinsic method, there is no guarantee of matching the healing rate to the damage rate. Moreover, the volume of the healing agent is limited and a healing event happens only once. After this, the healing agent is consumed and subsequent damages at the same site cannot be self-healed again. Adequate amounts of healing agents need to be released to entirely or partially fill the damaged areas. The incorporation of a higher concentration of microcapsules may affect the thermomechanical properties of the network. The storage of the healing agent can be enhanced by using microvascular self-healing materials (as a substitute for microcapsules), which store the healing agent in a network of capillaries or hollow channels, interconnected in one, two, or three dimensions until damage occurs triggering the self-healing. However, ensuring the proper release of the healing agent is more difficult compared to microcapsules. Also, the incorporation of such materials may affect the processability of the resin to a great extent.

Another extrinsic approach for self-healing is by incorporating a thermoplastic additive, which softens above its T_g and will flow into the cracks, thereby reconnecting the matrix material. The advantage of thermoplastic additive over liquid healing agent is that the healing process can be repeated a number of times for the same fracture surface with thermoplastic particles. It was reported that the use of thermoplastic linear poly(bisphenol-A-co-epichlorohydrin) resulted in a significant decrease in delamination area and matrix cracking can be eliminated by thermal healing [154,155]. It was demonstrated that the combination of an amorphous polymer above its T_g and nanoparticle reinforcements lead to a polymer system with both good healing efficiency and mechanical properties [156]. Luo et al. [157] prepared a thermally mendable thermoset epoxy resin by dispersing phase-separated poly(caprolactone) (PCL) in an epoxy matrix. Upon heating, the PCL melts and undergoes a thermal expansion and thereby provides self-healing attribute.

Though most of the works have been done on epoxy resins, microcapsule-based self-healing strategy has also been extended for other thermoset matrices. For example, Cho et al. [158] dispersed phase-separated droplets of hydroxyl-terminated poly dimethylsiloxane and poly diethoxysiloxane into a vinyl ester (VE) matrix to achieve a self-healing effect. Jadhav and coworker [159] reported the microencapsulation of linseed oil along with drier and corrosion inhibitor and used for development of polyurethane-based self-healing coatings. The cracks in the paint film could be successfully repaired by the release of linseed oil from microcapsules ruptured understimulated mechanical action, while corrosion inhibitor plays an important role to prevent corrosion in a scribed line region. They also reported phenol-formaldehyde microcapsules with linseed oil as an active agent for obtaining self-healing PU as well as epoxy coatings [160]. Suryanarayana et al. [161] successfully used linseed oil filled urea-formaldehyde resin microcapsules for healing of cracks generated in paint/coatings. Urea-formaldehyde/epoxy-based microcapsules were incorporated into an IPN of polyarylates-polysilicone and used as self-healing material to protect steel surface [162].

2.11.1.2 Intrinsic self-healing

We have discussed in the earlier section about extrinsic self-healing, which requires a stored healing agent. On the other hand, intrinsic self-healing materials do not involve a stored healing agent. The healing takes place through latent self-healing functionalities present in their polymeric structure. They are triggered by damage and driven by surface tension and elastic energy of the stress source or by an external stimulus such as heat, light, solvent, and others. Intrinsic self-healing has been investigated initially with thermoplastics such as amorphous, semicrystalline, block copolymers and subsequently extended to thermoset resins. The basic requirement is the presence of reversible dynamic covalent or noncovalent bonds. The concept is that these types of bond are able to reversibly associate and dissociate upon exposure to external stimuli like heat, light, microwave, IR irradiation, etc. Generally the bond strength of such reversible bonds is weak and hence tends to break first upon stress. However, on exposure of a suitable external stimulus, these bonds at the damaged areas can be reconstructed and thereby driving the polymer chains to contact, diffuse and reentangle above the glass transition temperature (T_g) of the polymer, and hence healing of the material is possible [163–166].

A number of chemical reactions have been utilized for the development of self-healing thermosets. For example, photo-induced [2 + 2]-cycloaddition reaction of 1,1,1-tris-(cinnamoyloxy methyl) ethane results high T_g transparent film with seal-healing ability [167]. Thermoreversible [4 + 2] Diels–Alder (DA) reaction is widely used for intrinsic self-healing. Li et al. [168] functionalized a novolac epoxy resin with furan moieties. Single polymer chains were crosslinked via N, N'-(4,4_-diphenylmethane) bismaleimide via a DA reaction. A crack on the surface of the polymer was treated with 130°C for 10 min for a retro-DA (r-DA) reaction to enable the crosslinking at 60°C (DA) again. Intrinsic self-healing also reported in polyketone (PK)/bismaleimide-based system where crosslinking and decrosslinking can be done via DA and r-DA reactions [169] as shown in Fig. 2.29. An aliphatic PK was modified polymers by using the Paal–Knorr reaction. The dicarbonyl arrangement of PK was converted to pyrrole groups bearing furan or amine active groups. Subsequently, the modified PK was covalently crosslinked through the DA reaction by using bismaleimide. DA reaction is also utilized to develop self-healing polyurethane composites for applications as phase change material [170]. Furan-modified dopamine particles were used as a photothermal filler as well as a crosslinker for the PEO-based PU. The crosslinking takes place via DA reaction. The PU composite is capable of converting efficiently the solar energy into heat. It exhibits near-infrared radiation-induced fast self-healing (within the 30 s) and offers up to 93% recovery in tensile strength.

Thio-disulfide reaction has also been used to achieve intrinsic self-healing. The disulfide linkage in the polymer can be reduced to a single thiol and healed via subsequent oxidation of the thiol regenerating disulfide linkages [171]. A good combination of adequate mechanical properties and healing at low temperatures can be achieved by the introduction of covalently bonded disulfide groups within a conventional epoxy-based network [172–174]. Recently Garcia and coworker [175] reported disulfide-linkage containing epoxy-based glass fiber-reinforced composite and demonstrated multiple healing of flexural

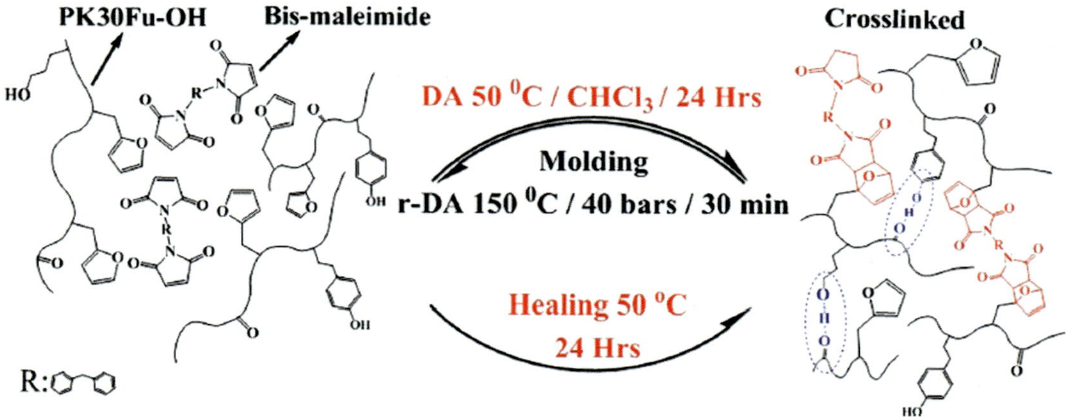

Figure 2.29
Schematic representation of DA/r-DA sequence and molding of functionalized polyketone PK-Fu-OH crosslinked with bismaleimide. In red reversible covalent interactions and blue hydrogen bonding (aliphatic or aromatic). Source: *Reprinted with permission from R. Araya-Hermosilla, G.M.R. Lima, P. Raffa, G. Fortunato, A. Pucci, E. Mario, et al., Eur. Polym. J. 81 (2016) 186, ©2016 Elsevier Publishers.*

and interlaminar fracture properties as well as the healing of low energy impact damage. The composites offer a combination of a decent stiffness (800–1200 MPa), moderate healing temperatures (70°C–85°C) below the curing temperature, good damping properties, and the possibility to process by conventional vacuum infusion. Introduction aliphatic of diselenide compounds to produce healable materials in which the exchange of diselenide moieties is triggered by the application of UV and visible light [176–178]. In comparison to disulfide bonds, diselenide bonds possess lower bond energy (diselenide bonds: 172 kJ mol^{-1}; disulfide bonds: 240 kJ mol^{-1}). This suggests that diselenide bonds can be more dynamic, and that exchange reaction might happen under much milder conditions, which could help to overcome the disadvantages of disulfide exchange reactions. Zhu and coworker [179] utilized aromatic diselenide compound show healing abilities without the need for any external intervention due to their ability to undergo diselenide metathesis at room temperature. In addition, they offer enhanced healing and reprocessing characteristics compared to disulfide-based polymers, as the diselenide exchange occurs faster than the disulfide exchange. Ying et al. [180] reported rational design of hindered urea bonds (urea with bulky substituent attached to its nitrogen) and used of them to make polyureas and poly(urethane-urea)s capable of catalyst-free dynamic property change and autonomous repairing at low temperature.

The examples of self-healing discussed above involve dynamic reversible chemical bonds. The self-healing can also be obtained by utilizing reversible noncovalent interactions [181,182] such as hydrogen bonding, π-π stacking, ionic interactions as shown [151] in Fig. 2.30. The noncovalent interactions function via the formation of supramolecular

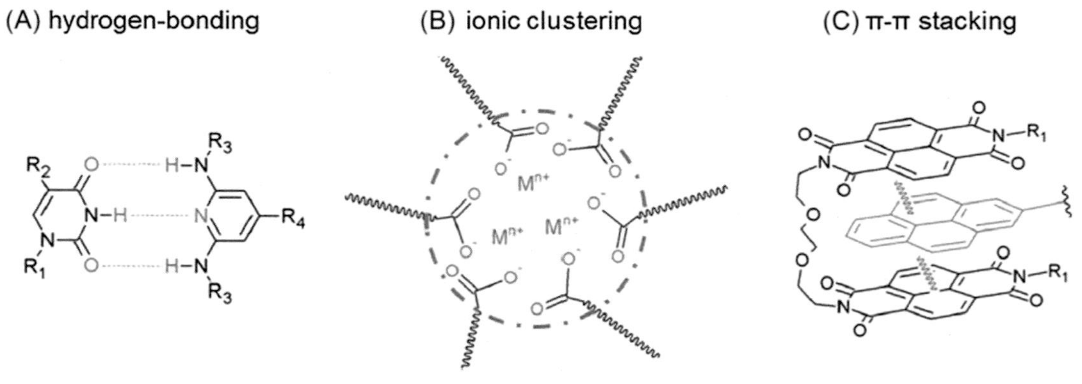

(A) hydrogen-bonding (B) ionic clustering (C) π-π stacking

Figure 2.30
Schematic noncovalent network formation strategies. (A) Hydrogen bonding between pyrimidine base residues. (A) Ionomers with anionic charged polymer chains and positive counter metal ions. (C) π–π stacking of π-electron-rich pyrenyl units with electron poor π systems. Source: *Reprinted with permission from K. Urdl, A. Kandelbauer, W. Kern, U. Müller, M. Thebault, E. Zikulnig-Rusch, Prog. Org. Coat. 104 (2017) 232–249, ©2017 Elsevier Publishers.*

networks. As soon as supramolecular bonds are present in a network and mechanical stress is applied, these weaker bonds will break during damage formation. However, the bond will subsequently re-form due to the dynamic character and thereby will provide a healing effect [183]. The recombination strongly depends initially, on the molecular dynamics of the individual polymer chains that enable certain mobility of the supramolecular groups and then the time period allowed for healing.

Host–guest interactions are another example of supramolecular binding, which can automatically form highly selective recognition between host and guest moieties. Such interactions have been utilized to develop self-healing hydrogels (low mechanical properties). Recently, Zhen Hua et al. [184] have extended the idea to thermoset resins and developed a vinyl ester (epoxy acrylate)-based polymeric system whose self-healing is driven by noncovalent host–guest interaction of the supramolecular complexation of azo (—N=N) and β-cyclodextrine (β-CD). Where vinyl β-CD acts as host moieties that form a stable inclusion complex with acrylamide azobenzene (AAAB). This bifunctional complex is incorporated into the epoxy crosslinking structure via copolymerization as a healable element. This results in the formation of a strong reversible and light-responsive switchable epoxy 3D network with host–guest interactions capable of exhibiting excellent self-healing properties. The healing efficiency reported is 63.3% on a restore of 9.3 MPa tensile strength. The mechanism of self-healing in the present case is presented in Fig. 2.31. Because of the light-responsive nature of vinyl β-CD and AAAB, the prior UV irradiation process makes β-CD exclude the AAAB from its cavity to disconnect the interaction between the host and guest molecules. This step ensures sufficient mobility of the

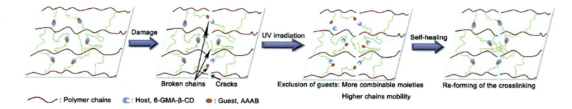

Figure 2.31
Schematic illustration of the self-healing mechanism based on host–guest chemistry. Source: Reprinted with permission from Z. Hua, Y. Liua., X. Xua, W. Yuana, L. Yanga, Q. Shaob, et al., Polymer 164 (2019) 79, ©2019 Elsevier Publishers.

host–guest moieties essential for self-healing. The vinyl ester network further gains mobility on heating up to above T_g, which supports a reflow of the polymer chains. Consequently, the dissociated host–guest interactions are able to spontaneously recombine and thereby providing a self-healing effect.

As we can see from the above discussion, networks with higher modulus require higher temperature for achieving a healing effect. It is well known that for structural application high strength/modulus of a thermoset network is absolutely necessary. Therefore it is desirable to develop a thermoset with high modulus and a healing approach which can function at a lower temperature. Fig. 2.32 summarizes [174] the major developments in the field by representing the elastic modulus versus the healing temperature of a wide range of intrinsic healing polymer matrices. The figure also includes vitrimers, which will be elaborated in subsequent sections. The figure clearly shows that the current challenge in self-healing polymers lies within the development of systems with a high modulus and low healing temperature in order to reach properties of interest that can compete with commercial high-performance polymers.

2.11.1.3 Shape-memory thermoset

Shape-memory thermosets (SMT) are a new class of smart materials, which can be deformed and fixed into a temporary shape and recover their original permanent shape only on exposure to external stimuli like heat, light, etc. [185–189]. We have seen when rubber is stretched by applying tensile force, it is elongated, and on the release of the force, it returns to its original size. This is not an example of the shape-memory effect because there is no control over the recovery. It should recover on demand (not automatically) for behaving like a smart material. Though a variety of stimuli can be used to trigger the recovery process, the thermally induced shape-memory effect is more common where the recovery takes place with respect to a certain critical temperature. The basic requirement for a polymer to display the shape-memory effect is the presence of a reversible switching segment plus a physically/chemically crosslinked network for the stability of the permanent

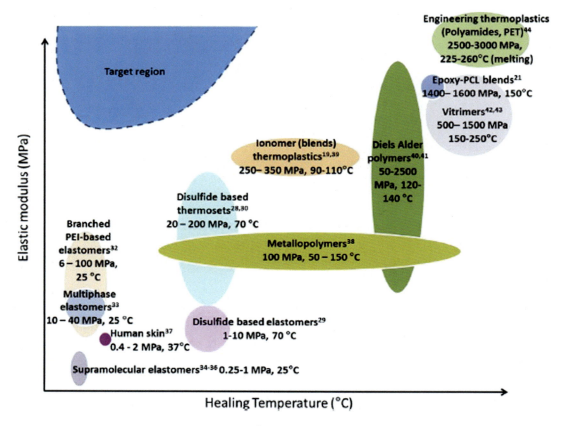

Figure 2.32
Overview of elastic modulus versus healing temperature of a wide range of intrinsic self-healing polymer matrices developed in the past decade. Source: *Reprinted with permission from W. Post, A. Cohades, V. Michaud, S. van der Zwaag, S.J. Garcia, Compos. Sci. Technol. 152 (2017) 85 (2017) 152, ©2017 Elsevier Publishers.*

shape. SMT can be designed by taking a polymer material, in which the polymer chains are able to fix a given deformation by cooling below a certain transition temperature. The transition temperature (T_s) can be a glass transition or a melting point. Melting point is preferred over glass transition as melting is comparatively sharper. Upon reheating above the T_s, the oriented chains in the network restore the random coil conformation resulting in a macroscopic recovery of the original shape. Such polymer systems consist of two segments/phases; one is reversible or switching phase for maintaining a transient shape and the other is a fixed one for recovering the original shape.

The SMTs are characterized (for describing shape-memory) by two important parameters namely strain recovery rate (R_r) and strain fixity rate (R_f) [185]. The strain recovery rate qualifies the ability of the material to memorize its permanent shape whereas strain fixity rate describes the ability of the switching segment to fix the mechanical deformation. These two parameters are determined from a thermomechanical test in which the sample is heated to a temperature above T_s and is deformed to a constant strain (ε_m) at a constant strain rate. While maintaining the strain at ε_m, the samples were cooled to a low temperature and unloaded. Upon removing the constraint undercooled condition, a substantial recovery of strain ε_u occurs. The sample was subsequently heated above T_s and kept at that temperature allowing the recovery to take place. This completes a thermomechanical cycle leaving a residual strain ε_p. The R_r and R_f for Nth cycle can be determined from the following formula:

$$R_r(N) = \frac{\varepsilon_m - \varepsilon_p(N)}{\varepsilon_m - \varepsilon_p(N-1)}, \qquad (2.44)$$

$$R_f(N) = \frac{\varepsilon_u(N)}{\varepsilon_m}. \qquad (2.45)$$

The most widely used shape-memory material is Ni–Ti alloy (Nitinol). Shape-memory alloys (SMA) exhibit outstanding properties such as small size and high strength and have found wide technical applications. However, they have obvious disadvantages such as high manufacturing cost, limited recoverable deformation and appreciable toxicity. SMT offers deformation to a much higher degree and a wider scope of varying mechanical properties compared to SMAs or ceramics, in addition to their inherent advantages of being cheap, lightweight and easy processability. In the case of SMA and shape-memory ceramics, maximum recoverable strains are 8% and <1%, respectively, whereas in case of SMT (e.g., PUs), the same can be up to 800% [185]. They also offer extra advantages due to the fact that they are biocompatible, nontoxic, and can be made biodegradable.

2.11.1.4 Polyurethane-based shape-memory polymer

PUs when suitably designed exhibit an excellent shape-memory effect [190,191]. Shape-memory polyurethanes (SMPU) are available commercially from Mitsubishi, Japan. The effect of processing on thermomechanical properties of the shape-memory polyurethanes has been investigated [192]. There was a significant variation in rubbery modulus for the sample cooled at different rates during the processing. Thus thermomechanical properties of these materials can be adjusted for various medical applications. The thermal and dynamic mechanical properties of thermosetting Mitsubishi PU materials are given in Table 2.9. The exact compositions of these SMPUs have not been published, but they seem to consist of a complex combination of short amorphous polyester and polyether soft segments, short diol extender, and aromatic urethane hard segments. The materials were synthesized by polymerizing alternatively hard segment (diphenylmethane diisocyanate) and soft segment (polyol) and various interactions among the segments led to the domain formation; hard

Table 2.9: Thermal and viscoelastic properties of DiAPLEX® (Mitsubishi) thermoset shape-memory PU.

Trade name	T_s (°C)	T_g^a (°C)	Tanδ^b peak temperature (°C)	E_g^b (Pa)	E_r^b (Pa)
MP3510	35	31	46.1	5.2E8	1.6E6
MP4510	45	42	54.1	7.0E8	1.7E6
MP5510	55	52	65.2	8.1E8	1.9E6

E_g and E_r are glassy and rubbery dynamic modulus.
[a]Determined by differential scanning calorimetry.
[b]Determined by dynamic mechanical analysis.

segment worked as a pivoting point for shape recovery and soft segment could mainly absorb external stress applied to the polymer.

The T_g of PU materials can be controlled in a wide range from −30°C to 100°C by using different kinds of urethane ingredients (diisocyanate, polyol, and chain extenders) and by adjusting their molar ratios [193,194]. In order to have effective shape-memory property, the glassy hard segments should maintain the shape through inter- or intrapolymeric chain attractions such as hydrogen bonding or dipole−dipole interaction, together with the physical crosslinking, but soft segments could freely absorb external stress by unfolding and extending their molecular chains. If the stress exceeds and breaks the interactions among hard segments, shape-memory will be lost and the original shape cannot be restored. Therefore precise control of the composition and structure of hard and soft segments is very important to satisfy the conditions required for various applications. In the segmented PUs, the hard segment (HS) acts as the physical crosslink and hard segment concentration should be sufficiently high (>20 wt.%) in order to generate a shape-memory effect [195]. Optimum shape-memory properties were achieved at 35−40 wt.% of HS concentration [196].

The shape-memory PUs can be classified into two categories namely amorphous and crystalline. The compositions and properties of different types of amorphous PUs reported in recent literature are summarized [197−201] in Table 2.10. Wang and Yuen [202] synthesized a series of thermoplastic PUs using aromatic chain extenders such as 4,4-bis(4-hydroxyhexoxy)-isopropylene or naphthoxy diethanol, in addition to 1,4-BD and reported an improvement in shape-memory properties as a result of introduction aromatic structure into the main chain. Yang et al. [203] compared the mechanical, dynamic mechanical, and shape-memory properties of PU block copolymers with planar shape HS (1, 6-diphenyl diisocyanate—PDI) and bent shape HS (MDI). The PDI-based PU shows superior property compared to MDI-based PU (Table 2.10) as a result of better interaction among hard segments due to the planar shape of PDI.

The compositions and properties of crystalline PU materials are presented in Table 2.11. Polycaprolactone (PCL) has been extensively used for the synthesis of PU with crystalline soft phase. PCL is first converted to PCL diol and the diol is used to synthesize the

Table 2.10: Compositions and properties of amorphous PUs.

Polyol Mw (g mole^{-1})	Isocyanate	Extender	Hard segment (wt.%)	T_g (°C)	% Recovery	Ref
PTMG 650	MDI	1,4-BD	31 to 86	−13 to 38	95	[197]
PTMG 2000	MDI	1,4-BD	12 to 50	–	90	[198]
PTMG	MDI	1,4-BD	20 to 50	−15 to 2	95	[196]
PTMG	MDI	1,4-BD	30 to 35	0 to −15	85	[202]
PTMG	PDI	1,4-BD	20 to 25	−1 to −17	96	[202]
PTMG 1000, 2000	MDI	1,4-BD + BES + BPE/ND	52 to 73	83 to107	>90	[201]
BHBP	TDI + HDI	1,3-BD	21 to 41	10 to 35		[199]
PTMG 1800	MDI	1,4-BD	30	−10	>80	[201]
PTMG 1000, 2000	MDI	1,4-BD	35	−10 to 30	>80	[200]

Table 2.11: Structure and properties of crystalline shape-memory PUs.

Hard segment	Soft segment (SS)	%SS	Switching temperature (°C)	% Recovery	Ref.
MDI/1,4-BD	PCL	70	44−55	82	[204]
MDI/1,4-BD	PCL/PVC blend (6/4)	38−62	18−48	92	[205]
HDI/4,4'-dihydroxybiphenyl	PCL	30−80	38−58	90	[206]
MDI/BEBP	PCL	57−70	41−50	80	[207]
MDI/DMPA	PCL	70	45	95	[208]

PU-based shape-memory polymers (SMPs) [204,209]. The PCL segment undergoes microphase separation and forms the soft switching segment. The recovery temperature, T_r (the temperature at which fast recovery takes place) can vary from 40°C to 60°C depending on the soft phase/hard phase composition and molecular weight of PCL. The relation between the shape-memory effect and molecular structure has been investigated and it was concluded that high crystallinity of the soft segment region at room temperature was a necessary prerequisite for segmented PUs to demonstrate shape-memory behavior. However, the crystallization of PCL segments is hindered by incorporating them into the multiblock copolymers. No crystallization was observed when PCL diol has a number average molecular weight less than 2000 g mole^{-1} and the optimum molecular weight in terms of shape-memory properties was found to be 5000−6000 g mole^{-1} [205].

The major issues related to the developments of such PU-based SMPs are to achieve maximum crystallization and stable hard segment domains. Various strategies like the incorporation of mesogenic segment [206] and ionic groups [207] have been adopted to enforce microphase separation (Table 2.11). Mesogenic segments were incorporated in the PU by partially replacing 1, 4-BD in PU formulation with 4, 4′-bis(2-hydroxyethoxy) biphenyl (BEBP) [206]. The improvement in solubility and shape fixity was observed as a result of the incorporation of mesogenic groups. The influence of ionic groups on the crystallization behavior of segmented PU using PCL diol of various molecular weights (4000, 20,200, 58,500, 71,600 g mole^{-1}) was also investigated [207]. Ionomers were synthesized by using dimethylol propionic acid (DMPA) in addition to 1,4-BD [208]. Crystallization studies indicate that crystallization rate increases as a result of the incorporation of ionic groups in the case of PU made from PCL-4000 whereas in the case of samples having a higher molecular weight of PCL the crystallization rate increases. Moreover, the ionomers show higher tensile strength, modulus, and fatigue resistance compared to the corresponding non ionomers due to the columbic forces between the ionic centers within the polymer backbone.

In case of PU with crystalline soft segments, the crosslinking (introduced in the structure) decreases the crystallization of the soft segment and upgrades the mechanical properties of the resulting crosslinked structure (Table 2.12). Buckley et al. [210] reported a novel thermosetting PU using 1,1,1-trimethylol propane as a crosslinker. Improvement in creep, increase in recovery temperature and recovery window, was observed due to the introduction of the crosslinking. Xu et al. [211] synthesized hybrid PU, crosslinked with Si-O-Si linkages, formed through hydrolysis and condensation of ethoxy silane groups. The Si-O-Si linkages not only act as the netpoints but also act as inorganic fillers for reinforcement. Ping et al. [212] reported PCL-based biodegradable polyurethane. Thermosetting PU with both shape-memory and hydrogel properties were also investigated [213,214]. They can be made by making graft copolymers of PU and other hydrophilic polymers, for example, polyacrylamides, polyacrylates or by introducing hydrophilic groups into the crosslinked PU backbone [213,214].

Although SMPUs have been used as smart materials in numerous applications, it remains a challenge to improve the mechanical properties and thermal stability of SMPUs. Researchers are actively designing new kinds of SMPUs which would allow the precise

Table 2.12: Effect of crosslinking on mechanical and shape-memory properties of crosslinked PUs.

Hard phase	Soft phase	Crosslinker	E' (GPa)	σ (MPa)	T_m (°C) Soft phase	% Recovery
MDI,1,4-BD, DMPA	PCL Mn ~ 4000	Nil	0.2	18	50.3	80
MDI,1,4-BD, DMPA	PCL Mn ~ 4000	Glycerin 6 wt.%	0.5	40	46.9	90

E' and σ are storage modulus and tensile strength respectively, at room temperature.

tuning of shape-memory properties. The incorporation of nanoparticles has been attempted to improve the thermomechanical properties of SMPU [215]. However, this is associated with a sacrifice in shape-memory properties, because fillers block the movement of the polymer chains and result in a delay in the shape recovery process. Details of nanocomposite technology will be discussed in Chapter 5. Fu et al. [216] successfully modified polyurethane (PU) by 1, 3, 5, 7-tetrahydroxy adamantine and found that the thermal stability of the PU was greatly improved. The PU frameworks were constructed by using adamantane as the core and PCL segments as star branches. The incorporation of adamantane tends to enhance the thermal stability of AdPU. The AdPU forms a microphase-separated structure composed of a semicrystalline soft phase and an amorphous hard phase. AFM images show that the aggregation of AdPU can form nano-sized crystal fibers by controlling the drying conditions. Finally, shape-memory tests indicate that the AdPU has thermal-induced shape-memory properties with 99% shape fixation and 85% shape recovery.

2.11.1.5 Epoxy-based shape-memory thermoset

The versatility and possibility of the wide scope of chemical modification for epoxy resins are discussed in Chapter 1. This wide scope for chemical modification of epoxy resin has been utilized to develop SMT with adjustable recovery temperature keeping various applications in mind. Xie et al. [217,218] fabricated SMT comprising epoxy resin and aliphatic amine curing agent where the transition temperature can be tuned from 6°C to 88°C. Zhang et al. [219] synthesized epoxy-based SMTs with T_g of 65°C–140°C by curing epoxy with methylhexahydrophthalic anhydride and reinforcing with multiwall carbon annotate.

The Tag of epoxy-based SMT can be attended by reducing crosslink density or by introducing flexible linkages [220]. The polyether amine-cured epoxies show better shape recovery compared to the amine-cured epoxy network. However, polyether amine-cured epoxy networks exhibit lower strength and modulus. The thermo oxidative stability of an epoxy-based SMT can be enhanced by blending with high-temperature polymers like cyanate ester and shape-memory properties can be further enhanced by introducing crystallizable segments in the network, which acts as a switching segment. For example, a network based on a blend of epoxy and cyanate ester, when coreacted with phenol end-capped poly(tetramethylene oxide) (PTOH), offers good shape-memory properties [221]. PTOH acts as a crystallizable switching segment. The origin of switching temperature is related to both meltings of PTMO and T_g of the thermoset matrix comprising oxazolidone and polycyanurate structures.

The strategy of incorporation of dynamic reversible linkages for the development of self-healing thermoset has been discussed in the previous section. The same strategy can also be used for the development of SMTs. In fact, such systems exhibit both self-healing and self-memory attributes. Ding et al. [222] reported a novel SMT comprising epoxy resin (EP) and chain-extended bismaleimide (CBMI), which offer shape-memory properties due to the presence of dynamic transesterification bonds. CMBI is prepared by reacting bismaleimide (BMI) with diaminodiphenyl sulfone (DDS). The epoxy/CMBI mixture is cured with glutaric anhydride (GA) to make SMT. The reaction scheme is shown in Fig. 2.33. The epoxy/CBMI-based SMTs can be deformed by an external force to different temporary shapes at temperature 150°C. The temporary shape of the samples can be fixed after cooling to room temperature. The % shape recovery and shape recovery time for EP/CBMI samples of different compositions are reported to be 93.9%−99.8% and 10−21 s, respectively. As CBMI concentration increases, the shape recovery time tends to increase. This can be explained by considering the fact that EP with a higher concentration of CBMI shows higher T_g, resulting in relatively slower segmental motion of polymer chains.

Let us now discuss mechanism of reconfiguration of polymer chains, which provides both shape-memory and self-healing attributes. The schematic of the self-reconfiguration mechanism is shown in Fig. 2.34. When the deformed polymer is continuously heated at a temperature up to the transesterification temperature ($T_v = 200°C$) as shown in Fig. 2.34C, the reversible transesterification reaction between ester group and hydroxyl group becomes activated in the presence of a catalyst (zinc (II) acetylacetonate) and a new ester group and a new hydroxyl group are produced until a dynamic equilibrium of transesterification is established. As a result, polymer networks are rearranged and reconfigured leading to the stress relaxation of the polymer chains [223]. Therefore a new geometrical configuration is obtained after the sample is cooled, and the shape change of the sample is permanent for involving no entropy change (Fig. 2.34D). The reconfigured sample can be temporarily deformed again by the external force at a temperature above T_g and retain the shape after cooling and unloading the force as shown in Fig. 2.34E. The two-deformed sample can recover its reconfigured shape when heated again up to a temperature close to or above T_g as shown in Fig. 2.34F.

2.11.1.6 Interpenetrating polymer networks-based shape-memory thermoset

As discussed above, there are a plethora of synthetic variables available at our disposal to prepare an IPN with tunable properties. IPN technology has been successfully utilized for the development of SMTs [121−124]. Ratna et al. [121] have developed SMT systems based on IPN of PEO and polymethacrylates. Methyl methacrylate (MMA) and 2-hydroxyethyl methacrylate (HEMA) were used separately as a monomer for the synthesis of IPNs. The IPNs were synthesized by polymerizing the methacrylate monomer in the immediate presence of PEO using benzoyl peroxide as an initiator and triethylene glycol dimethacrylate (TEGDM) as a crosslinker. The IPNs made using MMA and HEMA are designated as PEO/x-PMMA and PEO/x-PHEMA, respectively.

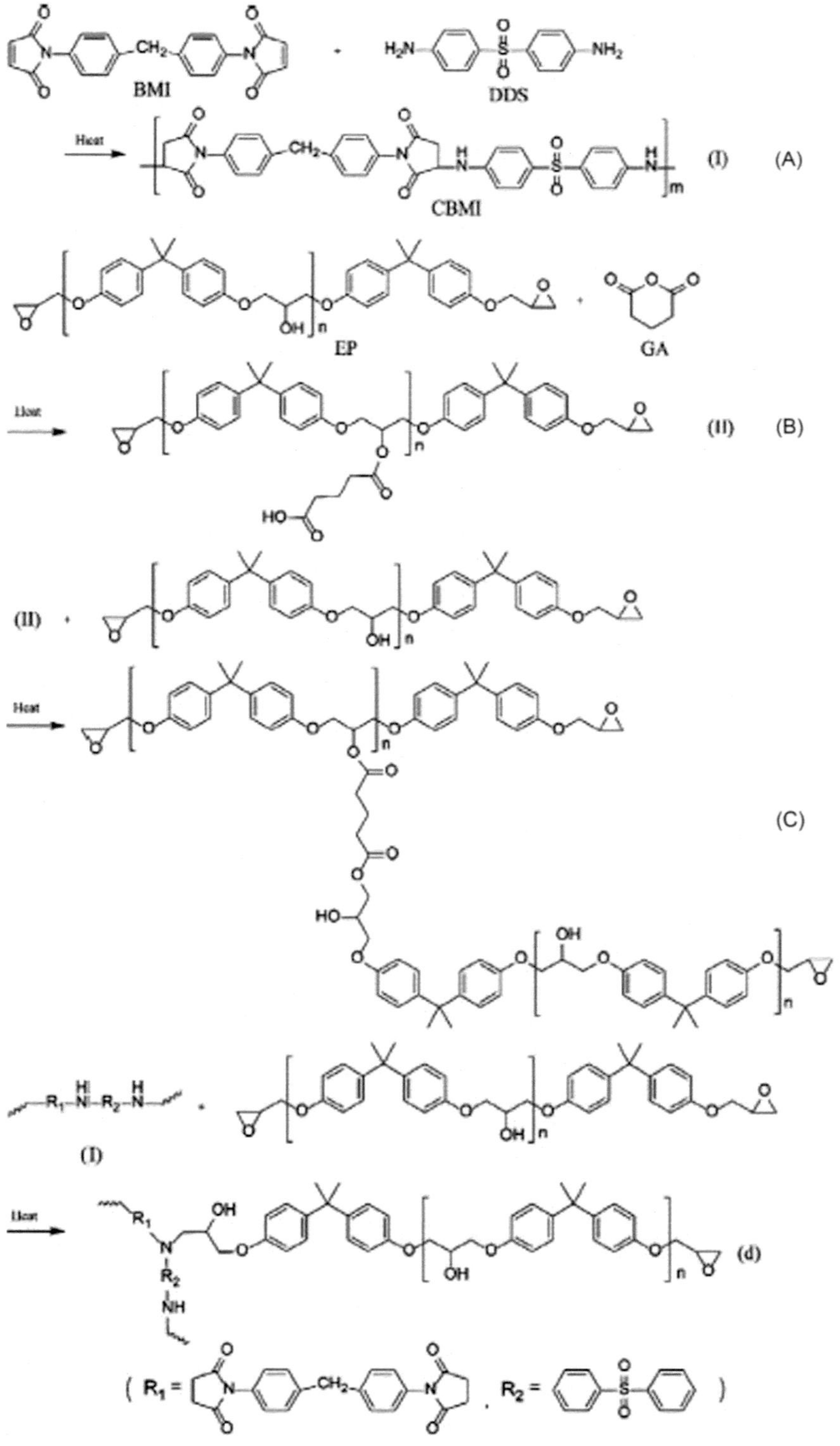

Figure 2.33
Reaction mechanisms during the synthesizing EP/CBMI polymers. Source: *Reprinted with permission from Z. Ding, L. Yuan, Q. Guan, A. Gu, G. Liang, Polymer 147 (2018) 170–182, ©2018 Elsevier Publishers.*

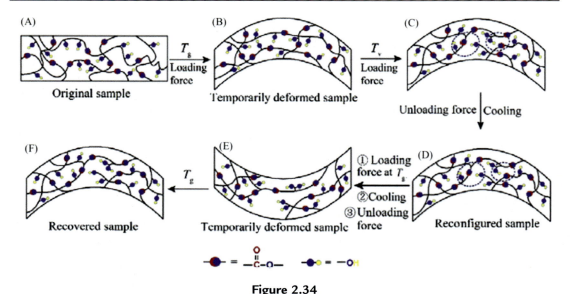

Figure 2.34
Schematic of shape reconfiguration mechanism. Source: *Reprinted with permission from Z. Ding, L. Yuan, Q. Guan, A. Gu, G. Liang, Polymer 147 (2018) 170–182, ©2018 Elsevier Publishers.*

PEO is miscible with the methacrylate monomers (MMA or HEMA) and TEGDM mixture at 60°C. However, with the advancement of the free radical-initiated crosslinking reaction of methacrylate/TEGDM system, the combinatorial entropy of mixing decreases. As a result, the crystalline part of PEO undergoes reaction-induced phase separation leading to the formation of a two-phase microstructure and the amorphous part of PEO remains dissolved with the methacrylate network. Thus the methacrylate network forms the continuous phase and the intermingled crystalline phase of PEO forms the dispersed phase, which acts as the switching phase due to the reversible process of melting and crystallization. The recovery process for a circular disc-shaped sample of IPN (PEO/PMMA: 70/30 w/w 4 wt.% crosslinkers) is shown in Fig. 2.35. The disc is first heated to about 70°C and deformed to a temporary shape (folded) and fixed the shape by cooling to room temperature under stressed conditions. The sample was then kept in an oven at 100°C. The photographs were taken at various intervals during the recovery process (Fig. 2.35). The recovery process clearly indicates almost 100% shape recovery. The shape recovery and fixity depend on the IPN composition and crosslink density, which can be controlled by just changing the crosslinker concentration. Shape fixity can be enhanced by increasing crosslinker concentration but shape recovery reduces beyond an optimum crosslinker concentration. Keeping the network the same, the recovery time and temperature depending upon the application can be developed by changing the switching segment. For example, the crystallization temperature of PEO can be decreased and tailored by using PEO/novolac blends [224,225].

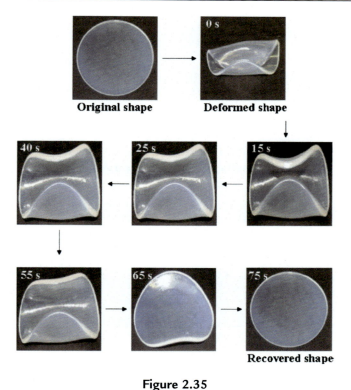

Figure 2.35
Recovery of a disk shaped sample of x-PMMA/PEO semi-IPN (x-PMMA/PEO = 68/32, 4 wt.% crosslinker) as a function of time at 100°C. Source: *Reprinted with permission from R. Debdatta, J. Karger-Kocsis, Polym. 52 (2011) 1063, ©2011 Elsevier Publishers.*

The PEO/x-PMMA IPNs are slightly translucent and become fully transparent at a temperature beyond 70°C due to the melting of the PEO crystals. This behavior is usual for any semicrystalline polymer system. The crystallites scatter light and reduce the transparency. When the crystals melt the transparency increases. It may be noted that PEO form compatible blend with PMMA. Interestingly, the PEO/x-PHEMA IPN transformed [124] from translucent to opaque when heated to a temperature beyond 70°C The transformation of translucent to opaque is reversible. For a thin sample (<3 mm thick), the reversibility can be demonstrated by simply blowing hot air using a drier as shown in Fig. 2.36. This behavior is quite different than that observed in a semicrystalline system discussed above. This unusual optical behavior can be explained by considering the fact that PEO is incompatible with x-PHEMA due to the polarity mismatch [124]. However, because of H-bonding between the hydroxyl groups of x-PHEMA with the ether group of PEO, the PEO crystallites remain sufficiently small to make IPN system translucent. When PEO melts, volume expansion takes place and the melt spreads the entire volume making the system opaque. When the sample is allowed to cool, it becomes translucent due to the

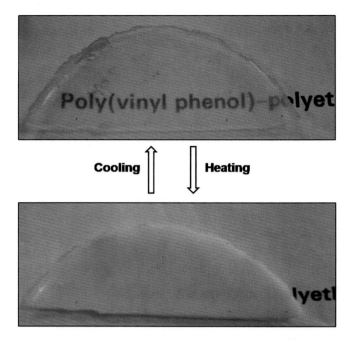

Figure 2.36
Demonstration of the reversible change in optical properties of the IPN. The sample is heated by blowing hot air using a drier. When the crystalline phase melts, the sample becomes opaque. Once the blower is stopped the sample starts getting cool and becomes translucent as a result of crystallization. Source: *Reprinted with permission from S.B. Jagtap, V.S. Rao, S. Barman, D. Ratna, Polymer 63 (2015) 41−51, ©2019 SCI Wiley Publishers.*

crystallization. Hence, in addition to shape-memory application, it can also find application as a smart optical window material.

Recently, Wang et al. [226] investigated epoxy/PU IPN for shape-memory applications. Hyperbranched polyurethane (HBPU) with s-triazine unit in the structure is synthesized by reacting polytetramethylene ether glycol (PTMEG) and 2,4-tolylene diisocyanate (TDI) with 1,3,5-tris(2-hydroxyethyl) isocyanurate. The HBPU/EP IPNs were synthesized through the grafting reaction of −NCO terminal group of HBPU prepolymer with oxirane ring of epoxy resin to form the network structure. EP/HBPU (80/20) IPN displays the best shape-memory effect with R_r and R_f more than 99%.

The shape-memory effect discussed above is the "one-way" type. How reversible or two-way shape-memory effect has also been reported. For example, Wu et al. [227] already proposed a reversible shape-memory polymer based on an IPN of poly(ε-caprolactone) (PCL) and poly(tetramethylene ether) glycol (PTMEG) with a potential application as artificial tendon. The crystalline PCL network is responsible for switching action.

The authors have used a "switch-spring" mechanism and provided a toolkit to easily convert existing shape-memory polymers toward two-way, reversible, shape-memory polymers. The PCL network offers shape shrinkage at high temperatures, through melt-induced compression, and PTMEG network was responsible for the elongation upon cooling by triggering the shape to expand.

2.11.1.7 Application of shape-memory thermoset

Potential applications for SMT exist in almost every area of daily life: from self-repairing autobodies to kitchen utensils, from switches to sensors, from intelligent packing to tools [228]. Other potential applications are drug delivery [229], biosensors, biomedical devices [230,231], microsystem components [232], and smart textile [233]. Since PU-based SMT can be made biodegradable, they can be used as short-term implants where removal by surgery can be avoided. Some important applications are discussed below in detail.

SMPs in general and SMTs, in particular, have tremendous applications in biology and medicine [234,235] especially for biomedical devices, which might permit new medical procedure. Because of the ability of shape to memorize a permanent shape that can be substantially different from an initial temporary phase, a bulky device could potentially be introduced into the body in a temporary shape like a string that could go through a small laparoscopic hole and then expanded on demand into a permanent shape at body temperature. As a result, major surgeries can be avoided.

Shape-memory PU has been proposed as a candidate for aneurysm coils [236]. An intracranial aneurysm can be a serious condition that can go undetected until the aneurysm ruptures, causing hemorrhaging within the subarachnoid space surrounding the brain. The typical treatment for large aneurysms is by remobilization using platinum coils. However, in about 15% of the cases treated by platinum coils, the aneurysm eventually reopens as a result of the bioinertness of platinum. One solution to this is to develop suitable materials with increased bioactivity like SMPs to use as coil implants.

Another example of a biomedical application is a microactuator made from thermosetting PU, which was used to remove blood vessel clots [237]. A microactuator, with a permanent shape of a cone-shape coil, can be elongated to a straight wire and fixed before surgery and delivered to an occlusion through a catheter. On triggering the shape recovery using an optical heating method, the original coil shape is recovered and the blood flow is restored. A concept of cold-hibernated elastic memory utilizing SMPs in open cellular structures was proposed for space-bound structural application [237]. The concept of cold-hibernated elastic memory can be extended to a variety of new applications like microfoldable vehicles, shape determination, and microtags [238]. Recent studies on shape-memory PU-based conductive composites using conducting polymer and carbonaceous nanomaterials show considerable promise for their application as electroactive and remote sensing actuators [239]. SMTs are also the potential materials for deployable structures related to aircrafts and spacecrafts. The deployable thin film structure in a satellite is to be folded using the hinges (made up of shape-memory material) to save storage space and required to be opened to their full span, once deployed into space by shape recovery of the hinges. The recovery process should be steady and well-controlled because fast recovery may vibrate the satellite in space. SMTs offer wide scope of design variations to meet such demands.

2.12 Vitrimers

As discussed above thermosets offer better dimensional stability, environmental stress cracking resistance, easier processing compared to thermoplastics. Thermoplastics are easily deteriorated by chemicals and mechanical stress, while the opposite is true for thermosets, which are insoluble and infusible. This makes thermosets an indispensable part of structural components in the automotive and aircraft industries. However, a major drawback of thermoset is that molding is not possible once the polymerization is completed due to the formation of irreversible linking by covalent bonds. This means once a product is made out of a thermoset, it cannot be reprocessed or recycled. Since they are not biodegradable, they cannot be destroyed biologically. The burning of thermosets causes environmental pollution. Thus the use of thermoset-based products poses a waste management issue that has been expressed by various social concerns for the environment. On the other hand, thermoplastics in which polymer chains are held together by weak interactions like van der Waals forces can be deformed reversibly above their glass transition temperature or their crystalline melting point and can be easily processed by extrusion, injection molding, and welding.

From the above discussion, it is clear that thermosets and thermoplastics have their own advantages and disadvantages. Considering the advantages and drawbacks of both thermoplastics and thermosets, the ideal polymer would be one, which is easy to process but held the chains together by strong covalent bonds. Vitrimers are a new class of polymers, which offer processabilities like a thermoplastic and mechanical property and chemical stability like a thermoset. The term was first coined by Leibler and coworkers in 2011 [240]. If we look at the chemistries of vitrimers, they consist of molecular, covalent networks, which can change their topology by thermally activated bond-exchange reactions. At high temperatures, they can flow like viscoelastic liquids. However, at low temperatures, the bond-exchange reactions are extremely slow, because of this reason the vitrimers behave like classical thermosets at this point.

Vitrimers are the first strong organic glass former and can be welded like silicon glasses or metals. Vitrimers possess permanent yet dynamic networks that rely on an associative exchange mechanism of the covalent bonds. This ensures that the network connectivity in thermally triggered vitrimers is preserved at all temperatures. As a result, vitrimers do not undergo a gel-to-sol transition and in principle remain insoluble. Unlike classical polymer melts, whose flow properties are largely dependent on friction between monomers, vitrimers become a viscoelastic fluid because of exchange reactions at high temperatures as well as monomer friction. These two processes have different activation energies, resulting in a wide range of viscosity variation. Moreover, because the exchange reactions follow Arrhenius' Law, the change of viscosity of vitrimers follows an Arrhenius relationship with the increase of temperature, differing greatly from conventional organic polymers.

An attractive chemical strategy to introduce plasticity in crosslinked polymer networks is offered by the introduction of exchangeable chemical bonds, leading to the formation of dynamic crosslinks. If chemical crosslinks can be efficiently and reliably exchanged between different positions of the organic polymer chains, macroscopic flow can be achieved without risking structural damage or permanent loss of material properties. Polymer networks containing such exchangeable bonds are also known as covalent adaptable networks (CANs). CANs may be further classified into two groups depending on their exchange mechanism. The first group of CANs makes use of a dissociative crosslink exchange mechanism. In this exchange, chemical bonds are first broken and then formed again at another place. The second group of CANs makes use of associative bond exchanges between polymer chains, in which the original crosslink is only broken when a new covalent bond to another position has been formed.

2.12.1 Synthesis of vitrimers

Leibler and coworkers [240] developed for the first time vitrimer (silica-like organic networks) using the well-established transesterification reaction of epoxy and fatty dicarboxylic or tricarboxylic acids. The synthesized networks have both hydroxyl and ester groups, which undergo exchange reactions (transesterifications) at high temperatures, resulting in the ability of stress relaxation and malleability of the material. On the other hand, the exchange reactions are suppressed to a great extent when the networks are cooled down, leading to behavior like a soft solid. This whole process is based only on exchange reactions, which is the main difference from that of thermoplastics. Following this pioneering work, several variations and alternative chemistries have been employed to develop new vitrimers.

Yang et al. [241] demonstrated that instead of heat, light can be used to process epoxy/acid-based transesterfication vitrimers by dispersing 1 w/w% of carbon nanotubes (CNTs) into the polymer matrix. CNTs absorb light of almost all wavelengths and transform the energy to heat, resulting in fast and precise local heating. CNT-impregnated vitrimers were successfully reshaped, healed and even welded with non-CNT vitrimers via irradiation with an infrared laser. The transesterfication vitrimers were also synthesized through isocyanate chemistry [242], using hydroxyl-terminated four-arm shaped polylactide and methylene diphenyl diisocyanate (MDI) monomers. A dioctyltin catalyst (Sn(oct)$_2$) was used for both the polymerization and the transesterfication reactions. The PU-based vitrimers made with isocyanate/alcohol ratios displays remarkably shorter relaxation times (50 s at 140°C) compared to epoxy vitrimer. Williams and coworkers [243] prepared networks from epoxidized soybean oil and citric acid in the presence of water, which are thus similar to Leibler's original system, using only renewable feedstock chemicals. Remarkably, these materials relax stresses and are healable, even without the addition of any catalyst. Since no catalyst is present, relaxation times are considerably higher, 5.5 versus 2.4 h for epoxy/acid resins with 5% Zn(OAc)$_2$ at 150°C. However, full stress relaxation could not be achieved in these epoxy/acid resins, probably due to the formation of irreversible crosslinks upon prolonged heating.

Prez and coworkers [244] explored vinylogous urethanes as a catalyst-free exchangeable group for developing vitrimers as an alternative to transesterfication vitrimers. From a thermodynamic point of view, the exchangeable carbon−nitrogen bond group is closely related to an amide or a urethane bond, because of the favorable electronic conjugation with the carbonyl function, mediated by the carbon−carbon double bond (cf. vinylogy principle). Vinylogous urethanes are thus stable toward hydrolysis and can even be formed quantitatively in water as a solvent. From a kinetic point of view, the vinylogous urethane moiety also possesses the reactivity of a typical Michael acceptor through a facile conjugated nucleophilic addition of an amine group, displacing the less stable carbon−carbon double bond, rather than the much stronger carbonyl double bonds. Drockenmuller and coworkers [245] reported new vitrimers based on transalkylation reactions in networks containing 1,2,3-triazolium salts, triazolines, and pendant alkyl halide chains. Such polyionic networks were synthesized via a one-pot process through a polyaddition of azide−alkyne monomers involving a thermal "nonclick" azide−alkyne Huisgen 1,3-dipolar cycloaddition and a simultaneous crosslinking step, using a bifunctional alkylating agent such as dibromo and diiodo alkanes or alkyl mesylates. The resulting networks showed typical Arrhenian stress relaxation with an experimental activation energy of 140 kJ mol^{-1} for the species with a bromide counter-ion. Short relaxation times ranging from 30 min at 130°C to a few seconds at 200°C were observed.

The olefin metathesis reaction (a very powerful tool for the formation of C−C bonds) has been used to prepare vitrimer based on crosslinked polybutadiene networks containing second-generation Grubbs catalysts [246]. This catalyst was used because of its superior stability toward air and moisture and its good functional group compatibility, as compared to more classical metathesis catalysts. Cross-metathesis is a good method for topology rearrangements via group exchange equilibria in networks, as no driving force is required. Note that cross-metathesis as such lacks an internal driving force.

Pepels et al. [247] reported dynamic networks based on base-catalyzed thiol-disulfide exchange reactions and could correlate the mechanical relaxation times with those of the model compounds, showing properties reminiscent of vitrimers, probably related to the associative exchange mechanism. However, as oxidation of free thiols readily occurs under air and free thiols are required for the exchange reaction, these materials show a marked deterioration of dynamic mechanical properties over time. Zhang and coworkers [248] exploited associative imine chemistry for the design of malleable and recyclable polymer networks. These crosslinked polyimine networks were prepared via the condensation of commercially available aldehydes and a mixture of di- and triamines in a solvent combination that further drives the reversible imine forming condensation, which only has a very small intrinsic thermodynamic driving force. The stress-relaxation behavior of the dried polyimine networks exhibited Arrhenius-like temperature dependence together with water-induced stress relaxation at room temperature, indicating vitrimer like properties. An overview of different vitrimer materials published recently [249] and their healing conditions are presented in Table 2.13.

Very recently, a concept to convert permanently crosslinked thermosets into vitrimer-type dynamic networks with a simple one-step method without depolymerization has been established and termed as "vitrimerization" [250]. The process is efficient at the same time economic, ecofriendly, and scalable. In this method, an anhydride-cured epoxy network is converted to a vitrimer by subjecting it to ball milling and treatment with a zinc acetate catalyst. The metal-carboxylate ligands are formed during ball milling and serve as junctions for the transesterification exchange reaction. The process is illustrated in Fig. 2.37. The vitrimer epoxy can be recycled and reprocessed. Vitrimer epoxy displays comparable properties with virgin epoxy and can be easily reprocessed into nanocomposites to further improve the properties for value-added products. The process can be extended to other thermoset systems encouraging more and more research in this area. Leibler and coworker [251] used metathesis of dioxaborolanes to prepare vitrimers from different polymers such as PMMA, polystyrene, high-density polyethylene, which can be processed multiple times by means of extrusion and injection molding. They exhibit superior chemical resistance and dimensional stability and can be assembled efficiently. The strategy can be extended to any polymer with carbon−carbon single bonds present in the backbone.

2.12.2 Application of vitrimers

Vitrimers can be used for all possible applications where conventional thermosets are used with the added advantage of self-healing attributes. A surfboard of vitrimers could be brought into a new shape, scratches on a car body could be healed and crosslinked plastic or synthetic rubber items could be welded. Vitrimers which are prepared from metathesis of

Table 2.13: Overview of different vitrimer materials and their healing conditions.

Mechanism	Material	Transition temp (T_g)	Healing conditions	Healing efficiency	Mechanical strength
Transestrification	Biobased triepoxy	187°C	220°C and slight pressure	90%	62.8–69.2 MPa[a]
	Epoxy with tertiary amine	130°C	160°C, 3 h	57%	0.461 MPa[a]
	Amino-capped aniline trimer liquid crystalline	58°C	IR laser	100%	20–22 MPa[b]
	Epoxy with polydopamine-modified gold nanospheres	42°C	Light, 120 min	90%–100%	25 MPa[b]
Transalkylation	Poly(1,2,3-triazolium ionic liquid)	−8°C to 23°C	—	—	15 MPa[f]
	Poly(thioether)	−20°C	160°C, 45 min	100%	80 MPa[f]
Disulfide	Epoxy with silica nanoparticles	151.5°C–155.5°C	180°C, 30–120 min	36.8%–78.1%	764 ± 8[c]
	Polyhydroxy urethane	66°C	150°C, 5–10 MPa for 30 min	65%	35 MPa[a]
Transamination	Epoxy with FRP	127°C	200°C, 100 bar for 5 min	—	242–292 MPa[d]
	Polycarbonate	19°C – 34°C	160°C, 5–10 MPa	64%–80%	7 ± 1 MPa[c]
	Poly(vinylogous urethane)	87°C	150°C, 30 min	100%	2 × 10³ MPa[e]
Trans-carbamoylation	Polyhydroxy urethane	54°C	160°C, 4 MPa for 8 h	76%	2.2 ± 0.4 GPa[e]
Imine	Polyimine	102°C	110°C for 30 min	—	49 MPa (dry sample) and 32 MPa (wet sample)

[a] Tensile strength.
[b] Stress–strain.
[c] Tensile modulus.
[d] Compression strength.
[e] Young modulus.
[f] Storage modulus.

Source: Reprinted with permission from B. Krishnakumar, R.V.S.P. Sanka, W.H. Binder, V. Parthasarthy, S. Rana, N. Karak, Chem. Eng. J. 385(2020) 123820, ©2020 Elsevier Publishers.

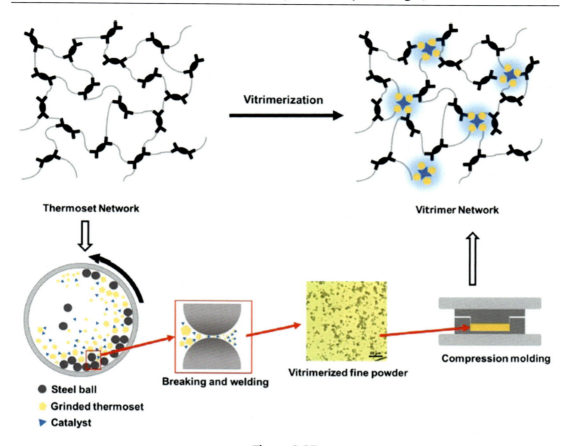

Figure 2.37
Illustration of the vitrimerization concept of converting the thermoset network into a vitrimer network via mechanochemical approach. Source: *Reprinted with permission from L. Yue, H. Guo, A. Kennedy, A. Patel, X. Gong, T. Ju, et al., ACS Macro Lett. 9 (2020) 836 – 842, ©2020 ACS Publishers.*

dioxaborolanes with different polymers that are commercially available can have both good processability and outstanding performance such as mechanical, thermal, and chemical resistance. The polymers that can be utilized in such methodology range from poly(methylmethacrylate), polystyrene, to polyethylene with high density and crosslinked robust structures, which makes this preparative method of vitrimers able to be applied to a wide range of industries. Recent NASA-funded work on reversible adhesives for in-space assembly has used a high-performance vitrimer system called aromatic thermosetting copolyester as the basis for coatings and composites reversibly bondable in the solid-state. Thus vitrimers are going to provide new possibilities for the assembly of large, complex structures for space exploration and development.

2.13 Metal−organic frameworks

Metal−organic frameworks (MOFs) are coordination polymers, consisting of metal ions/clusters and poydentate organic ligands [252−254]. Organic ligands are the groups that can donate multiple pairs of electrons to metal ions. The polydentate organic bridging ligands are coined as secondary building units (SBU) in MOFs chemistry. MOFs can be microporous, mesoporous as well as macroporous materials depending upon the size of organic ligand and metal ion. In most cases, the pores are stable during the elimination of guest molecules (often solvents) and can be refilled by others materials like gases and liquids. Because of this property, MOFs have been viewed as potential materials for various applications like the storage of gases like hydrogen and CO_2. Different combinations of metals and ligands have been used to synthesize a large number of MOFs with wide structural diversity and topologies with highly adjustable pore sizes and shape. Thus it is possible to design MOFs with tunable cavity formation and properties.

MOFs are characterized by an extraordinarily large surface area, thus have application in gas storage, separation and catalysis. They have well-arranged crystalline structure adjustable functionality and high porosity. Generally, they are resistant to structural collapse upon heating. MOFs can be theoretically designed into networks having finely tuned pore size, flexibility, and crystal structure. In recent years, MOFs have become a fast-growing research field. Although the development of molecular-based crystal engineering led coordination chemists to focus on the assembly of organic and inorganic building blocks at the beginning of 1990 [255], in order to form porous structure, the first MOF was synthesized in the late 1990s by Prof. Omar Yaghi and group [256]. So far, more than 20,000 different MOFs have been synthesized and characterized. The structural and chemical flexibility of MOFs has motivated extensive research and proposed their potential applications for gas storage, ion exchange, molecular separation, sensor, and heterogeneous catalysis.

The synthesis and design of MOFs have taken focus on the reticular chemistry (small net). In this approach, small net molecules are used as organic linkers. Flexible MOFs, also known as soft porous crystals (SPCs), are a subset of MOFs that possess both a highly ordered network and structural transformability. A subset of SPCs is the so-called elastic-layered MOFs (ELFs). ELFs are made up of metal vertex ions, bridging ligands, and charge-balancing counter-ions, which are connected into two-dimensional sheets that form a three-dimensional stacked structure. These types of MOFs have a latent porosity for the adsorption of gas molecules above a specific pressure (called "gate pressure") that results in an expansion of the interlayer spacing between the two-dimensional sheets and a corresponding jump in the adsorption isotherm that cannot be classified in accordance with the conventional IUPAC isotherm designation.

2.13.1 Synthesis of metal–organic framework

The synthetic scheme of three MOFs, for example, MOF-5, IRMOF-3, and PSM-IRMOF-3, is presented [257] in Fig. 2.38. MOF-5 and IRMOF-3 are generated through the linkage of 1,4-benzenedicarboxylate units through octahedral $[Zn_4O]^{6+}$ clusters. These linear and octahedral building blocks combine to form primitive-cubic lattices (Fig. 2.38). If we look at the structure of MOF-5 and IRMOF-3, then we can see the presence of an additional amino group on each of its carboxylate linkers in IRMOF-3. MOF-5 does not have any amino group. These types of structures are called isoreticular frameworks, which possess the same underlying structure, in this case, a primitive-cubic lattice, but have unique pore metrics and functionality. MOF-5 is prepared by heating solutions of 1,4-benzenedicarboxylic acid and zinc nitrate hexahydrate in dimethylformamide (DMF) at 80°C for 10 h. Similarly, IRMOF-3 can be prepared by heating solutions of 2-amino-1,4-benzenedicarboxylic acid and zinc nitrate hexahydrate in DMF at 100°C for 18 h. IRMOF-3 is further reacted with an anhydride to form PSM-IRMOF-3 having amide linkages.

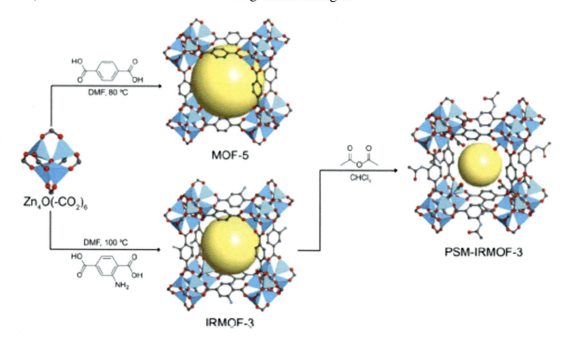

Figure 2.38

Synthesis of MOF-5 and IRMOF-3 and postsynthetic introduction of amide groups onto IRMOF-3 yielding PSM-IRMOF-3. In the diagram of the SBU on the left, the $-CO_2$ units originate from the carboxylate linkers and have been truncated to highlight the structure and coordination geometry of the cluster. Yellow spheres represent the frameworks' free spaces and have no chemical meaning. Atom colors: Zn, blue tetrahedra; C, gray; O, red; and N, blue. H atoms are omitted for clarity. Source: *Reprinted with permission from S.J. Lyle, R.W. Flaig, K.E. Cordova, O.M. Yagi, J. Chem. Edu. 95 (2018) 1512–1519, ©2018 ACS Publishers.*

There are two important synthetic methods used for the synthesis of MOFs, such as conventional methods and unconventional methods. In the conventional method, a solvent is used as medium and slow crystal growth is allowed. The method discussed above is an example of the solvothermal process. However, there are other approaches as well namely microwave-assistant method, sonochemical and mechanochemical synthesis, spray-drying synthesis, and electrochemical synthesis. The unconventional method is based on the concept of green chemistry. This process does not involve any solvent. Organic linkers and metal salts are ground together and then gently heated. This approach gives a high yield of product. The graphical presentation of the different synthetic procedures [258] has been mentioned in Fig. 2.39.

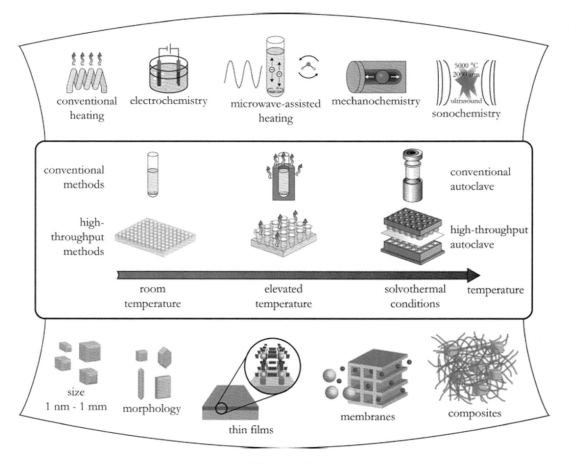

Figure 2.39
Overview of synthesis methods, possible reaction temperatures, and final reaction products in MOF synthesis. Source: *Reprinted with permission from N. Stock, S. Biswas, Chem. Rev. 112 (2012) 933–969, ©2012 ACS Publishers.*

2.13.2 Applications of metal–organic frameworks

In recent years, MOFs have been viewed as a potential candidate for many applications, for example, CO_2 adsorption, gas separation, catalysis and as electrode materials for supercapacitor and battery. MOFs have small and tunable pore size, better thermal and chemical stability owing to application for CO_2 adsorption. After some modification, MOFs show excellent selectivity for greenhouse gas. After saturation of MOF with CO_2, it can be removed by the temperature swing method. MOFs have a low cost of energy for regeneration. If we look at literature for the development of MOFs for CO_2 adsorption, various strategies have been adopted and observations are summarized below:

1. The selective adsorption efficiency of MOF increases with the introduction of basic groups in organic ligands. This is because Lewis acid CO_2 easily interacts with basic binding sites.
2. The carboxylates and amine-containing organic ligands are more applicable for the synthesis and designing of CO_2 capturing for MOFs.
3. Carbonized doped MOFs show better adsorption efficiency. This is a postsynthesis approach.
4. Amide functionalization can improve the storage capacity of MOF.
5. Doping of alkali metals in MOFs can improve the greenhouse gas adsorption capacity. Because unsaturated alkali metal sites can be bound CO_2 in MOFs.

MOFs can be used in the separation and purification of different gases [259] like H_2, SO_2, CH_4, etc. This is possible because pores can be tuned for different gases and functional groups with specific interactions can be incorporated by a suitable design strategy. Several MOFs are used as a thin film of semiconductors. It can be impregnated with grapheme for excellent results [260]. MOFs are novel candidates as electrode materials due to their salient features such as large surface area, three-dimensional porous architecture, permeability to foreign entities, structural tolerability, etc. In general, the pristine MOFs suffer from poor conductivity and not suitable for electrode applications. However, the conductivity issue can be resolved by making a composite with a suitable electronically conducting material. MOF-based electrodes are highly promising for flexible and wearable supercapacitors since they exhibit good energy and power densities [261]. Because of their inherently high surface area, MOFs have drawn wide attention as potential catalysts, offering a high density of catalyst sites in porous solids. The compositions of MOFs and the pore structures can strongly influence the catalytic performance [262].

2.14 Flame-retardant thermoset

Thermosets are made of flame retardants by incorporating suitable additives known as flame retardants (FR). For the various procedures of testing flame-retardant thermosets, the readers may refer to Chapter 6. FRs are known to function through two basic mechanisms of flame inhibition namely gas-phase mechanism and condensed-phase one. In the gas-phase mechanism of fire inhibition, the fire-retardant materials, when ignited, liberate nonflammable stable gases. These gases got mixed with the oxygen and fuel gases present above the surface. Thus they dilute the concentration of oxygen and flammable gases and also decrease the overall temperature of the surrounding polymer matrix. It also works through radical scavenging, thus reducing the count of reactive free radicals and energy feedback by the polymers. Halogenated compounds go through this mechanism during flame inhibition. Some of the fire-retardant compounds encourage the formation of an impervious blockade over the surface of the polymer. This barrier called char interrupts the escape of combustible gases and also inhibits the penetration of heat inside the bulk of the polymer. Compounds containing phosphorus, silicon and some inorganic borates show flame inhibition through condensed-phase mechanism. FRs can be divided into two major categories on the basis of their way of incorporation into the matrix such as additive type and reactive type. Additive flame-retardant chemicals are applied to a matrix material through blending with it. They do not react with each other and no bond formation takes place. Additive FRs known for their easy availability and easy processability, however, have problems like poor compatibility, leaching and reduced mechanical properties of the related thermoset networks. These problems can be resolved by using reactive type FR, which are incorporated into polymer matrix through chemical reactions. Reactive FRs require a much lower concentration to be added to a thermoset matrix compared to an additive FR for achieving flame-retardant properties. Different types of FRs are elaborated below with examples.

2.14.1 Inorganic flame retardants

Inorganic fillers are well-known flame retardants and smoke suppressants for polymers in general and thermosetting resins in particular. The main advantage of inorganic fillers is that they are nontoxic. However, the flame-retardant efficiency of inorganic fillers is so low that a very high loading of filler is required to be incorporated in order to achieve adequate flame retardancy. For example, aluminum trihydrate is the most commonly used inorganic flame retardant which is incorporated in amounts up to 300% of the resin level. These extremely high loadings pose a processing problem and deteriorate the mechanical properties of a composite substantially. The use of a combination of aluminum trihydrate with ammonium polyphosphate or red phosphorus significantly reduces the loading ($\sim 50\%$) to achieve optimum flame retardancy. Red phosphorous, an allotropic form of phosphorous, is not spontaneously flammable and is thermally stable up to 450°C. A silane treated grade of aluminum trihydrate (more expensive) is also available which is easier to incorporate into a thermoset resin system.

2.14.2 Halogen-containing flame retardants

Thermoset resins are made fire-resistant either by modifying the resin (or any component) with halogen or by using halogen-containing flame retardants. Halogenated polyols (brominated polyol, dibromo neopentyl glycol) are advantageously used for making flame-retardant PU. The polyols with aliphatically bound bromine like dibromo neopentyl glycol, tribromo neopentyl glycol are more effective than the polyols with aromatically bound bromine, which do not decompose at the lower temperature range of the burning of polymer. The polyester polyol containing hexachloroendomethylene tetrahydro phthalic acid structures are frequently used for the synthesis of flame-retardant PU. Chlorinated paraffins (CP) and decabromobiphenyl oxide (DB) [263] are generally blended with epoxy and polyester resins to improve flammability. However, the low molecular weight of CP or DB has a problem with migration. The migration problem can be solved by using brominated epoxy resins, which are mostly derived from diglycidyl ether of bisphenol-A (DGEBA) and tetrabromo bisphenol-A (TBBA) with a suitable catalyst [264]. Brominated epoxies of different grades are commercially available by Atul and Vantico. Halogen-containing compounds release hydrogen halide during decomposition, which interrupts the chain reaction. Due to the participation of halogen-containing compound in combustion the highly reactive ·OH and ·H radicals are replaced by the less reactive halogen X as shown below:

$$R\text{-}X + P\text{-}H \rightarrow H\text{-}X + R\text{-}P$$
$$H\text{-}X + H \rightarrow H_2 + X$$
$$H\text{-}X + OH \rightarrow H_2O + X$$

where RX and PH represent the flame retardant and the polymer, respectively. This strongly reduces the combustion rate. The effectiveness of halogens deceases in the order: $HI > HBr > HCl > HF$. Brominated and chlorinated organic compounds are generally used because iodides are thermally unstable at processing temperatures and the effectiveness of fluorides is too low. The choice depends on the type of polymer, for example, in relation to the behavior of the halogenated flame retardant under processing conditions (stability, melting, distribution) and/or the effect on properties and long-term stability of the resulting material.

Generally, halogen-containing flame retardants are used in combination with metal oxide-like antimony trioxide to achieve a synergistic effect. During fire, the antimony trioxide reacts with the chlorine of the flame retardant or polymer and forms antimony halides, which create a blanket of the gaseous layer. This acts as a gas barrier between the fuel gas and condensed phase. The reaction scheme is shown in Fig. 2.40. The problem in general with halogenated flame retardants is that during combustion the burning is associated with the release of toxic and corrosive gases, like hydrogen halides, which is a potential health hazard. Also, they can cause the severe degradation of the polymer chain to a combustible

$$R\text{-}X + P\text{-}H \longrightarrow H\text{-}X + R\text{-}P$$

$$Sb_2O_3 + 2HX \longrightarrow 2\,SbOX + H_2O$$

$$5\,SbOX \longrightarrow Sb_4O_5X_2 + SbX_3$$

$$4\,Sb_4O_5X_2 \longrightarrow 5Sb_3O_4X + SbX_3$$

$$3\,Sb_3O_4X \longrightarrow 4\,Sb_2O_3 + SbX_3$$

Figure 2.40
Reaction scheme for combustion of halogen-containing compound in presence of Sb_2O_3.

monomer or similar species [265]. Halogenated flame retardants not only affect humans but also animals and the environment. A few legislations were enacted to control the application of bromine-containing fire-retardant materials. For instance, the Waste Electrical and Electronic Equipment and the Restriction of Hazardous Substances came into act in European Union in 2003. In recent years, the research on the development of environment-friendly, so-called green flame retardants has received considerable attention. The need has not only been expressed by acute government regulations but equally and persuasively by various social concerns for the environment.

2.14.3 Phosphorous-containing flame retardant

Organophosphorous compounds have high flame-retardant efficiency in thermoset resins and also been found to generate less toxic gases and smoke compared to halogen-containing compounds [266]. Hence, the replacement of halogen-containing fire retardant by a phosphorous-containing one has a noteworthy benefit in terms of environmental protection. The mechanisms of flame retardancy have already been discussed in earlier sections. The presence of phosphorous helps to form a carbonaceous char or a barrier layer of polyphosphoric acid on the burning of the polymer in a condensed phase [265–267]. A chemical vapor-phase mechanism is found to be effective in those cases where the phosphorous-containing degradation products are capable of being vaporized at the temperature of the pyrolyzing surface. Triphenyl phosphine oxide and triphenyl phosphate have been shown to break down in the flame to small molecular species such as PO, HPO_2, PO_2, and P_2. The rate-controlling hydrogen atom concentration in the flame has been shown to reduce in presence of these species. Studies on the systems with a phosphorous atom in the chain as well as with the volatile phase species indicate that condensed-phase mechanism is more effective than vapor phase mechanism [268].

Very recently, Huo and coworker reported the mechanism (Fig. 2.41) for the flame-retardant performance of an epoxy resin cured with a liquid phosphaphenanthrene/imidazole-containing curing agent (DOPO-AI). DOPO-AI is prepared by reaction of

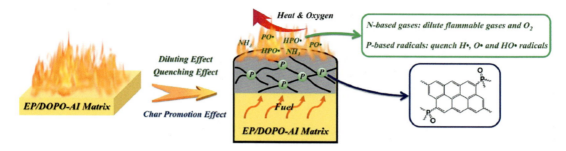

Figure 2.41
Illustration of the flame-retardant mechanism of EP/DOPO-AI thermosets. Source: *Reprinted with permission from S. Huo, S. Yang, J. Wang, J. Cheng, Q. Zhang, Y. Hu, G. Ding, Q. Zhang, P. Song, H. Wang, ACS Appl. Polym. Mater. 2 (2020) 3586−3575, ©2020 ACS Publisher.*

(9,10 dihydro-9-oxa-10-phosphaphenanthrene-10-oxide) (DOPO) and 1-(3-aminopropyl)-imidazole (AI). It was reported that DOPO-AI functions in both condensed and gaseous phases during combustion. In the condensed phase, P-based fragments from DOPO-AI promote the epoxy matrix to form a compact and intumescent char layer on the surface, which suppresses the heat release and smoke emission. In the gaseous phase, NH_3 and P-containing radicals generated by the degradation of DOPO-AI dilute the fuel and quench the H·, O·, and HO· radicals, which suppress the combustion reaction. Therefore it can be concluded that the significant enhancements in flame retardancy and smoke suppression are a result of both the condensed-phase and gaseous-phase effects of DOPO-AI.

The phosphorous-based flame-retardant system can be made by modifying the resin or curing agent to incorporate phosphorous into their chemical structure or by using reactive or nonreactive phosphorous-containing flame retardants. The reactive type flame retardants exhibit much better flame retardancy and overcome several drawbacks associated with physical blends of the epoxy and the flame retardants [270]. For phosphorous-containing epoxy resin and phosphorous-containing curing agents (amine, acid or anhydride) can be used to make the flame-retardant epoxy resin. A conventional epoxy can be modified with phenyl phosphate or a phosphorous-containing diol bis (3-hydroxyphenyl) phenyl phosphonate to prepare phosphorous-containing epoxies [271] as shown in Fig. 2.42. By the judicious selection of the curing agent of the epoxy, the fire-retardant property can be manipulated taking advantage of phosphorous-nitrogen synergism. Note that the synergistic property occurs probably due to the formation of P−N bonded intermediates, which are better phosphorylating agents than those of the related phosphorous compounds without nitrogen [268]. However, the introduction of phosphorous in the resin backbone or in the curing agents may affect the curing behavior and thermomechanical properties of the cured resin. It is necessary to study detailed kinetics and mechanical properties before selecting for a particular application.

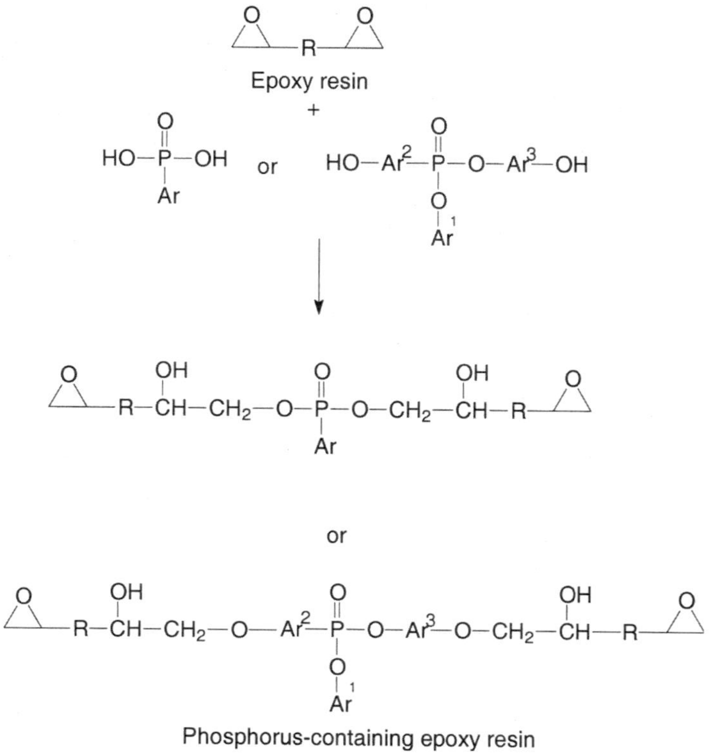

Figure 2.42
Synthesis of phosphorous-containing epoxy from conventional epoxy.

Hergenenrother et al. [272] synthesized the epoxy and the diamine curing agents containing phosphorous and evaluated the resins for composite applications. The optimized formulations showed excellent flame retardation with phosphorous content as low as 1.5 wt.%. The flame retardancy is achieved without any sacrifice in the properties due to the incorporation of phosphorous. Braun et al. [273] investigated the influence of the oxidation state of phosphorous on the decomposition and fire behavior of epoxy-based composites highlighting the potential for optimizing flame retardancy while maintaining the mechanical properties of epoxy/CF composites. They have used phosphene oxide, phosphinate, phosphonate, and phosphates (phosphorous content about 2.6 wt.%) and found that with increasing oxidation state of the phosphorous, additional charring was observed. So, the thermally stable residue increases whereas the flame inhibition, which plays an important role in the fire performance of the composites, decreases. The study of the decomposition behavior indicates that phosphorous-containing groups influence the decomposition of the epoxy, resulting in a clear multistep decomposition with mass losses between c. 15 and 20 wt.% in subsequent processes after the main decomposition step. The mass loss of the main decomposition process is reduced as a result of the incorporation of phosphorous.

A novel phosphorous-modified polysulfone was developed [274], which can act as a flame retardant-cum-impact modifier for an epoxy thermoset. Incorporating the modifier in an amine-cured difunctional epoxy could enhance both the toughness and the flame retardancy without a significant sacrifice in thermomechanical properties. Also, it was possible to generate a network of higher T_g under certain curing conditions due to the formation of an interlocked epoxy-thermoplastic network. In order to explain the fire property, decomposition behavior has been investigated. The possible decomposition pathway for DGEBA/DDS network is shown [274] in Fig. 2.43. The analysis of decomposed products indicates the formation of gases like SO_2, CO, CH_4, etc.

DOPO has been widely used to make a flame-retardant thermoset system. DOPO-based compounds have been shown [275,276] to induce a significant improvement in the flame

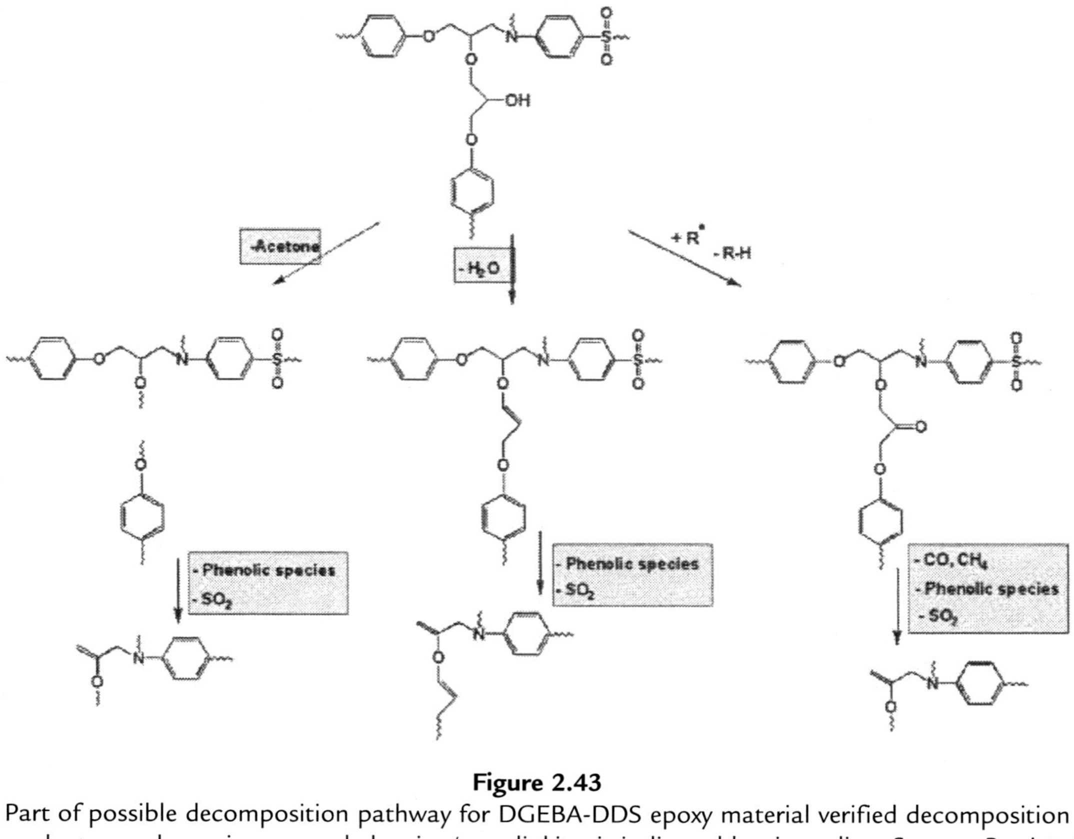

Figure 2.43
Part of possible decomposition pathway for DGEBA-DDS epoxy material verified decomposition products are shown in gray and charring/crosslinking is indicated by zigzag line. Source: *Reprinted with permission from U. Braun, U. Knoll, B. Schartel, T. Hoffman, D. Pospiech, J. Artner, et al., Macromol. Chem. Phys. 207 (2006) 1501, ©2006 Wiley-VCH Verlag GmbH & Co KgaA.*

retardancy of epoxies, while at the same time avoiding many disadvantages such as poor compatibility, migration of compound and release of toxic gases upon burning as compared to conventional flame retardants. DOPO can be used as such or it can be reacted with the resin or DOPO-based resin can be synthesized. For example, DOPO-based epoxy resin is synthesized by first reacting DOPO with p-benzoquinone to get hydroxyl derivative of DOPO. The DOPO derivative is further reacted with epiclorohydrin in presence of a base catalyst to prepare DOPO-based epoxy resins. The reaction scheme is shown in Fig. 2.44. The beauty of this system is that flame retardancy can be improved at a phosphorous content as low as 3 wt.%, thereby limiting the processing difficulties and the often severe degradation of the thermomechanical properties of the base resin.

A synergistic effect is reported to be observed when different flame-retardant groups or elements are integrated into one flame retardant. The synergistic flame-retardant systems constructed by two or more functional groups have been investigated with the aim to achieve higher flame-retardant efficiency while maintaining the good comprehensive performance of the resulting thermosets [277–282]. Zhang and coworker [283–285] used DOPO and polyhedral oligomeric silsesquioxanes (POSS) in combination to enhance the

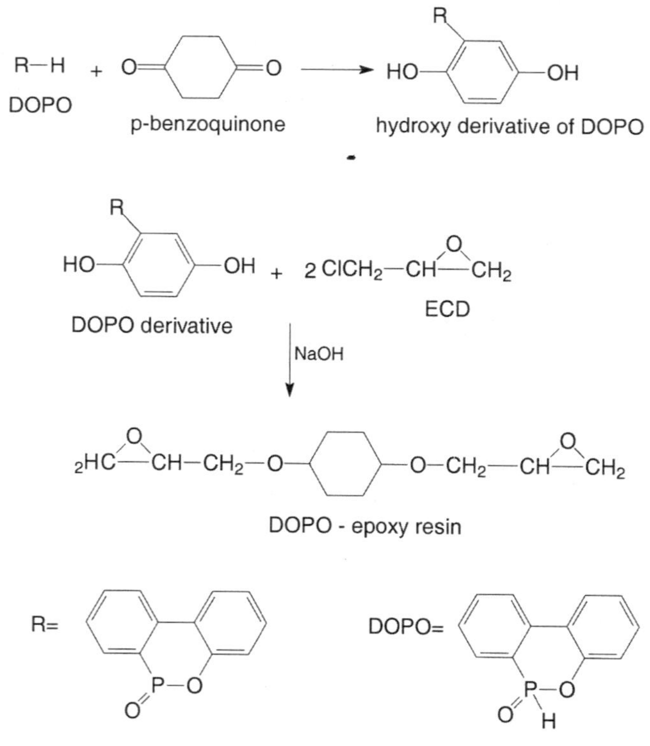

Figure 2.44
Synthesis of DOPO-based epoxy.

flame retardancy of the epoxy resin cured with m-phenylene diamine (MPDA) and demonstrated the synergistic effect of DOPO and POSS. It was observed that the pure EP burns rapidly after ignition and HRR reaches a sharp peak with a peak heat release rate (p-HRR) of 855 kW m^{-2}. The p-HRR of EP/5% DOPO is 731 kW m^{-2}. A similar observation is reported by others as discussed above. In the case of the EP/5% POSS sample, the p-HRR is 712 kW m^{-2} indicating the similar flame retarding effect of POSS. On the other hand, the p-HRR of EP/2.5%POSS/2.5%DOPO is 603 kW m^{-2}, which is much lower than that of EP/2.5% POSS/2.5% DOPO. This means that for the same total amount of flame retardants, the mixture of POSS and DOPO presents higher flame-retardant efficiency than that achieved using POSS or DOPO alone. This clearly indicates the P–Si synergistic effect.

The same group [285] has also successfully synthesized DOPO-containing POSS (DOPO-POSS) and used it as a flame retardant. A typical structure of DOPO-POSS is presented in Fig. 2.45. The synthesis of DOPO-POSS comprises two steps of the reaction. First DOPO is allowed to react with vinyl triethoxy silane (VTES) at 80°C for 10 h using azobisisobutyronitrile as a catalyst to prepare DOPO–VTES, which on hydrolytic condensation produce DOPO-POSS. The hydrolytic condensation was carried out in a methanol medium at 80°C for 24 h using hydrochloric acid as a catalyst. The DOPO-POSS has been incorporated into various epoxy systems cured with either aliphatic or aromatic amine and its effect on flame-retardant properties of the composites were investigated. The limiting oxygen index (LOI) values of epoxy resins cured with m-PDA increase obviously with the DOPO-POSS loading and vary in the range of 25%–30% for addition up to 2.5 wt.% DOPO-POSS. However, the LOI value reduces when 10 wt.% DOPO-POSS is added into the epoxy system cured with MPDA.

These results are quite different from the other studies, which show a linear increase of LOI with increasing the concentration of FR. The results of UL-94 test show a similar trend. This is explained by considering an interesting blowing-out extinguishing effect, which was

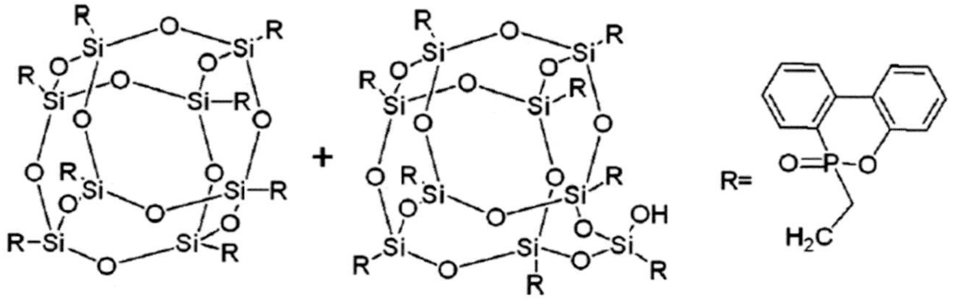

Figure 2.45
Typical chemical structures of DOPO-POSS molecules. Source: *Reprinted with permission from W. Zhang, X. Li, R. Yang, Polym. Degrad. Stab. 96 (2011) 2167–2173, ©2011 Elsevier Publishers.*

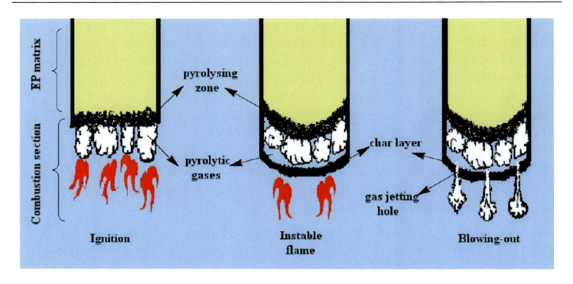

Figure 2.46
The model of the blowing-out effect. Source: *Reprinted with permission from W. Zhang, X. Li, L. Li, R. Yang, Polym. Degrad. Stab. 97 (2012) 1041–1048, ©2012 Elsevier Publishers.*

observed during UL-94 test. The model of the "blowing-out effect" is shown Fig. 2.46. The "blowing-out extinguishing effect" is that: "After the sample was ignited, it showed an unstable flame for several seconds; with the pyrolytic gaseous products jetting outward from the condensed-phase surface, the flame was extinguished, it looks like that the gas blew out the flame." A char layer survives and grows in the initial several seconds after ignition, yielding a rigid char-brick with the same dimensions as the original sample. The most attractive phenomenon is that jet of the pyrolytic gases from the holes on the char layer and flame hardly persists on the top of the airflow, which has separated from the surface of char layer. This is the so-called self-extinguishing effect through the pyrolytic property.

The effect of the nature of the curing agent on flame-retardant properties of epoxy/DOPO-POSS has also been investigated. It was reported that an aromatic amine (e.g., diaamino diphenyl sulfone, DDS, or MPDA) system offers better flame-retardant property than that of an aliphatic amine or amide-based one and the blowing-out effect not observed in the case of an aliphatic hardener-based system. For example, an aromatic amine (DDS) cured composite containing 2.5% DOPO-POSS offers UL-94 V_{-1} rating and an LOI of 27.1% whereas to achieve an UL-94 V_{II} rating, an aliphatic amine-cured epoxy composite requires 10 wt.% DOPO-POSS [286]. A novel metal (Ti)-embedded POSS was also reported; the same was further functionalized with DOPO through a feasible route. The FR when incorporated to an epoxy matrix, desirable LOI and char yield was achieved 1.5–6.0 wt.% loading. [287].

Very recently, Aurelio Bifulco et al. [288] reported the synergistic effect of melamine, silica and phosphorus by synthesizing silica−epoxy nanocomposites through a sol−gel method. Melamine and 3-(6-oxidodibenzo [1,2] oxaphosphinin-6-yl) propenamide (DA) were incorporated into an a cycloaliphatic amine-cured epoxy system in various concentration and fire properties of the modified epoxy networks were recorded. At only 2 wt.% of DA, V_{-0} rating was obtained and a remarkable reduction (48%−70%) in HRR was observed. In addition, the modulus of the network is also improved due to the combined effect of DA, melamine, and silica. Organic/inorganic hybrid materials have been synthesized by reacting DOPO with vinyl-trimethoxysilane (VTS) by a sol−gel technique [288]. The LOI value of the epoxy thermoset obtained after the addition of DOPO-VTS was 32 (LOI for pure epoxy, 22.5) and V_{-0} rating was obtained at 15 wt.% loadings into the resin. The char residue was 23.8 wt.% at 700°C as observed in the thermogravimetric analysis [289]. A P/N/S-based flame-retardant cocuring agent (PAD) containing phosphaphenanthrene, phosphonate, and benzothiazole functional groups was also reported [286]. The chemical structure and the mechanism by which PAD imparts flame-retardant property to a thermoset like epoxy presented in Fig. 2.47. As shown in the figure, phosphate and the related analogs were formed due to the degradation of PAD in advance. These compounds catalyzed the charring of EP matrix. On the other hand, the phosphorus radicals and nonflammable gas like NH_3 and CO_2 derived from the pyrolysis of PAD were wrapped in the thick char, which made the char become intumescent. The phosphorus radicals could quench the hydroxyl free radical and the nonflammable gas could dilute the flammable gas. On burning the volatile gases are produced and the number of gaseous products finally exceeded the intake capacity of the char layer and large amounts of gaseous products were released into the gaseous phase with a rush to extinguish the flame, namely the blowing-out effect as discussed above [285]. Epoxy/PAD/DDS composition [286] with a phosphorous content of 0.5% offers UL-94 V_{-0} rating with a value of LOI 33.6%. This is achieved with only a slight (3−10°C) sacrifice in T_g.

Xu et al. [290] reported a phosphorous-nitrogen synergistic effect using DOPO-containing Schiff-base flame retardant for epoxy resin cured with diaminodiphenyl methane (DDM). The modified epoxy thermoset exhibits excellent flame retardancy at the low phosphorus content, in which the V_{-0} grade of UL-94 classification with LOI of 39.7% is achieved when the phosphorus content is only 0.5 wt.%. Besides, T_g of the flame retarded epoxy thermoset remains relatively constant. Phosphorus element plays a leading role in the flame retardation due to PO. released from phosphorus-containing species during combustion, which exhibits the flame retardancy through the quenching mechanism of active radicals in the gaseous phase. Besides, the degradation of the phosphorus-containing group also produces phosphate and polyphosphate structure in the condensed phase, which promotes the configuration of dense carbon layer to prevent heat transfer. The presence of nitrogen elements generates a lot of nitrogen gases to dilute the concentration of oxygen necessary

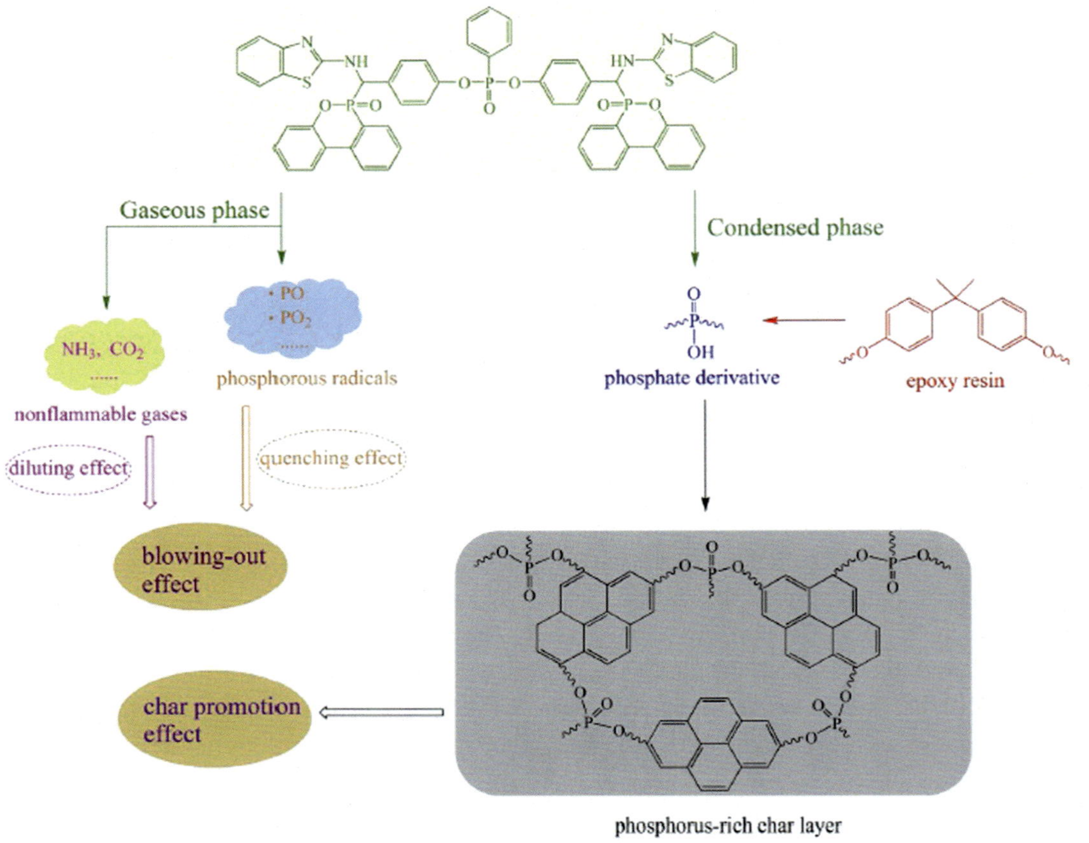

Figure 2.47
The flame-retardant mechanism of PAD on EP. Source: *Reprinted with permission from Q. Zhang, J. Wang, S. Yang, J. Cheng, G.D.Y. Hu, S. Huo, et al. Fire Safety J. 113 (2020) 102994, ©2020 Elsevier Publishers.*

for maintaining combustion. Therefore highly efficient flame retardancy can be obtained from the combination of these two major impacts.

Most of the reported flame-retardant compositions with UL-94 V_{-0} rating are based on aromatic amines because of inherently higher flame-retardant properties of aromatic amine-cured epoxy compared to aliphatic amine-cured ones. However, an ambient temperature curing system saves energy and very much required for the application where heat curing is practically not feasible. Wang et al. [291] reported a poly(DOPO substituted dihydroxy phenyl pentaerythritol diphosphonate)-based flame retardant and used it to make a room-temperature curable flame-retardant epoxy composition. They have achieved UL-94 V_{-0} rating in a room-temperature curable epoxy system by incorporating a 10 wt.% modifiers in a polyamide cured epoxy system (phosphorous content 1.7%). However, being a nonreactive modifier, it will likely have issues related to compatibility and durability.

Sen and coworker [292] have utilized phosphorous-containing epoxy resin (modified with DOPO) and phosphorous-containing silica filler to achieve UL-94 V_{-0} rating in a room-temperature curable epoxy system.

A novel DOPO-based oligomer indicated as PDAP was synthesized [293] via the nucleophilic addition reaction between DOPO and imine obtained from the condensation reaction of terephthalaldehyde and p-phenylenediamine. The video screenshots of pure epoxy (EP) and epoxy/PDAP composites with varying concentrations of PDAP (EP/PDAP4, EP/PDAP7, EP/PDAP10) during combustion are presented in Fig. 2.48. It is noteworthy that EP is ignited quickly, and then burns aggressively with serious flaming drips, indicating its highly flammable nature. In contrast, all composite sample exhibit auto-extinguishable features, and moreover, the average burning time of the PDAP-modified epoxies decreases with increasing PDAP content. This clearly suggests improved flame retardancy epoxy/PDAP systems. Upon close inspection of video screenshots, one can easily note that all the flaming flame-retardant epoxy samples seemingly blow out by the ejected pyrolytic gases from the surface of char layers. A similar phenomenon was also observed previously by Zhang et al. [283–285] in epoxy resin modified with DOPO-POSS during combustion. This interesting phenomenon was suggested by them and was named as "blowing-out effect."

As discussed above, DOPO has been extensively used for epoxy systems. It is also successfully used for other thermosets [294]. Zhou and coworker [295] reported a novel flame-retardant composition containing phosphorous and nitrogen in the molecule and utilized for the development of rosin-based rigid polyurethane foam. The flame-retardant properties are achieved without affecting the closed-cell content, shape of the cells and thermal conductivity. Duan et al. [296] synthesized a series of fire-retardant vinyl ester (VE) resins via the copolymerization of phosphorus/nitrogen-containing acrylates [1,3,5-triglycidyl isocyanurate (TGIC)−acrylic acid (AA)−DOPO] and the commercially available VE resin (901-VER). The low LOI value (19.5%) of 901-VER casting suggested poor fire resistance in the ordinary condition, whereas the flame-retardant performance improved significantly after the introduction of TGIC−AA−DOPO. The LOI value of the 40% TGIC−AA−DOPO/901-VER casting reached 31.7%; this was 62.6% higher than that of the neat 901-VER casting. Similar observations were reported for cone calorimetry and UL-94 tests as well.

A DOPO-containing benzoxazine was successfully prepared from phenol, aniline and DOPO by a three-step procedure [297]. The first step is the condensation of 2-hydroxybenzaldehyde with aniline, forming an intermediate imine. The second step is the addition of DOPO to the imine resulting in a secondary amine. The third step is the ring closure condensation leading to the formation of DOPO-containing benzoxazine. The flame-retardant benzoxazine resins were also prepared [298] by copolymerization of

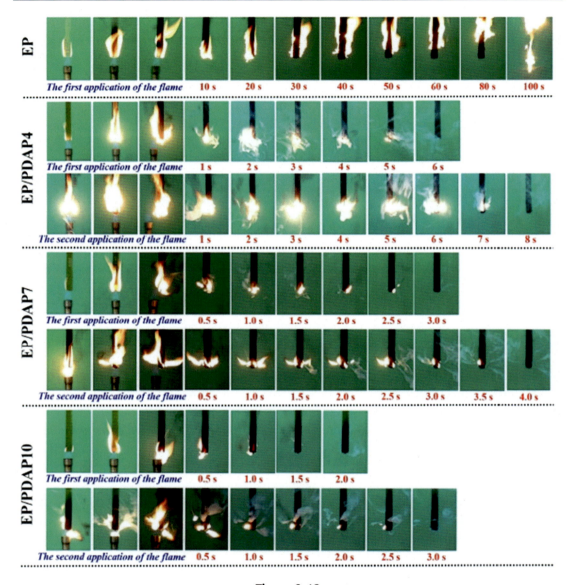

Figure 2.48
Video screenshots of pure epoxy EP, PDAP-modified epoxy thermosets with varying concentration of PDAP during the vertical burning test. Source: *Reprinted with permission from P. Wang, F. Yang, L. Li, Z. Cai, Polym. Degrad. Stab. 129 (2016) 156–167, ©2016 Elsevier Publishers.*

bisphenol-A-based benzoxazine (BA-a) and a phosphorous-containing phenolic derivative (DOPO-HPM), which display the highest LOI value of 32.6% and UL-94 V_{-0} rating. Chen et al. [299] reported a new cyanate ester (CE)-based resin system (DPDP/CE), which was developed by copolymerizing 10-(2,5-dihydroxyphenyl)-9,10-dihydro-9-oxa-10-phosphate phenanthrene-10-oxide (DPDP) and 2,2′-bis(4-cyanatophenyl) propane (CE). The modified CE resin requires a lower postcure temperature compared to the pure CE and offers lower water absorption and UL-94 V_{-0} rating. Note that CE resin has inherent flame-retardant property and shows UL-94 V_{-2} rating.

2.15 Additives for thermoset resin

2.15.1 Antioxidants

Thermoset resins (polymers in general) are susceptible to degradation through oxidation reactions during the processing at a high temperature or during thermal treatment and outdoor exposure. Oxidation of polymer resulted in changes in molecular structures causing a loss in mechanical properties (e.g., tensile, flexural, and impact strength) of the cured resin network and physical characteristics (e.g., gloss, finish, color, etc.) of the resin surface. The extent of oxidative degradation of polymer chains depends on the processing or service conditions and the chemical structure of the polymer. Aliphatic chains degrade more easily compared to the aromatic structure. Hence it is necessary to prevent this degradative reaction in order to increase the life expectancy of polymer or thermoset materials. The additives, which are used to protect the polymer materials from oxidative degradation, are called "antioxidants." β-carotene and α-tocophenol are the examples of natural antioxidants.

In order to know how the antioxidants function, it is necessary to know the mechanism of the oxidation process. The oxidation reaction is believed to proceed through a free radical mechanism [300–303]. The main feature of the oxidation mechanism is the initial peroxidation of the polymer substrate through the reaction of alkyl peroxyl radical (ROO.) with the substrate, which produces hydroperoxide (RCOOH). The peroxide is the major free radical generator for continuing free radical-initiated oxidation. The generation of free radicals and propagation of the oxidative degradation process is represented in Fig. 2.49.

Antioxidants interfere with the free radical oxidative cycle to inhibit or retard the oxidation mechanism. On the basis of the mechanisms by which antioxidants function, they can be classified into two categories namely primary or chain breaking (CB) antioxidants and secondary or preventive antioxidants. Chain-breaking antioxidants are of two types: chain-breaking donor (CB-D) antioxidants and chain-breaking acceptor (CB-A) antioxidants. CB-D antioxidants operate through stabilization reaction by reducing ROO. to ROOH (Fig. 2.50). The reaction is facilitated due to the higher stability of antioxidant free radicals (A.). These stable radicals do not participate in propagation reactions and get converted into nonradical products. Hindered phenol (e.g., 2.6-t-butyl 4 methyl phenol) and amines (2,2,6,6-tetramethyl piperidine) are good examples of CB-D antioxidants. Chain-breaking acceptor antioxidants perform by removing alkyl radicals from the chain propagating reactions and are effective under oxygen-deficient conditions [302,303]. Quinones are

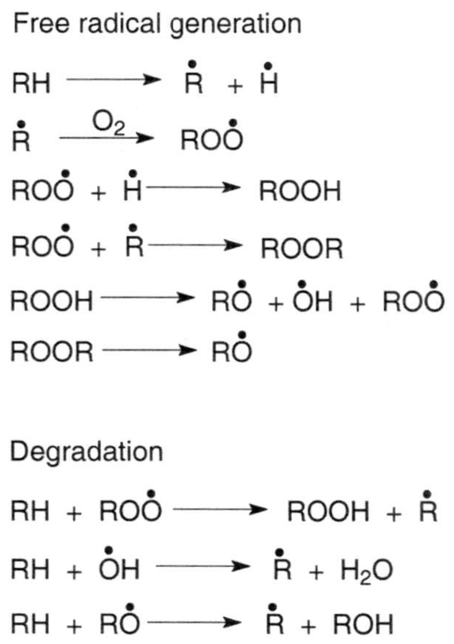

Free radical generation

$$RH \longrightarrow R^\bullet + H^\bullet$$
$$R^\bullet \xrightarrow{O_2} ROO^\bullet$$
$$ROO^\bullet + H^\bullet \longrightarrow ROOH$$
$$ROO^\bullet + R^\bullet \longrightarrow ROOR$$
$$ROOH \longrightarrow RO^\bullet + OH^\bullet + ROO^\bullet$$
$$ROOR \longrightarrow RO^\bullet$$

Degradation

$$RH + ROO^\bullet \longrightarrow ROOH + R^\bullet$$
$$RH + OH^\bullet \longrightarrow R^\bullet + H_2O$$
$$RH + RO^\bullet \longrightarrow R^\bullet + ROH$$

Figure 2.49
Free radical mechanism of oxidative degradation of polymers.

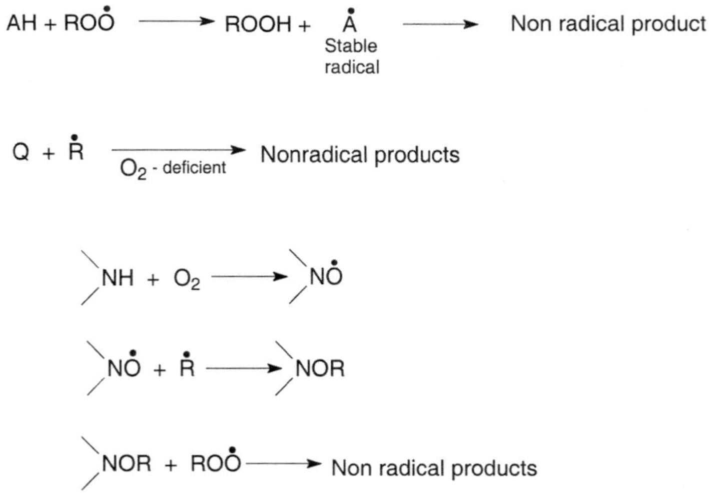

$$AH + ROO^\bullet \longrightarrow ROOH + A^\bullet \text{ (Stable radical)} \longrightarrow \text{Non radical product}$$

$$Q + R^\bullet \xrightarrow{O_2 \text{-deficient}} \text{Nonradical products}$$

$$\text{\textbackslash NH} + O_2 \longrightarrow \text{\textbackslash NO}^\bullet$$

$$\text{\textbackslash NO}^\bullet + R^\bullet \longrightarrow \text{\textbackslash NOR}$$

$$\text{\textbackslash NOR} + ROO^\bullet \longrightarrow \text{Non radical products}$$

Figure 2.50
Mechanism of antioxidant effect—formation of stable radicals or nonradical products.

important examples of CB-A antioxidants. Preventive or secondary antioxidants prevent the generation of free radicals and thereby reduce degradation [304,305]. They decompose the hydroperoxide into nonradical products. Phosphites and sulfide esters are classic examples of secondary antioxidants.

The use of both primary and secondary antioxidants results in a synergistic effect with respect to stabilization. The addition of a small number of such additives (0.05–0.5 wt.%) considerably increases the life of thermoset materials. Benzophenone and benzotriazoles compounds absorb ultraviolet (UV) radiation (referred to as UV absorber) and protect the polymer from the harmful degradative effect of UV light. Note that like heat the UV light initiates the free radical oxidation of the polymer chain. The UV absorbers convert the harmful UV radiation to comparatively harmless heat. In the case of 2-hydroxy benzophenone, the UV light is utilized for a reversible keto-enol transition [305]. While selecting an antioxidant for a particular thermoset matrix, it is necessary to consider the following aspects: (1) it should be compatible with the matrix, (2) it should not degrade during curing or postcuring, and (3) it should not exude out from the matrix after curing. However, the loss of antioxidants due to leaching or diffusion cannot be avoided for the conventional antioxidants discussed above [306–308]. In order to solve the migration problem, reactive antioxidants, which can be chemically anchored to the polymer backbones, have been explored [309–311].

2.15.2 Filler

Fillers are solid additives mostly of inorganic type, which are added to reduce the cost or to modify the physical properties (usually mechanical) of polymer materials. Fillers are mainly classified into two groups namely inert filler and reinforcing filler. Reinforcing filler can dramatically improve the mechanical properties of a polymer and are used for the development of composite materials. The composite materials will be discussed in greater detail in subsequent chapters. Inert fillers do not improve the mechanical properties significantly and are used to reduce the cost and other properties like viscosity, opacity, etc. Fillers impart hiding ability and help to reduce the shrinkage stress, which is very important for adhesive and coating applications. The curing of thermosetting resin is associated with a shrinkage (known as a cure shrinkage) which results in the development of residual stress in the matrix. This effect is more prominent for the resin with a high cure shrinkage as for example phenolic, polyester resins. Inert filler improves the rigidity at the cost of toughness and strength. The fillers have to be selected judiciously and their size and shape have to be manipulated accordingly, to keep such loss in strength and toughness to a minimum. Al-powder, calcium carbonate, china clay, talc, barium sulfate, etc. are commonly used as filler.

2.15.3 Blowing agent

Blowing agents are used to producing cellular products like foam. They are used mostly for phenolic and polyurethane resins. Chlorofluro carbon compounds (CFCs), namely, trichlorofluro methane (b.p., 23.8°C) 1,1,1-trifluorodichloroethane (b.p., 28.7°C), and

1,1-dichloro-1-fluoro ethane (b.p., 32°C) are the classic examples of blowing agents. A CFC compound is mixed with the polyol before the polymerization reaction. During the processing, the liquid CFCs change to gases and develop foam. The frothed foam fills cavities quickly and imparts uniform cells. Though CFCs are very successful as blowing agents, they are not environmentally safe. The concern over ozone layer depletion allegedly caused by the CFCs has prompted polymer technologists to search for alternative blowing agents.

The addition of a small amount of water produces carbon dioxide by reacting with an isocyanate compound. The produced carbon dioxide goes out and produces cellular structure. Other class of blowing agents for examples sodium bicarbonate, hydrazide derivatives, azodicarbonamide, which work on the principle that they decompose during the processing and produce gases (like carbon dioxide, nitrogen), which generate cellular structures. However, such blowing agents are to be chosen considering the reaction conditions and the decomposition temperature of the respective blowing agent. The processing temperature of the thermosetting resin system must be higher than the decomposition temperature of the blowing agent. For example, if sodium bicarbonate is selected as a blowing agent the processing temperature must be higher than 130°C and similarly, for azodicarbonamide the processing temperature should be higher than 200°C. Bicarbonates and sulfonylhydazides are used in unsaturated polyester resin where the metal catalysts used as a promoter. The catalyst reduces the decomposition temperature and ensures the generation of gases at a lower temperature (less than the processing temperature).

2.15.4 Coupling agent

The property of a polymer blend or a filled polymer system largely depends on the compatibility between the constituent polymers in the blends or between the polymer and filler in filled polymer systems. Sometimes other terminologies like "wetting," "interfacial adhesion" are used to mean compatibility. In the case of filled polymer systems, the incompatibility generates weak interfaces, leading to a drastic reduction in mechanical properties. Coupling agents are the compounds, which are used to improve the compatibility or wetting or interaction between the constituent polymers in the blends or between the polymer and filler in filled polymer systems. Coupling agents have been found to be very successful for improving the properties of thermoplastic blends [312]. However, in thermosetting resin systems, coupling agents are mostly used to improve the wetting of filler by the resin matrices. Silane compounds like trichlorovinyl silane, triethoxyvinyl silane, and 3-glycidoxypropyl-trimethoxy silane are the examples of coupling agents for thermosetting resins [313].

2.15.5 Surfactants

Surfactants are added to a thermoset resin system to promote the dispersion of fillers in the resin matrix. Recently, surfactants have been used to disperse carbon nanotubes in polymer matrices [314–316]. Surfactants are of two types neutral and ionic. Surfactants find lots of applications in coating industries for the development of water-based resin systems [317]. Surfactants are added to the phenolic or polyurethane foam formulation in which facilitates the formation of small bubbles. The size and uniformity of bubble formation result in a fine cell structure. A surfactant reduces the surface tension of resin formulations and provides an interface between the highly polar resin and the nonpolar blowing agent. Note that surfactant for a particular resin system has to be selected carefully so that it is compatible with the resin and resistant to the acidic or basic catalyst. When the foam develops, the surfactant protects the developing foam from collapse or rupture. However, it is important to optimize the concentration of surfactant as a function of foam properties in order to develop foam with desirable structure and properties.

2.15.6 Colorant

Thermoset resins can be colored by introducing a chromophore group or using a suitable additive (colorant). The first approach is expensive and the chemical modification may affect the other physical properties of the resin. The other method i.e. to use a colorant is comparatively easier and widely accepted. Colorants are generally divided into two classes such as pigment (insoluble colorant) and dyestuffs (soluble colorant). As mentioned for other additives, the colorant should be stable under the processing condition and it should have good covering power.

2.16 Processing of thermoset resin

2.16.1 Die casting

Die casting is a very common and simple method to fabricate an object from liquid thermoset resins. In this method, the liquid resin is thoroughly mixed with the curing agent and other additives using a glass rod or a mechanical stirrer. The resin mixture is then subjected to vacuum to remove the air bubbles and poured into a preheated mold. Mold can be made up of aluminum, glass, plaster of paris, etc. The resin takes the shape of the mold. The mold is kept in the oven to cure the resin for a specified time. Once the curing is complete, the mold is allowed to cool and the sample is removed. The resin shrinks during curing and allows the easy removal of the cured product. When a thin aluminum mold is

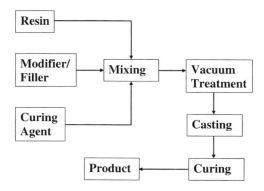

Figure 2.51
A flow diagram for a die casting process of thermoset resins.

used, the sample can be recovered by simply tearing off the aluminum foil. The sample is then postcured if necessary. A flow diagram for a die casting process is shown in Fig. 2.51.

2.16.2 Rotational casting

Die casting is successfully used to make a solid object. However, hollow objects cannot be produced by this method. Hollow objects like balls, dolls, etc. are made by rotational casting. The resin and curative mixture are placed in a hollow mold which is able to rotate simultaneously along the primary and secondary axes. The resin takes the shape of the mold. The shape gets fixed due to the curing of the resin under rotation at an elevated temperature. The mold is then allowed to cool and opened to remove the product.

2.16.3 Compression molding

The compression molding process is widely used to fabricate thermoset resin-based castings and composites. A typical mold used for compression molding is shown in Fig. 2.52. The mold consists of two halves: the upper or male and the lower or female. The lower half usually contains a cavity and the upper half has a projection that exactly fits into the cavity when the mold is closed. The gap between the projected upper half and the cavity in the lower one gives the shape of the molded articles. The material in the mold is subjected to heat and pressure simultaneously using a hydraulic press with a heating facility. Molding temperature can be up to 300°C and pressure up to 80 kg cm^{-2}. Depending on the thermal and rheological properties the pressure and temperature have to be adjusted. A sufficient amount of material has to be put in so that the cavity is full. As the mold closes down under pressure, the material is squeezed or compressed between the two halves and excess material

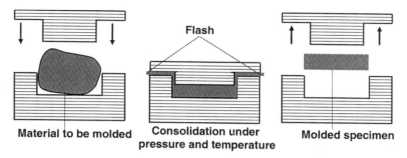

Figure 2.52
A typical mold used for compression molding of thermoset resins.

flows out of the mold as a thin film known as flash. The mold is cooled down and the product is removed.

2.16.4 RIM process

Reaction injection molding (RIM) is a high-productivity part-manufacturing processor process used for the rapid and automated production of large, thin, and complex shaped parts. The difference between RIM and ordinary injection molding (used for thermoplastic) is that RIM uses polymerization in the mold unlike the cooling to form a solid polymer in injection molding. Polyurethane is the most common system used for RIM [318].
A schematic representation of RIM process is shown in Fig. 2.53. In RIM process two or more low-viscosity reactant liquids (monomer, prepolymer or both) are accurately metered according to chemical stoichiometry and mixed at high pressure (200–300 bar), giving turbulent flow conditions in a hydraulically operated mixed. Unlike normal polymerization, which is initiated by heat, the polymerization in a RIM process is initiated by impingement mixing. Hence it is possible to activate the polymerization at a relatively low temperature (~40°C). The reactant mixture then flows with a low pressure under laminar conditions into the mold cavity. The principal advantage of RIM is the ability to process low-viscosity liquid reactants (typically 0.1–1.0 Pa s) using low temperature and pressures, particularly during the mold filling stage. This means that small-scale hydraulic equipment (lighter weight and lower cost mold) can be used. This facilitates the short production runs and prototype applications.

2.17 3D printing

2D printing, developed in the 15th century, which is known to create a revolutionary effect all over the world, has now been advanced and perfected to a higher level of sophistication to an additive process in which successive layers of materials are distributed to form a 3D

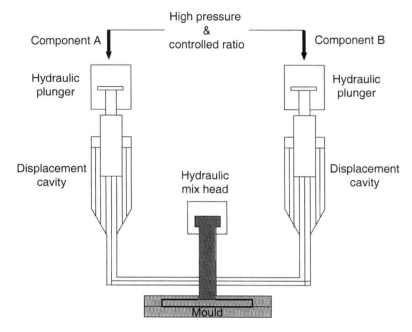

Figure 2.53
Schematic diagram of RIM process.

shape. This is known as rapid additive manufacturing or 3D printing. 3D printing was first described by Charles W. Hull in 1986. His manufacturing process involves sequential printing of thin layers of materials (which can be cured with ultraviolet radiation) to form a 3D structure. The process was named "stereolithography" which was later applied to create sacrificial resin molds for the formation of 3D scaffolds for biological materials. The first step of 3D printing is to create a 3D model of the object to be printed. This model can be designed with computer-aided design (CAD) software. The CAD file is then converted to a standard additive manufacturing file format. The file is then digitally sliced into layers. In the fourth step, the machine which is only controlled by a computer builds the model layer by layer. The layer thickness dictates the final quality and depends on the machine and process.

Recently, 3D printing has emerged as an advanced additive manufacturing technology that has drawn considerable research interest and has promoted significant innovations. It offers a great potential for a growing range of applications which include both civil and defense sectors namely microelectronics, biomedical engineering, microfluidic devices, soft robots etc. This is due to the fact that 3D printing possesses a strong ability to customize macropatterns and precise control of microstructure. It offers several technological advantages over conventional manufacturing processes. For example, 3D printing does not require any tooling, which is essential for conventional manufacturing techniques such as injection molding. The requirement of tooling in widely used manufacturing techniques mentioned above can be a barrier to production due to the high cost. In the case of 3D printing, various parts of a component can be sent digitally and printed in homes or locations near to consumers. As a result, the necessity of transport and related expenditure can be reduced. Thus 3D printing is more environmentally friendly compared to conventional manufacturing processes. Geometric limitations in conventional manufacturing methods can be overcome by using 3D printing which can produce models in hours, not in weeks or months. The requirement of resource and skilled manpower is much lesser in 3D printing compared to conventional model-making craftsmanship. 3D printing offers better customization particularly within the medical sector, where parts can be fully customized to the patient and their individual requirements. 3D printing involves more efficient use of material due to the exact production. Because the wastage of material due to overproduction (based on estimated demand) as in the case of the conventional manufacturing process, can be avoided. 3D printing is more attractive from the commercial point of view as well because the development time from concept to the manufacturing of parts can be minimized considerably. Different types of 3D printing are described below.

2.17.1 Types of 3D printing

2.17.1.1 Stereolithography

Stereolithography (SLA) utilizes a machine known as stereolithography. Unlike 2D printing in which the machine extrudes ink or some other liquid onto the surface, a stereolithography starts with an excess of liquid plastic, some of which is cured, or hardened, to form a solid object. It comprises four main parts: a tank that can be filled with liquid plastic (photopolymer), which is curable with UV radiation, a perforated platform that is lowered into the tank, an ultraviolet (UV) laser, and a computer controlling the platform. The UV-curable resin hardens instantly when the UV laser touches it, forming the first layer of the 3D-printed object. Initially a thin layer of photopolymer (usually between 50 and 100 µm) is exposed above the perforated platform. The UV laser hits the perforated platform of the object being printed. Once the initial layer of the object has hardened, the platform is lowered, exposing a new surface layer of liquid resin. The laser again traces a cross section of the object being printed, which instantly bonds to the hardened section beneath it. This process is repeated again and again until the entire object has been formed and is fully submerged in the tank. The platform is then raised to expose a three-dimensional object. After it is rinsed with a liquid solvent to remove the excess resin, the object is baked in an ultraviolet oven to further cure the thermoset resin. Objects made using stereolithography generally have smooth surfaces, but the quality of an object depends on the quality of the SLA machine used to print it.

2.17.1.2 Digital light processing

Digital light processing (DLP) is similar to SLA as discussed above and also utilizes a photocurable resin. One major difference between the two is the source of light. While SLA uses ultraviolet light, DLP makes use of a more traditional light source, usually arc lamps. This process results in pretty impressive printing speeds. When there's plenty of light, the resin is quick to harden in seconds. The 3D-printing process using DLP is much faster compared to the same using SLA. This is because in DLP, the entire layers are exposed at once and therefore DLP achieves quicker print times for most parts. The reason for its faster printing is because it exposes entire layers at once whereas, in SLA printing, a laser has to draw out each of these layers, which is a time-consuming process. The other merit of DLP technology is that it is robust and always produces high-resolution models using cheaper materials. Over the last few years, a lot of advancement has taken place in the field of DLP 3D-printing technology, such as projection microstereolithography providing micro-/sub-micrometer printing resolution, continuous liquid interface production enabling 100 times faster printing, and large-area projection microstereolithography creating 3D features spanning several orders of magnitude from nanometers to centimeters

2.17.1.3 Fused deposition model

Fused deposition model (FDM) rapid prototyping, in general, consists of an extrusion system, which has the ability to precisely control where the extrudate is laid in three-dimensional space in order to build a prototype with complex geometry. Basically, layers are built by extruding a small bead of material in a particular lay-down pattern such that the layers are covered with the beads. After a layer is completed, the height of the extrusion head is increased and the subsequent layers are built to construct the part. The quality of printed parts can be controlled by altering printing parameters, such as layer thickness, printing orientation, raster width, raster angle, and air gap. FDM printers also offer advantages, including low cost, high speed, and simplicity. It allows the deposition of diverse materials simultaneously by using multiple extrusion nozzles. Since different materials can be loaded at a time, multifunctional parts with the desired composition can be easily made. The viscosity of molten material plays an important role in FDP process. If the viscosity is very high it will be difficult to extrude the same. At the same time if the viscosity is very low structural support will be disturbed. Therefore the optimum melt viscosity of the polymer has to be used for processing. The molten viscosity should be high enough to provide structural support and low enough to enable extrusion. Also, complete removal of the support structure used during printing may be difficult.

2.17.1.4 Selective laser sintering technology

Selective laser sintering (SLS) technology (invented in the mid-1980s by an American businessman) is a rapid manufacturing technique, which uses high-power CO_2 lasers to fuse particles together. The laser sinters powdered materials (metallic, ceramics, or polymer materials like white nylon powder). For each successive laser scan, the build platform or bed is to be lowered. The unique feature of this process is that it repeats one layer at a time until it reaches the desired height of the object. During the building process, it does not need any additional support materials because the unsintered support from other powders surrounds and protects the model. The unsintered powders are manually removed after printing. SLS produces durable, high-precision parts, and it can use a wide range of materials. It is a perfect technology for fully functional, end-use parts, and prototypes. SLS is quite similar to SLA technology with regards to speed and quality. The main difference is with the materials, as SLS uses powdered substances, whereas SLA uses liquid resins. Therefore a wide variety of thermosetting resin can be used in SLA; however, they cannot be processed using SLS. This is attributed to the popularity of SLA over SLS.

2.17.1.5 Selective laser melting technology

Selective laser melting (SLM) is a comparatively newer 3D-printing technology and developed in 1995 by German scientists. Similar to SLA where UV laser is used, a high-powered laser beam is used in SLM to form 3D parts. During the printing process, the laser beam melts and fuses various metallic powders together. As the laser beam hits a thin layer of the material, it selectively joins or welds the particles together. After one complete print cycle, the printer adds a new layer of powered material to the previous one. The object is then lowered by the precise amount of the thickness of a single layer. When the print process is complete, someone will manually remove the unused powder from the object. The main difference between SLM and SLS is that SLM completely melts the powder, whereas in SLS, only partly melted or sintered powdered is used. In general, SLM end products tend to be stronger as they have fewer or no voids. A common use for SLM printing is with 3D parts that have complex structures, geometries and thin walls. The aerospace industry uses SLM 3D printing in some of its pioneering projects. These are typically those which focus on precise, durable, lightweight parts. SLM is quite widespread now among the aerospace and medical orthopedics industries. Those who invest in SLM 3D printers include researchers, universities, metal powder developers and others, who are keen to explore the full range and future potential of metal additive manufacturing.

2.17.1.6 Electron beam melting technology

Electron beam melting (EBM) technology is a comparatively new 3D-printing technique, which is similar to the SLM as powder beds are used in this case as well. They differ in the nature of power sources used for the printing process. The SLM approach above uses the high-powered laser in a chamber of noble, or inert gas. EBM, on the other hand, uses a powerful electron beam in a vacuum. It is mostly used for metal. The main attraction of this process is its high efficiency and ability to work without any auxiliary equipment for the

3D-printing process. The process offers lesser lead time and therefore faster delivery of the related products is possible using EBM.

2.17.1.7 Binder jetting technology

Binder jetting (BJ) technology also known as powder bed printing, inkjet 3D printing is first invented by researcher at the Massachusetts Institute of Technology, USA. Two types of materials are used to build objects namely powder-based material (usually gypsum) and a bonding agent. The role of the bonding agent is to act as a strong adhesive to attach the powder layers together. The printer nozzles extrude the binder in a liquid form similar to a regular 2D inkjet printer. After completing each layer, the build plate lowers slightly to allow for the next one. This process repeats until the object reaches its required height. Different types of materials such as plastics, ceramic, metals can be processed using this technology. One advantage of this process is that the printers allow one to print parts in full color by adding suitable color pigment. Lots of efforts are being done all over the globe for advancement and to perfect this process to a higher level of sophistication to enhance the resolution of BJ process.

2.17.2 3D printing of thermoplastic

If we look at the use of various rapid prototyping fabrication technologies, fused deposition modeling (FDM) has been extensively used for 3D printing of polymers. The research on 3D printing started with the use of thermoplastic as a feedstock. Most of the studies used a copolymer of acrylonitrile butadiene and styrene (ABS). Although FDM is an efficient technology compared to conventional polymer processing technologies, full-scale application of the same still remains a challenge because of the compatibility of presently available materials with rapid prototyping technologies [319,320]. To overcome this limitation, a good number of researchers have devoted to the study of the effect of various process parameters on the properties of the parts/components manufactured by 3D printing [321–323]. It was observed that the properties of parts prepared by FDM are a function of various process-related parameters and can be significantly improved with the proper adjustment such parameters. Since mechanical properties are important for functional parts, it is absolutely essential to study the influence of various process parameters on mechanical properties so that improvement can be made through judicious selection of the related process parameters.

The mechanical strength of a component processed by FDM primarily depends on five important control factors [324] such as layer thickness, part build orientation (the inclination of part in a build platform with respect to X, Y, Z axis), raster angle (direction of raster relative to the x-axis of build table), raster width (width of raster pattern used to fill interior regions of part curves), and raster to raster gap. A study made by Es Said et al.

[325] shows that raster orientation causes alignment of polymer molecules along the direction of deposition during fabrication and tensile, flexural and impact strength depends on orientation. Since semimolten filament is extruded from the nozzle tip and solidified in a chamber maintained at a certain temperature, a change of phase is likely to occur. As a result, volumetric shrinkage takes place resulting in weak interlayer bonding and high porosity leading to reduced load-bearing area. Ahn et al. [326] have pointed out that process parameters such as air gap and raster orientation significantly affect the tensile strength of FDM processed part as compared to other parameters like raster width, model temperature and color through experimental design and analysis. Khan et al. [327] have concluded that layer thickness, raster angle and air gap influence the elastic performance of the compliant FDM ABS prototype.

Sood et al. [328] reported a study that is focused on the assessment of mechanical properties, namely, tensile, flexural, and impact strength of part fabricated using FDM technology. As the relation between mechanical property and process parameters is difficult to establish, an attempt has been made to derive the empirical model between the processing parameters and mechanical properties using response surface methodology. In addition, the effect of each process parameter on the mechanical property is analyzed. In actual practice, the parts are subjected to various types of loadings and it is necessary that the fabricated part must withhold more than one mechanical property simultaneously. They have built up an empirical model by fitting a second-order polynomial and adopted a desirability function approach to optimize more than one response at a time.

It was observed that the number of layers in a part depends upon the layer thickness and part orientation. If the number of layers is more, it will result in a high-temperature gradient toward the bottom of the part. This will increase the diffusion between adjacent rasters and strength will improve. But high-temperature gradient is also responsible for distortion within the layers or between the layers. Moreover, an increase in the number of layers also increases the number of heating and cooling cycles and thus residual stress accumulation increases. This may result in distortion, interlayer cracking and part delamination or fabrication failure. Hence, the strength of the part will reduce. Small raster angles are not preferable as they will result in long rasters which will increase the stress accumulation along the direction of deposition resulting in more distortion and hence weak bonding. But small raster angle also means that rasters are inclined along the direction of loading and will offer more resistance thus strength will improve. Thick rasters result in stress accumulation along the width of part and have the same effect as the long rasters. But this stress accumulation results in high temperature near the boding surfaces which may improve the diffusion and may result in strong bond formation. Zero air gap will improve the diffusion between the adjacent rasters but may also decrease the heat dissipation as well as total bonding area.

2.17.3 3D printing of thermoset

Photocurable thermosets can be processed via 3D printing using a relatively complex and costlier printer than one generally used for 3D printing of thermoplastic (as discussed above). For example, vinyl ester resin (epoxy diacrylate) and epoxy resin were directly printed via photo-induced radical polymerization and UV light-induced cationic polymerization, respectively [329]. The printed thermoset show high modulus but brittle behavior because of low strain at break. In addition, the high cost of the cationic initiators restricts their use for practical applications. Lewis and coworkers [330,331] reported a direct-ink-write (DIW) approach so that thermally cured thermoset systems can be processed via 3D printing. In this approach, the ink containing curable resin is extruded through a nozzle attached to a syringe dispenser. It is necessary to add some thixotropic filler so that the part can retain its shape due to the recovery of high viscosity after deposition. The part is subsequently treated in an oven, for network formation to take place as a result of curing. Note that it is necessary to control the thermal curing process carefully or to provide a procuring at low temperature to avoid deflection of the part. Because the viscosity of part will decrease sharply at high temperature, which may lead to shape distortion.

The distortion issue can be resolved by using a UV-assisted DIW fabrication approach [332]. During the extrusion of the viscous photocurable ink at room temperature, crosslinking is initiated instantly as a result of UV exposure and the shape of the part gets fixed. Though this approach helps to address the distortion problem, 3D printing of thermoset is still associated with issues like clogging of nozzles, brittleness of printed networks and so on. Chen et al. [333] reported a new UV-assisted DIW technique via a two-stage cure. They used a mixture of a thermally curable epoxy resin and a UV-curable one and each layer is printed followed by ex situ UV curing instead of in situ curing. As a result of the use of ex situ curing, the clogging issue can be avoided. The flexible network formed by the UV-curable resin can hold the shape of the part very well even at an elevated temperature. In fact, it leads to the formation of an IPN which displays shape-memory properties. The flexibility of a 3D-printed network tends to increase with the increase in the concentration of photopolymer. A highly stretchable UV-curable PU elastomer was also prepared using DLP printing technique [334]. The DLP printable PU elastomer resin formulations were prepared by mixing a monofunctional monomer consisting of epoxy aliphatic acrylate (EAA), and a difunctional crosslinker consisting of aliphatic urethane diacrylate (AUD) diluted with 33 wt.% of isobornyl acrylate. The AUD oligomer comprises "soft domains" made up of aliphatic chains and "hard domains" made up of urethane units. The printed elastomers can be stretched by up to 1100%, which is more than five times the elongation at break of the commercial UV-curable elastomers and is comparable to that of the most stretchable silicon rubber.

Femmer et al. [335,336] demonstrated the 3D printing of silicone resin using the DLP approach where they achieved for the first time in the literature, where 100 μm resolution was attained. They also demonstrated the fabrication of complex architectures where a

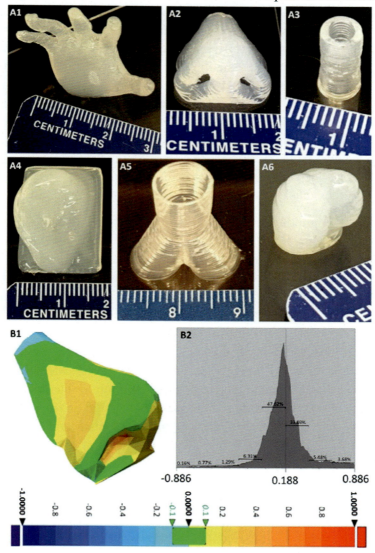

Figure 2.54
(A) 3D-printed organ models: (A1) human hand, (A2) nose, (A3) blood vessel, (A4) ear, (A5) branched structure, (A6) femoral head. (B) Fidelity analysis of printed nose geometry: (B1) heat maps and (B2) average percent frequency histogram representing surface deviations between printed and model geometries. Source: *Reprinted with permission from V. Ozbolat, M. Dey, B. Ayan, A. Povilianskas, M.C. Demirel, I. Ozbolat Madhuri, ACS Biomater. Sci. Eng. 4 (2) (2018) 682−693, ©2018 ACS Publisher.*

sacrificial mold was the first 3D printed followed by casting and removing the sacrificial mold thereafter. Ozbolar et al. [337] optimized the printing parameters for 3D printing of silicone and prepared various models in high fidelity as shown in Fig. 4A, including structural models of the human hand, nose, blood vessel, ear, a branched structure, and femoral heads. Characterization of the shape fidelity of the printed nose model was performed through surface deviation analysis by comparing the difference between the scanned and original models and presented in Fig. 2.54. A 78.45% of the surface was within the ± 100 μm tolerance threshold with 1.86% underprint and 19.68% overprint error. The average surface deviation of 189 μm was observed with 263 μm overprint deviation along the curved surface. Furthermore, 787-μm deviation was observed along the upper portion of the nostrils where the unsupported bridging occurred. Sharp edges of the model along the nasion resulted in an underprint error of −181 μm. These results display the ability to reproduce geometrically complex structures with relatively less accuracy along sharp edges and overhanging features.

As discussed above, 3D printing of thermoset mostly deals with photocurable or very fast curing resins like silicone. Unlike thermoplastic, where the general method for 3D printing is well established, the same for a thermoset resin is not yet established. The major challenges in the preparation of a thermoset-based composite using 3D printing are that thermosets mostly do not retain their shape until crosslinked and once they are crosslinked, they cannot be extruded, molded, and reshaped. Due to the long-term curing process involved for many thermoset resins, it is difficult to match the continuous layer-by-layer processing of 3D printing. Recently, Dong et al. [338] reported a novel versatile strategy for direct 3D printing of a variety of thermosets. They used a readily available extrusion-based technology using sacrificial sodium chloride particles as a removable thickener for printing. The carrier material (salt) can be easily combined with other additives such as organoclay, graphene and carbon nanotubes, etc., to fabricate high-performance and functional thermosetting composites. The strategy offers several degrees of such as particle size, nozzle sizes, extrusion temperature and design. The strategy has been successfully utilized to fabricate sophisticated construct with hierarchical and porous structures (e.g., nanopores) for applications ranging from regenerative medicines and sensors to catalyst support and energy absorbers.

References

[1] A. Gardziella, L.A. Pilato, A. Knop, Phenolic Resins, 2nd ed., Springer-Verlag, Heidelberg, 2000.
[2] H.L. Fisher, Chemistry of Natural and Synthetic Rubbers, Reinhold Publishing Corporation, New York, 1958.
[3] P.J. Flory, Principle of Polymer Chemistry, Cornell University Press, Ithaca, NY, 1953.
[4] W.H. Stockmayer, J. Chem. Phys. 12 (1944) 125.
[5] F.W. Billmeyer, Textbook of Polymer Science, 3rd ed., Wiley, New York, 1984.
[6] W.H. Carothers, Trans. Faraday Soc. 32 (1936) 39.

[7] S.D. Senturia, N.F.K. SheppardDusek (Eds.), Advances in Polymer Science, 80, Springer-Verlag, New York, 1986, pp. 1–48.
[8] J.B. Enns, J.K. Gillham, J. Appl. Polym. Sci. 28 (1983) 2567.
[9] S. Sourour, M.R. Kamal, Thermochem. Acta 14 (1976) 41.
[10] M.R. Dusi, C.A. May, C.J. Seferis, ACS Symp. Ser. 227 (1983) 301.
[11] L.M. Chiacchiarelli, J.M. Kenny, L. Torre, Thermochim. Acta 574 (2013) 88–97.
[12] F. Dimier, N. Sbirrazzuoli, B. Vergnes, M. Vincent, Polym. Eng. Sci. 44 (2004) 518–527.
[13] X. Sun, J. Toth, L.J. Lee, Polym. Eng. Sci. 37 (1997) 143–152.
[14] S.V. Muzumdar, L.J. Lee, Polym. Eng. Sci. 36 (7) (1996) 943.
[15] M.R. Kamal, S. Sourour, Polym. Eng. Sci. 13 (1973) 59–64.
[16] H.L. Friedman, J. Polym. Sci. C. Polym. Symp. 6 (1964) 183–195.
[17] T. Ozawa, Bull. Chem. Soc. Jpn. 38 (1965) 1881–1886.
[18] H.E.A. Kissinger, Chem 29 (1957) 1702–1706.
[19] B. Fernandez d'Arlas, L. Rueda, P.M. Stefani, K. de la Caba, I. Mondragon, Eceiza, Thermochim. Acta 459 (2007) 94–103.
[20] S. Vyazovkin, Isoconversional Kinetics of Thermally Stimulated Processes, Springer, Cham, Switzerland, 2015.
[21] S. Heinze, A.T. Echtermeyer, Appl. Polym. 8 (2018) 2227.
[22] H. Cai, P. Li, G. Sui, Y. Yu, G. Li, X. Yang, et al., Thermochim. Acta 473 (2008) 101–105.
[23] P. Zhang, S.A.A. Shah, F. Gao, H. Sun, Z. Cui, J. Cheng, et al., Thermochimica Acta 676 (2019) 130–138.
[24] C. Billotte, F.M. Bernard, Edu Ruiz, Eur. Polym. J. 49 (2013) 3548–3560.
[25] J.H. Flynn, L.A. Wall, J. Res. Nat. Bur. Stand. Part. A Phys. Chem. 70A (1996) 487–523.
[26] R. Jain, P. Kukreja, A.K. Narula, V. Chaudhary, J. Appl. Polym. Sci. 100 (2006) 3919–3925.
[27] S. Vyazovkin, A. Mitielu, N. Sbirrazzuoli, Macromol. Rapid Commun. 24 (2003) 1060–1065.
[28] S.B. Jagtap, V.S. Rao, S. Barman, D. Ratna, Polymer 63 (2015) 41–51.
[29] S.B. Jagtap, D. Ratna, J Appl. Polym. Sci. 134 (2017) 44595.
[30] M. Ghaemy, H. Behmadi, J. Therm. Anal. Calorim. 101 (2010) 1011–1018.
[31] J.S. Najat, J.K. Bashar, Eng Tech J. 31 (2013) 1659–1673.
[32] T.J. Horong, E.M. Woo, Angew. Macromol. Chem. 260 (1998) 31.
[33] C.S. Chern, G.W. Pohlein, Polym. Eng. Sci. 27 (1987) 782.
[34] I. Havlicek, K. Dusek, in: B. Sedlecek, J. Kahobek (Eds.), Crosslinked Epoxies, Walter de Gruyter, Berlin, 1987, p. 359.
[35] K. Potter, Resin transfer molding, 196, Chapman and Hall, London, 1997.
[36] M. Stanko, M. Stommel, Polymers 10 (2018) 698–713.
[37] O.G. Kravchenko, S.G. Kravchenko, R.B. Pipes, Composites, Part A 80 (2016) 72–81.
[38] S. Heinze, A.T. Echtermeyer, Appl. Sci. 8 (2018) 2227–2258.
[39] R.R. Hill, J.R. Shailesh, V. Muzumdar, L.J. Lee, Polym. Eng. Sci. 35 (1995) 852–861.
[40] C. Li, K. Potter, M.R. Wisnom, G. Stringer, Compos. Sci. Technol. 64 (2004) 55–64.
[41] M. Holst, K. Shanzlin, M. Wenzel, J. Xu, D. Lellinger, I. Alig, J. Polym. Sci. 43 (2005) 2314–2325.
[42] Y. Abou Msallem, F. Jacquemin, N. Boyard, A. Poitou, D. Delaunay, S. Chatel, Composites: Part. A 41 (2010) 108–123.
[43] Y. Nawab, N. Boyard, F. Jaquemin, J. Composite Mater. 48 (2014) 3191–3199.
[44] K. Shaker, Y. Nawab, A. Saouab, Int. J. Mater. Form. 13 (2020) 897–906.
[45] J. Pinoteck, S. Richter, S. Zschoche, K. Sahre, K.F. Arndt, Acta Polym. 49 (1998) 192.
[46] E. Haberstroh, W. Michaeli, E. Henze, J. Reinf. Plast. Comp. 21 (2002) 461.
[47] N. Boyard, M. Vayer, C. Sinturel, R. Erre, D. Delaunay, J. Appl. Polym. Sci. 88 (2003) 1258.
[48] M. Zarreli, A.A. Skordos, I.K. Patridge, Plast. Rubber Com. 31 (2002) 377.
[49] Y. Nawab, X. Tardif, N. Boyard, Compos. Sci. Technol. 73 (2012) 81–87.
[50] J. Ramos, N. Pagani, C.C. Riccardi, J. Borrajo, S.N. Goyanes, I. Mondragon, Polymer 46 (2005) 3323.
[51] T. Mesogitis, A. Skordos, A. Long, Compos. Sci. Technol. 110 (2015) 145–151.
[52] R. Subbiah, J. Tjong, S. Nayak, M. Sain, Can. J. Chem. Eng. 94 (2016) 1375–1380.

[53] D. Abliz, T. Artys, G. Ziegmann, J. Appl. Polym. Sci. 134 (2017) 45137.
[54] G. Struzziero, A. Skordos, Part. A Appl. Sci. Manuf. 93 (2017) 126–136.
[55] Z.-S. Guo, S.Y. Du, B.M. Zhang, Compos. Sci. Technol. 65 (2005) 517–523.
[56] X. Yan, J. Appl. Polym. Sci. 103 (2007) 2310–2319.
[57] J. Zhang, Y.C. Xu, P. Huang, Polym. Express Lett. 9 (2009) 534–541.
[58] T.A. Bogetti, J.W. Gillespie Jr, J. Compos. Mater. 25 (1991) 239–273.
[59] D. Ratna, Rapra Review Report, 2005 Smithers Rapra Technology, UK.
[60] Y.G. Lin, J. Gally, H. Sautereau, J.P. Pascault, in: B. Sedlecek, J. Kahobek (Eds.), Crosslinked Epoxies, Walter de Gruyter, Berlin, 1989, pp. 147–168.
[61] H.E. Adaboo, R.J. Williams, J. Appl. Polym. Sci. 29 (1982) 1327–1334.
[62] L.E. Nielsen, J. Macromol, Sci. Rev. Macromol. Chem., C3 ((1969) 69.
[63] J.P. Pascault, R.J.J. Williams, J. Polym. Sci. Part. B 28 (1990) 85.
[64] P.R. Couchman, Macromolecules 20 (1987) 1712.
[65] S. Montserrat, Polymer 36 (1995) 435.
[66] I. Havlicek, K. Dusek, in: B. Sedlecek, J. Kahovec (Eds.), Crosslinked Epoxies, de Gruyter, Berlin, 1987, pp. 417–424.
[67] R.F.T. Stepto (Ed.), Polymer Networks, Blackie Academic and Professional, New York, 1998.
[68] E. Ruiz, F. Trochu, J. Compos Mater. 39 (2005) 881–916.
[69] A.F. Baffa, J. Borrajo, J. Appl. Polym. Sci. 102 (2006) 6064.
[70] A.M. Atta, S.M. El-Saeed, R.K. Farag, React. Funct. Polym. 66 (2006) 1596–1608.
[71] A.B. Samui, B.C. Chakraborty, D. Ratna, Int. J. Plastic Technol. 8 (2004) 279–286.
[72] C.P. Reghunadhan Nair, Prog. Polym. Sci. 29 (2004) 401–498.
[73] D. Ratna, G.P. Simon, J. Appl. Polym. Sci. 117 (2010) 557–564.
[74] D. Ratna, J. Adhes, Sci. Technol. 17 (2003) 1655.
[75] V.V. Ginzburg, J. Bicerano, C.P. Christenson, A.K. Schrock, A.Z. Patashinski, J. Polym. Sci., Part. B: Polym. Phys. 16 (2007) 2123–2135.
[76] K. Aou, A.K. Schrock, V.V. Ginzburg, P.C. Price, Polymer 54 (2013) 5005–5015.
[77] M.A. Corcuera, L. Rueda, A. Saralegui, M.D. Martín, B.F. Arlas, I. Mondragon, et al., J. Appl. Polym. Sci. 122 (2011) 3677–3685.
[78] S.R. Williams, W. Wang, K.I. Winey, T.E. Long, Macromolecules 41 (2008) 9072.
[79] P.J. Halley, M.E. Mackay, Polym. Eng. Sci. 36 (1996) 593.
[80] D. Ratna, R. Varley, G.P. Simon, J. Appl. Polym. Sci. 92 (2004) 1604.
[81] X. Wang, J.K. Gillham, J. Appl. Polym. Sci. 45 (1992) 2127.
[82] R.J. Varley, D.G. Hawthorne, J. Hodgkin, G.P. Simon, J. Polym. Sci. Part B Polym. Phys. 35 (1997) 153.
[83] P.J. Flory, Principle of Polymer Chemistry, Cornell University Press, Ithaca, New York, 1975.
[84] D. Ratna, R. Varley, G.P. Simon, J. Appl. Polym. Sci. 89 (2003) 2339.
[85] H.E. Adabbo, R.J.J. Williams, J. Appl. Polym. Sci. 27 (1982) 1327.
[86] R.J.J. Williams, in: A. Whelan, J. Craft (Eds.), Development in Plastic Technology-2, Elsevier, Barking, England, 1985, p. 339.
[87] P.A. Oyanguren, R.J.J. Williams, J. Appl. Polym. Sci. 47 (1993) 1361.
[88] C. Carfagna, E. Amendola, M. Giamberini, Macromol. Chem. Phy. 195 (1994) 279.
[89] C. Carfagna, E. Amendola, M. Giamberini, Macromol. Chem. Phy. 195 (1994) 2307.
[90] C. Carfagna, E. Amendola, Giamberini. Macromol. Chem. Phy. 198 (1997) 3185.
[91] D. Wohrle, A. Hartwig, G. Schnurpfeil, A. Harder, H. Schroder, Polym. Adv. Technol. 11 (2000) 739.
[92] W. Mormann, M. Brocher, Macromol. Chem. Phy. 197 (1994) 1841.
[93] D. Ribera, A. Mantecon, A. Serra, Macromol. Chem. Phy. 202 (2001) 1658.
[94] D. Ribera, A. Mantecon, A. Serra, J. Polym. Sci. Part. A: Polym Chem. 40 (2002) 4344.
[95] S.M. Kelly, Liq. Cryst. 24 (1998) 71.
[96] J.I. Mallon, P.A. Adams, J. Polym. Sci. Part. A: Polym. Chem. 31 (1993) 2249.
[97] D.J. Broer, J. Lub, G.N. Mol, Macromolecules 26 (1993) 1244.

[98] K. Sadagopan, D. Ratna, A.B. Samui, Polym. Sci. Part. A: Polym. Chem. 41 (2003) 3375.
[99] W.F.A. Su, J. Polym. Sci. Part. A: Polym. Chem. 31 (1993) 3251.
[100] G.W. Calundann, Husam, A.A. Rasoul, H.K. Hall, Jr, US Patent, 4654412, 1987.
[101] A. Shiota, C.K. Ober, J. Polym. Sci. Part. A: Polym. Chem. 34 (1996) 1291.
[102] A. Shiota, C.K. Ober, Polymer 38 (1997) 5837.
[103] C. Carfagna, E. Amendola, Giamberini, Prog. Polym. Sci. 22 (1997) 1607.
[104] D.H. Choi, Y.K. Cha, Bull. Korean Chem. Soc. 23 (2002) 587.
[105] B. Koscienlny, A. Pfiatzmann, M. Fedtke, Polym. Bull. (Berl.) 32 (1994) 529.
[106] P. Castell, M. Galia, A. Serra, Macromol. Chem. Phy. 202 (2001) 1649.
[107] R. Balamurugan, P. Kannan, High. Perform. Polym. 21 (2009) 251–264.
[108] M. Iqbal, B. Norder, E. Mendes, T.J. Dingemans, J. Polym. Sci. Part. A: Polym Chem. 47 (2009) 1368–1380.
[109] D. Ratna, N.R. Manoj, L. Chandrasekhar, B.C. Chakraborty, Polym. Adv. Technol. 15 (2004) 583.
[110] B.C. Chakraborty, D. Ratna, ISBN: 978-0-12–819252-8Jan Polymers for Vibration damping Applications, Elsevier, USA, 2020pp. 1–327.
[111] M.L. Williams, R.F. Landel, J.D. Ferry, J. Am. Chem. Soc. 77 (1955) 3701–3707.
[112] J.D. Ferry, Viscoelastic Properties of Polymers, John Wiley, New York, 1980.
[113] L.H. Sperling, J.J. Fay, Polym. Adv. Technol. 49 (1991) 2.
[114] D. Ratna, Sprayable damping material for constrained layer damping (CLD) applications, in: Proceeding, International Conference on Recent Trend in Materials Science and Technology, held at ATF area, VSSC, Thiruvanthapuram, Kerala, India, on 29–31 October 2010, p. 32.
[115] D. Ratna, R. Khuswaha, S. Barman, B.C. Chakraborty, Indian Patent 300740 (2010).
[116] L.H. Sperling, in: G. Allen, J.C. Bevington (Eds.), Comprehensive Polymer Science, Vol. 6, Pergamon Press, New York, 1989.
[117] N.R. Manoj, R.D. Rout, P. Shivraman, D. Ratna, B.C. Chakraborty, J. Appl. Polym. Sci. 96 (2005) 4487.
[118] N.R. Manoj, D. Ratna, V. Dalvi, M. Patri, B.C. Chakraborty, P.C. Deb, Polym. Eng. Sci. 42 (2002) 1748.
[119] S. Zhang, Y. Feng, L. Zhang, J. Sun, X. Xu, Y. Xu, J. Polym. Sci. A Polym Chem. 45 (2007) 768–775.
[120] A.B. Samui, V. Dalvi, M. Patri, B.C. Chakraborty, J. Appl. Polym. Sci. 99 (2006) 2542.
[121] R. Debdatta, J. Karger-Kocsis, Polym. 52 (2011) 1063.
[122] S. Zhang, Y. Feng, L. Zhang, J. Sun, X. Xu, Y. Xu, J. Polym. Sci. A Polym Chem. 45 (2007) 768.
[123] G. Liu, X. Ding, Y. Cao, Z. Zheng, Y. Peng, Macromol. Rapid Commun. 26 (2005) 649.
[124] S.B. Jagtap, V. Dalvi, K. Shankar, D. Ratna, Polym. Int. 8 (2019) 812–817.
[125] M.S. Silverstein, Polymer 207 (2020) 122929.
[126] O. Gryshchuk, J. Karger-kocsis, J. Nanosci. Nanotechnol. 6 (2006) 345.
[127] S. Giaveri, P. Gronchi, A. Barzoni, Coatings 7 (2017) 213.
[128] H. Jin, Y. Zhang, C. Wang, Y. Sun, Z. Yuan, Y. Pan, et al., J. Therm. Anal. Calorim. 117 (2014) 773.
[129] Y. Zhang, D.J. Hourston, J. Appl. Polym. Sci. 69 (1998) 271.
[130] P. Yu, G. Li, L. Zhang, F. Zhao, S. Chen, A.I. Dmitriev, et al., Tribol. Int. 131 (2019) 454.
[131] S. Gao, L. Zhang, Q. Huang, J. Appl. Polym. Sci. 90 (2003) 1948.
[132] K.T. Lu, C.T. Liu, H.L. Lee, J. Appl. Polym. Sci. 89 (2003) 2157.
[133] J. Huang, L. Zhang, J. Appl. Polym. Sci. 86 (2002) 1799.
[134] L. Zhang, J. Huang, J. Appl. Polym. Sci. 81 (2001) 3251.
[135] X. Cao, L. Zhang, Biomacromolecules 6 (2005) 671.
[136] H. Yeganeh, P. Hojati-Talemi, Polym. Degrad. Stab. 92 (2007) 480.
[137] C. Meier, P. Parlevliet, M. Dring, High. Perform. Polym. 29 (2017) 556.
[138] C.-Y. Qu, W. Xiao, D. Wang, C. Liu, H. Li, Polym. Mater. Sci. Eng. 32 (2016) 83.
[139] R. Ballestero, B.M. Sundaram, H.V. Tippur, M.L., Express Polym. Lett. 10 (2016) 204.
[140] Y. Wen, J. Yan, J. Liu, Z. Wang, Express Polym. Lett. 11 (2017) 963.
[141] D. Ratna, V. Dalvi, S. Rath, S. Billa, ACS Appl. Polym. (Under Review). 3 (2021) 5073–5086.
[142] S.K. Sharma, P.K. Pujari, Prog. Polym. Sci. 75 (2017) 31–47.

[143] J. James, G.V. Thomas, P.M. Akhil, P.M.G. Nambissan, K. Nandakumar, S. Thomes, Phys. Chem. Chem. Phys 22 (2020) 18169–18182.
[144] S.R. White, N.R. Sottos, P.H. Geubelle, J.S. Moore, M.R. Kessler, Autonomichealing of polymer composites, Nature 409 (2001) 794.
[145] G.O. Wilson, M.M. Caruso, N.T. Reimer, S.R. White, N.R. Sottos, J.S. Moore, Chem. Mater. 20 (2008) 3288.
[146] J.M. Kamphaus, J.D. Rule, J.S. Moore, N.R. Sottos, S.R. White, J. R. Soc. Interface 5 (2008) 95.
[147] Y.C. Yuan, M.Z. Rong, M.Q. Zhang, B. Chen, G.C. Yang, X.M. Li, Macromolecules 41 (2008) 5197.
[148] M.R. Kessler, N.R. Sottos, S.R. White, Compos. Part. A: Appl. Sci. Manuf. 34 (2003) 743.
[149] A. Ghaemi, A. Philipp, A. Bauer, K. Last, A. Fery, S. Gekle, Chem. Eng. Sci. 142 (2016) 236.
[150] G.O. Wilson, J.S. Moore, S.R. White, N.R. Sottos, Adv. Funct. Mater. 18 (2008) 44.
[151] K. Urdl, A. Kandelbauer, W. Kern, U. Müller, M. Thebault, E. Zikulnig-Rusch, Prog. Org. Coat. 104 (2017) 232–249.
[152] M.M. Caruso, B.J. Blaiszik, S.R. White, N.R. Sottos, J.S. Moore, Adv. Funct. Mater. 18 (2008) 1898.
[153] H. Jin, C.L. Mangun, D.S. Stradley, J.S. Moore, N.R. Sottos, S.R. White, Polymer 53 (2012) 581.
[154] S.A. Hayes, F.R. Jones, K. Marshiya, W. Zhang, Compos. Part. A 38 (2007) 1116.
[155] S.A. Hayes, W. Zhang, M. Branthwaite, F.R. Jones, J. R., Soc. Interface 4 (2007) 381.
[156] S.J. Picken, S.D. Mookhoek, H.R. Fischer, S. van der Zwaag, Self Healing in Nanoparticle Reinforced Polymers and Other Polymer Systems, in Optimization of Polymer Nanocomposite Properties, Wiley-VCH Verlag GmbH & Co KGaA, Weinheim, Germany, 2010, pp. 261–278.
[157] X.F. Luo, R.Q. Ou, D.E. Eberly, A. Singhal, W. Viratyaporn, P.T. Mather, ACS Appl. Mater. Interfaces 1 (2009) 612.
[158] S.H. Cho, H.M. Andersson, S.R. White, N.R. Sottos, P.V. Braun, Adv. Mater. 18 (2006) 997.
[159] R.S. Jadhav, V. Mane, A.V. Bagle, D.G. Hundiwale, P.P. Mahulikar, G. Waghoo, Int. J. Ind. Chem. 31 (2013) 1.
[160] R.S. Jadhav, D.G. Hundiwale, P.P. Mahulikar, J. Appl. Polym. Sci. 119 (2011) 2911.
[161] C. Suryananarayana, K. Chowdoji Rao, D. Kumar, Prog. Org. Coat. 63 (2008) 72.
[162] S. Neema, M. Selvaraj, J. Raguraman, S. Ramu, J. Appl. Polym. Sci. 127 (2013) 740.
[163] S. Burattini, B.W. Greenland, D.H. Merino, W. Weng, J. Seppala, H.M. Colquhoun, et al., J. Am. Chem. Soc. 132 (2010) 12051.
[164] P. Zheng, T.J. McCarthy, J. Am. Chem. Soc. 134 (2012) 2024.
[165] M. Burnworth, L. Tang, J.R. Kumpfer, A.J. Duncan, F.L. Beyer, G.L. Fiore, et al., Nature 472 (2011) 334.
[166] L. Huang, N. Yi, Y. Wu, Y. Zhang, Q. Zhang, Y. Huang, et al., Adv. Mater. 25 (2013) 2224.
[167] C.M. Chung, Y.S. Roh, S.Y. Cho, J.G. Kim, Chem. Mater. 16 (2004) 3982.
[168] J. Li, G. Zhang, L. Deng, K. Jiang, S. Zhao, Y. Gao, et al., J. Appl. Polym. Sci. 132 (2015) 1.
[169] R. Araya-Hermosilla, G.M.R. Lima, P. Raffa, G. Fortunato, A. Pucci, E. Mario, et al., Eur. Polym. J. 81 (2016) 186.
[170] S. Yang, X. Du, S. Deng, J. Qiu, Z. Du, X. Cheng, et al., Chem. Eng. J. 398 (2020) 125654.
[171] N.V. Tsarevsky, K. Matyjaszewski, Macromolecules 35 (2002) 9009.
[172] J. Canadell, H. Goossens, B. Klumperman, Macromolecules 44 (2011) 2536.
[173] M. Abdolah Zadeh, S. van der Zwaag, S.J. García, Adv. Polym. Sci. 273 (2016) 185.
[174] M. Abdolahzadeh, A.C.C. Esteves, S. van der Zwaag, S.J. García, J. Polym. Sci. Part. A Polym. Chem. 52 (2014) 1953.
[175] W. Post, A. Cohades, V. Michaud, S. van der Zwaag, S.J. Garcia, Compos. Sci. Technol. 152 (2017) 85.
[176] S. Ji, W. Cao, Y. Yu, H. Xu, Angew. Chemie 53 (2014) 6781.
[177] S. Ji, W. Cao, Y. Yu, H. Xu, Adv. Mater. 27 (2015) 7740.
[178] X. Pan, F. Driessen, X. Zhu, F.E. DuPrez, Macro Lett. 6 (2017) 89.
[179] X. An, R.H. Aguirresarobe, L. Irusta, F. Ruipérez, J.M. Matxain, X. Pan, et al., Polym. Chem. 8 (2017) 3641–3646.
[180] H. Ying, Y. Zhang, J. Cheng, Nat. Commun. 5 (2014) 3218.
[181] F. Herbst, D. Döhler, P. Michael, W.H. Binder, Macromol. Rapid Commun. 34 (2013) 203.

[182] R.J. Varley, S. van der Zwaag, Polym. (Guildf.) 51 (2010) 679.
[183] S. Burattini, H.M. Colquhoun, J.D. Fox, D. Friedmann, B.W. Greenland, P.J.F. Harris, et al., Chem. Commun. 44 (2009) 6717.
[184] Z. Hua, Y. Liua, X. Xua, W. Yuana, L. Yanga, Q. Shaob, et al., Polymer 164 (2019) 79.
[185] D. Ratna, K. Kocsis, J. Mater. Sci. 43 (2008) 254.
[186] J. Delaey, P. Dubruel, S.V. Vlierberghe, Adv. Funct. Mater. 30 (2020) 1909047.
[187] A. Lendlein, Science 296 (2002) 1673.
[188] M. Behl, J. Zotzmann, A. Lendlein, Adv. Polym. Sci. 226 (2009) 1.
[189] T. Sauter, M. Heuchel, K. Kratz, A. Lendlein, Polym. Rev. 53 (2013) 6.
[190] M. Li, J. Chen, M. Shi, H. Zhang, P.X. Ma, B. Guo, Chem. Eng. J. 375 (2019) 121999.
[191] O. Castillo-Cruz, F. Avilés, R. Vargas-Coronado, J.V. Cauich-Rodríguez, L.H. Chan-Chan, V. Sessini, et al., Mater. Sci. Eng., C 102 (2019) 887.
[192] G. Baer, T.S. Wilson, D.L. Mathews, D.J. Maitland, J. Appl. Polym. Sci. 103 (2007) 3882.
[193] S. Mondal, J.L. Hu, J. Elast. Plast. 39 (2007) 81.
[194] T. Takahashi, N. Hayashi, S. Hayashi, J. Appl. Polym. Sci. 60 (1996) 1061.
[195] H.S. Park, J.W. Kim, S.H. Lee, B.K. Kim, J. Macromol. Sci. Phys. 43 (2004) 447.
[196] B.S. Lee, B.C. Chun, Y.C. Chung, K. Sul, J.W. Cho, Macromolecules 34 (2001) 6431.
[197] J.R. Lin, L.W. Chen, J. Appl. Polym. Sci. 69 (1998) 1563.
[198] J.R. Lin, L.W. Chen, J. Appl. Polym. Sci. 69 (1998) 1575.
[199] H.M. Jeong, S.Y. Lee, B.K. Kim, J. Mater. Sci. 35 (2000) 1579.
[200] B.C. Chun, T.K. Cho, Y.C. Chung, J. Appl. Polym. Sci. 103 (2007) 1435.
[201] J.W. Cho, Y.C. Jung, B.C. Chun, Y.C. Chung, J. Appl. Polym. Sci. 92 (2004) 2812.
[202] H.H. Wang, U. Yuen, J. Appl. Polym. Sci. 102 (2006) 607.
[203] J.H. Yang, B.C. Chun, Y.C. Chung, J.H. Cho, Polymer 44 (2003) 3251.
[204] M. Xu, F. Li, Chin, J. Polym. Sci. 17 (1999) 203.
[205] H.M. Jeong, B.K. Ahn, B.K. Kim, J. Polym. Sci. Polym. Phys. 38 (2000) 3009.
[206] H.M. Jeong, B.K. Ahn, B.K. Kim, Polym. Int. 49 (2000) 1714.
[207] H.M. Jeong, B.K. Kim, Y.J. Choi, Polymer 41 (2000) 1849.
[208] Y. Zhu, J.L. Hu, K.W. Yeung, Y.Q. Liu, H.M. Lien, J. Appl. Polym. Sci. 100 (2006) 4603.
[209] F.K. Li, X.Y. Zhang, M.T. Wang, D.J. Ma, M. Xu, X.L. Luo, et al., J. Appl. Polym. Sci. 64 (1997) 1511.
[210] C.P. Buckley, C. Prisacariu, A. Caraculacu, Polymer 48 (2007) 1388.
[211] J. Xu, W. Shi, W. Pang, Polymer 47 (2006) 457.
[212] P. Ping, W. Wang, X. Chen, X. Jing, Biomacromolecules 6 (2005) 587.
[213] W. Chen, C. Zhu, X. Gu, J. Appl. Polym. Sci. 84 (2002) 1504.
[214] B.K. Kim, S.H. Park, J. Polym. Sci. Part. A Polym. Chem. 37 (1999) 2703.
[215] V.D. Punethaa, Y.-M. Hab, Y.-O. Kimb, Y. ChaeJungb, J.W. Choa, Polymer 181 (2019) 121791.
[216] S. Fu, J. Zhu, F. Zou, X. Zeng, S. Chen, Lett. 229 (2018) 44–47.
[217] T. Xie, I.A. Rousseau, Polymer 50 (2009) 1852–1856.
[218] N. Zheng, G. Fang, Z. Cao, Q. Zhao, T. Xie, Polym. Chem. 6 (2015) 3046–3053.
[219] Y. Liu, J. Zhao, L. Zhao, W. Li, H. Zhang, X. Yu, et al., ACS Appl. Mater. Interfaces 8 (44) (2016) 311–320.
[220] T. Xie, I.A. Rousseau, Polymer 50 (2009) 1852.
[221] B. Rajendran, K.S.S. Kumar, R.S. Rajeev, C.P.R. Nair, Polym. Adv. Technol. 24 (2013) 623–629.
[222] Z. Ding, L. Yuan, Q. Guan, A. Gu, G. Liang, Polymer 147 (2018) 170–182.
[223] M. Capelot, M.M. Unterlass, F. Tournilhac, L. Leibler, ACS Macro Lett. 1 (2012) 789–792.
[224] T. Abraham, J. Karger-Kocsis, J. Appl. Polym. Sci. 108 (2008) 2156–2162.
[225] T. Abraham, J. Karger-Kocsis, Macromol. Chem. Phys. 209 (2008) 723–733.
[226] C. Wang, Y. Zhang, J. Li, Z. Yang, Q. Wang, T. Wang, et al., Eur. Polym. J. 123 (2020) 109393.
[227] Y. Wu, J. Hu, J. Han, Y. Zhu, H. Huang, J. Li, et al., J. Mater. Chem. A 2 (2014) 18816.
[228] A. Lendlein, S. Kelch, Angew. Chem. Int. (Ed.) 41 (2002) 2034.

[229] J. Han, G. Fei, G. Li, H. Xia, Macromol. Chem. Phys. 214 (2013) 1195.
[230] M.F. Metzger, T.S. Wilson, D. Schumann, D.L. Mathew, D.J. Maitland, Biomed. Microdevices 4 (2002) 89.
[231] H.M. Wache, D.J. Tartakowska, A. Hentrich, M.H. Wagner, J. Mater. Sci. Mater. Med. 14 (2003) 109.
[232] K. Gall, M. Mikulas, N. Munsi, M. Tupper, J. Intell. Mater. Syst. Structure 11 (2000) 877.
[233] M. Enomoto, K. Suehiro, Text. Res. J. 67 (1997) 601.
[234] R. Langer, D.A. Tirrell, Nature 428 (2004) 487.
[235] A. Lendlein, R. Langer, Science 296 (2002) 1673.
[236] J.M. Hampikian, B.C. Heaton, F.C. Tong, Z. Zhang, Mater. Sci. Eng. C. 26 (2006) 1373.
[237] J. Bontinck, P. Goverde, H. Schroë, J. Hendriks, L. Maene, F. Vermassen, J. Vascul. Surg. 64 (2016) 1311.
[238] S.J. Tey, W.M. Huang, W.M. Sokolowski, Smart Mater. Struct. 10 (2001) 321.
[239] W.M. Huang, C.W. Lee, H.P. Teo, J. Intel. Mater. Sys. Str. 17 (2006) 753.
[240] M. Capelot, D. Montarnal, F. Tournilhac, L. Leibler, J. Am. Chem. Soc. 134 (2012) 7664–7667.
[241] Y. Yang, Z. Pei, X. Zhang, L. Tao, Y. Wei, Y. Ji, Chem. Sci. 5 (2014) 3486–3492.
[242] J.P. Brutman, P.A. Delgado, M.A. Hillmyer, ACS Macro Lett. 3 (2014) 607–610.
[243] F.I. Altuna, V. Pettarin, R. Williams, Green. Chem. 15 (2013) 3360–3366.
[244] W. Denissen, G. Rivero, R. Nicolaÿ, L. Leibler, J.M. Winne, F.E. Du Prez, Adv. Funct. Mater. 25 (2015) 2451–2457.
[245] M.M. Obadia, B.P. Mudraboyina, A. Serghei, D. Montarnal, E. Drockenmuller, J. Am. Chem. Soc. 137 (2015) 6078–6083.
[246] Y.-X. Lu, F. Tournilhac, L. Leibler, Z. Guan, J. Am. Chem. Soc. 134 (2012) 8424–8427.
[247] M. Pepels, I. Filot, B. Klumperman, H. Goossens, Polym. Chem. 4 (2013) 4955–4965.
[248] P. Taynton, K. Yu, R.K. Shoemaker, Y. Jin, H.J. Qi, W. Zhang, Adv. Mater. 26 (2014) 3938–3942.
[249] B. Krishnakumar, R.V.S.P. Sanka, W.H. Binder, V. Parthasarthy, S. Rana, N. Karak, Chem. Eng. J. 385 (2020) 123820.
[250] L. Yue, H. Guo, A. Kennedy, A. Patel, X. Gong, T. Ju, et al., ACS Macro Lett. 9 (2020) 836–842.
[251] M. Rottger, T. domenech, R. Weegen, A. Breuillac, R. Nicolay, L. Leibler, Science 356 (2017) 62–65.
[252] P. Waller, F. Gándara, O.M. Yaghi, Chemistry of covalent organic frameworks, Acc. Chem. Res. 48 (2015) 3053–3063.
[253] J. Crane, K. Anderson, S. Conway, J. Chem. Educ. 92 (2015) 373–377.
[254] K. Sumida, J.J. Arnold, Chem. Educ. 88 (2011) 92–94.
[255] B.F. Hoskins, R. Robson, J. Am. Chem. Soc. 112 (1990) 1546.
[256] O.M. Yaghi, H. Li, C. Davis, D. Richardson, T.L. Groy, Acc. Chem. Res. 31 (1998) 474–484.
[257] S.J. Lyle, R.W. Flaig, K.E. Cordova, O.M. Yagi, J. Chem. Edu. 95 (2018) 1512–1519.
[258] N. Stock, S. Biswas, Chem. Rev. 112 (2012) 933–969.
[259] J. Hungerford, S. Bhatacharyha, U. Tumuluri, S. Nair, Z. Wu, K.S. Walton, J. Phys. Chem. C. 122 (2018) 23493–23500.
[260] S.S. Chui, Science 283 (1999) 1148–1150.
[261] J. Cherusseri, D. Pandey, K.S. Kumar, J. Thomas, L. Zhai, Nanoscale 12 (2020) 17649–17662.
[262] D. Yang, B.C. Gates, ACS Catal. 9 (3) (2019) 1779–1798.
[263] C.P. Yang, T.M. Lee, J. Appl. Polym. Sci. 34 (1987) 2733.
[264] S.W. Shalaby, E.M. Pearce, Int. Polym. Mater. 3 (1) (1974) 81.
[265] S.D. Shaw, A. Blum, R. Weber, K. Kannan, D. Rich, L.S. Birnbaum, Rev. environ. health, 25, 2010, p. 261.
[266] C.S. Wang, J.Y. Shieh, Eur. Polym. J. 36 (2000) 433.
[267] C.S. Wang, C.H. Lin, J. Appl. Polym. Sci. 75 (2000) 429.
[268] U. Braun, B. Schartel, J. Fire. Sci. 23 (2005) 5.
[269] S. Huo, S. Yang, J. Wang, J. Cheng, Q. Zhang, P. Song, H. Wang, ACS Appl. Polym. Mater. 2 (2020) 3575–3586.
[270] C.H. Lin, C.S. Wang, Polymer 42 (2001) 1869.
[271] P. Jain, V. Choudhury, I.K. Verma, J. Macromol. Sci. Polym. Rev. 42 (2) (2002) 139.

[272] P.M. Hergenrother, C.M. Thompson, J.G. Smith, J.W. Connell, J.A. Hinkley, R.E. Lyon, et al., Polymer 46 (14) (2005) 5012.
[273] U. Braun, A.I. Balabanovich, B. Schartel, U. Knoll, J. Artner, M. Ciesielski, et al., Polymer 47 (26) (2006) 8495.
[274] U. Braun, U. Knoll, B. Schartel, T. Hoffman, D. Pospiech, J. Artner, et al., Macromol. Chem. Phys. 207 (2006) 1501.
[275] X. Wang, Q. Zhang, Eur. Polym. J. 40 (2004) 385.
[276] T. Hoffmann, D. Pospiech, L. Hausler, H. Komber, D. Voigt, C. Harnisch, et al., Macromol. Chem. Phys. 206 (2005) 423.
[277] Y. Qiu, L.J. Qian, H.S. Feng, S.L. Jin, J.W. Hao, Macromolecules 51 (2018) 9992−10002.
[278] S. Tang, L.J. Qian, Y. Qiu, Y.P. Dong, Polym. Degrad. Stabil. 153 (2018) 210−219.
[279] L. Qian, Y. Qiu, J. Liu, F. Xin, Y. Chen, J. Appl. Polym. Sci. 131 (2014) 39709.
[280] S.Q. Huo, J. Wang, S. Yang, J.P. Wang, B. Zhang, B. Zhang, et al., Stabil 131 (2016) 106−113.
[281] P. Wang, L. Chen, H. Xiao, J. Anal. Appl. Pyrol. 139 (2019) 104−113.
[282] P. Wang, Z.S. Cai, Polym. Degrad. Stabil. 137 (2017) 138−150.
[283] W. Zhang, X. Li, L. Li, R. Yang, Polym. Degrad. Stab. 97 (2012) 1041−1048.
[284] W. Zhang, X. Li, R. Yang, Polym. Degrad. Stab. 96 (2011) 2167−2173.
[285] W.C. Zhang, X.M. Li, R.J. Yang, J. appl. Polym. Sci. 122 (2011) 3383.
[286] W. Zhang, X. Li, R. Yang, Polym. Degrad. Stab. 96 (97) (2011) 1314−1324.
[287] H. Wu, B. Zeng, J. Chen, T. Wu, Y. Li, Y. Liu, et al., Chem. Eng. J. 374 (2019) 1304−1316.
[288] A. Bifulco, D. Parida, K.A. Salmeia, R. Nazir, S. Lehner, R. Stampfli, et al., Mater. Des. 193 (2020) 0−8.
[289] X. Qian, L. Song, Y. Hu, R.K.K. Yuen, Thermochim. Acta 552 (2013) 87−97.
[290] W. Xu, A. Wirasaputra, S. Liu, Y. Yuan, J. Zhao, Polym. Degrad. Stab. 122 (2015) 44−51.
[291] X. Wang, L. Song, W. Xing, H. Lu, Y. Hu, Mater. Chem. Phys. 125 (3) (2011) 536−541.
[292] A.K. Vibha Shree, J.S. Sen, Gel. Sci. Tech. 85 (2018) 269−279.
[293] P. Wang, F. Yang, L. Li, Z. Cai, Polym. Degrad. Stab. 129 (2016) 156−167.
[294] J. Sag, D. Goedderz, P. Kukle, L. Grener, M. Doring, Molecules 24 (2019) 3746.
[295] M. Zhang, Z. Luo, J. Zhang, S. Chen, Y. Zhou, Polym. Degrad. Stab. 120 (2015) 427−434.
[296] H. Duan, S. Ji, T. Yin, X. Tao, Y. Chen, M. Huiru, J. Appl. Polym. Sci. 137 (2020) 47910.
[297] M. Spontó, G. Lligadas, J.C. Ronda, M. Galià, Polym. Degrad. Stab. 94 (2016) 1693−1699.
[298] S. Huo, J. Wang, S. Yang, Y. Tang, J. Appl. Polym. Sci. 133 (2016) 43403.
[299] X. Chen, G. Liang, A. Gu, L. Yuan, Ind. Eng. Chem. Res. 54 (2015) 1806.
[300] H. Zweifel, Macromol. Symp. 115 (1997) 181.
[301] P.P. Nicholas, A.M. Luxader, L.A. Brooks, P.A. Hammes, Antioxidants and antiozonants, in: M. Grayson (Ed.), Kirk-Othmer Encyclopedia of Chemical Technology, Wiley, New York, 1978, p. 128.
[302] P.P. Klemchuk, Antioxidants, Ullmann's Encyclopedia of Industrial Chemistry, Vol A3, Wiley-VCH Verlag GmbH & Co KgaA, Weinheim, 2002.
[303] S. Al-Malaika, Chap. 2 in: G. Scott (Ed.), Atmospheric Oxidation and Antioxidants, Vol 1, Elsevier Applied Science Publisher, London, 1993.
[304] J. Pospisil, Ch. 1 in: G. Scott (Ed.), Developments in Polymer Stabilization, Vol 1, Applied Science Publishers, London, 1979.
[305] R. Bagheri, K.B. Chakraborty, G. Scott, Polym. Deg. Stab. 4 (1982) 1.
[306] G. Gungumus, Ch. 3 in: R. Gachter, H. Muller (Eds.), Plastic Additives Handbook, 2nd ed), Hanser Publisher, Munich, Germany, 1997.
[307] N. Grassie (Ed.), Developments in Polymer Degradation, Applied Science Publisher, London, 1977.
[308] Ch. 4 J.W. Moison, J.W. Comyn (Eds.), Polymer Permeability, Applied Science Publisher, London, 1985.
[309] S. Al-Malaika, Polym. Plast. Technol. Eng. 29 (1990) 73.
[310] J. Pospisil, Ch. 6 in: P. Klemchuk, J. Pospisil (Eds.), Oxidation Inhibition of Organic Materials, Vol 1, CRC Press, Boca Raton, FL, 1996.

[311] S. Al-Malaika, Chemical Reactions on Polymer, in: J.L. Benham and J.F. Kinstle (Eds.), ACS Symposium Series 364, American Chemical Society, Washington, DC, Ch. 29, 1988.
[312] D.R. Paul, S. Newman, Polymer Blends, Vol. II, Academic Press, New York, 1978.
[313] K.L. Mittal (Ed.), Silane and Other Coupling Agents-2, VSP, Utrecht, The Netherlands, 2000.
[314] X. Gong, J. Liu, S. Baskaren, R.D. Voise, J.S. Young, Chem. Mater. 12 (2000) 1049.
[315] T. Fukushima, A. Kosaka, Y. Ishimura, T. Yamamoto, T. Takigawa, N. Ishii, et al., Science 300 (2003) 2072.
[316] P.V. Kodgire, A.R. Bhattacharyya, S. Bose, N. Gupta, A.R. Kulkarni, A. Misra, Chem. Phys. Lett. 432 (2006) 480.
[317] D. Ratna, A.K. Banthia, N.C. Maity, Ind. J. Chem. Tech. 2 (1995) 253.
[318] C.W. Macosko, RIM: Fundamentals of Reaction Injection Molding, Hanser, Munich, 1989.
[319] G.N. Levy, R. Schindel, J.P. Kruth, CIRP Ann-Manuf Technol. 52 (2) (2003) 589−609.
[320] A. Pilipovic, P. Raos, M. Šercer, Int. J. Adv. Manuf. Technol. 40 (2009) 105−115.
[321] U.K. Zaman, E. Boesch, A. Siadat, M. Rivette, A.A. Baqai, Int. J. Adv. Manuf. Technol. 101 (2019) 1215−1226.
[322] K. Chockalingama, N. Jawahara, U. Chandrasekarb, K.N. Ramanathana, J. Mater. Process. Technol. 208 (1019) (2008) 348−365.
[323] B. Vasudevarao, D.P. Natarajan, A. Razdan, H. Mark, Sensitivity of RP surface finish to process parameter variation, in: Proceedings of Solid Free Form Fabrication, Austin, TX, USA, 2000, pp. 252−258.
[324] F.D.M. Stratasvs, Vantage User Guide Version 1.1, Stratasvs;, Minnesota, 2004.
[325] Hh, E. Said, J. Os Foyos, R. Noorani, M. Mandelson, R. Marloth, et al., Mater. Manuf. Process. 15 (2000) 107−122.
[326] S.H. Ahn, M. Montero, D. Odell, S. Roundy, P.K. Wright, Rapid Prototyp. J. 8 (2002) 248−257.
[327] Z.A. Khan, B.H. Lee, J. Abdullah, J. Mater. Process. Technol. 169 (2005) 54−61.
[328] A.K. Sood, R.K. Ohdar, S.S. Mahapatra, Mater. Des. 31 (2010) 287−295.
[329] R. Yu, X. Yang, Y. Zhang, X. Zhao, X. Wu, T. Zhao, et al., ACS Appl. Mater. Inter. 9 (2017) 1820−1829.
[330] J.A. Lewis, Adv. Func. Mater. 16 (2006) 2193−2204.
[331] B.G. Compton, J.A. Lewis, Adv. Mater. 26 (2014) 5930−5935.
[332] L.L. Lebel, B. Aissa, M.A.E. Khakani, D. Therriault, Adv. Mater. 22 (2010) 592−596.
[333] K. Chen, X. Kuang, V.C.F. Li, G. Kang, H.J. Qi, Soft Matter 14 (2018) 1879−1886.
[334] D.K. Patel, A. Hosein Sakhaei, M. Lyani, B. Zhang, Q. Ge, S. Magdassi, Adv. Mater. 29 (2017) 1−7.
[335] T. Femmer, A.J.C. Kuehne, M. Wessling, Lab. Chip 14 (15) (2014) 2610.
[336] T. Femmer, A.J.C. Kuehne, J. Torres-Rendon, A. Walther, M. Wessling, J. Memb. Sci. 478 (2015) 12−18.
[337] V. Ozbolat, M. Dey, B. Ayan, A. Povilianskas, M.C. Demirel, I. Ozbolat Madhuri, ACS Biomater. Sci. Eng. 4 (2) (2018) 682−693.
[338] D. Lei, Y. Yang, Z. Liu, S. Chen, B. Song, A. Shen, et al., The Royal Society of Chemistry, 6, 2019, pp. 394−404.

CHAPTER 3

Toughened thermoset resins

Abbreviations

FRP	fiber-reinforced polymer
ABS	acrylonitrile−butadiene−styrene
PVC	poly(vinyl chloride)
PMMA	poly(methyl methacrylate)
PP	polypropylene
PC	polycarbonate
PPO	poly(phenylene oxide)
HIPS	high-impact polystyrene
PBT	poly(butylene terephthalate)
MLT	matrix ligament thickness
SEM	scanning electron microscope
CTBN	carboxyl-terminated copolymer of acrylonitrile and butadiene
HTBN	hydroxyl-terminated polybutadiene
VTBN	vinyl-terminated poly(butadiene−acrylonitrile)
ETBN	epoxy-terminated vinyl-terminated poly(butadiene−acrylonitrile)
HTE	hydroxyl-terminated polyether
HBP	hyperbranched polymers
VEUH	vinyl ester-urethane hybrid
LPA	low-profile additives
ATBN	amine-terminated copolymer of acrylonitrile and butadiene
PCF	P-cresol formaldehyde
PEO	polyethylene oxide
BMI	bismaleimide
PEP	poly(ethylene phthalate)
AT-PAEKI	allyl-terminated hyperbranched polyimide
MPA-BMI	modification of bisphenol-A bismaleimide
HBPEE	hyperbranched polyether epoxy
DGEBA	Diglycidyl bisphenol-A
CTPEGA	carboxyl-terminated poly(ethylene glycol) adipate
TPP	triphenylphosphine
ATPS	amine-terminated polysulfone
DDS	Diamino diphenyl sulphone
PDMS	Poly (dimethyl siloxane)
PCL	polycaprolactam
GMA	glycidyl methacrylate
PECH	polyepichlorohydrin
ESO	epoxidized soybean oil
CTPPGA	carboxyl-terminated poly(propylene glycol) adipate
PU	polyurethane

ATPEHA	amine-terminated poly(2-ethylhexyl acrylate)
CTPEHA	carboxyl-terminated poly(2-ethylhexyl acrylate)
DTDGA	dithiodiglycolic acid
ABCVA	4, 4′-azobis (4-cyanovaleric acid)
CRPEHA	Carboxyl-randomized poly (2-ethyl hexyl acrylate)
TGMDA	tetraglycidyl methylene dianiline
PEHA	poly(2-ethylhexyl acrylate)
DDS	diamino diphenyl sulfone
BA	butyl acrylate
VBGE	vinyl benzyl glycidyl ether
HBP	hyperbranched polymer
PR	powdered rubber
PES	poly(ether sulfone)
PEK	poly(ether ketones)
APN	amine-containing novolac-phthalonitrile
PEEK	Polyether ether ketone
PMS	N-phenylmaleimide–styrene
PHMS	N-phenylmaleimide-N-cyclohexylmaleimide styrene
PBT	poly(butylene terephthalate)
PVDF	poly(vinylidene fluoride)

3.1 Introduction

Thermosets, because of their crosslinked structure, offer many useful properties such as high strength, good dimensional stability at elevated temperature, very low creep, solvent and water resistance, etc. In order to get the desired thermomechanical properties, it is necessary to maintain a high crosslink density of the thermoset resin network. Unfortunately, these highly crosslinked glassy networks suffer from the problem of poor fracture resistance. Hence since the late sixties, efforts have been made to toughen the thermosets [1–10]. Although the use of plasticizers is a very successful route to improve the ductility of a brittle thermoplastic, it does not work well in the case of thermosets resins. This is because thermoset resins are characterized by a crosslinking reaction as a result of which, the modifier either exudes out from the matrix or undergoes a macrophase separation. Moreover, the accumulation of free liquid plasticizer molecules at the fiber surface can act as a weak boundary layer and cause a substantial decrease in the mechanical performance of a composite. A similar problem arises for adhesive applications as well. The general strategies used to toughen a thermoset resin are given below:

1. Introduction of flexible chain (e.g., ether linkages) into the network structure.
2. Compatible blending with a flexible/ductile polymer.
3. Reduction in crosslink density of the network.
4. Introduction of a suitable filler (rubber, thermoplastic, or rigid particles) as a second phase.

The first three approaches improve the toughness by flexibilizing the thermoset networks. The flexibilization is achieved by various methods like chemical modification of the resin, use of reactive diluents, long-chain curing agents, and chain extension of the resin. However, the improvement in toughness due to flexibilization using the first three strategies is associated with a significant sacrifice in T_g and mechanical properties. Mechanical properties of a thermoset resin can be adjusted by reinforcing them with high-strength inorganic fillers/fibers. However, a reduction in T_g as a result of modification (flexibilization) of the resin limits the applications of the related cured networks at a high service temperature. Because at a temperature higher than the T_g, the cured resin loses its dimensional stability and cannot serve its functions. For example, it cannot hold the fibers as a matrix in fiber-reinforced polymer (FRP) composites and cannot offer good bond strength as a structural adhesive.

The fourth strategy provides a solution to this problem in which the modifier is introduced as a separate phase so that it does not affect the bulk thermomechanical properties, and at the same time, improves the fracture properties of the thermoset network. This approach is called the second phase toughening, which offers a significant increase in toughness/impact strength without a significant sacrifice in thermomechanical properties especially the T_g. This is particularly required where the material has to experience high temperature, for example, aircraft application. As discussed in Chapter 1, Chemistry and General Applications of Thermoset Resins, an epoxy network with desirable mechanical properties and a T_g of about 240°C−250°C can be made by combining the conventional difunctional and multifunctional epoxies. If the T_g is reduced further due to modification, then epoxy formulation cannot satisfy the requirement for aircraft application ($T_g > 200°C$). Hence the technology of second phase toughening is extremely important for the development of advanced thermoset polymer materials. This can be achieved by using rubber, thermoplastic or rigid fillers. However, the most effective toughening is achieved by using a suitable rubbery filler (rubber toughening). Block copolymers and nanofillers are also used to toughen a thermoset resin, which will be elaborated in Chapter 5, Thermoset Nanocomposites.

In this chapter, the general principles of rubber toughening and the works on thermosetting resins will be reviewed. Though this book is dedicated to thermoset resins, toughening of thermoplastic [11−103] will also be briefly discussed for completeness and a better understanding of the subject. If we look at the literature on toughening of thermosets, we find that maximum number of publications a dealt with epoxy resins and toughening technology has been most successfully exploited for epoxy resins. This is because of the versatility, reactivity, and availability of epoxy resins as discussed in Chapter 1, Chemistry and General Applications of Thermoset Resins. Therefore the works on toughened epoxy will be discussed at the end after discussing the general principle and toughened thermosets other than epoxy.

3.1.1 Toughening of thermoplastics

The earliest and the most well-known rubber-toughened thermoplastics are high-impact polystyrene (HIPS) [11,12] and acrylonitrile—butadiene—styrene copolymer (ABS) [13,14]. Both the rubber-modified polymers possess greatly improved toughness as compared to the unmodified polystyrene (PS) and styrene-acrylonitrile copolymer. At an early age, styrene-butadiene rubber (SBR) was used as a toughening agent. Initially, butadiene rubber ($T_g \sim -85°C$) replaced SBR ($T_g \sim -50°C$) and increased its application for a wide range of temperatures. Other thermoplastics, which have been successfully toughened by the introduction of a rubbery second phase, include poly(vinyl chloride) (PVC) [15–17], poly(methyl methacrylate) (PMMA) [18–20], polypropylene (PP) [21–26], polycarbonate (PC) [27–30], poly(phenylene oxide) (PPO) [31–33], nylons [34–36], etc.

HIPS materials were prepared by a melt or latex blending of rubber and polystyrene during the early attempt of making rubber-toughened plastics. However, in these multiphase polymers, the rubber particles were only poorly bonded to the surrounding matrix. The interfacial adhesion problem has been addressed by polymerizing styrene in the presence of the rubber, which gives well-bonded particles due to the chemical grafting [37,38]. The graft copolymer, formed during polymerization, acts as a coupling agent and the ungrafted rubber acts as a toughening agent. The polystyrene produced due to the homopolymerization of styrene forms the matrix. Graft copolymerization via a bulk or bulk suspension process is currently the standard method for the manufacture of HIPS [9]. The HIPS materials prepared by the grafting route exhibit significantly higher toughness compared to those prepared by a melt blending. Similarly, ABS and the related polymers such as methacrylate-styrene-butadiene copolymers and toughened PMMA are prepared by techniques that lead to the grafting reactions [9]. In these materials, such reactions are typically induced by first preparing rubber latex by emulsion polymerization and then adding the monomer (e.g., styrene, acrylonitrile) to the reactor vessel and polymerizing them in the presence of rubber latex [9]. Hardward and Mann [39] have compared the properties of ABS polymers prepared by blending or by grafting. They have demonstrated that the direct blending route results in low particle/matrix adhesion and poor impact strength compared to the grafted polymers.

The grafting strategy cannot be readily adopted in some materials such as rubber-toughened PVC, nylon, etc. Hence they are prepared by the conventional blending route. The criteria for a rubber to be an effective toughening agent is that the rubber should possess a solubility parameter sufficiently different from that of the matrix polymer to ensure a fine second phase dispersion, but close enough to promote adequate adhesion of the particles with the matrix [9,40]. This requirement may limit the choice of rubber at the expense of mechanical properties. Block copolymers with hard and soft segments can overcome these limitations and broaden the scope of application of the toughened plastics. As a result, the block copolymers have been investigated extensively [41–49] to examine their suitability as toughening agents for a number of plastics, especially PS. Their soft segments provide the low modulus needed for the dispersed particles to act as effective stress concentrators while their hard segments can lead to the microdomains that act as physical crosslinks and can provide a mechanism of physical adhesion to the rigid matrix phase. It has been reported that a styrene-based block copolymer adheres well to PS and PPO, with which PS is miscible. The effectiveness of block copolymers as toughening agents depends strongly on their molecular architecture [33].

3.1.2 Mechanism of toughening of brittle polymer

When mechanical stress is applied to a brittle polymer, the polymer cannot dissipate the mechanical energy adequately. This inadequate energy dissipation causes overstressing resulting in the formation of cracks, which propagate rapidly through the glassy material leading to the catastrophic failure of the material. In order to increase its fracture toughness, it is required to provide some mechanisms for the dissipation of mechanical energy, while at the same time, limit the growth and breakdown of voids and crazes to prevent premature crack initiation. Depending on the inherent nature of the matrix polymer, the major energy dissipating mechanism may vary. Two important energy dissipating mechanisms, namely, shear yielding and crazing are discussed in subsequent sections.

3.1.2.1 Shear yielding

When a polymer starts to deform plastically under applied stress it is said to have yielded. Shear yielding consists of a change of shape without a significant change in volume and is the major energy absorbing mechanism for toughened pseudo-ductile polymers [27,50–67] such as PC, PVC, polyamides, etc. Although shear yielding operates in the unmodified polymers also, they do so only in highly localized regions around the crack tip. In rubber-toughened thermoplastics, each rubber particle initiates shear yielding and the same is propagated through the matrix and thus enhanced to a great extent. When a multiphase material is subjected to uniaxial stress, the necking, drawing, and orientation hardening occurred indicating shear deformations, and the matrix had to undergo localized plastic deformations around virtually every rubber particle [50].

In order to understand the role of rubber particles in enhancing the shear yielding mechanism, various micromechanisms have been proposed. Most convincing among these is the rubber cavitations, first identified by Haff et al. [51] and Donald and Kramer [52]. This micromechanism not only can explain the enhanced shear yielding in rubber-toughened polymers but also accounts for the stress whitening observed during the fracture of rubber-toughened polymers. Stress whitening is observed during the mechanical stretching of a polymer, which cannot be explained by the shear yielding mechanism without considering the rubber cavitations.

3.1.2.2 Rubber cavitation

In rubber-modified plastics, under triaxial tensile stresses, voiding can be initiated inside the rubber particles. Once the rubber particles are cavitated, the hydrostatic tension in the material is relieved, with the stress state in the thin ligaments of the matrix between the voids being converted from a triaxial to a more uniaxial tensile stress state. This new stress state is favorable for the initiation of shear bands. In other words, the role of rubber particles is to cavitate internally, thereby relieving the hydrostatic tension and initiating the ductile shear yielding mechanism [53–58]. The process of the rubber cavitation and the associated matrix shear yielding are found in rubber-modified PC [27,59–61], PVC [62–64], poly(butylene terephthalate) (PBT) [65], and nylon [66–68]. Depending on the nature of the matrix, rubber cavitation may initiate shear yielding or crazing. In the case of plastics having a high entangle density or a thermoset with very high crosslink density, crazing is suppressed [69], and shear yielding is initiated to a great extent. In the case of a very brittle polymer, though rubber cavitation can occur to a very large extent, this does not lead to matrix shear yielding [61]. That is why in a very brittle polymer the major energy dissipating mechanism is crazing initiated by rubber cavitation.

Recently, Dompus and Groeninckx [70,71] developed a criterion for internal cavitation. The condition has been treated as an energy balance between strain energy relieved by cavitation and surface energy required to create a new surface and given by the following equation:

$$U_{total} = U_{strain} + U_{surface} \\ = -\pi/12 K_r \Delta^2 d_o^3 + \gamma \pi \Delta^{2/3} d_o^2 < 0 \quad (3.1)$$

where K_r, Δ, d_o, and γ are the rubber bulk modulus, relative volume strain, rubber particle diameter, and surface energy per unit area, respectively.

Accordingly, rubber's cavitation depends on the elastic and molecular properties of rubber, rubber particle size, and the applied volume strain, which again depends on the difference in Poisson's ratio between the matrix and the rubber. Cavitation resistance increases with the decrease in particle size and the difference in Poisson's ratio [72]. From Eq. (4.1), it is clear that the bigger rubber particles cavitate more easily than the smaller rubber particles. Critical particle size for cavitation in the range of 100–200 nm is estimated [70,71]. The same critical particle size is found for effective toughening in rubber-toughened PVC [73,74], nylon [75,76], PMMA [77–79], and PC [53]. This explains the experimental fact that there is a little toughening effect when the particle size is reduced beyond 0.1 mm. Now a question may arise in the inquisitive readers' minds that if rubber cavitation is favored for bigger particles, why should we use sufficiently smaller rubber particles (\\10.42.4.54\bkuserworkarea\comp\48-els-che\ratna-rat-1633100\ch003 <5 mm) for an effective toughening? This will be addressed in the course of the discussion presented shortly.

3.1.2.3 Crazing

When triaxial stress is applied to a brittle polymer, since it cannot easily deform, microvoiding is initiated in the matrix. The localized yielded region, therefore, consists of an interpenetrating system of voids and polymer fibrils (1 mm diameter), called craze. A craze can grow indefinitely in length following the minor principal trajectories. The craze also grows in width, although craze opening is limited to between 0.5 and 1.0 mm [80]. Unlike a crack, a craze is capable of transmitting loads across the faces and this process of energy dissipation is called crazing. In rubber-toughened brittle plastics, crazing is the major energy dissipating mechanism [87–93] because the shear yielding cannot be initiated. From optical microscopy studies on HIPS, Bucknall and Smith [81] have concluded that the function of rubber particles is to initiate multiple crazes in the PS matrix. Subsequent studies [81–86] have reported the results on transmission electron microscope (TEM) analysis of thin sections of various rubber-modified polymers by first staining the unsaturated rubber phase with osmium tetraoxide. These results clearly confirmed that crazes frequently initiate from the rubber particles.

The role of rubber cavitation on the initiation of shear yielding has been discussed in Section 3.1.2.2. In brittle polymers, the cavitated rubber particles on relieving the stress cannot initiate shear deformation. Now the point is to know how rubber cavitation favors the crazing mechanism. Bucknall and coworkers [86] proposed that rubber particles have two separate, but equally important functions. First, under applied tensile stress, crazes are initiated at the points of maximum stress concentrations and grow approximately normal to the applied stress. Second, the rubber particles are craze terminators, preventing the growth of very large crazes. Thus the result is that a large number of small crazes are generated in contrast to the small number of large crazes formed in the same polymer in absence of rubber particles. The multiple crazing that occurs throughout a comparatively large volume of rubber-modified materials explains the high energy absorption in the fracture tests and exclusive stress whitening which accompanies deformation and failure.

3.1.3 Morphological aspects

Phase morphology is a key issue for effective rubber toughening. It is generally believed that there is an optimum rubber particle size, which is closely related to the nature of the postyield deformation mechanism of the matrix materials that can be triggered by rubber particles. Very brittle polymers, like PS, require relatively large particles of the order of several microns [52,94–96] as large particles are more effective in crazing, which is the major energy dissipating mechanism for brittle polymers [80–93]. If the rubber particle size is less than craze thickness then a growing craze may engulf the rubber particles smaller than the craze thickness rather than having its growth arrested. If the growth of crazes is not arrested then the crazes may combine and form a bigger craze, which may act as a defect and leads to the catastrophic failure of the material. Note that the rubber cavitation micromechanism is also more effective for bigger particles.

On the other hand, the more ductile polymers like PPO, polyamides, and PC, deform by shear yielding and require relatively smaller particles for toughening than the materials that craze [36,97,98]. When toughness value is plotted against particle size at a constant amount of rubber and interphase adhesion, a sharp brittle-tough transition occurs at a critical rubber particle diameter, d_c [99,100]. When the average particle size (d) is smaller than d_c, a blend is tough; when d is larger than d_c, the blend is brittle [35,100]. For example nylon 6 and nylon 66 are not effectively toughened unless the rubber particles are considerably smaller than 1 mm [101]. The optimum particle size for toughening PPO is well below that of PS [102] but probably somewhat larger than the optimum range for certain polyamides [101] or for styrene-acrylonitrile copolymers [103,104]. The critical particle size depends on the amount of rubber, that is, decreasing with decreasing amount of rubber. The lower limit of the optimum particle size range effective for rubber toughening can be explained in terms of reduction in the effectiveness of rubber cavitation with decreasing size and engulfment of smaller rubber particles by the growing crazes. The higher limit of the optimum range can be explained by considering the fact that the particles having bigger than the optimum size cannot serve as effective stress concentrators rather than act as defects. The existence of a higher limit of the optimum particle size range can also be very precisely explained by using the concept of matrix ligament thickness (MLT), which will be discussed in Section 3.3.3.3.

3.2 Toughening of thermosets

Unlike the modification of thermoplastics, which often requires only a simple physical blending of a particular elastomeric modifier, the toughening of thermoset by rubbery second phase is achieved almost exclusively through the chemistry and a challenge to the expertise of polymer chemist. A useful final product with improved durability and applicability is a major concern of materials scientists and engineers who are responsible for answering this challenge. There are two approaches for incorporating rubber into a thermoset network. One is the direct dispersion of preformed core–shell particles (CSPs) and the other approach is to blend with a liquid rubber, that is, liquid rubber toughening. Liquid rubber toughening has been more extensively investigated and exploited.

3.3 Liquid rubber toughening

In this method, liquid rubber is blended with a thermoset resin before curing [8,34,62,97,100,105–190]. Various liquid rubbers, used for toughening thermoset resins, are presented in Table 3.1. It may be noted that toughening effect of a particular modifier largely dependent on the molecular weight and functionality of a particular liquid rubber.

Table 3.1: List of liquid rubbers used for toughening of a thermoset.

Name of liquid rubber	Abbreviation/source/reference(s)
Carboxyl-terminated copolymer of butadiene and acrylonitrile	CTBN, Goodrich Company
Amine-terminated copolymer of butadiene and acrylonitrile	ATBN, Goodrich Company
Vinyl-terminated copolymer of butadiene and acrylonitrile	VTBN [185]
Hydroxyl-terminated polybutadiene	HTBN [185]
Hydroxyl-terminated polyether	THE [185]
Carboxyl-terminated poly(butyl acrylate)	[2,107]
Vernoria oil	[2,10,108]
Epoxidized soybean oil	ESO [2,10,11]
Epoxy-terminated siloxane oligomer	[2,10]
Poly(ethylene phthalate)	PEP [191]
Carboxyl-terminated poly(ethylene glycol) adipate	CTPEGA [2,10]
Carboxyl-terminated poly(propylene glycol) adipate	CTPPGA [2,10]
Epoxy-functionalized hyperbranched polymer	Perstrop, HBP (epoxy) [105]
Hydroxy-functionalized hyperbranched polymer	Perstrop, HBP (OH) [10,105,192,193]
Carboxyl-terminated poly(2-ethylhexyl acrylate)	CTPEHA [106,107]
Carboxyl-randomized poly(2-ethylhexyl acrylate)	CRPEHA [2,10]
Hydroxyl-terminated poly(propylene oxide-block-ethylene oxide)	[10,107]
N-phenylmaleimide—styrene copolymer	PMS [194]
Isocyanate end-capped polybutadiene	[195]

The criteria for selecting the rubber are as follows: (1) it must be a low-molecular-weight liquid to ensure easy miscibility with a thermoset resin, (2) it must have functional groups, capable of reacting with the base thermoset resin, and (3) it must have borderline miscibility with the thermoset resin. The borderline miscibility means that the rubber is initially miscible with the resin making a homogeneous solution and undergoes reaction-induced phase separation leading to the formation of a two-phase microstructure. Scanning electron microscope (SEM) photographs of a CTBN-toughened epoxy network is shown in Fig. 3.1, which clearly shows a uniform dispersion of spherical rubber particles in the epoxy matrix. Such rubber-modified resin networks with a two-phase microstructure show improved fracture toughness as the rubber particles, dispersed and bonded to the matrix act as the centers for dissipation of mechanical energy. The detailed mechanisms of toughening and roles of rubber particles will be discussed shortly. The principle of reaction-induced phase separation is discussed below considering the thermodynamics of mixing.

3.3.1 Reaction-induced phase separation

The thermodynamic condition for compatibility of a resin and a rubber is that the free energy change of mixing (ΔG_m) at constant pressure (P) and temperature (T) should be negative [105]:

$$(\Delta G_m)_{P,T} < 0 \tag{3.2}$$

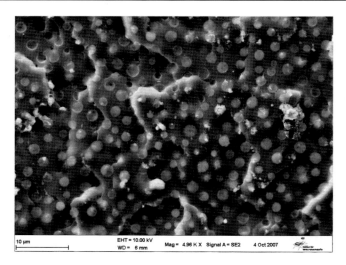

Figure 3.1
SEM photographs of a CTBN-toughened epoxy network showing a two-phase morphology.

Combining the Flory–Huggins equation and the Hildebrand equation, the free energy of mixing can be expressed as:

$$\Delta G_m/V = \phi_R \phi_r (\delta_R - \delta_r)^2 + RT(\phi_R/V_R . \ln \phi_R + \phi_r/V_r . \ln \phi_r) \qquad (3.3)$$

where ϕ_R and ϕ_r are the volume fractions, δ_e and δ_r are the solubility parameters, V_e and V_r are the molar volume of the resin and rubber, respectively, V is total volume, R is the universal gas constant, and T is temperature.

Since both ϕ_R and ϕ_r are fractions, the second term is always negative. The equation clearly explains why high temperature and low molar volume (molecular weight) favor miscibility as both the factor decrease the free energy of mixing (thermodynamic spontaneity). For a fixed resin/rubber composition, ΔG_m at constant temperature depends on δ_r, that is, the chemical nature of rubber and V_r, which is dependent on the molecular weight of the rubber. A toughening agent has to be designed (in terms of chemistry and molecular weight) in such a way that the free energy of mixing is marginally negative at the curing temperature. Then the rubber will be compatible with the resin before the curing, but with the advancement of curing reaction, V_e and V_r will increase due to the increase in molecular weight of the rubber and the resin and at a certain stage, ΔG_m will become positive. At that point, the rubber starts undergoing phase separation and it is called the cloud point. The process continues until the gelation point, where the phase separation is arrested due to the tremendous increase in viscosity. The final network is obtained after a heat treatment called postcuring. The process is schematically described in Fig. 3.2.

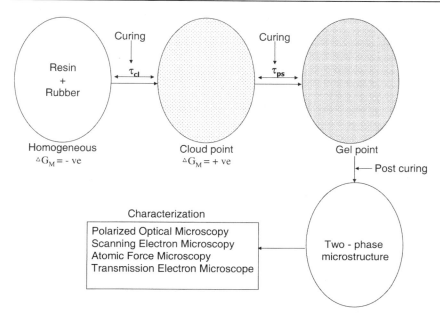

Figure 3.2
Description of reaction-induced phase separation in rubber-modified thermoset resin.

If $(\delta_e - \delta_r)$ is very low, then ΔG_m will be highly negative and entropy change (during curing) cannot make ΔG_m equal to negative, before gelation, leading to the formation of a single-phase morphology. Again if $(\delta_e - \delta_r)$ is very high, then ΔG_m will be positive at curing temperature and rubber will be immiscible at the initial stage itself, leading to a macro level phase separation. If the phase separation starts in a very early stage of curing then rubber particles will coalesce leading to the formation of bigger particles. Again if phase separation starts at a later stage of curing, the rubber may not get a chance to phase separate completely. Hence it is necessary to design the toughening agent for a particular thermoset considering the miscibility criteria. For example, carboxyl-terminated polybutadiene undergoes macrophase separation when blended with epoxy and is not a suitable toughening agent for epoxy. This is because the solubility parameter of butadiene is much lower than that of epoxy and it is immiscible with epoxy. However, a carboxyl-terminated copolymer of butadiene and acrylonitrile with 20–30 wt.% of acrylonitrile (polar acrylonitrile increases the solubility parameter) has close proximity of solubility parameter with an epoxy resin and is an effective toughening agent for epoxy. The copolymers are commercially produced by Goodrich Company and are known as trade Hycar CTBN. CTBN with higher acrylonitrile content has better miscibility with the epoxy resin in terms of solubility parameter and undergoes phase separation at a later stage of curing resulting in a lower amount of phase-separated rubber. CTBN with an acrylonitrile content of more than 30 wt.% results in a single-phase morphology and the modifier acts as a flexibilizer rather than a toughening agent.

Complete phase separation is always desirable as residual rubber remains miscible with the resin and reduces the T_g of the resin network. The other factor (apart from the nature of rubber), which plays an important role in the development of morphology is the cure condition. The morphology is controlled by the initial cure temperature, and postcuring conditions (though it affects the final properties) have no role in morphology development [106–108]. The time from cloud point to gelation is the effective phase separation time (t_{ps}). For complete phase separation, the t_{ps} has to be higher than the time required for diffusion of rubber from the resin medium (t_{diff}). The diffusivity is the controlling factor of phase separation if t_{diff} is greater than t_{ps}. The diffusivity of rubber in the resin medium (D_r) is considered to be proportional to the temperature and viscosity ratio through the Stokes–Einstein equation [109] as given below:

$$D_r = kT/6\pi R_r \, \eta_R \qquad (3.4)$$

where k is the Boltzmann constant, R_r is the radius of rubber adducts, h_R is the viscosity of the resin, and T is the absolute temperature. The characteristic time scale for diffusion in two dimensions is:

$$t_{diff} = L^2/2\, D_r \qquad (3.5)$$

A length scale (L) can be assigned from the average two-dimensional distance between domain centers obtained from micrographs of the cured specimen. Monternal et al. [107] reported that the system viscosity at cloud point (η_{cp}) is the most fundamental parameter rather than the t_{ps} and t_{diff} for morphology development in rubber-toughened thermoset. They correlated the system viscosity with the particle size ($\overline{D}$) of the dispersed phase. A straight line with a negative slope was obtained when $\ln \eta_{cp}$ is plotted against $\overline{D}$. The validity same is also verified by Verchere et al. [108] for epoxy/CTBN systems using two different types of CTBN. They reported that correlation was found for both the systems independently of the cure temperature and initial rubber concentration. Different systems show different fittings indicating that other factors like thermodynamic miscibility play an important role.

3.3.2 Mechanism of toughening of thermoset

A number of theories have been proposed to explain the toughening effect of rubber particles on the brittle thermoset matrix, based on the fractographic features and fracture properties of the rubber-toughened thermoset networks. Garg and Mai [110] proposed 13 different mechanisms to explain the impact behavior of rubber-toughened epoxy systems. The various mechanisms are (1) shear band formation near rubber particles, (2) fracture of rubber particles after cavitation, (3) rubber particle stretching, (4) rubber particle deboning, (5) tearing of rubber particles, (6) transparticle fracture, (7) debonding of hard particles,

(8) crack deflection by hard particles, (9) voiding and cavitation of rubber particles, (10) crazing, (11) plastic zone at craze tip, (12) diffuse shear yielding, and (13) shear band/craze interaction.

Among the various mechanisms mentioned above few important mechanisms are discussed below [8,34,62,97,100,111–185].

3.3.2.1 Rubber bridging and tearing

According to the bridging and tearing mechanism [111,112] when a crack opens up during its advancement, the rubber particles are stretched between the crack surfaces, behind the crack tip, and bridge the crack surfaces and break ultimately. Yee and Pearson [115] studied a highly crosslinked rubber-modified epoxy system and attributed the enhancement in toughness to the rubber-bridging mechanism. Similar evidence was reported by others [111–114]. The model is based on the assumption that crosslinked thermoset resins are truly brittle materials and incapable of any localized plastic deformation. Thus the mechanical energy is absorbed exclusively by the rubber particles, not the matrix. However, even in a highly crosslinked thermoset some plastic deformation always takes place. Since the absorbed energy is utilized to stretch and tear the rubber particles, the fracture energy can be estimated quantitatively as below:

$$G_{IC} = G_{ICe}(1 - V_p) + 4\Gamma V_p(1 - 6/\lambda^2 + \lambda + 4) \quad (3.6)$$

where V_p, G are the volume fraction and tear energy of the rubbery particles; G_{ICe}, G_{IC} are the fracture energy of the unmodified and rubber-modified thermoset matrix, respectively.

However, this theory has several drawbacks. It can only explain the modest increase in toughness and cannot explain the dramatic increase in fracture energy reported in many rubber-toughened epoxy systems. It may be noted that there is a lot of discrepancies between the reported data for improvement in toughness for thermoset resins. This is due to the sensitivity of toughening effect to the curing condition and inherent matrix ductility [116,117]. The effect of G_{IC} on test temperature and strain rate cannot be explained using the above equation as the dependence of G is exactly the reverse of G_{IC}. Moreover, the theory cannot explain the stress whitening observed in fracture surface analysis. Hence this mechanism cannot be considered as a major energy dissipating mechanism for highly crosslinked thermoset resins.

3.3.2.2 Crazing

A second proposal to explain the toughening of a thermoset network involves conventional crazes being initiated and growing in the thermoset matrix. Crazing has already been discussed in Section 3.1.2.3 and in some rubber-toughened thermoplastics such as HIPS; multiple crazing as stress concentrators around the rubber particles has been established as the major toughening mechanism [87–93]. Donald and Kramer [118,119] showed that in thermoplastic there is a transition from crazing and shear yielding as the length of the polymer chain between physical entanglements, decreases below ~ 20 nm. Thus crazing is less probable in highly crosslinked thermoset resins. Van den Boogaat [120] observed crazes at the tips of moving cracks in partially cured epoxy. In fully cured resins, the crazes were not found [118].

Sultan and McGarry [121,122] identified the microcavitation at the crack tip as crazing and the proposed mechanism has been used to explain the toughening behavior of CTBN-modified epoxy networks. Lilley and Holloway [123] have reported seeing crazes in the vicinity of the crack tips in some epoxy resins although the features they showed could have been microcracks. Morgan et al. [124–126] have shown craze-like entities in transmission electron micrographs obtained from strained films of epoxy resins. They claimed that the presence of crazes results from the existence of a local region of low crosslink density. However, the crazes are relatively short (~ 1 mm) and they appear to be bridged by a drawn polymer. Hence no concrete evidence whatsoever has been found for crazing for toughened thermoset resins [127,128].

3.3.2.3 Shear yielding and crazing

As discussed in the earlier sections, shear yielding is the most probable and effective mechanism for toughening a crosslinked thermoset system. However, always there exists a possibility of two or more mechanisms acting simultaneously in a complex multiphase system. Bucknall and coworkers proposed [9] that massive crazing and shear flow are the two energy absorbing mechanism in rubber-modified epoxies. They used tensile creep dilatometry and determined the volume strain ($\Delta V/V_0$) to study the contribution of shear yielding and crazing on toughening of a thermoset resin [129,130]. This study is based on the assumption that the shear yielding creates no volume change, whereas the crazing is associated with a volume change. Therefore the slope of the volume strain ($\Delta V/V_0$) against longitudinal strain curve should be zero for a pure shear yielding process and equals to unity for a pure crazing process. Intermediate slopes would be produced by the proportionate mixing of the two processes. Hence by assuming the rule-of-mixture, it is possible to determine the contributions of the two mechanisms. Bucknall and coworkers investigated rubber-modified epoxies using this method and found that both the volume strain and toughness increased with rubber content. From this observation, they concluded that a combination of shear yielding and massive crazing was the toughening mechanism for thermoset resins.

However, volume dilation cannot be unambiguously and exclusively attributed to crazing; voiding can cause volume dilation [62,97,131]. No other supporting evidence to corroborate the volumetric data has been reported. In fact, several authors have cast doubt on the existence of crazes in a tightly crosslinked glassy polymer [111,112,132,133] despite reports of having present [123,127,128]. Hence it can be concluded that even if crazes exist in the rubber-modified thermoset networks, their contribution in improving the toughness of highly crosslinked thermoset networks is not convincing.

3.3.2.4 Cavitation and shear yielding

A direct relationship between the toughness and plastic zone size was reported by Bascom and coworkers [134,135] in rubber-toughened epoxy adhesives. They proposed that cavitation caused by the triaxial tension ahead the crack tip increases the size of the plastic zone and that plastic flow of the epoxy matrix and elongation of the particles also contribute to the toughness. The spectroscopic evidence for cavitation of rubber particles has been reported for various rubber-modified thermoset systems [136–138]. The SEM photograph for the fracture surface of ETBN-modified VE resin is shown in Fig. 3.3, which clearly shows cavitation of rubber particles [185]. Cavitations in the particle or in the matrix/particle interface and plastic shear yielding of the thermoset matrix are the microdeformation mechanisms occurring at the crack tip that dissipate energy and produce the toughening effect. This was further supported by the fact that the modified VE matrix, which shows cavitation, resulted in higher toughness [185] compared to the matrix where no cavitation was observed as will be discussed critically in Section 3.4.

The fractographic features of the stress whitened region indicate a number of deformation processes. The hole formation suggests that there has been a dilational deformation because holes are larger in diameter than the undeformed particles seen in the rapid crack growth region [62,97,131]. This dilational deformation is consistent with the triaxial stress field that is known to be present at the crack tip [138]. These stresses lead to the rapture of the

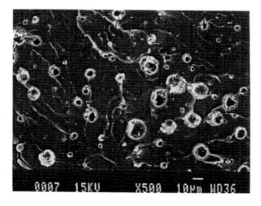

Figure 3.3
SEM photograph for the compact tension fracture surface of 4 phr (parts per hundred grams of resin) ETBN-modified VE resin illustrating cavitated particles. Propagation was from left to right. The specimen was stained with osmic acid. Source: *Reprinted with permission from J.S. Ullett, R.P. Chartoff, Polym. Eng. Sci. 35 (1995) 1086, © 1995 John Wiley and Sons Publishers.*

rubber particles, which then relax back into the enlarged cavity and give the appearance of holes. Results from several different experiments support this idea, but the most convincing is that of Kinloch et al. [139]. They exposed the whitened region of the fracture surface to an organic fluid to swell the rubber. The SEM then revealed that the hole had become hillocks. Moreover, a number of workers [140–142] have made attempts to capture any rubber particle that may have been ejected from the holes during crack growth. But none of the efforts have produced any evidence that particles are ejected. Thus this result suggests that microvoids are produced during crack growth by cavitation of rubber particles.

The cavitation is followed by the onset of the shear localization process, which would not have taken place in the net resin under the same conditions. The rubber cavitation and associated shear yielding have been discussed in Section 3.1.2.2 for rubber-toughened thermoplastics. The difference in elastic modulus of a glassy thermoset and rubber is much higher than the corresponding differences in their bulk modulus. Hence at the crack tip where the hydrostatic tensile component is large, the magnitude of the concentrated deviatoric stress in the vicinity of rubber particles is insufficient to promote shear yielding. The internal cavitation of the rubber particles relieves the plain strain constrain by effectively reducing the bulk modulus. After an effective reduction, the magnitude of the concentrated deviatoric stresses becomes sufficient for shear yielding. The voids left behind by the cavitated rubber particles act further as stress concentrators. Theoretical model calculations show that they reduce the octahedral shear stress to yield [143] and they grow and promote the formation of the plastic zone between voids [144]. Shear band formation is enhanced in such a voidy solid [145,146].

All these theoretical predictions are consistent with the results observed by Yee and Person [117,128], which clearly show shear bands between the cavitated particles in rubber-toughened epoxy materials. They also found that ability of CTBN rubber to toughen epoxy is closely related to CTBN rubber cavitation, which is seen as a thick dark circle within the rubber particles in the optical microscopy. The need for internal cavitation of rubber particles has been questioned by others [147,148], but those researchers have only considered the case of uniaxial tension, which has been shown to be quite different than the case of triaxial stress state at the crack tip.

Li et al. [149] have studied the fracture behavior of unmodified and CTBN-modified epoxies under hydrostatic pressure. They found that when rubber cavitation is suppressed by superimposed hydrostatic pressure, the fracture toughness of CTBN-modified epoxy is no higher than that of the unmodified thermoset resin network. This implies that stress concentration by rubber particles alone will not necessarily induce massive shear yielding and increase the fracture toughness. Hence rubber cavitation is very important to the toughening of rubber-modified epoxies; without cavitation, these rubber particles can still cause stress concentration but they are not effective in toughening. From the above discussion, it is clear that though all the four mechanisms have got some support from experimental observations, the most accepted mechanism is the rubber cavitation and associated shear yielding.

According to Evans et al. [150,151], no single mechanism can explain the toughening effect of the thermoset network by rubber particles. They proposed a synergistic model, which takes all the possible mechanisms including rubber bridging, void growth, and shear banding into consideration. The synergistic relationship between the rubber-bridging mechanism and the other two mechanisms was based on the assumption that the fracture energies increase proportionately with the size of the process zone. The validity of such relation is yet to be accurately validated. Yee [152] has cast doubt on the existence of synergism in the toughening mechanism observed in rubber-modified plastics. Huang [153,154] calculated the contribution of different mechanisms and concluded that the localized shear yielding, which is initiated by rubbery particles is a major toughening mechanism. Plastic void growth is another main toughening mechanism at higher test temperatures and the contribution increases sharply with rising temperature. The rubber-bridging mechanism can also play a role when the ability of the matrix to undergo plastic deformation is suppressed.

3.3.3 Microstructural features

Microstructural features of the dispersed rubbery phase have a major effect on the toughness of the multiphase thermoset network. The morphological parameters are dependent on the nature of the resin, toughening agent, and the cure condition. Various morphological parameters and the molecular parameters, which control the morphology, are listed in Table 3.2. It is very difficult to study the effect of individual parameters on toughening effect as the parameters are interrelated. In many studies, several crucial features have been changed simultaneously. Moreover, the parameters are interrelated and it is difficult to change one parameter without affecting others. For example, if we want to

Table 3.2: Parameters influencing rubber toughening.

Molecular parameters		Morphological parameters	Processing parameters
Epoxy	**Rubber**		
Matrix ductility	Polarity	Rubber volume fraction	Initial cure temperature
Functionality	Molecular weight	Particle size	Postcure temperature and time
Molecular weight	Functionality	Particle size distribution	—
Curing agent (type and concentration)	Concentration	Matrix ligament thickness	—
Viscosity	Viscosity	Particle to matrix adhesion	—

study the effect of the functionality of rubber by changing the functionality, this changes the solubility parameter difference and affects the other morphological parameters. However, attempts have been made here to correlate the fracture behavior and microstructural features of the toughened networks based on the available results in the literature.

3.3.3.1 Volume fraction

The fracture energy of toughened thermoset resins increases with an increase in the concentration of elastomer, up to a point where it becomes difficult to obtain the desired particle−matrix morphology. Bucknall and Yoshii [155] have determined the fracture energy of rubber-modified epoxy resins, all containing 8.7% w/w of CTBN rubber but with various rubbery phase volume fractions, achieved by varying the curing condition employed. They varied the rubber volume fraction from 0 to 0.2 and reported a linear relationship between fracture energy and volume fraction of rubber for different epoxy resin hardener systems using a CTBN rubber. Different rubber volume fractions for a modified epoxy system with the same concentration of rubber were achieved by changing the cure conditions. As commented earlier the morphology of a rubber-toughened thermoset system is tremendously influenced by the curing conditions.

However, Kunz and coworkers [113,156] found (for both CTBN- and ATBN-modified epoxies) that once a volume fraction of about 0.1 had been achieved in the polymer, further increase in volume fraction resulted in only a minor increase in the value of fracture energy. The apparent discrepancy can be attributed to the use of matrices with different inherent ductility, curing conditions and test conditions employed [157,158]. It may be noted that fracture energy in rubber-toughened epoxy is dependent on the test temperature and the test rate [158−160]. The fracture energy increases with an increase in the test temperature and a decrease in the test rate below the T_g.

Kinloch and Hunston [157,161] and others used the concept of time−temperature superpositioning and separated the effect of changing the volume fraction and the properties of the matrix phase. No definite relationship between fracture energy and volume fraction of rubber was observed. The maximum value of volume fraction that can be achieved is about 0.2−0.3. Attempts to produce higher volume fraction result in phase inversion and loss of mechanical fracture properties. Though in most of the studies [162,163], phase inversion is reported at a volume fraction higher than 20%, in some recent studies, inversion is reported [164,165] at a very low volume fraction (∼2%) of rubber.

3.3.3.2 Particle size

Several studies have suggested that particle size may influence the extent and type of micromechanism and so affect the measured toughness. Sultan and McGarry [122,123] proposed that large particles (0.1−1 mm in diameter) cause crazing and are five times more effective in toughening the thermoset than small particles (about 0.01 mm in diameter), which initiates shear yielding. However, evidence for crazing is not convincing as discussed in Section 3.3.2.2. Furthermore, volume fraction was not determined in these studies.

Kunz et al. [111,112] have reported that large (40 mm diameter) particles are not efficient in providing an increase in fracture energy as smaller (1 mm) particles. Many authors [128,129,166,167] reported that fracture energy of rubber-toughened thermoset is independent on the particle size. However, the average particle size was only varied between 0.5 and 5 mm. Pearson and Yee [168] have studied the effect of particle size on the fracture properties of rubber-toughened epoxy networks having a particle size ranging from 0.2 to 200 mm. They found that large (100-mm-diameter particles) are not as effective in providing a toughening effect whereas small, 0.1 mm diameter particles appear to be the most efficient and provides over a 10-fold increase in fracture energy. Hence it can be concluded that for an effective toughening of a thermoset network, the rubber particle size should be in the range of 0.1–10 mm.

The origin of this size dependence arises from the role played by the particles, which is governed by the size of the process zone. Large rubber particles lying outside the process zone are only able to act as bridging particles, which provides only a modest increase in fracture energy. Small rubber particles, which lie in the process zone, are forced to cavitate by the large hydrostatic stress component that exists in the process zone and contributes to the increase in fracture energy.

3.3.3.3 Matrix ligament thickness

Wu [34,100,169] proposed that the matrix ligament thickness (MLT), that is, surface to surface interparticle distance is a more fundamental parameter. For an effective toughening, the average matrix ligament thickness (t) should be less than that of the critical value (t_c) where brittle to tough transition occurs. The t_c is independent of rubber volume fraction, particle size, and characteristics of the matrix alone at a given test temperature and rate of deformation. For blends with dispersed spherical particles, the t_c can be related to the rubber particle size and rubber volume fraction ϕ_r by the following equation [34]:

$$\tau_c = d_o[k(\pi/6\phi_r)^{1/3} - 1] \tag{3.7}$$

where k is geometric constant and d_o is particle diameter. t_c is reported to be 0.30 mm for nylon/rubber blend.

The existence of critical matrix ligament thickness for an effective rubber toughening can be explained [34,71,170] in the light of two basic mechanisms, namely, rubber cavitation followed by the formation of shear bands and crazing. The low MLT maintains the connectivity of the yielding process, which then propagates over the entire deformation zone and makes the blend tough. This happens when t is less than t_c. In the cases where crazing is the major energy dissipating mechanism, the high MLT causes the formation of secondary crazes at the highly stressed region of the ligament, which then propagates rapidly leading to the catastrophic failure of the materials.

Using the critical MLT concept, the effect of particle size on the toughness of modified thermoset resin can be explained. If we consider the rubber cavitation mechanism, which is accepted as a major energy dissipating mechanism in toughened thermoset, then bigger particles should perform better, because the cavitation resistance decreases with an increase in particle size. Recall Eq. (3.1), where the power of particle size (d_0) is more for the negative term. However, particles should be smaller enough to satisfy the critical MLT condition. For a given volume fraction, the critical MLT is achieved by decreasing the particle size and by improving the dispersion. For most of the systems, this concept works very well, a decrease in particle size corresponding to a lower brittle-tough transition temperature. However, very small particles (<0.1 mm) are not able to cavitate and therefore do not release the hydrostatic tension in the material to promote ductile shear yielding [171]. As a result, it has been found in a number of systems that there exists a minimum particle size below which the brittle to tough transition no longer shifts to lower temperatures.

Other researchers [156,157] have claimed that the optimal particle size actually involves a bimodal distribution of small and large particles. They reported that rubber-modified networks having a bimodal distribution of particles exhibit higher fracture toughness than that of the matrix having a unimodal distribution of particles. Bimodal distributions were achieved by using BPA. BPA acts as a chain extender and significantly reduces the crosslink density. It is well known that toughenability of a matrix increases with a decrease in crosslink density. Hence these claims have neglected the effect of crosslink density, which has been shown to be quite substantial [8,172–182]. Moreover, Yee and Pearson [168] have disproved the effect of bimodal particle distribution on several morphological scales. They have not found any synergistic effect combining 0.2 mm particles with 2 mm particles and 2 mm particles with 100 mm particles.

3.3.3.4 Interfacial adhesion

Matrix–rubber particle adhesion is an important parameter for rubber toughening. For an effective rubber toughening, the rubber particles must be well bonded to the thermoset matrix. The poor intrinsic adhesion across the particle–matrix interface causes premature particles debonding leading to the catastrophic failure of the materials. Nearly all the studies [2–10,183] reported in the literature have been concerned with reactive rubbers as toughening agents, which result in dispersed particles having interfacial chemical bonds as a consequence of chemical reactivity.

However, there are some reports that have denied to accept the role of interfacial adhesion on toughening of thermoset resins. Lavita et al. [181] reported that nonreactive rubber can toughen BPA-modified epoxy. However, the mechanism has not been discussed critically. Huang et al. [184] have shown that when the second phase consists of micron-size rubber particles, the interfacial bonding has only a modest effect on the fracture properties of blends.

3.4 Toughening of vinyl ester resin

The various strategies for toughening thermoset resins in general as discussed in earlier sections have been utilized to toughen VE resins. Incorporation of flexible units into the dimethacrylate backbone by reacting a carboxyl-terminated butadiene−acrylonitrile liquid rubber (CTBN) with the VE resin system has been reported [185−187]. The CTBN is chemically bonded through terminal ester groups with 2,2-bis (4-hydroxyphenyl) propane units during manufacture [188]. The incorporation of such flexible units into the crosslinked structure improves the fracture toughness at the cost of thermomechanical properties. In order to achieve improved fracture toughness without a significant sacrifice in thermomechanical properties, blends or interpenetrating networks with two-phase morphology have been investigated. Rubber toughening of VE resins can be achieved by two different methods. The first method is to blend with the preformed CSPs. Additions of core−shell rubber particles have been found to increase the toughness of vinyl ester resin-based fiber-reinforced composite systems [189]. The second method is liquid rubber toughening discussed shortly.

3.4.1 Liquid rubber toughening

As discussed in earlier sections, a thermoset toughener is required to be soluble with the resin at the beginning and undergoes phase separation during the curing reaction leading to the formation of a two-phase microstructure. However, most of the commercially available or developed modifier (for epoxy resin) suffers from the problem of poor miscibility with the VE resins. The poor compatibility between the liquid rubber and the resin tended to result in large particles with poor interfacial adhesion [190,196]. Pham and Barchill [188] reported a particle size of 40 mm using a hydroxyl-terminated polybutadiene (HTBN) rubber. Suitable modification of the tougheners [50] or VE resin [51] and the use of compatibilizer [52] have been reported for the successful toughening of VE resin.

The initial compatibility can be improved by the use of rubber with the reactive functional groups to enable the rubber molecules to react with the resin to produce an adduct. This strategy can significantly promote the compatibility of the rubber leading to a blend with smaller particle size. Rubber particles with sizes up to 0.1 μm have been reported by using a vinyl-terminated poly(butadiene−acrylonitrile) rubber (VTBN) [197]. Pham and Burchill [188] modified HTBN by reaction with a diisocyanate and various alcohols to increase the compatibility between the rubbery phase and the VE resin.

Ullett and Chartoff [185] investigated fracture behavior of VE resins toughened with a variety of rubbers such as CTBN, VTBN, epoxy-terminated vinyl-terminated poly (butadiene—acrylonitrile) rubber (ETBN), hydroxyl-terminated polyether (HTE), etc. They analyzed the morphology and reported the presence of single-phase particles using VTBN and two-phase particles using ETBN and CTBN. Two-phase particles containing systems led to a greater increase in toughness than those containing single-phase particles. Examination of the fracture surfaces of these systems revealed that the toughening mechanisms for the systems with the two-phase particles were cavitation and shear yielding as evident from the SEM photograph for the compact tension fracture surface of 4 phr (parts per hundred grams of resin) ETBN-modified VE resin (Fig. 3.3). On the other hand, only crack bridging and particle deformation were observed for the systems with single-phase particles [185]. For example, SEM photographs for compact tension fracture surface of 4 phr VTBN modified VE resin is presented in Fig. 3.4. It can be seen that the particles appeared to have been stretched. Dreerman et al. [196] investigated the effect of modification with VTBN and ETBN on the fracture properties of VE resin. Among VTBN and ETBN, ETBN has been found to be the most effective in increasing toughness. They also observed the internal structure in the ETBN particles and suggested that it implies the existence of a dual-phase structure in the rubber particles. The toughening effect depends on several factors such as type of rubber [185], rubber volume fraction [186,196], the morphology of the rubber particles [185], the cure regime [185], the testing rate [188,196], and the compatibility or chemistry of the rubber/matrix system [188].

The results reported on fracture properties of rubber-toughened VE resin are contradictory. For example, Liao et al. [197] observed impact toughness to increase with additions of VTBN, while Dreerman et al. [196] observed no improvement in toughness using VTBN. This can be explained in terms of the difference in the size of rubber particles generated during reaction-induced phase separation in the two different systems investigated by the two research groups. Liao et al. [197] found very small particle sizes (up to 10 nm) and Dreerman et al. [196] observed particle sizes greater than 1 mm. They suggested that the very small particle size obtained was due to the gelation of VE resins at an early stage in the reaction, which prevented aggregation of the rubber molecules. The origin of discrepancies in toughness results can also be explained by considering the fact that fracture behavior is sensitive to the test conditions and inherent matrix properties, especially ductility. The pure resin material was found [188,197] to fail catastrophically using the three-point bend with a single notch, but when tested using compact tension sample geometry [185] or the much slower double-torsion method [188] stick/slip or stable crack propagation was observed. Dreerman et al. [196] and Ullett and Chartoff [185] found that additions of rubber changed the fracture mode from catastrophic failure to stick/slip. While Pham and Burchill [188] observed no change in the failure mode with the addition of rubber using the three points bend with a single notch, all samples failed catastrophically.

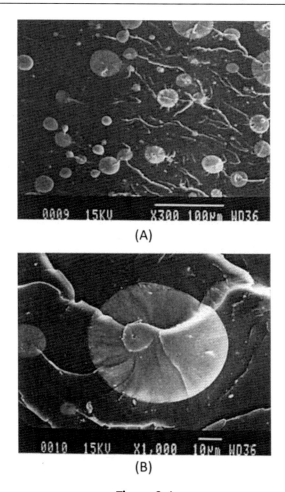

Figure 3.4
SEM photographs for compact tension fracture surface of 4 phr VTBNX-modified VE resin showing (A) transition from fast fracture to crack-initiation region and (B) close-up of one particle. Propagation was from left to right. The specimen was stained with osmic acid. *Source: Reprinted with permission from J.S. Ullett, R.P. Chartoff, Polym. Eng. Sci. 35 (1995) 1086, © 1995 John Wiley and Sons Publishers.*

Toughness values determined in Pham and Burchill's study by the single-edge-notch three-point bend technique were lower than those determined by a double-torsion technique. This observation highlights the influence of test conditions on measurements of fracture toughness. Karger-Kocsis and coworkers [192,193] have investigated the reactive hyperbranched polymers (HBPs) for toughening vinyl ester-urethane hybrid (VEUH) resin [192,193]. The toughening effect is influenced by the architecture and terminal groups (vinyl/hydroxyl ratio). Less compact star-like polymers with long, flexible arms with high vinyl functionality resulted in the highest improvement in toughness. Such HBPs copolymerize with VE resin and styrene and allow the polyaddition reactions (with polyisocyanate) leading to the formation of a regular crosslinked network structure. About four times increase in fracture toughness was reported by adding 20 wt.% of HBPs with long, flexible arms with high vinyl functionality they also investigated the interpenetrating structure of VE resin and epoxy [198]. They studied various combinations of IPN systems using different types of epoxies and reported significant improvement in toughness using an aliphatic epoxy system.

3.5 Modification of UPE resin

The curing of UPE resin is associated with large volume shrinkage (7%–10%) [199]. This is attributed to the extensive intramolecular or cyclization reactions of UPE molecules and the formation of compact microgel structures during cure [200]. The cure shrinkage induces a series of defects in the final product such as sink mark, surface waviness, cracks, etc. This creates challenges in fabricating a high-quality surface and maintaining the dimensional stability of a part made of UPE resin. The addition of nonreactive additives like thermoplastic polyurethane, polymethyl methacrylate (PMMA), poly(vinyl acetate) (PVAc), etc. offers a unique solution to this problem. These additives are called "low-profile additives" (LPAs) [201–206].

Depending on the chemical structure and molecular weight, LPA can be compatible or incompatible with the uncured UPE resin. However, curing both compatible and incompatible LPA lead to the formation of a two-phase microstructure. This reaction-induced phase separation is favored by the reduction in combinatorial entropy as a result of curing. Detailed thermodynamic analysis of such reaction-induced phase separation in thermoset has already been discussed in section. The phase separation is mostly described through a nucleation and growth process [207]. Li and Lee [208] proposed a phase separation mechanism through spinodal decomposition for the low-temperature cure of UPE–PVAc systems. It was pointed out that the time between the onset of phase separation and gelation is the phase separation time, which determines the final morphology of the modified UPE resin network. Boyard et al. [209] confirmed the spinodal decomposition mechanism in a similar system by using small-angle laser light scattering.

At the early stages of curing, microgels of polyesters are formed because of the strong tendency toward intramolecular cyclization (a characteristic of chain-growth polymerization). With further progress in curing, LPA and the unreacted styrene separate from these microgels in a matrix, which surrounds and isolates the microgels. Subsequently, the microgel contacts each other leading to the formation of a cocontinuous structure. Bulliard et al. [210] carried out a detailed SEM analysis to study the characteristics morphologies of UPE/PVAc (LPA) system. The SEM photographs for the modifier UPE system (cured for 90 min) with different concentrations of PVAc are presented in Fig. 3.5.

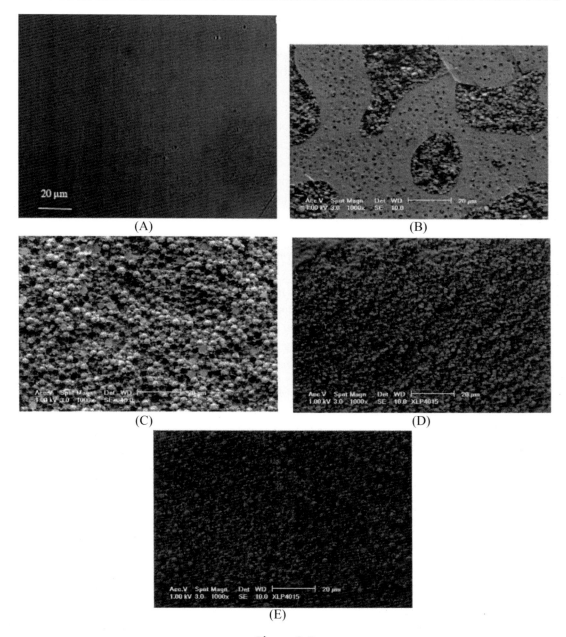

Figure 3.5
SEM photographs for the modifier UPE system (cured for 90 min) with different concentration of PVAc: (A) 0 wt.%, (B) 5 wt.%, (C) 6 wt.%, (D) 11 wt.%, and (E) 14 wt.%. All samples were etched in 2-butanone, except (A). Source: *Reprinted with permission from X. Bulliard, V. Michaud, J.-A.E. Manson, J. Appl. Polym. Sci. 102 (2006) 3877,* © *2006 John Wiley and Sons Publishers.*

The polished cured UPE resin exhibits a single-phase morphology. The addition of PVAc induces a two-phase structure. At a low concentration in PVAc (Fig. 3.5B), the PVAc-enriched phase forms domains of about 20 mm in diameter dispersed in the continuous polyester phase. The PVAc-enriched phase is composed of a globular structure. Small inclusions of PVAc are visible in the continuous polyester phase. Increasing the concentration over 5 wt.% led to a transition to a morphology, which is composed of fine polyester globules of a few microns in size that seem to be connected to each other. They correspond to the polyester microgels. This connected globule structure implies a two-phase morphology of interconnected domains of a polyester-rich phase regularly dispersed in a matrix enriched in PVAc. This morphology is referred to as a cocontinuous one. For the particular concentration of 14 wt.% of PVAc, this interconnected globule morphology implies a step of coarsening and growth of the structure between phase separation and gelation. The structure that was shown to be cocontinuous after phase separation would tend to form spherical domains of polyester because of the differences in the interfacial tension between the phases, as reported in epoxy-based systems, and because of intramolecular cyclization of the polyester molecules. The morphology is consisting of three elements: (1) microgel particles, (2) LPA layer, and (3) lightly styrene crosslinked polyester chain and polystyrene chain between the LPA-covered microgel particles between the LPA-covered microgel. The overall morphology is represented by a model, which is a parallel combination of the three elements mentioned above. The mechanical behavior can be approximately represented by the Takayanagi models [211,212].

Several studies have elaborated on the origin of the function of LPA. The formation of microcracks/microvoids has been assigned as the origin of volume shrinkage control of UPE resins by nonreactive LPA [213–216]. Pattison et al. [199,213] have proposed that as crosslinking of LPA-containing UPE resins proceeds, strain due to polymerization develops in the system, particularly at the interface of the LPA phase and crosslinked UPE phase. This strain can increase to a point that stress cracking propagates through the week LPA phase. In this process, the strain is relieved leading to the generation of microcrack and mocrovoid or both. The decrease in volume due to cure shrinkage is compensated by the space generated by microcracks or microvoids.

Reactive LPAs like PVAc-b-PMMA, PVAc-b-polystyrene with many peroxide linkages along the skeleton of LPA are also developed. The reactive LPAs, in addition to microcracking, reduce the inherent cure shrinkage of the resin and thereby decrease overall volume shrinkages. Dong et al. [216] investigated a series of reactive LPAs and found that volume shrinkage control cannot be explained in terms of microcrack or microvoid formation. If microvoiding and microcracking is the sole mechanism of volume shrinkage control then a system with a smaller volume fraction of microvoids would lead to a higher volume shrinkage which was not found to be the case. Thus besides microvoid formation, the intrinsic polymerization shrinkage is also important in determining the volume shrinkage control during the cure of UPE/LPA systems.

UPE resins are blended with several materials like rubber, thermoplastics to improve their impact strength and fracture properties. Both solid and liquid rubbers are dispersed in the resin for enhancement of toughness. The commercial toughening agents are carboxyl-terminated copolymer of acrylonitrile and butadiene (CTBN) and amine-terminated copolymer of acrylonitrile and butadiene (ATBN). Various grades of CTBN and ATBN of different molecular weights and acrylonitrile contents are available. The toughening effect of an elastomer additive depends on phase separation to provide the final morphology and to control the rubber particle size and volume fraction [217]. The toughening effect is often modest due to the low chemical reactivity of the rubber toward polyester end groups. In order to initiate chemical bonding, condensation of hydroxyl or carboxyl-terminated liquid rubbers and the polyester reactants have been tried out, which result in polyesters containing rubber segments in the main chain [218]. Other examples of chemically reactive rubbers are methacrylate end-capped carboxyl-terminated nitrile rubber [219] or isocyanate end-capped polybutadiene [195] or epoxy-terminated nitrile rubber [182]. Block copolymers of UPE with polyurethanes, polyureas, polysiloxanes, polyimides, polyoxazolines, or polyglycols have also been reported [220]. Attempts have been made to modify UPE resins with dicyclopentadiene [221], bismaleimide [222], poly(-caprolactone)-perfluoro polyethers [223], and epoxy [224].

IPN strategy has also been used for such modifications. Details of IPNs have been discussed in Chapter 2, Properties and Processing of Thermoset Resin. Alcoholic and carboxyl end groups of the resin can be blocked by isocyanates that form covalent urethane bonds [225,226]. Chou and Lee [225] studied the morphology–kinetics–rheology relationship of polyurethane (PU)-unsaturated polyester interpenetrating polymer networks (IPNs). They found that the PU-rich phase formed the dispersed domains in the continuous UP-rich phase. The interaction between the two reactive systems may give rise to a synergistic effect on the properties of IPN. The interactions have been attributed to an increase in crosslink density because of the interpenetration of network chains [227]. Other specific interactions like grafting [228] and interactions between oppositely charged groups [229] between the two networks are possible reasons for property enhancement. Cherian et al. [230] investigated IPNs of polyester and polyurethane (PU) prepolymers based on various polyols like hydroxyl-terminated natural rubber (HTNR), hydroxyl-terminated polybutadiene (HTPB), PEG, and castor oil (containing the glyceride ester of ricinoleic acid). Urethane linkages are formed by reaction between terminal hydroxyl and carboxyl groups of UPR and isocyanate groups of a PU prepolymer. The schematic representation of PU-UPR IPN is shown in Fig. 3.6. The effect of incorporation of PU on the strength and toughness properties of IPNs is shown in Figs. 3.7 and 3.8, respectively. It is evident that both the tensile strength and toughness increase as a result of the incorporation of PU in the IPN form and maximum property enhancement were achieved in the case of HTBP-based PU.

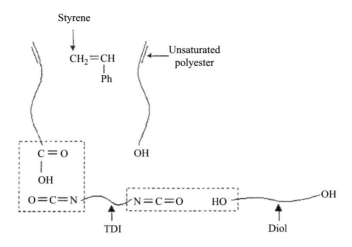

Figure 3.6
The schematic representation for the formation of PU-UPE IPN. Source: *Reprinted with permission from A.B. Cherian, B.T. Abraham, E.T. Thachil, J. Appl. Polym. Sci. 100 (2006) 449, © 2006 John Wiley and Sons Publishers.*

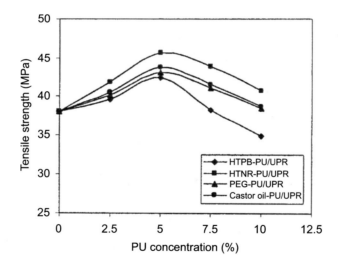

Figure 3.7
Tensile strength of PU-UPE IPN versus PU concentration. Source: *Reprinted with permission from A.B. Cherian, B.T. Abraham, E.T. Thachil, J. Appl. Polym. Sci. 100 (2006) 449, © 2006 John Wiley and Sons Publishers.*

3.6 Toughening of phenolic resin

The brittleness of phenolic resin can be reduced using derivatives of phenol (such as dihydroxy diphenyl ether) in the main reaction with formaldehyde [231]. The toughness increases due to the incorporation of ether linkages in the network. Novolac resin has been

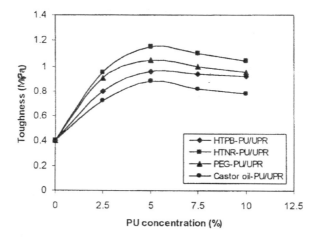

Figure 3.8
Toughness of PU-UPE IPN versus PU concentration. The effect of incorporation of PU on toughness properties of PU-UPE IPNs. Source: Reprinted with permission from A.B. Cherian, B.T. Abraham, E.T. Thachil, J. Appl. Polym. Sci. 100 (2006) 449, © 2006 John Wiley and Sons Publishers.

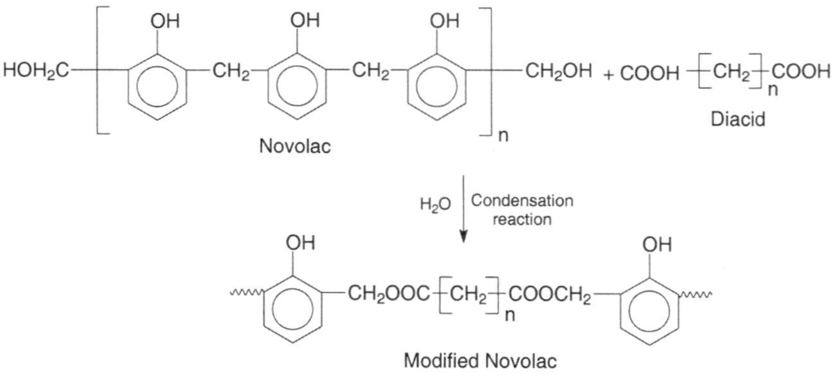

Figure 3.9
Chemical modification of novolac type phenolic resin by a diacid.

modified with flexible carboxylic acid like adipic acid [232] as shown in Fig. 3.9. Such modified novolac resins on curing with HMTA produce tougher networks due to the introduction of flexible aliphatic chains and reduction in crosslink density of the networks. Investigation on various dicarboxylic acids and their concentrations, 0.0007 moles of adipic acid was established as an optimum level of modification. An increase in strain from 3% to 7% and an increase in compression strength from 99 to 142 MPa are achieved as a result of the modification.

The blending of rubber has been tried out for toughening phenolic resin. Phenolic-nitrile rubber (NBR) blend was developed in the early 1950s, which has found wide applications

[233] such as structural adhesives, O-rings, gaskets, etc. However, it was reported that when NBR is mixed with resole and cured, they phase-separate into undesirably large domains with low adhesion between the phases [234]. A discussed in earlier sections, for an effective toughening effect the domain size the dispersed phase and interfacial adhesion are extremely important. The addition of p-cresol formaldehyde (PCF) into the phenolic/NBR blends resulted in a reduction in the domain size of the dispersed phase and an improvement in mechanical properties [235]. PCF resin has an intermediate polarity compared to NBR and resole and can react faster with NBR. Therefore PCF molecules are likely to be concentrated at the phenolic/NBR interface and act as an external compatibilizing agent [236]. Thus compatibility and chemical bonding between NBR and phenolic resin is improved leading to the enhancement in properties. The other materials used as toughening agents of phenolic resin include elastomers such as natural rubber and nitrile rubber [237,238], reactive liquid polymers [239], thermoplastics such as polysulfone, polyamide, and polyethylene oxide [240,241]. Gietl et al. [242] investigated the efficiency of various types of toughening agents like CTBN, ATBN, and core−shell rubber particles. They found the best performance for ATBN rubber.

Ratna et al. [243] modified resole with polyethylene oxide (PEO). Resole/PEO blends are prepared by a solution cast method. Generally, blending of a thermoplastic with a highly crosslinked thermoset is associated with a phase separation (either micro or macro) due to the substantial decrease in combinatorial entropy of mixing as a result of curing, which makes free energy of mixing positive before gelation. However, in the present case, PEO forms a compatible blend with the resole and T_g of the blends are slightly higher than that theoretically calculated using Fox equation. This is attributed to the strong H-bonding interactions between the -OH groups of resole and ether chain of PEO as confirmed FTIR analysis [191]. The flexural properties of the pure resole and resole/PEO blends are given in Table 3.3. The results show the typical behavior of a compatible blend, that is, the flexural strength decreased and the flexural strain increased with increasing PEO concentration. We can see that a wide variation in mechanical properties is possible by using blending approach and the blends can be selected accordingly depending upon the applications where the materials are going to be used.

Table 3.3: Flexural properties of PF resole/PEO blends.

Blend composition resole/PEO (w/w)	Flexural strength (MPa)	Flexural modulus (MPa)	Flexural strain (%)
100/0	97 ± 4	4097 ± 25	1.9 ± 0.2
85/15	72.1 ± 3.5	3122 ± 30	2.4 ± 0.1
70/30	38.3 ± 3.5	1238 ± 20	3.9 ± 0.3
55/45	11.3 ± 1.5	185 ± 10	10.4 ± 0.5

Source: *Reprinted with permission from D. Ratna, J. Karger-Kocsis, J. Appl. Polym. Sci. 127 (2013) 4039−4043, © 1995 John Wiley and Sons Publishers.*

3.7 Toughening of PI, BMI, and CE resins

Rubber toughening of high-temperature resins like polyimide (PI), BMI, and cyanate ester (CE) resins is not very successful as the addition of rubber reduces the thermal stability of the resin drastically. The toughness of such resins is improved by either changing the chemical structure or by blending with engineering thermoplastics or other thermoset resins. The toughness of PI and BMI resins are mostly improved by using the amines containing ether linkages and chain extension strategy as discussed in Chapter 1, Chemistry and General Applications of Thermoset Resins. BMI and CE resins can be blended with the epoxy to improve the processability and toughness. However, such modifications are associated with a reduction in water absorption owing to the strong polarity of epoxy groups [194]. Cyanate esters have been successfully modified by BMIs. However, the modified resins are extremely sensitive to curing conditions and it is very difficult to guarantee the good quality of the cured product.

Poly(ethylene phthalate) (PEP) has been successfully utilized to toughen BMI and CE resins [244]. However, the low water absorptivity of the parent resin was significantly deteriorated. Iijima et al. [245] used N-phenylmaleimide–styrene copolymer (PMS) for toughening BMI and CE resins. The effect of modification of CE resin with PMS (M_w = 129,000 g mol^{-1}) on thermomechanical properties and fracture toughness of the modified network is presented in Fig. 3.10. The fracture toughness increases remarkably at 9–12 wt.% PMS loading and decreases thereafter. The increase in toughness is associated with a modest reduction in flexural strength and slight increase in moduli. The thermal properties of the modified resins are comparable to the same of the unmodified resins. Yang et al. [246] modified CE resins using an epoxidized polysiloxane, produced by the reaction between an epoxy resin and amidosiloxane. Epoxidation of the siloxane with epoxy improves the miscibility with CE resin and thus the macrophase separation can be avoided. The modified resin offers better processability compared to the unmodified resin. A significant improvement in impact strength was achieved while maintaining the desirable flexural strength. For example the unmodified CE resin (cured) exhibits an Izod impact strength of 5.4 kJ m^{-2} whereas the modified resin with 18 wt.% polysiloxane, exhibits an impact strength of 9.7 kJ m^{-2}.

Mather and coworker [247] reported modification of bisphenol-A bismaleimide resin (MPA-BMI) with an allyl-terminated hyperbranched polyimide (AT-PAEKI). Synthesis of hyperbranched polymers have already been discussed in Chapter 1, Chemistry and General Applications of Thermoset Resins. HBP-based toughening agents have been extensively used for toughening epoxy resin, which will be elaborated in subsequent sections. It was observed that MPA-BMI/AT-PAEKI form miscible blends up to 16 wt.% of AT-PAEKI loading. A modest toughening effect, that is, enhancement (K_{1C}) from 0.48 to 0.55 MPa m^{-1} was reported. Due to lack of heterogeneity in the modified network, the

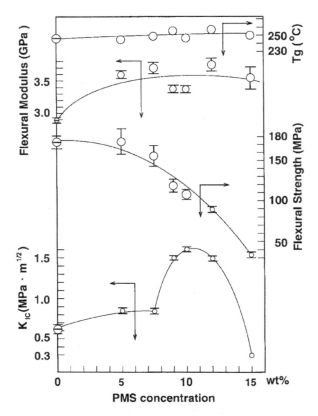

Figure 3.10
Physical properties of the modified cyanate ester resins as a function of PMS (N-phenylmaleimide–styrene copolymer) concentration ⊖, control and O, PMS (M_w 129,000 or 133,000). Source: *Reprinted with permission from T. Iijima, T. Maeda, M. Tomoi, J. Appl. Polym. Sci. 74 (1999) 2931, © 1999 John Wiley and Sons Publishers.*

toughening mechanisms as discussed above no longer available in the present case resulting in only a modest improvement in toughness.

Wang et al. [248] used hyperbranched polyether epoxy (HBPEE) to toughen benzoxazine resin (Bz). HBP with both the rigid and flexible moieties are known to provide good performance in toughening a thermoset and HBPEEs is designed accordingly having both the flexible aliphatic chains and rigid aromatic structure. The HBPEE is incorporated into bezoxazine using a solution blending method. Mechanical properties of pure epoxy and modified epoxy networks with varying concentration of HBPEE of three different molecular weights 2872 g mole^{-1} (HBPEE-1), 3885 g mole^{-1} (HBPEE-2), 4938 g mole^{-1} (HBPEE-3) are presented in Fig. 3.11. We can see from the figure that the impact strength of all the modified network tend to increase with increasing HBPEE concentration attains a maxima and decreases thereafter. The modified network containing 10 wt.% HBPEE-2 shows

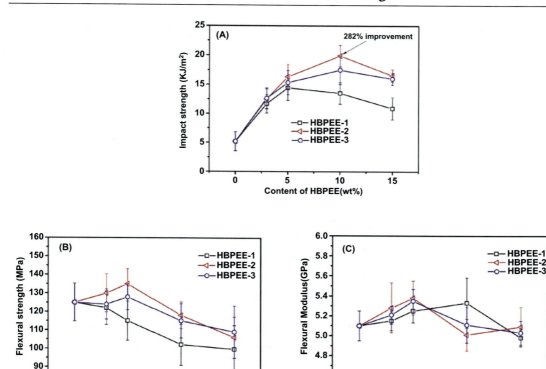

Figure 3.11
Effects of HBPEE loading on (A) impact strength, (B) flexural strength, and (C) flexural modulus of cured hybrids; error bars represent ± 1 standard deviation. Source: *Reprinted with permission from X. Wang, L. Zong, J. Han, J. Wang, C. Liu, and X. Jian, Polymer 121 (2017) 217–227, © Elsevier publisher.*

remarkable impact strength showing about 280% increase compared to pure bezoxazine. If we look at flexural property, for HBPEE-1 the flexural strength decreases for all the concentrations. However, in the case of HBPEE-2 and HBPEE-3, flexural strength increases with HBPEE loading up to 5 wt.%. The increase in flexural property is due to increase in crosslink density of the networks. Hence considering both the impact and flexural properties, it can concluded that Bz/HBPEE-2 (5 wt.%) is the balanced modified benzoxazine formulation.

3.8 Toughened epoxy resins

In the previous sections, toughened thermoset resins have been discussed in general. Among the commercially available thermosetting resins, epoxy resins has been most extensively studied and toughening technology has been exploited in the field of adhesive and fiber-reinforced composites. This is due to the inherent ductility of the cured epoxy resins and versatile epoxy resin chemistry. In this section, toughened epoxy systems will be elaborated.

3.9 Chemical modification

Unlike thermoplastics, the nonreactive liquid plasticizers are not successful in improving the toughness and bond strength of epoxy resins. This is because after the formation of network, as a result of curing, the modifier either exudes out from the matrix or undergoes macrophase separation. Moreover, the accumulation of free liquid plasticizer molecules at the substrate surface can act as a weak boundary layer in the adhesive joints leading to a substantial decrease in adhesive joint strength [249,250]. Reactive diluents, which are basically monoepoxide compounds, are used as plasticizers for epoxy. The use of a long-chain hardener is reported in the literature as a solution to this problem of low bond strength [251,252] as discussed in this chapter.

Ratna and colleagues [253–255] have reported that chemical modification of DGEBA with carboxyl-terminated poly(ethylene glycol) adipate (CTPEGA). CTPEGA was prepared by the condensation polymerization of polyethylene glycol (PEG) with an excess amount of adipic acid using *p*-toluene sulfonic acid as a catalyst. The epoxy resin was prereacted with CTPEGA in presence of triphenylphosphine (TPP) as a catalyst till no carboxyl group was found by the titration. The reaction is basically a carboxyl–epoxy esterification reaction, as proposed by Romanchick et al. [256]. The product was epoxy end-capped PEG adipate, which can be cured in the same way as epoxy. A large excess of epoxy resin was used for end-capping the CTPEGA to essentially prevent further polymerization. The reaction is illustrated in Fig. 3.12. The modified epoxy can be cured in the same way as epoxy leading to the formation of a crosslinked network. The stress–strain diagram for the modified networks containing varying concentrations of CTPEGA [252] is shown in Fig. 3.13. From the figure, it is clear that as the concentration of CTPEGA increases, the tensile strength decreases, and % elongation at break increases slowly. This can be attributed to the increase in flexibility of the matrix as a result of the incorporation of flexible ether linkage (−O−) into the matrix. Moreover, due to the prereaction of epoxy with CTPEGA, the molecular weight between crosslinks increases, which enhances the flexibility of the matrix. CTPEGA can be further reacted with DDS to make an amine-terminated polysulfone (ATPS) [253] as shown in Fig. 3.14. ATPS can be used as hardener-cum-flexibilizer for epoxy resins [254].

Though chemical modification as discussed above resulted in a significant improvement in flexibility, the major downside of this approach is that it is associated with a significant decrease in modulus and T_g as shown in Table 3.4. This problem can be overcome by using

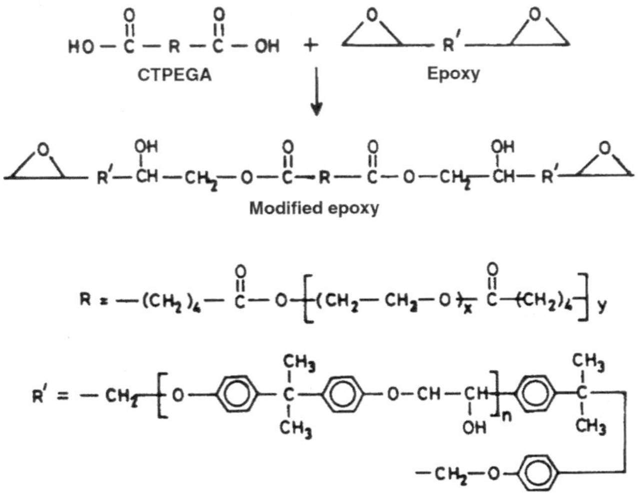

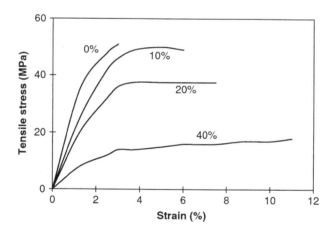

Figure 3.12
Chemical modification of epoxy resin with carboxyl-terminated poly(ethylene glycol) adipate (CTPEGA). Source: *Reprinted with permission from D. Ratna, A.B. Samui, B.C. Chakraborty, Polym. Int. 53 (2004) 1882, © 2004 John Wiley and Sons Publishers.*

Figure 3.13
The stress–strain diagrams for the modified epoxy networks containing varying concentrations of CTPEGA. Source: *Reprinted with permission from D. Ratna, A.B. Samui, B.C. Chakraborty, Polym. Int. 53 (2004) 1882, © 2004 John Wiley and Sons Publishers.*

the second phase toughening strategy as discussed in Chapter 4, Thermoset Composites, in general. The various toughening agents reported will be discussed under four subsections, namely, liquid rubber toughening, CSP toughening, thermoplastic toughening, and rigid particle toughening.

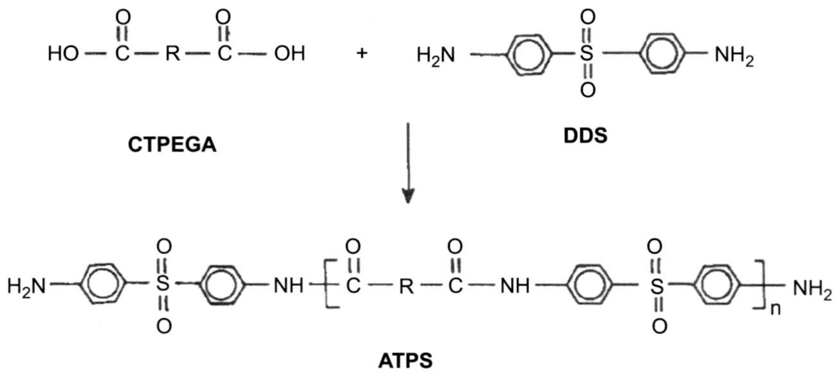

Figure 3.14
Synthesis of amine-terminated polysulfone (ATPS). Source: *Reprinted with permission from D. Ratna, M. Patri, B.C. Chakraborty, P.C. Deb, J. Appl. Polym. Sci. 65 (1997) 901, © 1997 John Wiley and Sons Publishers.*

Table 3.4: Effect of chemical modification on thermomechanical properties of epoxy.

Concentration of CTPEGA (phr)	Tensile strength (MPa)	Flexural strength (MPa)	Impact strength (J m^{-1})	T_g (°C)
0	54	135	19	115
10	48	120	26	90
20	39	98	32	75
30	28	50	27	65
40	18	20	24	50

3.10 Rubber toughening

In rubber toughening, an elastomeric modifier is incorporated into the epoxy matrix as a second phase as discussed above and that is why improvement in toughness is achieved without a significant sacrifice in T_g and mechanical properties. The toughening agents reported so far can be classified as commercial, other rubber-based, acrylate-based, and hyperbranched polymer-based modifiers. The liquid rubber-based toughening agent will be discussed in the subsequent sections.

3.10.1 Commercial toughening agents

The commercial toughening agents used widely are carboxyl-terminated copolymer of butadiene and acrylonitrile (CTBN) and amine-terminated copolymer of butadiene and acrylonitrile (ATBN) from Goodrich Company. A multifunctional liquid rubber (nitrile-diene-acrylamide terpolymer) for toughening epoxy composites and coatings was developed by the Wolverine Gasket division of Eagle-Picher Industries, Inkster Mich, United States [257]. A new additive that increases the strength and toughness of amine cured epoxy resins has been commercialized by Unirayal Ltd. of Elmira, Canada. NASA has discovered [258] that fiber-reinforced epoxy composites can be made tougher by incorporating bromine-containing additive, which resulted in a substantial increase in flexural and impact strength. The addition of a small amount of CTBN or ATBN further improved these properties. McGarry [259] from B. F. Goodrich Company first initiated the work on toughening of epoxy resins with CTBN and extensive works have been done subsequently by various research groups including that of the author of this book [106,110,260,261]. The principle of reaction-induced phase separation and mechanism of toughening formation of two-phase microstructure have already been discussed in Section 3.3. The rubber particles in most of the reports remain as uncrosslinked. Zhou et al. [262] reported a novel method that involves a precrosslinking process of rubber to improve the interfacial adhesion between the liquid rubber and the matrix. In this method, before addition of curing agent, an initiator is added to the liquid rubber–epoxy mixture to initiate crosslinking reaction of liquid rubber. Subsequently, curing agent was added to form the thermoset network. A comparison of mechanical properties [263] of epoxy network cured with piperidine (Epoxy-P), CTBN-modified epoxy network made in traditional way without precrosslinking (Tra-C) and the same made with precrosslinking (Cro-C) is shown in Fig. 3.15. It is clear from the figure that CTBN is able to toughen epoxy effectively and the new approach resulted in about 20% increase in impact strength compared to the conventional blending method.

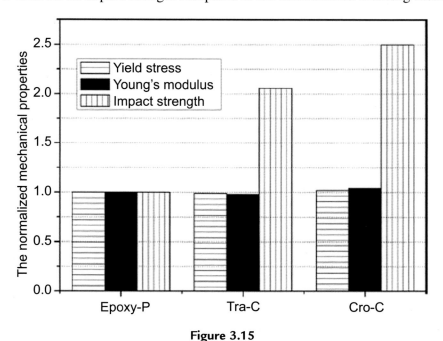

Figure 3.15
Normalized mechanical properties of pure epoxy, as well as traditional and precrosslinked CTBN/epoxy blends. Source: *Reprinted with permission from H.S. Zhou, X.X. Song, S. Xu, J. Appl. Polym. Sci. 131 (2014) 41110, © 2014, John Wiley and Sons Publishers.*

3.10.2 Rubber-based toughening agents

Apart from carboxyl- and amine-terminated nitrile rubbers, the rubbers having other end groups like amine [264–266], mercaptan [178,267], hydroxyl [268], epoxy [269,270], etc. have also been examined. Riew et al. [271] showed that ATBN could be added to the amine curatives, which could then be mixed with the epoxy resin. Unfortunately, ATBNs are not universally soluble in amine curatives. But among the reactive ended nitrile rubbers, the CTBN gives the best performance. The better adhesive strength of the CTBN-modified epoxy in comparison epoxy system modified with NBR with other reactive groups can be attributed to the better adhesion of CTBN-modified epoxy with the substrate due to the presence of carboxyl groups.

Cunliff et al. [272,273] in a series of papers described the generation of rubber-modified epoxy prepared by anionic polymerization of monomers such as isoprene, butadiene, and acrylonitrile. Their method ensures the reaction of anionic functional rubber with epoxy, which then guarantees the reaction of epoxy end-capped rubber to the matrix. They found all the rubbers increased the shear strength of the epoxy-based adhesive system. However, they have not carried out a detailed fracture analysis. The main problem with their techniques was the instability of the rubber particles, which tended to coalesce upon standing. Meek [116] investigated the use of preformed ABS powder as a toughening agent. He found that the fracture energy value surpassed the same for CTBN-modified system.

Rezaiford et al. [274] reported that significant toughness enhancement could be achieved by using PMMA-grafted natural rubber instead of CTBN. Carboxyl-terminated polyisobutylene [275] and polysulfide [276,277] rubbers are also reported as an effective modifier for DGEBA resin using different types of amine hardeners. Mizutani [278] reported an anhydride-cured transparent toughened epoxy network using hydroxyl-terminated liquid polychloroprene rubber as a toughening agent. The modified networks containing 0–10 vol.% of liquid rubber lead to a completely phase-separated microstructure. However, the modified networks are transparent, because the refractive index of rubber is comparable with that of the epoxy.

Siloxane rubbers, because of their attractive properties like high chain flexibility, extremely low T_g ($\sim -100°C$), low surface tension and surface energy, and hydrophobic behavior, have been used for toughening epoxy resins. Poly(dimethylsiloxane) oligomer (PDMS) is not compatible with epoxy resin as pure PDMS has a calculated solubility parameter of about 7.6 $(cal/mole)^{0.5}$, which is much lower than that of epoxy resin ($d_s = 7.4-7.8$) [279]. Due to its extreme incompatibility with epoxy resin, it is either macroscopically immiscible or exudes out from the crosslinked matrix during the curing process. So pure PDMS has little use as a toughening agent for epoxy. 2-Aminopropyl siloxane is somewhat better and studied as a modifier for epoxy [280]. Yet they are not highly miscible either.

The mechanical properties of the siloxane-modified epoxy network depend on the method of modification. For example, the modulus of silicone-modified epoxy is lower when modification is carried out in toluene solution compared to the modified networks made in melt [280,281].

The compatibility is increased by copolymerizing dimethylsiloxane with diphenylsiloxane or dimethyl fluoro propyl siloxane [281]. Such copolymers having controlled structures can be successfully used as toughening agents [282,283]. Konzol et al. [284] reported that the miscibility of PDMS can be increased by introducing polycaprolactam (PCL) block into PDMS. Hydroxyl-terminated block copolymers having basic structure $(PCL)_2$-b-PDMS-b-$(PCL)_2$ with a controlled composition leads to a phase-separated system with uniformly dispersed elastomeric particles with size in the order of magnitude of 20 nm. Incorporation of about 5 wt.% of such modifier resulted in a twofold increase in toughness. Tog et al. [285–287] showed that the miscibility of PDMS with epoxy can be improved by grafting the PDMS with glycidyl methacrylate (GMA). The graft segments of PMMA and PGMA on the siloxane particles surface enhance their compatibility with the epoxy and ensure homogeneous distribution of particles leading to an effective toughening effect.

Riffle et al. [288,289] developed epoxy-terminated siloxane oligomer and used it successfully for the toughening of epoxy resin. The epoxy-terminated oligomer was further reacted with piperazine to get secondary amine group end-capped oligomer, which can be used to cure and modify epoxy resins. Silicone-epoxy block copolymers have also been investigated [290] as toughening agents for epoxy. Lanzetta et al. [291] modified epoxy resin by hydroxyl-terminated block copolymer of PDMS and polyoxyethylene (PDMS-co-PEO) elastomer and functionalized saturated polybutene. These modified networks exhibit better oxidative stability but do not produce similar effects reported in the literature for CTBN and ATBN on the mechanical properties of epoxy.

Polyepichlorohydrin (PECH) is reported as a toughening agent for epoxy in both patents [292,293] and published literature [294]. The structure of PECH, with its pendent chloromethyl groups and terminal hydroxyl groups, gives great flexibility in the variation of interactions that could be introduced either by prereaction or curing reaction. The degree of toughening depends on the molecular weight of the PECH and on the curing temperature. Optimum properties were achieved with PECH of the highest molecular weight of 3400 g mole^{-1}. The PECH-modified epoxy exhibits similar thermomechanical and hot-wet properties as CTBN-modified epoxy.

Epoxy groups containing triglyceride oils, namely, vernoria oil [295,296], epoxidized soybean oil [297,298], and castor oil [299,300] have been utilized to toughen epoxy resins. Direct mixing of epoxidized oil and epoxy resin followed by curing of the mixture does not lead to a two-phase microstructure and offers only plasticizing effect. The two-phase microstructure was obtained by using a prepolymer of epoxidized soybean oil (ESO) and amine hardener instead of pure ESO. The high molecular weight of the prepolymer compared to pure ESO reduces its compatibility with the epoxy and favors the phase separation [297,298]. The optimum modifier concentration was found to be 10−20 phr. Above 30 phr of rubber concentration, phase inversion takes place.

Samajima et al. [301] described a poly(ether ester)-based modifier made from dimethyl terephthalate, butanediol, and poly(tetramethylene ether) glycol. They have demonstrated that the elastomer also phase-separate and increases flexural strength of epoxy over that of unmodified epoxy. Carboxyl-terminated poly(propylene glycol) adipate (CTPPGA) has been used as liquid rubber for the development of ambient temperature-curing epoxy adhesive [302]. The effect of molecular weight of liquid rubber and concentration of rubber added on the adhesive strength has been investigated. It was reported that for an effective toughening, the molecular weight must be above 5000 g mole^{-1} and the addition of 15 wt.% of such CTPPGA resulted in a threefold increase in lap shear strength and fivefold increase in T-peel strength.

Polyurethane (PU) oligomer has been investigated as a modifier for epoxy resins by Wang and Chen [303,304]. They developed hydroxyl-, amine-, and anhydride-terminated PU prepolymer by reacting the isocyanate-ended PU oligomers with bisphenol-A, DDS, and benzophenone tetracarboxylic anhydride respectively. Evaluating the oligomers having different reactive groups, they found that hydroxyl-terminated oligomers offer the best toughening effect. The incorporation of about 20 wt.% of this oligomer into the epoxy matrix resulted in a fivefold increase in fracture energy compared to the unmodified epoxy.

Laeuger et al. [305,306] reported the toughening of anhydride-cured DGEBA resin by using diol and bis(4-hydroxybenzoate)-terminated poly(tetrahydrofuran) liquid rubber. The morphology development occurs via spinodal decomposition and monitored by TEM and small-angle light scattering. The liquid rubbers react with DGEBA to form segmented liquid rubbers with poly(tetrahydrofuran) and poly(hydroxy ether) segments. The bis (4-hydroxy benzoate)-terminated poly(tetrahydrofuran) gives better performance compared to the diol-terminated one as far as mechanical and fracture properties are concerned. Modification with 20−30 phr of poly(propylene carbonate) resulted [307] in 30% improvement of lap shear strength and 75% increase in impact strength.

3.10.3 Acrylate-based toughening agents

The commercially available liquid rubber-toughened epoxy (CTBN, ATBN) often shows outstanding fracture properties and the technology is exploited in the field of engineering adhesives [10]. However, since the butadiene component of the elastomer contains unsaturation, it would appear to be a site for premature thermal and/or oxidative instability and such modified resins are not suitable for application at high temperature. One would imagine that excessive crosslinking could take place with time, which would detract from otherwise desirable improvements accomplished with these structures. Second, there is some limitation in its use due to the possibility of the presence of traces of free acrylonitrile, which is carcinogenic. Hence considerable efforts have been made to develop saturated liquid rubbers alternative to CTBN [2,308].

Functionalized acrylic oligomers can be envisaged to exhibit considerable utility as toughening agents for epoxy due to the following reasons: (1) The T_g of various acrylates are well below room temperature ($\sim 30°C$). Moreover, some of these acrylates are known for their high damping peak, which in turn may be helpful for in toughening of epoxy resin from the mechanistic point of view. (2) Even at room temperature low-molecular-weight acrylate oligomers which are elastomeric in nature are viscous liquid. (3) Low-molecular-weight acrylates are compatible with DGEBA resin as the difference in solubility parameter of certain acrylates are relatively small. (4) Due to lack of unsaturation, the oxidative stability and aging resistance of acrylates are much superior compared to the unsaturated rubbers like CTBN. (5) Acrylate monomer can be conveniently polymerized or copolymerized by conventional technology.

3.10.3.1 Synthesis of functionalized acrylate rubbers

Ratna et al. [298,309–314] have extensively reported the acrylate-based liquid rubbers for toughening epoxy resins. Various acrylate-based modifiers have been synthesized, namely, carboxyl-terminated poly(2-ethylhexyl acrylate) (CTPEHA), carboxyl-randomized poly(2-ethylhexyl acrylate) (CRPEHA), amine-terminated poly(2-ethylhexyl acrylate) (ATPEHA), and poly(2-ethylhexyl acrylate) oligomer with terminal and pendant reactive groups (carboxyl or amine). CTPEHA or ATPEHA can be prepared by bulk polymerization of EHA using carboxyl- or amine-ended initiator and chain transfer agent. A typical recipe and synthetic process for CTPEHA are briefly described below.

CTPEHA of varying molecular weights were synthesized [298,309,310] by bulk polymerization, using 4, 4′-Azobis (4-cyanovaleric acid) (ABCVA) (Aldrich) as a free radical initiator and dithiodiglycolic acid (DTDGA) as a chain transfer agent. The reaction was carried out in a three-necked reaction flask (500 mL) fitted with a stirrer, a thermometer pocket, and a gas inlet. Approximately, 100 g (0.54 moles) of EHA monomer was taken into the reaction flask and was rapidly brought to the desired temperature. After the system was well purged with nitrogen gas, 3.0 g of ABCVA (2 mol%) and the required amount of DTDGA (2.3–11.3 g, 2–10 mol%) were added and the reaction was allowed to occur for 1 h with stirring. The mixture was then diluted with toluene (200 mL) and cooled to room temperature and kept overnight. The unreacted ABCVA and DTDGA precipitated out and were removed by filtration. Unreacted monomer and the solvent were removed under vacuum on a rotary evaporator until a constant weight was obtained.

The initiator ABCVA, on dissociation, generates carboxyl-ended free radicals, which initiate monomer molecules to produce the growing macro radicals. These growing macroradicals evidently react with DTDGA through cleavage of the sulfur−sulfur bond to form carboxyl end-capped oligomers. This reaction will reduce the probability of chain termination of growing acrylate radicals by combination and disproportionation. A representation of the tentative reaction mechanism is shown in Fig. 3.16. Since acrylates are known to terminate through a combination of free radicals rather than disproportionation, the functionality of the liquid rubber should be theoretically 2. However, the above-mentioned process generates liquid rubber with functionality in the range of 1.6−1.9 eq mole^{-1} (Table 3.5). The deviation of the value of functionality from the theoretical value of 2 can be attributed to the chain transfer to monomer. In order to keep the functionality close to 2, it is necessary to synthesize the liquid rubbers by bulk polymerization. In the case of solution polymerization, the solvent competes with the functionalized chain transfer agent in the chain transfer process and significantly reduces the functionality. Solution polymerization produces almost monofunctional resin as shown in Table 3.3.

3.10.3.2 Acrylate-modified epoxy

The acrylate modification of epoxy resins is carried out in two ways, namely, one-stage method and two-stage method. In the one-stage method, the modifier is directly blended with the resin/hardener mixture. In a two-stage process, the modifier is prereacted with the epoxy resin and the prereacted resin is further cured with a suitable hardener. For the modification of high-temperature-curing epoxy systems, any of the above-mentioned processes can be adopted. In the case of room temperature-curing epoxy systems, a two-stage modification method is recommended because generally, the modifiers react with the epoxy at a much slower rate compared to a curing agent. For example, when a carboxyl-functionalized modifier is directly blended with a room temperature-curing epoxy system (resin + hardener) the modifier cannot form a chemical bond with the epoxy as epoxy−carboxyl reaction does not take place at room temperature. Hence in order to initiate chemical bonding of the modifier with the epoxy, it is necessary to adopt the prereact method for the modification of room temperature-curing epoxy system.

The modified resin in general shows a higher gel point compared to the unmodified resin due to the lower reactivity of the unmodified resin as a result of chain extension and dilution effect. The effects of CTPEHA modification (using a two-stage method) on gel time and T_{peak} (temperature corresponding to the exothermic peak obtained in DSC) are shown in Table 3.6. As expected both the gel time and T_{peak} increase with increasing concentration of CTPEHA. However, when CTPEHA modification is carried out by the one-stage method (i.e., by direct blending of CTPEHA with epoxy without any prereaction), the T_{peak} was found to decrease. This can be explained by considering the accelerating

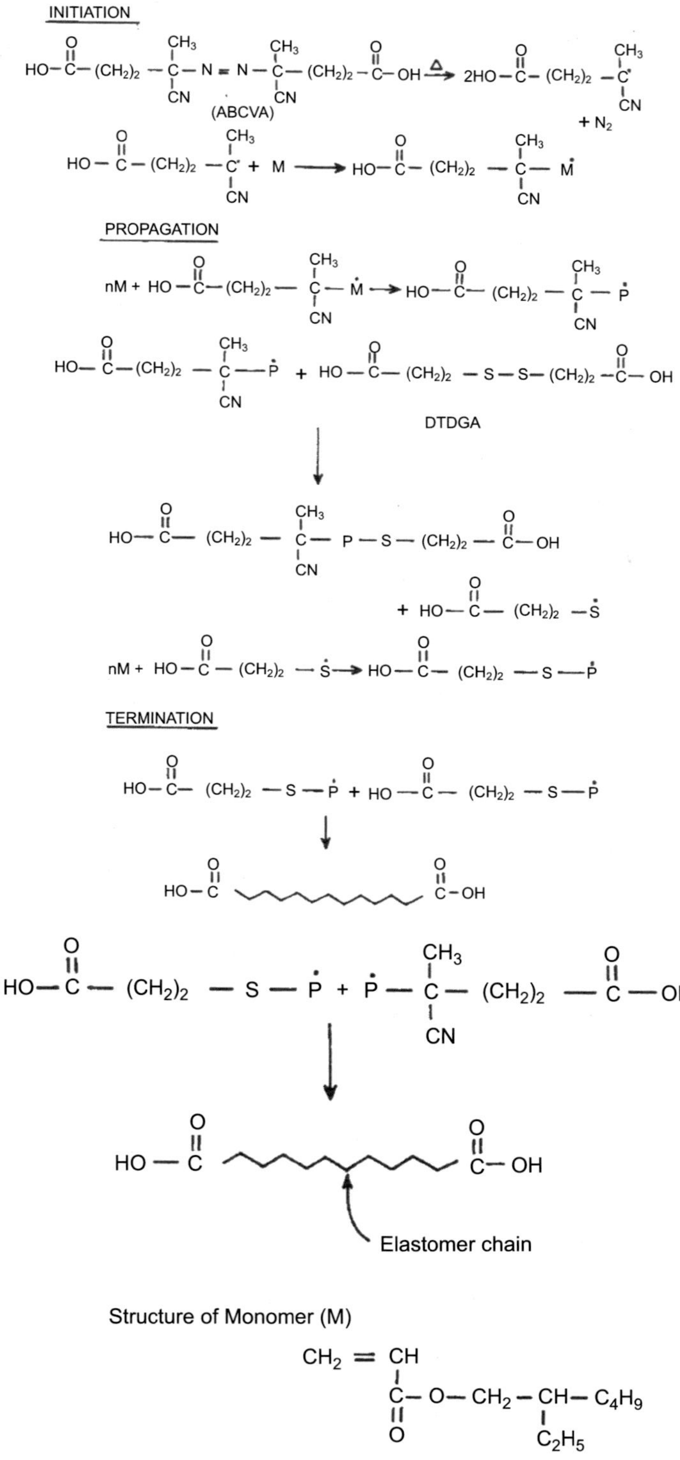

INITIATION

$$HO-\underset{O}{\overset{\parallel}{C}}-(CH_2)_2-\underset{CN}{\overset{CH_3}{\underset{|}{C}}}-N=N-\underset{CN}{\overset{CH_3}{\underset{|}{C}}}-(CH_2)_2-\underset{O}{\overset{\parallel}{C}}-OH \xrightarrow{\Delta} 2HO-\underset{O}{\overset{\parallel}{C}}-(CH_2)_2-\underset{CN}{\overset{CH_3}{\underset{|}{C}}}\cdot + N_2$$

(ABCVA)

$$HO-\underset{O}{\overset{\parallel}{C}}-(CH_2)_2-\underset{CN}{\overset{CH_3}{\underset{|}{C}}}\cdot + M \longrightarrow HO-\underset{O}{\overset{\parallel}{C}}-(CH_2)_2-\underset{CN}{\overset{CH_3}{\underset{|}{C}}}-\dot{M}$$

PROPAGATION

$$nM + HO-\underset{O}{\overset{\parallel}{C}}-(CH_2)_2-\underset{CN}{\overset{CH_3}{\underset{|}{C}}}-\dot{M} \longrightarrow HO-\underset{O}{\overset{\parallel}{C}}-(CH_2)_2-\underset{CN}{\overset{CH_3}{\underset{|}{C}}}-\dot{P}$$

$$HO-\underset{O}{\overset{\parallel}{C}}-(CH_2)_2-\underset{CN}{\overset{CH_3}{\underset{|}{C}}}-\dot{P} + HO-\underset{O}{\overset{\parallel}{C}}-(CH_2)_2-S-S-(CH_2)_2-\underset{O}{\overset{\parallel}{C}}-OH$$

DTDGA

↓

$$HO-\underset{O}{\overset{\parallel}{C}}-(CH_2)_2-\underset{CN}{\overset{CH_3}{\underset{|}{C}}}-P-S-(CH_2)_2-\underset{O}{\overset{\parallel}{C}}-OH$$

$$+ HO-\underset{O}{\overset{\parallel}{C}}-(CH_2)_2-\dot{S}$$

$$nM + HO-\underset{O}{\overset{\parallel}{C}}-(CH_2)_2-\dot{S} \longrightarrow HO-\underset{O}{\overset{\parallel}{C}}-(CH_2)_2-S-\dot{P}$$

TERMINATION

$$HO-\underset{O}{\overset{\parallel}{C}}-(CH_2)_2-S-\dot{P} + HO-\underset{O}{\overset{\parallel}{C}}-(CH_2)_2-S-\dot{P}$$

↓

$$HO-\underset{O}{\overset{\parallel}{C}}\sim\sim\sim\sim\sim\sim\underset{O}{\overset{\parallel}{C}}-OH$$

$$HO-\underset{O}{\overset{\parallel}{C}}-(CH_2)_2-S-\dot{P} + \dot{P}-\underset{CN}{\overset{CH_3}{\underset{|}{C}}}-(CH_2)_2-\underset{O}{\overset{\parallel}{C}}-OH$$

↓

$$HO-\underset{O}{\overset{\parallel}{C}}\sim\sim\sim\sim\sim\sim\underset{O}{\overset{\parallel}{C}}-OH$$

↑ Elastomer chain

Structure of Monomer (M)

$$CH_2=\underset{\underset{O}{\overset{\parallel}{C}}-O-CH_2-\underset{C_2H_5}{\overset{|}{CH}}-C_4H_9}{\overset{CH}{\underset{|}{}}}$$

(Continued)

Table 3.5: Physicochemical properties of acrylate rubbers.

Solvent	DTDGA concentration mole %	$\overline{M}_n^a$	$\overline{M}_n^b$	$\overline{M}_n^b/\overline{M}_n^b$	Functionality (f) eq mole^{-1}
Nil	2	9500	8000	7.8	1.8
Nil	5	7000	4700	8.2	1.7
Nil	10	3600	2400	7.3	1.9
Dioxan	2	2360	1900	2.3	1.2
THF	2	2700	2200	2.7	1.3

^{a}Measured by VPO.
^{b}Measured by GPC.
Source: Reprinted with permission from D. Ratna, A.K. Banthia, P.C. Deb, J. Appl. Polym. Sci. 80 (2001) 1792.

Table 3.6: Effect of CTPEHA modification on gel time, T_{peak}, and T_g of modified epoxy cured with DETDA.

Concentration of CTPEHA (phr)	Gel time (min)	T_{peak} (°C)	Viscosity (Pa s)	T_g^a (°C)
0	43	195	2.8	203
5	48	205	9.1	198, −60
10	51	210	14.8	196, −60
15	53	215	18.1	185, −55
20	61	217	23.2	174, −58

^{a}Values determined from DSC analysis of cured network using the same cure schedule. The two T_g values refer to the epoxy-rich phase (high value) and CTPEHA-rich phase (low value).
Source: Reprinted with permission from D. Ratna, Polymer 42 (2001) 4209, © 2001, Elsevier Publisher.

effect of carboxyl groups of CTPEHA on epoxy—amine reaction. The behavior can be interpreted in terms of the intermolecular transition state of epoxy—amine reaction. According to this mechanism, strong hydrogen bonding species such as acids and alcohols stabilize the transition state and strongly accelerate the reaction as discussed in Chapter 4, Chemistry and General Applications of Thermoset Resins.

The acrylate rubbers are initially miscible with the epoxy resin and undergo phase separation during the curing leading to the formation of a two-phase microstructure. Such a two-phase microstructure with uniformly dispersed rubber particles often offers outstanding fracture properties as the rubber particles dissipate the mechanical energy by various mechanisms as discussed in Chapter 4, Thermoset Composites. The effect of addition CTPEHA on the impact strength of the modified epoxy is shown in Fig. 3.17. The impact

Figure 3.16
A representation of tentative reaction mechanism for formation of carboxyl-terminated poly(2-ethylhexyl acrylate) (CTPEHA) from free radical polymerization of EHA in presence of 4, 4′-azobis (4-cyanovaleric acid) (ABCVA) as a free radical initiator and dithiodiglycolic acid (DTDGA) as a chain transfer agent (P˙ = nM˙). Source: Reprinted with permission from D. Ratna, A.K. Banthia, P.C. Deb, J. Appl. Polym. Sci. 80 (2001) 1792, © 2000, John Wiley and Sons Publishers.

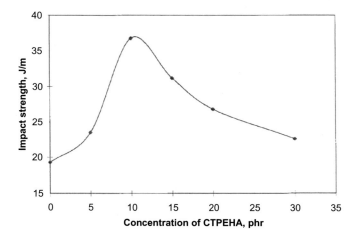

Figure 3.17
Effect of CTPEHA content on impact strength of modified epoxy network. Source: *Reprinted with permission from D. Ratna, A.K. Banthia, P.C. Deb, J. Appl. Polym. Sci. 80 (2001) 1792,* © *2001, John Wiley and Sons Publishers.*

strength of modified epoxy increases with increasing CTPEHA concentration attains a maximum at a concentration of 10–15 phr and decreases thereafter. The decrease in impact strength beyond an optimum concentration is attributed to the deviation from uniformly dispersed morphology as a result of agglomeration.

The morphology depends on the functionality and molecular weight of the rubber. Initial cure temperature also plays an important role in deciding the morphology development for a particular rubber-modified system. The postcuring temperature has no role in morphology development because the phase separation is arrested at gelation due to a tremendous increase in viscosity. The theoretical reasons for the effect of these parameters have already been discussed in Section 3.3.1. Since the impact property of the modified network depends on the resulting morphology, adjustment of initial cure temperature is equally important as the selection of correct liquid rubber. For example, CTPEHA-modified epoxy (containing 10 phr CTPEHA and cured with DETDA) shows a significant enhancement in impact properties when the resin is cured at 140°C compared to the unmodified epoxy. On the other hand, when the same system is cured at 180°C, the cured resin displays much lower impact strength. The results can be explained by considering the fact that at a higher temperature the curing reaction is so fast that the gelation is reached immediately and the dissolved rubber does not get sufficient time to undergo phase separation. The SEM photographs for fracture surfaces of the modified epoxy/DETDA systems cured at 160°C and 180°C are shown in Fig. 3.18. The modified system cured at 180°C shows almost single-phase morphology with very little phase-separated rubber. This explains the poor impact performance of the epoxy system cured at 180°C as it has been already explained in

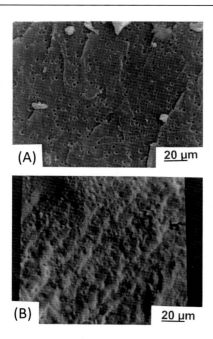

Figure 3.18
The SEM photographs for fracture surfaces of the CTPEHA-modified epoxy/DETDA systems cured at (A) 140°C and (B) 180°C. *Source: Reprinted with permission from D. Ratna, Polymer 42 (2001) 4209, © 2001 Elsevier Publisher.*

Section 3.3.2 that phase-separated rubbers particles initiate the shear yielding by various mechanisms and effectively enhance the toughness of the matrix whereas the dissolved rubbed plasticizes the matrix and cause a nominal increase in toughness.

Thus a complete phase separation can be achieved for such acrylate liquid rubbers by increasing the molecular weight of the liquid rubber and decreasing the cure temperature within the processing window. The T_g values for CTPEHA (M_n = 3600 g mole^{-1}) modified epoxy (cured with DETDA) are presented [311] in Table 3.4. The modified DSC analysis shows two T_gs: one for epoxy network and the other is due to phase-separated CTPEHA. Since CTPEHA forms chemical bonds with the epoxy, it is expected that the T_g of phase-separated CTPEHA will be higher than the T_g of pure CTPEHA. However, it was observed [289,314] that the T_g of phase-separated CTPEHA (−45°C) is lower than pure CTPEHA (−60°C). This can be explained by considering thermal shrinkage stress [289]. Triaxial thermal shrinkage stress develops in the dispersed domain phase on cooling through epoxy T_g because the coefficient of thermal expansion of the rubbery state is larger than that of the glassy state. The triaxial stress increases the free volume of the rubbery CTPEHA particles and depresses the T_g of the precipitated CTPEHA. It is also evident from Table 3.7 that up to 10 phr CTPEHA concentration, there is no significant change in epoxy which suggests almost a complete phase separation.

An increase in molecular weight of liquid rubber favors phase separation. Hence to ensure complete phase separation, it is necessary to use liquid rubber with sufficiently high molecular weight. However, beyond a certain molecular weight, the liquid rubber undergoes phase separation at a very early stage leading to agglomeration and macrophase separation as shown in Fig. 3.19. The fracture surface of modified epoxy with CTPEHA having molecular weight 10,000 g mole^{-1} shows agglomerated particles having a size greater than 10 μm. The optimum molecular weight reported being about 7000 g mole^{-1} for CTPEHA [311,312] and 3000 for CRPEHA g mole^{-1} [312].

Table 3.7: A comparison of toughening effects of various modifiers for epoxy matrix.

Liquid rubber	Optimum concentration	Curing agent	Improvement in property	Reference (s)
CTBN	15 phr	HY 960[a]	400% T-peel strength	[252]
CTBN	15 phr	Piperidine	100% T-peel strength	[250,251]
CTPPGA	15 phr	HY 960[a]	400% T-peel strength	[302]
CTPEHA	10 phr	HY 951[b]	100% impact strength	[310]
ESR	20 wt.%	HY 960[a]	100% Lap shear strength	[297]
PECH	10 phr	Pipyridine	66% fracture toughness K_{IC}	[294]
PU prepolymer	Not established	DDS	400% fracture energy G_{IC}	[303,304]
BA, VBGE, St terpolymer	20 wt.%	DDS	80% fracture energy G_{IC}	[315,316]
HBP	5 wt.%	Isophoronediamine	600% G_{IC}	[317,318]
HBP	15 wt.%	DETDA	300% impact strength	[319]

[a]Tris-2,4,6-(N, N-dimethylamino methyl) phenol.
[b]Trimethylene tetramine.

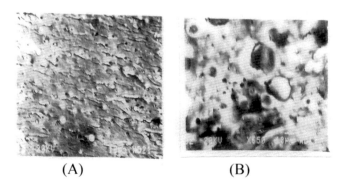

Figure 3.19
SEM microphotographs of the fracture surfaces of TETA-cured epoxy network modified with CTPEHA of different molecular weights: (A) $\overline{M}_n$ = 3600 g mole^{-1} and (B) $\overline{M}_n$ = 9500 g mole^{-1}.
Source: Reprinted with permission from D. Ratna, A.K. Banthia, P.C. Deb, J. Appl. Polym. Sci. 78 (4) (2000) 716, © 1995 John Wiley and Sons Publishers.

In order to study the effect of the functionality on toughening effect, various acrylate rubbers having only pendant carboxyl group and both the terminal and pendant groups have been investigated. The acrylate rubber with pendant carboxyl groups (i.e., CRPEHA) can be prepared by copolymerizing EHA and acrylic acid. The advantage of CRPEHA over CTPEHA is that CRPEHA can be made by solution polymerization, which offers better control over polymerization reaction compared to the bulk polymerization (method of preparation for CTPEHA). Note that the synthesis of CTPEHA using solution polymerization produces almost monofunctional resins (Table 3.4). The impact strength of the liquid acrylate rubber-modified epoxy networks as a function of carboxyl functionality of the liquid rubber is shown in Fig. 3.20. It was found that impact strength increases with an increase in the carboxyl functionality, attains a maximum, and then decreases [313]. It appears that an increase in chemical interaction of the liquid rubber and the epoxy matrix leads to the initial increase in impact strength of the matrix.

However, the increase in toughness cannot be attributed to only the increase in rubber particle to matrix adhesion as the thermal analysis results indicate that the T_g value of the modified networks (all containing the same amount of rubber) decreases with an increase in

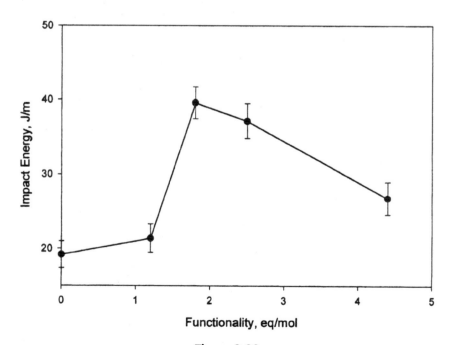

Figure 3.20
Effect of carboxyl functionality of carboxyl-randomized poly(2-ethylhexyl acrylate) (CRPEHA) on impact strength of modified epoxy network. Source: *Reprinted with permission from D. Ratna, G.P. Simon, Polymer 42 (2001) 7739–7747,* © *2001 Elsevier Publisher.*

functionality of the resin. The depression of epoxy T_g in the rubber-modified system arises due to residual rubber which remains dissolved in the epoxy matrix. Hence for a particular composition, the amount of dissolved rubber increases with an increase in functionality of the rubber. The dissolved rubber increases the ductility of the epoxy matrix. This has twofold effects on toughness; one is the improvement of toughness due to plasticizing effect and another is the increase in the effectiveness of the toughening process imparted by the rubber particles as demonstrated by several studies. The toughening mechanism highlights the role of inherent ductility of the matrix in influencing the toughness of the multiphase network. For example, a decrease in the stress needed for localized shear yielding should obviously assist in increasing the toughness if all other microstructural features are unchanged [314]. Rubber toughening of the brittle tetraglycidyl methylene dianiline (TGMDA)-based networks have resulted in minimum improvement of fracture energy in contrast to the comparatively ductile DGEBA-based networks [320,321]. Lavita et al. [181] has plotted the fracture energy of toughened thermoset versus the fracture energy of the unmodified epoxy, which represents an amplification factor of 10. TGMDA-based resin formulation cured with pipyridine (PIP) is more amenable to toughening by the addition of CTBN than the DDS-hardened system because of the lower ductility of the latter [321]. The fracture energy of TGMDA-PIP is four times greater than that of TGMDA-DDS system but the fracture energy of CTBN-modified (20 phr) TGMDA-PIP system is 24 times greater than that of the toughened TGMDA-DDS system [322,323]. This is because the rubber-rich particles as stress concentrators induce plastic deformation of the highly brittle matrix to a far lesser extent. Hence both the increase in rubber particle to matrix adhesion and the flexibilizing effect of dissolved rubber contribute to the improvement in toughness.

A decrease in toughness above an optimum value of the functionality of the liquid rubber can be explained in terms of solubility parameter and chemical interactions. The solubility parameter of epoxy is higher than acrylate rubber (PEHA). With the introduction of carboxyl functionality, the solubility parameter of the acrylate oligomer increases. When the functionality reaches an optimum value, the solubility parameter difference between the epoxy and acrylate rubber $(\delta_e - \delta_r)$ becomes less than 0.1. As a result, the decrease in combinatorial entropy cannot compensate for the enthalpy term [recall Eq. (4.3), Chapter 4: Thermoset Composites] and the free energy change of mixing does not become positive before gelation. Hence the rubber with higher functionality cannot phase separate and leads to a single-phase morphology. This explains the decrease in impact strength of acrylate-modified epoxy when the functionality goes beyond an optimum value. The nominal enhancement in impact strength results due to plasticizing effect of the dissolved rubber. In addition, the liquid rubber having higher functionality undergoes an extensive chemical reaction with the epoxy, which further makes the phase separation more difficult.

Iijima et al. [315,316] used an acrylic elastomer containing pendent epoxy groups to reduce the brittleness of DGEBA resin cured with diamino diphenyl sulfone (DDS).

The elastomers were prepared by copolymerization of butyl acrylate (BA), vinyl benzyl glycidyl ether (VBGE), and styrene (St), or acrylonitrile. The addition of 20% of the terpolymer (74 mole % BA, 18 mole% VBGE, and 8 mole % St) resulted in an 80% increase in fracture energy of the cured resin at the slight expense of its mechanical properties.

3.10.4 Hyperbranched polymer-based toughening agents

The modification of epoxy with linear elastomers, as discussed in previous sections, is associated with a considerable increase in viscosity, which is disadvantageous from a processing point of view. The problem can be overcome by using dendritic hyperbranched polymer (HBP)-based toughening agents [317,319]. Due to the compact, three-dimensional (3D) structure of dendritic polymers, these molecules mimic the hydrodynamic volume of spheres in solution or melts and flow easily past each other under applied stress. This results in low melt viscosity, even at high molecular weights, due to a lack of restrictive interchain entanglements [109]. Indeed, dendritic polymers have been shown to exhibit melt and solution viscosities that are an order of magnitude lower than their linear analogs of similar molecular weight. The high density of functional terminal groups on dendritic polymers also offer the potential for tailoring their compatibility with the epoxy either through conversion of dendritic polymer end groups to chemically suitable moieties or through in situ reaction to form covalently bound networks. These two properties low viscosity and tailorable compatibility make HBPs excellent candidates as flow additives that could act simultaneously as toughening agents. These polymers are commercially available, for example, Bolton. Schematic representation of hydroxyl- and epoxy-functionalized HBP is shown in Fig. 3.21.

Boogh et al. [318,324] first investigated the effect of incorporation of HBP on fracture properties of epoxy resins. They reported an increase in critical strain energy release rate (G_{IC}) by a factor of 6 as a result of addition of 5 wt.% of epoxy-functionalized HBP in the epoxy matrix. Ratna et al. [319,325−330] studied extensively the blends of epoxies (difunctional and multifunctional) and HBP with various functionalities like epoxy, hydroxyl, etc. A significant toughening effect has been demonstrated. A comparison of toughening effect of various toughening agents is presented in Table 3.6. Epoxy-functionalized HBP is not miscible with epoxy at room temperature but becomes miscible at higher temperatures (100°C). With the advancement of curing reaction, the HBP undergoes phase separation leading to the formation of a two-phase microstructure as shown by optical microscopic analysis (Fig. 3.22). It was confirmed from dynamic mechanical analysis and microscopic studies that the HBP phase separates completely up to 15 wt.% of HBP concentration. The loss factor versus temperature plots for epoxy/HBP (epoxy-functionalized) are shown in Fig. 3.23. The plot clearly shows two relaxation peaks: one is at high temperature for epoxy and the other is at low temperature for the HBP. There is no change in epoxy T_g up to 15 wt.% of HBP suggesting a complete phase separation.

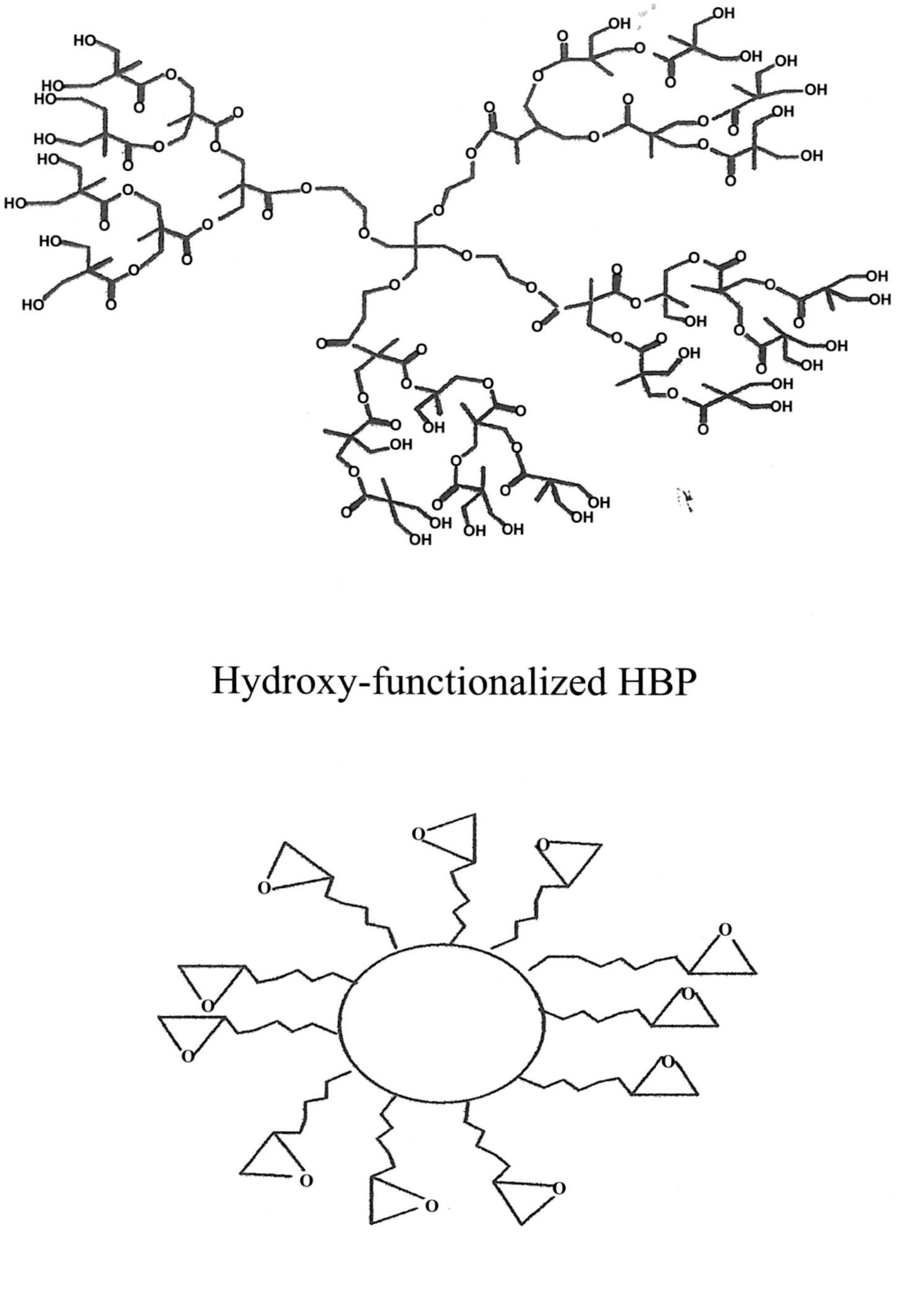

Hydroxy-functionalized HBP

Epoxy-functionalized HBP

Figure 3.21
Chemical structures of hydroxyl and epoxy-functionalized hyperbranched polymers (HBPs).

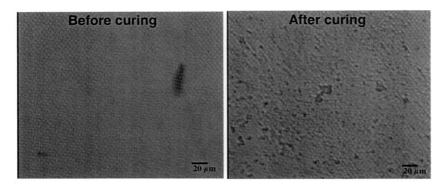

Figure 3.22
Optical microscopy photograph of epoxy/HBP/DETDA mixture (A) before curing and (B) after curing. Source: *Reprinted with permission from D. Ratna, Composites A39 (2008) 462, © 2008 Elsevier Publisher.*

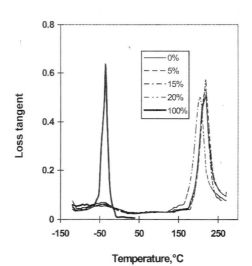

Figure 3.23
Loss tangent versus temperature plots of epoxy, epoxy-functionalized HBP and epoxy/HBP blends cured with DETDA. Source: *Reprinted with permission from D. Ratna, R. Varley, G.P. Simon, J. Appl. Polym. Sci. 89 (2003) 2339, © 2003 Springer publisher.*

The effect of the addition of epoxy-functionalized and hydroxyl-functionalized HBP on gelation at 160°C for epoxy/HBP blends is shown in Fig. 3.24. The figure indicates that the addition of epoxy-functionalized HBP has hardly any effect on gelation and vitrification, although HBP is known to cure at a slower rate compared to that of epoxy resin. This behavior is different from that observed in the case of other liquid rubbers like CTPEHA, discussed earlier, where a significant increase in gel time was observed. In those cases, the delay has been attributed to the viscosity effect, which retards the movement of reactive

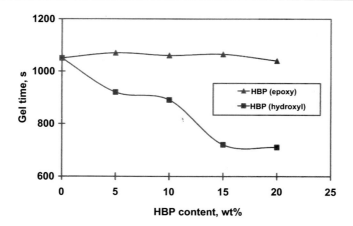

Figure 3.24
Effect of addition of hydroxyl and epoxy-functionalized hyperbranched polymers (HBPs) on the gel time of modified epoxy.

molecules. The blending of HBP with epoxy, however, does not lead to any significant change in viscosity and hence the reaction rate remains unaffected. The viscosity of HBP at 25°C (15 Pa s^{-1}) is about 3 times lower than that of the acrylate-based liquid rubber CTPEHA (42 Pa s^{-1}). On the other hand, addition of hydroxyl-functionalized HBP was found to decrease the gel time of epoxy/HBP blends (Fig. 3.24). This indicates that hydroxyl-functionalized HBP blends undergo curing faster than epoxy alone. This can be explained by considering the accelerating effect of -OH group of HBP on epoxy−amine reaction as discussed in Chapter 1, Chemistry and General Applications of Thermoset Resins.

The work discussed above is involved with commercial HBP. A new class of epoxy-ended hyperbranched polyether (HBPEE) with aromatic skeletons was also synthesized and used as a toughening agent for epoxy [331−333]. The HBPEE was prepared through a one-step proton transfer polymerization [332,333]. Compared with the neat DGEBA, the modified epoxy networks showed excellent balanced mechanical properties at 5 wt.% HBPEE loading. In the case of epoxy networks modified with commercial HBP, the glass transition temperature (T_g) was found to decrease. However, in the present case, the T_g is reported to be increased due to increase in crosslink density, introduction of rigid skeletons and the molecular-scale cavities brought by the reactive HBPEE. The modified networks tend to form homogeneous system due to extensive chemical interaction with the matrix because of the reactivity of HBPEE. The improvement in toughness is due to fibril formation as evident from SEM analysis. Similar observation was also reported when the same is used to toughen benzoxazine resins as discussed in Section 3.7. Hyperbranched polyethers containing cyclic carbonate groups have also been successfully used [334] for toughening epoxy resins. Brocks et al. [335] synthesized hyperbranched poly(butylene adipate) (HPBA) polymer and evaluated as a toughening agent for epoxy resin. They have demonstrated a significant toughening effect better than that of a commercial dendritic polyol (HPOH). SEM analysis clearly shows the fracture mode changing from brittle to ductile as a result of addition of modifiers, which was more evident for blends of HPBA than HPOH.

3.11 Core−shell particle toughening

Liquid rubber toughening is associated with a reaction-induced phase separation process during which the rubber (initially miscible with epoxy) undergoes phase separation and generates a morphology with a two-phase microstructure. The phase separation process depends upon the formulation, processing and curing conditions. Incomplete phase separation can result in a significant lowering of glass transition temperature (T_g). Moreover, the rubber phase that separates during the cure is difficult to control and may result in uneven particle size. The differences in morphology and volume of the separated phase affect the mechanical performance of the product. The factors that affect the fracture toughness of the modified epoxy such as morphology, particle size, and composition are interdependent; hence it is very difficult to study the effect of an individual parameter. These problems can be minimized by using insoluble preformed particles directly. Since the size, morphology and composition, shell thickness, and crosslink density of the rubbery cores can be controlled separately by employing emulsion polymerization techniques, the effects of various parameters on the toughening of epoxies can be investigated individually. The control of the particle parameters by emulsion polymerization has been extensively studied, and various efficient technologies have been developed [336−339]. Monodispersed latex particles with a diameter from submicron to micron range can be prepared by sequential seeded emulsion polymerization [340], sequential swelling process [341], and dispersion polymerization [336,337]. Cohesive strength, which is influenced by the crosslink density of the rubber phase, can be controlled by the conversion of polymerization and the amount of crosslinking agent [339]. Interfacial architecture can be controlled by changing the following parameters: (1) thickness of the shell which depends on the ratio of the shell−core materials and polymerization mechanism; (2) chemical bonding and physical interaction between particles and matrix, which can be enhanced by introducing functional groups onto the surface of the shell; (3) grafting between the shell and core; and (4) molecular weight of shell materials. Various morphologies of the composite such as core−shell, occluded, or multilayer can be achieved through two or multiple stage emulsion polymerization [342]. Typically, core−shell morphology latex particles can be made by semicontinuous process under a monomer "starved" condition [343].

The preformed particles are incorporated into the epoxy matrix by simple mechanical mixing. The dispersibility of the particles can be improved by (1) introducing crosslink into the shell or (2) using comonomer like acrylonitrile or glycidyl methacrylate (GMA), which increases the interfacial adhesion by polar or chemical interaction [344,345]. Quan et al. [346] reported that in the case of poly(butadiene-co-styrene) core and poly(methyl methacrylate) (PMMA) shell particles, the cluster size reduces from 3–5 mm to 1–3 mm as a result of the use of 5 wt.% crosslinker (divinylbenzene). They also found that the cluster size could be further reduced to 1–2 mm by using MMA-AN or MMA-GMA copolymer as shell composition.

Lin et al. [347,348] have prepared reactive CSPs with butyl acrylate as a core and MMA-GMA copolymer as a shell and used them as toughening agents for DGEBA epoxy. They found that shell crosslinked CSPs had a higher toughening effect than core crosslinked CSPs. They also reported a tremendous effect of comonomer like GMA on the toughening effect. This seemed to contradict Sue's observation that the chemical bonding of CSP to the epoxy matrix did not significantly contribute to the toughening performance.

Kinloch and coworker [349] investigated the toughening effect of preformed polysiloxane core–shell rubber (S-CSR) particles in an anhydride-cured epoxy polymer. Because of very low T_g of polysiloxane (100°C), the toughened networks are expected to offer good fracture toughness at subambient temperatures unlike commercial rubber which are effective at very low temperatures. The fracture properties of the modified epoxy networks with varying concentrations (2–20 wt.%) of S-CSR measured at different temperatures (20°C, −55°C, −80°C, −109°C). It is observed that the variation of fracture toughness of unmodified epoxy with the temperature is insignificant considering the standard deviation. However, for modified epoxy networks, there are significant variations in temperature. At ambient temperature, the fracture energy of 20 wt.% S-CSR containing network is 947 J m^{-2} whereas the same values measured at −109°C is 481 J m^{-2}, Thus the toughening efficiency of S-CSR decreases with decreasing temperature as expected due to decrease in cavitation of rubber particles. However, the value of fracture energy at −100°C is still 50% higher compared to unmodified epoxy. Therefore it is very noteworthy that the addition of the S-CSR particles can significantly toughen the epoxy polymer even at a very low temperature close to −100°C. SEM studies indicates that rubber is cavitated even at −109°C but the size of cavities is reduced with decreasing temperature. The SEM microphotographs for 20 wt.% S-CSR containing epoxy tested at −55°C and −109°C are shown in Fig. 3.25. Gong et al. [350] used preformed 3-D powdered rubber (PR) to toughen the DGEBA epoxy resins. K_{IC} and G_{IC} values of modified epoxy networks containing 3 wt.% and 6 wt.% of PR are 1 MPa m$^{0.5}$, 3.1 MPa m$^{0.5}$ and 307 J m^{-2}, 568 J m^{-2}, respectively. Both the K_{IC} and G_{IC} values for the modified epoxies are considerably higher than those of pure epoxy (0.51 MPa m$^{0.5}$, 77 J m^{-2}) indicating significant toughening effect. The toughening effect is achieved with a simultaneous enhancement in tensile strength and a modest decrease in elastic modulus.

3.12 Thermoplastic toughening

As discussed in the earlier sections, rubber toughening resulted in a dramatic improvement in the impact properties of the cured epoxy. However, the presence of the rubber phase

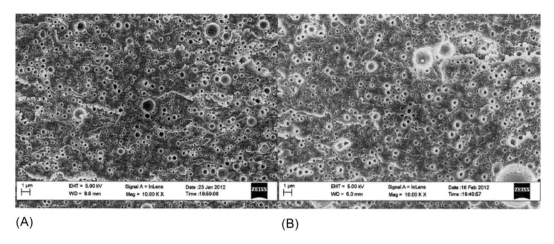

Figure 3.25
FEG-SEM images of the fracture surface of the 20 wt.% S-CSR particle-modified epoxy polymer tested at (A) 55°C and (B) 109°C. Source: *Reprinted with permission from J. Chen, A.J. Kinloch, S. Sprenger, A.C. Taylor, Polymer 54 (2013) 4276–4289,* © *2013, Elsevier Publisher.*

does somewhat decrease the modulus and thermal stability of the materials and increase the tendency of water absorption with an accompanying loss of properties at elevated temperatures. Moreover, the reactive rubbers like carboxyl-terminated copolymer of butadiene and acrylonitrile (CTBN) have been reported as an ineffective modifier for a highly crosslinked system based on epoxy having a functionality of more than 2 (i.e., 3 or 4). This is because the rubber-rich particles, as stress concentrators, induce the plastic deformation of a highly crosslinked matrix to a far less extent and dissipation of fracture energy by the enlargement of the deformation zone can hardly be attained [322,323]. The search for an alternative method to rubber toughening led to the development of thermoplastic-toughened epoxy. The thermoplastics that have been reported as efficient modifiers for epoxy resin can be classified as (1) engineering thermoplastic, (2) amorphous thermoplastics, and (3) crystalline thermoplastics.

3.12.1 Engineering thermoplastics

Tough and ductile engineering thermoplastics have been experimented with for modifying epoxy resin [351–354]. The chemical structures of few commonly used thermoplastic modifiers are given in Fig. 3.26. Because of their high modulus and high T_g of engineering thermoplastics, the modified epoxy resin will reach or even exceed the corresponding values for the unmodified resin. Unlike the rubber toughening where a significant reduction in stiffness and modulus was observed at ambient temperature, in the case of thermoplastic-toughened epoxy reduction in stiffness becomes significant only at a temperature near to the T_g of the thermoplastic moiety [355].

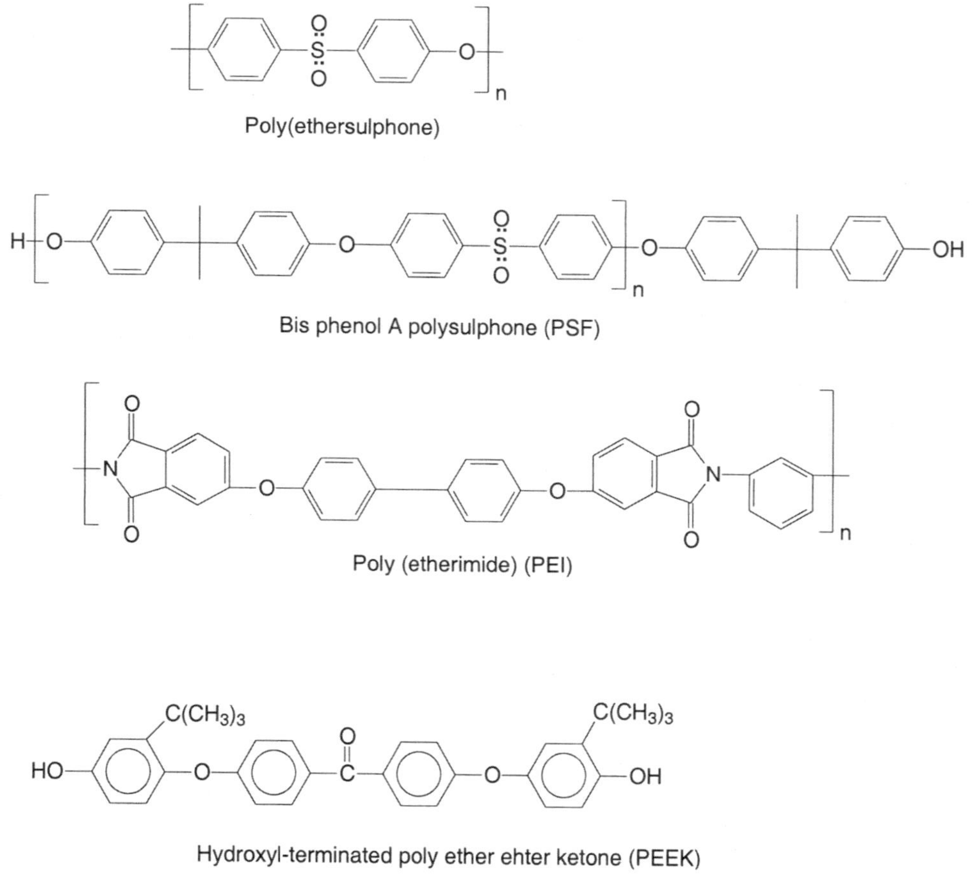

Figure 3.26
Chemical structures of some thermoplastic modifier used to toughen epoxy resins.

The earliest studies on the addition of thermoplastics into the epoxy resins were by Bucknall and Patridge [356] using poly(ether sulfone) (PES) and by Diamant and Moulton [357] using poly(ether imide) (PEI), by the physical blending of epoxy and respective thermoplastics using methylene dichloride as a solvent. Using trifunctional (Ciba Giegy, MY 051) and tetrafunctional epoxy (Ciba Giegy, MY 720), they have reported little improvement in fracture energy of the modified epoxy networks cured by dicyanamide (Dicy) and 4,4′-diamino diphenyl sulfone (DDS). Additionally, Bucknall and Patridge [358] found that the fracture energies and fatigue properties of fiber-reinforced composites were relatively insensitive to the presence of the thermoplastic. Part of the explanation for the ineffectiveness of PES as a toughening agent in this study may derive from the morphology of the systems under observation as Bucknall and Patridge reported that they did not find any phase separation in three of the systems studied, namely, PES/MY720/Dicy, PES/MY720/DDS, PES/MY720/ERL 0570/DDS. However, the phase separation was observed in the systems, namely, PES/ERL 0510/Dicy and PES/MY 720/ERL 0510/Dicy. They concluded that the type and concentration of resin and hardener influenced the morphology of the PES-rich phase and that the addition of PES has no significant effect upon the creep and fracture toughness of epoxy resin irrespective of the phase separation and morphology.

Raghva [359,360] studied the influence of microstructure on the mechanical properties of a modified epoxy system derived from a tetrafunctional epoxy and an aromatic anhydride. Characterizing the resulting morphology, he found that Victex 100 P (PES with a molecular weight $M_w = 145,000$) gave a higher percentage of large round particles (1–5 μm) as compared to the agglomerated particles by Victex 100 P (PES with molecular weight $M_w = 85,000$). He reported that there is only a marginal improvement in the fracture toughness for both the blend systems over the unmodified epoxy network at and below room temperature but a significant improvement in fracture toughness was observed at higher test temperature. He concluded that the lack of improvement of fracture toughness observed in these systems may be due to the fact that cured epoxy resin's high crosslink density inhibited the primary toughening mechanism, namely, the formation of shear bands.

Hedrick et al. [361,362] considered poor interfacial adhesion to be the main reason for the inability of commercial thermoplastics (nonreactive) in improving the toughness of epoxy resin. They used phenolic —OH-ended bisphenol-A-based PES and amine-terminated PES oligomers as toughening agents and claimed that this approach resulted in a remarkable increase in fracture energy. The theory is similar to the liquid rubber toughening. The thermoplastic modifier having reactive ends reacts with the epoxy resin. The presence of the excess epoxy resin essentially produces epoxy end-capped thermoplastic, modifier preventing further polymerization. Initially, the thermoplastic is compatible with the epoxy resin but as molecular weight increases due to the curing reaction, the homogeneous mixture starts phase separating by a spinodal decomposition resulting in the development of a two-phase microstructure.

The thermodynamic feasibility of phase separation (positive ΔG_m) at a certain stage of the curing reaction can be attributed to the substantial decrease of combinatorial entropy of mixing (ΔS_m) due to the increase in molecular weight. The extent of phase separation and hence the resulting morphology largely depend on the system viscosity and gel period. When the phase separation proceeds, the periodic distance increases simultaneously. At the same time, the phase connectivity will be interrupted by the increase of interfacial tension, resulting in dispersed droplets. Once the interruption has taken place, the dispersed droplets grow in size without changing their loci since they are dispersed in a matrix of high T_g. Further, the droplets contact one another to yield a globular structure [363,364]. A nucleation and growth mechanism has been used to describe the morphology of epoxy/thermoplastic blends. Using a suitable model, it is possible to predict the fraction, composition, and average radius of the dispersed phase segregated during the thermoset polymerization, based on the thermodynamic consideration as discussed in Chapter 4, Thermoset Composites, for rubber-modified epoxy systems and the constitutive equations of the rate of nucleation, coalescence, and growth. Another origin of phase separation is spinodal decomposition, which is usually assigned to the textures displaying some degree of connectivity, known as cocontinuous structure. Unlike rubber-modified epoxy, thermoplastic-modified epoxy systems mostly display cocontinuous morphology.

The popularity of PES as a toughening agent of epoxy is due to their good oxidative, thermal and hydrolytic stability as well as good mechanical properties [365,366]. Once the -OH-terminated bisphenol-A (BPA)-based PES was found to be successful for toughening the epoxy matrix, efforts have been concentrated on amine-terminated oligomers of PES, which can be synthesized by adding a stoichiometric amount of amino phenol as the end-capping agent [367,368]. Pac et al. [369] have used PES with pendant amino groups as the modifier for epoxy resin. Like amine-terminated PES, they can be used as such or after modification with maleic anhydride. It has been found that with an increase in $-NH_2$ content, the toughness increases initially, passes through a maximum, and then decreases. A similar observation was reported by Lin et al. [370] while investigating PEI as a modifier. The initial increase in fracture toughness with an increase in $-NH_2$ content is due to the increase in interfacial adhesion between the epoxy matrix and the dispersed PES particles, which prevents the debonding of particles. The decrease in fracture energy after an optimum $-NH_2$ concentration can be attributed to the higher miscibility of PES containing higher $-NH_2$ content with epoxy resin leading to a single-phase morphology of the modified network, which is undesirable for toughening. Other engineering thermoplastics have also been examined and found to be effective, for example, polyetherimide [371−374], poly(ether ketones) (PEK) [375,376], poly(phenylene oxide) [377−381], and liquid crystalline polymer [380,381].

Nair and coworker [382] investigated toughening of a blend of epoxy and amine-containing novolac-phthalonitrile (APN) resins with hydroxyl-terminated polyether ether ketone (PEEK) and successfully used for high-temperature adhesive applications. The adhesive property of APN, APN/epoxy blend, and APN/epoxy blend toughened with PEEK is shown in Fig. 3.27. We can see that adhesive strength is improved due to blending and further improved due to modification with PEEK. PEEK is initially miscible with the APN/epoxy blend but undergoes phase separation during curing leading to the formation of two-phase microstructure, which imparts toughening effect and enhances the adhesive strength. The toughening effect is accomplished without affecting the thermal stability required for high-temperature adhesive applications. Details mechanism of thermoplastic-toughened epoxy will be discussed in subsequent sections.

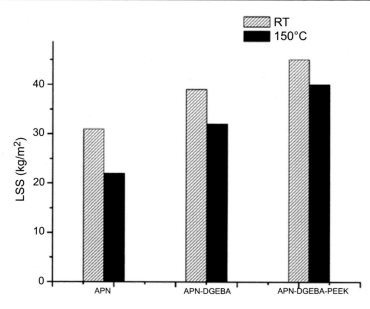

Figure 3.27
Lap shear strength for APN, APN/epoxy, and APN/epoxy/PEEK blends. Source: *Reprinted with permission from D. Augustine, K.P. Vijayalakshmi, R. Sadhana, D. Mathew, C.P. Reghunadhan Nair, Polymer 55 (2014) 6006–6016, © 2014, Elsevier Publisher.*

3.12.2 Amorphous thermoplastics

Aromatic polyesters were investigated to examine their suitability as a toughening agent for epoxy and found to be effective for a highly crosslinked system. Note that liquid rubber is less efficient in toughening highly crosslinked systems. Iijima et al. [383,384] synthesized a series of polyesters by the reaction of 1,2-ethane diol and aromatic dicarboxylic acid. The aromatic dicarboxylic acid and the derivatives contained phthalic anhydride, isophthalic acid, dimethyl terepthalate and dimethyl 2,6-napthalene dicarboxylate. These polyesters were reported to be soluble with epoxy without using any solvent at the initial stage but during the curing reaction undergo phase separation leading to the formation of a two-phase microstructure.

The T_gs of polyesters is somewhat higher than room temperature (intermediate between reactive liquid rubber and engineering thermoplastics). Their effectiveness was hardly influenced by the polyester structure and depended on polyester molecular weight and content. Maximum fracture toughness was obtained by 20 wt.% addition of polyester. Fracture resistance gradually increases with an increase in molecular weight of polyester passes through a maximum and then starts decreasing. The optimum molecular weight is dependent on the nature of the polyester.

PMS copolymer (PMS) and N-phenylmaleimide-N-cyclohexylmaleimide styrene terpolymer (PHMS) were also reported [385,386] to be effective modifiers for DGEBA resin. PMS was also an effective modifier for liquid amino cresol type trifunctional epoxy resin [387]. These copolymers can be prepared by radical polymerization and hence the high molecular weight could be easily achieved. On the other hand, it is rather difficult to get engineering thermoplastics with high molecular weight as they are generally prepared by polycondensation. In polycondensation, the molecular weight of the product is sensitive to stoichiometry and hence synthesis of high-molecular-weight polymer requires maintenance of purity level, which requires stringent purification. The studies show that the most suitable composition for modification of epoxy resin was the inclusion of 10 wt.% of PMS with weight average molecular weight (M_w) of 344,000 g mole^{-1}. The copolymers of still higher molecular weight are difficult to handle because of an increase in viscosity of the uncured epoxy mixture. The disadvantage of this particular system is that the fracture resistance was achieved at the cost of medium deterioration of flexural strength. This can be attributed to the poor interfacial adhesion between epoxy and PMS due to the absence of reactive groups in PMS and PHMS. The decrease in flexural strength was suppressed to some extent by introducing pendent hydroxyl phenyl groups using hydroxyl-containing monomer like hydroxyl styrene, N-(p-hydroxyl) phenylmaleimide [388].

Jyotishkumar et al. [389] reported epoxy resin cured with 4,4′-diaminodiphenyl sulfone (DDS) and modified with poly(acrylonitrile−butadiene−styrene) (ABS). They have studies the rheological property and cure shrinkage. It was observed that the cure blends exhibit lower shrinkage compared to pure epoxy and the cure shrinkage decreased linearly up to 6.9 wt.% ABS concentration. However, the mechanical properties of cured networks were not reported.

3.12.3 Crystalline thermoplastics

Crystalline thermoplastics have also been utilized in modifying resin. Among the crystalline thermoplastics, poly(ethylene oxide) (PEO) is most extensively investigated [390−393]. The presence of a polyether chain helps to impart flexibility. The -OH groups of PEO react with epoxy at elevated temperature and form a compatible blend with single-phase morphology or two-phase microstructure depending on the molecular weight of PEO and curing condition. Kim and Robertson [394] reported the toughening by particulate inclusions of three crystalline polymers such as poly(butylene terepthalate) (PBT), nylon 6, poly(vinylidene fluoride) (PVDF) of a crosslinked epoxy that cannot be toughened significantly by liquid rubber. This improvement in toughness was achieved without any loss of either Young's modulus or yield strength. Nicholson and Robertson [395] reported a systematic exploration of the relationship between thermal history, morphology and mechanical properties of PBT-epoxy blend. They found that 5 wt.% of a thermoplastic PBT was able to increase the fracture energy (G_{IC}) of a brittle anhydride-cured epoxy from 180 to 2000 J m^{-2} with proper control over the resulting morphology. Exceptionally higher toughening ability of PBT in comparison to nylon 6 and PVDF can be attributed to phase transformation of PBT at the crack tip [394,395].

3.12.4 Morphology and microstructural aspects

Unlike to rubber-toughened epoxy polymers, the relationship between the microstructure and toughness of thermoplastic-toughened epoxy polymers is not fully understood. A number of studies have been initiated for fracture properties and fractography of thermoplastic-toughened epoxy system in order to establish the definitive relationships between the microstructure and fracture properties. However, as discussed in detail below, the complex nature of materials precludes straightforward interpretations.

In contrast to the liquid rubber-modified epoxy systems, which tend to form a simple particulate structure, the microstructure of the thermoplastic-toughened epoxy network change from one form to the other one with an increase in thermoplastic concentration [396,397]. Initially, at a low concentration of thermoplastic in the epoxy/thermoplastic blend, the thermoplastic gets miscible in the epoxy matrix and generates a single-phase morphology. On a subsequent increase in thermoplastic concentration, the phase separation occurs leading to the development of a two-phase microstructure. The microstructure changes to particulate, cocontinuous, and finally to a phase inverted one [398,399]. In the sample containing particulate structure, the particle size increases with an increase in modifier concentration. Min et al. [398,399] identified a partially phase inverted structure (intermediate between cocontinuous and phase inverted structure) in their phenolic hydroxyl-ended polysulfone ($M_n = 10,000$ g mole^{-1}) modified epoxy system at 15 wt.% modifier concentration.

The microstructure of the modified epoxy networks largely depends on the cure condition. In the case of a multistep cured system, the final microstructure depends on initial cure temperature and postcuring condition has no role in microstructure development as phase separation gets arrested at gelation. An increase in initial cure temperature causes an increase in particle size due to a decrease in the system viscosity. These observations are similar to that observed in a rubber-toughened epoxy system. Yamanka et al. [368] observed a phase inverted cocontinuous structure by decreasing the cure temperature in their PES-modified epoxy system. They reported that a decrease in the cure temperature slowed down the rate of phase separation (based on the spinodal decomposition process) without significantly reducing the rate of the chemical reaction. This arrested the phase separation at an early stage of phase separation and finally resulted in an interconnected globular epoxy particle in a cocontinuous modifier-rich matrix. Using a PES with a reactive functional group, Mackinnon et al. [400] reported that phase separation was needed in order to record a significant increase in toughness but no discontinuities in the plot of toughness (G_{IC} or K_{IC}) versus the concentration of added thermoplastic were observed as microstructure changes from one form to another. Wilkinson et al. [401] have used a reactively terminated polysulfone to modify thermosetting BMI resins and reported that there is no correlation between toughness and various microstructures.

Considering other engineering thermoplastic Bucknall and Gilbert [378] found that phase separation was needed in order to record a significant increase in toughness of epoxy polymers modified with PEI but again did not get any discontinuities in the toughness versus the concentration of added thermoplastic as the microstructure changes from one to another. Hourtson and Lane [402] investigated the toughening of a trifunctional epoxy cured with DDS using PEI (Ultem-1000) as a modifier and reported that fracture toughness increases significantly at 15 wt.% of modifier concentration when phase inversion starts. On the other hand, Murakami et al. [372] concluded that a morphology-possessing particle of PEI-rich phase is sufficient for toughening. Kinloch et al. [403] studied the relationship between microstructure, and fracture properties of high-temperature-curing epoxy systems have been studied using PES modifier having reactive end groups and varying the concentration from 0 to 140 phr. The toughness of the system does not begin to increase until phase separation of the thermoplastic copolymer occurs (at about 8 phr). Up to this point, no significant increase in fracture toughness was observed. As the concentration of PES is further increased, the toughness value increases steadily. They reported a greater increase in toughness when the level of added thermoplastic exceeds about 55 phr and this is approximately the concentration at which the cocontinuous thermoplastic phase is first observed. However, they did not get any discontinuities as microstructure changes from particulate to cocontinuous to phase inverted one. Therefore it appears that phase separation is required in order to achieve a significant increase in the toughness of thermoplastic-modified epoxy polymers, which is in broad agreement with the observation reached by other authors [372,403].

3.12.5 Mechanism of toughening

In the case of rubber-toughened epoxy, it is well established that shear yielding by rubber cavitations and voiding is the main mechanism of toughening. In the case of thermoplastic-toughened epoxy, it is still contradictory. Kim and Brown [404] investigated the fracture properties of glassy oligomer (undisclosed composition)-modified epoxy system and suggested that plastic yielding of epoxy-rich and thermoplastic-rich phases were the active toughening mechanism. Bucknall and Gilbert [378] and Hedrick et al. [327] have concluded that ductile tearing in the thermoplastic-rich phase is the major toughening mechanism, but they did not observe any plastic yielding of the thermoplastic-rich phase. According to these views, the first stage of fracture is the brittle fracture of the epoxy-rich phase, leaving the more ductile thermoplastic-rich phase bridging behind the advancing crack front. Both the groups of workers supported their suggestion by observing the plastic drawing and ductile failure of the thermoplastic-rich phase on the fracture surfaces of the materials. Iijima et al. [361] also supported the ductile drawing and tearing mechanism studying the PMS copolymer-toughened epoxy system. However, other authors have not observed any significant ductile tearing of the thermoplastic-rich phase in their studies. Diliello et al. [381] claimed that for a TGDDM/DDS system toughened with PEI, the toughening mechanism was due to the plastic deformation of the PEI phase while it was part of the dispersed phase.

3.12.6 Effect of matrix crosslink density

In previous sections, it has been discussed that rubber toughening is extremely successful in the modification of difunctional epoxy for toughness improvement. However, a tough rubber modifier can dramatically increase the fracture toughness of difunctional epoxy they cannot toughen the multifunctional epoxy significantly. Hence it is interesting to know the performance of thermoplastic modifiers in this respect. In order to establish this comparative performance, various types of thermoplastics have been investigated using both the difunctional epoxy (DGEBA) as well as the highly crosslinked epoxies like triglycidyl amino phenol (TGAP) and tetraglycidyl diaminodiphenyl methane (TGDDM) [405,406]. Fracture energy (K_{IC}) versus PEI content plots for various epoxy systems [407] are shown in Fig. 3.28. It is clear that the toughenability of using a thermoplastic modifier is better in the case of highly crosslinked epoxies compared to the conventional difunctional epoxy. This is the advantage of thermoplastic toughening compared to rubber toughening, where toughenability decreases with increasing crosslink density and rubber modification is virtually ineffective for the highly crosslinked epoxy systems. Thus thermoplastic modifiers can toughen difunctional epoxies moderately (to a lesser extent compared to rubbery modifiers) as well as multifunctional functional epoxy systems to a significant extent (better than rubbery modifiers).

3.13 Rigid particle toughening of epoxy

The third approach generally taken to improve crack resistance of epoxy resin is the incorporation of rigid inorganic filler into the glassy epoxy matrix. Among the various rigid fillers, glass beads and alumina [408–410] have been investigated most extensively. Kinloch [411] examined epoxy resin-toughened with CTBN and glass beads. It was observed that these hybrid particle composite showed greater toughness than the

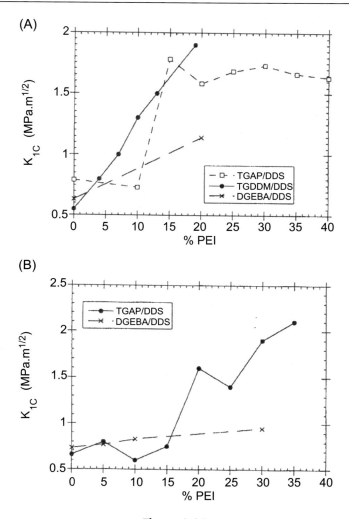

Figure 3.28
Fracture energy (K_{IC}) versus weight percent of thermoplastic modifier plots for difunctional and multifunctional epoxy systems: (A) PEI and (B) PES. *Source: Reprinted with permission from J.H. Hodgkin, G.P. Simon, R. Varley, Polym. Adv. Technol. 9 (1998) 3–10, © 1995 John Wiley and Sons Publishers.*

composition containing CTBN alone. A similar observation was reported by Geisler and Kelly [412] using alumina particles. Lange and Redford [413] used aluminum trihydrade to modify epoxy networks and found an enhancement in toughness which depended on volume fraction and particle size of the rigid filler. Molony et al. [414,415] reported a considerable improvement in fracture toughness using angular-shaped silica particles ranging from 60 to 300 mm in diameter. Cured epoxy resin filled with such silica particles in size from <1 mm to ~100 mm were used as packaging materials for integrated circuits.

Nakumura et al. [416] used angular-shaped silica, prepared by crushing fused natural raw silica. Studying the particles of six different sizes ranging from 2 to 47 mm, they found that fracture toughness increases with an increase in particle size. Nano- and microsilica particles have also been successfully used [417,418].

The energy dissipating mechanism for rigid filler-toughened epoxy is believed to be the crack pinning. According to this mechanism when a sharp crack approaches a layer of rigid particles, the crack front is pinned, perhaps due to the development of sufficient adhesion between the particles and the matrix. The crack is then bowed creating a new surface area before fracture. However, unlike rubbery filler that can cavitate, get stretched, and remain well bonded due to the chemical interaction to sustain the imposed load, the rigid filler cannot deform or cavitate and easily get debonded from the matrix leading to the catastrophic failure of the material. It is difficult also to achieve a good dispersion of such materials in the epoxy matrices. That is why the rigid particles cannot enhance the toughness of the matrix to a great extent like rubbery filler and in most cases achieving and reproducing the dispersion of such filler remains a problem.

Recently, Meng and Wang [419] described the crack propagation process for a rigid-particle-toughened thermoset system and reported an improved crack-bridging model using critical stress intensity factor to represent the fracture toughness of the composite. As mentioned above, the particle−matrix adhesion is much lesser in such composites compared to a rubber-toughened system where rubber particles remain chemically bonded with the thermoset matrix. Under tension loads, crack propagation generally originates from microcracks in the matrix, as shown in Fig. 3.29. As the load is increased, the microcracks are formed (due to the expansion as well as coalescence of microcracks) and propagate in the matrix. When the crack front encounters rigid spherical particles, the crack bows out and is blunted at discrete locations along the front, while the crack continues to propagate in the matrix. The rigid particles present in the matrix provide a hindrance to the propagation of crack and bridge the crack over the crack faces [420,421]. Once the additional load or energy overcomes the interfacial adhesion strength between the particle and the matrix, the crack face gradually opens up because of interfacial debonding. As a result, expansion of the area of the particle-bridging zone near the crack tip takes place. When the particles are pulled out completely, hemispherical holes are generated leading to an increase in roughness of the fracture surface [422]. This is similar to the fiber−matrix debonding process observed in the fiber-reinforced polymer (FRP) composites [423]. However, particle−matrix debonding is not associated with any postdebonding friction stage unlike in the case of FRP composites, due to the spherical shape of the rigid particles. It is evident from the above description that matrix cracking, particle debonding and bridging are the main energy dissipating mechanisms, which predominantly operate in a rigid-particle-toughened thermoset system.

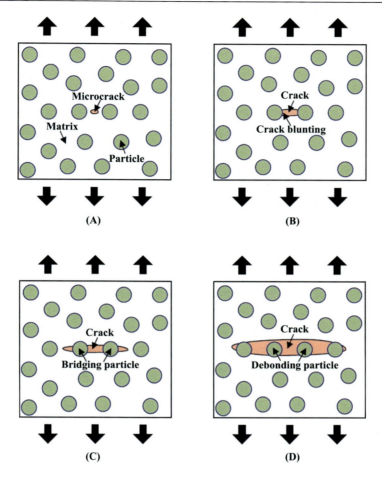

Figure 3.29
Schematic of crack propagation in rigid microparticle-reinforced polymer composites: (A) Initiating matrix microcrack, (B) matrix crack propagation, (C) crack propagation with particle bridging, and (D) steady-state crack propagation. Source: *Reprinted with permission from Q. Meng, T. Wang, Eng. Fract. Mech. 211 (2019) 291–302, © 2019, Elsevier Publisher.*

3.14 Summary and conclusion

Thermoset resins are widely used as casting, coating, adhesive, and matrices for FRP composites in various structural and functional applications as discussed in Chapter 1, Chemistry and General Applications of Thermoset Resins. However, poor fracture toughness of thermosets (much lesser compared to engineering thermoplastics) limits their application, especially in structural applications. The toughness can be improved by chemically modifying the resin with a flexible modifier or by reducing the crosslink density (long-chain hardener and epoxy with high epoxy equivalent). However, this approach is associated with a significant deterioration in thermomechanical properties. The improvement in toughness without a significant deterioration in thermomechanical properties can be achieved by incorporating a suitable filler (rubber, thermoplastic, or rigid fillers) as a second phase. The addition of rubbery filler often resulted in a dramatic increase in fracture toughness of difunctional epoxy, which can be attributed to the energy dissipating mechanism, namely, rubber cavitation followed by shear yielding. The rubber toughening is generally achieved by blending liquid rubber, which is initially miscible with the epoxy and undergoes phase separation with the advancement of the curing reaction leading to the formation of a two-phase microstructure. The toughening effect depends on the morphological parameters, namely, particle size, particle size distribution, matrix ligament thickness, particle to matrix adhesion, which are controlled by molecular parameters such as molecular weight and functionality of the rubber and epoxy equivalent of the resin. Initial cure temperature also plays an important role in deciding the morphology. However, rubbery fillers are not effective in toughening multifunctional epoxy with higher crosslink density. Multifunctional epoxy resins are successfully toughened by using thermoplastic-based toughening agents. Rigid fillers like alumina and glass beads are also reported to enhance the epoxy network by matrix cracking, particle debonding, and bridging but they are not as effective as rubbery fillers. Block copolymers have been viewed as potential candidates for toughening of thermoset resins, which will be discussed in Chapter 5, Thermoset Nanocomposites, because they self-assemble in a thermoset matrix, leading to the formation of a nanostructure morphology.

References

[1] A.J. Kinloch, R.J. Young, Fracture Behavior of Polymers, Applied Science, London, 1983.
[2] D. Ratna, Rapra Rev. Rep. 16 (2005) 1.
[3] S.N. Tong, D.S. Chen, T.K. Kwei, Polym. Eng. Sci. 25 (1985) 54.
[4] P.J. Madec, E. Marecha, J. Polym. Sci. Polym. Chem. (Ed.) 16 (1978) 3165.
[5] S.N. Tong, C. Chen, T.K. Peter, Rubber Toughened Plastics, in: C.K. Riew (Ed.), Advances in Chemistry Series, 222, American Chemical Society, Washington, DC, 1989, p. 375.
[6] J.M. Brown, S. Srinivasan, A. Rau, T.C. Word, J.E. MaGrath, A.C. Loos, et al., Polymer 37 (1996) 1691.
[7] J. Borrajo, C.C. Riccardi, R.J.J. Williams, Z.Q. Cao, J.P. Pascault, Polymer 36 (1995) 3541.
[8] A.J. Kinloch, S.J. Shaw, D.A. Tod, Rubber Modified Thermoset Resins, in: C.K. Riew, J.K. Gillham (Eds.), Advances in Chemistry Series, 208, American Chemical Society, Washington, DC, 1984, p. 101.
[9] C.B. Bucknall, Toughend plastics, Applied Science Publishers Ltd., London, 1977.
[10] D. Ratna, J. Adhes. Sci. Technol. 17 (2003) 1655.
[11] H. Keskkula, A.E. Platt, R.F. Boyer, 2nd ed., Encyclopedia of Chemical Technology, 19, Wiley Interscience, New York, 1969, p. 85.
[12] J.L. Amos, Polym. Engg. Sci. 14 (1974) 1.
[13] L.E. Daly, US Patent 2439202 (1948).
[14] C.W. Childers, C.F. Fisk, US Patent 2773 (1958).
[15] W.A. Semon, G.A. Stahl, History of Polymer Science and Technology, 1, Wiley Interscience, New York, 1985, p. 338.

[16] D.M. Kulich, P.D. Kelley, J.E. Pace, Encyclopedia of Polymer Science and Technology, 1, John Wiley and sons, New York, 1985, p. 338.
[17] A.W. Carlson, T.A. Zones, J.L. Martin, Modified. Plast. 44 (1957) 155.
[18] R.G. Bauer, R.M. Pierson, W.C. Mast, N.C. Blatso, L. Shepherd, Mullticomponent Polymer Systems, Advances in Chemistry Series, 199, American Chemical Society, Washington, DC, 1971, p. 251.
[19] B.D. Gesner, J. Appl. Polym. Sci. 11 (1967) 2499.
[20] C.J.G. Plumer, P. Bengnelin, H.H. Kausch, Polymer 37 (1996) 7.
[21] B.Z. Jang, D.R. Ulhmann, J.B. Vandersande, J. Appl. Polym. Sci. 30 (1985) 2485.
[22] A.K. Gupta, S.N. Purwar, J. Appl. Polym. Sci. 30 (1984) 1799.
[23] A.K. Gupta, S.N. Purwar, J. Appl. Polym. Sci. 31 (1984) 535.
[24] A.K. Gupta, K.R. Srinivasan, P. Krishnakumar, J. Appl. Polym. Sci. 42 (1991) 2595.
[25] C.B. Bucknall, C.J. Page, J. Mater. Sci. 17 (1982) 808.
[26] F.C. Stehling, T. Huff, C.S. Speed, G. Wissler, J. Appl. Polym. Sci. 26 (1981) 2693.
[27] A.F. Yee, J. Mater. Sci. 12 (1977) 757.
[28] A.F. Yee, W.V. Olszewski, S. Miller, in: R.D. Deanin, A.M. Crugnala (Eds.), Toughness and Brittleness of Plastics, Academic Press, New York, 1976, p. 97.
[29] J. Karger-Kocsis, I. Csikai, Polym. Engg. Sci. 27 (1987) 241.
[30] C.K. Riew, R.W. Smith, Rubber toughened plastics, in: C.K. Riew (Ed.), Advances in Chemistry Series, 222, American Chemical Society, Washington, DC, 1989, p. 225.
[31] H.E. Bair, Polym. Engg. Sci. 10 (1970) 247.
[32] M. Kramer, Appl. Polym. Symp. 15 (1971) 227.
[33] I. Park, H. Keskula, D.R. Paul, J. Appl. Polym. Sci. 45 (1992) 1313.
[34] S. Wu, J. Appl. Polym. Sci. 35 (1988) 549.
[35] S. Wu, Polymer 26 (1955) 1855.
[36] R.J.M. Borggreve, R.J. Gaymans, J. Schuijer, J.F. Ingen Housz, Polymer 28 (1987) 1489.
[37] A. Echte, Rubber Toughened Plastics, in: C.K. Riew (Ed.), Advances in Chemistry Series, 222, American Chemical Society, Washington, DC, 1989, p. 15.
[38] G.E. Molau, H. Keskkula, J. Polym. Sci. P A 1 (4) (1966) 1595.
[39] R.N. Haward, J. Mann, Proc. R. Soc. A 282 (1964) 120.
[40] A.J. Kinloch, J. Mater. Sci. 15 (1980) 2141.
[41] H. Keskkula, in: K. Solc (Ed.), Polymer Compatibility and Incompatibility, Harwood Academic Publishers, London, 1982, pp. 323–354.
[42] N.R. Legge, S. Devison, H.E. De La Mare, G. Halden, M.K. Martin, Am. Chem. Soc. Symp. Ser. 285 (1985) 175.
[43] J.R. Campbell, S.Y. Hobbs, T.J. Shea, D.J. Smith, Topics in Polymer Science, in: B.M. Culbertson (Ed.), Multiphase Macromolecular Systems, 6, Plenum, New York, 1990, p. 439.
[44] S.L. Aggarwall, Polymer 17 (1976) 938.
[45] A.L. Bull, G. Molden, J. Elastom. Plast. 9 (1977) 281.
[46] A.F. Yee, J. Diamant, Polym. Prep. 9 (1978) 92.
[47] W.J. Coumans, D. Heikens, S.D. Sjoerdsma, Polymer 21 (1980) 103.
[48] S.L. Aggarwall, R.A. Livigni, Polym. Engg. Sci. 45 (1992) 1313.
[49] D. Trifonova, S. Vasileva, J. Mater. Sci. 27 (1992) 3657.
[50] S. Newman, S. Strella, J. Appl. Polym. Sci. 9 (1965) 2297.
[51] F. Haff, H. Breuer, J. Stabenow, J. Macromol. Sci. Phys. B-14 (1977) 387.
[52] A.M. Donald, E.J. Kramer, J. Mater. Sci. 17 (1982) 1765.
[53] G. Signa, P. Lomellini, M. Merlotti, J. Appl. Polym. Sci. 37 (1989) 1527.
[54] F.A. Flexman, Polym. Eng. Sci. 19 (1979) 564.
[55] S.Y. Hobbs, R.C. Bopp, V.H. Watkins, Polym. Eng. Sci. 23 (1983) 380.
[56] G. Weber, J. Schoeps, Angew. Makromol. Chem. 136 (1985) 45.
[57] F. Speroni, E. Castoldi, P. Fabri, T. Casirangi, J. Mater. Sci. 24 (1989) 2165.

[58] T. Kunori, P.H. Geil, J. Macromol. Sci. Phys. B18 (1980) 135.
[59] D.S. Parker, H.J. Sue, J. Huang, A.F. Yee, Polymer 31 (1985) 2267.
[60] F.C. Chang, J.S. Wu, L.H. Chu, J. Appl. Polym. Sci. 44 (1992) 491.
[61] R.A. Bubech, D.J. Buckley, E.J. Kramer, H.R. Brown, J. Mater. Sci. 26 (1991) 49.
[62] H. Breuer, F. Haff, J. Stabenow, J. Macromol. Sci. Phys. 14 (1977) 387.
[63] F. Haff, H. Breuer, A. Echte, B.J. Schmitt, J. Stabenow, J. Sci. Ind. Res. 40 (1981) 659.
[64] A. Tse, E. Shin, A. Hittner, E. Baer, R. Laakso, J. Mater. Sci. 26 (1991) 2823.
[65] D.J. Hourston, S. Lane, H.X. Zhang, Polymer 32 (1991) 2215.
[66] D. Dijkstra, PhD thesis, University of Twente, The Netherlands (1993).
[67] R.J.M. Borggreve, R.J. Gaymans, J. Schuijer, Polymer 30 (1989) 71.
[68] R.J.M. Borggreve, R.J. Gaymans, H.M. Eichenwald, Polymer 30 (1989) 78.
[69] E.J. Kramer, L.L. Berger, Adv. Polym. Sci. 91 (1990) 1.
[70] D. Tompas, G. Groeninckx, Polymer 35 (1994) 4743.
[71] D. Tompas, G. Groeninckx, Polymer 35 (1994) 4751.
[72] Y. Huang, A.J. Kinloch, Polymer 33 (1992) 5338.
[73] M. Morton, M. Cizmecioglu, R. Lhila, Adv. Chem. Ser. 206 (1984) 221.
[74] D.L. Dunkelberger, E.P. Doughterty, J. Vinyl Technol. 12 (1990) 212.
[75] A.J. Oostenbrink, K. Dijkstra, S. Wiegersma, A.V.D. Wal, R.J. Gaymans, PRI International Conference on Deformation, Yield and Fracture of Polymers, Cambridge (April 1990).
[76] A.J. Oshinski, H. Keskula, D.R. Paul, Polymer 33 (1992) 268.
[77] S. Wu, Polym. Int. 29 (1992) 229.
[78] C. Wrotecki, P. Heim, P. Gaillard, Polym. Eng. Sci. 31 (1991) 213.
[79] H. Liang, W. Jiang, J. Zhang, J. Bingzheng, J. Appl. Polym. Sci. 59 (1996) 505.
[80] E.J. Kramer, L.L. Berger, Advances in Polymer Science, 91/92, in: H.H. Kausch (Ed.), Crazing in Polymers, 2, Springer-Verlag, Berlin, 1990, p. 1.
[81] C.B. Bucknall, R.R. Smith, Polymer 6 (1965) 437.
[82] M. Matuo, Polymer 7 (1966) 421.
[83] R.P. Kambour, D.R. Russel, Polymer 12 (1971) 237.
[84] M. Matuo, A. Ueda, Y. Kondon, Polym. Eng. Sci. 10 (1970) 253.
[85] J.D. Moore, Polymer 12 (1971) 478.
[86] C.B. Bucknall, D. Clayton, W.E. Keast, J. Mater. Sci. 7 (1972) 1443.
[87] M. Matsuo, C. Nozaki, Y. Jyo, Polym. Eng. Sci. 9 (1969) 197.
[88] P. Beahan, A. Thomas, M. Bevis, J. Mater. Sci. 11 (1976) 1207.
[89] K. Kato, J. Electron. Micros. 14 (1968) 220.
[90] K. Kato, Polym. Eng. Sci. 7 (1967) 38.
[91] H.H. Kausch, Polymer Fracture, 2nd ed., Springer Verlag, Berlin, 1987.
[92] I. Narisawa, M. Ishikawa, H. Ogata, J. Mater. Sci. 15 (1980) 2059.
[93] G.H. Michler, Colloid Polym. Sci. 267 (1989) 377.
[94] R.N. Haward, C.B. Bucknall, Pure Appl. Chem. 46 (1976) 227.
[95] I.V. Yannas, R.R. Luise, J. Macromol. Sci. Phys. B-21 (1982) 443.
[96] H. Keskkula, M. Schwarz, D.R. Paul, Polymer 27 (1986) 211.
[97] M.A. Maxwell, A.F. Yee, Polym. Eng. Sci. 21 (1981) 205.
[98] S.T. Wellinghoff, E. Baer, J. Appl. Polym. Sci. 22 (1978) 2025.
[99] S. Wu, J. Polym. Sci. B. Polym. Phys. 21 (1983) 699.
[100] S. Wu, A. Mongalina, Polymer 31 (1990) 972.
[101] R.J.M. Borggreve, R.J. Gaymans, A.R. Luttmer, Makromol. Chem. Macrromol. Symp. 16 (1988) 195.
[102] G.D. Cooper, G.F. Lee, A.K. Katchman, C.P. Sannk, Mater. Technol. 12 (1981). Spring.
[103] C.B. Bucknall, I.C. Drinkwater, J. Mater. Sci. 8 (1973) 1800.
[104] R.R. Durst, R.M. Griffith, A.J. Arbenic, W.J. Van Essen, Adv. Chem. Ser. 154 (1976) 239.
[105] D. Ratna, R. Varley, G.P. Simon, J. Appl. Polym. Sci. 89 (2003) 2339.

[106] D. Ratna, A.K. Banthia, Macromol. Res. 12 (2004) 11.
[107] S. Manternal, J.P. Pascault, H. Sautereau, Rubber Toughened Plastics, in: C.K. Riew (Ed.), Advances in Chemistry Series, 222, American Chemical Society, Washington, DC, 1989, p. 193.
[108] D. Verchere, J.P. Pascault, H. Sautereau, S.M. Moschair, C.C. Riccardi, R.J.J. Williams, J. Appl. Polym. Sci. 42 (1991) 701.
[109] P.J. Flory, Principles of Polymer Chemistry, Cornell University Press, Ithaca, NY, 1975.
[110] A.C. Garg, Y.W. Mai, Comp. Sci. Technol. 31 (1988) 179.
[111] S. Kunz-Douglass, P.W.R. Beamont, M.F. Ashby, J. Mater. Sci. 15 (1980) 1109.
[112] S. Kunz-Douglass, P.W.R. Beamont, J. Mater. Sci. 16 (1981) 3141.
[113] J.A. Sayre, S.C. Kunz, R.A. Assink, Rubber Modified Thermoset Resins, in: C.K. Riew, J.K. Gillham (Eds.), Advances in Chemistry Series, 208, American Chemical Society, Washington, DC, 1984, p. 215.
[114] A.N. Gent, Science and Technology of Rubber, Academic Press, New York, 1978, p. 419.
[115] R.A. Pearson, A.F. Yee, J. Mater. Sci. 24 (1989) 2571.
[116] A.C. Meeks, Polymer 15 (1974) 675.
[117] A.C. Soldatos, A.S. Burhans, Advances in Chemistry Series, 99, American Chemical Society, Washington, DC, 1971, p. 531.
[118] A.M. Donald, E.J. Kramer, J. Mater. Sci. 17 (1982) 1871.
[119] E.J. Kramer, Advances in Polymer Science, in: H.H. Kausch (Ed.), Advances in Polymer Science Series No. 52/53, 52/53, Springer-Verlag, Berlin, 1983. Chapter 1.
[120] A. Van den Boogaart, in: Z.R.N. Haward (Ed.), Physical Basis of Yield in Glassy Polymers, Institute of Physics, London, 1979.
[121] J.N. Sultan, R.C. Liable, F.J. McGarry, Polym. Symp. 16 (1971) 127.
[122] J.N. Sultan, F.J. McGarry, Polym. Eng. Sci. 13 (1973) 29.
[123] J. Lilley, D.G. Halloway, Phil. Mag. 28 (1973) 215.
[124] R.J. Morgan, J.E. O' Neel, J. Mater. Sci. 12 (1982) 295.
[125] R.J. Morgan, E.T. Mones, W.J. Steele, Polymer 23 (1982) 295.
[126] R.J. Morgan, J.E. O' Neel, D.B. Miller, J. Mater. Sci. 14 (1979) 109.
[127] A.F. Yee, R.A. Pearson, J. Mater. Sci. 21 (1986) 2462.
[128] R.A. Pearson, A.F. Yee, J. Mater. Sci. 21 (1986) 2475.
[129] C.B. Bucknall, F.F.P. Cote, I.K. Partdridge, J. Mater. Sci. 21 (1986) 301.
[130] C.B. Bucknall, T. Yoshii, Yield and fracture of polymers, in: Proceedings of the International Conference on Deformation, Plastic and Rubber Institute, Cambridge, 1976, p. 131.
[131] R. Ramsteiner, Polymer 20 (1979) 839.
[132] J. Mijovik, J.A. Koutsky, Polymer 20 (1979) 1095.
[133] R.J. Young, in: E.H. Andrews (Ed.), Development of Polymer Fracture, vol. 1, Applied Science, London, 1979, p. 183.
[134] W.D. Bascom, D.L. Hunston, Proceedings of the International Conference on Toughening of Plastics, Plastic and Rubber Institute, London, 1978, p. 22.1.
[135] W.D. Bascom, D.L. Hunston, Rubber Toughened Plastics, in: C.K. Riew (Ed.), Advances in Chemistry Series, 222, American Chemical Society, Washington, DC, 1989, p. 135.
[136] A.J. Kinloch, S.J. Shaw, D.A. Tod, D.L. Hunston, Polymer 24 (1983) 1355.
[137] A.J. Kinloch, D.L. Hunston, J. Mater. Sci. Lett. 5 (1986) 909.
[138] A.J. Kinloch, Rubber Toughened Plastics, in: C.K. Riew (Ed.), Advances in Chemistry Series, 222, American Chemical Society, Washington, DC, 1989, p. 67.
[139] A.J. Kinloch, S.J. Shaw, D.A. Tod, D.L. Hunston, Polymer 24 (1983) 1341.
[140] D.L. Hunston, J.L. Bitner, J.L. Rushford, J. Oroshnik, W.S. Rose, Elast. Plast. 12 (1980) 133.
[141] C.C. Chau, J.C.M. Lii, J. Mater. Sci. 16 (1981) 1858.
[142] D.R. Uhlmann, Faraday Disc 68 (1979) 87.
[143] A.L. Gurson, J. Eng. Mater. Tech. Trans. ASME 99 (1977) 2.
[144] A. Needleman, J. Appl. Mech. Trans. ASME 39 (1972) 964.

[145] H. Yamamoto, Int. J. Fract. 14 (1978) 347.
[146] V. Tvergaard, Int. J. Fract. 17 (1981) 389.
[147] M.E.J. Dekkers, S.Y. Hobbs, V.H. Watkins, J. Mater. Sci. 23 (1988) 1225.
[148] F.J. Guild, R.J. Young, J. Mater. Sci. 24 (1989) 2454.
[149] D. Li, A.F. Yee, I.W. Chen, S.C. Chang, K. Takahashi, J. Mater. Sci. 29 (1994) 2205.
[150] A.G. Evans, Z.B. Ahmad, D.G. Gilbert, P.W.R. Beaumont, Acta Metall. 34 (1986) 79.
[151] K.T. Faber, A.G. Evans, Acta Metall. 31 (1983) 565.
[152] A.F. Yee, R.A. Pearson, J. Mater. Sci. 26 (1991) 2273.
[153] Y. Huang, A.J. Kinloch, J. Mater. Sci. 27 (1992) 2763.
[154] Y. Huang, A.J. Kinloch, J. Mater. Sci. Lett. 11 (1992) 484.
[155] C.B. Bucknall, T. Yoshii, Br. Polym. J. 10 (1978) 53.
[156] S.C. Kunz, J.A. Sayre, R.A. Assink, Polymer 23 (1982) 1897.
[157] A.J. Kinloch, D.L. Huntston, J. Mater. Sci. Lett. 6 (1987) 137.
[158] A.J. Kinloch, G.A. Kodokian, M.B. Jamarani, J. Mater. Sci. 22 (1987) 4111.
[159] B.J. Cardwell, A.F. Yee, Polymer 34 (8) (1993) 1695.
[160] H. Huange, A.J. Kinloch, J. Adhes. 41 (1993) 5.
[161] D.L. Hunston, A.J. Kinloch, S.J. Shaw, S.S. Wang, in: K.L. Mittal (Ed.), Adhesive Joints, Plenum, NY, 1984, p. 789.
[162] K. Fulin, U.L. Chung, J. Mater. Sci. 29 (1994) 1198.
[163] A.R. Siebert, Rubber Modified Thermoset Resins, in: C.K. Riew, J.K. Gillham (Eds.), Advances in Chemistry Series, 208, American Chemical Society, Washington, DC, 1984, p. 179.
[164] A. Lowe, O.H. Kwon, Y.H. Mai, Polymer 37 (4) (1996) 565.
[165] H.R. Daghyani, L. Ye, Y.W. Mai, J.S. Wu, J. Mater. Sci. Lett. 13 (18) (1994) 1330.
[166] L.C. Chen, J.K. Gillham, S.J. Kinloch, S.J. Shaw, Rubber Modified Thermoset Resins, in: C.K. Riew, J.K. Gillham (Eds.), Advances in Chemistry Series, 208, American Chemical Society, Washington, DC, 1984, p. 261.
[167] Y.B. Zeng, L.Z. Zang, W.Z. Peng, Q. Yu, J. Appl. Polym. Sci. 42 (1991) 1905.
[168] R.A. Pearson, A.F. Yee, J. Mater. Sci. 26 (1991) 3828.
[169] A. Mongolina, S. Wu, Polymer 29 (1988) 2170.
[170] S.D. Sjoerdsma, Polym. Commun. 30 (1989) 106.
[171] D. Tompas, G. Groeninckx, Polymer 35 (1994) 4760.
[172] J.P. Bell, J. Appl. Polym. Sci. 14 (1970) 1901.
[173] J.P. Bell, J. Polym. Sci. A-2 (6) (1987) 13.
[174] J.D. Lemay, B.J. Swetlin, F.N. Kelley, Characterization of highly cross-linked polymer, in: S.S. Labana, R.A. Dickie (Eds.), ACS Symposium Series 243, Washington, DC, American Chemical Society, 1984, p. 165.
[175] S.L. Kim, M.D. Skibo, J.A. Manson, R.W. Hertzberg, J. Janisjewski, Polym. Eng. Sci. 18 (1978) 1093.
[176] B.W. Cherry, K.W. Thomson, J. Mater. Sci. 16 (1981) 1913.
[177] R.A. Pearson, A.F. Yee, Polym. Mater. Sci. Eng. Prepr. 186 (1983) 316.
[178] G. Levita, A. Marchetti, E. Butta, Polymer 26 (1985) 1110.
[179] G. Levita, A. Marchetti, A. Lazzeri, V. Frosini, Polym. Compos. 8 (1987) 141.
[180] G. Lavita, S. Depetris, A. Marchetti, A. Lazzeri, J. Mater. Sci. 26 (1991) 2348.
[181] G. Levita, Rubber Toughened Plastics, in: C.K. Riew (Ed.), Advances in Chemistry Series, 222, American Chemical Society, Washington, DC, 1989, p. 93.
[182] G.A. Crosbie, M.G. Philips, J. Mater. Sci. 20 (1985) 182.
[183] K.A. Hodd, Rapra Rev. Rep. 38 (2) (1990) 2.
[184] Y. Huang, A.J. Kinloch, R. Bertsch, A.R. Siebert, Advances in Chemistry Series, No. 233, C.K. Riew, A.J. Kinloch (Eds.) (1993), p. 189.
[185] J.S. Ullett, R.P. Chartoff, Polym. Eng. Sci. 35 (1995) 1086.
[186] J.S. Ullett, R.P. Chartoff, Polym. Mater. Sci. Eng. 70 (1993) 100.

[187] A.R. Siebert, C.D. Guiley, A.J. Kinloch, M. Fernando, E.P.L. Heijnsbrock, in: C.K. Riew, A.J. Kinloch (Eds.), Toughened Plastics II: Novel Approaches in Science and Engineering, American Chemical Society, Washington, DC, 1996, pp. 151–160.
[188] S. Phas, P.J. Burhcill, Polymer 36 (1995) 3279.
[189] P.J. Burchill, A. Kootsooks A, M. Lau, J. Mater. Sci. 36 (2001) 4239.
[190] P.J. Pearce, A.R. Siebert, D.R. Egan, C.D. Guiley, R.S. Drake, J. Adhes. 49 (1995) 245.
[191] D. Ratna, T. Abraham, J. Karger-Kocsis, Macromol. Chem. Phys. 209 (2008) 723–733.
[192] J. Karger-Kocsis, J. Frohlich, O. Gryshchuk, H. Kautz, H. Frey, R. Mulhaupt, Polymer 45 (4) (2004) 1185.
[193] O. Gryshchuk, N. Jost, J. Karger-Kocsis, Polymer 43 (17) (2002) 4763.
[194] G.Z. Liang, M.X. Zhang, J. Appl. Polym. Sci. 85 (2002) 2377.
[195] M. Malinconico, E. Martuscelli, G. Ragosta, M.G. Volpi, Inter. J. Polym. Mat. 11 (1987) 317.
[196] E. Dreerman, M. Narkis, A. Siegmann, R. Joseph, H. Dodiuk, A.T. Dibenedetto, J. Appl. Polym. Sci. 72 (1999) 647.
[197] F.H. Liao, N.J. Chu, S.N. Tong, J. Appl. Polym. Sci. 35 (1988) 797.
[198] J. Karger-Kocsis, O. Gryshchuk, N. Jost, J. Appl. Polym. Sci. 88 (8) (2003) 2124.
[199] V.A. Pattison, R.R. Hindersinn, W.T. Schwartz, J. Appl. Polym. Sci. 18 (1974) 2763.
[200] Y.S. Yang, L.J. Lee, Polymer 29 (1988) 1793.
[201] K.E. Atkins, in: H.G. Kia (Ed.), Sheet Molding Compounds: Science and Technology, Hanser, New York, 1993. Chapter 4.
[202] L. Suspene, D. Fourquier, Y.S. Yang, Polymer 32 (1991) 1593.
[203] C.B. Bucknall, I.K. Partridge, M.J. Phillips, Polymer 32 (1991) 638.
[204] Y.J. Huang, C.M. Liang, Polymer 37 (1996) 401.
[205] Y.J. Huang, C.J. Chu, J.P. Dong, J. Appl. Polym. Sci. 78 (2000) 543.
[206] Y.J. Huang, T.S. Chen, J.G. Huang, F.H. Li, J. Appl. Polym. Sci. 89 (2003) 3336.
[207] M. Kinkelaar, B. Wang, L. Lee, J. Polym. 35 (1994) 3011.
[208] W. Li, J. Lee, Polymer 41 (2000) 697.
[209] N. Boyard, M. Vayer, C. Sinturel, R. Erre, J. Appl. Polym. Sci. 95 (2005) 1459.
[210] X. Bulliard, V. Michaud, J.-A.E. Manson, J. Appl. Polym. Sci. 102 (2006) 3877.
[211] M. Takayanagi, K. Imada, T. Kojiyama, J. Polym. Sci. Pt C. 15 (1966) 263.
[212] Y.J. Huang, J.C. Horing, Polymer 39 (1998) 3683.
[213] V.A. Pattison, R.R. Hindersinn, W.T. Schwartz, J. Appl. Polym. Sci. 19 (1975) 3045.
[214] Z. Zhang, S. Zhu, Polymer 41 (2000) 3861.
[215] J.P. Dong, J.G. Huang, F.H. Lee, J.W. Roan, Y.J. Huang, J. Appl. Polym. Sci 91 (2004) 3388.
[216] J.P. Dong, S.G. Chiu, M.W. Hsu, Y.J. Huang, Appl. Polym. Sci. 100 (2006) 967.
[217] L. Suspene, Y.Y. Show, P.J. Pierre, Rubber Toughened Plastics, in: C.K. Riew, A.J. Kinloch (Eds.), Advances in Chemistry Series 233, American Chemical Society, Washington, DC, 1993.
[218] V.M. Rosa, J. Karger-kocsis, M.I. Felisberti, J. Appl. Polym. Sci. 81 (2001) 3280.
[219] F.J. McGarry, E.H. Rowe, C.K. Riew, Polym. Eng. Sci. 18 (1978) 2.
[220] C.K. Riew, Rubber Toughened Plastics, Advances in Chemistry Series 230, American Chemical Society, Washington, DC, 1989.
[221] H.T. Chiu, S.C. Chen, J. Polym. Res. 8 (2001) 183.
[222] B. Gawdzik, T. Matynia, E. Chmielewska, J. Appl. Polym. Sci. 82 (2001) 2003.
[223] M. Messori, M. Toselli, F. Pilati, C. Tonelli, Polymer 42 (2001) 9877.
[224] Z.G. Shaker, R.M. Browne, H.A. Stretz, P.E. Cassidy, M.T. Blanda, J. Appl. Polym. Sci. 84 (2002) 2283.
[225] Y.C. Chou, L. Lee, J. Polym. Eng. Sci. 35 (1995) 976.
[226] L. Valette, C.-P. Hsu, Polymer 40 (1999) 2059.
[227] K.C. Frisch, D. Klempner, H.L. Frisch, Polym. Eng. Sci. 22 (1982) 1143.
[228] P.R. Scarito, L.H. Sperling, Polym. Eng. Sci. 19 (1979) 297.
[229] J.K. Yeo, L.H. Sperling, D.A. Thomas, J. Appl. Polym. Sci. 26 (1983) 3283.
[230] A.B. Cherian, B.T. Abraham, E.T. Thachil, J. Appl. Polym. Sci. 100 (2006) 449.

[231] Y. Takase, M. Nishi, T. Yamagishi, Y. Nakamoto, Polym. Prepr. 54 (2005) 397.
[232] T.B. Iyim, J. Appl. Polym. Sci. 106 (2007) 46.
[233] N.G. DeLollis, Adhesives, Adherend, Adhesion, Kienger Publishing Company Inc., New York, 1980, p. 105.
[234] P.S. Achary, G. Chockalingam, R. Ramaswamy, J. Adhes. Sci. Technol. 6 (1988) 415.
[235] P.S. Achary, R. Ramaswamy, J. Appl. Polym. Sci. 69 (1998) 1187.
[236] D.R. Paul, S. Newman, Polymer Blends, II, Academic Press, New York, 1978, p. 9.
[237] P.S. Achary, G. Chockalingam, R. Ramaswamy, J. Appl. Polym. Sci. 69 (1998) 1187.
[238] C.C. Kaynak, O. Agatay, Polym. Test. 25 (2006) 296.
[239] G. Camino, E. Alba, P. Buonfico, J. Appl. Polym. Sci. 82 (2001) 1346.
[240] J.T. Carter, Plast. Rubber Compos. Process. Appl. 16 (1991) 157.
[241] H. Wu, C.C. Ma, J. Lin, J. Appl. Polym. Sci. 63 (1997) 911.
[242] T. Gietl, H. Lengsfeld, V. Altstadt, J. Mater. Sci. 41 (2006) 8226.
[243] D. Ratna, J. Karger-Kocsis, J. Appl. Polym. Sci. 127 (2013) 4039–4043.
[244] T. Iijima, H. Shiono, W. Fukuda, M. Tomoi, J. Appl. Polym. Sci. 65 (1997) 1349.
[245] T. Iijima, T. Maeda, M. Tomoi, J. Appl. Polym. Sci. 74 (1999) 2931.
[246] C. Yang, A. Gu, H. Song, Z. Fang, L. Tong, J. Appl. Polym. Sci. 105 (2007) 2020.
[247] H. Qin, P.T. Mather, J.-B. Baek, L.-S. tan, J. Polym. 47 (2006) 2813–2821.
[248] X. Wang, L. Zong, J. Han, J. Wang, C. Liu, X. Jian, Polymer 121 (2017) 217–227.
[249] C.A. May, G.Y. Tanka, Epoxy Resin Chemistry and Technology, Marcel Dekker, New York, 1973.
[250] D. Ratna, A.K. Banthia, Polym. Int. 49 (2000) 281.
[251] R.S. Bouer, Epoxy Resin Chemistry, Advances in Chemistry Series, 114, American Chemical Society, Washington, DC, 1979.
[252] P.S. Achary, P.B. Latha, R. Ramaswamy, J. Appl. Polym. Sci. 41 (1990) 151.
[253] D. Ratna, M. Patri, B.C. Chakraborty, C. Madhavan, P.C. Deb, Bull. Mater. Sci. 18 (1995) 1013–1019.
[254] D. Ratna, A.B. Samui, B.C. Chakraborty, Polym. Int. 53 (2004) 1882.
[255] D. Ratna, B.C. Chakraborty, P.C. Deb, J. Polym. Mater. 14 (1997) 185.
[256] D. Ratna, M. Patri, B.C. Chakraborty, P.C. Deb, J. Appl. Polym. Sci. 65 (1997) 901.
[257] W.A. Romanchick, J.E. Sohn, J.F. Gaibel, Epoxy Resins Chemistry II, in: R.S. Bauer (Ed.), Advances in Chemistry Series, 221, American Chemical Society, Washington, DC, 1983, p. 93.
[258] Wolverine Gasket Co., Plastic Technol. 34 (1988) 13.
[259] F.J. McGarry, Proc. R. Soc. A 319 (1970) 59.
[260] R. Bagheri, B.T. Marouf, R.A. Pearson, J. Macromol. Sci. C 49 (2009) 201.
[261] K.P. Unnikrishnan, E.T. Thachil, Des. Monomers. Polym. 9 (2006) 129.
[262] H.S. Zhou, S.A. Xu, Mater. Lett. 121 (2014) 238.
[263] H.S. Zhou, X.X. Song, S. Xu, J. Appl. Polym. Sci. 131 (2014) 41110.
[264] Composites and Adhesive Newsletter, NASA, 7 (1990), p. 4.
[265] S. Koushal, Kishore, J. Mater. Sci. Lett. 11 (1992) 86.
[266] S. Koushal, K. Tankala, R.M.V.G.K. Rao, Kishore, J. Mater. Sci. 26 (1991) 6293.
[267] T.T. Wang, H.M. Zupko, J. Appl. Polym. Sci. 15 (1981) 1211.
[268] K. Nakao, K. Yamanka, J. Adhes. 37 (1992) 15.
[269] C.K. Riew, Rubber Chem. Technol. 58 (1985) 622.
[270] E.H. Rowe, A.R. Siebert, R.S. Darke, Modified. Plast. 47 (1970) 110.
[271] C.K. Riew, E.H. Rowe, A.R. Siebert, Toughness and Brittleness of Plastics, in: R.D. Deanin, A.M. Crugnala (Eds.), Advances in Chemistry Series, 154, American Chemical Society, Washington, DC, 1976, p. 326.
[272] C.K. Riew, Rubber Chem. Technol. 54 (1981) 374.
[273] A.V. Cunliffe, M.B. Huglin, P.J. Pearce, D.H. Rechards, Polymer 16 (1975) 622.
[274] A.H. Rezaiford, K.A. Hodd, D.A. Tod, J.M. Barton, Int. J. Adhes. Adhes. 14 (1994) 153.
[275] R. Slysh, Epoxy Resins, Advances in Chemistry Series, 92, American Chemical Society, Washington, DC, 1970, p. 108.

[276] T.J. Kemp, A. Wilford, O.W. Howarth, T.C.P. Lee, Polymer 33 (1992) 1860.
[277] R.A. Peter, T.J. Logan, Adhesive Age. 17 (1975).
[278] K. Mizutani, J. Mater. Sci. 28 (1993) 2178.
[279] E.M. Yorkgitis, et al., Rubber modified thermoset resins, in: C.K. Riew, J.K. Gillham (Eds.), Advances in Chemistry Series, 208, American Chemical Society, Washington, DC, 1984, p. 137.
[280] E.M. Yorkgitis, N.S. Eiss, C. Tran, G.L. Wilkes, J. McGrath, Adv. Polym. Sci. 72 (1985) 79.
[281] J.L. Hedrick, B. Haider, T. Russel, D.C. Hofer, Macromolecules 21 (1988) 1967.
[282] P. Liu, L. He, J. Song, X. Liang, H. Ding, J. Appl. Polym. Sci. 109 (2008) 1105.
[283] Y. Zheng, K. Chonung, X. Jin, P. Wei, P. Jiang, J. Appl. Polym. Sci. 107 (2008) 3127.
[284] L. Konczol, W. Doll, V. Buchholz, R. Mulhaupt, J. Appl. Polym. Sci. 54 (1994) 815.
[285] J.D. Tong, R.K. Bai, Y.F. Zou, C.Y. Pan, S. Ichimura, J. Appl. Polym. Sci. 52 (1994) 1373.
[286] J.D. Tong, R.K. Bai, C.Y. Pan, E.J. Goethals, J. Appl. Polym. Sci. 57 (1995) 895.
[287] J.D. Tong, R.K. Bai, Y.F. Zou, C.Y. Pan, S. Ichimura, Macromol. Rep. A 30 (1993) 407.
[288] J. Riffle, A.K. Banthia, D.C. Webster, J.E. McGrath, Org. Coat. Plast. Chem. 42 (1980) 122.
[289] J. Riffle, R.G. Freelin, A.K. Banthia, J.E. McGrath, J. Macromol. Sci. Chem. 15 (1981) 967.
[290] M. Ochi, H. Yamana, Kobunsi Ronbunghu 49 (1992) 499.
[291] N. Lanzetta, P. Laurienzo, M. Malinconico, E. Martuscelli, G. Rongosta, M.G. Valpe, J. Mater. Sci. 27 (1992) 786.
[292] C.K. Riew, A.R. Siebert, E.H. Rowe, US Patent 3387 (1976).
[293] K. Sakamoto, US Patent 3972 (1976).
[294] M.B. Jackson, L.N. Edmond, R.J. Varley, P.G. Warden, J. Appl. Polym. Sci. 48 (1993) 1259.
[295] I. Frischinger, S.K. Dirlikov, Polym. Mater. Sci. Eng. 65 (1991) 178.
[296] I. Frischinger, S.K. Dirlikov, Polym. Mater. Sci. Eng. 70 (1993) 1.
[297] D. Ratna, A.K. Banthia, J. Adhes. Sci. Technol. 14 (2000) 15.
[298] D. Ratna, Polym. Int. 50 (2001) 179.
[299] S.K. Dirlikov, I. Frischinger, M.S. Islam, T.J. Lepkowski, in: C.G. Gebelain (Ed.), Biotechnology and Polymers, Plenum Press, New York, 1991.
[300] R.A. Ruseckaite, D.P. Fasce, R.J.J. Williams, Polym. Int. 30 (3) (1993) 297.
[301] H. Samejima, T. Fukuzawa, H. Tada, M. Sage, Ind. Eng. Chem. Pro. Res. Div. 22 (1983) 10.
[302] P.S. Achary, C. Gouri, R. Ramamurty, J. Appl. Polym. Sci. 42 (1991) 743.
[303] H.H. Wang, J.C. Chen, J. Appl. Polym. Sci. 57 (1995) 671.
[304] J.M. Zeilingki, M.S. Vratsanos, J.H. Laurer, R.J. Spoutek, Polymer 37 (1996) 75.
[305] J. Laeuger, P. Albert, J. Kressler, W. Grouski, R. Mulhaupt, ACTA Polym. 46 (1) (1995) 68.
[306] J. Laeuger, P. Albert, J. Kressler, W. Grouski, R. Mulhaupt, ACTA Polym. 46 (1) (1995) 74.
[307] J. Wang, Y. Huang, G. Cong, Gongneng Gaofenzi Xuebao 7 (2) (1994) 181.
[308] D. Ratna, Intl. J. Plastic Technol. 8 (2004) 279.
[309] D. Ratna, A.K. Banthia, P.C. Deb, J. Appl. Polym. Sci. 80 (2001) 1792.
[310] D. Ratna, A.K. Banthia, P.C. Deb, J. Appl. Polym. Sci. 78 (4) (2000) 716.
[311] D. Ratna, Polymer 42 (2001) 4209.
[312] D. Ratna, A.K. Banthia, Polym. Int. 49 (2000) 309.
[313] D. Ratna, G.P. Simon, Polymer 42 (2001) 7739–7747.
[314] S. Kar, D. Gupta, A.K. Banthia, D. Ratna, Polym. Int. 52 (2003) 1332–1338.
[315] T. Iijima, M. Tomoi, J. Yamasaki, H. Kakiuchi, Eur. Polym. J. 26 (1990) 145.
[316] M. Tomoi, H. Yamazaki, H. Akada, H. Kakiuchi, Angew. Makromolek. Chem. 163 (1988) 63.
[317] R. Mezzenga, L. Boogh, J.A.E. Manson, Compos. Sci. Technol. 61 (2001) 787.
[318] L. Boogh, B. Pettersson, E.J. Manson, Polymer 40 (1999) 2249.
[319] D. Ratna, R. Varley, R.K. Singh, G.P. Simon, J. Mater. Sci. 38 (2003) 147.
[320] J. Diamant, R.J. Moulton, SAMPA Q16 (1984) 13.
[321] T. Iijima, N. Yoshioka, M. Tomoi, Eur. Polym. J. 28 (1992) 573.
[322] P.J. Pearce, B.C. Ennis, C.E.M. Morris, Polym. Comm. 29 (1988) 93.

[323] P.J. Pearce, C.E.M. Morris, B.C. Ennis, Polymer 37 (1996) 1137.
[324] R. Mezzenga, L. Boogh, E.J. Manson, J. Polym. Sci. Polym. Phy. 38 (2000) 1883.
[325] D. Ratna, G.P. Simon, J. Polym. Mater. 19 (2002) 349.
[326] D. Ratna, G.P. Simon, Polym. Eng. Sci. 41 (2001) 1815.
[327] D. Ratna, G.P. Simon, Polymer 42 (2001) 8833.
[328] D. Ratna, O. Becker, R. Krishnamurthy, G.P. Simon, R. Varley, Polymer 44 (2003) 7449.
[329] D. Ratna, R. Varley, G.P. Simon, J. Appl. Polym. Sci. 92 (2004) 1604.
[330] D. Ratna, Composites A39 (2008) 462.
[331] L. Luo, Y. Meng, T. Qiu, X. Li, J. Appl. Polym. Sci. 130 (2013) 1064–1073.
[332] J. Lv, Y. Meng, L. He, X. Li, H. Wang, Chin. J. Polym. Sci. 30 (4) (2012) 493–502.
[333] G.M. Roud, A.K. Mahanty, M. Misra, ACS Sustain. Chem. Eng. 5 (2017) 9528–9541.
[334] P.G. Parzuchowski, M. Kizlinska, G. Rokicki, Polymer 48 (2007) 1857–1865.
[335] T. Brocks, L. Ascione, V. Ambrogi, M. Cioffi, P. Persico, J. Mater. Res. 30 (2015) 869–878.
[336] J. Ugelstad, P.C. Mork, K.H. Kaggurud, T. Ellingsen, A. Berge, Adv. Colloid Interface Sci. 13 (1980) 101.
[337] K.E.J. Barret, Disperson Polymerization in Organic Media, Wiley, New York, 1975.
[338] S. Shen, E.D. Sudol, M.S. El-Aasser, J. Polym. Sci. Pt A Polym. Chem. 32 (1993) 1393.
[339] M.P. Markel, V.L. Dimonie, M.S. El-Aasser, J.W. Vanderhoff, J. Polym. Sci. Pt A Polym. Chem. 25 (1987) 1219.
[340] H.J. Sue, Polym. Eng. Sci. 31 (1991) 270.
[341] A. Maazouz, H. Santerean, J.F. Genard, Polym. Bull. 33 (1994) 67.
[342] J.L. Jhonson, H. Hassander, L.H. Jansson, B. Tornwell, Macromolecule 24 (1991) 126.
[343] S. Lee, A. Rudin, Polymer Letex preparation, characterization and applications, in: E.S. Daniels, E.D. Sudol, M.S. El-Aasser (Eds.), Symposium on ACS Series, 492 (1992), p. 234.
[344] K.-F. Lin, Y.-D. Shieh, Polym. Mater. Sci. Eng. 76 (1997) 358.
[345] H.J. Sue, E.L. Garcia-Maitin, N.A. Orchard, J. Polym. Sci. Polym. Phy. (Ed.) 31 (1993) 595.
[346] J.Y. Qian, R.A. Pearson, V.L. Dimonie, M.S. El-Aasser, J. Appl. Polym. Sci. 58 (1995) 139.
[347] K.-F. Lin, Y.-D. Shieh, J. Appl. Polym. Sci. 70 (1998) 2313.
[348] K.-F. Lin, Y.-D. Shieh, J. Appl. Polym. Sci. 69 (1998) 2069.
[349] J. Chen, A.J. Kinloch, S. Sprenger, A.C. Taylor, Polymer 54 (2013) 4276–4289.
[350] L. Gong, L. Zhao, L.C. Tang, H.Y. Liu, Y.W. Mai, Compos. Sci. Technol. 121 (2015) 104–114.
[351] K. Mimura, H. Ito, H. Fujuka, Polymer 41 (2000) 4451.
[352] I. Martinez, M.D. Martin, A. Eceiza, P. Oyanguren, I. Mondragon, Polymer 41 (2000) 1027.
[353] I. Mandragon, M.A. Corcuera, M.D. Martin, A. Valea, M. Franco, V. Bellenguer, Polym. Int. 47 (1998) 152.
[354] R.J. Varley, D.G. Hawthorne, J. Hodgkin, G.P. Simon, J. Polym. Sci. Pt B Polym. Phys. 35 (1997) 153.
[355] B.S. Kim, T. Ciba, T. Inoue, Polymer 34 (1993) 2809.
[356] C.B. Bucknall, I.K. Patridge, Polymer 24 (1983) 639.
[357] J. Diamant, R.J. Moulson, 29th National SAMPE Sympoium, April 3–5, 1994, p. 422.
[358] C.B. Bucknall, I.K. Patridge, Polym. Eng. Sci. 26 (1986) 54.
[359] R.S. Raghava, 29th SAMPE Symposium, Covina, California, April 3–5, 1984, p. 1384.
[360] R.S. Raghava, J. Polym. Sci. Pt B. Polym. Phys. 26 (1988) 65.
[361] J.L. Hedrick, I. Yilgor, G.L. Wilkes, J.E. McGrath, Polym. Bull. 13 (1985) 201.
[362] J.L. Hedrick, M.J. Jureck, I. Yilgor, J.E. McGrath, Polym. Prepr. 26 (1985) 293.
[363] K. Yamanka, T. Inoue, Polymer 30 (1989) 662.
[364] C. Girodet, E. Espuche, H. Sautereau, B. Chabert, R. Ganga, E. Valot, J. Mater. Sci. 31 (1996) 2997.
[365] T.E. Attwood, P.C. Dawson, J.L. Freeman, J.B. Rose, P.A. Staniland, Polymer 22 (1981) 1096.
[366] P.M. Hergenother, B.J. Jensen, S.J. Havens, Polymer 25 (1988) 357.
[367] J.L. Hedrick, D.K. Mahanty, B.C. Johnson, R. Biswanathon, J.A. Hinkley, J.E. McGrath, J. Polym. Sci. Pt A 23 (1986) 287.
[368] J.L. Hedrick, I. Yilgor, M. Jureck, J.C. Hedrick, G.L. Wilkes, J.E. McGrath, Polymer 32 (1991) 2020.
[369] S.J. Pac, G.D. Lyle, R. Mercier, J.E. Mcgrath, Polymer 34 (1993) 885.

[370] S.J. Li, B.L. Hsu, S.Z.D. Cheng, Polym. Mat. Sci. Eng. 70 (1994) 51.
[371] S.C. Kim, H.R. Brown, J. Mater. Sci. 22 (1987) 2509.
[372] A. Murakami, D. Saunders, K. Osihi, T. Yoshiki, M. Saitoo, O. Watanabe, et al., J. Adhes. 39 (1992) 227.
[373] W.G. Liu, T.L. Yu, W.Y. Chen, J.C. Chen, J. Appl. Polym. Sci. 61 (1996) 2423.
[374] C.B. Bucknall, A.H. Gilbert, Polymer 30 (1989) 213.
[375] B.Z. Zang, J.Y. Lian, L.R. Hwang, W.K. Shih, J. Rienf. Plast. 8 (1989) 312.
[376] T. Iijima, T. Tochimoto, M. Tomoi, J. Appl. Polym. Sci. 43 (1991) 1685.
[377] R.A. Pearson, A.F. Yee, Polymer 34 (1993) 3568.
[378] R.W. Venderbosch, J.G.L. Nelissen, H.G.H. Miijer, P.J. Lemstra, Makromol. Chem. Macromol. Symp. 75 (1993) 73.
[379] A.F. Yee, R.A. Pearson, Polym. Mater. Sci. Eng. 63 (1990) 311.
[380] L.N. Carfangna, E. Amendola, C. Carfagna Jr, A.G. Filippov, J. Appl. Polym. Sci. 44 (1992) 1465.
[381] K. Sodagapan, D. Ratna, A.B. Samui, J. Polym. Sci. Pt A Polym. Chem. 41 (2003) 3375−3383.
[382] D. Augustine, K.P. Vijayalakshmi, R. Sadhana, D. Mathew, C.P. Reghunadhan Nair, Polymer 55 (2014) 6006−6016.
[383] T. Iijima, N. Arai, W. Fukuda, M. Tomoi, Eur. Polym. J. 31 (1995) 275.
[384] T. Iijima, T. Tochimoto, M. Tomoi, H. Kakiuchi, J. Appl. Polym. Sci. 43 (1991) 463.
[385] T. Iijima, N. Arai, K. Kakematsu, W. Fukuda, M. Tomoi, Eur. Polym. J. 28 (1992) 1539.
[386] T. Iijima, K. Sato, W. Fukuda, M. Tomoi, J. Appl. Polym. Sci. 48 (1993) 1859.
[387] T. Iijima, S. Miura, W. Fukuda, M. Tomoi, Eur. Polym. J. 29 (1993) 1103.
[388] T. Iijima, S. Mirua, W. Fukuda, M. Tomoi, J. Appl. Polym. Sci. 57 (1995) 819.
[389] P. Jyotishkumar, J. Pionteck, C. Ozdilek, P. Moldenaers, U. Cvelbar, M. Mozeticd, et al., Soft Mater. 7 (2011) 7248.
[390] X. Luo, S. Zeng, Z. Naibin, M. Dezhu, Polymer 35 (1994) 2619.
[391] Q. Guo, X. Peng, Z. Wang, Polymer 32 (1991) 53.
[392] Q. Guo, X. Peng, Z. Wang, Polymer Polym. Bull. 21 (1989) 593.
[393] Q. Guo, Polymer 34 (1993) 70.
[394] J. Kim, R.A. Robertson, J. Mater. Sci. 27 (1992) 3000.
[395] M.E. Nichols, R.E. Robertson, J. Mater. Sci. 29 (1994) 5916.
[396] A.M. Romano, F. Garbassi, R. Bragila, J. Appl. Polym. Sci. 52 (1994) 1775.
[397] C.D. Wingerd, C.L. Beatty, J. Appl. Polym. Sci. 41 (1990) 2539.
[398] B.G. Min, Z.H. Stachurski, J.H. Hodgkin, J. Appl. Polym. Sci 50 (1993) 1511.
[399] B.G. Min, J.H. Hodgkin, Z.H. Stachurski, J. Appl. Polym. Sci 50 (1993) 1065.
[400] A.J. Mackinnon, S.D. Jenkins, P.T. McGrail, R.A. Patrick, Macromolecules 25 (1992) 3492.
[401] S.P. Wilkinson, S.C. Liptac, P.A. Wood, J.E. McGrath, T.C. Ward Proceedings of the 23rd International SAMPE Conference, Kiamesha, NY, USA, 1991, p. 201.
[402] D.J. Hourston, S. Lane, Polymer 33 (1992) 1379.
[403] A.J. Kinloch, M.L. Yuen, S.D. Jenkins, J. Mater. Sci. 29 (1994) 3781.
[404] C. Kim, H.R. Brown, J. Mater. Sci. 22 (1987) 2589.
[405] V. Diliellow, E. Martuscelli, P. Musto, G. Ragosta, G. Scarinzi, Angew. Makromol. Chem. 213 (1993) 93.
[406] S. Horiuchi, A.C. Street, T. Ougizawa, T. Kitano, Polymer 35 (1994) 5283.
[407] J.H. Hodgkin, G.P. Simon, R. Varley, Polym. Adv. Technol. 9 (1998) 3−10.
[408] J. Spanoudakis, R.J. Young, J. Mater. Sci. 19 (1984) 473.
[409] J. Spanoudakis, R.J. Young, J. Mater. Sci. 19 (1984) 487.
[410] R.J. Young, D.L. Maxwell, A.J. Kinloch, J. Mater. Sci. 21 (1986) 380.
[411] A.J. Kinloch, D.L. Maxwell, R.J. Young, J. Mater. Sci. 20 (1985) 3797.
[412] B. Geisler, F.N. Kelley, J. Appl. Polym. Sci. 54 (1994) 177.
[413] F.F. Lange, K.C. Reinford, J. Mater. Sci. 6 (1971) 1197.
[414] A.C. Moloney, H.H. Kausch, T. Kaiser, H.R. Beer, J. Mater. Sci. 18 (1983) 208.
[415] A.C. Moloney, H.H. Kausch, T. Kaiser, H.R. Beer, J. Mater. Sci. 19 (1984) 1125.

[416] Y. Nakamura, M. Yamaguchi, A. Kitayma, M. Ocubo, T. Matsumoto, Polymer 32 (1991) 3221.
[417] S.Y. Fu, X.Q. Feng, B. Lauke, Y.W. Mai, Compos. Pt B-Eng. 39 (2008) 933−961.
[418] W.J. Cantwell, J.W. Smith, H.H. Kausch, T. Kaiser, J. Mater. Sci. 25 (1990) 633−648.
[419] Q. Meng, T. Wang, Eng. Fract. Mech. 211 (2019) 291−302.
[420] R. Kitey, H.V. Tippur, Acta Mater. 53 (2005) 1153−1165.
[421] R. Kitey, H.V. Tippur, Acta Mater. 53 (2005) 1167−1178.
[422] K.C. Jajam, H.V. Tippur, Compos. Pt B-Eng 43 (8) (2012) 3467−3481.
[423] Q. Meng, Z. Wang, Arch. Appl. Mech. 85 (2015) 745−759.

CHAPTER 4

Thermoset composites

Abbreviations

FRP	fiber-reinforced plastic
PMC	polymer−matrix composite
MMC	metal−matrix composite
CMC	ceramic−matrix composite
RTM	resin transfer molding
VARTM	vacuum-assisted resin transfer molding
FRC	fiber-reinforced composite
SRIM	structural RIM
RFI	resin film infusion
ASTM	American Society for Testing and Materials
SBS	short-beam shear
DMA	dynamic mechanical analysis
TGAP	trifunctional epoxy resin
GFRP	glass fiber-reinforced plastic
UPE	unsaturated polyester
GFRT	glass fiber-reinforced thermoset
CFRT	carbon fiber-reinforced thermoset
BFRT	basalt fiber-reinforced thermoset composite
KFRT	Kevlar fiber-reinforced thermoset
RIM	reaction injection molding
CFRP	carbon fiber-reinforced plastic
TGDDM	tetraglycidylether of 4,4′ diaminodiphenyl methane
DGEBA	diglycidyl ether of bisphenol-A
ILSS	interlaminar shear stress
TGAP	triglycidyl p-aminophenol
LCM	liquid composites molding
CFRC	carbon fiber-reinforced composite
CNT	carbon nanotube
CNF	carbon nanofiber
CVD	chemical vapor deposition
DCB	double cantilever beam
CFRC	carbon fiber-reinforced composite
CVD	chemical vapor deposition
CSRP	core−shell rubber particle
SEM	scanning electron microscope
PEBA	poly(ether-block-amide)
CTPEGA	carboxyl-terminated poly(ethylene glycol)
DGEBA	diglycidyl ether of bisphenol-A
HBP	hyperbranched polymer

CSRP	core−shell rubber particle
CF	carbon fiber
RIFT	resin infusion flexible tooling

4.1 Introduction

A composite is defined as a combination of two or more materials with a distinguishable interface. On the basis of nature of matrices, composites can be classified into four major categories: polymer−matrix composite (PMC), metal−matrix composite (MMC), ceramic−matrix composite (CMC), and carbon−matrix composite or carbon−carbon composites. PMCs can be processed at a much lower temperature, compared to MMC and CMC. Over the last three decades, the use of PMCs, especially fiber-reinforced plastic (FRP) composites, has increased tremendously and this dramatic growth is expected to continue in the future. The composites possess many useful properties like high specific stiffness and strength, dimensional stability, adequate electrical properties, and excellent corrosion resistance. The implications are easy transportability, high payload for vehicle, low stress for rotating parts, and high ranges for rockets and missiles, which make them attractive for both the civil and defense applications. For a given design load, the weight component is much lower if manufactured with a composite than with a metal. The high specific stiffness and strength of PMCs offered the potential for reduced fuel consumption and increased range with passenger aircraft and enhance the stealth performance for military aircrafts [1,2].

Depending on the types of polymer matrices, PMCs are classified as thermoset resin composites and thermoplastic resin composites. The composite industries are dominated by thermoset resins. This is because of their availability, relative ease of processing, lower cost of capital equipment for processing, and low material cost. Since thermosetting resins are available in oligomeric or monomeric low-viscosity liquid forms, they have excellent flow properties to facilitate resin impregnation of fiber bundles and proper wetting of the fiber surface by the resin. They are characterized by a crosslinking reaction or curing, which converts those into a three-dimensional (3D) network form (insoluble, infusible). Because of the crosslinked structure, thermoset composites offer better creep properties and environmental stress cracking resistance compared to many thermoplastics, for example, polycarbonate [3,4].

Thermoset composites can be broadly grouped into fiber-reinforced plastic (FRP) composites, particulate composites and nanocomposites [5,6] depending on the reinforcement used as shown in Table 4.1. Different forms of reinforcements used for making thermoset composites are given in Fig. 4.1. Thermoset-based FRPs particularly offer very high strength and opportunity to tailor the material properties through the control of fiber and matrix combinations and processing conditions. The FRP composite and

Table 4.1: Various types of thermoset composites and the manufacturing processes.

Composites	Reinforcement	Process
Fiber-reinforced plastic composite	Glass fiber, carbon fiber, kevlar fiber, besalt fiber	Wet lay-up and compression molding, prepreg lay-up with vacuum bagging and autoclave curing, filament winding with oven curing, pultrusion, resin transfer molding (RTM), vacuum bag molding, liquid composite molding (LCM), structural reaction engineering molding
Particulate microcomposite	Silica, carbon black, calcium carbonate, glass bead, glass balloons, silicon carbide	Mechanical mixing and casting, compression molding, matched-die molding
Nanocomposite	Nanosilica, nanocalcium carbonate, nanotitania, nanoclay, carbon nanofibers, carbon nanotubes	Mechanical mixing and sonication followed by casting or compression molding

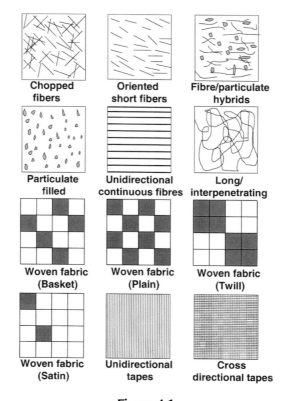

Figure 4.1
Schematic representations of various types of reinforcements used to make thermoset-based composites.

particulate composites are microcomposites because reinforcements sizes are at the microlevel. In recent years, considerable attention has been made on the development of nanocomposites where the enforcement size reduces to nanolevel. The nanocomposites will be discussed in Chapter 5, Thermoset Nanocomposites.

4.1.1 Constituents of FRP composites

FRP composites are made of two major constituents namely the thermosetting resin matrix and the fiber reinforcement. Unlike the blends, in composites both the constituents namely fibers and matrix grossly retain their identities and, at the same time, result in properties that can never be achieved with either of the constituents acting alone. The fibers are usually of high strength and rigidity and predominantly responsible for load-bearing capacity of composites. The role of matrix resin is to keep the fibers in desired location and orientation. The fibers must be separated from each other to avoid mutual abrasion during the deformation of the composites. The load applied into the composite is distributed into the fibers through the matrix. Since fibers are mostly brittle, the resin is the source of toughness for a composite. However, as it will be discussed in subsequent sections impact property (toughness) cannot be simply predicted from the constituents because the toughness originates from the various complex mechanisms.

The thermosetting resins used as matrices in FRP composites have been elaborated in Chapter 2, Properties and Processing of Thermoset Resin. The other constituent (fiber) and the interface between the matrix and the fiber are to be discussed. A wide variety of fibers are available for use in composites. The physical properties of the available fibers are presented in Table 4.2. Most commonly used fibers are glass, carbon, and Kevlar fibers. Each kind of fibers can have a different shape, size, and orientation as shown in Fig. 4.1. A wide variety of composites can be made using the same matrix, which may differ in respect to type of fiber, amount of fiber, fiber length, fiber orientation, etc. Fiber hybridization can further be exploited to tailor the properties of a composite. On the basis of length, fibers are designated in two ways, namely, short fiber and long continuous fibers. Continuous fiber-reinforced composites contain reinforcements having lengths much greater than their cross-sectional dimensions. A composite is considered to be discontinuous or short-fiber composites if its properties vary with the fiber length. On the other hand, when the length of the fiber is such that any further increase in length does not result in further increase in elastic modulus or strength of the composite, the composite is considered to be continuous fiber-reinforced. Most continuous fiber-reinforced composites contain fibers, which are comparable in length to the overall dimension of the composite part. Thus continuous fiber-reinforced composites offer higher strength compared to short-fiber-based composites. Hence short-fiber composites are used for secondary structural applications whereas continuous fiber composites are utilized in primary structural applications and considered as

Table 4.2: Physical properties of commercially available man-made fibers and natural fibers used for making thermoset composites.

Material fibers	Tensile modulus (GPa)	Tensile strength (GPa)	Density (g cc^{-1})	% Strain at break	Specific strength (kNm kg^{-1})
E-glass	72.4	3.5	2.54	2.30	1380
S-glass	85.5	4.6	2.48		1850
Carbon (high modulus)	390	2.1	1.90	0.6	1100
Carbon (high strength)	240	2.9	1.77	1.1	1640
Besalt	85	4.8	2.7	3.1	1780
Boron	385	2.8	2.63	–	1100
Silicon carbide	400	3.5	3.5	–	1000
Silica	72.4	5.8	2.19	–	2650
PBT (heat treated)	331	4.2	1.58	–	2650
Kevlar-49	130	2.8	1.45	2.25	1870
Kevlar-29	6	2.8	1.44		1800
Spectra-900	117	2.6	0.97		2800
Spectra-1000	172	3	0.97	2.33	3200
Polyester	<18	0.8	1.38	15	600
Silicon nitride	380	11	3.2	–	3440
Sisal	7–13	0.5	1.41	4.5	350
coconut	19–26	1.4	1.42		980
Jute	10–30	0.6	1.5	–	400

high-performance engineering materials. The mechanical properties of a composite also depend on the orientation of the fiber. The maximum mechanical performance is achieved when all the fibers are oriented in the fiber axis direction.

4.2 Composite interface

The properties of a composite not only depend on the nature of constituents, that is, matrix and fiber, but also largely depend on the interface. The high strength of fiber can be realized in composites only by designing the interface judiciously. Otherwise, failure may take place in the week interface. An interface is a boundary demarcating the distinct phases namely matrix and fiber. At the interface there can be chemical and physical interactions between the fiber and matrix and also there can be voids, adsorbed gases and surface chemical groups. Bascom and Drzal [7] have cited a number of possible polymer fiber interactions, namely, selective adsorption of matrix components, conformational effects, penetration of polymer molecules into the fiber surface, diffusion of low molecular weight component from the fiber, and catalytic effect of fiber surface on resin curing [8]. A possible representation of interface is given in Fig. 4.2. Another term is often used to

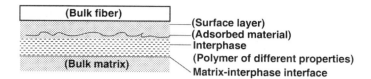

Figure 4.2
A possible representation of interface in fiber-reinforced thermoset composites.

indicate the region where the fiber and resin are chemically and/or mechanically combined is called an interphase. Depending on the processing condition the chemical reactions, the generated volumetric changes and stresses will be different at the interface. Thus the resultant interface will be very complex.

For a good interfacial strength, there should be a good wetting of the fiber surface by the resin. In this regards, it is necessary to discuss the surface tension and contact angle and their relation with wetting of the fiber by the resin. Better wetting resulted in a better interfacial bonding between the fibers and the resin matrix, leading to a better performance of the composites.

4.2.1 Surface tension and contact angle

The molecules on the surface of a liquid are always in a state of unsaturation of molecular forces. Unlike a molecule in the bulk, which is attracted by other molecules from all directions, the molecules on the surface are pulled inward by other molecules deeper inside the liquid and are not attracted as intensely by the molecules in the neighboring medium (be it vacuum, vapor, or air). That is why they have a net force acting on it pulling it toward the interior. The existence of the force is manifested as the formation of "convex meniscus" on a liquid since the relaxation time of the liquid is in the order of microsecond. This is known as surface tension. It is defined as the force along a line of unit length, where the force is parallel to the surface but perpendicular to the line. The value of surface tension increases with increasing polarity of the liquid. Various methods of for determination of surface tension can be found in the related literature [9]. The surface molecules in a solid also face the similar situations but the atoms or molecules in a solid cannot rearrange themselves spontaneously like in a liquid (higher relaxation time) and that is why the surface appears to be unaffected by any disturbances.

When a liquid is placed on a solid surface then depending on the nature of surfaces, the liquid will spontaneously form a droplet or spread out into a film. If a droplet is formed then there will be a value of contact angle, which is the angle, which the tangent to the surface makes with the solid surface. The contact angle can be correlated with the surface tension values of the corresponding solid and liquid in the equilibrium condition as follows:

$$\gamma_S = \gamma_{LS} + \gamma_L \cos\Theta \tag{4.1}$$

where γ_S, γ_L, and γ_s are the surface tensions of solid and liquid and interfacial surface tension. Adhesion between the solid fiber and liquid resin can be thermodynamically defined as the "work of adhesion" (W_A), which is the energy required for generation of two new surfaces and elimination of an interface.

$$W_A = \gamma_L + \gamma_S - \gamma_{LS} \qquad (4.2)$$

Combining Eqs. (4.1) and (4.2), we get

$$W_A = \gamma_L(1 + \cos\Theta) \qquad (4.3)$$

When $\Theta = 0$, W_A is two times the surface tension of the liquid.

As Bascom and Drzal [9] have pointed out, this force should not be confused with the work of disrupting an adhesively bonded interface between the reinforcing fiber and the matrix resin, which includes many substantial energy-absorbing and dissipating mechanisms. Although these expressions apply to the equilibrium condition, they can be utilized to understand the wetting process in resin-reinforcement systems of composites. If Θ is greater than 90 degrees, then the liquid is a nonwetting one for that particular solid. If Θ is greater than 90 degrees, then the process is called "wetting." The lower the value of Θ, higher is the extent of "wetting." When Θ is equal to zero, droplet does not form and the liquid is said to be spreading.

4.2.2 Surface treatment of fibers

The design strategy of interface will change depending on the nature of the matrix as well as the nature of fiber. Reinforcing fibers are never used in the state they are manufactured. They are used after subjecting them to suitable physical or chemical treatments.

4.2.2.1 Glass fiber

The surfaces of the freshly prepared glass fiber absorb water very quickly. This results in the formation of hydroxilated surface. If exposure to moisture is continued, critical size flaws are generated due to absorption of water. This process considerably reduces the strength of the glass fibers [10,11]. The intensity water absorption and surface oxidation can vary in intensity depending on the composition of glass fiber. Thus the glass surface must be protected from the chemical attack of water. Organofunctional silanes are widely used for the surface protection purpose. Silane molecules having three hydrolysable groups and one functional group capable of reacting with the matrix are usually applied to the glass surface. The silanol groups react with surface hydroxyl groups and form stable siloxane bonds, which are not attacked by water. In commercial glass fibers, the formulation, which is applied for surface protection is called "sizes," which is a mixture of a suitable silane coupling agent, binder, antistatic agent, lubricants, and other ingredients. An approximately

0.1-μm-thick coating is applied. Thus the "sizing" surface treatment not only provides the corrosion protection to the fiber but also provides protection for the fiber surface from surface damage during handling and aids in the infiltration of the matrix into the fiber tows during processing of composites using the fibers.

4.2.2.2 Carbon fiber

Unlike glass fiber surfaces, carbon fiber surfaces are not so reactive. Majority of carbon fiber surfaces consist of graphite planes, which are very stable. However, the surface generated during carbonization or graphitization is not suitable to withstand the high shear required for composite applications. During chemical treatment, the native fiber surface is etched away and reactive functional groups are produced on the surface. Thus a fresh surface capable of withstanding much higher shear loading coupled with the formation of reactive groups lead to much improved compatibility and bonding of carbon fibers with the matrix. The edges and corner of the planes and the resulting crystallites are only susceptible to chemical reaction and utilized for surface modification.

A variety of surface modification techniques, which can effectively promote interfacial bonding, have been reported in the literature [12–16] as summarized in Fig. 4.3. The modification techniques can be broadly classified into two categories, namely, oxidative methods and nonoxidative methods. In oxidative methods, the fiber is oxidized with either liquid oxidizing agents such as concentrated nitric acid or by gases like air, oxygen, carbon dioxide, ozone, etc. Oxidation can also be carried out using an electrochemical technique using the electrolytes like caustic soda, phosphoric acid, potassium dichromate, potassium

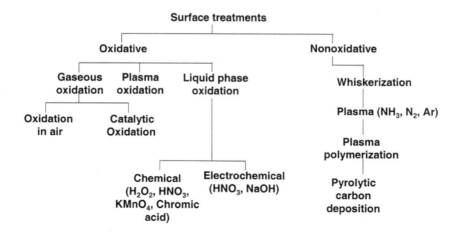

Figure 4.3
Various surface treatment techniques for modification of fibers to improve interfacial adhesion in fiber-reinforced thermoset composites.

permanganate, etc. In this process, fibers are moved continuously through an electrolytic bath. The fibers act as anode. The electrochemical techniques offer fast and uniform surface modification. The oxidation of fiber surface is associated with the generation of functional groups (namely, carbonyl, carboxyl, hydroxyl, etc.) on the fiber surfaces. The implantation of polar organic groups significantly enhances the compatibility and interfacial bonding between the reinforcing fiber and the matrix resin.

In a nonoxidative process, the fiber surface is whiskerized by the deposition of whiskers, which provides the sites for mechanical anchorage and significantly increases the interfacial adhesion. Plasma treatment with ammonia produces amine groups on the surface and effectively increases the interfacial bonding for epoxy-based composites due to the possibility of chemical reaction between the epoxy groups of matrix resin with the amine groups of the fibers. Plasma polymerization process provides a polymeric layer on the fiber surface and thereby promotes the compatibility. It was observed that fibers after subjecting to plasma coating treatment exhibit higher tensile strength compared to the untreated carbon fibers. This indicates that the plasma coating probably healed some of the surface flaws of the fiber. Thus depending on the nature of the matrices, plasma treatment promotes interfacial bonding by various mechanisms such as removal of native weak surfaces and contaminants, increasing the surface roughness of the fibers, which enhance mechanical locking and implantation of specific functional groups, which can react with the polymer matrix.

Various other methods of modification (mentioned in Fig. 4.3) can also be utilized. For the intermediate modulus fibers, only 20% of the fiber surface contains reactive functional groups. Further increase in surface reactivity usually results in a loss of strength due to the generation flaws in the fiber. In order to protect the treated surface from damage during handling, a thin layer of polymer coating (which is essentially same with the matrix to be used for the fabrication of composite) of thickness of about 0.1 μm is given.

4.2.2.3 Polymeric fiber

The polymer-based reinforcing fibers are not sensitive to oxidation or mechanical damage due to handling like glass or carbon fibers. They do not require "sizing" as necessary for glass fiber and "finishing" as necessary for carbon fiber. However, the surfaces of polymeric fibers are low energy surfaces and require some surface treatment to improve wettability. The various surface preparation techniques shown in Fig. 4.3 are also partially applicable for the modification of polymeric fibers. Like in the case of other fibers, in polymeric fibers as well, the surface treatment improves interfacial adhesion between the matrix and the fibers. This can be attributed to the removal of low energy contaminants from the fiber surface, increase in surface roughness, and implantation of reactive groups capable of reacting with the matrix [17,18].

4.3 Processing of composites

The final properties of a composite material depend not only on the nature of the matrix and fiber but also, to a great extent, on processing of such materials. The properties of composites made from the same resin and reinforcements may vary widely depending on the processing technique and processing parameters used for their fabrication. It has already been demonstrated in the last decades that attractive properties of composites could be realized reasonably. The composite materials have found application both in low-tech application like toys, household appliances, as well as high-tech applications like aerospace and biomedical applications. The growing awareness and increasing confidence in composite materials prompted the concerned scientists to develop/perfect the existing processing techniques and to introduce newer techniques. With the growth of industries and requirements, the mass production of composites has increased. The mass production demands for reduced production time, quality control, and greater repeatability. While selecting a particular processing technique, apart from the final properties of composites, the cost and volume production have to be considered.

In this section, various processing techniques used for the fabrication of thermoset fiber-reinforced composites are discussed. A comparison of advantages and limitations of various processes are presented in Table 4.3.

Table 4.3: Comparison of various processing techniques.

Process	Advantage	Limitation	Preferred matrix
Contact molding	Low tooling cost, simple process, no shape restriction	Labor-intensive, high volatile emission, slow production rate	Epoxy, polyester, vinyl ester
Compression molding	Low cycle time, low volatile emission, structural integrity	High tooling cost, average finish	Epoxy, polyester, vinyl ester, polyimide, cyanate ester
RTM	Variety of reinforcement, closed mold, good tolerance	Medium production time	Epoxy
RIM	Fast cycle automation	Resin selectivity	Polyurethanes
Pultrusion	Continuous process, low tooling cost, easy to handle	Joining difficulty, shape restriction	Polyester
Filament winding	Continuous process, control of fiber orientation, automation	Expensive tooling, slow production rate, shape restriction	Epoxy, polyester
Resin film infusion	High-quality product, reproducibility, no need of impregnation, minimum material wastage	Difficult to adopt for resin other than epoxy	Epoxy
Prepreg molding	High-quality product, reproducibility, no need of impregnation, minimum material wastage	High cost	Epoxy, bismaleimides, polyimide

4.3.1 Contact molding

Contact molding is a simple and well-known method for fabrication of fiber-reinforced thermoset composites. Thermoset resins, which cure at room temperature like unsaturated polyester, vinyl ester, epoxy, etc., are used to make glass, jute, or carbon fiber-reinforced composites using contact molding. A wide variety of structures can be fabricated using this technique without any limitation in respect to the size and complexity of the shape of the structure. However, the process is very labor-intensive and time consuming and hence it is used only for a short run or one-off.

The process requires only one mold or molding tool. The mold can be made from wood, plaster of paris, or metal. The wood and plaster of paris molds are used for one-time molding whereas a metal mold is ideal for longer runs. To begin with, a release agent (e.g., waxes, polyvinyl alcohol, etc.) is applied onto the surface of the mold. The release agent prevents the final fabricated product from sticking to the mold. This is followed by application of a gel-coat by using a spray gun or a painting brush. A gel-coat is generally about 0.5-mm-thick resin-rich layer, which provides a smooth, hard surface and prevents the fiber reinforcements from appearing on the surface. The gel-coat can be provided with a particular color on request depending on the application using a suitable colorant. The mold is then coated with a resin layer and presized reinforcement is then laid over the resin layer. Alternate layers of resin and fiber reinforcements are applied until the required thickness is build up. The fibers are continually rolled to ensure proper wetting of the fibers, expulsion of air, and consolidation of the materials. The glass content of 25–30 wt.% can be achieved for random glass mat-based composites while the same up to 50 wt.% can be achieved using fine glass fabric.

In order to increase the production rate, spray-up technique is used instead of a hand lay-up technique. This technique employs a multiple headed spray gun. The resin mixture and chopped fibers are discharged simultaneously by using the spay gun on the surface of the mold where they get deposited to a uniform thickness. Sheets, storage vessels, auto body parts, boat hulls, truck bodies, and building components are made using this process.

4.3.2 Wet lay-up molding

In this method, the resin and fiber can be combined by usual hand lay-up or spray-up techniques. The reinforcement (rovings, mat, and fabric) is cut into the desired shape using a knife/scissor. The reinforcement of desired shape are kept on a clean mold (with mold release agent applied on it) and impregnated with liquid resin and hardener mixture. Depending on the design of the products, the reinforcements are applied layer by layer and laid on the mold. The mold is then kept inside a compression mold press and allowed to cure under pressure. The process of compression molding of unreinforced resin is already discussed in Chapter 2, Properties and Processing of Thermoset Resin. The same procedure is followed for composite as well. Fiber-reinforced composites can be fabricated in a similar way. Glass fiber content up to 70 wt.% or 50 vol.% can be achieved for the fiber composites processed by this technique. Press molding offers obvious advantages like fast cycle time, good finish, structural integrity, and high reinforcement content. Since the process is not automated and human error is involved, the materials made using this method suffer from the lack of consistency and quality.

4.3.3 Resin transfer molding

Resin transfer molding (RTM) offers fabrication of a composite structure with dimensional tolerances and good surface finish. At the same time, the process is less costly compared to an autoclave molding. The popularity of this process has increased in recent years due to the increased use of resin transfer molded composites in automotive and aerospace industries. A schematic diagram of RTM process is shown in Fig. 4.4. This process consists of filling a mold cavity by injecting a resin through one, or several, points depending on the size of the component. The reinforcements are placed in the interior of the mold previously, before closing and locking it firmly. The closed-mold pressure injection system allows faster gelling as compared to contact molded parts. Use of closed mold enables to achieve good tolerances and control of emission of volatiles. The basic difference between RTM and compression molding is that in RTM, the resin is inserted into the mold, which contains the reinforcement whereas in compression molding, both reinforcement and resin remain in the mold, where heat and pressure are applied.

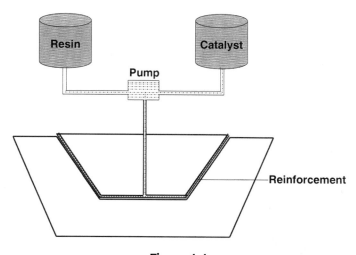

Figure 4.4
A schematic diagram of resin transfer molding (RTM).

In this process, the reinforcement (rovings, mat, and fabric) is cut into the desired shape using a knife/scissor or pattern cutter and shaped into a preform by laying up individual layers of fabrics with the correct orientation. The preform is normally modified by using a fast curing resin or a small amount of thermoplastic binder is added on the surface of the reinforcement, which softens on heating and helps to set the reinforcement in shape. For high-performance applications, the preform is compressed under pressure using a matched lightweight tooling. Once the preform is ready, it is inserted into the mold and mold is closed. Resin is injected into the mold cavity. The resin flows under pressure through the reinforcement preform and impregnate the fibers by expelling the air in the cavity. The injection of resin is continued till the resin starts flowing from the vent areas of the mold. The preform must not extend beyond the pinch-off area in the mold to keep the mold close safely. The mold is kept in the curing temperature for several minutes or hours depending on the resin systems. When the curing process ends the item is removed from the mold and subjected to a postcuring treatment to complete the residual curing reaction of the resin. RTM offers rapid production, ease to vary the reinforcement and use of inclusion when required, quality product and low volatile emission. The example of components made by this process include propeller blades, panels, compressor casings, bumper beam, missile and helicopter component, satellite disks, etc. The distinct advantages of RTM are given below:

1. RTM offers good surface finish on both sides of moldings.
2. RTM allows to make items with close dimensional tolerances.
3. RTM offers fabrication of composites with high fiber volume fraction up to 65%.
4. RTM involves low pressure injection.
5. RTM involves low tooling cost compared to other sophisticated molding process.
6. RTM offers fabrication of both the high and low thickness laminates.
7. RTM involves low material wastage.
8. RTM offers processing with low volatile emission.
9. RTM has the ability to mold complex structural and hollow shapes.

4.3.4 Vacuum-assisted resin transfer molding

Vacuum-assisted resin transfer molding (VARTM) and advanced version of RTM in which environmental pressure is utilized to provide the compressive pressure against the fiber preform the assembly. In the RTM process, the positive injection pressure is applied to process the fiber-reinforced composite (FRC) and the injection pressure is significantly higher than the environmental pressure. The molds used for RTM should have sufficiently high to overcome mold deflection particularly during manufacturing of large FRC parts. On the other hand, in VARTM, the pressure difference between the vacuum pressure and the environmental pressure is utilized to infuse the resin into the preform and compress the preform to secure them against the mold. In fact, a VARTM mold looks similar to the open mold used in a hand lay-up process and can be constructed easily with much larger dimensions than an RTM mold. It offers flexibility to change the mold as per the design. The VARTM process is a closed-mold process with reduced volatile organic compounds emission offers the benefits of high quality, repeatability, and clean handling of an RTM process as well as the advantages of flexibility and scalability of an open-mold wet lay-up processing. VARTM processes are widely used in the marine, energy, infrastructure building, aerospace, and defense industries. Many variations of VARTM have also been developed recently for manufacturing more complex composite parts with improved quality and lower cost [19].

In this process, the dry fiber preform is laid upon a clean mold surface. As mentioned for other methods, a suitable mold release agent has to be applied on the mold surface before keeping the preform on the mold surface. The preform is then covered with a peel ply. On the top of the peel ply, the flow distribution medium layer is applied as per requirement. The flow distribution medium layer can help to enhance the resin infusion speed and is commonly used for manufacturing large composite laminate parts. The flow distribution medium layer is connected to the resin injection port. The resin injection port is placed on one end of the flow distribution medium layer. A helical open tube (like a helical spring shape) or an omega-shaped tube can also be used as a resin injection line source, which serves the purpose of promoting fast resin supply in the helical or omega tube and simultaneously infusing the resin into the flow distribution medium layer. The vent port is to be placed on the top of the peel ply above the preform. It may be noted that the vent must be placed slightly away from the flow distribution medium layer. One can then apply a double-sided tacky tape for adhering to the mold surface and the vacuum bag together surrounding the preform assembly. Vent tube and the injection tube are connected to the vent port and the injection port, respectively. One should close the injection port and open the vacuum port to apply the vacuum inside the bagged preform assembly. At this point, a debulking process can be adopted by cyclically compressing and relaxing the preform to better compact the fiber preform. After performing the debulking process, the resin is put on the reservoir and the injection port is opened to allow the resin to be drawn into the vacuum bagged fiber preform assembly. Once the curing of resin is completed, one can turn off the vacuum and demold the composite part from the mold. The advantages of VARTM have already mentioned and are widely used in making product for civil and naval applications. However, this has some limitations as well. For example, filled resins are difficult to infuse due to their higher viscosity. As discussed earlier, various microfiller and nanofillers are used to modify the properties of thermoset resins. Second, vacuum bag, flow distribution medium, peel ply, sealing tape, and resin tubing may not be reusable. A lower compressive pressure is usually applied on the fiber preform in a VARTM compared to a wet lay-up method, which can limit the fiber volume fraction of the composite part.

4.3.5 Reaction injection molding

Reaction injection molding (RIM) is a relatively new process and it can be used for the processing of unfilled resin as well as fiber-reinforced composites. The process has already been discussed in Chapter 2, Properties and Processing of Thermoset Resin. The process is similar to RTM (discussed above) with some variation in mold release and reinforcement sizing to optimize the resin chemistry with the process. The low-viscosity reactant systems facilitate composite materials production, the so-called structural RIM (SRIM) composites [20,21] in which continuous fiber reinforcement mats are placed in mold cavities prior to the injection. Capital investment and operational costs in RIM are therefore much less than those for conventional injection molding. Normally polymerization of a monomer is initiated by heat. However, in the RIM process, the polymerization is initiated by impingement mixing (not by heat). Hence it is possible to activate polymerization at relatively low temperature. Unlike RTM, in RIM the mold-fill times are very low (~ 1 s) and cycle times of less than 60 s are typical. The process is used for the rapid and automated production of large, thin and complex-shaped parts.

4.3.6 Pultrusion

Pultrusion is a simple technique that is employed mostly for the fabrication of products based on unsaturated polyester resins (or other resins like epoxy, vinyl ester) and continuous strand mat (fiber or fabric). A pultrusion machine consists of creel for supplying fiber, a resin tank, forming dies, machined die with a temperature control facility, puller, and saw for cutting the product from the continuous system. The continuous strand mats are passed through a bath containing a mixture of resin and curatives. In the resin bath, the fiber is passed through a series of rods to remove entrapped air and the excess resin to ensure complete wetting of the fiber. The resin-impregnated fibers or fabrics were then pulled through a series of forming dies. The final die is heated to cure the resin system in order to produce a rigid composite structure. Depending on the tool cross-section, products of different profiles can be made. The cured profile is continuously pulled out of the die, which provides the driving force for the impregnation of fibers to be forced through the die. The process is called pultrusion because the raw materials are pulled through the combining, shaping, and curing operations.

The process has a similarity with extrusion, which is generally used for the processing of thermoplastics. The difference is that in extrusion, the thermoplastic is forced through the die from inside by a rotating screw; on the other hand, in pultrusion, the thermoset material is pulled through the die orifice from the outer side. Both solid and hollow products with high stiffness and strength can be made using this technique. Typical products made by pultrusion are ladders, stanchions, pipes, aerial booms, building parts. Pultrusion offers the advantages like high specific strength of the product, easy handling, and a variety of product profiles. The disadvantages are expensive tooling and difficulty in joining the products. A description of pultrusion process, pultruded profile, and fiber—matrix architecture is shown [22] in Fig. 4.5. The FRP composite materials made using the pultrusion process mentioned above have unidirectional fiber—matrix architecture and therefore a product of virtually any shapes that can be manufactured.). Pultruded FRP composites have a relatively higher fiber volume ratio (60%—75%) compared to the laminates made by the wet lay-up technique discussed above (up to 50 vol.%). The mechanical strength is known to increase with fiber volume ratio and hence high-strength FRPs for civil infrastructure can be made by pultrusion. However, for such applications analysis of long-term performance is extremely important. The reader may refer to a recent review published by Liu et al. [23] on various experimental studies investigating the

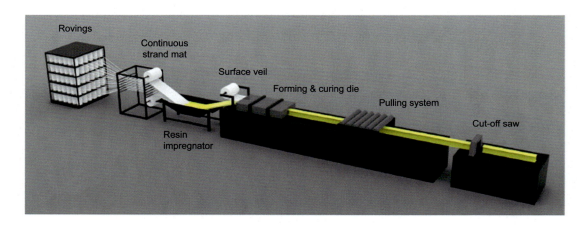

(A) Pultrusion process

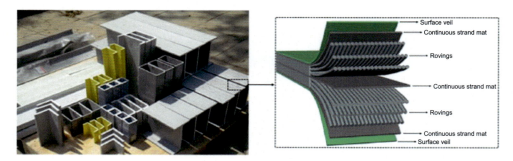

(B) Pultruded profiles and fiber-matrix architecture

Figure 4.5
Typical pultruded FRP profiles. Source: *Reprinted with permission from T.Q. Liu, X. Liu, P. Feng, Compos. Pt. B 191 (2020) 107958, ©2020, Elsevier Publishing Company.*

mechanical performance of pultruded FRP composites subjected to long-term environmental effects, including water/moisture, alkaline solutions, acidic solutions, low/high temperatures, ultraviolet radiation, freeze−thaw cycles, wet−dry cycles, and in situ environments.

4.3.7 Filament winding

Filament winding is an important continuous processing technique, which is widely used in defense sectors for the fabrication of structures like pressure vessels, driveshaft, and radomes. The basic difference between filament winding and pultrusion is that pultrusion involves the placement of fiber in the longitudinal direction, whereas filament winding has reinforcement in hoop direction. This involves laying down of resin-impregnated reinforcement in the form of rovings or tapes onto a mandrel. A continuous length of roving or tape of fibers is passed through a bath containing resin and curatives. Before winding over a mandrel the excess resin is squeezed out. The winding process is continued till the desired thickness is achieved. The mandrel is then subjected to initial cure according to the recommended cure schedule of the matrix resin. The component is then pulled off the mandrel for postcuring treatment. Depending on the shape of the mandrel, components of different sizes and shapes can be produced. By adjusting the manner and direction of fiber lay down, various patterns like helical, hoop, or polar can be achieved. Generally, the mandrel is rotated continuously in one direction and the fiber source reciprocates parallel to it. A synchronization of rotation and reciprocation is necessary to provide total cover of the mandrel surface. The process offers the possibility to tailor the orientation of reinforcement to match the direction and magnitude of applied loads. A successful development of a product using a filament winding process requires a judicious selection of winding type, mandrel design, and winding equipment. The component with complex shapes can be made by using a multiaxis winding spindle. The process produces high-quality components and it can be automated to produce high-volume products.

4.3.8 Resin film infusion

Resin film infusion (RFI) makes use of thin resin film of the thermoset resins instead of liquid resin used in a hand-lay/wet lay-up or VARTM process. RFI method has been successfully used so far for epoxy resin only. A suitable mixture of solid and liquid epoxy and hardener can be cast into a partially cured tacky film, which can be used to make composite. Nowadays automatic machines for making epoxy-based resin films are available in many industries. However, it is difficult to make resin film with highly filled thermoset resin system. Both the reinforcement and the resin film are cut to the desired dimension and placed over a clean mold one after one. It has to be ensured that the resin film just sticks to the fabric. The process is repeated to yield sandwiches of resin films with two fabric layers on both sides. Such sandwiches are placed on one another to build the desired thickness as per the design of the product. The sandwich stack over the mold is then cured by vacuum bagging technique as used for VARTM and discussed in the previous sections. After completing the heating cycle, oven is switched off and the job is allowed to cool slowly to room temperature before demolding. A thermocouple is placed on top of the job to continuously monitor the temperature. On heating, the resin films melt and infuse amidst the fabric plies. The cure cycle and extent of fiber impregnation depends upon the resin permeability of the fiber mat. Hence it is necessary measure the permeability to optimize the manufacturing of a composite product using RFI. In fact the permeability measurement in one of the crucial requirement for design of composite product made by RFI method. fiber optic sensors are widely used for permeability measurement [24]. The fiber optic sensor allows to identify in situ the flow properties during real manufacturing process, which can be used to simulate the RFI process. A typical experimental tool [24] used to characterize carbon braided fabric is shown in Fig. 4.6. We can see in the figure seven braided carbon fibers are placed on the top of resin films. An aluminum perforated plate is

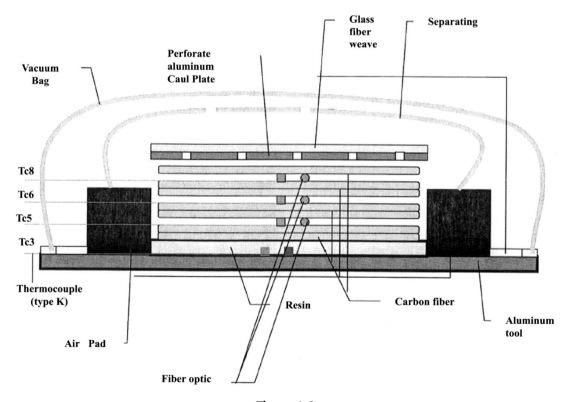

Figure 4.6
RFI element with thermocouple and fiber optic. Source: *Reprinted with permission from V. Antonucci, M. Giordano, L. Nicoalis, A. Cusano, A. Calabro, S. Insera, et al., J. Mater. Proc. Technol. 143 (2003) 687−692, ©2003 Elsevier Publishing Company.*

used to absorb the excess resin. A nylon separating layer is introduced to prevent the escape of resin and to prevent the assembly to come in contact the vacuum bag.

4.3.9 Prepreg molding

4.3.9.1 Prepreg

Fibers that are prepregnated with matrix resin and the resin is cured to B-stage (tacky semisolid form) are called prepreg. Depending on the nature of fibers, prepregs can be of various types such as unidirectional prepreg, woven fabric prepreg, multidirectional tape-prepreg, tow prepregs, etc. On the basis of applications, prepregs are grouped into two categories, namely, general purpose prepregs and high-performance prepregs. The high-performance prepregs are used for aerospace applications. Solvent impregnation is the common process for manufacturing of prepreg tapes. The fabric or a ribbon is passed through a bath containing resin solution. After the fabric getting absorbed with the resin solution, the fabric is passed through a series nip-roller. During this process the excess resin is squeezed off and wetting of the fiber is ensured. The rollers are designed keeping in mind the requirements of fiber to resin ratio. The objective to use solvent is to reduce the viscosity of resin mixture. Lower viscosity promotes better impregnation. The impregnated ribbons or fabrics are then passed through a heated drying chamber for the removal of solvents. The dried fabrics are then partially cured to attain a state of restricted flow but adequate tack. Such prepregs are then covered with plastic film from both side, wound on a drum, and stored in a refrigerator preferably at less than $-15°C$.

4.3.9.2 Molding of prepregs

Prepregs are ready-made materials for molding with optimized composition for specific purposes and hence by using prepregs, manufacturers can avoid cumbersome process of stocking of resins, hardeners and reinforcements [25]. The process is simple because the resin impregnation is excluded as a variable in contrast to other process like contact or compression molding where resin impregnation is an important parameter. The process offers good alignment of reinforcements, minimum material wastage, and excellent quality products.

The prepreg sheets are cut to a specific shape as per requirement with a template. A gel-coat is applied on the mold surface. As soon as the gel-coat is partially cured, the prepreg layers are laid on to top of one another in appropriate orientation. Prepregs-based composites can be made either by vacuum bag molding or by application of pressure in an autoclave (autoclave molding). A typical vacuum bag configuration is shown in Fig. 4.7. It consists of various release layer and air bleed layers to make the surface nonsticking and to absorb the excess resin coming out from the prepregs, respectively. The setup can be used for application of vacuum or the entire setup can be put in an autoclave and cured the composite under heat and pressure.

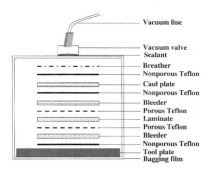

Figure 4.7
A typical vacuum bag configuration used for processing of prepregs.

The prepreg molding offers the highest level of uniformity and excellent properties of the composites [26]. Prepreg-molded composites are extensively used in aerospace industries, where the ultimate mechanical properties and lightness are the major consideration rather than production rate and cost. The composites made by autoclave molding exhibit better properties compared to those prepared by vacuum bag molding. This is because autoclave can exert more pressure than that is generally achieved in vacuum bag molding and thus results greater packing-up and void elimination.

4.4 Analysis and testing of composites

Analysis and testing of composites include the quality control of matrix resins and reinforcements, which have been discussed in earlier sections. The composition of a composite, that is, resin content and fiber contents, should be accurately determined for proper quality control.

4.4.1 Determination of glass content

The glass content for the composite samples was determined by a resin burn-off test according to American Society for Testing and Materials (ASTM) D3171-99. The wt.% glass (x) was determined from the following formula:

$$x = \frac{w}{w_o} \times 100 \tag{4.4}$$

where w_o and w are the initial weight and weight after the matrix burn-off, respectively.

The vol.% glass (y) was determined from the formula given below:

$$y = \frac{x}{x + (100 - x)\frac{d_g}{d_e}} \tag{4.5}$$

where d_g and d_e are the densities of glass and epoxy, respectively.

4.4.2 Mechanical testing of composites

Composites are tested in similar ways as discussed in Chapter 1, Chemistry and General Applications of Thermoset Resins, for unreinforced resins. In some cases, different ASTM standards are used. The ASTM standards used for testing composites are presented in Table 4.4. For example, tensile testing of a composite is carried out using ASTM-D3039. Uncured resins are mostly tested with dumbbell-shaped specimens whereas composites are mostly tested with a rectangular specimen. Since composites require a much higher load to break compared to the corresponding casting, tabbed specimens are used as shown in Fig. 4.8.

4.4.3 Interlaminar shear stress

Interlaminar shear stress (ILSS) of a thermoset composite is generally determined using a short-beam shear (SBS) test (ASTM-D2344) and a double-notched shear test (ASTM-D3846-99). In addition, a 4-point bending test (ASTM-D790) was also used for ILSS of laminates. A review of analytical and experimental data reveals that the 3-point or 4-point SBS test at best represents a quality control test and cannot correctly provide the actual ILSS. Because many times, failure does not occur through an interlaminar shear, instead by indentation and/or flexure. In the case of compressive testing, the loading heads cause indentation deformation and concentration of compressive and transverse shear stresses. These stresses either individually or in combination exceed the material strength before the interlaminar failure occurs. Thus it is extremely difficult to pinpoint the failure load that is used in the calculation of ILSS [27]. Defects or waviness of textile fabric composites further deteriorate the compression strength [28] and cause compression failure on the loading side. Short [29] also developed a method with the sandwiched specimen to measure ILSS of graphite-epoxy composite.

Table 4.4: ASTM standards for various tests of thermoset composites.

Test methods	ASTM standard	Test methods	ASTM standard
Three-point and four-point flexural test	D 790M	Density measurement test	D 3800
Short-beam test	D 2344	Interlaminar shear test	D 3846
Impact test	D 3029	In-plane shear test	D 4255
Tensile test	D 3039	Shear test	D 5379
Constituent content test	D 3171	Interlaminar fracture toughness test	D 5528
Compressive test	D 3410	Bearing response test	D 5961
In-plane shear test	D 3518	Damage resistance test	D 6264

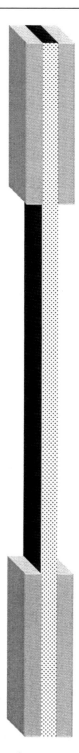

Figure 4.8
A representation of tabbed specimens used for tensile testing of fiber-reinforced composites.

4.5 Prediction of composite strength and rigidity

A theoretical analysis for predicting the functional properties of a material is very important to materials scientists in order to design materials for a particular application. This is more important in multiphase and multicomponent systems like a composite. Such theoretical analysis can be utilized to save a lot of time for experiments that are otherwise necessary for designing material for a particular application. Extensive works have been carried out to predict the rigidity and strength of a polymer composite material from the corresponding constituent properties and their proportions [30–33]. Note that such predictions are only valid for a composite with a fairly good interface. That means processing parameters are to be taken care of in order to validate the prediction.

Generally in multicomponent systems, a statistical approach is used for the prediction of a particular property. The statistical approach requires knowledge of the distribution of the individual phases. In a composite, it is very difficult to ascertain the distribution of individual phases. Hence a statistical approach cannot be used to predict the mechanical properties of a composite. That is why mostly a two-phase model in which average stresses and strains are considered to exist in each of the phases is used. If we assume that fiber and matrix experience equal strain then a parallel model can be used to predict the elastic modulus under longitudinal loading and the composite modulus can be correlated with the modulus of the constituents by the following equation:

$$Y_c = Y_f V_f + Y_m V_m \tag{4.6}$$

$$V_f + V_m = 1 \tag{4.7}$$

where Y_c is the composite modulus, Y_f is the fiber modulus, Y_m is the matrix modulus, V_f is the fiber volume fraction, and V_m is the matrix volume fraction. Typically, Y_f is much higher than Y_m and V_f is equal or greater than V_m; hence the above equation can be approximated as below:

$$Y_c = Y_f V_f \tag{4.8}$$

This suggests that the longitudinal modulus of a unidirectional composite is practically dictated by the axial modulus of the fibers. On the other hand, when the load is applied in transverse direction or in case shear loading the parallel model mentioned above cannot be applied. Rather the matrix and fiber can be considered to act in series. Assuming iso-stress condition where both the matrix and fiber carry the same load, the transverse modulus of the composite can be expressed by the following equations.

$$Y_c = \frac{Y_m T_f}{Y_m V_f + Y_f V_m} \tag{4.9}$$

$$\frac{1}{Y_c} = \frac{V_f}{Y_f} + \frac{V_m}{Y_f} \tag{4.10}$$

Similarly, under shear loading, the shear modulus (G_c) of a composite can be expressed as below:

$$\frac{1}{G_c} = \frac{V_f}{G_f} + \frac{V_m}{G_m} \quad (4.11)$$

where G_c, G_f, and G_m are the shear modulus of the composite, fiber, and matrix, respectively.

Hirsch [34] proposed a model considering the stress transfer between fiber and matrix. The model is in fact a combination of parallel and series model and can be expressed as below:

$$Y_c = x(Y_f V_f + Y_m V_m) + (1-x)\frac{Y_m Y_f}{Y_m V_f + Y_f V_m} \quad (4.12)$$

where x is a parameter, which determines the stress transfer between the fiber and the matrix. It is always assumed that x is determined mainly by the fiber orientation, fiber length, and stress amplification effect at the fiber ends.

Combining Eqs. (4.6) and (4.7), we get

$$Y_c = (Y_f - Y_m)V_f + Y_m \quad (4.13)$$

This is the well-known rule-of-mixture, which describes a rather idealized situation and can predict the modulus only for continuous fiber-reinforced composites where there is sufficient stress transfer from the matrix to the fiber. However, generally short fibers are much shorter than the specimen length. For short fibers, we must consider the matrix–fiber stress transfer. When the matrix is under stress, the maximum stress transferred to the fiber is described by the interfacial stress transfer (τ). The stress transfer depends on the fiber length (l), so that at some critical length, l_c, the stress transferred is large enough to break the fiber. The stress transferred to the fiber builds up to its maximum value (σ_f that which causes breakage) over a distance l_c from the end of the fiber. This means that the long fibers carry the load more efficiently than short fibers.

In other words, it can be stated that the short fibers offer lower effective modulus for reinforcement purposes compared to continuous fibers. This was first considered by Cox [35] who showed that for aligned fibers, the composite modulus is given by the following equation:

$$Y_c = (\eta_l Y_f - Y_m)V_f + Y_m \quad (4.14)$$

where η_l is the length efficiency factor, which is described by the following equations:

$$\eta_l = 1 - \frac{\mathrm{Tanh}(a.l/D)}{a.l/D} \quad (4.15)$$

$$a = \sqrt{\frac{-3Y_m}{2Y_f \ln V_f}} \quad (4.16)$$

From Eq. (4.15), it is clear that as the value of aspect ratio (l/D) increases the value of η_1 (<1) increases and approaches to unity for a high value of aspect ratio ($l/D > 50$). This indicates that fillers with a higher aspect ratio offer a better reinforcing effect compared to the ones with a lower aspect ratio.

Another assumption for the equation based on the rule-of-mixture is that all the fibers are aligned. In the case of short fibers, this assumption does not hold well. Hence it is necessary to consider the orientation effect. Thus for nonaligned, short fibers, the composite modulus is given by the following equation:

$$Y_c = (\eta_0 \eta_1 Y_f - Y_m) V_f + Y_m \tag{4.17}$$

where η_0 is the orientation efficiency factor. This has values of $\eta_0 = 1$ for aligned fibers, $\eta_0 = 3/8$ for fibers aligned in-plane and $\eta_0 = 1/5$ for randomly oriented fibers. At a low V_f, the measure of reinforcement, which takes into account of both the magnitude of the stiffness increase and the amount of fiber, is represented by the following equation:

$$\frac{dY_C}{dV_f} \approx \eta_0 \eta_1 Y_f - Y_m \tag{4.18}$$

A similar calculation can be used to derive an equation for composite strength. For very long aligned fibers, the composite strength is described as given below:

$$\sigma_C = (\sigma_f - \sigma_m) V_f + \sigma_m \tag{4.19}$$

where σ_C, σ_f, and σ_m are the tensile strength of the composite, fiber, and matrix, respectively. The equation shows a similar trend as was found for the modulus. That is the reinforcement is reduced as the fiber length is decreased. For medium length fibers ($l > l_c$), the composite strength can be expressed by the following equation:

$$\sigma_C = (\eta_s \sigma_f - \sigma_m) V_f + \sigma_m \tag{4.20}$$

where η_s is the strength efficiency factor and given by $\eta_s = (1 - l_c/2\,l)$. Note that when l is greater than the critical fiber length (l_c), there is good load transfer between the fiber and the matrix and the fiber breaking takes place at higher applied stress. However, if l is lower than the critical fiber length (l_c), the matrix cannot effectively transfer the load to the fibers and hence fibers do not experience enough load for breaking. As a result, failure takes place through fiber pull-out from the matrix. Under this situation, the strength can be expressed as follows:

$$\sigma_C = (\tau l/D - \sigma_m) V_f + \sigma_m \tag{4.21}$$

Another common model is that developed by Halpin and Tsai [36,37], which may be used for calculating the in-plane stiffness of a single, transversely isotropic sheet having all the fibers unidirectionally aligned in the plane of the sheet. The values of moduli calculated from the equations agree reasonably well with the experimental values for a variety of reinforcement geometries including fibers, flakes and ribbons. One of the advantages of Halpin and Tsai equations is that they cover both the particulate reinforced case (which can be considered as fiber with aspect ratio 1) as well as the continuous fiber-reinforced case where the aspect ratio of the fiber tends to infinity. For aligned fiber composites, the Halpin–Tsai model gives the composite modulus to be

$$Y_C = Y_m \frac{(1 - \xi \eta V_f)}{(1 - \eta V_f)} \quad (4.22)$$

where $\xi = 2l/D$ and $\eta = \frac{Y_f/Y_m - 1}{Y_f/Y_m + 1}$

For randomly oriented composites, the expression becomes slightly more complicated as presented below:

$$\frac{Y_C}{Y_m} = \frac{3}{8}\left[\frac{1 + \xi \eta_L V_f}{1 - \eta_L V_f}\right] + \frac{5}{8}\left[\frac{1 + 2\eta_T V_f}{1 - \eta_T V_f}\right] \quad (4.23)$$

where $\eta_L = \frac{Y_f/Y_m - 1}{Y_f/Y_m + \xi}$ and $\eta_T = \frac{Y_f/Y_m - 1}{Y_f/Y_m + 2}$

The Halpin–Tsai equation can explain the experimental results fairly well at a low volume fraction of fiber. However, the modulus of a composite, calculated using the Halpin–Tsai equation often shows a lower value in comparison to the experimental data. At a low fiber volume fraction, the change in modulus as a function of fiber concentration can be expressed by the following equation:

$$\frac{dY_C}{dV_f} \approx \frac{3}{8} Y_m \eta_L (\xi + 1) + \frac{15}{8} Y_m \quad (4.24)$$

The equation can be successfully used to predict the mechanical properties (strength and modulus) of a composite up to a certain volume fraction of fibers. At higher fiber volume fractions the results show a negative deviation.

4.6 Thermomechanical properties of thermoset composites

4.6.1 Thermal properties

The thermal properties of a composite are mainly governed by the resin as the inorganic fibers are highly heat resistant. The T_g of the matrix is very important for the performance of a composite. At a temperature higher than the T_g of the matrix, the composite loses its dimensional stability and cannot be used for load-bearing applications any more. Hence the knowledge of the T_g of the matrix in the composite is essential from the application point of view. The T_g of a matrix in the form of a composite is generally measured my dynamic mechanical analysis (DMA), though both the differential scanning calorimetry and DMA

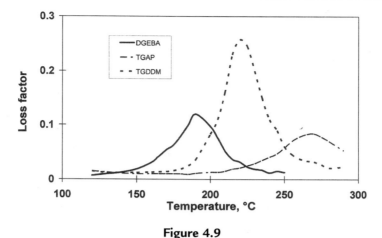

Figure 4.9
Loss factor versus temperature plots of glass fiber-reinforced composites based on epoxies of different functionalities as matrices.

are normally used to determine the T_g of a pure matrix resin [38]. The loss factor versus temperature plots of glass fiber-reinforced composites using epoxies of different functionalities as matrices are shown in Fig. 4.9.

Comparing these data with the same for unreinforced resin (presented in Table 3.4), it can be observed that a composite sample shows a relaxation peak at a temperature lower than that obtained in the case of the corresponding bulk cured resin. For example, the tanδ peak temperature (T_g) for a trifunctional epoxy resin (TGAP)-based glass fiber-reinforced plastic (GFRP) composite was 268°C whereas the value for the unreinforced cured epoxy was 284°C [39]. Theocaris and Papanicolaou [40] reported that the temperature of tanδ peak of an epoxy resin was significantly higher (25°C) in the bulk resin than that in a GFRP composite containing the same resin. Ghosh et al. [41,42] reported similar effects in the case of jute fiber-reinforced composites. This can be explained by considering the effect of organosilane coating, which is used for commercial fibers in order to increase their adhesion with the matrix. The organosilane coating with unreactive organic groups leads to an interface with many unrestrained or free end groups, which result in a reduction in the crosslink density of the polymer network in the interface region [43]. The plasticized region then yields decreased internal friction and thereby causes a reduction in relaxation temperature.

4.6.2 Mechanical properties

In contrast to the thermal properties, the mechanical properties of a composite depend mostly on the type, content, and orientation of the fiber used as the reinforcement. The mechanical properties of glass fiber-reinforced unsaturated polyester (UPE) composites [44]

Table 4.5: Properties of glass fiber-reinforced UPE resin.

Fiber type	Glass content (%)	Tensile strength (MPa)	Tensile modulus (GPa)	Flexural strength (MPa)	Flexural modulus (GPa)
NIL	0	59	2.4	88	3.9
Chopped-stretched mat	50	288	16.7	197	14.5
Roving fabric	60	314	19.5	317	15.0
Woven glass fabric	70	331	25.8	403	17.4
Unidirectional roving fabric	70	611	32.5	403	29.4

are shown in Table 4.5. Since mechanical strength of a composite is mostly contributed by the fibers due to much higher strength of inorganic fibers compared to any thermoset resin. Thus the mechanical strength of a composite increases with increase in fiber content up to an optimum concentration. Increasing the fiber concentration beyond the optimum concentration resulted in the formation of flaws inside the composite leading to a drastic reduction in mechanical properties. As a result, other properties of the composites also get affected. For example, it is observed that increasing the fiber volume ratio V_f could lead to increased water uptake [45] in FRP materials since a greater number of voids exist at higher fiber volume ratios. The diffused water inside the composite leads to delamination and catastrophic failure of the structural parts. FRPs with higher water uptake are not recommended for many application, for example, marine applications. Because such materials degrade easily in marine environments.

Hence processing of flawless composites using a minimum amount of resin is a real challenge. As evident from Table 4.5, the fiber glass orientation has a great influence on mechanical properties of glass-reinforced composites. For a constant fiber content, the unidirectional composites showed much higher tensile properties and flexural modulus when tested in the fiber direction. Properties of unidirectional composites are highly anisotropic. In the fiber direction they show the maximum tensile strength whereas when the measurement is carried out in a direction transverse to the fiber direction, the tensile strength of the composite tend to almost same as the strength of the unreinforced matrix resin. It is also reported that fiber orientation play an important role in deciding the degradation behavior [46,47] when exposed to moisture, high temperatures, and freeze−thaw and wet−dry cycles. Before discussing the effect of resin systems on the mechanical properties of composites, the average mechanical properties of glass fiber-reinforced thermoset (GFRT), carbon fiber-reinforced thermoset (CFRT), basalt fiber-reinforced thermoset composite (BFRT), and kevlar fiber-reinforced thermoset (KFRT) are presented in Table 4.6 for better understanding of the readers. The properties of steel are also given in the table for comparison. We can see that thermoset-based composites display higher specific strength compared to steel.

Table 4.6: Comparison of average mechanical properties of fiber-reinforced thermoset composites and steel.

Materials composite	Density (g cc^{-1})	Tensile strength (MPa)	Tensile modulus (GPa)	% Elongation at break	Specific strength (kNm kg^{-1})
GFRP	1.8	400	25	2	220
CFRP	1.4	700	90	1	500
KFRC	1.2	1200	30	3	830
BFRC	1.9	900	40	1	470
Steel	7.8	500	200	7	60

Table 4.7: Mechanical properties of FRP composites based on difunctional and multifunctional epoxies.

Composition of composites	Flexural strength (MPa)	Flexural modulus (GPa)	Impact strength (J m^{-1})
TGDDM/GF	350	20	700
TGAP/GF	330	19	700
TGDDM + TGAP/GF (50:50)	440	21	900
DGEBA/GF	400	19	950
DGEBA/CF	730	53	600
TGDDM + TGAP/CF (50:50)	780	55	600

Ratna and coworkers [48] evaluated the GFRP and carbon fiber-reinforced plastic (CFRP) composites based on the difunctional and multifunctional epoxies for their mechanical properties. The results are shown in Table 4.7. Readers can find the detailed chemistry and properties of various epoxy resins in Chapter 1, Chemistry and General Applications of Thermoset Resins. It is expected that the tetrafunctional epoxy resin (TGDDM)-based composites would exhibit superior properties compared to the difunctional epoxy (DGEBA)-based composites made using a similar method. However, it was found that the mechanical properties of TGDDM-based composites are inferior compared to those based on difunctional epoxy. The anomalous behavior in case of composites might be due to very high viscosity of tetrafunctional epoxy (Table 3.1), which hinders impregnation with the fibers [49]. This can be explained in terms of ILSS, which is considered as the bonding strength between the fiber, and the resin. The ILSS value of a TGDDM-based composite was 20 MPa, compared to 32 MPa for a DGEBA-based composites. The lower ILSS value for TGDDM-based composite clearly supports the phenomenon of poor resin impregnation due to the higher viscosity of TGDDM.

The trifunctional epoxy (TGAP)-based composites also do not show good mechanical properties because of its highly brittle nature. This can be attributed to the inherent brittleness of the resin due to very low molecular weight between crosslinks. However, when TGDDM is blended with TGAP, the viscosity decreases and the mechanical properties of the composites improve due the betterment of impregnation of resin mixture into the fiber. The mechanical properties of the FRP composites based on the multifunctional epoxy blend (TGDDM/TGAP = 50/50 w/w) are superior to the same based on difunctional epoxy resin (Table 4.7). The improvement in impregnation as a result of blending is reflected from the increase in ILSS from 20 to 35 MPa. It may be noted that the results mentioned above have been reported for composites made by a hand lay-up followed by compression molding. This discrepancy may not arise if sophisticated processing like autoclave molding is used.

In recent years, lots of works [22,50−60] have been reported on FRP composites using natural fibers like sisal, jute, coconut, etc. because of the environment-friendly nature of such composites. The natural fibers though preferable from an environmental point of view, their mechanical properties are much inferior compared to inorganic fiber like carbon and glass (see Table 4.2). Since the mechanical properties of an FRP composite is contributed mostly by the reinforcing fiber, the natural fiber-based composites also exhibit inferior mechanical properties compared to inorganic fiber-based composites. Moreover, unlike inorganic fibers, which do not absorb water, the natural fibers tend to absorb water, inducing a greater extent of fiber swelling and rupture [61−64]. Therefore natural fiber composites are not recommended for marine applications.

The mechanical properties of natural fiber-based composites can be enhanced by incorporating glass fiber partially [51]. The effect of fiber length on mechanical properties of composites based on polyester resin and sisal fiber was investigated. It was reported that the composites made with 2 cm fiber length exhibits better mechanical properties compared to the same made with fibers of 1 or 3 cm length. This indicates that there exists an optimum fiber length beyond which impregnation of resin becomes difficult. Gowthami et al. [52] investigated the effect of the incorporation of silica on the mechanical properties of sisal fiber-reinforced polyester composites. It was reported that the tensile strength and tensile modulus of composites with silica are 1.5 and 1.08 times greater than that of the composite without silica respectively. Similarly, the impact strength of the composite with silica were1.36 and 1.8 times greater than that of composites without silica, respectively. The mechanical properties of sisal, jute, and glass fiber-reinforced polyester composites were investigated by Ramesh et al. [53]. They observed that the addition of glass fiber into jute fiber composite resulted in maximum tensile strength. In the same way, they have observed that jute and sisal mixture composites sample is capable of having maximum flexural strength and maximum impact strength was obtained for the sisal fiber composite. The effect of weaving patterns and random orientation of banana, kenaf, and banana/kenaf fiber on the mechanical properties of the related polyester resin-based FRC was studied [54]. Composites were prepared using a hand lay-up method with two different weaving patterns, namely, plain and twill type. Out of the two weaving patterns used for the fabrication of composites, the plain type showed improved tensile properties compared to the twill type in all the fabricated composites. Moreover, the plain-woven hybrid composites display higher mechanical strength compared to the randomly oriented composites. This indicates the distribution of load transfer along the fiber direction resulted in a minimum stress development at the interface of composites.

4.7 Toughened composites

As discussed earlier, thermoset composites offer many interesting properties and received wide applications in various fields. However, the thermoset-based FRP composites are known to be highly susceptible to internal damage caused by a low-velocity impact, which may lead to severe safety and reliability problems. Hence, the improvement of damage tolerance of FRP composites by enhancing their impact strength has been the subject of considerable research interest [63–71].

The impact response of a composite is a complex phenomenon unlike the mechanical strength measured at a small strain rate. As discussed in earlier sections, the modulus of a composite material can be predicted from the same values for the constituents, that is, matrix and fiber. In many cases, even simple rule-mixture can fairly explain the experimental results. However, no such theoretical analysis can be successfully utilized for the assessment of impact properties of a composite. When an impact load is subjected to a composite structure, it absorbs the impact energy by two types of energy-absorbing or energy-dissipating mechanisms. The first is an elastic nonfailure response due to the rearrangement of the reinforcement and other is energy absorption through creation of damaged areas. Matrix–fiber delamination is one of the important energy absorption mechanisms and the increase in absorbed energy is linearly related to an increase in delaminated area [63–65]. Fiber breakage, fiber pull-out, and fiber debonding are other important energy-dissipating mechanisms. Because of involvement of these energy-dissipating mechanisms, a composite exhibits much higher toughness than that expected from the toughness of individual components. For example, the impact strength of cured epoxy is about 20 $J\ m^{-1}$, the impact strength of glass is still less; however, the impact strength of an epoxy–glass fiber composite is about 900 $J\ m^{-1}$, which is about 50 times higher than the value for the tougher constituent.

There are several methods to enhance the toughness of FRP composites; examples of these are matrix toughening, interleaving (insertion of interlaminer "interleaf" layers), short-fiber reinforcement, and stitching [66–69]. The impact strength of a thermoset composite can be improved by using high-strain fiber or fiber hybridization [70,71]. However, high-strain fibers often show lower modulus and hence cannot satisfy the requirement for dimensional stability in high-performance engineering applications [72]. This approach is particularly exploited for the application of composites under high-incident-energy condition as for example ballistic applications [73].

Interleaving involves incorporation of a tough resin layer in the interface, which enhances the plastic deformation at the interface that is otherwise highly elastic. Wimolkiat and Bell [74] electropolymerized a high-temperature thermoplastic (3-carboxy phenyl maleimide−styrene copolymer) interphase onto graphite fiber and evaluated the fiber-reinforced epoxy composites. The improvement in critical strain energy factor of about 100% and notched impact strength of about 60% were achieved while maintaining the interlaminar shear strength at around the same value as for controlled composite. Other thermoplastic and liquid rubbers have also been successfully utilized as interleaving materials [75−77]. However, use of such fillers in high concentration (5−20 wt.%) required for achieving the desired toughness is associated with a significant deterioration in modulus and glass transition temperature. Another issue is that an excessive interlaminar toughness between plies may reduce the in-plane and out-of-plane permeability of textile reinforcements, which will affect resin impregnation characteristics and decrease the process efficiency of liquid composites molding (LCM) processes.

Recently Zhao and coworkers [78] used polyamide-6 (PA6) and PA6/graphene oxide (GO)-based interlaminar toughening takifiers in order improve the toughness of textile-reinforced epoxy composites. In addition to enhancement in interlaminar properties, the process efficiency is also simultaneously enhanced during the manufacture of via a LCM process. PA6/GO composite was made by in situ polymerization method [79]. The PA6 and PA6/GO composite-based takifier powder for the CF3031 woven fabrics was prepared by ethanolic solution precipitation method. The takifier powders were first dispersed in ethanol homogenously via ultrasonication and the suspension was sprayed uniformly on one side of the woven fabric. The ethanol is removed by drying at 70°C. The laminates were made after application of takifier and the fracture properties of the laminates made using PA6- and PA6/GO-based takifier are presented in Table 4.8. We can observe that use of PA6-based takifier resulted in a significant enhancement in both first and second mode fracture energy up to 5 wt.% of takifier concentration. Addition of graphene in the takifier improves the G_{IC} values further. A takifier with 5.0 wt.% of PA6/GO-(1 wt.%) offer the highest value of G_{IC} (0.753 kJ m^{-2}). If we compare this with the value of pure epoxy-based laminate (0.417 kJ m^{-2}), then improvement is more than 80%.

The toughness of carbon fiber-reinforced composite (CFRC) has been significantly improved by using the partially cured CNT/epoxy CNF/epoxy as interleaving materials [80−83]. Similarly, 3D graphene foams, synthesized by different methods, have also been exploited in recent years [84,85]. However, the processing GFs using the chemical vapor deposition (CVD) technique requires a complicated procedure and expensive facilities, which restricts the use of such materials for the manufacturing of composites for

Table 4.8: Mechanical properties of the control composite laminate and composite laminates toughened with PA6, /GO.0.5, and PA6/GO.1.0 takifiers at varying loadings.

Specimens	G_{IC} (kJ m^{-2})	G_{IIC} (kJ m^{-2})
Control	0.147 ± 0.027	1.243 ± 0.097
C (PA6)-2.5 wt.%	0.533 ± 0.043	1.347 ± 0.079
C (PA6)-5.0 wt.%	0.619 ± 0.071	1.482 ± 0.102
C (PA6)-10.0 wt.%	0.467 ± 0.041	0.941 ± 0.124
C (PA6/GO-0.5)-2.5 wt.%	0.587 ± 0.047	1.529 ± 0.120
C (PA6/GO-0.5)-5.0 wt.%	0.679 ± 0.042	1.557 ± 0.106
C (PA6/GO-0.5)-10.0 wt.%	0.524 ± 0.032	1.206 ± 0.021
C (PA6/GO-1.0)-2.5 wt.%	0.661 ± 0.062	1.693 ± 0.105
C (PA6/GO-1.0)-5.0 wt.%	0.753 ± 0.035	1.568 ± 0.039
C(PA6/GO-1.0)-10.0 wt.%	0.552 ± 0.015	1.344 ± 0.087

Source: Reprinted with permission from X. Zhao, W. Chen, X. Han, Y. Zhao, S. Du, Compos. Sci. Technol. 191 (2020) 108094, ©2020, Elsevier.

commercial applications. Du et al. [86] prepared graphene sheets by a simple thermal reduction method in air and used as interleaved materials for epoxy/CF composite and evaluated the mode I delamination resistance of interleaved CF/E composites. A schematic for fabrication of graphene interleaved CF/E composites and the double cantilever beam (DCB) geometry of specimen used for mode I interlaminar fracture toughness test are presented in Fig. 4.10. It can be seen that laminates interleaved by graphene sheets (0.5 and 1.0 wt.%) display improved G_{IC} values compared with the control made without use of interleaving. Beyond 1 wt.% concentration of GO in interleave, the fracture energy tends to decrease, which can be attributed to the serious agglomeration of graphene sheets at high loadings in the epoxy matrix. Addition of 1.0 wt.% graphene/epoxy interleaves in epoxy/CF composites leads to a 145% increase in fracture energy. Table 4.9 compares this result with the same reported by other research groups. It is obvious that the G_{IC} of graphene/epoxy interleaves developed in this work [86] exhibits the highest increase in mode I fracture energy compared to other methods using interlayers of SiC whiskers [87], short carbon fiber tissue [88], CNTs [80–82,89–95] and carbon black [96], and GO/epoxy [97].

Very recently, a scalable method in interleaving is reported [98] in which CNT fiber fabric veils with highly porous structure synthesized by CVD method are directly affixed to the surface of carbon fiber prepreg without any treatment. The CNT fibers were continuously drawn out at the exit of the furnace and directly deposited onto the surface of a piece of unidirectional carbon fiber prepreg (HexPly AS4/8552). Schematic of direct spinning of CNT fibers from the gas phase by CVD is shown in Fig. 4.11. It is observed that when pure CNT veil without any resin is used as interleave, only a 12% increase in G_{IIC} the fracture energy is found to decrease. On the other, when the same is used with the epoxy resin (AS4/8552), a 88% enhancement in G_{IIC} is observed. The interleave materials based on self-healing polymers have also been investigated. Kostopoulos and coworkers [99] used

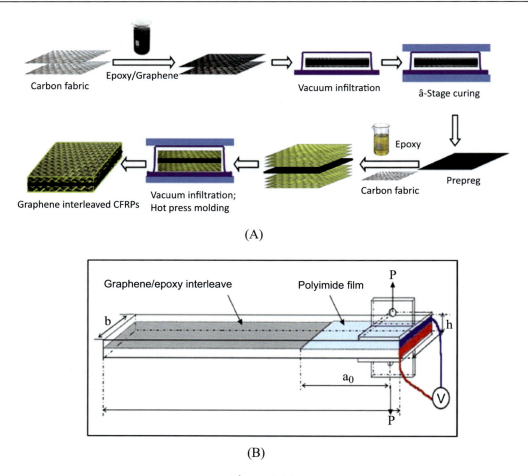

Figure 4.10
(A) Schematic for fabrication of graphene interleaved CF/E composites. (B) DCB geometry for mode I interlaminar fracture toughness test. Source: *Reprinted with permission from X. Du, H. Zhou, W. Sun, H.-Y. Liu, G. Zhou, Y.-W. Mai, et al., Compos. Sci. Technol. 140 (2017) 123–133, ©2017, Elsevier Publishing Company.*

flexible supramolecular polymers (SPs) comprising ureiodypyrimidone hydrogen bonding unit (UPy) moieties as interleave additives into conventional unidirectional (UD) CFRC to make a tough hybrid composite with self-healing attributes. They have demonstrated dramatic enhancement in fracture energy (almost one order of magnitude higher) as result of interleaving with the SP. The G_{IC} values of the interleaved hybrid CFRC and the reference composite are 0.33 and 5.45 kJ m^{-2}, respectively. This can be explained by considering the bridging mechanism that occurs during the crack propagation, due to the presence of the SP-based interleaves (SPSH01) in between the upper and lower crack surfaces of the sample as shown in Fig. 4.12. The interleaved CFRC also exhibit

Table 4.9: Improvements in mode I interlaminar fracture energy of CF/E composites with interleaves modified by different nanofillers.

Nanofiller dispersion techniques in interleaves/interlayer	Mode I interlaminar fracture energy improvement	Reference
Interlayer with SiC whiskers	16.1%	[39]
Interlayer with CB/epoxy composite (15 g m^{-2})	88.6%	[7]
Interlayer with CNF powder (20 g m^{-2})	26%	[16]
Interlayer with CNT paste (20 g m^{-2})	20%	[17]
Interlayer with CNT/epoxy composite (0.5 wt.%)	−5%	[14]
Interlayer with CNT/epoxy composite (0.3 wt.%)	48%	[12]
Interlayer with sprayed CNT (1.32 g m^{-2})	32%	[10]
Interlayer with sprayed CNT (0.047 wt.%)	21%	[8]
Interlayer with CVD grown CNT arrays	43% (for IM7 carbon fabrics) 36% (for AS4 carbon fabrics)	[11]
Interlayer with short carbon fiber tissue	99% (for fiber length of 0.8 mm) 63% (for fiber length of 4.3 mm)	[40]
Interlayer with carbon tissue or CNT-modified fiber tissue	−5% for carbon fiber tissue 35% for CNT-modified tissue	[15]
Growth of CNTs on CF by CVD	46%	[13]
Growth of CNTs on CF by flame	67%	[9]
Interlayer with GO/E composite (2 g m^{-2})	108%	[6]
Interlayer with graphene/epoxy composite (1 wt.% to 1 g m^{-2})	145%	This work

Source: Reprinted with permission from X. Du, H. Zhou, W. Sun, H.-Y. Liu, G. Zhou, Y.-W. Mai, et al., Compos. Sci. Technol. 140 (2017) 123−133, ©2017, Elsevier Publisher.

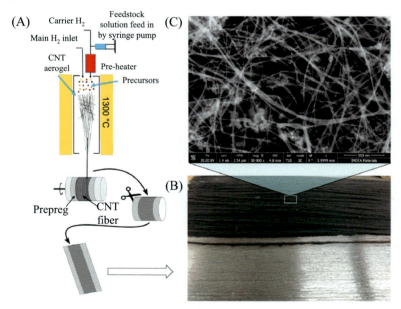

Figure 4.11
(A) Schematic of direct spinning of CNT fibers from the gas phase by FFCVD; (B) unidirectional carbon AS4/8552 prepreg with fluffy CNT veils deposited on the surface; and (C) high-resolution SEM picture of as-produced CNT fiber veil. Source: Reprinted with permission from Y. Ou, C. González, J.J. Vilatela, Compos. Pt B Eng., 201 (2020) 108372, ©2020, Elsevier Publishing Company.

Figure 4.12
Illustration of the extended bridging phenomenon of the SPSH01 interleaf between the upper and the lower adjacent surfaces during the mode I testing. Source: *Reprinted with permission from V. Kostopoulos, A. Kotrotsos, S. Tsantzalis, P. Tsokanas, T. Loutas, A.W. Bosman, Compos. Sci. Technol. 128 (2016) 84–93, ©2016, Elsevier Publishing Company.*

self-healing attribute; however, the G_{IC} values are reduced with increasing number of healing cycles. This is attributed to an uneven distribution of the SPSH01 film over the debonded surfaces during the healing cycle at 100°C, due to inhomogeneous polymer coverage of the surface resulting from the fracturing events. It may be noted that the characteristic values for G_{IC} after the seventh healing cycle are still much higher compared to the values obtained for the unfractured reference material.

Short-fiber reinforcement is another effective technique for making toughened composites. Unlike interleaving, it does not rely on the generation of plastic deformation for consumption of mechanical energy. Short-fiber reinforcement enhances the toughness of a composite by increasing the energy consumption for a given damage area using semielastic failure mechanism of fiber pull-out or fiber breakage. Generally, short-fiber loading ranges from 1 to 4 wt.%. They can be deposited parallel to the ply surface or they can be oriented translaminarly or through the thickness of the ply. The short-fiber reinforcement resulted in increase in toughness by 100%–300% [100]. SEM analysis indicates the generation of increased levels of fiber bridging, crack deviation, and crack bifurcation due to the addition of short fibers [101]. Electrospun nanofibers have also been used for toughening of fiber-reinforced thermoset composites [102]. The toughening effect of the nanofibers is directly dependent on the effects of nanofiber bridging of the crack and yielding of the nanofibers [103]. Since yielding of nanofiber is involved, the fracture properties are influenced by the inherent properties of the nanofibers. The fracture behavior of nanofiber-toughened composites is also dependent on the interfacial adhesion between the matrix and the nanofiber on its turn influenced by the adhesive interactions between matrix, which has also been shown by Beckermann and Pickering [103]. It was reported that the crack path of the delamination should frequently pass through the nanofiber-toughened zone in order for the nanofibers being able to exert their toughening effect [104–106]. For a composite comprising same matrix and fiber, the toughening effect of nanofiber may vary with the orientation of long fibers present in the composite. Because depending on the orientation of fiber, the toughening mechanism may vary. For example, in the case of multidirectional interfaces, the fracture toughness tends to increase with increasing interface angle. Recently, Meireman et al. [107] used poly(ether-block-amide) (PEBA)-based nanofibers to toughen epoxy-based FRC with dissimilar interfaces in terms materials and fiber orientations. PEBA nanofibers were directly deposited on the glass or carbon fiber ply at 2.5 g m^{-2} using an electrospinning machine with multinozzle design. This study is particularly relevant for practical applications, where dissimilar interfaces form a damage-prone boundary between composite plies having different mechanical strength/toughness. A significant enhancement in fracture toughness was achieved, which is attributed to the plastic deformation of the nanofibers, nanofiber bridging effects, and the frequent crossings of the crack path through the nanofiber-toughened interlaminar zone. Additionally, in case of multistructural FRC, augmented occurrence of crack bifurcations toward adjacent nontoughened interlayers also contributes toward toughening effect as shown in Fig. 4.13.

The impact strength and damage tolerance of a composite can also be significantly enhanced by the introduction of z-direction fibers (3D fibers) or stitching. Cox et al. [108] investigated tensile behavior of graphite/epoxy composites with 3D woven interlock reinforcements and reported the contributions of various mechanisms of fracture. The through-thickness reinforcement has the ability to maintain the transverse damage almost constant over the depth of the composite. It also helps in crack bridging [109–112]. As far as the mode I failure is concerned, the stitching yarns carry most of the load at the crack tip, reducing the stress intensity in the surrounding matrix. The stitching provides a crack closure force and thus increases the load required to propagate the crack through the matrix. It was reported that bridging effect was most prominent in case of 90/90 degree interface and negligible for 0/0 degree interfaces. Hence a stitched composite displays higher interlaminar fracture toughness and offers a better damage resistance. The damage area was found to be significantly reduced due to the incorporation of stitching for a constant impact loading. In contrast to the composites made from 2D fabrics, which fail catastrophically, in stitched composites, the failure process usually proceeds gradually through a combination of various mechanisms like matrix microcracking, fiber–matrix debonding, fiber breakage, stress distribution, and delamination.

At the same time unlike interleaving where energy absorption is reduced, the stitching has been shown to improve the energy absorption [113–115]. The nature of stitch and yarn plays an important role for determining the behavior of composite materials under impact

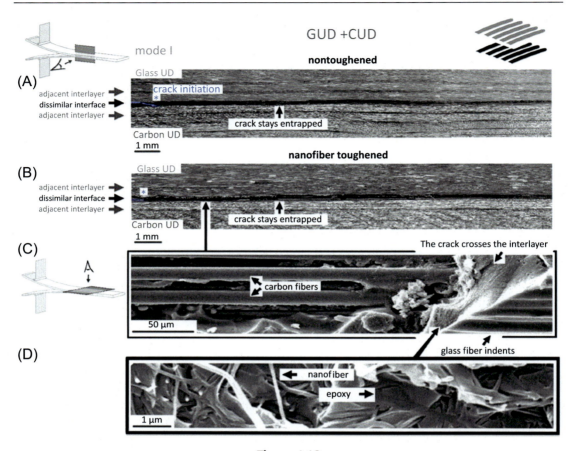

Figure 4.13
The lateral cross section of a mode I tested unidirectional multimaterial composite (GUD-CUD) without nanofibers (A) shows that the crack propagates evenly within the unidirectional interface. In case of a nanofiber-toughened composite (B), the crack alternates between both sides of the interlayer and repeatedly crosses the nanofiber-toughened interlaminar region (C) that results in plastic deformations of the nanofibers (D) and a high IFT increase. A similar fracture development is noted for the equivalent similar GFRP (S3) and CFRP. Source: *Reprinted with permission from T. Meireman, L. Daelemans, E.V. Verre, W.V. Paepegem, K.D. Clerck, Mater. Des. 195 (2020) 109050, ©2020, Elsevier Publishing Company.*

condition. The knot strength controls the fiber's ability to crimp without breaking while the denier indicates the volume percent of stitching fiber and also degree of in-plane fiber disruption. There also exists an optimum value for stitch pitch (number of stitches per unit length) like weaving of fiber in other direction. The number of stitches per unit length should be sufficient to be effective in suppressing the extent of damage induced by the impact. However, it is necessary to consider another issue that the resin impregnation becomes more and more difficult with increasing the value of stitch pitch. For achieving good mechanical properties of a composite, resin impregnation is very much essential. Hence the 3D fibers have to be designed accordingly. Also it is sometimes necessary to modify a thermoset resin formulation by introducing diluents to make it less viscous in order to ensure the sufficient resin impregnation. Again use of too much quantity of diluent may affect the properties of the composites. Hence it is necessary to optimize the amount of diluent to be added.

4.7.1 Resin toughening

Using a particular reinforcement, low-velocity impact resistance (desirable for structural applications) of a particulate or FRP composite can be improved to a great extent by increasing the toughness of the matrix. The various methods for toughening thermoset resins have been discussed in this and next chapters. Few such examples will be discussed here to highlight their potential for composite applications.

Ratna et al. [116] evaluated GFRP composites (glass content almost 45 vol.%) based on carboxyl-terminated poly(ethylene glycol) (CTPEGA)-modified epoxy cured with an ambient temperature curing amine hardener. The readers can find the details of CTPEGA modification in Chapter 3, Toughened Thermoset Resin. The flexural stress versus strain curves for the GRFP composites (containing 0–40 phr of CTPEGA) are shown in Fig. 4.14. The knee on the stress–strain curves and the stiffness represented by the lower portion of the curve are identical for all the FRP samples. This indicates that the flexural modulus of a composite is mostly dominated by the fiber. As expected, a significant increase in flexural strain and toughness was observed without abruptly changing the strength and modulus up to 20 phr of CTPEGA concentration. Beyond a concentration of 20 phr, an abrupt decrease in strength and modulus takes place. The area under the curve gives an idea about toughness of the material. The increase in toughness (as a result of modification) of the

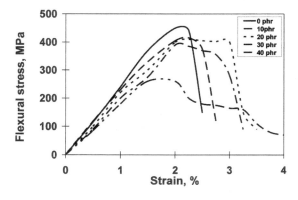

Figure 4.14
The flexural stress versus strain curves for the glass fiber-reinforced CTPEGA-modified epoxy-based composites (containing 0–40 phr of CTPEGA). Source: *Reprinted with permission from D. Ratna, T.K. Chongdar, B.C. Chakraborty, Polym. Compos. 25 (2004) 165, ©2004, John Wiley and Sons Publishers.*

Table 4.10: Impact and ILSS data for various toughened DGEBA epoxy composites.

Modifier	Concentration of modifier (wt.%)	Impact strength (J m^{-1} GFRP)	Impact strength (J m^{-1} CFRP)	ILSS (MPa GFRP)
CTPEGA	0	950 ± 50	500 ± 25	32.5
CTPEGA	10	1120 ± 40	–	29.2
CTPEGA	20	1350 ± 45	–	26.4
HBP	0	900 ± 40	540 ± 30	–
HBP	10	1150 ± 45	510 ± 25	–
HBP	15	1325 ± 35	500 ± 20	–

composites is much lower compared to that observed in the unreinforced matrix (reported in Fig. 5.2). The poor translation of resin toughness into GFRP composites can be attributed to fiber constraints, suppressing the inelastic resin deformation at the crack tips [117,118].

Impact strength and ILSS data of various composites based on epoxy (DGEBA)- and CTPEGA-modified epoxy are presented in Table 4.10. It can be seen from the figure that the composites based on CTPEGA-modified epoxy show higher impact strength compared to that based on unmodified epoxy, and its value passes through a maximum (20 phr CTPEGA concentration) with increasing CTPEGA concentration. The ILSS of the unmodified epoxy-based composite is higher than that of a composite made using 20% CTPEGA-modified epoxy, although the impact strength of the modified epoxy-based was much higher compared to the unmodified epoxy-based composite. This indicated that the reduction in ILSS is associated with an increase in impact strength up to a certain value of ILSS. This is attributed to the fact that the strong adhesion can limit the energy absorption mechanisms like fiber debonding, pull-out, and postdebonding friction during the fracture process leading to a brittle failure of the composites [119,120]. However, a satisfactory adhesion is required to avoid debonding before the failure [120,121]. Hence it is not necessary for a composite to have very high ILSS in order to exhibit very good overall mechanical properties. A moderate ILSS value (need to be optimized) results in a balanced strength and toughness properties.

As discussed in Chapter 5, Thermoset Nanocomposites, CTPEGA forms a compatible blend with the epoxy (DGEBA) and improves the impact strength by enhancing the plastic deformation. Such modification always reduces the T_g and restricts the use of the related composites for high-temperature applications. In order to solve this problem, a second-phase toughening strategy is used. The modifier is selected in such a way that it is miscible with the thermoset resin, initially, and while processing of the composite, it undergoes a reaction-induced phase separation leading to the formation of a two-phase microstructure. Since the modifier is introduced in a separate phase, it does affect the T_g of the composites significantly. Hence such composites can be used for high-temperature applications. Note that a composite when used at a temperature higher than the T_g of the matrix, its dimensional stability is lost and it remains no longer suitable for structural applications. The composites made with such modified resin systems exhibit considerably higher impact strength and damage tolerance due to increase in the toughness of the matrix resin. The toughening of resins has been elaborately discussed in Chapter 5, Thermoset Nanocomposites. In this section, the properties of the composites based on the modified two-phase resin systems will be discussed taking an epoxy/hyperbranched polymer (HBP) blend system as a reference.

Ratna [122] has studied epoxy/HBP-based GFRP and CFRP composites. Fig. 4.15 compares the effects of incorporation of HBP on the T_g for castings and composites. It was observed that the T_g of epoxy/blend in the composite form is lower than the same epoxy/blend in the casting form. The difference originates from the interaction of epoxy with organosilane coating of glass fiber as discussed in Section 4.2.2. Apart from the difference in T_g of epoxy or blend in different form (reinforced or unreinforced), what is more interesting is that the effects of incorporation of HBP on T_g are different in castings and in composites. In case of castings, there is no change in the T_g of epoxy/HBP blends up to 15 wt.% of HBP, whereas in case of composites, the reduction in epoxy T_g was observed as a result of incorporation of HBP for all the compositions. The decrease in T_g for cured rubber-modified epoxy systems arises from the incomplete phase separation and plasticization phenomenon caused by the dissolved rubber that has been noted in varied rubber-modified epoxy formulations [123,124]. Hence the quantity of dissolved HBP for a particular blend is more in the composite form than in the casting form. This indicates that the presence of

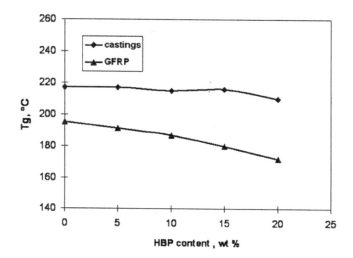

Figure 4.15
Effects of incorporation of HBP on the T_g of modified epoxy both in castings and glass fiber-reinforced composite form. Source: *Reprinted with permission from D. Ratna, Compos. Pt A Appl. Sci. Manuf. 39 (2008) 462, ©2008, Elsevier Publishing Company.*

fibers initiates partial miscibility of the HBP with the epoxy matrix due to the chemical interaction of the epoxy/HBP blend with the fiber at the interface. The thermodynamic consideration for explaining the above observation is given below.

In case of composite, the free-energy change of the system including three components, epoxy, HBP, and fiber, can be described as

$$\Delta G_m = \Delta G_{EF} + \Delta G_{HF} - \Delta G_{EH} \qquad (4.25)$$

where ΔG_{EH} is the free energy of mixing between epoxy and HBP and ΔG_{EF} and ΔG_{HF} are the free energy of interaction of the two components with the fiber surface. Since both the components are strongly adsorbed to fiber surface, ΔG_m will always be negative and hence equilibrium phase diagram is shifted to a higher compatibility. This explains why epoxy/HBP blend shows more compatibility in composites compared to that in casting.

SEM photographs for the fracture surfaces of epoxy/HBP blend (15 wt.% HBP) in various forms such as castings, GFRP, and CFRP composites are shown in Fig. 4.16. In the case of castings, the HBP particles are uniformly distributed throughout the matrix as already discussed in Chapter 5, Thermoset Nanocomposites. It is interesting to note that in composites, the particles are comparatively bigger and the uniformity in particle size distribution is lost to a certain extent. Both the GFRP and CFRP composites exhibit similar morphologies. This indicates that the fibers simply act as nucleating agents and there is not much effect of the nature of the fiber on the resulting morphology. However, the two types of composites discussed above differ drastically in their impact behavior. Since epoxy/HBP blends are tougher than the unmodified epoxy alone, it is expected that the composites based on epoxy/HBP blend would be tougher than the unmodified epoxy-based composites. As expected, the GFRP composites offer a significant improvement in impact strength as shown in Table 4.10. On the other hand, no such toughening effect is observed in case of CFRP composites (Table 4.10). The fracture property of CFRP composites is reported [125,126] to be governed by the interfacial adhesion and internal stress development during cure, which can be manipulated by material and process tailoring. Thus these aspects need to be analyzed critically. However, the apparent difference in behavior of HBP-modified epoxy-based GFRP and CFRP composites prepared under the same conditions can be explained in the light of the inherent properties of the fibers (Table 4.2). Carbon fibers are much more brittle (tensile strain = 1%) compared to glass fiber (tensile elongation = 2.5%) [118]. This is also reflected in the impact behavior of GFRP and CFRP composites. The impact strength of unmodified epoxy-based CFRP composite is 500 J m^{-1} compared to 950 J m^{-1} for corresponding GFRP composite. It is well established that dispersed rubber particle enhances the toughness of the epoxy system by cavitations of rubber particle followed by shear yielding and their effectiveness decreases with increase in the rigidity of the system [127,128]. Highly brittle carbon fiber imposes restriction on the induction of

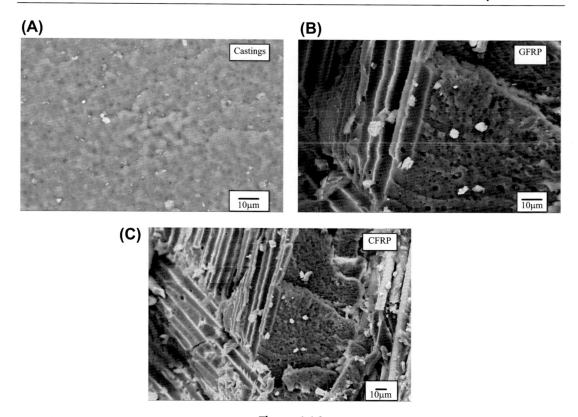

Figure 4.16
(A) SEM photographs for the fracture surfaces of epoxy/HBP blend (15 wt.% HBP) in castings form. (B) SEM photographs for the fracture surfaces of epoxy/HBP blend (15 wt.% HBP) in glass fiber-reinforced plastic (GFRP) form. (C) SEM photographs for the fracture surfaces of epoxy/HBP blend (15 wt.% HBP) in carbon fiber-reinforced plastic (CFRP) form. Source: *(A, B) reprinted with permission from D. Ratna, Compos. Pt A Appl. Sci. Manuf. 39 (2008) 462, ©2008 Elsevier Publishing Company.*

plastic deformation by the rubber particle. In fact failure of the CFRP composites occurs due to fiber breaking.

Kinloch and coworkers [129–131] investigated the effect incorporation of nanosilica (NS) on the toughness of FRC under both static and fatigue loading conditions. They have reported that silica nanoparticles can be easily dispersed epoxy matrix and The FRC made with NS-modified epoxy display higher fracture energy compared to pure epoxy-based FRC. The modulus and thermal properties remain unaltered due after NS is incorporated in the composite structure. An improved fatigue life and reduced fatigue crack growth have also been observed [132] for the FRC based on NS-modified epoxy matrix. The main advantage of NS over other resin modifier is that the particles do not increase the viscosity of the resin significantly and do not exhibit thixotropic properties. This makes NS-modified epoxies suitable for their processing by resin infusion manufacturing method [133]. The suitability of such modified epoxy as a new type of matrix for manufacturing FRCs using injection technology (or liquid composite molding techniques) has also been reported [133]. NSs are known to toughen a thermoset resin by various mechanisms like debonding

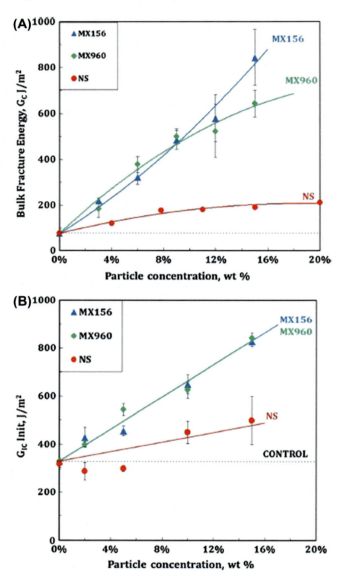

Figure 4.17
Fracture energy versus particle concentration (A) bulk G_{IC} and (B) composite G_{IC} (for interpretation of the references of color in this figure legend, the reader is referred to the web version of this article.) Source: *Reprinted with permission from S.A. Ngah, A.C. Taylor, Composites Pt A Appl. Sci. Manuf. 80 (2016) 292–303, ©2016, Elsevier Publishing Company.*

followed by plastic void growth, and shear deformation in the polymer matrix. In addition to these toughening mechanisms, fiber bridging increases the toughness further in case of FRCs. However, the addition of silica nanoparticles may affect the level of fiber to matrix adhesion. It has been reported that fiber−matrix interface of glass fiber-reinforced composite (GFRC) with a NS-modified thermoset matrix was very weak, leading to a nearly entire interfacial failure.

Taylor and coworkers [134] compared the properties GFRC made using NS-modified and core−shell rubber particle (CSRP)-modified epoxies. Two types of CSRPs were used: Kane Ace MX156 and MX960 (from Kaneka, Belgium). The modified resins are infused into 8-ply glass fiber preforms, arranged as a quasi-isotropic lay-up of $[+45/-45/90/0]_s$ to give laminates with a central 0/0-ply interface. An additional two layers of carbon fiber (CF) preforms were added either side of the GF preform stacks. The laminates were made by resin infusion flexible tooling (RIFT) method under vacuum. The plot bulk and the composite fracture energies as a function of particle concentration are presented in Fig. 4.17. It is evident that toughness increases linearly with particle concentration for both NS and CSRP. The linear increase is steeper for CSRP than that for NS. This is expected due to higher ability of CSRP in toughening a thermoset matrix than NS. If we concentrate on the translation of bulk toughness into the composite, then we can see for lesser bulk toughness, the extent of transfer of toughness from bulk to composite is higher. For a brittle matrix, complete transfer of bulk toughness into FRC takes place. The higher composite toughness can be explained by considering the additional toughening mechanisms like fiber bridging, which are not present in the bulk. In composites with tougher matrices, there is only a partial transfer of matrix to composite toughness. This can be attributed to the fact that the presence of the fibers considerably restricts the plastic zone size and limits the toughening mechanisms, which fully operate in the bulk.

References

[1] L.T. Drzal, Epoxy Resin and Composites IISer. No. 75 in: K. Dusek (Ed.), Adv. Polym. Sci, Springer-Verlag, Berlin, 1986.
[2] D.H. Middleton (Ed.), Composite Materials in Aircraft Structure, Longman, New York, 1990.
[3] A.A. Bakker, R. Jones, R. Callinan, J. Compos. Struct. 15 (1985) 154.
[4] B.Z. Jang, Sci. Eng. Compos. Mater. 2 (1) (1991) 29.
[5] D. Ratna, Rapra Rev. Rep. 16 (5) (2005) 1.
[6] M. Okamoto, Rapra Rev. Rep. 14 (2003) 1−166.
[7] W.D. Bascom, L.T. Drzal, NASA Contract Report 4084 (1987), N87-25434.
[8] P.S. Chua, Polym. Compos. 8 (1987) 308.
[9] A.J. Walter, Phys. Educ. 7 (8) (1972) 491.
[10] E.A. Plueddmann, Silicate Coupling Agents, Plenum Press, New York, 1982.
[11] C. Shen, L.S. Liu, J.T. Yeh, Polym. Polym. Compos. 21 (1999) 21.
[12] L.T. Drzal, M. Rich, P. Lloyd, J. Adhes. 16 (1983) 1.
[13] K.J. Hook, R.,K. Agrawal, L.T. Drzal, J. Adhes. 32 (1990) 157.

[14] T. Takahagi, A. Ishitani, in: H. Ishida, G. Kumar (Eds.), Molecular Characterization of Composite Interfaces, Plenum Press, New York, 1985.
[15] H. Zhang, Z. Zhang, B. Claudia, Compos. Sci. Technol. 64 (2004) 2021.
[16] Y. Yang, F. He, M. Wang, B. Zhang, J. Mater. Sci. 33 (1998) 3651.
[17] D. Ratna, R. Kushwaha, V. Dalvi, N.R. Manoj, A.B. Samui, B.C. Chakraborty, J. Adhes. Sci. Technol. 22 (1) (2008) 93.
[18] D.S. Bag, V.A. Kumar, S. Maiti, J. Appl. Polym. Sci. 71 (1999) 1041.
[19] A.T. Bhatt, P.P. Gohil, IOP Conf. Ser. Mater. Sci. Eng. 1004 (2020) 012001.
[20] C.W. Macosko, RIM, Fundamentals of RIM, Hanser Publisher, New York, 1989.
[21] E. Haberstroh, I. Kleha, Macromol. Mater. Eng. 289 (2000) 108.
[22] T.Q. Liu, X. Liu, P. Feng, Compos. Pt B Eng. 191 (2020) 107958.
[23] T. Liu, X. Liu, P. Feng, Compos. Pt B Eng. 191 (2020) 107958.
[24] V. Antonucci, M. Giordano, L. Nicoalis, A. Cusano, A. Calabro, S. Insera, et al., J. Mater. Proc. Technol. 143 (2003) 687–692.
[25] S.G. Advanie, M. Sozer, Compr. Compos. Mater. 2 (2000) 807–844.
[26] S. Park, D. Lee, J. Song, Int. J. Precis. Eng. Manuf. 19 (2018) 441–446.
[27] J.M. Whitney, Interlaminar Response of Composite Materials, in: N.J. Pagano (Ed.), Composite Materials Series, vol. 5, Elsevier, 1989, pp. 161–250.
[28] T.C. Emehel, K.N. Shivakumar, J. Reinf. Plast. Compos. 16 (1997) 86.
[29] S.R. Short, Composites 26 (1995) 431.
[30] C.L. Tucker, E. Liang, Compos. Sci. Technol. 59 (1999) 655.
[31] W.D. Callister, Materials Science and Engineering, an Introduction, Wiley, New York, 2003.
[32] A. Kelly, W.R. Tyson, J. Mech. Phys. Solids 13 (1965) 329.
[33] G.P. Carman, K.L. Reifsnider, Compos. Sci. Technol. 43 (1992) 137.
[34] T.J. Hirsch, Modulus of elasticity of concrete affected by elastic moduli of cement paste matrix and aggregate, in Proceedings of the American Concrete Institute, 59 (1962) 427.
[35] H.L. Cox, J. Appl. Phys. 3 (1) (1952) 72.
[36] B.W. Rosen, Fibre Composite Materials, American Society for Metals, Metals Park, 1965, p. 58.
[37] L.E. Nielson, Mechanical Properties of Polymers and Composites, 2, Marcel Dekker, New York, 1974.
[38] T. Murayama, Dynamic Mechanical Analysis of Polymeric Materials, Elsevier Science Publishing Co., Amsterdam, 1978.
[39] D. Ratna, T.K. Chongder, B.C. Chakraborty, Polym. Int. 49 (2000) 815.
[40] P.S. Theocaris, G.C. Papanicolaou, Coll. Poly. Sci. 258 (1980) 1044.
[41] P. Ghosh, N.R. Bose, J. Appl. Polym. Sci. 58 (1995) 2177.
[42] P. Ghosh, N.R. Bose, B.C. Mitra, S. Das, J. Appl. Polym. Sci. 65 (1997) 2467.
[43] V.B. Gupta, L.T. Drzal, C.Y.C. Lee, M.J. Rich, J. Macromol. Sci. Phys. B-23 (1985) 435.
[44] M. Grayson, D. Eckroth, third ed., Encyclopedia of Chemical Technology, vol. 18, John Wiley and Sons, 1982, p. 575.
[45] M. Akay, S.K.A. Mun, A. Stanley, Compos. Sci. Technol. 57 (1997) 565–571.
[46] R.E. Allred, J. Compos. Mater. 15 (2) (1981) 100–116.
[47] R.E. Allred, J. Compos. Mater 15 (2) (1981) 117–132.
[48] A.B. Samui, B.C. Chakraborty, D. Ratna, Int. J. Plastic Technol. 8 (2004) 279.
[49] I.M. Low, P. Schmidt, J. Lane, M. McGrath, J. Appl. Polym. Sci. 54 (1994) 2191.
[50] G.C. Sanjeevamurthy, G. Rangasrinivas, S. Manu, Int. J. Eng. Res. Appl. (IJERA) 2 (2012) 615–619.
[51] K.P. Ashik, R.S. Sharma, J. Min. Mater. Charac. Eng. 3 (2015) 420–426.
[52] A. Gowthami, K. Ramanaiah, A.V.R. Prasad, K.H.C. Reddy, K.M. Rao, J. Mater. Environ. Sci. 4 (2012) 199–204.
[53] M. Ramesh, K. Palanikumar, K.H. Reddy, Compos. Pt B Eng. 48 (2013) 1–9.
[54] A. Alavudeen, N. Rajini, S. Karthikeyan, M. Thiruchitrambalam, N. Venkateshwaren, Mater. Des. 66 (2015) 246–257.

[55] K. Agarwal, S.K. Kuchipudi, B. Girard, M. Houser, J. Compos. Mater. 52 (2018) 3173–3318.
[56] R.B. Ashok, C.V. Srinivasa, B. Basavaraj, Sci. Technol. Mater. 30 (2018) 120–130.
[57] A. Alavudeena, N. Rajinia, S. Karthikeyana, M. Thiruchitrambalamb, N. Venkateshwaren, Mater. Des. 66 (2015) 246–257.
[58] K.P. Ashik, R.S. Sharma, Miner. Mater. Char. Eng. 3 (2015) 59743–59749.
[59] I. Shakir, A. Siefaldeen, O. Kamalaldin, Mater. Today Proc. V. 43 (2021) 1003–1008.
[60] D. Bolcu, M.M. Stănescu, Polymers 12 (2020) (1649).
[61] O. Faruk, A.K. Bledzki, H.P. Fink, M. Sain, Progr. Polym. Sci. 37 (2012) 1552–1596.
[62] D.B. Dittenber, H.V.S. Ganga Rao, Compos. Pt A Appl. Sci. Manuf. 43 (2012) 1419–1429.
[63] J.P. Slow, V.P.W. Shim, J. Compos. Mater. 32 (1998) 1178.
[64] A. Kessler, A.K. Bledzki, Polym. Compos. 20 (1999) 269.
[65] A.A.J.M. Peijs, S.W. Venderbosch, P.J. Lemstra, Composites 25 (1990) 522.
[66] W.J. Cantwell, J. Morton, Composites 22 (1991) 347.
[67] I.K. Patridge, D.D.R. Cartie, Compos. Pt A Appl. Sci. Manuf. 36 (2005) 55.
[68] M.V. Hosur, U.K. Vaidya, C. Ulven, J.S. Tuskegee, Compos. Struct. 64 (2004) 455.
[69] J.C. Chen, C.K. Lu, C.H. Chiu, H. Chin, Composites 25 (1994) 251.
[70] R.C.L. Dutra, B.G. Soares, E.A. Campose, J.L.G. Silva, Polymer 41 (2000) 3841.
[71] M.S. Sohn, X.Z. Hu, Compos. Sci. Technol. 58 (1998) 211.
[72] D.F. Adams, R.S. Zimmerman, SAMPE J. 22 (1986) 10.
[73] R.L. Ellis, F. Lalande, H. Jia, C.A. Rogers, J. Reinf. Plast. Comos. 17 (1998) 147.
[74] A.S. Wimolkiatisak, J.P. Bell, J. Appl. Polym. Sci. 46 (1992) 1899.
[75] S. Kunz, J. Sayre, R. Assink, Polymer 23 (13) (1982) 1897–1906.
[76] J. Ma, M.S. Mo, X.S. Du, S.R. Dai, J. Polym. Sci. 110 (2008) 304–312.
[77] Z.J. Thompson, M.A. Hillmyer, J. Liu, H.-J. Sue, M. Dettloff, F.S. Bates, Macromolecules 42 (2009) 2333–2335.
[78] X. Zhao, W. Chen, X. Han, Y. Zhao, S. Du, Compos. Sci. Technol. 191 (2020) 108094.
[79] X. Zhao, Y. Li, W. Chen, S. Li, Y. Zhao, S. Du, Compos. Sci. Technol. 171 (2019) 180–189.
[80] S.H. Lee, H. Kim, S. Hang, S.K. Cheong, Compos. Sci. Technol. 73 (2012) 1–8.
[81] Y. Li, N. Hori, M. Arai, N. Hu, Y. Liu, H. Fukunaga, Compos. Pt A Appl. Sci. Manuf. 40 (2009) 2004–2012.
[82] M. Arai, N. Yukihiro, K.-I. Sugimoto, M. Endo, Compos. Sci. Technol. 68 (2008) 516–525.
[83] S.U. Khan, J.K. Kim, Carbon 50 (2012) 5265–5277.
[84] J. Jia, X. Du, C. Chen, X. Sun, Y.-W. Mai, J.-K. Kim, Carbon 95 (2015) 978–986.
[85] J. Jia, X. Sun, X. Lin, X. Shen, Y.-W. Mai, J.-K. Kim, ACS Nano 8 (2014) 5774–5783.
[86] X. Du, H. Zhou, W. Sun, H.-Y. Liu, G. Zhou, Y.-W. Mai, et al., Compos. Sci. Technol. 140 (2017) 123–133.
[87] W. Wang, Y. Takao, T. Matsubara, H. Kim, Compos. Sci. Technol. 62 (2002) 767–774.
[88] F. Xu, X. Du, H.Y. Liu, Y.-W. Mai, X.J. Wang, Compos. Sin. 32 (2015) 1784–1790.
[89] H. Zhang, Y. Liu, M. Kuwata, E. Bilotti, T. Peijs, Compos. Pt A Appl. Sci. Manuf. 70 (2015) 102–110.
[90] X.S. Du, H.-Y. Liu, F. Xu, Y. Zeng, Y.-W. Mai, Sci. Technol. 101 (2014) 159–166.
[91] S.C. Joshi, V. Dikshit, J. Compos. Mater. 46 (2011) 665–675.
[92] E.J. Garcia, B.L. Wardle, A.J. Hart, Compos. Pt A Appl. Sci. Manuf. 39 (2008) 1065–1070.
[93] V. Mirjalili, R. Ramachandramoorthy, P. Hubert, Carbon 79 (2014) 413–423.
[94] K.L. Kepple, G.P. Sanborn, P.A. Lacasse, K.M. Gruenberg, W.J. Ready, Carbon 46 (2008) 2026–2033.
[95] R.J. Sager, P.J. Klein, D.C. Davis, D.C. Lagoudas, G.L. Warren, H.J. Sue, J. Appl. Polym. Sci. 121 (2011) 2394–2405.
[96] H. Ning, Y. Li, J. Li, N. Hu, Y. Liu, L. Wu, et al., Compos. Pt A Appl. Sci. Manuf. 68 (2015) 226–234.
[97] H. Ning, J. Li, N. Hu, C. Yan, Y. Liu, L. Wu, et al., Carbon 91 (2015) 224–233.
[98] Y. Ou, C. González, J.J. Vilatela, Compos. Pt B Eng. 201 (2020) 108372.

[99] V. Kostopoulos, A. Kotrotsos, S. Tsantzalis, P. Tsokanas, T. Loutas, A.W. Bosman, Compos. Sci. Technol. 128 (2016) 84–93.
[100] L. Walker, X. Hu, Scr. Mater. 41 (1999) 575.
[101] L. Walker, M.S. Sohn, X. Hu, Compos. Pt A Appl. Sci. Manuf. 33 (2002) 893.
[102] R. Palazzetti, A. Zucchelli, Compos. Struct. 182 (2017) 711–727.
[103] G.W. Beckermann, K.L. Pickering, Compos. Pt A Appl. Sci. Manuf. 72 (2015) 11–21.
[104] L. Daelemans, N. Kizildag, W. Van Paepegem, R.D. Dagmar, K. De Clerck, Compos. Sci. Technol. 175 (2019) 143–150.
[105] T. Meireman, L. Daelemans, S. Rijckaert, H. Rahier, W. Van Paepegem, K. De Clerck, Compos. Sci. Technol. 193 (2020) 108–126.
[106] S. van der Heijden, L. Daelemans, K. De Bruycker, R. Simal, I. De Baere, K. De Clerck, et al., Compos. Struct. 159 (2017) 12–20.
[107] T. Meireman, L. Daelemans, E.V. Verre, W.V. Paepegem, K.D. Clerck, Mater. Des. 195 (2020) 109050.
[108] B.N. Cox, M.S. Dadkhan, W.L. Morris, Compos. Pt A Appl. Sci. Manuf. 27 (1996) 447.
[109] D. Shu, Y. Mai, Compos. Sci. Technol. 47 (1) (1993) 25.
[110] D. Shu, Y. Mai, Compos. Sci. Technol. 49 (1993) 165.
[111] L.K. Jain, Y.W. Mai, Compos. Sci. Technol. 51 (1994) 331.
[112] W.C. Chung, B.Z. Zang, T.C. Chang, L.R. Hwang, R.C. Wilcox, Mater. Sci. Eng. A112 (1989) 157.
[113] S.T. Jenq, S.L. Sheu, Compos. Struct. 25 (1993) 427.
[114] P.J. Hogg, Mater. Technol 8 (1993) 51.
[115] K. Dransfield, D.C. Baillie, Y.W. Mai, Compos. Sci. Technol. 50 (1994) 305.
[116] D. Ratna, T.K. Chongdar, B.C. Chakraborty, Polym. Compos. 25 (2004) 165.
[117] J.M. Scott, D.C. Phillips, J. Mater. Sci. 10 (1995) 551.
[118] W.D. Bascom, J.L. Bitner, R.J. Moulton, A.R. Siebert, Composite 11 (1980) 9.
[119] P. Marshall, J. Price, Composites 22 (1991) 53.
[120] B. Zinger, S. Shkolnik, H. Hoecker, Polymer 30 (1989) 628.
[121] A.S. Crasto, S.H. Own, R.V. Subramaniam, Polym. Compos. 9 (1) (1988) 78.
[122] D. Ratna, Compos. A 39 (2008) 462.
[123] M. Murli, D. Ratna, A.B. Samui, B.C. Chakraborty, J. Appl. Polym. Sci. 103 (2007) 1723.
[124] S. Kar, D. Gupta, A.K. Banthia, D. Ratna, Polym. Int. 52 (2003) 1332.
[125] Y. Eom, L. Boogh, V. Michaud, J.-A.E. Månson, Polym. Compos. 23 (2002) 1044.
[126] J. Verry, Y. Winkler, V. Michaud, J.-A.E. Månson, Compos. Sci. Technol. 65 (2005) 1527.
[127] G. Levita, S. Petris, A. Marchetti, A. Lazzer, J. Mater Sci 26 (1991) 2348.
[128] D. Ratna, A.K. Banthia, Macromol. Res. 12 (2004) 11.
[129] B.B. Johnsen, A.J. Kinloch, R.D. Mohammed, A.C. Taylor, S. Sprenger, Polymer 48 (2) (2007) 530–541.
[130] T.H. Hsieh, A.J. Kinloch, K. Masania, J. Sohn Lee, A.C. Taylor, S. Sprenger, J. Mater. Sci. 45 (5) (2010) 1193–1210.
[131] T.H. Hsieh, A.J. Kinloch, K. Masania, A.C. Taylor, S. Sprenger, Polymer 51 (2010) 6284–6294.
[132] C. Manjunatha, A. Taylor, A. Kinloch, S. Sprenger, J. Mater. Sci. 44 (2009) 4487–4490.
[133] T. Mahrholz, J. Stängle, M. Sinapius, Compos. Pt A Appl. Sci. Manuf. 40 (2009) 235–243.
[134] S.A. Ngah, A.C. Taylor, Compos. Pt A Appl. Sci. Manuf. 80 (2016) 292–303.

CHAPTER 5

Thermoset nanocomposites

Abbreviations

POSS	polyhedral oligomeric silsesquioxanes
TEOS	tetraethoxysilane
PCN	polymer—clay nanocomposite
CNT	carbon nanotube
NMS	nanoscale mesoporous silica
CEC	cation exchange capacity
DMA	dimethyl acetamide
WAX	wide-angle X-ray diffraction
TEM	transmission electron microscopy
XRD	X-ray Diffraction
DETDA	diethyl toluene diamine
SAX	small-angle X-ray scattering
SEM	scanning electron microscope
AFM	atomic force microscope
NMR	nuclear magnetic resonance
PP	polypropylene
DDS	diamino diphenyl sulfone
TDA	tetradecyl amine
HDA	hexadecyl amine
ODA	octadecyl amine
POSS	polyhedral oligomeric silsesquioxanes
PCL	poly (ε-caprolactone)
NMS	nanoscale mesoporous silica
APTMS	3-aminopropyltrimethoxysilane
ODPA	(4,4′-oxydipthalic anhydride).
BAPP	2-bis-[4-(4-aminophenoxy) phenyl] propane
DA	Diels—Alder
PEP	poly (ethylene-alt-propylene)
PEO	poly(ethylene oxide)
MA	methyl acrylate
GMA	glycidyl methacrylate
OP	poly(ethylene-alt-propylene)-b-poly(ethylene oxide)
BCP	block copolymer
MAM	triblock copolymer poly(methyl methacrylate)-b-poly(butyl acrylate)-b-poly(methyl methacrylate)
SWCNT	single-wall carbon nanotubes
MWCNT	multiwall carbon nanotubes
FCNT	fluorinated carbon nanotubes
APTS	3-aminopropyltriethoxysilane
HNAA	half-neutralized salt of adipic acid

SAHA	sodium salts of 6-aminohexanoic acid
MAC	microwave-absorbing coating
HBAU	hyperbranched alkyd urethane
WVP	water vapor permeability
GONS	prepared graphene oxide nanosheets
GO	graphene oxide
MDI	4,4′-methylenediphenyl diisocyanate
PTMG	poly(tetramethylene glycol)
SGO	sulfonic graphene oxide
SNP	silver nanoparticles
HBUA	hyperbranched urethane alkyd
HBA	hyperbranched alkyd
HBP	hyperbranched polymer
CSR	core−shell rubber
UPE	unsaturated polyester
LNR	liquid natural rubber
LPA	low-profile additives
IPN	interpenetrating network
MFC	microfibrillated cellulose
CRGO	chemically reduced graphene oxide
HRR	heat release rate
ATH	aluminum trihydate
PHRR	peak heat release rate
PPA	polyphosphoric acid
SiC	silicon carbide
CI	carbonyl iron

5.1 Introduction

In Chapter 4, Thermoset Composites, particulate and fiber-reinforced thermoset composites have been discussed in detail. Those are examples of microcomposites because the reinforcement sizes are in the microscale. Wood is a classic example of a natural composite where cellulose fibers act as a reinforcement in a lignin matrix. The oldest example of manmade composite material is concrete where macrosized steel rods are used for reinforcement. In the due course of time, materials scientists have reduced the sizes of reinforcements from the macro- to the microscale. This is because a reinforcement with a smaller size offers a better opportunity to control their orientations and interactions with the matrix. Since such interactions play an important role in determining the overall mechanical properties of a composite, the composite property can be manipulated to a great extent [1] by reducing the size of the reinforcement. That is why, today, microcomposites have formed the basis of multicrore industries. The recent trend in composite research is to further reduce the reinforcement size up to the nanoscale. The composites made with nanometer-sized reinforcements are called nanocomposites. In a nanocomposite, it is not necessary for the reinforcement to be nanosized in all dimensions but at least in one dimension, it should be in the nanoscale (1−100 nm). For example, carbon nanotube has a diameter in nanoscale but length in microscale. Though the materials with sizes ranging from 1 to 100 nm are defined as nanomaterials, to realize the actual nanoeffect, the size should be less than 20 nm. The unique properties of nanomaterials arise due to a tremendous increase in surface area when a conventional material is converted to a nanomaterial. For example, the surface area of nanotubes of a diameter of 1 nm is 1000 times higher than the same volume of short fibers of a diameter of 1 μm.

Though nanocomposites or nanostructured inorganic organic hybrid materials are considered as a novel class of materials, they are available naturally. Nature has long demonstrated the propensity to synthesize organic–inorganic hybrids as exemplified by several ubiquitous materials such as bones, abalone, shells of mollusks, and marine sponges. For example, nacre (available in abalone shell) is composed of 99 vol.% aragonite ($CaCO_3$) and 1 vol.% biopolymer. Nacre is 2 times harder and 1000 times tougher than its main component aragonite [2]. Such a remarkable enhancement in properties due to the presence of only 1 vol.% of biopolymer is the result of a highly controlled structure, which nature produces by a process called biomineralization. It is very difficult for mankind to match such a controlled structure. So similar to many other technologies, nanocomposite technology is also an outcome of man's attempt to mimic nature.

In this context, it is relevant to mention an example of old manmade hybrid material, which was produced without a proper understanding of nanoscience [3]. This is a painting called "Maya blue" available in Bunampark, Mexico. The painting is a wonderful discovery, which retains its intense blue color even after experiencing the harsh jungle atmosphere for the several centuries. A careful analysis of the painting indicates that the coloring pigment is indigo (not so stable) and it is encapsulated within the channels of clay minerals palygorskite. Hence the material is nanostructured in nature, which is responsible for the remarkable stability offered by the painting despite constituting a coloring pigment (indigo), which itself is not so stable.

5.2 Thermoset nanocomposites

In the last two decades, extensive works have been done to develop various types of organic–inorganic hybrid materials. Various routes for synthesis of such hybrid nanocomposites are presented [3] in Fig. 5.1. One successful approach to achieve such nanocomposites is the in situ polymerization of metal alkoxides in organic matrices [4–6]. Inorganic components, especially silica, have been formed by the hydrolysis and condensation of a mononuclear precursor such as tetraethoxysilane (TEOS) in many polymer systems. Due to the loss of volatile by products formed in the hydrolysis/condensation reaction, it is difficult to control sample shrinkage after molding. Self-assembly is another important method for the synthesis of nanocomposites. This is mostly carried out through block copolymerization, which may lead to a hierarchical structure [7,8]. The third approach is to intercalate the polymer materials in a layered silicate, which leads to

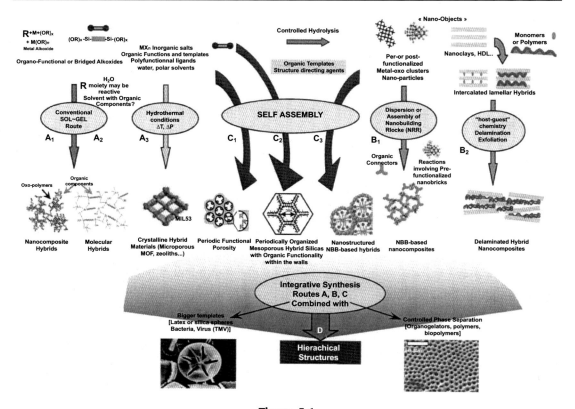

Figure 5.1
Scheme of main chemical routes for the synthesis of organic–inorganic hybrids. Source: *Reprinted from C. Sanchez, P. Julian, P. Pellerille, M. Popall, J. Mater. Chem. 15 (2005) 3559, ©2005 The Royal Society of Chemistry.*

the formation of polymer/clay nanocomposites (PCN). Nanocomposites can also be made by directly dispersing nanomaterials like carbon nanotube (CNT), nanoscale mesoporous silica (NMS), and polyhedral oligomeric silsesquioxane (POSS) in a thermoset matrix.

The main advantage of a nanocomposite is the property tradeoffs associated with conventional composites. For example, nanomodification can improve the stiffness without sacrificing toughness, can enhance barrier properties without sacrificing transparency, and offers flame retardancy without deteriorating mechanical property and color. The advantages of nanocomposites over conventional composites are given below:

1. Efficient reinforcement with minimal loss of ductility and impact strength.
2. Much lower weight penalty compared to filled polymers and conventional polymer composite systems, as nanocomposites contain a very little amount of inorganic fillers.
3. Outstanding barrier properties without requiring a multilayered design.
4. Improved heat stability and flame resistance characteristics.

5. Improved chemical and weather resistance properties.
6. Improved biodegradability for a pristine biodegradable polymer.
7. Retained transparency for a pristine transparent polymer.

In nanocomposites, a significant mechanical reinforcement is achieved by the incorporation of only a low level of clay (<5%) in contrast to microcomposites where a high level of filler incorporation is required for even a modest improvement in properties. Rubbery modifiers and rigid fillers are very often used to adjust the mechanical properties of a polymer for various applications [9]. The superiority of a nanofiller over conventional modifiers lies in the fact that nanomodification offers an improvement in mechanical properties in all aspects. The addition of a conventional filler in a polymer matrix often leads moderate increase in modulus at the cost of strength and elongation at break. However, nanofiller, if dispersed properly, can lead to increase in all the three properties modulus, strength, and % elongation (impact strength). The impermeable inorganic nanolayers like nanoclay, CNT, etc. mandate a tortuous pathway for a permeant to transverse the nanocomposite. The enhanced barrier characteristics, chemical resistance, and reduced solvent uptake of the polymer nanocomposites make them suitable for various applications. The nanocomposites especially polymer/clay nanocomposites show flame-retardant properties, which will be discussed in detail shortly. Key evidence for the resilience of nanocomposites to combustion includes the following:

1. An increase in heat distortion temperature.
2. An increase in peak heat release temperature.
3. A change in char structure.
4. A decrease in char permeability.
5. A decrease in the rate of mass loss.
6. A tendency to self-extinguish.

5.2.1 Thermoset/clay nanocomposites (PCN)

Polymer/clay nanocomposites are a new class of nanocomposite that makes use of clay materials, which are cheap and well-known filler for polymer materials. The research on polymer–clay intercalation has been reported before the 1980s [10]. However, these works were not taken in the history of polymer/clay nanocomposites as these did not result in a dramatic improvement in the physical and engineering properties of the polymers. The researchers at Toyota, Japan, demonstrated for the first time that clay (the so-called filler) can do miracles in 1993 [11,12]. While searching for lightweight material for automotive applications, they successfully developed a nylon-6/clay nanocomposite, which results in a dramatic improvement in properties compared to the pristine polymer. Subsequently, the technique was extended to thermoset resins leading to the formation of thermoset nanocomposites.

5.2.1.1 Principle of polymer/clay nanocomposite formation

The model structure of layered silicates such as montmorillonite [13] is shown in Fig. 5.2. The crystal lattice of the single 0.95-nm-thick layer consists of two tetrahedral sheets fused to one octahedral sheet of either aluminum or magnesium hydroxide (2:1 layer). These 2:1 layers are not electrostatically neutral. The excess layer charge, caused by isomorphous substitutions of Si^{4+} for Al^{3+} in the tetrahedral lattice and Al^{3+} for Mg^{2+} in the octahedral sheet, is balanced by interlayer cations, which are commonly Na^+, Ca^{2+}, or Mg^{2+} ions. The number of sites of the isomorphous substitution determines the surface charge density and hence significantly influences the surface and colloidal properties of the layered silicate [14]. The charge per unit cell is thus a significant parameter necessary to describe the layered silicates. The intermediate value for the charge per unit cell of smectites ($x \approx 0.25-0.6$) compared to talc ($x \approx 0$) or mica ($x \approx 1-2$) enables cation exchange and gallery swelling for this group of layered silicate, making them suitable for the formation of thermoset nanocomposites [15,16]. The negative surface charge determines the cation exchange capacity (CEC) (meq/100 g), which is the key factor to the organic surface modification. The crystallites are composed of as many as 100 individual layers stacking together and they are known as tactoids. The primary particles consist of compact face-to-face stacking or low-angle intergrowth of individual tactoids. Finally, weak agglomerations of these primary particles form the powders, which are agglomerates. Conventionally, these are called "particles." It is necessary to break such agglomerates in order to make a nanocomposite. The strong hydrophilic nature of the clay surface results in high interfacial tension with organic materials, making the layered silicate difficult to intercalate and disperse homogeneously in a polymer matrix. That is why in situ polymerization or blending of a polymer with clay leads to a conventional microcomposite with a particle size of about 5–15 μm [17] as shown in Fig. 5.3.

The fundamental principle behind the formation of a PCN is that the monomer is able to move in and react within the interlayer galleries. In order to facilitate the organic material

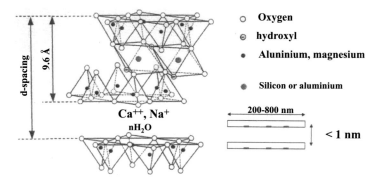

Figure 5.2
Model structure for montmorilonite clay.

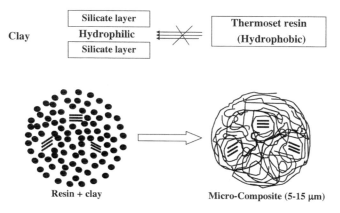

Figure 5.3
Blending of a thermoset resin with pristine clay.

to penetrate into the gallery space, alkyl ammonium cations are introduced into the clay by a simple ion-exchange method as the alkali cations residing in the clay gallery are exchangeable [18]. A typical laboratory method for the modification of clay is described below. At first, Na^+ and/or NH_4^+ exchanged clays are obtained by the reaction of aqueous NaCl and or NH_4Cl solution (~1.0 M) with the pristine clay at elevated temperature (70°C–80°C) for a specific period of time (~24 h). The resulted products are washed with distilled water until no chloride ion is detected while adding a dilute $AgNO_3$ (~0.1 N) solution. Then it is air-dried. This hydrophilic exchanged clay is then used to get organophilic alkyl ammonium exchanged clays by the following procedure. At first the neutral alkyl amine ($CH_3(CH_2)_{n-1}NH_2$, $n = 4-8$) is dissolved in ethyl alcohol and added to 1.0 N aqueous HX solution (X = Cl or Br) with constant stirring at about 60°C to get alkyl ammonium halide salts. Sometimes alkyl diamines or amino acids are also used. The Na^+/NH_4^+ exchanged clay is dispersed thoroughly into the alkyl ammonium halide salt solution under the vigorous stirring condition at 60°C–70°C for 3–6 h. The precipitate formed is isolated by filtration and washed several times with warm deionized water and water–alcohol mixture until no trace of chloride is detected while treating with dilute $AgNO_3$ solution. The treated clay is dried at 80°C in a vacuum oven until it was dried properly. The drying can also be done by freeze-drying process after dispersion of the modified clays in deionized water.

This simple method offers a wide scope of modification of clay with different types of onium ions as presented in Fig. 5.4. Such modification serves two purposes: it introduces a hydrophobic character into the clay gallery and also increases the d-spacing. The modified clay is termed organoclay or nanoclay. The hydrophobicity of the clay gallery increases with an increase in the number of carbon atoms present in the alkyl ammonium cation. However, it is difficult to introduce an alkyl ammonium cation with a large number of

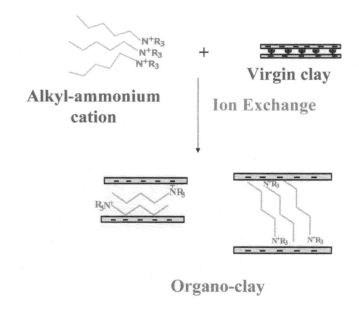

Figure 5.4
Modification of clay with alkyl ammonium cation by ion-exchange method.

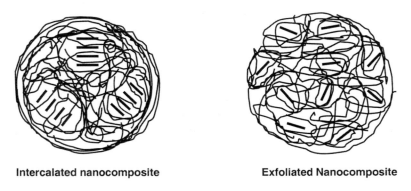

Figure 5.5
Schematic representations of intercalated and exfoliated nanocomposites.

carbon atoms. Generally, the number of carbon atom should be eight or above to generate the required hydrophobicity. Blending of a polymer or in situ polymerization with a suitably designed organoclay leads to the formation of a nanocomposite.

In general, two idealized polymer layered silicate structures are possible: intercalated and exfoliated. Schematic representations of conventional, intercalated, and exfoliated or delaminated nanocomposites are shown in Fig. 5.5. In an intercalated nanocomposite, the insertion of polymer into the clay structure occurs in a crystallographically regular fashion and a few molecular layers of polymer typically occupy the gallery region. Consequently, the properties of an intercalated composite usually resemble those of the ceramic host. In contrast, in an exfoliated nanocomposite, the individual 1-nm-thick clay layers are dispersed in a continuous polymer matrix and segregated from one another by average distances that depend on the clay loading. Hence, an exfoliated nanocomposite has a monolithic structure with properties related to those of the starting polymer.

5.2.1.2 Methods of nanocomposite synthesis

The process for synthesizing a polymer nanocomposite by various researchers can be classified into four categories, namely, monomer intercalation (in situ polymerization), covulcanization, common solvent, and polymer melt blending. Rubber-based nanocomposites are made by a covulcanization method where the nanoclay is dispersed in the rubber directly or using some liquid rubber and vulcanized [19]. In a melt mixing technique, the intercalation or exfoliation of clay particles occurs during the melting of polymer either statically or under shear. The efficiency of intercalation using this method may not be as good as that of the in situ polymerization technique. However, the approach can be applied by the polymer processing industry to produce nanocomposites based on traditional polymer processing techniques, such as extrusion and injection molding [20,21]. This process also eliminates the use of hazardous solvents used in the solution technique. Therefore this technology has played an important role in proliferating the application of PCNs.

The common solvent technique is based on a solvent system in which the polymer or prepolymer is soluble and the silicate layers are swellable. The layered silicate is first swollen in a solvent, such as water, chloroform, or toluene. When the polymer and layered silicate solutions are mixed, the polymer chains intercalate and displace the solvent within the interlayer of the silicate. Upon solvent removal, the intercalated structure remains, resulting in the formation of PCNs. Polyimide-based nanocomposites are made by using a common solvent like dimethyl acetamide (DMAc) [22].

Nanocomposites based on the other thermosetting resins like epoxy, polyurethane, unsaturated polyester resins, etc. are made by in situ polymerization method and the resulting morphology of the nanocomposite depends on various factors, for example, layered silicate charge density, the nature of interlayer exchanged ion, cure condition, nature of resin, curing agent, and their composition [23–25]. In situ polymerization of layered silicate nanocomposite was first reported for the synthesis of nylon-6 polymer nanocomposites [11,12]. In this method, the organoclay is initially swollen by the liquid monomer (ε-caprolactam) enabling polymer formation outside and inside the interlayer galleries. The layered silicate gallery surface was pretreated with 12-aminolauric acid, which takes part with the ε-caprolactam in the reaction. In some of the early studies, the layered silicate was added directly to the resin/hardener mixture [26,27]. However, the

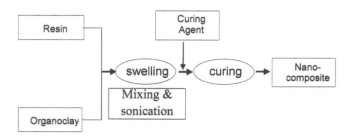

Figure 5.6
A flowchart describing the synthesis of a thermoset/clay nanocomposite.

more established methods of thermoset nanocomposite formation [28] include the preintercalation of the layered silicate by swelling in the resin for a period of time prior to the curing followed by the addition of a suitable curing agent and crosslinking. A flowchart describing the synthesis of a thermoset/clay nanocomposite is shown in Fig. 5.6. Sonication is used in many cases to break the tactoids. Another processing techniques that have been examined include the use of a three-roll mill to impart high shear forces into the system [29]. For such processing, the clay is added to the epoxy, which initially becomes more viscous and opaque, attaining clarity after shearing,

5.2.1.3 Characterization of PCN

The key question regarding the structure of a thermoset/clay nanocomposite system is whether a true nanocomposite has formed or not. If not, the material is comparable to a conventional filled micro composite. Generally, wide-angle X-ray diffraction (WAX) analysis and transmission electron microscopy (TEM) are used to elucidate the structure of a nanocomposite. Due to its easiness and availability, WAX is most commonly used to establish the nanocomposite structure [30,31]. By monitoring the position, shape, and intensity of the basal reflections from the distributed silicate layers, the nanocomposite structure (intercalated or exfoliated) may be identified. The X-ray technique is often applied to identify nanocomposite structures through Bragg's relation, which is given below:

$$\lambda = 2d\sin\theta \qquad (5.1)$$

where λ is the wavelength of the X-ray used ($\lambda = 1.5405$ Å), d is the spacing between specific diffraction lattice planes, and θ is the measured diffraction angle. The increase in d as a result of intercalation will be reflected through a decrease in θ as λ is constant. Thus any shift of peak in the X-ray spectrum of the composite is an indication of the formation of an intercalated nanocomposite. If a crystalline order is manifested as a peak corresponding to the same basal spacing as in the pristine silicate, we have a conventional composite material. An increase in the basal spacing indicates the formation an intercalated composite. An extensive layer separation associated with the delamination of the original

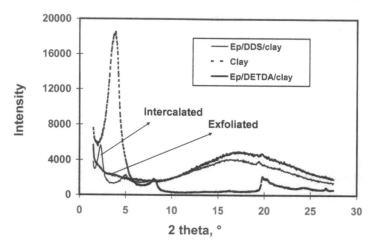

Figure 5.7
XRD plots of organoclay (modified with octadecyl ammonium cation), epoxy/DDS/clay (intercalated) and epoxy/DETDA/clay (exfoliated) nanocomposites. Source: *Reprinted from D. Ratna, N.R. Manoj, R.K. Singh Raman, R. Varley, G.P. Simon, Polym. Int. 52 (2003) 1403, ©2003 John Wiley and Sons Publishers.*

silicate layers in the polymer matrix results in the eventual disappearance of any coherent X-ray diffraction from the distributed silicate layers. Hence the absence of any corresponding peak is an indication of delamination of the clay platelets leading to the formation of an exfoliated nanocomposite. The typical X-ray diffraction (XRD) plots of clay and epoxy-based intercalated and exfoliated nanocomposites [32] are shown in Fig. 5.7. The clay shows d_{100} peak at $2\theta = 3.9$, which corresponds to a d-spacing of 2.26 nm. This indicates that modification of clay with organic ions, not only makes the clay surface hydrophobic but also results in a tremendous increase in d-space (d-space for untreated clay <1 nm), which facilitates the penetration of liquid resin into the interlayer galleries. In the intercalated nanocomposite, the peak shifted to a position at $2\theta = 3.9$, corresponding to the d-spacing of 4.42 nm. The absence of d_{100} peak in the other nanocomposite indicates extensive delamination leading to the formation of an exfoliated nanocomposite. The origin of the resultant of different morphology for the epoxy systems will be discussed later.

Although WAX offers a convenient method to determine the interlayer spacing of the silicate layers in the original layered silicates and in the intercalated nanocomposites (<8 nm), it cannot provide sufficient information about the spatial distribution of the silicate layers or any structural inhomogeneities in nanocomposites. Further complicacy arises due to the fact that some layered silicates initially do not exhibit well-defined basal reflections. This makes it difficult to investigate the peak broadening and change in intensity systematically. The elucidation of the exact structure of nanocomposite can be done by transmission electron microscopy (TEM), which allows a qualitative understanding of the internal structure, spatial distribution of the various phases, and views of the defect structure through direct visualization. It may be noted that very thin (<70 nm) and uniform samples are necessary for getting informative TEM pictures. A TEM picture of epoxy/diethyl toluene diamine (DETDA)-based nanocomposite [30] made using an octadecyl ammonium cation modified montmorillonite clay is shown in Fig. 5.8. XRD analysis of the same nanocomposite system (Fig. 5.7) indicates exfoliation. However, as evident from the TEM image that the tactoids are still present. This indicates the morphology falls in between the intercalated and exfoliated nanostructure, though XRD shows exfoliation. The advantage of XRD is that it provides the changes in layer spacing quantitatively unlike TEM, which gives only qualitative information. However, when the layer spacing exceeds 8–9 nm, the XRD feature weakens and remains no longer useful. Hence it is recommended that both WAX and TEM should be applied as complementary tools to characterize and describe the morphology of a nanocomposite. Small-angle X-ray scattering (SAX) is used

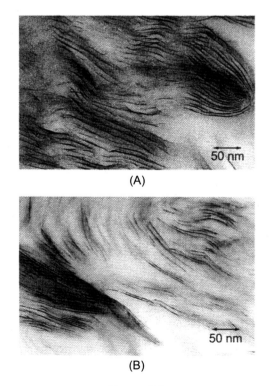

Figure 5.8
TEM micrographs of epoxy/DETDA/clay (modified with octadecyl ammonium cation) nanocomposite cured at (A) 100°C and (B) 160°C. Source: *Reprinted with permission from O. Becker, Y.B. Cheng, R.J. Verlay, G.P. Simon, Macromolecule 36 (2003) 1616, ©2003 American Chemical Society Publishers.*

to measure crystallite sizes [33]. The XRD analysis can also be extended to determine three-dimensional orientation as demonstrated by Bafna et al. [34] and Sinha Ray [31]. Nowadays, small-angle neutron scattering and dynamic light scattering are being used [35] to understand the underlying structural aspects of the silicate layers as well as the conformations of polymers in layered silicate-based nanocomposites.

High-resolution scanning electron microscope (SEM) and atomic force microscope (AFM) are increasingly becoming tools for characterizing nanocomposites. However, for a more clear picture TEM is necessary. The interlayer distance determined from AFM is generally higher than that obtained from WAX [36]. This can be explained by considering the fact that the AFM tip may deplete or distort the flexible silicate layers dispersed in a rigid thermoset matrix. Since all these measurements involve only a very small area, care must be taken to not to neglect the characterization in larger-scale inhomogeneities. Such inhomogeneities could not be detected at a very high magnification. Hence it is recommended to analyze a specimen microscopically in both the higher and lower magnifications.

VanderHart et al. [37] used solid-state nuclear magnetic resonance (NMR) (^{1}H and ^{13}C) spectroscopy for the first time as a tool to study morphology, surface chemistry, and to a very limited extent, the dynamics of exfoliated PCNs. This method uses the reduction in the spin–spin relaxation time, $^TH_1$, of a nanocomposite when compared with the neat system, as an indicator for the organoclay layer separation. It was shown that the paramagnetic Fe^{3+} ions in the crystal lattice of the montmorillonite provide an additional relaxation mechanism of the protons. The additional relaxation depends on average $Fe^{3+}-^1H$ distance, which is determined by clay concentration and dispersion of clay in the matrix.

5.2.1.4 Controlling factors for nanocomposite formation

The thermodynamic spontaneity of the intercalation process is determined by an interplay of entropic and enthalpic factors [38,39]. The confinement of the polymer inside the interlayer results in a decrease in the overall entropy of the polymer chains. However, the entropic penalty may be compensated by the increased conformational freedom of the tethered surfactant chains in a less confined environment, as the layers separate out. Even then, there is an overall decrease in entropic factor as the increase in gallery height is very small. Thus the degree of layer separation depends on the establishment of very favorable polymer surface interaction to overcome the penalty of polymer confinement. There are two different types of interactions, unfavorable apolar interaction and favorable polar interaction, which originates from the Lewis acid/Lewis base character of the layered silicates. Thus depending on the interactions and entropic factors three possible equilibrium states are generated namely immiscible (conventional), intercalated, and exfoliated.

The first condition of penetration of polymer/monomer into the clay galleries is the polarity matching between the clay gallery and the polymer or monomer. The modification of clay with alkyl ammonium cations increases the hydrophobicity of the galleries to match the polarity with the polymer/monomer to be intercalated. The hydrophobicity increases with an increase in the number of carbon atoms in the alkyl ammonium cation. It may be noted that incorporation of alkyl ammonium cation becomes more difficult with increase in the number of carbon atoms in the cation. Lan et al. [40] used a series of alkyl ammonium ions, $CH_3(CH_2)_{n-1}NH_3^+$, exchanged montmorillonite with $n = 4, 8, 10, 12, 16$, and 18, to study the intercalation of epoxy and observed that the number of carbon atoms in the alkyl ammonium ion greatly affects clay expansion before cure [40]. It was concluded that an alkyl amine cation with number of carbon atoms more than 8 was necessary for the successful formation of a nanocomposite. Zilg et al. [41] carried out a similar study using a fluorhectorite clay and reported that the number of carbon atoms in the alkyl ammonium cation had to exceed six to promote intercalation or exfoliation. A chain length of higher than eight units did not further improve the exfoliation process. Wang and Pinnavaia [42] studied a series of primary to quaternary onium (octadecyl ammonium) ion-modified clays and observed that primary and secondary onium ion-modified clays produces exfoliated nanocomposites in an aromatic amine-cured epoxy system, while the tertiary and quaternary ion-modified clays generate only intercalated one. This effect has been attributed to the higher acidity and stronger Brönsted-acid catalytic effect of the primary and secondary onium ions on the intergallery epoxy-amine curing reaction.

For a highly hydrophobic polymers like polypropylene (PP), the modification of clay gallery with an onium ion having the highest possible number of carbon atoms cannot introduce sufficient hydrophobicity to match the polarity of the clay gallery with PP. Modification polypropylene is necessary to match the polarity of the clay gallery with PP. PP nanocomposites have been successfully made by modifying PP with maleic anhydride [43]. Ratna et al. [32] investigated intercalation of epoxy/DDS and epoxy/DETDA systems using an octadecyl ammonium ion-modified clay. Under the same processing conditions, the epoxy/DDS system produces an intercalated nanocomposite and epoxy/DETDA system produces an exfoliated nanocomposite as evident from the XRD analysis shown in Fig. 5.7. The difference in results due to the change in curing agent may be attributed to a higher polarity of DDS compared to DETDA. Moreover, DETDA is liquid and offers lower viscosity of the epoxy/hardener mixture compared to DDS, which is a solid. The lower prepolymer viscosity favors penetration of polymer molecules into the clay gallery, which leads to the exfoliation. This indicated that exfoliation depended on the nature of thermoset system used especially on the viscosity of the resin/hardener system.

High processing temperatures have been used to improve intercalation. It was found that higher cure temperatures offered better exfoliation of the organosilicate in the epoxy matrix [44,45]. This improved exfoliation was mainly ascribed to the higher molecular mobility and diffusion rate of the resin and hardener into the clay galleries, leading to an improved balance between inter- and extragallery reaction rate. Simon and coworkers [30]

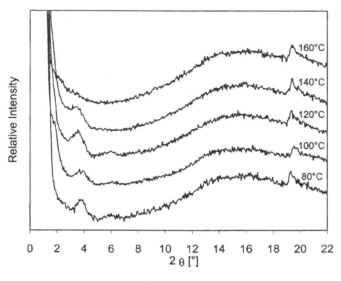

Figure 5.9
XRD plots of an epoxy/DETDA/clay (modified with octadecyl ammonium cation) nanocomposite cured at different temperature profile. Source: *Reprinted with permission from O. Becker, Y.B. Cheng, R. J. Verlay, G.P. Simon, Macromolecule 36 (2003) 1616, ©2003 American Chemical Society Publishers.*

investigated the effect of curing temperature on the intercalation of epoxy by preparing composites at different temperatures and characterizing them by XRD analysis. Fig. 5.9 shows the XRD plots of nanocomposites prepared using different curing temperatures. It is evident that d_{001} peak shifted toward lower value (indicative of an increase in d-spacing) with increasing curing temperature. This clearly indicated that an increase in cure temperature favors intercalation. However, since the window of processing temperatures is limited by the side reactions, thermal degradations, the variation of cure temperature alone may not be sufficient to form a fully dispersed true nanocomposite.

When an organoclay is dispersed in a thermoset resin mixture, the resin mixture penetrates into the gallery and the curing reaction occurs both inside the gallery and outside the gallery, which is called intragallery and extragalery reactions, respectively. For expansion of the clay gallery and delamination, the intragallery reaction should be slightly higher than extragalery reaction so that clay platelets are pushed apart. In case of epoxy resin, the alkyl ammonium cations (present in the galleries) catalyze the epoxy/amine reaction. Hence the intragallery reaction is faster than the extragallery reaction. For other thermosets, this can be achieved by attaching a reactive moiety with the alkyl ammonium cation. For example, a schematic drawing of synthesis of polyimide/clay nanocomposite using nonreactive and reactive modifier [46] is shown in Fig. 5.10. In the first case (A), the modifier contains NH_3^+ and $-COOH$ group. NH_3^+ replaces Na^+ and forms an ionic bond with the silicate layers and the $-COOH$ group does not react with amic acid (precursor of PI). In second

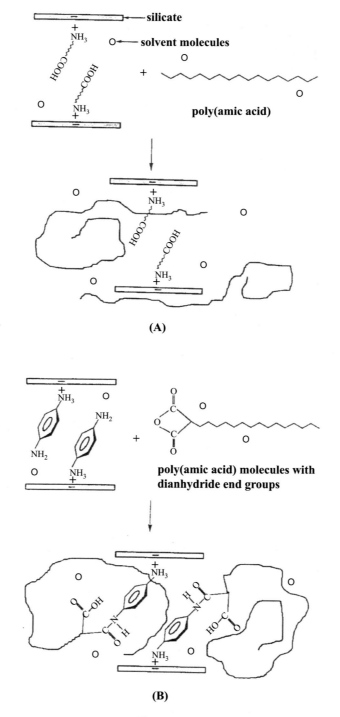

Figure 5.10
Schematic drawing of montmorilonite intercalated by poly (amic acid) and (A) nonoreactive swelling agent containing one functional group NH_3^+ (reversible) and (B) reactive swelling agent having double functional groups NH_3^+ and NH_2 (irreversible). Source: *Reprinted with permission from H.L. Tyan, Y. Liu, K. Wei, Chem. Mater. 11 (1999) 1942, ©1999 American Chemical Society Publishers.*

case (B), the modifier contains two $-NH_2$ groups: the protonated one forms ionic bond with the silicate layers and the other $-NH_2$ group can react with the dianhydride end group of polyamic acid as shown in Fig. 5.10. The incorporation of a reactive moiety helps the clay platelets to push apart leading to a higher extent of intercalation or exfoliation. Lan et al. [40] emphasized the importance of the balance between intergallery and extragallery reaction rates, as well as the accessibility of the resin and hardener mixture to the clay galleries for the exfoliation process. It is reported that the monomers penetrate and swell the silicate layers until a thermodynamic equilibrium is reached between the polar resin molecules or resin/hardener blend, and the high surface energy of the silicate layers. Actually further increases in the distance between organoclay platelets require the driving force in terms of elastic forces generated due to curing reaction or homopolymerization to overcome the attractive electric forces between the negative charge of the silicate layers and the counterbalancing cations in the galleries. Decreasing polarity during reaction of the resin in the galleries displaces the equilibrium and encourages further monomer to diffuse into, and react within, the silicate galleries. Yebassa et al. [47] investigated the effect of processing parameters on the morphology of vinyl ester resin-based nanocomposites and observed that shearing and sonication during the cure improve the exfoliation process. They also reported that treatment of clay with a reactive amine and addition of comonomer like styrene favor exfoliation.

It has been demonstrated by in situ SAX studies of an epoxy nanocomposite that the d-spacing increases with the advancement of curing reaction [48]. Only mixing of resin with the clay did not result in a significant increase in d-spacing. Chen et al. [49] divided the interlayer expansion mechanism into three stages. First stage is the initial interlayer expansion due to resin and hardener intercalation into the silicate galleries. The second stage is the interlayer expansion state where the interlayer spacing steadily increases due to intergallery polymerization. The third stage of interlayer expansion is characterized by a decreased interlayer expansion rate. In some cases, a slight decrease in interlayer spacing could be observed before the cessation of gallery change, due to restrictions on further extragallery change because of gelation.

Park and Jana suggested [50] the possible mechanism of nanoclay exfoliation in epoxy clay nanocomposites. They considered various opposing forces, which simultaneously act when a thermoset resin cures in presence of clay as shown in Fig. 5.11. They theorized that the primary force behind clay layer exfoliation in this nanocomposite system is elastic force. The elastic force exerted by the crosslinked epoxy molecules inside the clay galleries pushed out the outermost clay layers from the tactoids against the opposing forces arising from electrostatic and van der Waals attractions. The study suggested further that a complete exfoliation of clay structure can be produced if the ratio of shear modulus to complex viscosity is maintained in such a way so that the elastic forces inside the galleries outweigh the viscous forces offered by the extragallery resin. The authors also studied the

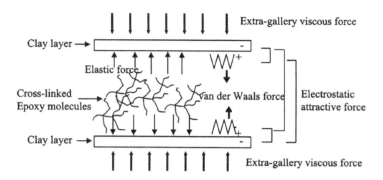

Figure 5.11
Schematic illustration of forces acting on a pair of clay layers during intercalation and exfoliation.
Source: *Reprinted with permission from J. Park, S.C. Jana, Macromolecules 36 (2003) 8391, ©2003 American Chemical Society.*

effects of quaternary ammonium ions in epoxy clay nanocomposite system on the curing rate, catalytic and plasticization actions. The study indicated that the degree of exfoliation strongly depends on the ratio of storage modulus of intragallery crosslinking epoxy molecules. The plasticization effect, which lowers the storage modulus, was found to be present in both aliphatic and aromatic epoxy and also a strong function of the nature of curing agent. However, in aromatic system, plasticization effect is less compared to the corresponding aliphatic one and thus there is a difference in exfoliation behavior of the nanocomposite. The aliphatic epoxy shows better exfoliation.

The use of a low reactive aromatic diamine as a curing agent favors exfoliation by retarding the crosslinking reaction in the extragallery region as reported by Kong and Park [51]. By analyzing the real-time small-angle XRD patterns, they suggested that the exfoliation occurred in a stepwise manner. The first step is the intercalation of epoxy into the intergallery region of the alkyl chain of clay. The second increase is by the clay layer folding, translating and expanding from the self-polymerization of epoxy with the catalytic effect of protonated alkyl amine and the third increase originates from the crosslinking of epoxy resin with the alkyl amine and the hardener in the intergallery region.

Ratna et al. [52] reported epoxy/clay nanocomposite cured with polyether amine and used a strategy of manipulation of extragallery and intragallery reactions to achieve good dispersion. Keeping the extragallery reaction the same they varied the intragallery reaction by changing the molecular weight of polyether amine modifier. Note that the reactivity of a polyether amine decreases with increasing molecular weight of the polyether chain. The detailed kinetics of polyeheramine cured epoxy systems are presented in Chapter 2, Properties and Processing of Thermoset Resin. The pristine clay (cloisite Na^+) is modified with amines having polyether chains of different molecular weights (300, 600, and

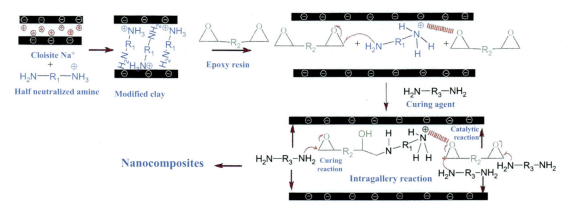

Figure 5.12
A schematic representation of curing reaction of epoxy/clay nanocomposites. Source: *Reprinted with permission from S.B. Jagtap, V.S. Rao, S. Barman, D. Ratna, Polymer, 63 (2015) 41–51, ©2015 Elsevier Publishers.*

900 g mole^{-1}: J-300, J-600, and J-900) to get three modified clay designated as CL-300, CL-600, and CL-900. Abovementioned three modified clays were separately incorporated in epoxy/J-900 network to make three composites (epoxy/J-900/CL-300, epoxy/J-900/CL-600, and epoxy/J-900/CL-900). A schematic representation of curing reaction of epoxy/clay nanocomposite is shown in Fig. 5.12. The TEM microphotographs of epoxy/J-900/CL-300, epoxy/J-900/CL-600, and epoxy/J-900/CL-900 are shown in Fig. 5.13. The presence of silicate nanolayers are observed in all the three nanocomposites. However, it was observed that as the intragallery reaction rate (R_{intrag}) increases, the level of clay dispersion increases. For preparation of all the three epoxy/clay nanocomposites, the epoxy/J-900 system has been used as a matrix. Therefore the extragallery reaction remains same only the intragallery reaction tend to vary. Since extragallery reaction is same in all the cases, it can be stated that the level of dispersion increases with increasing the difference between rate of intragallery and extragallery reaction (R_{extrag}), that is, ($R_{intrag} - R_{extrag}$). This helps in developing the elastic force, which pushed the clay platelets apart leading to a better dispersion.

5.2.1.5 Properties of PCN

Messersmith and Giannelis [53] carried out the dynamic mechanical characterization of an anhydride-cured epoxy-based nanocomposites and reported 58% increase in glassy storage modulus due to the incorporation of 4 vol.% of layered silicate. Ratna et al. [32] reported a similar observation using an aromatic amine-cured epoxy. The dynamic modulus versus temperature plots for the DETDA-cured epoxy system and the corresponding nanocomposites [32] are shown in Fig. 5.14. The storage modulus of the nanocomposite is

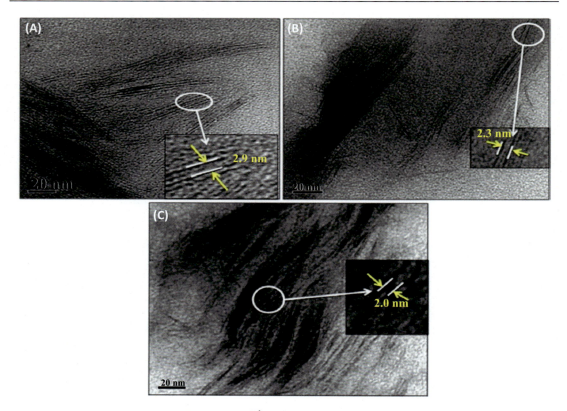

Figure 5.13
The TEM microphotographs of the three nanocomposites made with clays modified with amines of different polyether chain length. (A) EP/J-900/CL-900 [Rintrag /Rextrag ≈ 1], (B) EP/J-900/CL-600 [Rintrag /Rextrag = x > 1], and (C) EP/J-900/CL-300 [Rintrag /Rextrag = y > x]. Source: Reprinted with permission from S.B. Jagtap, V.S. Rao, S. Barman, D. Ratna, Polymer, 63 (2015) 41–51, ©2015, Elsevier publishers.

significantly higher compared to the neat epoxy. About 50% increase in storage modulus was achieved as a result of incorporation of 5 wt.% of clay into the glassy epoxy matrix. Although neither crosslinking density nor the modulus of such exfoliated silicate layers is known, this behavior can be attributed to the extraordinarily large aspect ratio of exfoliated silicate layers, which is frequently assumed to exceed 1000. The increase in modulus is more prominent in rubbery state, which shows about 300% improvement. The nanocomposites show higher T_g compared to the neat epoxy network. This can be explained by considering the confinement of polymer chain as a result of intercalation into the interlayer gallery of the clay. However, in case of highly crosslinked thermoset systems, a decrease in T_g was reported as a result of the incorporation of organoclay [30]. The probable reasons are the reduction in crosslink density due to the addition of nanoclay as happened in glass fiber-reinforced composites or the plasticization of the crosslinked matrix

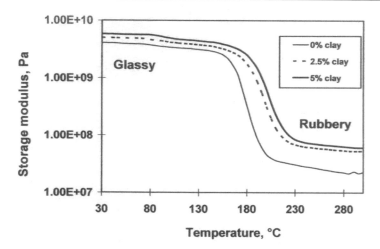

Figure 5.14
The dynamic modulus versus temperature plots for DETDA-cured epoxy system and the corresponding nanocomposites containing 2.5 and 5 wt.% clay (modified with octadecyl ammonium cation). Source: *Reprinted from D. Ratna, N.R. Manoj, R.K. Singh Raman, R. Varley, G.P. Simon, Polym. Int. 52 (2003) 1403, ©2003 John Wiley and Sons Publishers.*

by the unreacted resins/hardener present in the network. Note that it is difficult to pinpoint the exact reason due to the complexity of the cure reaction and possible side reaction involved especially when the resin is cured at more than 150°C.

Zilg et al. [54] characterized the tensile properties of anhydride-cured epoxies using various organoclays (modification with different alkyl ammonium cations). All systems investigated exhibited an increase in Young's modulus, although in several cases, decreases in tensile strength and failure strain were observed. It is thought that the loss in tensile strength might be related to an inhomogeneous network density due to the different cure rates of the intergallery and extragallery reactions or due to the internal stresses developed in the material. Pinnavaia and coworkers [55] found that m-phenyl diamine-cured epoxy (high T_g)-based PCNs showed only a marginal improvement of tensile strength and modulus compared to the pristine epoxy network. On the other hand, a significant reinforcement was observed for polyether amine-cured epoxies (low T_g) [27]. Ratna et al. [56,57] studied the reinforcing effect of a octadecyl ammonium cation modified clay on polyether amine cured (low T_g) epoxy systems. The reinforcing effect as evident from the plots of tensile strength and % elongation at break as a function clay concentration for a polyether amine-cured rubbery epoxy is shown in Fig. 5.15. The incorporation of 15 wt.% clay into a rubbery epoxy resulted in about 800% increase in tensile strength. It was also noted that the value of % improvement in tensile strength increases with increasing % elongation of the matrix. This can be explained by considering the phenomenon that under tensile loading, the clay platelets undergo orientation toward the applied load. Higher flexibility of the matrix

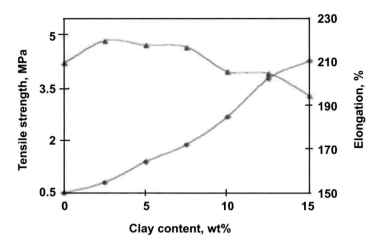

Figure 5.15
Effect of clay loading on tensile properties of rubbery epoxy network: tensile strength (-■- ■) and % elongation at break (-▲-▲-). Source: *Reprinted from D. Ratna, B.C. Chakraborty, H. Dutta, A. K. Banthia, Polym. Eng. Sci. 46 (2006) 1667, ©2006 John Wiley and Sons Publishers.*

facilitates the orientation of clay platelets resulting in a better strengthening effect. As in this case the room temperature is above T_g, the elasticity of the matrix increased thus some amount of applied stress will be transferred to the clay platelets.

Jagtap et al. [58] synthesized different types of organoclay by modifying closite-Na^+ (southern clay product) separately with alkyl amine of different chain lengths (e.g., tetradecyl, hexadecyl, and octadecyl amines). The nanoclays are designated as closite-tetradecyl amine (TDA), closite-hexadecyl amine (HDA), and closite-octadecyl amine (ODA). They have compared the reinforcing effects of the synthesized nanoclays as well as two commercial clay (Closite 10 A, 20 A, southern clay product) using a rubbery epoxy (DGEBA/J-900) as a matrix. XRD analysis indicated that the *d*-spacing gradually increases with increasing number of carbon atoms of the amine used for modification. The *d*-spacing of closite-Na^+, closite-TDA, closite-HDA, and closite-ODA are 12.65, 16.53, 17.70, and 18.62 nm, respectively. The increase in *d*-spacing helps in the intercalation of a thermosetting resin into the gallery space. The incorporation of long-chain cations into the gallery of the clay imparts hydrophobic character into the gallery resulting in polarity matching with the intercalating resin. The stress–strain diagram of a rubbery epoxy and the related nanocomposites are shown in Fig. 5.16. It may be noted that a significant improvement in tensile modulus and strength was observed for addition of 3 wt.% of organoclay. This is a unique advantage of nanocomposites where considerable reinforcement is achieved with a lower concentration of filler (<5 wt.%). Note that conventional fillers are required to be added in much higher concentration (>20 wt.%) for a modest improvement in tensile strength. In the figure, we can also see the comparison of the performance of two commercial nanoclays (Closite 10 A,

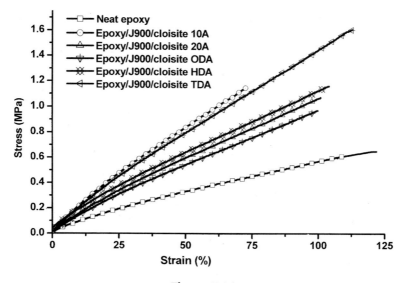

Figure 5.16
The stress–strain diagram of rubbery epoxy and the related nanocomposites. Source: *Reprinted with permission from S.B. Jagtap, V. Srinivasa Rao, D. Ratna, J. Rein. Plastic Compo. 32 (2013) 183, ©2013 Sage Publishers.*

20 A, southern clay product) and three synthesized nanoclays (closite-TDA, closite-HDA, and closite-ODA). The reinforcing ability found to increase with the increasing number of carbon atoms present in the modifier. The octadecyl amine-modified clay shows the best performance. This can be explained by considering the higher *d*-spacing of closite-ODA as discussed above. It is also observed that both the tensile strength and % elongation were found to increase unlike in the case of conventional filler where a modest improvement in tensile strength is associated with a significant decrease in % elongation.

Mechanical properties of polyurethane (PU)/clay nanocomposites have also been investigated [59–62]. The tensile strength of the PU/clay (closite 30B) nanocomposites increased with an increasing nanoclay content up to 3 wt.%. The tensile strength of PU-3 (PU/clay nanocomposite with 3 wt.% clay) nanocomposite sample (15.8 MPa) was more than two times higher than that of the pristine PU (7.1 MPa). This is due to the wide dispersion of clay nanolayers within the polymer matrix in which the induced forces could be spread uniformly within the polymer structure. However, the tensile strength of the nanocomposites decreased by the further increasing the concentration of nanoclay. The tensile strength of 7 wt.% clay containing nanocomposite was reported to be 7.6 MPa. The observed decline in tensile strength is attributed to incomplete distribution of nanoclay particles through the polymer matrix resulting in low particle surface area and high local stress concentration. On the other hand, Zhang et al. [63] reported a continuous increase in the tensile strength of PU polymer by the addition of 0.5–7 wt.% Cloisite 10 A.

The difference between the observed trends is due to the difference between the types of the used nanoclay. The maximum tensile strength obtained by means of Cloisite 10 A particles (7 wt.%) was relatively equal to that achieved using 3 wt.% of Cloisite 30B, demonstrating the effective performance of the latter nanoparticle.

The PCNs also characterized by a reduced permeability [64,65] with respect to moisture, solvents, etc. The permeability of a nanocomposite can be expressed in terms of clay concentration and aspect ratio as given below:

$$P_{\text{nanocomposite}} = \frac{(1 - \phi)P_{\text{matrix}}}{1 + \alpha\phi/2} \tag{5.2}$$

where P is the permeability, α is the volume fraction, and ϕ is the aspect ratio of the layered silicate used. The permeability of a nanocomposite can be enhanced by using a higher concentration of layered silicate with a higher aspect ratio. Note that the clay concentration should not exceed the optimum concentration beyond which layered silicates are agglomerated. That means the layered silicate should be well dispersed. The reduced permeability in the case of nanocomposite can be explained by considering a "tortuous path" concept. Because of the presence of impervious silicate layers, a permeant has to follow a tortuous path compared to a smooth one in the pure polymer as shown in Fig. 5.17.

Shah and Gupta [66] investigated moisture diffusion through vinyl ester-based nanocomposites using two different organoclays and showed that the clays were effective in reducing the diffusion coefficient and permeability. However, they did not get any correlation between morphology and water diffusion properties. Similar results were reported by Ratna et al. for nanocomposites of vinyl ester and clay (Nanocore I-30) [67]. The same group has also carried out [56] the swelling study of rubbery epoxy-based nanocomposites in chloroform to assess the solvent resistance. The solvent uptake in terms of weight gain as a function of time is shown in Fig. 5.18. It was observed that the rubbery

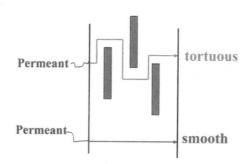

Figure 5.17
A tortuous model for permeability of thermoset/clay nanocomposites.

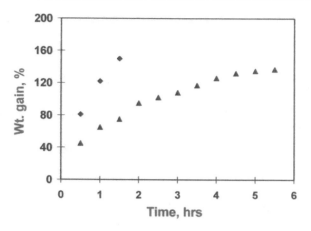

Figure 5.18
Swelling study (in chloroform) of a rubbery epoxy (-■-■-) and 15 wt.% clay containing nanocomposite (-▲-▲-). Source: *Reprinted with permission from D. Ratna, B.C. Chakraborty, H. Dutta, A.K. Banthia, Polym. Eng. Sci. 46 (2006) 1667, ©2006, John Wiley and Sons Publishers.*

epoxy disintegrates into pieces after exposure of about 2 h, whereas the nanocomposites survive the exposure up to 6 h without any damage. This clearly indicates that the solvent and water resistance of thermoset systems can be reduced significantly and the application of thermosets can be broadened.

5.2.2 POSS and silica-based nanocomposites

Polyhedral oligomeric silsesquioxanes (POSS) are a class of cage-like molecules that have been investigated recently as effective nanofillers for thermoset resins [68–72]. A typical POSS monomer possesses the structure of a cube (octameric framework) with eight organic corner groups. Depending on the nature of R group, POSS monomer can be nonfunctional (no reactive group), monofunctional (one reactive group), and polyfunctional (more than one reactive group). Polymerization of a multifunctional POSS monomer produces cage-like molecules with a general formula $[RSiO_{3/2}]_n$ where n is an even number in the range of 6–12 and R can be various types of organic groups, one or more of which is/are reactive or polymerizable. POSS materials are derived from the hydrolysis and condensation of trifunctional organosilanes [59]. POSS can be dispersed in a thermoset matrix to make POSS-based nanocomposites and the morphology depends on the degree of reaction between the POSS and the thermoset resin. Williams and coworkers [68] investigated the effect of incorporation of POSS on the dynamic mechanical properties and static compression properties of epoxy/POSS nanocomposites and reported an increase in dynamic glassy modulus from 2.8 to 3.43 GPa and uniaxial compression strength (at 25°C) from 80 to 100 MPa due to the incorporation of 50 wt.% POSS. The studies on modification of epoxy resins by a series of octasilsesquioxanes (with a variety of R groups such as aminophenyl and dimethyl siloxylpropylgylcidyl ether groups) indicated [61,62] that the dynamical mechanical properties, fracture toughness, and the thermal stability of the epoxy hybrids were closely dependent on the types of R group, tether structure between epoxy matrices, and the defects in silsesquioxanes cages.

Ni and Zheng [71] designed a novel amphiphilic macromolecule, that is, POSS-capped poly (ε-caprolactone) (PCL), which has both the organic (PCL) and inorganic (silsesquioxane cage structure) portions. POSS-capped PCL was synthesized via ring-opening polymerization of ε- caprolactone using 3-hydroxypropylheptaphenyl POSS as the initiator. The PCL subchain is miscible with the epoxy whereas octaphenyl-POSS is immiscible with the epoxy resin before and after the curing reaction. POSS−PCL forms a homogeneous solution with epoxy/ 4,4'-bis methylenebis (2-chloroaniline) (MOCA) system suggesting that no phase separation takes place before the curing. During the curing, reaction-induced phase separation takes place via self-assembly leading to the formation of a nanostructured epoxy as shown in Fig. 5.19. The hydrophobic character of the surface increases with an increase in the concentration of POSS as evident from the static contact angle measurements because the inorganic portion gets enriched on the surface. The contact angle of the cured epoxy/MOCA system with respect to water was 85.2 degrees whereas the same for a nanocomposite containing 40 wt.% POSS-capped PCL was 96.2 degrees.

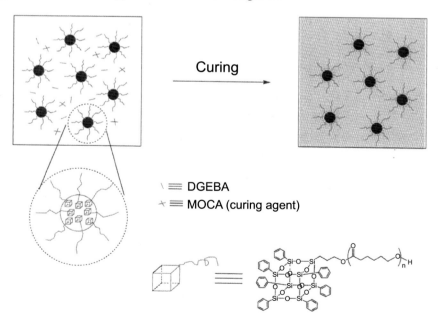

Figure 5.19
Preparation of POSS-based nanostructured epoxy thermoset. Source: *Reprinted with permission from Y. Ni, S. Zheng, Macromolecules, 40 (2007) 7009, ©2007, American Chemical Society Publishers.*

A reactive filler or modifier can be incorporated into a thermoset resin in two different ways: it can be directly blended and cured (chemical bonding occurs during curing) or the modifier/filler is first prereacted with the thermoset resin followed by curing, which is known as a prereact method. The effects of the direct blending and prereact methods on the morphology of the modified networks have been elaborately discussed for the liquid rubber-modified epoxy system in this chapter and in Chapter 4, Thermoset Composites. However, for thermoset/POSS composites, most of the studies have used the direct blending method. Recently Zucchi et al. [72] analyzed the effect of a prereaction for an aromatic amine-cured epoxy system using two types of monofunctional POSS, namely, glycidyloxypropyl-heptaisobutyl POSS (iBu-GlyPOSS) and glycidyloxypropyl-heptaphenyl POSS (iPh-GlyPOSS). The difference between the two POSS is the nature of the inert group; iBu-GlyPOSS contains seven isobutyl groups whereas iPh-GlyPOSS contains seven phenyl groups. The presence of phenyl groups causes stronger POSS−POSS interactions and that is why iPh-GlyPOSS is not compatible with the epoxy/amine mixture on direct blending. However, when iPh-GlyPOSS is prereacted with the epoxy, it makes a homogeneous solution, which on curing produces a two-phase nanostructure. Thus epoxy/iPh-GlyPOSS composites can only be made by a prereact method. On the other hand, the presence of iBu groups weakens the POSS−POSS interactions, which makes it possible to process epoxy/iBu-GlyPOSS composites by both a direct blending and a prereact method. Note that isobutyl group is less compatible with the epoxy and the hardener, which are aromatic in nature. This factor favors the phase separation of POSS from the epoxy matrix. The SEM photographs for the fracture surfaces of epoxy/iBu-GlyPOSS composites made without prereaction and with prereaction are shown [72] in Fig. 5.20. The composites made with a prereaction show a globular morphology with a micron-size POSS-rich phase. On the other hand, when the same composite is made after prereacting iBu-GlyPOSS with the epoxy followed by curing, irregular clusters (composed by a set of small individual domains) are formed. Thus prereaction produced a significant change in the morphology of the nanocomposites. This is because in the case of nonreacted POSS the reaction-induced phase separation takes place rapidly in the early stage of the pregel region, which produces a globular morphology. The prereaction enhances the compatibility of POSS with the epoxy; as a result, the phase separation takes place slowly at the later stage of the pregel region leading to the formation of clusters. Due to its better compatibility with the epoxy/MOCA system, iPh-GlyPOSS does not undergo any reaction-induced phase separation and mostly remains dissolved in the matrix resulting in a reduction in T_g of the nanocomposites.

As evident from the above discussion, POSS can effectively enhance the properties of a thermoset. However, the high cost of POSS limits its application for various applications. Recently, nanoscale mesoporous silica (NMS) has been viewed as a suitable candidate for reinforcement of thermoset resin because of its excellent thermal and mechanical stability and inherent air void inside the nanochannels. The major advantage of NMS over POSS is

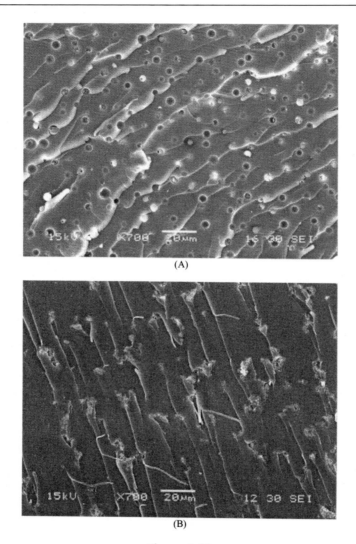

Figure 5.20
SEM micrographs of final morphologies generated in an epoxy network modified with iBu-GlyPOSS: (A) nonreacted iBu-GlyPOSS and (B) prereacted iBu-GlyPOSS. Source: *Reprinted with permission from I.A. Zucchi, M.J. Galante, R.J.J. Williams, E. Franchini, J. Gally, J.F. Gerard, Macromolecules, 40 (2007) 1274, ©2007 American Chemical Society Publishers.*

that NMS is much cheaper compared to POSS. NMS can be easily dispersed in a polar resin through polar interactions and covalent bonding. Recently, Cheng et al. [73] have successfully dispersed NMS in polyimide (PI) through a chemical bonding approach. The outline of synthesis of PI/NMS nanocomposite is shown in Fig. 5.21. NMS containing a large number of silanol groups inside the mesopore channels and on the surface was reacted with 3-aminopropyltrimethoxysilane (APTMS) to prepare an amino-functionalized NMS

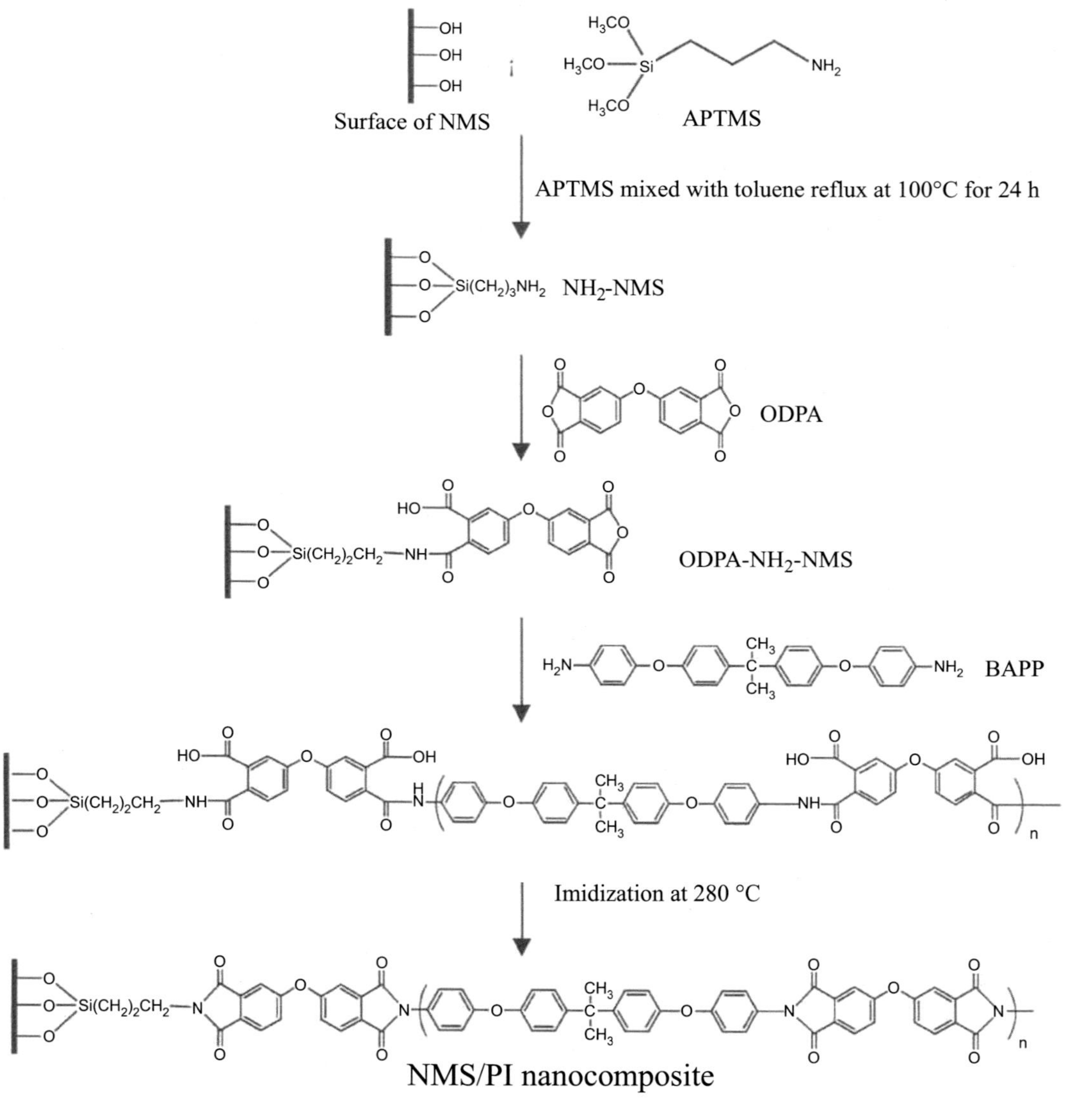

Figure 5.21
Synthesis of NMS/polyimide nanocomposite. Source: *Reprinted with permission from C. Cheng, H. Cheng, P. Cheng, Y. Lee, Macromolecules 39 (2006) 7583, ©2006 American Chemical Society Publishers.*

(Fig. 5.21). The modification involves the reaction between the hydrolysable Si(OMe)$_3$ moiety of APTMS and silanol groups of NMS. The amine-functionalized NMS is then converted to anhydride-functionalized one by reacting the same with ODPA (4,4'-oxydipthalic anhydride). Further reaction with 2-bis-[4-(4-aminophenoxy) phenyl] propane (BAPP) followed by imidization produces the PI/NMS nanocomposite. NMS-based nanocomposites offer better compatibility compared to other nanocomposites and show a lower dielectric constant than the corresponding pure polyimide. The covalent functionalization route discussed above resulted in good dispersion of NMS in the PI matrix. The TEM images of PI/NMS nanocomposites containing different concentrations of NMS are shown in Fig. 5.22. It can be observed that the nanocomposite displays an

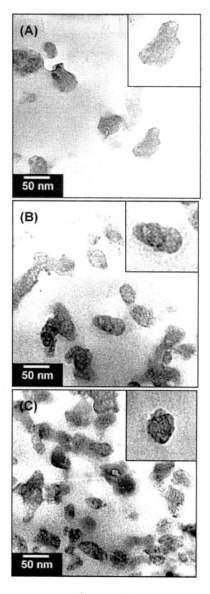

Figure 5.22
The TEM images of a PI/NMS nanocomposite containing (A) 1 wt.%, (B) 3 wt.%, and (C) 10 wt.% of amino-functionalized NMS. Source: *Reprinted with permission from C. Cheng, H. Cheng, P. Cheng, Y. Lee, Macromolecules, 39 (2006) 7583, ©2006 American Chemical Society Publishers.*

effective dispersion even at a higher concentration of NMS (10 wt.%). Note that it is very difficult to maintain a uniform dispersion of nanomaterials beyond 5 wt.% of filler loading. The chemical interaction of PI chains NMS through the amino group is the driving force to attain such good dispersion.

Recently, Zhang and coworker [74] reported a comparison of toughening effect among EP/POSS nanocomposites with different functional groups and factors, which influence the toughening of the related nanocomposites. POSS with three kinds of different functional groups containing octavinyl-POSS (OV-POSS), methacryl-POSS (M-POSS), and dodecaphenyl-POSS (DP-POSS) were incorporated into epoxy resin separately by a solution blending method to prepare the EP/POSS nanocomposites. The three nanocomposite samples made by incorporating 1 wt.% of OV-POSS, DP-POSS, and M-POSS exhibit maximum values of impact strength 22.60, 24.90, and 29.61 kJ m^{-2} and the minimum values of storage modulus 2.88, 2.25, and 2.14 GPa, respectively. This indicates that EP/M-POSS nanocomposites show the best performance. This can be attributed to the presence of intermolecular hydrogen bonds between the hydroxyl group the epoxy and the carbonyl group of the M-POSS. The dispersion of DP-POSS in the epoxy matrix is better than that of OV-POSS, which is attributed to the $\pi-\pi$ accumulation effect between the para-substituted benzene ring in the polymer and the single-substituted benzene ring in the DP-POSS.

POSS has also been used to develop a self-healing thermoset [75]. The details of self-healing thermosets have been discussed in Chapter 2, Properties and Processing of Thermoset Resin. A thermally mendable POSS nanocomposite has been prepared by directly crosslinking of POSS that brings eight furan-end functional groups with bismaleimide molecules via a Diels−Alder (DA) reaction. The POSS nanocomposite is stiff and transparent at room temperature but can heal the cracks thermally and the healing has excellent repeatability. When a crack is generated, it can be healed by heating the material to cleave the DA bonds between furan-end functionalized POSS and bismaleimide at above 94°C. At a lower temperature, the furan and maleimide moieties reconnect and provide self-healing property. This process is reversible and can be used to mend the internal cracks of the material.

5.2.3 Block copolymer-based nanocomposite

Block copolymers show an interesting property of self-assembling on polymeric length scale (nanostructure) into well-defined ordered nanostructure when blended with a suitable polymer [76−90]. The resulting morphology depends on various factors like molecular weight, block symmetry and interblock repulsive strength developed as a result of contact of dissimilar copolymer blocks. Amphiphilic block copolymers of poly(ethylene-alt-propylene) (PEP) and poly(ethylene oxide) (PEO) have been successfully used to generate self-assembled nanostructures with epoxy. Grubbs et al. [78] investigated both unreactive and reactive amphiphilic block copolymers to prepare epoxy-based nanocomposites. They have used a block copolymer, which consists of two blocks: one block is a random copolymer of methyl acrylate (MA) and glycidyl methacrylate (GMA) and the other one is polyisoprene. The epoxy groups of GMA can react with the amine group of the curing agent and participate in the curing reaction. Initially, MA/GMA copolymer block is solvated in epoxy and phase separation occurs due to the desolvation as a result of curing. The principle of such reaction-induced phase separation for epoxy/liquid rubber system has already been discussed in Chapter 3, Toughened Thermoset Resin. A comparison of phase separation processes of nonreactive, for example, poly (ethylethylene)-b-poly (ethylene oxide) (OP) and reactive block copolymers as mentioned above is shown in Fig. 5.23. Garate et al. [79] worked on controlling the morphology of self-assembled block copolymers (BCP) in the epoxy matrix using a reaction-induced microphase separation mechanism. For this purpose, they have designed an epoxidized poly-(styrene-b-isoprene-b-styrene) block copolymer incorporating amine-reactive functionalities (eSIS-AEP) in the epoxidized block as a modifier for an epoxy system. Oxirane ring-opening reaction of an epoxidized poly(styrene-bisoprene-b-styrene) (eSIS) block copolymer was performed using 1-(2-aminoethyl) piperazine as a nucleophile to produce a novel block copolymer (eSIS-AEP) capable of reacting toward diglycidyl ether of bisphenol A (DGEBA) during the curing reaction of the epoxy system. The incorporation of eSIS-AEP to the epoxy system as a modifier led to the formation of a nanostructured material with sphere-like nanodomains that were formed before curing by self-assembling of poly styrene) blocks as shown in Fig. 5.24.

As mentioned above, block copolymers may assemble into the matrix leading to the formation of nanostructured morphologies. The assembly type of block copolymers (BCP) depends upon a lot of factors like block length, the molecular weight of BCP, the interaction

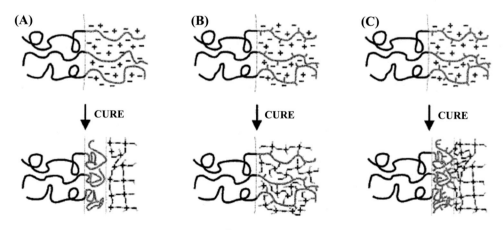

Figure 5.23
Comparison of microphase separation processes during cure of nonreactive OP copolymers (A) and reactive copolymers: copolymer cures within epoxy matrix (B) or copolymer cures interfacially after expulsion from epoxy phase (C). Source: *Reprinted with permission from R.B. Grubbs, J.M. Dean, M.E. Broz, F.S. Bates Macromolecules 33 (2000) 9522, ©2000 American Chemical Society Publishers.*

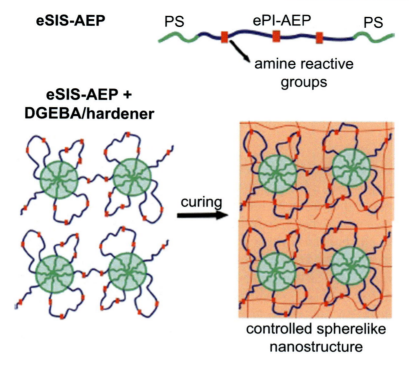

Figure 5.24
Schematic illustration of the formation of sphere-like nanodomains in the eSIS-AEP + DGEBA/hardener system. Source: *Reprinted with permission from H. Garate, S. Goyanes, N.B. D'Accorso, Macromolecules, 47 (2014) 7416—7423, ©2014 American Chemical Society Publishers.*

between BCP and matrix, and the concentration of the BCP in the matrix. Three different types of morphologies have been reported formed in the BCP modifies epoxy systems, such as spherical, worm-like, and vesicle type. Generally, the symmetric block copolymers assemble into spheres and asymmetric copolymers assemble either into a worm-like or vesicle morphology. The toughening effect is found to be different for different types of morphology of BCP. The advantage of BCP is that they toughen epoxy resin without sacrificing the modulus and T_g of the epoxy network. Moreover, the concentration required for toughening is also much lower compared to conventional toughening agents like liquid rubber. In spherical type micelle, it forms a core—shell type morphology, in which the rubbery part of the block forms the core and the epoxy-loving part forms the shell. When impact load is applied, the stress concentrates on the rubbery core, which leads to the cavitation of the core and subsequently causing plastic deformation of the surrounding matrix. This absorbs the energy of the impact and thereby imparts toughening effect to the brittle thermoset.

Redline et al. [80] modified the epoxy with 0%—6% of poly(ethylene-alt-poly)-b-(ethylene oxide) (PEP-PEO) and polystyrene-b-poly(ethylene oxide) (PS-PEO) diblock copolymers.

For the preparation of PEP-PEO diblock copolymer, the first isoprene was anionically polymerized and hydrogenated to give the structure of PEP. Further ethylene oxide was polymerized to give PEP-PEO type block copolymer. The resultant molecular weight (Mn) of corresponding PEP-PEO and PS-PEO was 27,000 and 56,500 g mole^{-1}, respectively. Both the PEP-PEO and PS-PEO form well-dispersed, spherical micelles with approximately 25 nm diameters in all concentrations and impart toughening effect (four times increase in G_{IC} for epoxy/PEP-PEO and two times increase in G_{IC} for epoxy/PS-PEO) without a significant sacrifice in T_g.

Liu and coworkers [81] studied the effect of amphiphilic PEP-PEO block copolymer on fracture properties of DGEBA-based epoxy resin. The composite samples were prepared by solvent casting method using acetone as a solvent. After mixing, the solvent was removed by a dynamic vacuum from room temperature to 130°C; thereafter, the catalyst was added to the mixture and cured at 200°C for 2 h. The epoxy-loving portion, that is, PEO blocks forms the shell structure while PEP forms the core structure. They have achieved spherical as well as worm-like morphologies by optimizing the molecular weight and PEO part of PEP-POP and crosslinking density. It may be noted that creating worm-like micelle structures is not straightforward because the composition window for successfully controlling and stabilizing this morphology is very narrow. In TEM micrograph (Fig. 5.25), we can see the tubular structure of diameter in the range of 10−15 nm. The incorporation of about 5 wt.% BCT resulted in more than 100% increase in fracture toughness and flexibility (% elongation at break) without any sacrifice in glass transition temperature. The author attributed these results to a combination of several toughening mechanisms like limited matrix shear yielding, voiding, or debonding at the interface between the phases, crack tip blunting, crack bridging, and viscoelastic energy dissipation.

Dean et al. [82] made a wide variety of PEO-based block copolymers using three polymerization techniques: anionic polymerization, anionic polymerization followed by polymer modification, and living free-radical polymerization. Block copolymers were incorporated into the epoxy system by solvent blending technique as discussed above. The block copolymers are reported to self-assemble into vesicles, spherical micelles, and worm-like micelles. It was reported that the self-assembly of block copolymer depends upon the symmetry of the block, interaction with the matrix and to some extent to the structure of the block repeat units. When the weight fraction of PEO (W_{PEO}) in the block copolymer is less than 0.37, then the formation of vesicles is favored. If W_{PEO} is greater than 0.42, a spherical micelle is formed and in between the spherical micelle and vesicles there lies a narrow region of W_{PEO}, in which worm-like micelle can be seen. All the modified epoxy compositions exhibit higher fracture toughness compared to pure epoxy. However, the composition with vesicle morphology was reported to be superior to the formulations with spherical and worm-like micelles in improving fracture toughness.

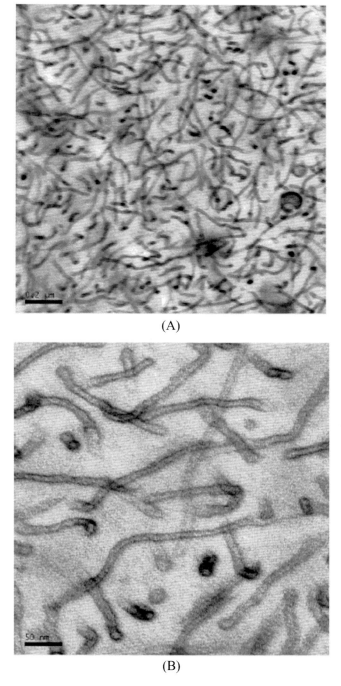

Figure 5.25
TEM micrographs of BCP worm-like micelle-modified epoxy at (A) low magnification and (B) high magnification. Source: *Reprinted with permission from J. Liu, Z.J. Thompson, H.J. Sue, F.S. Bates, M.A. Hillmyer, M. Dettloff, H. Pham, Macromolecules, 43(2010), 7238, ©2010 American Chemical Society Publishers.*

Pang et al. [83] studied the effect of poly(ethylene-alt-propylene)-b-poly(ethylene oxide) (OP) block copolymers on the adhesive strength of epoxy resin. The block copolymers of varying molecular weight were synthesized using anionic polymerization. The symmetric block copolymers (OP-S1, OP-S2) formed spherical micelle while the asymmetric copolymers (OP-V) with shorter poly (ethylene oxide) blocks formed small bilayer vesicles in the epoxy resin. The mechanical properties of neat epoxy networks and 5 wt.% BCP-modified networks are shown in Table 5.1. For all the BCP-modified epoxies studied, there is at least a threefold increase in K_{IC} and an eightfold increase in the G_{IC} compared to the neat epoxy. With a BCP self-assembling with spherical morphology, the adhesive strength increased up to nearly 46%. By contrast, a 28% reduction in adhesion strength was observed in the modified epoxies with bilayer vesicle morphology. Cavitation and void formation is reported to be the main toughening mechanism for increment in fracture toughness and adhesive strength.

Acrylic block copolymers such as poly(styrene-b-butadiene-b-methyl methacylate) and poly(methyl methacylate-b-butylacrylate-b-methyl methacrylate) were also investigated [84] as a toughening agent for epoxy resins. Depending upon the composition and ratio of block copolymers, worm-like and spherical micelles were obtained. The toughness of the thermoset increased with the addition of the copolymers without affecting the T_g of the epoxy thermoset.

Tao and coworkers [85] used the triblock copolymer poly(methyl methacrylate)-b-poly(butyl acrylate)-b-poly(methyl methacrylate) (MAM) to modify bisphenol A epoxy resin to improve its toughness. The good compatibility of MAM with epoxy resin was established by dissolution experiments, which show that MAM was completely dissolved in epoxy in 15 min at 150°C. Compared to the neat epoxy, the modified epoxy with 10 phr MAM shows a 91.5% increment in K_{1c} and 83.5% increase in the impact strength. However, the improvement in impact strength is associated with a decrease in tensile strength and modulus. This has been attributed to a decrease in crosslink density.

Bajpai A and coauthors [86] have modified the bisphenol F-based epoxy with the two different self-assembling block copolymers: (1) constituting of a center block of poly(butyl acrylate) and two side blocks of poly(methyl) methacrylate (BCP 1) and (2) poly(ethylene

Table 5.1: Bulk fracture and tensile properties of modified epoxies.

Sample	K_{IC} (MPa m$^{-1/2}$)	G_k (J m^{-2})	E (GPa)	σ_y (MPa)
Neat	0.68 ± 0.05	135 ± 18	3.02 ± 0.06	71.5 ± 04
OP-S1	1.94 ± 0.02	1090 ± 30	3.06 ± 0.04	62.4 ± 0.5
OP-S2	2.06 ± 0.09	1260 ± 120	2.97 ± 0.06	61.7 ± 0.6
OP-V	2.07 ± 0.14	1280 ± 180	2.96 ± 0.05	60.4 ± 0.4

Source: *Reprinted with permission from V. Pang, Z.J. Thompson, G.D. Joly, F.S. Bates, L.F. Francis, ACS Appl. Polym. Mater. 2 (2020), 464–474.*

oxide)-b-poly(butylene oxide) (PEO-PBO) diblock copolymer (BCP 2). The loading levels of BCP1 and BCP2 are varied from 1 to 12 wt.%. The maximum fracture toughness and fracture energy values obtained were $K_{IC} = 2.75$ MPa m$^{-1/2}$ and $G_{IC} = 2.37$ kJ m^{-2} with 10 wt.% addition of BCP 1, which are 366% and 227% higher in comparison to the neat epoxy system. Similarly, for BCP2-modified systems, the maximum value of $K_{IC} = 1.65$ MPa m$^{-1/2}$ and $G_{IC} = 1.10$ kJ m^{-2} was obtained for epoxy modified with 12 wt.% of BCP2. The toughening effect is associated with a decrease in tensile strength and modulus.

Li et al. [87] modified the epoxy thermoset with amphiphilic poly dimethyl siloxane-block-poly (ε-caprolactone) (PDMS-PCL) diblock copolymer and investigated their self-assembled morphologies. The author focuses on the spherical and worm-like morphologies and their effect on the toughness values. PDMS-PCL diblock polymer was synthesized using a ring-opening polymerization of ε-CL in presence of PDMS. They synthesized four different PDMS-PCL diblock copolymers with varying concentrations of PCL. The copolymers with a lower concentration of PCL resulted in spherical morphology and the size of nanoparticles increases with increasing concentration of block copolymer in the epoxy matrix. For the copolymers with a higher concentration of PCL, a combination of spherical and short worm-like nanostructures are observed. Unlike other block copolymers discussed above, PCL-based one required incorporation of higher concentration (20 wt.%) of BCP for significant enhancement of fracture toughness. However, beyond 5 wt.% of BCP the T_g of the modified network tend to decrease.

The block copolymer-modified epoxies discussed above is summarized in Table 5.2 so that the reader gets the information at a glance.

5.2.4 CNT-based nanocomposites

Since the discovery of carbon nanotubes (CNT) by Ijima in 1991 [91] and investigation of its properties by others subsequently, considerable attention has been made to carbon nanotube-based nanocomposites [92]. Carbon nanotubes are seamlessly rolled sheets of a hexagonal array of carbon atoms with a diameter ranging from few Angstroms to several tens of nanometer across. These nanometer-sized tubes exist in two forms: single-wall carbon nanotubes (SWCNT) in which the tube is formed from only a single layer of carbon atoms and multiwall carbon nanotubes (MWCNT), in which the tube consists of several layers of coaxial carbon tubes. A graphite sheet can be rolled up in many ways to form an SWCNT. The rolling action disturbs the symmetry of the planar system and an axial direction is imposed. Thus various types of tubes are formed namely metallic, semimetallic, or semiconducting. The exceptionally high strength (10–50 GPa) and elastic modulus (~ 1 TPa) of these tubes render them ideal candidates as ultrastrong reinforcement for composites. In addition, CNTs exhibit extremely high electrical and thermal conductivity.

Table 5.2: A comparison of various block copolymer-modified epoxy systems: fracture property and morphology.

Type of epoxy system	Modifier system	% Loading	K_{IC} (MPa$\sqrt{m}$)	G_{IC} (J m^{-2})	T_g (°C)	Morphology	Reference
Diglycidyl ether of bisphenol A	PEP-PEO and PS-PEO	0–4	1.85	~820	83	Nanosized spherical micelle	[80]
Diglycidyl eher of bisphenol A	PEP-PEO	0–5	1.79	–	152	Worm-like micelle	[81]
Poly (bisphenol A-co epichlorohydrin)	PEO-PI, PEO-PEP and PEO-PB	0–5	–	~530	–	Vesicle	[82]
Diglycidyl ether of bisphenol A	PDMS-PCL	0–40	3.83	–	~175	Worm-like morphology	[87]
Diglycidyl ether of Bisphenol A	SBM, MAM	0–10	3	4130	97	Worm-like morphology	[84]
Diglycidyl ether of bisphenol A	PEP-PEO	0–5	2.07	1280	–	Vesicles	[83]
Diglycidyl ether of bisphenol A	MAM	0–20	1.39	–	~130	Spherical micelles	[85]
Diglycidyl ether of bisphenol F	MAM, PEO-PBO	0–10	2.75	2370	142	Spherical micelle	[86]

Due to their one-dimensional nature, charge careers can travel through nanotubes without scattering. As a result, nanotubes can carry very large current densities (up to 100 MA cm^{-2}). Hence unlike layered silicate, they can be used to make conducting composites. The related composites offer much interesting applications like sensors, actuators, electrolytes, and conductive composites.

The technology of polymer/CNT nanocomposites is less matured compared to the contemporary technology of PCN. The probable reasons are difficulty in the synthesis of CNTs on large scale and their dispersion in a thermoset matrix unlike layered silicates, which are easily available and the related intercalation chemistry is well understood. Typically, CNTs tend to agglomerate as bundles in solvents or in the host resin and if dispersed, reagglomerate soon thereafter due to electrostatic attraction. Uniform dispersion within the polymer matrix and improved nanotube/matrix, wetting and adhesion are critical issues in the processing of these nanocomposites. Slipping of nanotubes when they are assembled in ropes significantly affects the elastic properties of the composite. In addition to slipping of tubes that are not bonded to the matrix in a composite, the aggregates of nanotube ropes effectively reduce the aspect ratio of the reinforcement [91]. As established for microfiber-reinforced composite, matrix−CNT interfacial bonding is a critical parameter, which controls the efficiency of stress transfer from CNT to the matrix and the efficiency of stress transfer, in turn, dictates the mechanical properties of the composites. Since the magnitude of CNT strength is very high (almost 10 times higher than typical carbon fiber), a very high interfacial shear strength may be required for more efficient strengthening of polymers with CNTs. Thus the current interest in using CNTs and anticipated potential applications for CNT-reinforced polymer composites demands a better understanding of CNT−matrix interfacial characteristics.

As discussed for PCN, processing of thermoset/CNT nanocomposites [93−100] also involves the use of ultrasonication, high shear mixing, etc. Initial academic research on polymer/CNT nanocomposites has been focused on single-wall carbon nanotubes (SWCNT) because of their simpler structure. Zhu et al. [101] treated SWCNT with dicarboxylic acid peroxide to generate alkyl carboxyl groups attached to SWCNT. The treated SWCNTS are further reacted with a diamine to make amine-functionalized SWCNT. These amino groups (covalently attached to SWCNT) can readily react with epoxy resin along with the added curing agent. Thus a crosslinked structure with covalent bonds between nanotubes and epoxy matrix is formed. The addition of 1 wt.% of such modified nanotube resulted in about 25% increase in tensile strength and 30% increase in both modulus and % elongation. This behavior is quite different from the one observed when conventional microfillers are incorporated into polymer matrix, that is, elongation at failure forms drastically when particulate or short fibers are incorporated into a polymer matrix. It may be noted that the addition of 1 wt.% of pristine SWNT in epoxy under similar conditions resulted in little enhancement in mechanical properties. This clearly demonstrates that through suitable modification, it is possible to realize the real nanoeffect in CNT-reinforced composites. Miyagawa et al. [102] have used fluorinated single-wall CNT for making anhydride-cured epoxy nanocomposite by sonication method as the fluorine atoms present in the nanotubes help to disrupt the Van der Waals forces between the nanotubes leading to homogeneous dispersion. However, during sonication, the fluorine atoms from free radical and result in partial breakage of the epoxy rings.

Khare and coworker [103] prepared the epoxy-based nanocomposites reinforced with amido-amine-functionalized single-wall carbon nanotubes (FCNTs) to elucidate the role of the matrix−filler interphase in the enhancement of mechanical and thermal properties of these nanocomposites. The presence of amido-amine groups not only help to achieve a good dispersion but they also form covalent bonds with the crosslinked epoxy matrices. They have also prepared a model structure of epoxy/FCNT nanocomposite from the simulation studies and compared it with the same for epoxy/pristine CNT-based system. A significant sacrifice in T_g (66K) is observed in the case of an epoxy/pristine CNT nanocomposite due to the formation of the compressible interphase region caused by weak matrix − filler interactions [104]. The strong chemical interaction of FCNT with matrix reduces the compressibility of the interphase and the nanocomposite exhibits T_g comparable to the cured epoxy. Additionally, the modification with FCNT resulted in a 50% enhancement in elastic modulus, a 12% increase in thermal conductivity compared to the neat crosslinked epoxy.

Although SWNT has a simpler structure and offers effective reinforcement to a thermoset matrix-like epoxy, the extremely high cost of SWCNT considerably restricts the commercialization of the related composites. On the other hand, the production of multiwall carbon nanotube (MWCNT) has already been scaled up by the industry and MWCNTs are available in sufficient quantities and at a reasonable cost. Therefore the research interest has been diverted toward the technological proliferation of MWCNT-based systems. Zhou et al. [105] incorporated 0–0.4 wt.% of MWCNT in the epoxy matrix by sonication technique and evaluated their mechanical properties. Improvement in property in all respect was observed up to 0.3 wt.% of the nanotube. Beyond this concentration, the strength and elongation decrease though the modulus keeps increasing. As a result of the addition of 0.3 wt.% nanotubes, the tensile strength increases from 93.5 MPa to 121 MPa and % elongation increases from 4.0 wt.% to 7.5 wt.%. Schadler et al. [106] evaluated the mechanical properties of MWCNT-epoxy nanocomposites in both tension and compression. They reported better reinforcement in terms of improvement in modulus (3.63–4.5 GPa) in compression than that in tension (3.1–3.71 GPa). This indicates a better load transfer in compression than in tension. Raman spectroscopic analysis indicates that the Raman peak position (indicative of strain in the carbon–carbon bonds under loading) shifts significantly under compression but not in tension. This can be explained by considering the fact that the load transfer in compression can be thought of as a hydrostatic pressure effect while the stress transfer in tension depends on the matrix to nanotube bonding. As an explanation to this behavior, it is proposed that during load transfer from matrix to MWCNT only the outer layers are stressed in tension whereas all the layers respond in compression. Very recently, Shen et al. [107] studied the tensile behavior of a randomly distributed epoxy/CNT nanocomposite with high CNT loading by experimental and finite element analysis (FEA). The EP/CNT nanocomposites with CNT loading of 22~25 wt.% are prepared by the resin film infusion process. The obtained tensile strength and modulus of the as-prepared EP/CNT nanocomposites are increased by 141.7% and 175.3%, respectively, compared to the pristine EP. When the CNT/EP prepreg (uncured) is prestretched with a tensile deformation of 2.5% to improve the alignment of CNTs, the yielded tensile strength and modulus of the cured composite are further improved by 6.3% and 10.8%, respectively. FEA result reveals that CNTs are the main stress carriers, and the load-carrying capacity of CNTs is closely correlative to their alignments. The maximum stress level undertaken by CNTs is 23-fold that of the EP matrix. Moreover, CNTs would affect their nearby stress fields within EP at a distance of 50–60 nm, which indicates that stress can be effectively transferred from EP matrix to CNTs.

Various strategies have been adopted to address the issue of achieving good dispersion and interfacial adhesion of MWCNT in a thermoset matrix. Surface oxidation of MWCNT by acid treatments, which generates polar groups on the surfaces of MWCNTs, has been proven as an effective way to improve the dispersion [108,109] of CNT. The oxygen functionality generated on the MWCNT surfaces creates electrostatic repulsion that prevents the reagglomeration of nanotubes. Once such polar groups (-OH, -COOH) are introduced into CNTs, the surface can be tailored to make them suitable for dispersion in various polymer matrices by covalent grafting methods using both "graft to" [110,111] and "graft from" [112,113] approaches. The "graft to" approach is to graft polymer chains onto the nanotube surface by direct esterification or amidation reactions and the "graft from" approach involves the growth of polymer chains from an initiator covalently immobilized on the nanotube surface (surface-initiated polymerization). Coleman et al. [114] published an exhaustive review on the covalent functionalization of CNTs and various aspects of polymer/CNT composites. Yaping et al. [115] functionalized MWCNT with amino groups by treating with an ethanol solution of triethylenetetramine and used the amino-functionalized CNTs for reinforcing epoxy resin. They reported an optimum MWCNT concentration of 0.6% where about a 100% increase in flexural strength was achieved. Yang and coworker [116] investigated the cure behavior and thermal stability of MWCNT/epoxy nanocomposites. MWCNTs are subjected to acid treatment and further functionalized by ethylenediamine/dicyclo hexylcarbodiimide/tetrahydrofuran solution. It was reported that MWCNTs have the retardation effect on the cure reaction of epoxy resin, while the functional groups on the surface of amine-modified MWCNTs tend to accelerate the cure reactions.

Recently Yaghoubi et al. [117] carried out surface modification of commercially available hydroxylated MWCNT (OH-MWCNTs) (Neutrino Corporation, Tehran, Iran) by silanization with 3-aminopropyltriethoxysilane (APTS) and dipodal silane (DSi). They have incorporated the silane-modified MWCNTs into a PU system to make PU/MWCNT nanocomposite for rigid foam applications. Uniform dispersion of MWCNT in PU matrix has been reported up to 1.5 wt.% of MWCT concentration and the nanocomposite-based foam offers higher cell density compared to pure PU-based foam prepared using the same technique. The elastic modulus and tensile strength of the PU matrix are enhanced by about 53% and 20%, respectively, as a result of the incorporation of 1.5% MWCNT into the PU matrix.

The covalent functionalization, although widely used for the preparation of polymer/MWCNT nanocomposite as discussed above, suffers from several limitations. First, the process involves handling concentrated nitric acid, which may pose a potential health hazard. Further removal of acid after treatment is a cumbersome task. The acid goes with the effluent, which causes environmental pollution. Second, the surface oxidation creates structural defects, onto the surfaces of CNTs, which disturb the π electron cloud and deteriorate the intrinsic properties of CNT. An alternate method is to use a surfactant, which interacts with the MWCNT physically and improves dispersion [118,119]. The advantage of this strategy is that it is possible to achieve homogeneous dispersion without disturbing the π electron clouds of CNT as opposed to chemical functionalization. Both the neutral and ionic modifiers have been reported for improving the dispersion of nanotube in the thermoset matrices.

Gong et al. [120] used surfactants as a wetting agent to improve the dispersion and showed about a 30% increase in dynamic storage modulus and an increase in T_g from 63°C to 88°C, due to the addition of 1 wt.% of nanotubes. However, effective load transfer from the matrix to the nanotube is only possible when there is a chemical interaction between the nanotube and matrix so that the nanotube forms an integral part of the matrix. Balogi et al. [121] investigated the effect of three structurally different and commercially available neutral surfactants, namely, BYK-110, Tween-80, and Nonidet-P40 on the dispersion of MWCNT in an epoxy matrix. It was observed that the addition of 0.1 wt.% unmodified MWCNT hardly shows any improvement in mechanical properties due to the poor dispersion of unmodified MWCNT in the epoxy matrix as confirmed by the TEM analysis. The improvement of dispersion was demonstrated for all the three surfactants and the performance of Nonidet-P40 was found to be the best in achieving good dispersion of MWCNTs in the epoxy matrix and improving the thermomechanical properties of the related nanocomposites. This can be explained by considering the $\pi-\pi$ interaction between the MWCNT and Nonidet-P40 and H-bonding interaction between the -OH groups of Nonidet-P40 and oxirane ring of epoxy resin as shown in Fig. 5.26.

Ratna and coworker [122,123] investigated rubbery epoxy/MWCNT nanocomposites using an organic acid salt-assisted dispersion technique. They have used half-neutralized salt of organic dicarboxylic acid or aminocarboxylic acid. A half-neutralized sodium salt of an organic diacid salt contains a carboxylate ion and a free carboxylic acid group. The sodium ion attached to the carboxylate ion interacts with the π electron cloud of MWCNT [119,124] and the carboxyl group can form a chemical bond with the epoxy. The organic diacid salts with a varying number of carbon atoms have been investigated and the half-neutralized salt of adipic acid (HNAA) has been found to be most effective. Fig. 5.27 shows the TEM pictures of unmodified MWCNT and HNAA-modified MWCNT. It may be noted that unmodified MWCNTs are highly agglomerated and bundled. On the other hand, after modification, the nanotubes are exfoliated due to cation$-\pi$ interactions between the MWCNTs and HNAA. In fact, we can see the individual tubes in the TEM photograph of the modified MWCNT. The unmodified MWCNTs, when dispersed in tetrahydrofuran, settle down slowly on storage, whereas the modified MWCNT forms a stable dispersion due to modification of the surface with HNAA. The photograph of sonicated mixtures (modified MWCNT in THF) was taken [123] after storage of 1 month and presented in Fig. 5.27. The modified MWCNTs form a stable suspension, whereas the unmodified MWCNTs settle down. For preparing the HNAA-modified MWCNT-based nanocomposites, the modified MWCNT is first prereacted with the epoxy so that the free -COOH groups of HNAA get linked with the epoxy resin. When the modified epoxy is cured with a polyether amine hardener to produce nanocomposites, the nanotubes become an integral part of the

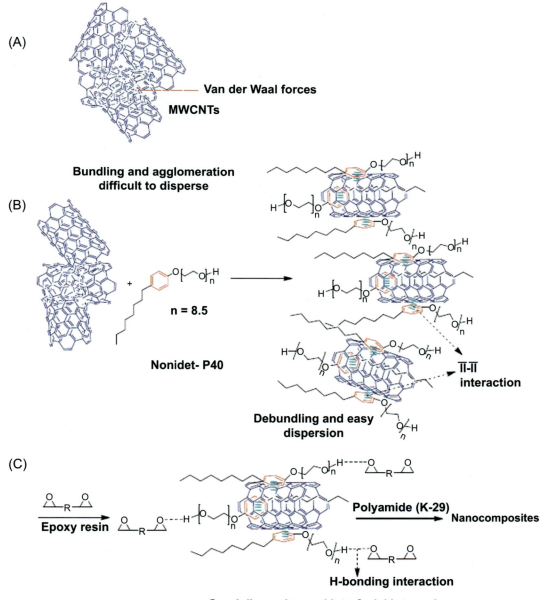

Figure 5.26
Schematic representation of π−π interaction between Nonidet-P40 and MWCNTs and H-bonding between Nonidet-P40 and the epoxy matrix. Source: *Reprinted with permission from R.B. Naik, S.B. Jagtap, R.S. Naik, N.G. Malvankar, D. Ratna, Prog. Org. Coat. 77 (2014) 1883, ©2014 Elsevier Publishers.*

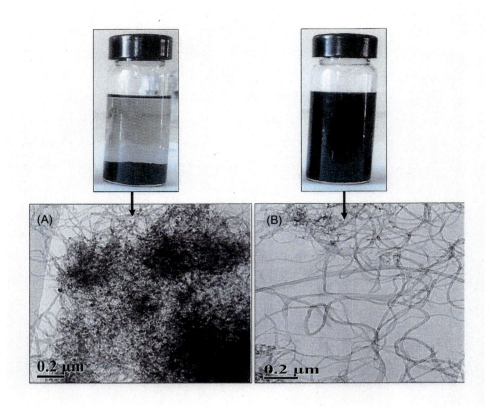

Figure 5.27
TEM images of the (A) unmodified MWCNTs and (B) HNAA-modified MWCNTs (1:1). Source: *Reprinted with permission from S.B. Jagtap, D. Ratna, J. Appl. Polym. Sci. 130 (2013) 2610, ©2013 John Wiley and Sons Publishers.*

crosslinked structure through the chemical bonding between HNAA and the epoxy resin. The reaction scheme for prereaction and curing showing cation–π interactions and chemical bonding is presented in Fig. 5.28.

The effect of cation–π interactions is also reflected in the morphology and mechanical properties of the nanocomposites. Fig. 5.29 shows the TEM microphotographs of the EP/MWCNT (0.5 wt.%), EP/MWCNT (0.5 wt.%)/HNAA, and EP/MWCNT (1 wt.%)/HNAA composites. The EP/MWCNT (0.5 wt.%) composite showed agglomerated MWCNTs because of the high degree of entanglement (Fig. 5.29A). On the other hand, the EP/MWCNT (0.5 wt.%)/HNAA composite showed well-dispersed MWCNTs (Fig. 5.29B). This indicated that the cation–p interaction between HNAA and the MWCNTs helped to break the agglomeration of the MWCNTs. Furthermore, at higher concentrations (1 wt.%) of MWCNTs, it was difficult to prevent the agglomeration, as shown in Fig. 5.29C.

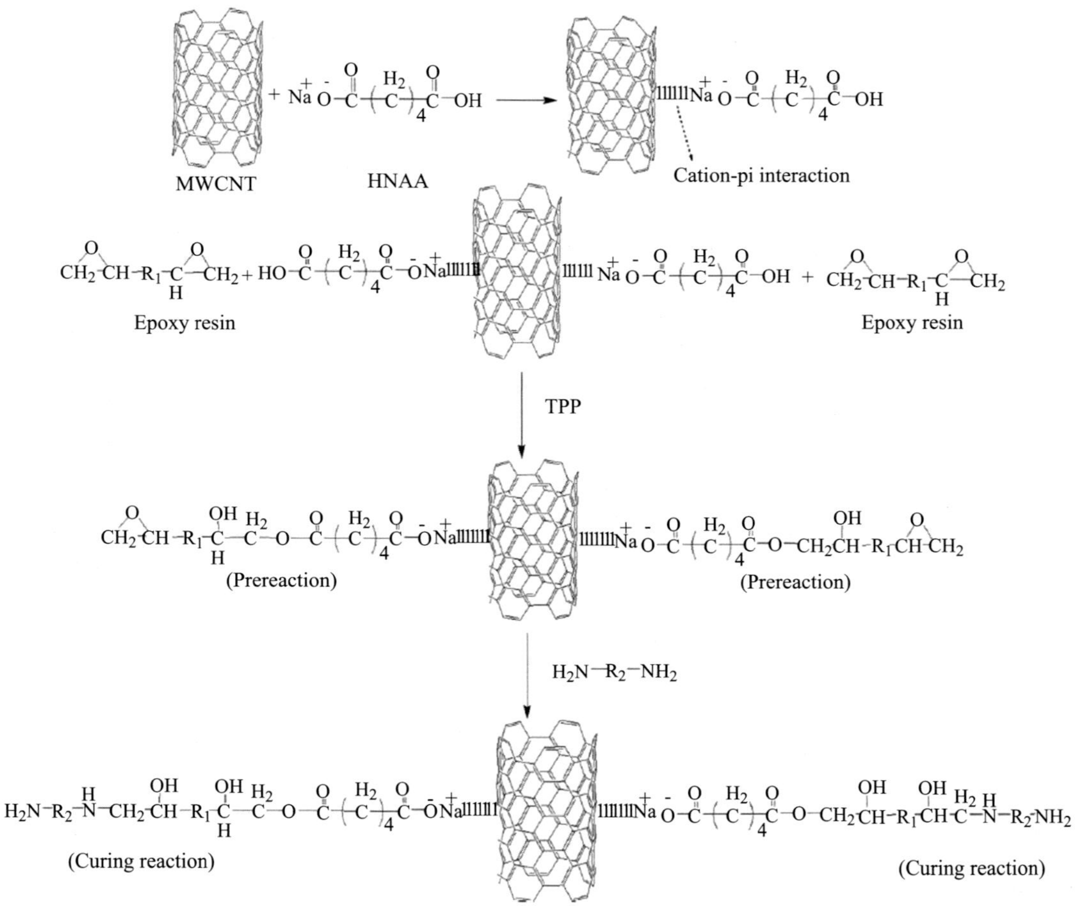

Figure 5.28
Representation of the cation–π interactions of HNAA the MWCNTs, chemical bonding between the HNAA and epoxy resin, and formation of the epoxy network. Source: *Reprinted with permission from S.B. Jagtap, D. Ratna, J. Appl. Polym. Sci. 130 (2013) 2610, ©2013 John Wiley and Sons Publishers.*

This explains why the tensile properties decreased beyond 0.5 wt.% of the HNAA-modified MWCNTs. The incorporation of 0.5 wt.% HNAA-modified MWCNTs showed 70% and 190% improvements in the tensile strength and tensile modulus, respectively.

A similar reinforcing effect was reported using another modifier, that is, sodium salts of 6-aminohexanoic acid (SAHA). In this case, the amine group reacts with epoxy resin. Since the amine group is equally reactive like the amine hardener, prereaction is not required. In the case of HNAA (discussed above), the carboxyl group does not react with the epoxy at room temperature; therefore a prereaction step was conducted to ensure its chemical bonding with the epoxy. The mechanical properties of various epoxy-based nanocomposites [122] made with varying concentrations of SAHA-modified MWCNT are shown in

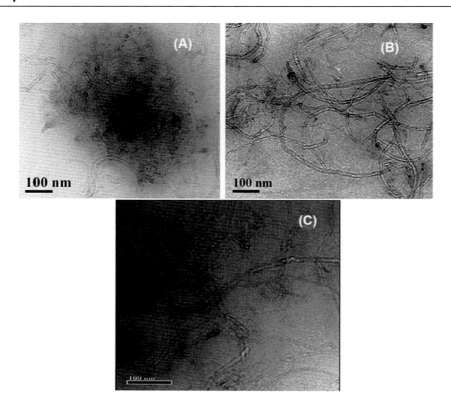

Figure 5.29
TEM images of the epoxy/MWCNT composites: (A) EP/MWCNT (0.5 wt.%), (B) EP/MWCNT (0.5 wt.%)/HNAA, and (C) EP/MWCNT (1 wt.%)/HNAA. Source: *Reprinted with permission from S.B. Jagtap, D. Ratna, J. Appl. Polym. Sci. 130 (2013) 2610, ©2013 John Wiley and Sons Publishers.*

Table 5.3. The nanocomposites display higher tensile strength and modulus compared to the pure epoxy network. The addition of optimum concentration of modified CNT (1 wt.%) resulted in around 90% and 264% improvement in the tensile strength and tensile modulus, respectively. The tensile strength decreases with a further increase in concentration due to the agglomeration of MWCNT.

As discussed above, both the modifiers HNAA and SAHA help in unbundling of the nanotubes due to the cation–π interaction. As a piece of evidence for the existence of cation–π interaction FTIR spectra of SAHA, SAHA-modified MWCNT, neat epoxy, and epoxy/MWCNT(1 wt.%)/SAHA composite samples are presented (Fig. 5.30) and discussed here. The SAHA shows a characteristic peak at 1563 cm^{-1} corresponds to C=O stretching of the carboxylate group. The peak appears in a much lower wavenumber than that observed in the carboxylic acid group (1710 cm^{-1}). This is because the negative charge on the oxygen atom of carboxylate ion undergoes dislocation and reduces the bond strength of

Table 5.3: Tensile properties of epoxy/MWCNT/SAHA (1:1.5) composite having varying concentration (wt.%) of MWCNT.

Samples (wt.% SAHA-modified MWCNT)	Tensile strength (MPa)	Tensile modulus (MPa)	Elongation at breaks (%)
0	0.61 ± 0.02	0.61 ± 0.03	121 ± 7
0.1	0.66 ± 0.02	0.98 ± 0.03	102 ± 6
0.25	0.74 ± 0.04	1.00 ± 0.04	90 ± 3
0.5	1.03 ± 0.04	1.65 ± 0.05	75 ± 4
0.75	1.09 ± 0.02	1.85 ± 0.07	64 ± 2
1	1.16 ± 0.05	2.22 ± 0.05	63 ± 3
1.5	0.96 ± 0.04	1.42 ± 0.06	58 ± 5
2	0.81 ± 0.03	1.40 ± 0.02	50 ± 2

Source: *Reprinted with permission from S.B. Jagtap, D. Ratna, J. Appl. Polym. Sci. 130 (2013) 2610, ©2013 John Wiley and Sons Publishers.*

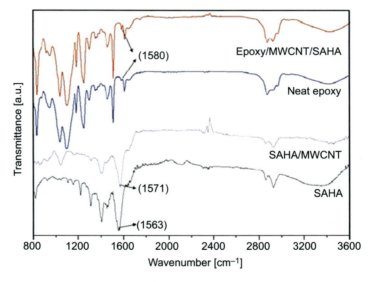

Figure 5.30
FTIR spectra for SAHA, SAHA-modified MWCNT, neat epoxy, and epoxy/MWCNT(1 wt.%)/SAHA composite. Source: *Reprinted with permission from S. B. Jagtap, D. Ratna, Poly. Lett. 7 (2013) 329, ©2013 Online Publications.*

C = O group and results in a reduction in the wavenumber. The C = O stretching peak of the carboxylate group is shifted from 1563 to 1571 cm^{-1} for SAHA-modified MWCNT. This shift of carboxylate ion peak is attributed to the strong electrostatic interaction between the cation (Na$^+$) with the π electron clouds of MWCNT (known as cation−π interaction). This interaction reduces the effectiveness of dislocation of negative charge on oxygen atom resulting in an increase in the wavenumber of carboxylate C = O peak. The epoxy/MWCNT (0.5 wt.%)/SAHA composite also exhibits a shift of the peak of the corresponding C=O group of the carboxylate ion from 1563 to 1580 cm^{-1}. It can be noted that the less intense peak at 1580 cm^{-1} appear also in the case of the neat epoxy, which corresponds to Ar−C=C−H stretching. The same peak obtained for epoxy/MWCNT (0.5 wt.%)/SAHA composite has stronger intensity. This can be attributed to the overlapping of carboxylate ion peak and Ar−C=C−H group stretching in this region.

Recently a new approach has been reported in the literature for making high strength thermoset/CNT nanocomposites. This is to directly induce the intertube crosslinking of CNTs by using high temperature, high pressure, electronic impact, or oxidizing agents [125,126]. This approach is associated with a possibility of damage of CNT structure in such harsh conditions [127]. Damage of CNTs may cause deterioration of mechanical performance of the related composites. Very recently, a facile method for synthesizing high-strength CNT/EP composite films has been demonstrated [128]. The schematics of the fabrication process of CL-CNT/EP films are shown in Fig. 5.31. The raw CNT film was prepared by a spray pyrolysis method and the film in a stretched state, immersed in into the diazonium salt solution at 60°C for the crosslinking reaction to take place. The optimum crosslinking time was found to be 2 h. The crosslinked CNTs film was infiltrated with epoxy resin to obtain a high-strength composite. The tensile strength of such composites is claimed to be 2.5 GPa.

In recent years, a lot of works have been done to develop thermoset/CNT composites for microwave-absorbing applications useful for both the civil and defense sector [129−133]. Such materials are necessary to avoid detection by enemy radar as well as for protecting communication and safety systems employed in aircrafts, ships and automobiles against electromagnetic interference. Dong et al. [130] investigated the influence of three commercial MWCNT materials with various diameters and length-to-diameter ratios on the X-band microwave absorption of epoxy nanocomposites with CNT contents from 0.125 to 2 wt.%, prepared by two dispersion methods, that is, a surfactant-assisted solution mixing method and a mixing via a ball milling. The laser diffraction particle size and TEM analysis showed that both methods produced good dispersions at the microscopic level of CNTs. The MWCNT with the largest aspect ratio resulted in composites with the best X-band microwave absorption performance. MWCNTs have been doped with various metal oxides to improve the dispersion as well as microwave-absorbing capacity [129,134].

Recently, an epoxy-based microwave-absorbing coating (MAC) containing carbon nanotube (CNT), silicon carbide (SiC), and carbonyl iron (CI) particles has been reported [135]. The effect of a single layer containing one type of filler, a single layer containing several types of fillers, and multilayer structures on mentioned key parameters are investigated. The multilayer MACs included nine layers in which each of the layers contained one type of the fillers (CNT, CI, and SiC) and three neat resin interlayers 2 mm in total thickness compared to other samples and provided a maximum return loss value of 17.6 dB at 08.50 GHz and

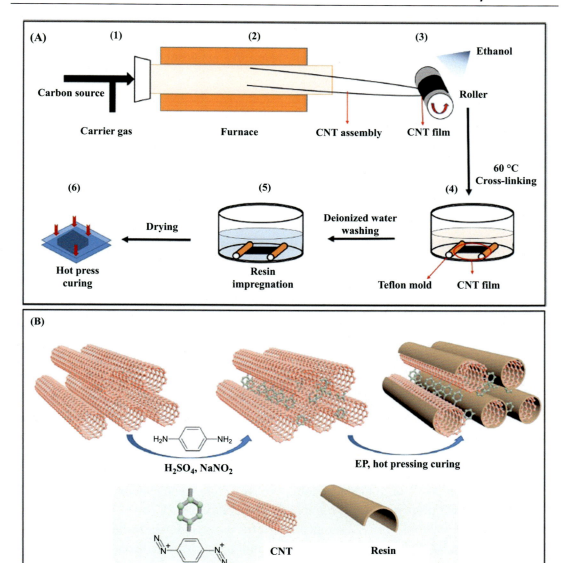

Figure 5.31
Schematics of the fabrication process of a CNT/EP composite film. (A) Schematic of experimental set-up, including (1) the injection of the reaction solution, (2) the formation of a hollow cylinder-like CNT assembly, (3) the preparation of a CNT film, (4) the crosslinking reaction, (5) the preparation of a CNT/EP film, and (6) the hot pressing of the film. (B) Reaction mechanism diagram for the preparation of the composite film. Source: *Reprinted with permission from J.C. Long, H. Zhan, G. Wu, Y. Zhang, J.N. Wang, Composites: Part A 146 (2021) 106409, ©2021 Elsevier Publishers.*

an absorption bandwidth (RL < 5 dB) of 16 GHz (2–18 GHz) with a density of 1.61 g cm^{-3}. So, it is simply possible to obtain an efficient MAC due to suitable microwave-absorbing filler distribution in multilayer coatings for any respective application.

5.2.5 Graphene-based nanocomposite

Another carbonaceous nanomaterial, which has drawn considerable attention in recent years, is graphene, which was first invented by Andre Geim and Konstatin Novoselov in 2004 [136]. Graphene nanosheets (2D) were discovered from graphitic crystal (3D) by splitting the crystal till individual planes were obtained. Graphene exhibits many outstanding properties such as high surface area (2630–2965 m^2 g^{-1}), high thermal conductivity ($\sim$5000 W m^{-1} K^{-1}), excellent electronic transport properties ($\sim$10^6 S cm^{-1}) and superior mechanical properties ($\sim$100 GPa) [137–139]. If we compare its properties with CNT, which is discussed above then we can see the electrical and thermal conductivities of graphene are higher than that of CNT (5000 S cm^{-1} and 3000 W m^{-1} K^{-1}). Because of the interesting properties of graphene, a lot of studies have been reported on nanocomposites of graphene and different types of thermosetting resins in recent years [140–145]. As discussed above for thermoset/MWCNT composites, the issues of dispersion and interfacial adhesion are applicable to graphene-based nanocomposite as well.

Before we discuss the studies conducted to resolve these issues specifically let us discuss the processing of such nanocomposites. It may be noted that because of higher surface area nanomaterials are more difficult to disperse in a polymer matrix compared to conventional fillers. Therefore it is necessary to used techniques like sonication, ball milling, etc. It is reported that different dispersion routes with and without a ball mill procedure caused different dispersion levels of graphene resulting in epoxy/graphene nanocomposites of different mechanical properties. Epoxy/graphene nanocomposite (0.2 wt.% graphenes) made by only sonication exhibits only a 24% increase in fracture toughness. When the same composite is made by sonication followed by ball milling, it displays better performance with a 52% increase in fracture toughness [146]. It is also recommended [147] that graphene should be first dispersed in a solvent (e.g., acetone, tetrahydrofuran) with vigorous mechanical stirring or sonication before incorporating into the identified polymer matrix.

Various noncovalent interactions and covalent functionalization-based strategies have been used to improve the dispersion and interfacial adhesion of graphene as well as the final properties of the related nanocomposites. For example, the author's research group has used [121,148] a nonionic surfactant "Nonidet P-40," which contains a benzene ring, a long aliphatic chain and poly (ethylene oxide) chain to disperse graphene in HBAU and epoxy matrix. It was demonstrated that dispersion of graphene in HBAU resin is significantly improved due to the addition of 0.5 wt.% "Nonidet P-40." This is attributed to $\pi-\pi$

interaction between π electrons of Nonidet-P40 with π electron clouds of graphene. The interfacial adhesion is improved as a result of H-bonding interaction between ether group of Nonider-P40 with —OH groups of HBUA matrix. This is responsible for reinforcing the effect of graphene. The addition of 0.25% graphene resulted in an increase in tensile strength from 7 MPa for pure HBAU to 9.5 MPa for the nanocomposite. The effect of incorporation of various carbonaceous nanofiller [148] of different geometries, for example, carbon black, MWCNT, graphene on water vapor permeability of HBUA are investigated and data for 0.25 wt.% filler containing nanocomposites are presented in Fig. 5.32. We can see the nanocomposite displays lower water vapor permeability (WVP) compared to pure HBUA. This is because the filler provides a tortuous path for the diffusion of water vapor and reduces the permeability. The lower water vapor permeability is necessary for better corrosion resistance of the coating. Among all the composites graphene-based nanocomposite offer the best performance, that is, the lowest WVP. This HBUA/graphene composite offers better barrier properties than HBUA/MWCNT composite, due to the 2D platelets like the structure of graphene. Ionic modifier like HNAA and SAHA used to modify MWCNT has also been extended for graphene [149].

Another effective approach to address the abovementioned issues is to construct the chemical bonding between the GO sheets and polymer matrix, which results in a significant increase in the mechanical and thermal properties of the related composites. It was demonstrated effective load transfer [150] between GO nanoplatelets and polyurethane (PU)

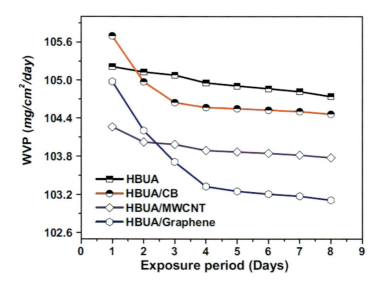

Figure 5.32
WVP versus exposure period plots of the neat HBUA, HBUA/CB, HBUA/MWCNTs, and HBUA/graphene composite coatings. Source: *Reprinted with permission from R.B. Naik, S.B. Jagtap, D. Ratna, Progr. Org. Coat. 87 (2015) 28–35, ©2015 Elsevier Publishers.*

matrix due to the formation of chemical bonding between the GO nanoplatelets and the hard segment of the PU. The nanocomposites with chemical interactions are prepared by functionalization of graphene and in situ polymerization in presence of functionalized graphene [151,152]. Li et al. [153] prepared graphene oxide nanosheets (GONS) functionalized PU and epoxy resin using the process as shown in Fig. 5.33. First, graphene oxide (GO) was produced following an improved Hummers' method [154]. The second step is in situ polymerization of 4,4′-methylenediphenyl diisocyanate (MDI) and poly (tetramethylene glycol) (PTMG) in the presence of GO nanosheets. In the third step, required amount of epoxy and 3,3′-dichloro-4,4′-diamino diphenylmethane (MOCA) were added and the mixture is cured to get the nanocomposite. The chemical bonding helped in

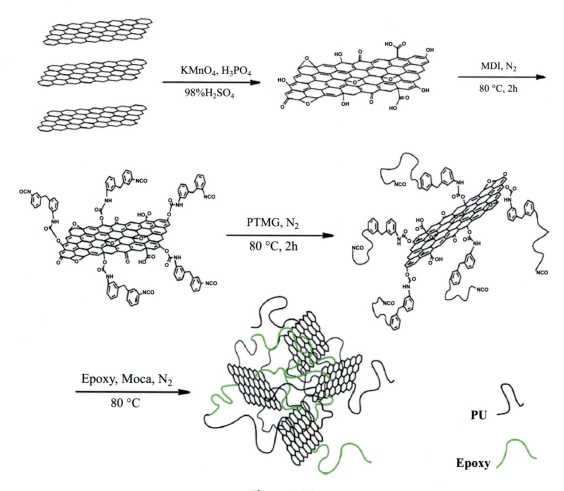

Figure 5.33
Synthetic route of PU/GO/EP nanocomposites. Source: *Reprinted with permission from Y. Li, D. Pan, S. Chen, Q. Wang, G. Pan, T. Wang., Mater. Design, 47 (2013) 850, ©2013, Elsevier Publishers.*

achieving good dispersion of graphene nanoplates and improvement of mechanical properties of the related nanocomposites. Incorporation of a small amount of 0.066 wt.% of GONS, the tensile modulus of PU/GO/EP nanocomposites increased from 218 MPa to 257 MPa. In addition, the tensile strength and % elongation at the break of the PU/GO/EP nanocomposites increased by more than 52% and 103%, respectively.

Amine-functionalized graphene was reacted with isocyanate and chemically bonded to PU through -NH-CO- covalent interface [155]. SEM analysis indicates a good dispersion of GO in the PU matrix with the average thickness of the GO strip being less than 50 nm. The well-dispersed GO in the PU matrix resulted in a 700% enhancement in Young's modulus and an approximate 50°C increment in decomposition temperature with 4% GO loadings.

The covalent interaction between graphene and the polymer matrix with functional groups other than ester/amide/urethane (as discussed above) has also been investigated. GO is modified with 3-amino triethoxysilane FG [156] and carried out in situ polymerization with PU and epoxy system to get PU/EP/GO nanocomposites. Scanning electron microscopy (SEM) analysis indicates the presence of a smooth silk-like wrinkle texture, which is the normal characteristic of a single layer GO.

Li et al. [157] reported nanocomposites based on epoxy and inorganic sulfonic graphene oxide (SGO). They have successfully established chemical bonding between epoxy and SGO via curing reaction and achieved a 113% improvement in the impact strength as a result of the incorporation of 1 wt.% SGO. Impact performance was found to be much higher for nanocomposites made using graphene oxide functionalized with peroxy radicals [158]. In this case, a 240% increase in fracture energy is reported due to the incorporation of 0.2 wt.% of functionalized GO. The dramatic increase in the fracture energy can be explained by considering the undisturbed sp^2 structure of graphene when functionalized with peroxy radicals that lead to a load transfer throughout the graphene hexagonal structure.

Nanocomposites based on bio-based thermoset resins and graphene have also been investigated. It may be noted that bio-based resins show many inferior mechanical properties compared to petroleum-based thermosets. Hence reinforcing such bio-based resins is very much relevant from an application point of view [159]. Baruah and Karak [160] reported nanocomposites of graphene and synthesized from phytochemical tannic acid. A 263% improvement toughness has been reported due to the addition of 0.5 wt.% GO into the bio-based epoxy.

Recently, GO sheets have been decorated with metal oxide (one-dimensional) and incorporated in various thermoset matrices to produce nanocomposites with enhanced mechanical, thermal, and anticorrosion activity [161]. Incorporation of such decorated GO in an alkyd resin is reported to improve the level of dispersion and alkyd-nanofillers bond strength and durability [162]. This indicates that the decoration of GO is a good strategy to address the inherent problem of dispersion of GO. A similar observation was reported by using epoxy as a matrix [163]. A close-packed GO layered structure by nano-Al_2O_3 has been successfully exfoliated in the epoxy matrix to generate a hybrid epoxy nanocomposite with higher corrosion resistance compared to the epoxy/GO composite.

Selim et al. [164] reported a ternary nanocomposite based on hyperbranched linseed oil-based alkyd, GO, and β-MnO_2 nanorods and used for corrosion-resistant coating applications. β-MnO_2 was prepared through a solvothermal procedure followed by thermal annealing without a catalyst. The hybrid ternary nanocomposite coating was prepared [164] using a single-step chemical deposition methodology as shown in Fig. 5.34. Up to 2.5%

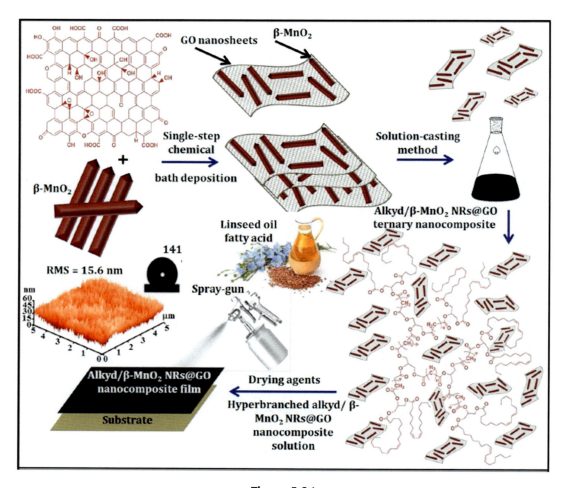

Figure 5.34
Film formation of hyperbranched alkyd/β−MnO_2 NRs@GO nanocomposite coating film. Source: Reprinted with permission from M.S. Selim, Z. Hao, P. Mo, Y. Jiang, H. Ou, Colloids Surf. A 601 (2020) 125057, ©2020 Elsevier Publishers.

loading of β-MnO₂ decorated GO, uniform dispersion of nanoparticles is achieved. The higher water contact angle of the nanocomposite coating (141 degrees) compared to the pure alkyd (98 degrees) indicated higher hydrophobicity of the nanocomposite coating. This explains the higher corrosion resistance of the nanocomposite coating compared to pure alkyd.

5.2.6 Silver nanoparticle-based nanocomposites

Silver nanoparticle (SNP)-based nanocomposites have been investigated in recent years for antimicrobial coating applications. SNP possesses considerably higher antibacterial activity compared to other nanomaterial-based biocides due to its ability to act as a reservoir for the continuous release of silver ions [165–175]. The continuous release of silver ions has been established by measuring its ionic conductivity [175]. The ionic conductivity tends to increase with the increasing concentration of silver nanoparticles dispersed in an acrylic polymer. The main attraction of SNPs is that it offers high toxicity for microorganisms and low toxicity for human beings [167] unlike conventional biocides, which are equally toxic to human beings as well. The SNPs can be incorporated into the polymer matrix in two different ways. One is a direct dispersion of SNPs into a polymer matrix, which is known as the ex situ approach. The second one is the generation of SNPs in a polymer matrix in situ. The dispersion studies indicate that a uniform dispersion of SNPs is very difficult if we adopt ex situ approach because of easy agglomeration of the nanoparticles. This results in a decrease in the antimicrobial activity of the related nanocomposites [174]. On the other hand, an in situ approach can provide a good dispersion of SNPs in a variety of polymer matrices. For example, Osman et al. [174] successfully utilized an in situ approach to generate SMPs by reducing silver nitrate. They have also demonstrated that nanocomposite samples exhibited an antibacterial effect against both the Gram-positive and Gram-negative bacteria. An antimicrobial coating is reported by Uday et al. [176], which is based on a highly branched polyester/clay/silver hybrid nanocomposite via reduction of silver salts by dimethylformamide at room temperature. They have demonstrated the formation of well-dispersed SNPs within the clay gallery with an average size of 15 nm. The influence of incorporation of SNPs on thermal and antibacterial properties of a waterborne polyurethane coating has also been reported [177]. It was reported a significant growth reduction of *Escherichia coli* and *Staphylococcus aureus* bacteria even at a very low loading (200 ppm) of SNP.

Ratna and Naik [178] developed a green antimicrobial coating based on SNP and hyperbranched urethane alkyd (HBUA). The properties of HBUA have already [179,180] been presented in Chapter 2, Properties and Processing of Thermoset Resin. HBUA requires a much lower amount of solvent for the preparation of the related coatings than that of a commercial alkyd resin. Therefore this development is associated with a significant reduction in VOC and therefore is a step toward green technology. Another reason for mentioning this technology as a "green" one is that it is associated with the generation of SNPs from silver benzoate without using any toxic reducing agents like dimethyl sulfoxide or hydrazine, which are used conventionally to generate SNPs. A similar process was also adopted for the preparation of oil-based composite of SNPs [181]. They reported that silver benzoate undergoes ligand exchange with the fatty acids, which causes the metal ions to dissolve in the oil and subsequently reduced the free radicals to form SNPs. In the case of HBUA, the free radicals produced during the curing of HBUA reduce the silver ion leading to the formation of SNPs as shown in Fig. 5.35. When the HBA resin dispersed with silver benzoate is cured with an isocyanate hardener, initially, a transparent coating is formed, which transformed into brown colored one after curing for about 6 h as a result of generation of SNPs. TEM analysis indicates a uniform distribution of SNPs in the HBUA matrix with average particle sizes in the range of 20–40 nm. This can be attributed to the secondary interactions between the polar group (-OH) present in the HBA resin and the SNPs, which resulted in good adsorption of HBA resins on the surfaces of SNPs. The embedded resin prevents the growth of agglomeration of SNPs. Antimicrobial tests were carried out in the laboratory against *Serratia marcescens* bacterial strain using agar-overlay and disk diffusion methods, which indicated an excellent performance of the developed HBUA/SNP nanocomposite coating. The details of antimicrobial tests will be discussed in Chapter 6, Characterization, Performance Evaluation, and Life Time Analysis of Thermoset Resin. *S. marcescens* is a Gram-negative rod-shaped bacteria in the family of Enterobacteriaceae and it produces a nondiffusible red pigment called prodigiosin.

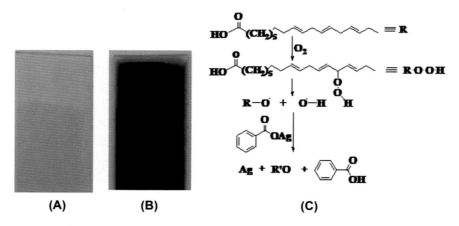

Figure 5.35
HBUA/SNP nanocomposite-based antimicrobial coating. (A) Photograph of the coating applied on glass panel before curing. (B) Photograph of the coating applied on glass panel after curing. (C) Mechanism of formation of SNP during curing. Source: *Adopted with permission from R.B. Naik, D. Ratna, J. Coat. Technol. Res. 12 (2015) 1073–1083, ©2015 Springer Publishers.*

Therefore the formation of red color indicates the growth of bacteria as observed in the Agar-overlay method using HBUA-based coating without the addition of SNPs. However, no red color was observed when SNPs (0.5% Ag) are incorporated into the resin. The absence of red color indicates that bacterial growth was completely inhibited.

5.2.7 Nanoreinforcement and toughening

Thermoset nanocomposites with well-dispersed layered silicate or CNTs offer improved thermal stability and elastic modulus as discussed in the earlier sections. However, the results on the effect of nanomodification on fracture toughness are contradictory. Some studies reported a reduction in toughness due to the addition of nanofiller [182] and other studies indicated a reduction in toughened as a result of nanomodification [183]. Thermoset resins are as such brittle and often required to be toughened using a suitable modifier. Thermoset resins can effectively be toughened by the modification with rubbers as discussed in Chapter 3, Toughened Thermoset Resin. However, rubber modification is accompanied by a substantial decrease in modulus. Hence the conventional toughening strategy can be coupled with the nanoreinforcement strategy to make a really strong and tough material. The ternary blending strategy (simultaneous nanoreinforcement and toughening) is schematically represented in Fig. 5.36. This requires the control of both the nanostructure (dispersion of nanofiller) and microstructure (dispersion of rubber). A number of studies have been carried out to examine the outcome of the abovementioned strategy [184,185]. The works done so far are reviewed below.

The chemorheology and morphology of a ternary system consisting of an epoxy resin, CTBN liquid rubber, and an organoclay modified with octadecyl ammonium cation have been investigated [186]. Analysis of both the nanostructure and microstructure indicates that although the liquid rubber intercalates the layered silicate alone, it does not do so to the same extent as the epoxy resin. Indeed, in the ternary blend, the clay remained in the epoxy-rich phase and the rubber phase-separated into fine particles. Flexural testing indicated that the clay was able to compensate for some of the modulus sacrificed as a result of rubber addition. The toughness was found to increase with the addition of clay alone and, of course, rubber alone. However, no synergistic toughening effect was observed. Frölich et al. [187] investigated a system in which epoxy was mixed with hydroxyl-terminated poly (propylene oxide-block-ethylene oxide) as the rubber, with a synthetic fluorohectorite clay treated with bis(2-hydroxyethyl) methyl tallow alkyl ammonium ions. The clay was first blended with the rubber, before being dispersed into the reactive epoxy mixture. Modification of the rubber allowed variation in miscibility and differing morphologies and properties. When the rubber was miscible, it imparts only plasticizing effect and the intercalated clay led to the improved toughness. If the rubber is sufficiently modified, such as with methyl stearate, then both the nanostructure and

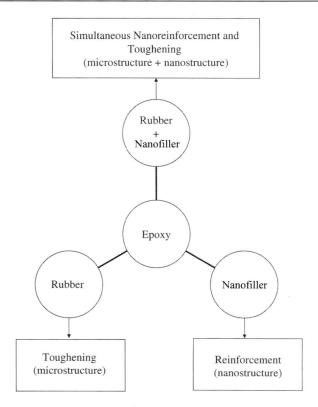

Figure 5.36
Schematic representations of simultaneous nanoreinforcement and toughening of thermoset resins.

microstructure coexist. The TEM pictures of various samples are shown in Fig. 5.37. The phase-separated morphology leads to a significant increase in toughness, with only a modest decrease in modulus.

Ratna et al. [188] investigated a ternary system containing epoxy, epoxy-functionalized hyperbranched polymer (HBP), and an octadecyl ammonium ion-modified nanoclay. Investigation of the morphologies of these materials showed that the epoxy nanocomposite had a well-dispersed structure with an average d-spacing of 8–9 nm. However, analysis and low magnification indicate the presence of tactoids. The thermomechanical properties of ternary systems with varying compositions are given in Table 5.4. It is clear that the incorporation of clay is able to compensate for the reduction in modulus as a result of rubber modification. However, the impact strength of the ternary system was found to be lower than epoxy/HBP blend alone. This can be explained by considering the fact the reaction-induced phase separation of HBP is disturbed to some extent due to the presence of clay as evident from SEM photographs shown in Fig. 5.38. The toughening effect in a liquid rubber-modified system largely depends on the microstructure, which is likely to be

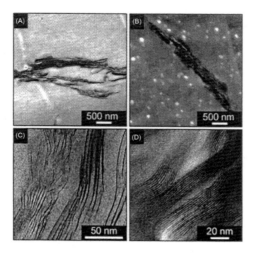

Figure 5.37
Transmission electron micrographs of epoxy hybrid nanocomposites. Images shown are examples for composites with (A) dissolved PPO in epoxy matrix, (B) phase-separated PPO-stearate spheres in epoxy matrix, (C) PPO-modified epoxy/clay nanocomposite, and (D) PPO-stearate-modified epoxy/clay nanocomposite. Source: *Reprinted with permission from J. Frohlich, R. Thomann, R. Mulhaupt, Macromolecule, 36 (2003), 7205,* ©*2003 American Chemical Society Publishers.*

Table 5.4: Properties of ternary blends of epoxy, epoxy-functionalized HBP, and nanoclay.

Blend composition epoxy/HBP/clay	Flexural strength (MPa)	Flexural modulus (MPa)	Impact strength (J m^{-1})	T_g (°C)
100/0/0	112	2920	740	192
100/15/0	109	2510	2250	182
100/0/5	146	4090	1060	206
100/15/5	135	3630	1540	192

Source: *Reprinted with permission from D. Ratna, O. Becker, R. Krishnamurty, G.P. Simon, R. Varley, Polymer 44 (24) (2003) 7449,* ©*2003 Elsevier Publisher.*

affected by the presence of a third component. Hence this issue has to be addressed properly for the development of strong and tough three component materials consisting of a thermoset resin, a liquid rubber, and a nanofiller.

Choi et al. [189] investigated a similar epoxy-based ternary system where they used a core−shell rubber (CSR) modifier, in which no reaction-induced phase separation is involved. They used cubic silsesquioxane (cubes typically 1.2−1.5 nm diameter) as a nanofiller. The stress−strain plots of the ternary systems containing different concentrations of CSR are shown in Fig. 5.39. It is clear from the figure that the toughness (area under the stress−strain plot) increases gradually with an increase in CSR concentration while the

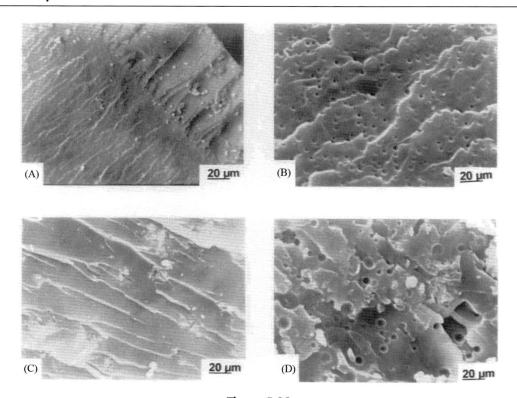

Figure 5.38
SEM micrographs for fracture surfaces of (A) neat epoxy, (B) epoxy/HBP, (C) epoxy/clay, and (D) epoxy/HBP/clay. Source: *Reprinted with permission from D. Ratna, O. Becker, R. Krishnamurty, G.P. Simon, R. Varley, Polymer, 44 (2003) 7449, ©2003 Elsevier Publisher.*

modulus remained constant. A similar observation was reported for the vinyl ester resin-based ternary system [188]. Hernandez et al. [190] investigated thermoplastic-modified epoxy with clay. The effect of clay dispersion and distribution on the fracture toughness of the thermoplastic/epoxy systems was evaluated. They found that thermoplastic particles and the silicate layers were dispersed throughout the continuous epoxy phase. A homogenous exfoliated structure resulted in less improvement in the fracture toughness compared to a heterogeneous intercalated/exfoliated structure. The presence of aggregates increased the crack propagation path in this ternary system, resulting in an improvement in toughness. Epoxy/thermoplastic/clay ternary systems have also been investigated using polyether polyols [191] and hydroxyl-terminated polyether ether ketone [192]. The addition of nanoclay led to an increase in the tensile modulus and flexural modulus in the systems. The value of the fracture toughness in the thermoplastic/clay/epoxy system was greater than pure epoxy; this improvement was caused by crack path deflection and debonding mechanisms.

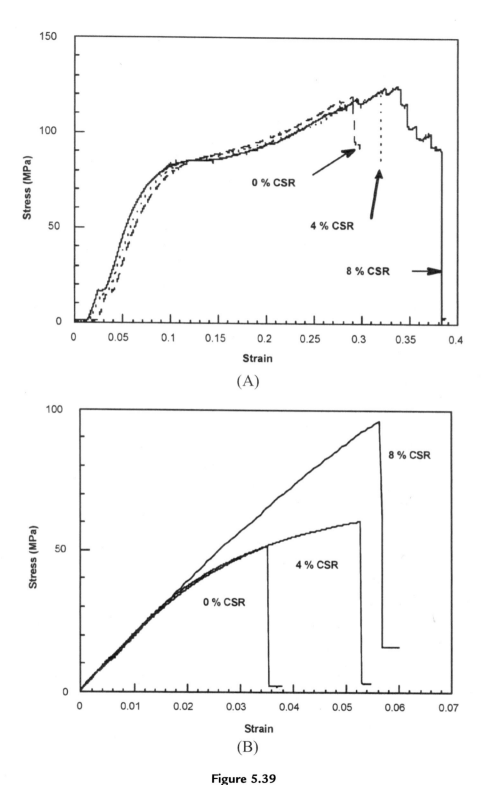

Figure 5.39
Room temperature mechanical tests of the ternary systems (epoxy + cubic silsesquioxane + CSR) containing different concentration of CSR (A) compression and (B) tension. Note that the strain scales for (A) and (B) are different. Source: *Reprinted with permission from J. Choi, A.F. Yee, R.M. Laine, Macromolecule, 37 (2004) 3267, ©2004 American Chemical Society Publishers.*

Ishak et al. [193] reported that the mechanical properties of unsaturated polyester (UPE)/rubber such as UPE/liquid natural rubber (LNR) can be improved by applying nanoreinforcements. The tensile strength, stiffness, and impact energy were slightly increased by the incorporation of clay into the UPE/LNR system. UPE/PU/clay system was also investigated in which polyester molecules are modified by forming a segmental copolymer with a rubber attached to the end group of UPE [194]. In this ternary system, a synergistic effect of clay and rubbery segment resulted in an improvement in toughness and flexural strength of the cured UPR; however, the modulus was still reduced.

A number of studies have been published related to UPE-based ternary systems looking at the effect of thermoplastic/clay combination on volume shrinkage. As discussed in Chapter 1, Chemistry and General Applications of Thermoset Resins, and Chapter 3, Toughened Thermoset Resin, thermoplastic additives have been widely used as low-profile additives (LPAs) to control the volume shrinkage of UPE. Note that the volume shrinkage of UPE as a result of curing is much higher than that of the epoxy resins and the inherent high cure shrinkage [194] is a major issue in regards to the processing of such materials. The studies showed that a combination of a nanoclay and LPA has a synergistic effect on the control of shrinkage [195,196]. Xu and Lee [195] claimed that almost all the silicate layers are distributed inside the LPA-rich phase, leading to faster reaction and earlier gelation of the LPA-rich phase, which is favorable for earlier stress cracking. A liquid LPA phase can release the stress arising from polymerization shrinkage of the system. Therefore stress-induced cracks are not formed in the LPA-rich phase or in the LPA/UPE interface until gelation happens in the LPA-rich phase. This local cracking leads to volume expansion, which can compensate for polymerization shrinkage. In addition, the presence of nanoclay compensates for the loss of some properties, caused by LPA, such as storage and flexural modulus.

Apart from the blending of thermoplastic and rubber as a thermoset to improve the toughness property, various toughened interpenetrating network (IPN) systems have also been reported. The reader may refer chapter for details of IPN technology. Karger-Kocsis et al. [197,198] reported vinyl ester (VE)/epoxy interpenetrating networks (VE/EP IPN), which shows a considerable higher toughness compared to the pristine VE resin. However, IPN materials exhibit low modulus and glass transition temperature. With the motivation to improve the modulus and T_g, the same group has investigated [197,198] VE/EP/clay nanocomposites using closite 30B and another synthetic flurohectorite clay. The nanocomposites were made by adding the treated clays to the premixed equally proportioned VE/EP resin mixture. The epoxy resin appeared to encapsulate the clays, which showed some level of intercalation. The fracture energy of the nanocomposite system containing 5 wt.% clay was found to be two times higher compared to the same for the unreinforced resin. The fracture energy decreased drastically with a further increase in clay concentration, which was attributed to a fundamental change in the failure mode.

The increase in toughness at a lower concentration of clay was not a nanophenomenon, rather due to the toughening effects of the softer interphase between encapsulated clay and the matrix. The course nonintercalated silicate particles (adhered to the bulk by a soft interface) act as a stress concentrator and initiate extensive shear deformation.

Yao and coworker [199] reported simultaneous nanoreinforcement and toughening of PU. They have successfully synthesized a series of PU/microfibrillated cellulose (MFC) nanocomposites via in situ polymerization. PU was covalently grafted onto the MFC by particular association with the hard segments. The nanocomposites offer simultaneous improvement of strength and flexibility. The tensile strength and elongation at break of the nanocomposite containing only 1 wt.% MFC were increased by 4.5-fold and 1.8-fold compared with that of neat PU, respectively.

Gong et al. [200] investigated ternary composite comprising epoxy matrix, performed 3D-powdered rubber (PR), and 2D chemically reduced graphene oxide (CRGO). Five samples were made by using a different combination of PR (3 and 6 wt.%) and CRGO (0.5 wt.%). The mechanical properties of the binary and ternary composites and pure epoxy network were compared in Fig. 5.40. We can observe that both the binary and ternary composites show higher toughness than the pure epoxy network. As expected, CRGO is more effective in enhancing elastic modulus than rubber and rubber is more effective in toughening than CRGO. Though both the PR and CRGO offer toughening effect, no synergistic effect was observed. Rather the toughening effect of PR is decreased in presence of CRGO. The composite containing 6 wt.% PR exhibits K_{1c} and G_{IC} values 1.31 MPa m^{05} and 568.3 J m^{-2}, respectively. The corresponding values are reduced to 1.06 and 343.5 J m^{-2} when 0.5% CRGO is incorporated into the same. This can be explained by considering the mechanism of toughening involved for various nanofillers as shown in Fig. 5.41. The toughening mechanisms on which rubber particles operate are rubber cavitation and shear yielding. These are inhibited or impeded in presence of CRGO sheets resulting in a reduction in toughening efficiency.

5.2.8 Nanotechnology and flammability

A thermosetting resin composition is required to be fire retardant to ensure reliability and human safety. Flame retardancy is a major concern in the applications, where fire can occur, namely, aircraft cabin, ships, submarines, offshore drilling platforms, and rail carriages. A number of studies have been carried out to examine the effect of high temperature or fire on the load-bearing properties of polymer laminates and sandwich structures [201,202]. It was concluded that thermal softening of the polymer matrix and reinforcement, creep, and decomposition of the polymer matrix deteriorate the tension properties, whereas matrix softening and delamination cracking are responsible for the reduction of compression properties. Among the thermoset resin discussed in previous

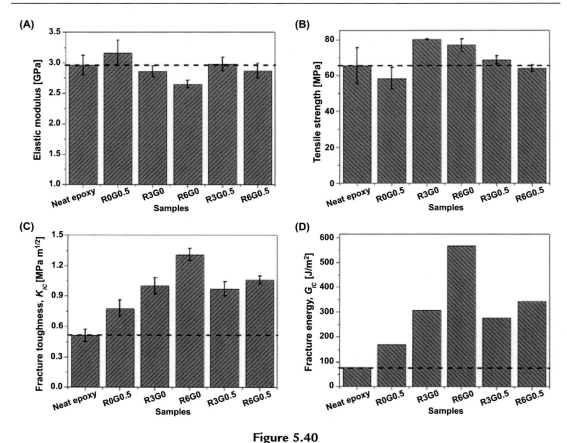

Figure 5.40
Mechanical properties of neat epoxy and epoxy composites: (A) elastic modulus, (B) tensile strength, (C) fracture toughness, and (D) fracture energy. Source: *Reprinted with permission from L. Gong, L. Zhao, L.C. Tang, H.Y. Liu, Y.W. Mai Compos. Sci. Technol. 121 (2015) 104, ©2015 Elsevier Publishers.*

sections, phenolic, polyimides, and amino resins are inherently flame retardant. Epoxies, PUs, UPEs are flammable and are to be modified in order to reduce their flammability for fire-sensitive applications. Such modification can be done by either chemically modifying the resin or by incorporating additives called flame retardants. Before discussing the role of nanotechnology on the fire retardancy of thermoset resin, it is necessary to know about the basics of flame retardancy and conventional flame retardants. The reader may refer to Chapter 2, Properties and Processing of Thermoset Resin, for the required information. The relevant testing for assessment of flammability and smoke behavior of the thermoset resins will be discussed in Chapter 6, Characterization, Performance Evaluation, and Life Time Analysis of Thermoset Resin.

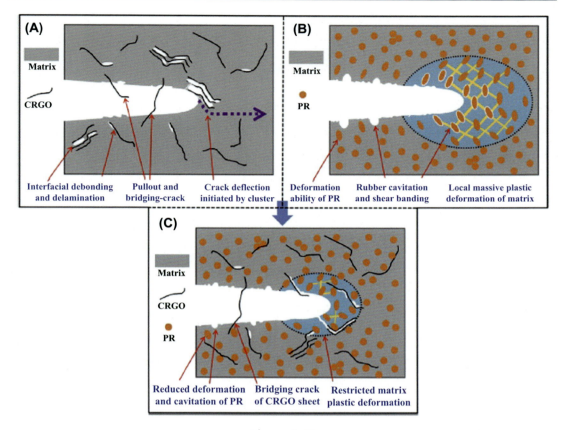

Figure 5.41
Schematic diagrams of toughening mechanisms within the damage zone around the crack tip for the binary and ternary composites: (A) epoxy/CRGO, (B) epoxy/PR, and (C) epoxy/CRGO/PR hybrid composites. *Source: Reprinted with permission from L. Gong, L. Zhao, L.C. Tang, H.Y. Liu, Y.W. Mai Compos. Sci. Technol. 121 (2015) 104, ©2015 Elsevier Publishers.*

5.2.9 Nanoclay-based flame retardant

Polymer/clay nanocomposites have been examined extensively for their flammability characteristics [203,204]. It was reported that if the samples were examined using a cone calorimeter, the peak heat release rate and mass loss rate tend to get reduced. For example, the cone calorimeter profile of a PU system and its nanocomposite as investigated by Berta et al. [205] are shown in Fig. 5.42. The figure compares the weight loss, mass loss rate, and heat release rate (HRR) of the nanocomposite and the corresponding PU matrix. It is evident that the incorporation of 2.5 wt.% of organoclay resulted in a significant decrease in weight loss, mass loss rate, and heat release profile. The nanocomposite displays a reduction in HRR of 43% and the total heat released is reduced to more than 80% in

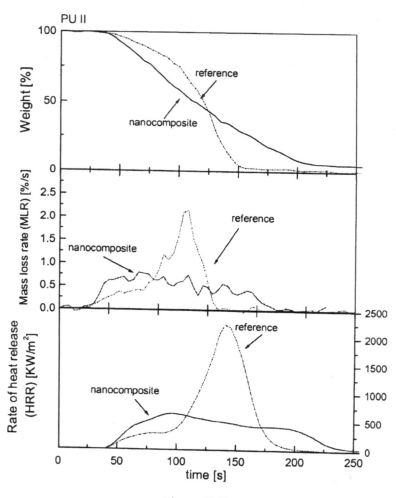

Figure 5.42
Cone calorimeter profiles of PU and PU/clay nanocomposite. Source: *Reprinted with permission from M. Berta, C. Lindsay, G. Pans, G. Camino, Polym. Degrad. Stab. 91 (2006)1179, ©2006 Elsevier Publisher.*

comparison to the PU matrix. It is not clear which type of nanocomposite (intercalated or exfoliated) improves the combustion behavior more. This is because it is difficult to prepare exfoliated nanocomposites without changing other parameters, which affect the flammability of a polymer. Moreover, in most cases, the morphology of a polymer/clay nanocomposite system falls in between the intercalated and exfoliated nanostructure. It may be noted that the time of ignition values of the nanocomposites was similar or slightly lower compared to the corresponding matrix. This is probably due to the early decomposition of the organic modifier on the clay surface, which catalyzes the polymer degradation.

The investigation on the mechanisms of thermal stabilization and flammability reduction in nanocomposites reveals that the addition of organoclays into a thermoset resin can substantially aid flame retardancy by encouraging the formation of a carbonaceous char in the condensed phase [206,207]. The nanoscale dispersed lamellae of clay, either intercalated or exfoliated in the polymer matrix, all enhance the formation of charring upon burning [208]. After pyrolysis, the nanocomposite forms a char with a multilayered carbonaceous silicate structure. The exfoliated clay layers first collapse into an intercalated structure, which is transformed into a multilayered carbonaceous silicate structure later [208,209]. The carbonaceous char builds up on the surface during burning and insulates the underlying materials. This limits the passage of degradation products from the matrix, which supports the continuous fueling of the fire [210]. In the case of the presence of char forming additives, the nanoclay reinforces the char and makes it more impervious. It was reported that the formation of a hard crust on burning of the nanocomposite, which can be observed on the surface of the residue. A TEM picture of the crust at the border between polymer and the residue as reported by Hartwig et al. [211] is presented in Fig. 5.43. It can be seen that the silicate is concentrated close to the interface of the residue and that particles are the main component of the residue. The particles contain silicon and aluminum, which shows that they are originated from clay. However, the layered structure of the clay is destroyed in the residue.

Hartwig et al. [211] examined the cone calorimetric analysis of three different epoxy samples: unmodified epoxy resin (E + T), layered silicate epoxy resin nanocomposite with 4.7 wt.% ammonium bentonite (E + T + TMA), and layered silicate epoxy resin nanocomposite with 4.7 wt.% phosphonium bentonite (E + T + TPP). The comparison of heat release rates and total heat release for the three epoxy samples is presented in Fig. 5.44. The external heat flux of 70 kW m^{-2} was maintained for all the samples. The unmodified resin shows a sharp peak whereas the nanocomposites show plateau-like behavior. The drastic reduction in peak heat release rate shows a typical improvement in fire resistance. The phosphonium bentonite-based nanocomposite shows superior flame resistance behavior compared to the ammonium bentonite-based nanocomposite. The advantages of organoclay over conventional flame retardants are manifolds: no generation of toxic gases (truly green flame retardant) and no discoloration. The use of organoclay improves the mechanical properties, unlike conventional flame retardants, which usually cause a deterioration in the thermomechanical properties [22–26]. However, it is usually found that such clay additions are not sufficiently effective to be classified as flame retardant. The clay-modified polymer compositions perform poorly when tested by the industrially significant UL-94 standard with respect to extinction time [212]. The failure is attributed to the adsorbed onium salt present in the clay (making it organophilic), which increases the early ignition. Also at high temperature, the onium ions present in the galleries of the clay undergo decomposition destroying the nanocomposite structure.

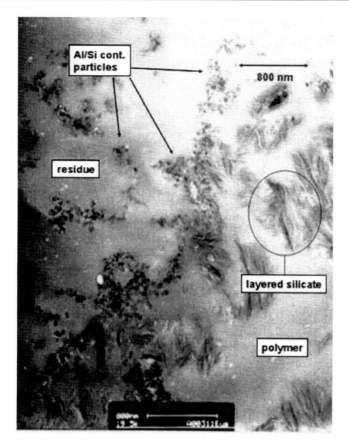

Figure 5.43
TEM image of the ignited polymer sample containing the sodium bentonite in the region between the decomposed and the intact polymer. Source: *Reprinted with permission from A. Hartwig, D. Putz, B. Schartel, M. Bartholmai, M.W. Josties, Macromol. Chem. Phys. 204 (2003) 2247, ©2003 Wiley-VCH Verlag GmbH & Co KgaA Publishers.*

5.2.10 Combination organoclay and other flame retardants

It has been established by recent studies that the presence of a nanofiller (organoclay) enhances the compatibility of an incompatible blend. For example, scanning transmission X-ray microscopy images [213] of PS/PMMA (30/70) and PS/PMMA/Closite 30B (27/63/10) taken after annealing at 190°C are presented in Fig. 5.45. It was observed that the size of the dispersed particles decreases due to the addition of clay into PS/PMMA blend indicating the improvement in the compatibility. The results have been attributed to the barrier effect or specific interaction of both the polymers with the clay, which resulted in a change in free energy of mixing. The modern flame retardants are basically organic molecules with halogen or phosphorous groups and are not well dispersed in the polymer system. Hence it can be

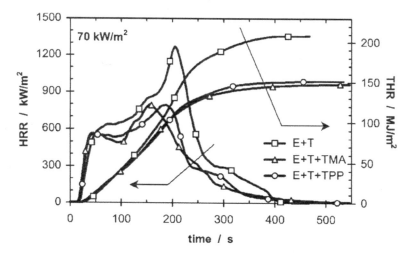

Figure 5.44
Cone calorimetric analysis of three different epoxy samples: unmodified epoxy resin (E + T), layered silicate epoxy resin nanocomposite with 4.7 wt.% ammonium bentonite (E + T + TMA), and layered silicate epoxy resin nanocomposite with 4.7 wt.% phosphonium bentonite (E + T + TPP). Source: *Reprinted with permission from A. Hartwig, D. Putz, B. Schartel, M. Bartholmai, M.W. Josties, Macromol. Chem. Phys. 204 (2003) 2247, ©2003 Wiley-VCH Verlag GmbH & Co KgaA Publishers.*

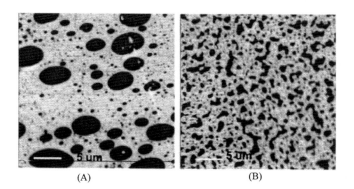

Figure 5.45
Scanning transmission X-ray microscopy images of PS/PMMA (30/70) and PS/PMMA/Closite 30B (27/63/10) taken after annealing at 190°C. Source: *Reprinted with permission from M. Si, T. Araki, H. Ade, A.L.D. Kilcoyne, R. Fisher, J.C. Sokolov, M.H. Rafailovich, Macromolecules, 39 (2006) 4793, ©2006 American Chemical Society Publishers.*

postulated that the presence of clay might enhance the compatibility between the polymer and the fire retardant resulting in an improvement in fire properties. Thus by using a small amount of clay, it is possible to significantly reduce the amount of conventional fire retardant additives, required for optimum flame retardancy. Conventional flame retardants have always some detrimental effects on the mechanical properties of the polymer system; hence by using nanoclay, such detrimental effects can be minimized and compensated. Synergistic effects of inorganic flame retardants (aluminum trihydrate) and phosphorous-based flame retardants with nanoclay were reported for thermoset resins like epoxy [214] and vinyl ester [215]. TEM and pyrolysis gas chromatography-mass spectrometry (GC-Ms) measurements of serially burned samples indicate that the clays play various roles in quenching the flame, such as promoting the char formation, improving the dispersion of fire retardants, and catalyzing the chain reaction for dissociation of halogenated compounds. However, Simon et al. [216] observed no synergistic effect while studying a tetrafunctional epoxy/clay nanocomposite and a phosphorous-containing flame retardant. The synergistic effect of nanoclay and conventional fire retardant (decabromodiphenyl oxide plus antimony trioxide) was also investigated for thermoplastic-based nanocomposite systems, for example, polypropylene-g-maleic anhydride system [208]. The synergy between the brominated fire retardant and antimony trioxide was found in the nanocomposite system whereas no synergy was observed in the virgin polymer matrix.

PU-based microcellular nanocomposite foams have been prepared by incorporating 10 wt.% organoclay into the PU matrix via a microcellular injection molding (MIM) method [217]. The XRD and TEM studies indicated that the layered nanoclays in the nanocomposite foams achieved a well-dispersed morphology within the cell walls due to the MIM foaming process. At the same time, the nanocomposite-based foam was able to maintain its foam porosity, original shape, and microcellular structure due to the synergistic effects of its microcellular structure and the fully dispersed nanoclays. The incorporation of 10 wt.% nanoclay resulted in a 27% decrease in heat release capacity (HRC) (measured by cone calorimeter) and 18 times increase in char yield. The flame retardant properties were attributed to a combination of two flame retardant mechanisms—the thermal barrier effect of the nanoclay layers and the microcellular self-intumescing system in the nanocomposite foams.

Bonati and coworker [218] used inorganic filler like aluminum trihydate (ATH) and asphalt binders nanomodified with organoclay in combination for development material used in tunnel structure. They have reported that the simultaneous use of ATH and nanocomposite asphalt binder gives a synergic effect that significantly changes and improves the fire behavior of the mixtures. Mishra and Vaidyanathan [219] fabricated clay-infused recycled carpet composite using a vacuum-assisted resin transfer molding (VARTM) process. It was demonstrated from flame retardancy tests that the presence of clay in the polymer network decreases the ignition time of the carpet composites.

5.2.11 Nanocarbon-based flame retardants

The use of carbonaceous nanomaterials, especially CNTs and graphenes, as a reinforcement for thermoset resins have been discussed in Sections 5.2.4 and 5.2.5, respectively, which have demonstrated unprecedented performance at ultralow fractions. In this section, a review on carbon-family material-based flame retardants and their comparison in terms of performance with conventional flame retardants and synergy effects while used in a combination will be presented. In this context, the readers may refer to two recent reviews published in 2017 and 2021 [219,220], which have elaborated the subject elegantly, highlighting the current state-of-the-art flame retarding technologies based on recently published work that used graphene, carbon nanotubes, and carbon black for multifunctional, flame retarding polymer nanocomposites.

When nanofillers such as graphene or carbon nanotubes are incorporated into the thermoset resin matrix as flame-retardant additives, the major flame retarding mechanism is through char layer formation. Undoubtedly, the low filler-to-filler distance and high interface area of these carbon nanomaterials promote the formation of compact, thick char layers. This nanoreinforced thick char acts as a thermal barrier coating and insulates the polymeric product from the heat feedstocks [221]. As a result, heat transfer to combustible gasses is reduced leading to slow pyrolysis of the material. The char layer provides a tortuous path to the volatiles and the gas phase thereby reducing the gas permeability [222]. Because of reduced permeability, O_2 cannot reach and come in contact with the underneath semipyrolyzed polymer. The flammable organic volatiles get adsorbed on the surface of the carbonaceous nanomaterials, which hinders the release of such gases during combustion. The oxygen-containing groups undergo decomposition and dehydration at low temperatures and consequently reduce the pyrolysis temperature and dilute the oxygen supply. Moreover, the small filler-to-filler distance and the large characteristic interfacial area of the nanomaterials are beneficial to form a thermally conductive three-dimensional network throughout the polymer. The formation of such a conductive network helps to relocate heat produced as a result of combustion from the hotter region to the cooler region leading to an increase in the time to ignition. Due to the formation of a 3D network, the viscosity of the melt tends to increase tremendously, which prevents dripping and reduces the risk of flame spread to the surrounding environment.

Lee and coworkers [223] incorporated nanoclay and MWCNT in combination into epoxy resins and investigated the fire-retardant property. The nanocomposites were made by varying concentrations of nanoclay from 6–10 wt.%. 1 wt.% of MWCNT was added into each of the formulations containing nanoclay and investigated the fire-retardant property. MWCNTs tend to reduce the degradation rate of the epoxy complex and increased the char yield, and the limiting oxygen index in the first-order relationship with char yield by working as a heat barrier and a thermal insulator. The addition of graphene into a thermoset is reported to reduce the PHRR determined from the cone calorimetry experiment [224,225]. From the mechanistic point of view, it is clear that better dispersion of CNT or graphene in a thermoset matrix is going to provide superior flame retardant property. In Section 5.2.5, we have discussed that decoration of GO with metal oxide is an effective approach to improve the dispersion of GO in a thermoset matrix and to enhance the mechanical and functional properties of the related nanocomposites. Hence it is expected that metal oxide decorated GO-based composite will show better flame-retardant properties compared to thermoset/GO composite. In this context, relevant work related to metal oxide decorated GO is elaborated as below.

Kalai et al. [226] synthesized three types of metal nanoparticles decorated graphene hybrids such as Ag@rGO, Cu2O@rGO, and TiO2@rGO, using a green and efficient microwave irradiation technique. The modified GOs were incorporated separately into epoxy resin to make EP/GO nanocomposites. HRR measured from the cone calorimetry experiments for all the three nanocomposites and the pure epoxy are presented in Fig. 5.46. One can observe in the figure, the nanocomposites display significantly lower HRR compared to pure EP. The pristine EP burnt fiercely after ignition and reached a PHRR of 1286 kW m^{-2}. The PHRR value was reduced to 849 kW m^{-2} for the EP composite with 2 wt.% rGO, which exhibited a 34% reduction in comparison to that of the pristine polymer. As previously reported, graphene acted as insulating barriers that slowed down the escape of volatile decomposition products during combustion. By comparison, either Ag@rGO/EP or TiO2@rGO/EP displayed a comparable PHRR value to rGO/EP, but Cu2O@rGO/EP showed a further decreased PHRR value of 643 kW m^{-2}. If we calculate the ratio of PHHR, tPHHR ( ), which is considered as an indication of fire spread behavior, then we find a reduction of 11.2 to 4.4 kW m^{-2} s^{-1} due to the incorporation of 2 wt.% Cu2O@rGO. This indicates a significant suppression of fire hazards. The pure epoxy showed a LOI value of 21.0%, which indicates its highly flammable nature. As expected

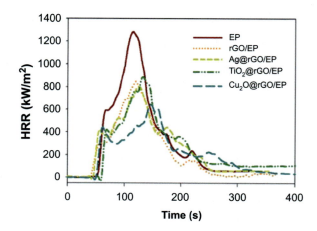

Figure 5.46
The HRR versus time profiles for the tested samples. (For interpretation of the references to color in this figure legend, the reader is referred to the web version of this article.) Source: *Reprinted with permission from E.N. Kalali, W. Guo, X. Wang, W. Xing, L. Song, Y. Hu, Compos. Part A 129 (2020) 105694, ©2020 Elsevier Publishers.*

and it does not show any classification in the UL-94 vertical burning test. When 2 wt.% of rGO is incorporated into the epoxy, the LOI value was increased to 23%, but still showed no rating in the UL-94 burning test. Incorporation of Ag@rGO or TiO2@rGO resulted in a slight improvement in LOI value with UL-94 V-1 rating in both cases. They achieved UL-94 V-1 rating in the vertical burning test. The epoxy composite modified with Cu2O@rGO exhibits a comparatively much better flame-retardant property, that is, an LOI value of 26.0% and UL-94 V-0 rating.

The use of graphene in combination with conventional flame retardants has also been investigated [227]. The presence of graphene can significantly reduce the optimal concentration of conventional flame retardant by providing a synergistic effect. For example, the addition of 8 wt.% of polyphosphoric acid (PPA) in epoxy resin caused a reduction of PHRR from 1730 to 1124 kW m^{-2}. When the same epoxy resin is blended with 2 wt.% PPA-modified graphenes, the resulting network displays a PHHR of 1154 kW m^{-2}. This means the addition of 2 wt.% PPA-modified graphene offers similar flame retardant properties achieved by the addition of 8 wt.% PPA. Hence the overall concentration of flame retardant additive can be reduced by using graphene or other carbon-based nanomaterials. It may be noted that the addition of a conventional FR additive may reduce the mechanical properties of the modified network significantly. Using this concept, an epoxy formulation [228] comprising 28 wt.% ammonium polyphosphate and 2 wt.% graphenes was developed which shows a V0 rating in the UL-94 burning test.

5.2.12 Application of nanocomposites

Nanocomposites have potential applications in various industries, namely, automotive, packaging, aerospace, electronics, biomedical, and defense. Various applications are briefly presented in Table 5.5. The first successful use of polymer/clay nanocomposites was in the automotive industry. Researchers at Toyota could reduce the weight of automobile parts by about 40% by using nanomaterials compared to conventional materials. Subsequently, the technology was adopted by other automotive companies. Thus the use of nanocomposites in automotive industries is likely to reduce fuel consumption and the related carbon dioxide emissions to a great extent.

The unique mechanical and barrier properties of nanocomposites justify their potential applications in structural and packaging applications. Very thin nanocomposite films can be used as wrapping materials for vegetables and fruits for long storage due to their inherent very low permeability to oxygen and moisture. Specific examples include packaging of processed meats, cheese, confectionery, and boil-in-the-bag foods. Since the nanocomposite film with much lower thickness compared to pristine polymer film can serve the packaging purpose, the solid waste and burden of the soldier can be reduced significantly. Thus nanocomposite has potential applications in defense technology. The low gas permeability

Table 5.5: Applications of polymer nanocomposites.

Field of application	Area/advantages
Automotive	Timing belt cover, engine cover, door handles, bumper, etc.
Food packaging	1. Enhancement of shelf life of food
	2. Reduction of solid waste
	3. Reduction in overall logistic burden of war fighter
Fuel tanks	1. Low fuel transmission
	2. Material for fuel tank and fuel line component of vehicles
Aerospace	1. Low level of incorporation
	2. Less weight penalty
Plastic film industries	Enhancement of transparency and haze reduction
Naval	Reduced water diffusion coupled with flame retardancy
Barrier coating	Reduced permeability
Medical	Dental filling material, artificial intestine
Electrical and electronic industries	Electrodes, sensors, gas separation membrane, fuel cell membrane, solid polymer electrolytes for batteries, and supercapacitors
Smart devices	Remote-sensing actuators
General	Impellers, blades for vacuum cleaner, power hoods, and covers for electronic equipment

of nanocomposites can be exploited for the development of gas separation membranes, where restricted and preferential permeability of gases is the prime requirement. Nanotechnology can be used for the development of high-performance proton-exchange membrane fuel cells (PEMFC). Fuel cells are an unconventional energy source, provide energy without environmental pollution, and are going to be the future energy sources. The role of the membrane is to allow proton transport from the anode to the cathode and to act as a separating barrier between hydrogen and oxygen. Presently, Nafion (DuPonT) is widely used as a membrane in PEMFC. However, it is permeable to methanol, very expensive, and limited to only low-temperature applications. The use of PEMFC membrane at a higher temperature (80°C) is required in order to increase the yield of the cell. By using nanomaterials like layered silicate or CNT, it is possible to control the properties like moisture retention, chemical stability, and proton conductivity of a conventional thermoset-based membrane so that they can compete with Nafion in terms of exchange property and at the same time offer better thermal stability [229].

Since thermoset nanocomposites show much lower water permeability compared to the corresponding matrix polymer, they have potential applications in the Navy. The electronic instruments (e.g., transducer) have to work underwater for navigation and ranging purposes. It is necessary to encapsulate such instruments with a suitable polymeric material. The nanocomposite-based encapsulating materials can offer longer life because of their lower water permeability. The thermoset resin and nanoclay combination can be integrated into glass or carbon fiber-reinforced composites for the development of FRP structures for underwater applications. Thus by utilizing nanotechnology the life of materials can be significantly extended for underwater applications.

Polymer nanocomposites have potential applications in the field of vibration damping. Vibration damping is necessary to avoid undesirable consequences such as unpleasant motions, noise, and dynamic stresses. Polymers (viscoelastic materials) are generally used to damp the vibration. For effective vibration damping, a material should have both high modulus and loss factor. When a polymer is blended with conventional fillers, the modulus increases moderately along with a drastic reduction in loss factor. By using nanomaterials, the modulus can be improved considerably without a significant sacrifice or even increase in loss factor. Hence such nanocomposite materials are going to find wide applications in the field of vibration damping. Rubbery epoxy/clay nanocomposite has been successfully utilized for the development of vibration damping coating for naval applications [230].

Thermoset/CNT or graphene nanocomposites can be used as conducting composites for various applications. One such example is antistatic coating. Polymers because of insulating nature often find problem arising due to the storage of static electricity. This problem can be solved by incorporating a small number of CNTs. Thermoset/CNT composites are the future materials for aerospace applications because of their antistatic property and radiation resistance. Shape memory polymers (SMP)/CNT composites have potential applications in remote-sensing actuators. PU-based SMPs have already been discussed in Chapter 2, Properties and Processing of Thermoset Resin. Most of the SMPs reported so far exhibit a thermally induced shape memory effect. Heat is a very good stimulus but it is very difficult to control from the remote. The SMP/CNT nanocomposite system is expected to offer electrically induced actuation, which can be controlled remotely, via Joule heating (current is passed through the conductive percolative network of the nanotubes within the SMP system) [231].

Recent research works on conducting polymer-based nanocomposite and liquid crystalline polymer-based nanocomposites show the promise for electronic applications. Conductive polymer-based nanocomposites present remarkable electrochemical behavior, attracting interest in applications that include modified electrodes, biosensors, solid-state batteries, smart windows, and other electrochemical devices. Polypyrrole (Ppy)—clay nanocomposite exhibits redox chemistry approximately 1 V more negative than the chemically synthesized Ppy. Although the capacitances of the materials are similar, the charge—discharge plateau is also shifted approximately 1 V. These observations could be related to the close interaction of the polymer with anionic clay layers, stabilizing the Ppy cationic form. The results reported recently on the electrochemical performance of conducting polymer/graphene oxide nanocomposite for application in flexible supercapacitor devices show considerable promise [232]. The conductivity of polyethylene oxide-based electrolytes (containing Li ions) can be significantly enhanced by incorporating a small amount of clay. The PEO-based electrolyte shows lower ionic conductivity due to the high crystallinity of PEO and the formation of ion pairs (ion-pair effect). The small anions get attached with cation strongly and move as ion-pair under electrical field, which significantly reduces the effective cation conductivity. Modification of PEO-based electrolyte with 7.5% organoclay resulted in an increase in conductivity by several times [233]. Such a dramatic change in conductivity due to the incorporation of clay can be explained by two reasons: First, the clay increases the amorphous content of PEO, which participates in the conduction of ions. Second, when Li ion enters into the galleries of clay, the silicate layers act as anion and under the electric field, only the Li ion moves causing an increase in effective cation conductivity.

Thermoset-based liquid crystalline materials have been discussed in Chapter 2, Properties and Processing of Thermoset Resin. When an electric field is applied, the LCs orient themselves in the direction of the field and once the electric field is turned off, the reformation takes place immediately. When clay is dispersed in LC matrix, the clay platelets anchor the LCs and prevent reform into the original random distribution. Such systems can find applications like adjustable light-controlling glass, erasable optical storage battery, thermooptical sensor, optical shutter, etc. The same strategy can be used in nonlinear optical material to increase the relaxation time of the aligned dipoles. NLO polymers possess dipoles, which can be aligned by the application of an electric field at a temperature higher than T_g of the polymer. The dipole arrangement can be fixed by cooling the polymer to room temperature, keeping the electric field on. Such poled system exhibits an interesting property of second harmonic generation. When radiation having a frequency of w is passed through the material, the radiation is converted into one with a frequency of $2w$. Thus such materials have potential use for frequency modulation and many other interesting applications. However, the diploes in a polymer with moderate T_g relax very fast leading to a loss in efficiency of the second harmonic generation. The incorporation of clay is expected to reduce the relaxations of the dipoles in the same way as in the LC systems discussed above.

In recent years, thermoset-based nanocomposite made with conducting nanofillers like carbon nanotubes, graphene have been proposed for the development of microwave-absorbing coating, which will find application to avoid detection of strategic platforms (airborne and marine) by enemy radar (radar stealth technology) [129–135]. They can also be used as electromagnetic interference (EMI) shielding for protecting communication and safety systems employed in aircrafts, ships and automobiles [234]. The rubbery thermoset nanocomposite-based EMI shielding offers advantages in terms of flexibility over metallic EMI shielding materials. Such materials can be utilized to enhance the efficiency of electronic equipment used for various applications.

5.2.13 Summary and outlook

The last two decades have witnessed extensive research and development in the field of thermoset nanocomposites. The physics and chemistry of intercalation of clay (layered silicate) in thermoset resin have been understood to a great extent through the synthesis of fully exfoliated thermoset-based nanocomposites still remains as a challenge. More researches are necessary to study the feasibility of integrating the nanoreinforced thermoset resins into the fiber-reinforced plastics to develop a composite structure not only with an improved strength but with improved water and solvent resistance. Such composites will have tremendous potential for use in the civil and defense sectors. It has been demonstrated that the reinforcing effect is much more effective in a rubbery matrix rather than in a highly brittle glassy matrix. The research will continue in this field focusing more on using low T_g thermoset matrices, and the thermosets with liquid crystalline and nonlinear optical properties.

Organoclays offer fire retardancy without any environmental hazard ("green" flame retardants). Note that the use of conventional flame retardants (FRs) is associated with environmental hazards, deterioration in mechanical properties, and discoloration. However, nanoclays are not themselves alone sufficiently effective. They can be used in combination with other FRs to achieve the required fire properties with a minimum amount of FR and to compensate for the degradation in thermomechanical properties of the composites occurring due to the addition of FRs. Chemically reactive phosphorous-containing flame retardants, in addition to organoclays, are going to be the future materials for the development of high-performance flame retardant thermoset composites. Note that phosphorous-containing FRs are reported to be better in terms of environmental protection compared to halogen-based ones. In addition, phosphorous-containing modifiers can simultaneously act as a flame retardant and as an impact modifier for thermoset.

Thermoset/clay technology has been exploited for the development of products for various civil and defense applications to a greater extent than thermoset/MWCNT or graphene nanocomposites. Hence the future works on thermoset nanocomposites is going to consider CNTs and grapheme in a bigger way. Since the magnitude of CNT strength is very high (almost 10 times higher than that of typical carbon fiber), very high interfacial shear strength may be required for a more efficient strengthening of polymers with CNTs. Recent strategies of using functionalized CNTs, graphenes, and cation$-\angle$ interactions show a considerable promise and are likely to be the subjects of major focus in the near future [103,117,145].

The toughening strategy can be coupled with a suitable nanoreinforcement by using a nanofiller like nanoclay and CNT to produce strong and tough composites. The control of both microstructures and nanostructures is required for achieving the optimum properties. In a liquid rubber-toughened system, the presence of nanomaterials disturbs the reaction-induced phase separation resulting in a change in microstructure as well as the toughening mechanism [184,185,235]. Thus extensive studies are needed to look into the various aspects for optimization of both the microstructures and nanostructures. Integration of such hybrid nanomaterials into fiber-reinforced thermoset composite is going to be focused on the future for the development of strong and tough structural composites for various civil and defense applications.

References

[1] B.M. Novak, Adv. Mater. 5 (1993) 422.
[2] P. Westbrock, F. Marin, Nature 394 (1998) 256.
[3] C. Sanchez, P. Julian, P. Pellerille, M. Popall, J. Mater. Chem. 15 (2005) 3559.
[4] G. Philipp, H. Schimdt, J. Non-Cryst. Solids 80 (1984) 283.
[5] G. Philipp, H. Schimdt, J. Non-Cryst. Solids 82 (1986) 31.
[6] M. Kakimoto, Y. Iyoku, A. Morikawa, H. Yamaguchi, Y. Imai, Polym. Prepr. 35 (1994) 393.
[7] S. Mann, S.L. Burkett, S.A. Davis, C.E. Fowler, N.H. Mendelson, S.D. Sims, et al., Chem. Mater. 9 (1997) 2300.
[8] T.P. Lodge, Macomol. Chem. Phys. 204 (2004) 265.
[9] D. Ratna, A.K. Banthia, Mocromol. Res. 12 (2004) 11.
[10] D.J. Greenland, J. Colloid Sci. 18 (1963) 647.
[11] A. Usuki, Y. Kojima, M. Kawasumi, A. Okoda, Y. Fukushima, T. Kurauchi, et al., J. Mater. Res. 8 (1993) 1179.
[12] Y. Kojima, A. Usuki, M. Kawasumi, A. Okada, T. Kurauchi, O. Kamigaito, J. Polym. Sci. Polym. Chem. 31 (1993) 983.
[13] M. Zanetti, S. Lomakin, G. Camino, Macromol. Mater. Eng. 279 (2000) 1.
[14] B.K.G. Theng, The Chemistry of Clay-Organic Reaction, Wiley, New York, 1974.
[15] E.P. Giannelis, Adv. Mater. 8 (1) (1996) 29.
[16] M. Kawasumi, J. Polym. Sci. Polym. Chem. 42 (3) (2004) 819.
[17] D. Ratna, J. Polym. Mater. 19 (2) (2002) 143.
[18] N. Karak, J. Polym. Mater. 23 (1) (2006) 1.
[19] S. Verghese, K.G. Gatos, A.A. Apostolov, J. Karger-Kocsis, J. Appl. Polym. Sci. 92 (2004) 543.
[20] A. Usuki, H. Hasegawa, M. Kato, Adv. Polym. Sci. 179 (2005) 135.
[21] Y. Tang, Y. Hu, S. Wang, J. Gui, Z. Chen, W. Fan, Polym. Adv. Technol. 14 (2003) 733.
[22] P.C. LeBaron, Z. Wang, T.J. Pinnavaia, Appl. Clay Sci. 15 (1) (1999) 11.
[23] D. Ratna, Rapra Rev. Rep. 16 (5) (2005) 1.
[24] N.A. Salahuddin, Polym. Adv. Technol. 5 (5) (2004) 251.
[25] T. Lan, P.D. Kaviratna, T.J. Pinnavia, Chem. Mater. 6 (1994) 573.
[26] Z. Wang, T. Lan, T.J. Pinnavaia, Chem. Mater. 8 (1996) 2200.
[27] T. Lan, T.J. Pinnavia, Chem. Mater. 6 (1994) 2216.
[28] J.M. Brown, D. Curliss, R.A. Vaia, Chem. Mater. 12 (2000) 3376.
[29] A. Yasmin, J.L. Abot, I.M. Daniel, Scr. Materiala 49 (1) (2003) 81.
[30] O. Becker, Y.B. Cheng, R.J. Verlay, G.P. Simon, Macromolecules 36 (2003) 1616.
[31] S. Sinha Ray, M. Okamoto, Prog. Polym. Sci. 28 (2003) 1539.
[32] D. Ratna, N.R. Manoj, R.K. Singh Raman, R. Varley, G.P. Simon, Polym. Int. 52 (9) (2003) 1403.
[33] I.J. Mathias, R.D. Davis, W.L. Jarrett, Macromolecules 32 (1999) 7958.
[34] A. Bafna, G. Beaucage, F. Mirrabella, S. Mehta, Polymer 44 (2003) 1103.
[35] T. McNally, W.R. Murphy, Y.L. Chun, R.J. Turner, G.P. Brennan, Polymer 44 (2003) 2761.
[36] C. Chen, D. Curliss, Nanotechnology 14 (2003) 643.
[37] D.L. VanderHart, A. Asano, J.W. Gilman, Macromolecules 34 (2001) 3819.
[38] R.A. Vaia, K.D. Jandt, E.J. Kramer, E.P. Giannelis, Macromolecules 28 (1995) 8080.
[39] Y. Ke, J. Lü, X. Yi, J. Zhao, Z. Qi, J. Appl. Polym. Sci. 78 (2000) 808.
[40] T. Lan, P.D. Kaviratna, T.J. Pinnavaia, Chem. Mater. 7 (1995) 2144.

[41] C. Zilg, R. Thomann, J. Finter, R. Mülhaupt, Macromol. Mater. Eng. 280 (2000) 41.
[42] Z. Wang, T.J. Pinnavaia, Chem. Mater. 10 (1998) 1820.
[43] M. Okamoto, P.H. Nam, P. Maiti, T. Kotaka, N. Hasegawa, A. Usuki, Nano Lett. 1 (6) (2001) 295.
[44] T.B. Tole, D.P. Anderson, Compos. Sci. Technol. 62 (2002) 1033.
[45] X. Kornmann, R. Thomann, R. Mulhaupt, J. Finter, L.A. Berglund, Polym. Eng. Sci. 42 (2002) 1815.
[46] H.L. Tyan, Y. Liu, K. Wei, Chem. Mater. 11 (7) (1999) 1942.
[47] D. Yebassa, S. Balakrishnan, E. Feresenbet, D. Raghavan, P.R. Start, S.D. Hudson, J. Polym. Sci. Polym. Chem. 42 (2004) 1310.
[48] C. Chen, D. Curliss, J. Appl. Polym. Sci. 90 (2003) 2276.
[49] J.S. Chen, M.D. Poliks, C.K. Ober, Y. Zhang, U. Wiesner, E.P. Giannelis, Polymer 43 (2002) 4895.
[50] J. Park, S.C. Jana, Macromolecules 36 (22) (2003) 8391.
[51] D. Kong, C.E. Park, Chem. Mater. 15 (2003) 419.
[52] S.B. Jagtap, V.S. Rao, S. Barman, D. Ratna, Polymer 63 (2015) 41−51.
[53] P.D. Messersmith, E.P. Giannelis, Chem. Mater. 6 (1994) 1719.
[54] C. Zilg, R. Mulhaupt, J. Finter, Macromol. Chem. Phys. 200 (1999) 661.
[55] T. Lan, P.D. Kaviratna, T.J. Pinnavia, J. Phys. Chem. Solids 57 (1996) 1005.
[56] D. Ratna, B.C. Chakraborty, H. Dutta, A.K. Banthia, Polym. Eng. Sci. 46 (12) (2006) 1667.
[57] D. Ratna, Int. J. Nanotechnol. Appl. 5 (4) (2011) 373−381.
[58] S.B. Jagtap, V. Srinivasa Rao, D. Ratna, J. Rein. Plastic Compo 32 (2013) 183.
[59] M.A. El-Fattah, A.M. El Saeed, M.M. Dardir, et al., Prog. Org. Coat. 89 (2015) 212−219.
[60] S. Ashhari, A.A. Sarabi, Polym. Compos. 38 (6) (2017) 1167−1174.
[61] J. Quagliano, J. Bocchio, IOP Conf. Ser.: Mater. Sci. Eng. 64 (2014) 012035.
[62] M. Kosari, S. Mousavian, S. Razavi, S. Ahmadi, M. Izadipanah, Plastic Rubber Compos. 46 (2017) 413−420.
[63] X. Zhang, R. Xu, Z. Wu, et al., Polym. Int. 52 (5) (2003) 790−794.
[64] O. Becker, R.J. Varley, G.P. Simon, Eur. Polym. J. 40 (1) (2004) 187.
[65] K. Yano, Y. Kojima, A. Usuki, A. Okada, T. Kurachi, O. Kamigaito, J. Polym. Sci. Pt A Polym. Chem. 31 (1993) 2493.
[66] A.P. Shah, R.K. Gupta, H.V.S. Gangarao, C.E. Powell, Polym. Eng. Sci. 42 (2002) 1852.
[67] D. Ratna, S. Khan, S. Barman, B.C. Chakraborty, Open. Macromol. J. 6 (2012) 59−67.
[68] D.P. Fasce, R.J.J. Williams, F. Mchin, J.P. Pascault, M.F. Llauro, R. Ptiaud, Macromolecules 32 (15) (1999) 4757.
[69] D.P. Fasce, R.J.J. Williams, R. Balsells, Y. Ishikawa, H. Nanami, Macromolecules 34 (11) (2001) 3534.
[70] S.A. Pellice, D.P. Fasce, R.J.J. Williams, J. Polym. Sci. Polym. Phys. 41 (13) (2003) 1451.
[71] Y. Ni, S. Zheng, Macromolecules 40 (19) (2007) 7009.
[72] I.A. Zucchi, M.J. Galante, R.J.J. Williams, E. Franchini, J. Gally, J.F. Gerard, Macromolecules 40 (4) (2007) 1274.
[73] C. Cheng, H. Cheng, P. Cheng, Y. Lee, Macromolecules 39 (2006) 7583.
[74] C. Zhang, T. Li, H. Song, Y. Han, H. Su, Y. Wang, et al., J. Photopolym. Sci. Technol. 30 (2017) 25−31.
[75] Z. Xu, Y. Zhao, X. Wang, T. Lin, Chem. Commun. 49 (2013) 6755−6757.
[76] M.W. Matsen, F.S. Bates, Macromolecules 29 (1996) 1092.
[77] P.M. Lipic, F.S. Bates, M.A. Hillmyer, J. Am. Chem. Soc. 120 (1998) 8963.
[78] R.B. Grubbs, J.M. Dean, M.E. Broz, F.S. Bates, Macromolecules 33 (2000) 9522.
[79] H. Garate, S. Goyanes, N.B. D'Accorso, Macromolecules 47 (2014) 7416−7423.
[80] E.M. Redline, C. Declet-Perez, F.S. Bates, L.F. Francis, Polymer 55 (16) (2014) 4172−4181.
[81] J. Liu, Z.J. Thompson, H.J. Sue, F.S. Bates, M.A. Hillmyer, M. Dettloff, et al., Macromolecules 43 (17) (2010) 7238−7243.
[82] J.M. Dean, R.B. Grubbs, W. Saad, R.F. Cook, F.S. Bates, J. Polym. Sci. Pt B Polym. Phys. 41 (20) (2003) 2444−2456.

[83] V. Pang, Z.J. Thompson, G.D. Joly, F.S. Bates, L.F. Francis, ACS Appl. Polym. Mater. 2 (2020) 464–474.
[84] P. Gerard, N.P. Boupat, T. Fine, L. Gervat, J.P. Pascault, Macromolecular Symposia, 256, WILEY-VCH Verlag, Weinheim, 2007, pp. 55–64.
[85] L. Tao, Z. Sun, W. Min, H. Ou, L. Qi, M. Yu, RSC Adv. 10 (3) (2020) 1603–1612.
[86] A. Bajpai, B. Wetzel, J. Compos. Sci. 3 (3) (2019) 68.
[87] M. Li, Z. Heng, Y. Chen, H. Zou, M. Liang, Ind. Eng. Chem. Res. 57 (39) (2018) 13036–13047.
[88] C. Declet-Perez, L.F. Francis, F.S. Bates, Macromolecules 48 (2015) 3672–3684.
[89] F. Nian, J. Ou, Q. Yong, Y. Zhao, H. Pang, B. Liao, J. Macromol. Sci. Pt A Pure Appl. Chem. 55 (7) (2018) 533–543.
[90] T. Li, M.J. Heinzer, L.F. Francis, F.S. Bates, J. Polym. Sci. Pt B Polym. Phys. 54 (2016) 189–204.
[91] S. Ijima, Nature 354 (1) (1991) 56.
[92] E.T. Thostenson, Z. Ren, T. Chou, Compos. Sci. Technol. 61 (13) (2001) 1899.
[93] M. Moniruzzaman, K.I. Winey, Macromolecules 39 (16) (2006) 5194–5205.
[94] F.H. Gojny, M.H.G. Wichmann, U. Köpke, B. Fiedler, K. Schulte, Comp. Sci. Technol. 64 (2004) 2363–2371.
[95] A. Kausar, I. Rafique, B. Muhammad, Polym. Plastic Tech. & Engg 55 (2016) 1167–1191.
[96] M. Bhattacharya, Materials 9 (2016) 262.
[97] I. Tarashi, A. Fereidoon, S. Paszkiewicz, Z. Roslaniec, Compos. Interfaces 25 (2018) 275–286.
[98] M. Treacy, T.W. Ebbesen, J.M. Gibson, Nature 381 (6584) (1996) 678–680.
[99] R. Andrews, M.C. Weisenberger, Curr. Opin. Solid. State Mater. Sci. 8 (1) (2004) 31.
[100] P. Puglia, L. Valentini, J.M. Kenny, J. Appl. Polym. Sci. 88 (2003) 452.
[101] J. Zhu, H. Peng, F.R.- Macious, J.L. Margrave, V.N. Khabashesku, A.M. Imam, et al., Adv. Func. Mater. 14 (2004) 643.
[102] H. Miyagawa, L.T. Drzal, Polymer 45 (15) (2004) 5163.
[103] K.S. Khare, R. Khare, J. Phys. Chem. B 117 (240) (2013) 7444–7454.
[104] K.S. Khare, FardinKhabaz, R. Khare, ACS Appl. Mater. Interfaces 6 (9) (2014) 6098–6110.
[105] Y.X. Zhou, P.X. Wu, Z.Y. Cheng, J. Ingram, S. Jeelani, Express Polym. Lett. 2 (1) (2008) 40.
[106] L.S. Schadler, S.C. Giannaris, P.M. Ajayan, Appl. Phy. Lett. 73 (1998) 3842.
[107] L. Shen, L. Liu, Z. Wu, Adv. Eng. Mater. 22 (4) (2020) 1900895.
[108] Y. Lin, M.J. Meziani, Y.-P.J. Sun, Mater. Chem. 17 (2007) 1143.
[109] A. Yaghoubi, A. Ramazani, Curr. Org. Chem. 22 (2018) 1505–1522.
[110] D. Bhaskaran, J.R. Dunlap, J.W. Mays, M.S. Bratcher, Macromol. Rapid Commun. 26 (2005) 481.
[111] B. Zhao, H. Hu, A. Yu, D. Perea, R.C. Haddon, J. Am. Chem. Soc. 127 (2005) 8197.
[112] C.Y. Hong, Y.Z. You, C.Y. Pan, Chem. Mater. 17 (2005) 2247.
[113] A.M. Shanmugharaj, J.H. Bae, R.R. Nayak, S.H. Ryu, J. Polym. Sci. Pt A Polym. Chem. 45 (2007) 460.
[114] J.N. Coleman, U. Khan, W.J. Blau, Y.K. Gun'ko, Carbon 44 (2006) 1642.
[115] Z. Yaping, Z. Aibo, C. Qinghua, Z. Jiaoxia, N. Rongchang, Mater. Sci. Eng. A 435 (1) (2006) 145.
[116] K. Yang, MingyuanGu, Y. Jin, J. Appl. Polym. Sci. 110 (5) (2008) 2980–2988.
[117] A. Yaghoubi, M. Nikje, Compos. Pt A Appl. Sci. Manufac. 109 (2018) 338–344.
[118] F.H. Gojny, J. Nastalczyk, Z. Roslaniec, K. Schulte, Chem. Phys. Lett. 370 (2003) 820–824.
[119] P.V. Kodgire, A.R. Bhattacharyya, S. Bose, N. Gupta, A.R. Kulkarni, A. Mishra, Chem. Phys. Lett. 432 (2006) 480–485.
[120] X. Gong, J. Liu, S. Baskaren, R.D. Voise, J.S. Young, Chem. Mater. 12 (2000) 1049.
[121] R.B. Naik, S.B. Jagtap, R.S. Naik, N.G. Malvankar, D. Ratna, Prog. org. Coat. 77 (2014) 1883–1889.
[122] S.B. Jagtap, D. Ratna, Express Poly Lett. 7 (2013) 329–339.
[123] S.B. Jagtap, D. Ratna, J. Appl. Polym. Sci. 130 (2013) 2610–2618.
[124] D. Ratna, T. Abraham, S. Siengchin, J. Karger-Kocsis, J. Polym. Sci. Pt B Polym. Phys. 47 (2009) 1156–1165.
[125] M. Schirowski, G. Abellán, E. Nuin, J. Pampel, C. Dolle, V. Wedler, et al., J. Am. Chem. Soc. 140 (2018) 3352–3360.

[126] T. Filleter, R. Bernal, S. Li, H.D. Espinosa, Adv. Mater. 23 (2011) 2855−2860.
[127] W. Jian, D. Lau, Compos. Sci. Technol. 191 (2020) 108076.
[128] J.C. Long, H. Zhan, G. Wu, Y. Zhang, J.N. Wang, Compos. Pt A Appl. Sci. Manuf. 146 (2021) 106409.
[129] M. Green, X. Chen, J. Materiomics 5 (2019) 503−541.
[130] B.D. Che, et al., Chem. Cent. J. 9 (1) (2015) 1−13.
[131] Y. Qing, et al., Carbon 48 (14) (2010) 4074−4080.
[132] I.H. Nwigboji, et al., Diam. Relat. Mater. 52 (2015) 66−71.
[133] N. Li, et al., Nano Lett. 6 (6) (2006) 1141−1145.
[134] T. Zhang, B. Zhong, J.Q. Yang, X.X. Huang, G. Wen, Ceram. Int. 41 (81) (2015) 163−170.
[135] G. Firouz, S.M. Dehaghi, H. Mahdavi, J. Coat. Technol. Res. 17 (2020) 650.
[136] K.S. Novoselov, A.K. Geim, S.V. Morozov, D. Jiang, Y. Zhang, S.V. Dubonos, Science 306 (2004) 666.
[137] K.S. Novoselov, A.K. Geim, S.V. Morozov, D. Jiang, M.I. Katsnelson, I.V. Grigorieva, Nature 438 (2005) 197.
[138] J.J. Mack, L.M. Viculis, A. Ali, R. Luoh, G. Yang, H.T. Hahn, et al., Adv. Mater. 17 (2005) 77.
[139] K.S. Subrahmanyam, S.R.C. Vivekchand, A. Govindaraj, C.N.R. Rao, J. Mater. Chem. 18 (2008) 1517.
[140] N.B. Thalib, S.N.H. Mustapha, C.K. Feng, R. Mustapha, Rev. Chem. Eng. 36 (5) (2020) 623−652.
[141] S. He, N.D. Petkovich, K. Liu, Y. Qian, C.W. Macosko, A. Stein, Polymer 110 (2017) 149−157.
[142] H. Zhang, H. Porwal, Z. Huang, E. Bilotti, T. Peijs, Compos. Pt A Appl. Sci. Manuf. 95 (2017) 229−236.
[143] M.T. Le, S.C. Huang, Materials 8 (2015) 5526−5536.
[144] N. Alexopoulos, Z. Paragkamian, P. Poulin, S.K. Kourkoulis, Compos. Sci. Technol. 150 (2017) 194−204.
[145] Y.T. Park, Y. Qian, C. Chan, T. Suh, M.G. Nejhad, C.W. Macosko, et al., Adv. Funct. Mater. 25 (4) (2015) 575−585.
[146] L.C. Tang, Y.J. Wan, D. Yan, Y.B. Pei, L. Zhao, Y.B. Li, et al., Carbon 60 (2013) 16−27.
[147] C. Bora, P. Gogoi, S. Baglari, S.K. Dolui, J. Appl. Polym. Sci. 129 (2013) 3432−3438.
[148] R.B. Naik, S.B. Jagtap, D. Ratna, Prog. Org. Coat. 87 (1) (2014) 28−35.
[149] S.B. Jagtap, R. Khushwaha, D. Ratna, RSC Adv. 5 (2015) 30555−30563.
[150] D.Y. Cai, K. Yusoh, M. Song, Nanotechnology 20 (2009) 085712.
[151] C. Bao, Y. Guo, L. Song, Y. Kan, X. Qian, Y. Hu, J. Mater. Chem. 21 (2011) 13290−13298.
[152] X. Wang, Y.A. Hu, L. Song, H.Y. Yang, W.Y. Xing, H.D. Lu, J. Mater. Chem. 21 (2011) 4222−4227.
[153] Y. Li, D. Pan, S. Chen, Q. Wang, G. Pan, T. Wang, Mater. Des. 47 (2013) 850−856.
[154] D.C. Marcano, D.V. Kosynkin, J.M. Berlin, A. Sinitskii, Z.Z. Sun, A. Slesarev, et al., ACS Nano 4 (2010) 4806−4814.
[155] D. Cai, J. Jin, K. Yusoh, R. Rafiq, M. Song, Compos. Sci. Technol. 72 (2012) 702−707.
[156] L. Zhang, H. Jiao, H. Jiu, J. Chang, S. Zhang, Y. Zhao, Compos. Pt A Appl. Sci. Manuf. 90 (2016) 286−295.
[157] W. Li, H. Li, X. Yang, W. Feng, J. Compos. Mater. 51 (2016) 1097−1208.
[158] S. Chhetri, N. Chandra, P. Samanta, N. Chandra, T. Kuila, Polym. Test. 63 (2017) 1−11.
[159] O. Eksik, A. Maiorana, S. Spinella, A. Krishnamurthy, S. Weiss, R.A. Gross, et al., ACS Sustain. Chem. Eng. 4 (2016) 1715−1721.
[160] P. Baruah, N. Karak, Polym. Degrad. Stab. 129 (2016) 26−33.
[161] A.A. Nazeer, E. Al-Hetlani, M.O. Amin, T. Quinones-Ruiz, I.K. Lednev, Chem. Eng. J. 361 (2019) 485−498.
[162] M.S. Selim, S.A. El-Safty, N.A. Fatthallah, M.A. Shenashen, Prog. Org. Coat. 121 (2018) 160−172.
[163] Z. Yu, H. Di, Y. Ma, L. Lv, Y. Pan, C. Zhang, et al., Appl. Surf. Sci. 351 (2015) 986−996.
[164] M.S. Selim, Z. Hao, P. Mo, Y. Jiang, H. Ou, Colloids Surf. A 601 (2020) 125057.
[165] M.Y. Mamaghani, M. Pishvaei, B. Kaffashi, J. Macromol. Res. 19 (2011) 243.
[166] Z. Xiu, Q. Zhang, H.L. Puppala, V.L. Colvin, P.J.J. Alvarez, Nano Lett. 12 (2012) 4271.
[167] A.S. Nair, N.P. Binoy, R. Ramakrishna, T.R.R. Kurup, L.W. Chan, C.H. Goh, et al., Appl. Mater. Interfaces 11 (2009) 2413.
[168] M. George, M. Jolocam, P. Nnamuyomba, H.C. Song, Nanoscien. Nanotech. 1 (2011) 40.

[169] Y. Jo, B.H. Kim, G. Jung, e-Xtra 93 (2009) 1037.
[170] R. Varshney, A.N. Mishra, S. Bhadauria, M.S. Gaur, J. Nanometer. Biostruct. 4 (2009) 349.
[171] P.J. Rivero, A.U.J. Goicoechea, C.R. Zamarreno, F.J. Arregui, I.R. Matlas, Nanoscale Res. Lett. 6 (2011) 305.
[172] M.K. De, M. Botes, T.E. Cloete, Nano. 6 (2011) 395.
[173] G.A. Sotiriou, S.E. Pratsinis, Environ. Sci. Technol. 44 (2010) 5649.
[174] E. Osman, A.E. Tuncer, Y. Yusuf, J. Macromol. Sci. Pt A Pure Appl. Chem. 45 (2008) 698.
[175] A. Pica, D. Ficai, C. Guran, Rev. Chim. (Buchar.) 63 (2012) 459−462.
[176] K. Uday, N. Karak, M. Mandal, Prog. Org. Coat. 68 (2010) 265.
[177] M. Akbarian, M.E. Olya, M. Ataeefard, M. Mahdavian, Prog. Org. Coat. 75 (2012) 344.
[178] R.B. Naik, D. Ratna, J. Coat. Technol. Res. 12 (2015) 1073−1083.
[179] R.B. Naik, D. Ratna, S.K. Singh, Prog. Org. Coat. 77 (2014) 369−379.
[180] R.B. Naik, D. Ratna, S.K. Singh, Prog. Org. Coat. 77 (2013) 369.
[181] A. Kumar, V.P. Kumar, P.M. Ajayan, G. John, Nat. Mater. 7 (3) (2008) 236−241.
[182] F.H. Gojny, M.H.G. Wichmann, U. Kopke, B. Fiedler, K. Schulte, Comp. Sci. Technol. 64 (2004) 2363.
[183] Y.T. Vu, G.S. Rajan, J.E. Mark, C.L. Myers, Polym. Int. 53 (2004) 1071.
[184] A. Surendran, J. Jay, S. Anas, S. Thomas, Soft Matter 16 (2020) 3363−3377.
[185] P. Vijayan, D. Pugilia, A. Jose, M. Kanny, S. Thomas, Mater. Sci. Eng. R. 116 (2017) 1−29.
[186] O. Becker, G.P. Simon, Adv. Polym. Sci. 179 (1) (2005) 29.
[187] J. Frohlich, R. Thomann, R. Mulhaupt, Macromolecule 36 (2003) 7205.
[188] D. Ratna, O. Becker, R. Krishnamurty, G.P. Simon, R. Varley, Polymer 44 (24) (2003) 7449.
[189] J. Choi, A.F. Yee, R.M. Laine, Macromolecule 34 (2004) 3267.
[190] M. Herandez, B. Sixou, J. Duchet, H. Sautereau, Polymer 48 (2007) 4075−4086.
[191] I. Isik, U. Yilmazer, G. Bayram, Polymer 44 (2003) 6371−6377.
[192] A. Asif, V.L. Roa, K.N. Ninan, Polym. Adv. Technol. 22 (2011) 437−447.
[193] A. Ishak, M.H. Farihan, J. Reinforced Plast. Compos. 29 (2010) 2834−2841.
[194] C. Wang, Z. Pan, L. Zhang, D. Yang, W. Li, Polym. Adv. Technol. 17 (2006) 528−533.
[195] L. Xu, L.J. Lee, Polymer 45 (2004) 7325−7334.
[196] M.H. Beheshty, M. Vafayan, M. Poorabdollah, Polym. Compos. 30 (2009) 629−638.
[197] J. Karger-Kocsis, O. Gryshchuk, N. Jost, J. Appl. Polym. Sci. 88 (2003) 2124.
[198] J. Karger-Kocsis, O. Gryshchuk, J. Frohlich, R. Mullhaupt, Compos. Sci. Technol. 63 (2003) 2045.
[199] X. Yao, X. Qi, Y. He, D. Tan, F. Chen, Q. Fu, ACS Appl. Mater. Interfaces 6 (2014) 2497−2507.
[200] L. Gong, L. Zhao, L.C. Tang, H.Y. Liu, Y.W. Mai, Compos. Sci. Technol. 121 (2015) 104−114.
[201] T. Keller, C. Tracy, A. Zhou, Compos. Pt A Appl. Sci. Manuf. 37 (2006) 1286.
[202] J.V. Bausano, J.J. Lesko, S.W. Case, Compos. Pt A Appl. Sci. Manuf. 37 (2006) 1092.
[203] Y. Hu, S. Wang, Z. Ling, Y. Zhuang, Z. Chen, W. Fan, Macromol. Mater. Eng. 288 (2003) 272.
[204] J.W. Gilman, Appl. Clay Sci. 15 (1999) 31.
[205] M. Berta, C. Lindsay, G. Pans, G. Camino, Polym. Degrad. Stab. 91 (2006) 1179.
[206] J.W. Gilman, C.L. Jackson, A.B. Morgan, R. Harris, E. Manias, E.P. Giannelis, et al., Chem. Mater. 12 (2000) 1866.
[207] S. Bourbigot, E. Devaux, X. Flambard, Polym. Degrad. Stab. 75 (2002) 397.
[208] M. Zanetti, G. Camino, D. Canavese, A.B. Morgan, C.A. Wilkie, Chem. Mater. 14 (2002) 189.
[209] J. Innes, A. Innes, Rapra Rev. Rep. 14 (2003) 1.
[210] A.E. Grand, C.A. Wilkie, Fire Retardency in Polymeric Materials, Marcel Dekker, Inc, 2000.
[211] A. Hartwig, D. Putz, B. Schartel, M. Bartholmai, M.W. Josties, Macromol. Chem. Phys. 204 (2003) 2247.
[212] A.B. Morgan, Polym. Adv. Technol. 17 (2006) 206.
[213] M. Si, T. Araki, H. Ade, A.L.D. Kilcoyne, R. Fisher, J.C. Sokolov, et al., Macromolecules 39 (2006) 4793.
[214] Y. Wang, Q. Zhang, Q. Fu, Macomol. Rapid. Commun. 24 (2003) 231.
[215] B. Schartel, U. Knoll, A. Hartwig, D. Puetz, Polym. Adv. Technol. 17 (2006) 218.
[216] G. Chigwada, D.D. Jiang, C.A. Wilkie, Proceedings of Fall Meeting 2004, ACS Polymeric Materials: Science and Engineering, Philadelphia (2004), 156−157.

[217] T. Geng, J. Han, C. Liu, C. Shen, L. Turng, H.E. Yang, Polym. Compos. 39 (2018) 1429–1440.
[218] A. Bonati, F. Merusi, G. Polacco, G. Bochicchio, B. Tessadri, G. Polacco, et al., Constr. Build. Mater. 47 (2013) 990–1000.
[219] X. Wang, E.N. Kalali, J.-T. Wan, D.-Y. Wang, Prog. Polym. Sci. 69 (2017) 22–46.
[220] S. Araby, B. Philips, Q. Meng, J. Ma, T. Laoui, C.H. Wang, Compos. Pt B 212 (2021) 108675.
[221] B. Dittrich, K.-A. Wartig, D. Hofmann, R. Mülhaupt, B. Schartel, Polym. Degrad. Stabil. 98 (2013) 1495–1505.
[222] M. Soheilmoghaddam, H. Adelnia, H.C. Bidsorkhi, G. Sharifzadeh, M.U. Wahit, N.I. Akos, et al., Macromol. Mater. Eng. 302 (2017) 1600260.
[223] S.K. Lee, B.C. Bai, J.S. Im, Y.-s Lee, J. Indust. Engg. Vhem. 16 (2010) 891–895.
[224] X. Wang, S. Zhou, W. Xing, B. Yu, X. Feng, L. Song, et al., J. Mater. Chem. 13 (2013) 4383–4390.
[225] R. Wang, D. Zhuo, Z. Weng, L. Wu, X. Cheng, Y. Zhou, et al., J. Mater. Chem. 18 (2015) 9826–9836.
[226] E.N. Kalali, W. Guo, X. Wang, W. Xing, L. Song, Y. Hu, Compos. Appl. Sci. Manuf. 129 (2020) 105694.
[227] X. Wang, W. Xing, X. Feng, B. Yu, L. Song, Y. Hu, Polym. Chem. 5 (2014) 1145–1154.
[228] T.-P. Ye, S.-F. Liao, Y. Zhang, M.-J. Chen, Y. Xiao, X.-Y. Liu, et al., Compos. B Eng. 175 (2019) 107189.
[229] S. Rao, V.R. Hande, S.M. Sawant, S.K. Rath, K. Sudarshan, D. Ratna, et al., ACS Appl. Mater. Interfaces 11 (2019) 37013–37025.
[230] D. Ratna, R. Khuswaha, S. Barman, B.C. Chakraborty, Indian Patent 300740.
[231] T. Liu, T. Zhou, Y. Yao, F. Zhang, J. Leng, Compos. A 100 (2017) 20–30.
[232] S. Dhibar, S. Malik, ACS Appl. Mater. Interfaces 12 (2020) 54053–54067.
[233] D. Ratna, S. Divekar, P. Sivaraman, A.B. Samui, B.C. Chakraborty, Polym. Int. 56 (2007) 900.
[234] A. Kumar, R. Anant, K. Kumar, S. Kumar, R. Kumar, RSC Adv. 6 (2016) 113405–113414.
[235] B.F. Jogi, M. Kulkarni, P.K. Bramhakar, D. Ratna, Proc. Mater. Sci. 5 (2014) (2014) 787–794.

CHAPTER 6

Characterization, performance evaluation and lifetime analysis of thermoset resin

Abbreviations

ATR	Attenuated total reflectance
NMR	Nuclear magnetic resonance
PALS	Positron annihilation lifetime spectroscopy
DNA	Deoxyribonuclic acid
GPC	Gel permeation chromatography
SEM	Scanning electron microscope
EDX	Energy-dispersive X-ray
TEM	Transmission electron microscope
AFM	Atomic force microscopy
PSD	Power spectral density
XRD	X-ray diffraction
DSC	Differential scanning calorimetry
DMA	Dynamic mechanical analysis
TGA	Thermogravimetric analysis
TTS	Time−temperature superposition
WLF	Williams−Landel−Ferry
UV	Ultraviolet
ASTM	American Society for Testing and Materials
UTM	Universal testing machine
LVDT	Linear variable differential transformer
EMY	Effective Medium Theory
TMA	Thermomechanical analysis
LCTE	Linear coefficient of thermal expansion
HDT	Heat deflection temperature
LOI	Limiting oxygen index
HPR	Heat release rate
MLR	Mass loss rate
SEA	Specific extinction area
CLD	Constrained layer damping
TAST	Thick adherent shear test
CT	Compact tension
SENB	Single edge-notched bending
CTBN	Carboxyl-terminated copolymer of butadiene and acrylonitrile
ATBN	Amine-terminated copolymer of butadiene and acrylonitrile
CTPEHA	Carboxyl-terminated poly (ethylene glycol adipate)
HTPB	Hydroxyl-terminated polybutadiene

FFT	Fast Fourier transform
FRF	Frequency response function
FWO	Flynn–Wall–Ozawa
KAS	Kissinger–Akahira–Sunose
LVDT	linear variable differential transformer
PALS	Positron annihilation lifetime spectroscopy
SEM	Scanning electron microscope
TEM	Transmission electron microscopy
NMR	Nuclear magnetic resosnance spectroscopy
EDX	Energy-dispersive X-ray
XRD	X-ray diffraction
GPC	Gel permeation chromatography
LCTE	linear coefficient of thermal expansion
UTM	Universal testing machine
VPO	Vapor pressure osmometry
1DPSD	One-dimensional power spectral density
LCTE	Linear coefficient of thermal expansion
UTS	Ultimate tensile strength

6.1 Analysis and Characterization

Thermoset resins are oligomers (low molecular weight) with reactive functional groups, which undergo crosslinking with or without a crosslinker and form three-dimensional networks. Thorough characterizations and analysis of the uncured resins are necessary for proper quality control. At the same time, effective exploitation of these resins requires analysis of cure characteristics and characterization of the cured networks. In this section, the general methods/techniques used for the characterization of thermoset resins and composites will be discussed.

6.1.1 Titration

Titration is an easy method to estimate the concentration of functional groups present in a thermoset resin. A small amount (accurately weighed) sample is dissolved in a suitable nonreactive solvent and titrated with a suitable standard solution. For example, carboxyl group and epoxy groups are titrated by a standard alcoholic solution of potassium hydroxide and acetic acid solution of hydrobromic acid, respectively. Unsaturation in the structure is estimated by reacting the material with iodine and titrating back the excess iodine using an aqueous solution of sodium thiosulfate.

6.1.1.1 Acid value

Acid value provides a quantitative estimation of carboxy or sulfonic groups present in the resin. This is used for quality control of thermoset resins like unsatuarted polyester and vinyl ester resins. The acid value is determined as per ASTM D 1639-90 by titrating the resin sample against standard alcoholic KOH using potassium hydrogen phthalate as a primary standard. Acid value (Z) was calculated using the following equation:

$$Z = \frac{56.1 \times A \times 0.1}{B} \times 100 \tag{6.1}$$

where A is the volume of KOH solution in milliliter and B is the weight of the sample in gram.

6.1.1.2 Hydroxyl value

Hydroxyl value is used for quality control of thermoset resins especially polyurethanes and is determined as per the procedure described in IS 354, 1987. About 1.0 g of resin is dissolved in 25 mL of toluene. 5 mL of acetylating mixture, obtained by mixing acetic anhydride and distilled pyridine at a ratio of 70:30 (w/w), is added to the reaction mixture and the contents were refluxed for 24 h. Afterward, 5 mL of water is added through the condenser and heating is continued for 10 min. The reaction mixture is allowed to cool and the condenser is washed with 15 mL of n-butanol. Unreacted acetic anhydride is determined by titrating with standard (approx. 0.5 N) alcoholic KOH solution using phenolphthalein as an indicator. The hydroxyl (OH) value determination was carried out in duplicate. A similar experiment was also carried out for blank.

$$\text{Hydroxyl value}(A) = \frac{(B - S) \times 56.1 \times N}{W} \tag{6.2}$$

where B is volume of KOH (mL) consumed for blank, S is volume of KOH (mL) consumed for sample, N is normality of KOH solution, and W is weight of the sample (g).

6.1.1.3 Determination of epoxy equivalent

Epoxy resins are characterized by the presence of two or more epoxy groups in the structure. In order to cure an epoxy resin with a suitable hardener, accurate estimation of epoxy groups is absolutely necessary. Epoxy equivalent is defined as the amount of resin that contains one mole of epoxy. Epoxy equivalent of an epoxy sample is determined by a standard titration method [1] using hydrogen bromide solution in acetic acid. The method is briefly described below.

About 100 mg of anhydrous sodium carbonate, 20 mL of chlorobenzene, and 10 mL of glacial acetic acid are taken in an Erlenmeyer flask and stirred magnetically till all the carbonate dissolved. Four drops of 0.1% solution of crystal violet in acetic acid are added. The purple-colored solution is then titrated with hydrogen bromide solution, prepared by diluting about 6 mL of the commercial reagent with 250 mL of acetic acid. The titer value is noted at the blue-green endpoint. About 300 mg of epoxy sample was titrated in a similar way. A blank titration was also carried out without any sample.

The epoxy equivalent (Z) is calculated from the following formula:

$$Z = W \times 53 \times V/(V_s - V_0) \times W_p \tag{6.3}$$

where W, W_P are the weight (g) of sodium carbonate taken and the same for the sample respectively, V and V_s are the volumes of HBr/CH$_3$COOH solution required for titration of Na$_2$CO$_3$ and the same for the sample, respectively in mL, and V_0 is the blank titer value in mL.

6.1.1.4 Solubility parameter

This is an important parameter to assess the compatibility of a thermoset resin with an additive, modifier or any other resin. It may be noted that a number of modifiers and additives are required to be added with a thermoset resin to qualify the resin for diverse applications. The solubility parameter (δ) is defined as a square root of cohesive energy density [2]. Cohesive energy density is the amount of energy needed to completely remove unit volume of molecules from their neighbor to infinity, which is equal to the enthalpy of vaporization divided by the molar volume. Thus the solubility parameter of small compounds or solvents can be determined easily from the corresponding heat of vaporization data available in the literature. However, macromolecules mostly degrade before vaporization; that is why the solubility parameter of polymer is determined indirectly by examining their solubility or swelling efficiency in solvents using the "like dissolves like" principle. Theoretically, the solubility parameter can be determined by Van Kravelen's group contribution method [3]. Hansen [4] considered the contribution of three different interaction forces namely London dispersion force, polar force and H-bonding interaction and introduced the concept of three-dimensional solubility parameters namely δ_p, δ_p, δ_h corresponding to the three different interactions. The overall solubility parameter can be expressed as follows

$$\delta = \sqrt{\delta_d^2 + \delta_p^2 + \delta_h^2} \tag{6.4}$$

Solubility parameters are determined by Hansen's iteration method from the three-dimensional solubility parameters of the solvents in which the polymer is miscible [4,5]. In this method, first, solubility of the polymer in various solvent is examined. The plots of the three-dimensional solubility parameters of the solvents (available in the literature [3]) give a three-dimensional spherical space, called Hansen's space. Now if the distance between any two points is measured by a computation method then the straight line connecting the two points situated at the longest distance will represent the diameter of the sphere and the center of the sphere will represent the three-dimensional solubility parameter of the polymer.

6.1.2 Infrared spectroscopy

Infrared (IR) spectroscopic analysis is used as a primary tool to characterize a chemical compound, monomer or polymer. The atoms in a molecule are considered to undergo various motions like stretching, bending, rotating, etc. The energy associated with a change in vibration level of a particular chemical bond corresponds to the energy of IR radiation (wavenumber 200–4000 cm^{-1}). Depending on the nature of bonds, they absorb at various frequencies. By analyzing two monochromic beams passing through a sample and a reference the absorption can easily be detected. Nowadays, instead of using a monochromatic beam, the polychromatic beam, after passage through a sample can be analyzed by means of a scanning Michelson interferometer. A spectrum is reconstructed from the information contained in the interferogram using a mathematical process known as Fourier transformation. A particular chemical bond will have a characteristic IR absorption frequency, which may change slightly due to interaction with other bonds. Thus from the interpretation of the IR spectra, it is possible to know whether certain functional groups are present or not in the monomer/polymer. Also, the presence of subtle interaction like H-bonding can be detected. The liquid samples are analyzed as thin-film cast in a sodium chloride cell, which is transparent at wave number up to 650 cm^{-1}. For frequencies under 600 cm^{-1}, a polyethylene cell can be used. The solid samples are analyzed in the pellet form. The specimens are generally prepared by adding approximately 1 wt.% of the sample to dry KBr powder and pressed into a disc of about 15 mm in diameter and 1.5–2 mm in thickness. For thick or highly filled samples, analysis is carried out by the technique known as attenuated total reflectance (ATR). In this method, the angle of incidence is adjusted so as to attain total internal reflection. When a material is placed in contact with a reflecting substrate the beam will lose energy at the frequency characteristic to the groups present in the sample. Fourier-transform infrared spectroscopy (FTIR) is extensively used to characterize thermoset resins. For example the −OH and −CO functional groups are characterized by FTIR analysis from their characteristic peaks at 3350–3500 and 1730 cm^{-1}, respectively. Oxirane ring of epoxy resin is identified from its characteristics peak at 910 cm^{-1} [6].

6.1.3 Nuclear magnetic resonance spectroscopy

Nuclear magnetic resonance (NMR) spectroscopic analysis is a well-known powerful technique for elucidation and characterization of the structure of thermoset resins. The advantage of NMR spectroscopy is that it is an extremely sensitive technique for the chemical environment of specific nuclei, which allows the precise characterization of the chemical structure of a thermoset resin. In NMR analysis, the sample is subjected to a magnetic field and radio frequency (R/F) field. Under the magnetic field, the ½ spin nuclei like ^{1}H, ^{13}C, etc. dissociate into two energy states and the nuclei undergo Larmor precessional motion about the field of direction. When the frequency of the R/F field matches the precessional frequency, the resonance takes place. The resonance frequency for a particular nucleus changes with its chemical environment. This is expressed by the chemical shift expressed in relation to the resonance of tetramethylsilane as the zero of the reference. The uncured thermoset resin can be characterized by a solution NMR. For example, the molecular weight of novolac can be determined from ^{1}H NMR analysis. The ratio of the intensity of aliphatic to aromatic proton (r) is used to determine the molecular weight from the following equations [7]:

$$r = \frac{2n-2}{3n+2} \tag{6.5}$$

$$M_w = 106n + 200 \tag{6.6}$$

Here n is the number of phenolic units present in the chain. The methylene bridges are characterized by ^{13}C NMR spectroscopic analysis. The characteristics peaks for o, o′, o, p, and p p′ methylene carbons are 30, 35, and 40 ppm. The free phenol is characterized by peaks at 115 and 120 ppm. The free phenol content of a resin sample can also be determined by dissolving the sample in acetone and gas chromatographic analysis of the solution using m-cresol as an internal standard. The cured resin can only be characterized by solid-state analysis. However, contrary to a solution NMR where spectra usually consist of a series of very sharp lines due to an averaging of all the anisotropic interactions by the molecular motion in the solution, for the solid samples very broad peaks are observed due to the anisotropic interactions between the nuclei. This makes the characterization of cured thermoset networks extremely difficult.

6.1.4 Positron annihilation lifetime spectroscopy

Positron annihilation lifetime spectroscopy (PALS) is normally applied to determine the free volume properties of a cured thermoset. The theory and methodology of PALS [8,9] is briefly described below. The positron, an antiparticle of an electron, is used to investigate the free volume between polymer chains. The birth of the positron can be detected by the release of a gamma-ray of characteristic energy, which occurs approximately 3 ps after positron emission when the ^{22}Na decays to ^{22}Ne. Once inside the polymer materials, the positron forms one of the two possible types of positroniums, an ortho-positronium or a para-positronium, obtained by pairing with an electron abstracted from the polymer environment. The decay spectra are obtained by the "death" event of the positron, parapositronium, or orthopositronium species. By appropriate curve fitting, the lifetimes of the various species and their intensity can be determined. The lifetime of orthopositronium (τ_3) and intensity (I_3) have been found to be indicative of the free volume in the polymer system since this is where the relevant species become localized. τ_3 is related to the size of the free volume sites and I_3 to their number concentration.

6.1.5 Crosslinked density

Crosslink density is an important parameter for a thermosetting system for tailor making the properties of the crosslinked network. As the crosslink density increases the hardness, modulus, mechanical strength and chemical resistance increase; on the other hand, % elongation at break, impact strength, toughness, resiliency, low-temperature flex tend to decrease. For example, when natural rubber is lightly crosslinked (low crosslink density) the network displays a highly flexible and resilient material. When the same rubber is highly crosslinked it generates a brittle material. Crosslinked density (X_c) is expressed as the number of crosslinks per unit volume or molecular weight between the two crosslink points (M_c). The methods for the determination of crosslink density are discussed below.

X_c can be determined from the swelling study using Flory–Rehner theory [10]. Unlike thermoplastics which dissolve in a solvent having close proximity of solubility parameters, the crosslinked network swells by absorbing solvent. As more and more solvent is adsorbed by the polymer network, the network expands progressively. The driving force toward swelling is the increase in entropy of the mixing of solvent with the polymer. However, during the swelling process, the network chains are forced to attain more elongated, less probable configurations. As a result, like pulling a spring from both ends, a decrease in chain configurational entropy is produced by swelling. This effect reduces the entropy and opposes the swelling. Considering the forces arising from the entropy effect as discussed above and the enthalpy component, that is, heat of mixing the equilibrium condition according to Flory–Rehner theory can be expressed as below:

$$\ln(1 - v_p) + v_p + \chi v_p^2 + V_s X_c \left[v_p^{1/3} - \frac{v_p}{2} \right] = 0 \tag{6.7}$$

$$\text{Hence, } X_c = \frac{\ln(1 - v_p) + v_p + \chi v_p^2}{V_s \left[v_p^{1/3} - \frac{v_p}{2} \right]} \tag{6.8}$$

$$\overline{M_c} = \frac{d_p V_s \left[v_p^{1/3} - \frac{v_p}{2} \right]}{\ln(1 - v_p) + v_p + \chi v_p^2} \tag{6.9}$$

$$\chi = k + \frac{V_s}{RT}(\delta_s - \delta_p)^2 \tag{6.10}$$

where d_p is the Density of the cured resin, V_s is the Molar volume of solvent, v_p is the Volume fraction of polymer in the swollen network, X_c is the Crosslink density, M_c is the Molecular weight between two crosslink point, X is the Flory–Huggins interaction parameter, δ_s is the Solubility parameter of solvent, δ_p is the Solubility parameter of resin, R is the Universal gas constant, T is the Temperature ($= 295K$), K is the Constant specific to solvent.

An estimation of the crosslink density of a thermoset network can be obtained from the storage modulus values in the rubbery plateau region. In principle, the crosslink density of a cured thermoset network could be calculated from the theory of rubber elasticity. The shear modulus G of a crosslinked rubbery network is given by [11]:

$$G = \frac{r_1^2}{r_f^2} \frac{dRT}{M_c} \left(1 - \frac{2M_c}{M_c}\right) \quad (6.11)$$

where d is the density, R is the universal gas constant, T is the absolute temperature, M_c is the molecular weight between crosslinks, M_n is the chain backbone molecular weight, and r_1^2/r_f^2 is the ratio of the mean square end to end distance of the polymer chain in the sample to the same quantity in a randomly coiled chain. The ratio is often assumed to be unity. For a highly crosslinked system, M_c/M_n is negligible and can be neglected. Hence the Eq. (1.34) can be written as

$$M_c = \frac{dRT}{G} = \frac{3dRT}{E'} \quad (6.12)$$

Thus M_c can be determined from the dynamic storage modulus value at a temperature at least 30°C higher than T_g. The details of the determination of dynamic modulus using dynamic mechanical analysis are discussed in Section 1.9.6.3.

6.1.6 Molecular weight distribution

Unlike simple molecules, polymer molecules are large in size. The physical parameter typically used to describe the polymer size is the molecular weight. With the exception of few naturally available polymers like protein, deoxyribonuclic acid (DNA), etc., a polymer sample consists of chains of varying lengths. That is why in the case of polymers, we talk about average-molecular weight, not the absolute molecular weight.

The various averages used commonly for molecular weights are number-average molecular weight (ΣM_n), weight-average molecular weight (ΣM_w), z-average molecular weight (ΣM_z), viscosity-average molecular weight (ΣM_v). ΣM_n is calculated like any other numerical average by dividing the sum of individual molecular weight values, $M_i N_i$, by the number of molecules, N_i.

$$M_n = \frac{\sum_{i=1}^{i=\infty} M_i N_i}{\sum_{i=1}^{i=\infty} N_i} \quad (6.13)$$

Any measurement that leads to the determination of number of molecules, functional groups or particles that are present in a given weight of samples allows the calculation of ΣM_n. Most thermodynamic properties are related to the number of particles present and thus are dependent on ΣM_n. Colligative properties dependent on the number of particles present are obviously related to (ΣM_n). ΣM_n values are independent of molecular sizes and are highly sensitive to the presence of small molecules in the mixtures.

The weight average molecular weight, ΣM_w is the second moment or second power average as shown mathematically:

$$M_w = \frac{\sum_{i=1}^{i=\infty} M^2{}_i N_i}{\sum_{i=1}^{i=\infty} M_i N_i} \qquad (6.14)$$

M_w is determined from the experiments, in which each molecule or chain makes a contribution to the measured results. The average is more dependent on the number of heavier molecules than is the number-average molecular weight, which is dependent simply on the total number of particles. Bulk properties associated with large deformations such as viscosity and toughness are particularly related to ΣM_w values. It is determined by the light scattering and ultracentrifugation techniques.

Melt elasticity is more closely dependent on z-average molecular weight, ΣM_z, which can also be determined by the ultracentrifugation technique. It is the third moment or third power average and is shown mathematically as:

$$M_z = \frac{\sum_{i=1}^{i=\infty} M^3{}_i N_i}{\sum_{i=1}^{i=\infty} M^2{}_i N_i} \qquad (6.15)$$

Although $z + 1$ and higher-average molecular weights can be calculated, the major interests for all practical purposes lie on ΣM_n, ΣM_w, and ΣM_z. Because ΣM_w is always greater than ΣM_n except in a monodisperse system, the ratio $\Sigma M_w : \Sigma M_n$ is a measure of polydispersity and is called the polydispersity index. The most probable distribution for polydisperse polymers produced by condensation techniques is a polydispersity index of 2.0. Thus for a polymer mixture, which is heterogeneous with respect to the molecular weight, $\Sigma M_z > \Sigma M_w > \Sigma M_n$. As the heterogeneity decreases, the various molecular weight values converge until for homogeneous mixtures where $\Sigma M_n = \Sigma M_w = \Sigma M_n$. Various methods for the determination of molecular weights are discussed below.

6.1.6.1 Viscometry

Viscometry is most widely used method for the characterization of polymer molecular weight because it provides the easiest and rapid means of obtaining molecular weight-related data and requires only a minimum amount of instrumentation. The molecular weight thus obtained is called viscosity-average molecular weight, ΣM_v. The most obvious characteristic of polymer solution is their high viscosity, even when the quantity of the added polymer is very small. Viscometry does not yield an absolute value of the molecular weight; hence, one must calibrate the viscometry results with the values obtained for the same polymer and solvent by using an absolute technique such as light scattering photometry.

6.1.6.2 End group analysis

The end group analysis is a chemical method used for calculating the number-average molecular weight of a polymer sample whose molecules contain reactive functional groups at one end or both ends. The number-average molecular weight can be expressed as below:

$$\sum M_n = \frac{F}{X} \tag{6.16}$$

where F is the functionality (eq mol^{-1}) and X is the equivalent of functional groups present.

6.1.6.3 Vapor pressure osmometry

Vapor pressure osmometry (VPO) is based on the principle (Rault's law) that vapor pressure of the solution is lower than vapor pressure of the solvent. Solvent and solution drop kept in two thermister beads placed in a thermostated chamber saturated with solvent vapor. Solvent condenses on the solution bead and temperature increases due to liberation of latent heat of fusion. The increase in temperature was noted. The temperature change can be correlated with molecular weight as given below:

$$\frac{\Delta T}{c} = k_s \left(\frac{1}{M_n} + A_2 c \right) \tag{6.17}$$

where ΔT is the increase in temperature, c is the concentration of the solution and A_2 is a second virial coefficient. K_s is a constant determined by calibrating the instrument with a substance of known molecular weight. Thus $\Delta T/c$ versus c plot will be a straight line. The number-average molecular weight and the second virial coefficient can be easily determined from the slope and intercept of the straight line, respectively.

6.1.6.4 Membrane osmometry

In this technique, osmotic pressure is measured for a polymer solution of various concentrations. The equation relating osmotic pressure (Π) and molecular weight and concentration can be written as follows [1,10]:

$$\frac{\Pi}{c} = \frac{RT}{M_n} + \frac{A_2}{RTM_n} c \tag{6.18}$$

where R is the universal gas constant, T is the temperature, c is the concentration of the solution, and A_2 is a second virial coefficient. ΣM_n can be determined from the slope (RT/ΣM_n) of the Π/c versus c plot (straight line).

6.1.6.5 Light scattering

The principal method for determining the weight-average molecular weight is light scattering although small-angle neutron scattering is now becoming important, especially in the bulk state. The radiation is said to be scattered when an object begins to approach the wavelength of radiation in size, down to atomic dimension. Many esthetic phenomena of nature, such as the rainbow, the blue of the sky and the sea, are in one way or another connected to the scattering of light, which occurs when a light beam encounters suspended matter. It is possible to determine the molecular weight of a polymer using a light scattering phenomenon because the polymer molecule in a solution can scatter the beam of light like a suspended matter.

The basic equation used for molecular weight and size can be written as [1,10]:

$$\frac{kc}{\Delta \tau} = \frac{Hc}{R(\theta)} = \frac{1}{M_w P_\theta} + 2A_2 c \tag{6.19}$$

where $\Delta \tau$ is the excess turbidity of the solution over that of pure solvent, c is the concentration of the solution, P_θ is the particle scattering factor, and A_2 and K, H are second virial coefficient and light-scattering calibration constants, respectively.

$$P_\theta = \left[1 - \frac{16\pi^2 R_g^2 \sin^2 \frac{\theta}{2}}{3\lambda^2} \right] \tag{6.20}$$

Hence the key equation at the limit of zero angle and zero concentration, respectively, relating the light scattering intensity to the weight average molecular weight, ΣM_w and the z-average radius of gyration (R_g) may be written as:

$$\left(\frac{Hc}{R(\theta)} \right)_{\theta=0} = \frac{1}{M_w} [1 + 2A_2 c] \tag{6.21}$$

$$\left(\frac{Hc}{R(\theta)} \right)_{c=0} = \frac{1}{M_w} \left[1 + \frac{16\pi^2 R_g^2 \sin^2 \frac{\theta}{2}}{3\lambda^2} \right] \tag{6.22}$$

From Eqs. (6.21) and (6.22), it is clear that by plotting $Hc/R(\theta)$ versus c and $Hc/R(\theta)$ versus $\sin^2 \theta$, the weight-average molecular weight and radius of gyration can be determined.

6.1.6.6 Gel permeation chromatography

Gel permeation chromatography (GPC) is a powerful technique and an instrument that is exclusively used for the characterization of molecular weight and molecular weight distribution of the polymer. GPC makes use of the size exclusion principle. In this technique, a dilute polymer solution is passed through a gel (usually made of crosslinked styrene-divinyl benzene copolymer) containing fine spherical beads. The gel contains voids as well very small pores (diameter 50–100 Å). First, the pure solvent is passed through the column so that all the pores get filled with the solvent. Afterward, when a solution is allowed to pass through the column, the smaller molecules diffuse into the pores due to the existence of a concentration gradient. The bigger molecules cannot penetrate into all the pores and elute faster than the smaller molecules. As the mobile phase passes, the porous particle the separation between the smaller and larger molecules becomes greater. By calibrating the column with a polymer solution of known molecular weight (polystyrene standard), it is possible to determine the molecular weight of an unknown polymer sample. The data are collected as the height of chromatograph versus retention counts. The resultant data are then analyzed, usually by software to yield various molecular weight averages. Thus the entire molecular weight distribution is determined in one experiment. When a thermoset resin is synthesized, the resin may vary with respect to branching, molecular weight distribution. Hence a thorough analysis and characterizations are necessary for proper quality control. Molecular weight and molecular weight distribution are analyzed by GPC.

6.1.7 Morphological characterization

6.1.7.1 Scanning electron microscope

Scanning electron microscope (SEM) is widely used to study the morphology (microstructure and nanostructure) of thermoset resin systems. SEM produces high-resolution images of a sample surface. Mostly cryogenically fractured surfaces with a conductive coating of gold or carbon are used for analysis. The coating makes the resin surface (insulating) conducting and prevents charging problems. SEM images have a characteristic 3D appearance and are therefore used to characterize surface structure (homogeneity, phase morphology) and mode of failure (ductile or brittle). The principle of SEM is that the primary electrons coming from the source strike the surface structure and they are inelastically scattered by the atoms in the sample. The electrons emitted are detected to produce an image. Besides the emitted electrons, X-rays are also produced by the interaction of electrons with the sample which is used to detect the presence of metallic elements by energy-dispersive X-ray (EDX) spectroscopy. The SEM can provide resolution up to 1 nm and hence is used for characterization of nanofiller thermoset or nanocomposites, which will be discussed in subsequent chapters.

6.1.7.2 Transmission electron microscope

A transmission electron microscope (TEM) is also used to study morphology especially to study the nanoscale dispersion of inorganic fillers or other polymeric components in the thermoset resin. Modern high-resolution TEM can provide resolution up to 0.1 nm. In TEM images are produced by focusing a beam of electrons onto a very thin specimen. The transmitted electrons carry the information about the inner structure, which is captured to produce an image. Hence the specimen thickness should be very low (<80 nm) to ensure transmittance. There are several problems for the characterization of polymer samples by TEM. First, the preparation of such a thin sample is a time-consuming process and difficult for highly flexible resin. Second, thermoset resin samples often produce poor contrast in TEM analysis due to a weak interaction of electrons with the resin sample. This problem can be partially solved by using stains such as heavy metal compounds. The dense electron clouds of the heavy metal atoms interact strongly with the electron beam. Third, sometimes the resin decomposes in the electron beam due to the increase in temperature as a result of electron beam exposure. In order to avoid the decomposition of resins cryogenic microscopy (cryo-TEM), which keeps the sample in liquid nitrogen, is used.

6.1.7.3 Atomic force microscopy

Atomic force microscopy (AFM) is another powerful tool, which has been used in recent years for the characterization of the morphology of modified thermosets and thermoset-based composites [12,13]. The development of this technique has helped to image surface topography on a nanometric scale [14–16]. AFM has several advantages over TEM and SEM such as AFM gives a three-dimensional surface profile of the sample and it does not need a high vacuum, conducting coating (for SEM), or exposure of high voltage electron beam (like SEM and TEM), which may degrade the surface of the resin sample. In AFM a cantilever with a sharp tip at its end (typically composed of silicon or silicon nitride) with tip sizes in the order of nanometer is used. The probe is brought into close proximity of a sample surface. The distance between the tip and the surface has to be selected with proper care otherwise the tip may get damaged due to the collision of the tip with the sample surface. The Van der Waal forces between the tip and the sample lead to a deflection of the cantilever. Typically, the deflection is measured by applying a laser beam, which is reflected from the top of the cantilever into an array of photodiodes. The sample is mounted on a piezoelectric holder in all three space directions (x and y for scanning the sample, z for maintaining a constant force). Scanning in an $x-y$ plan represents the topography of the sample. Contamination of the probe or the substrate surface is a common problem in AFM analysis. Hence special care has to be taken for sample preparation and analysis.

Another unique feature of AFM is that it allows estimating the adhesion of the filler with matrix in a filled thermoset system. The adhesion strength can be determined using a probe with a specific spring constant by calibrating the cantilever. The changes in surface topology can be determined quantitatively by the root mean square (RMS) roughness calculation (R_{rms}). This is used in a range of applications, particularly where the surfaces involved show a degree of randomness. Since the roughness analysis is based on the vertical axis, that is, z-axis, the RMS height deviations can be expressed as

$$R_{\text{rms}} = \sqrt{\frac{\sum (Z_i)^2}{n}} \quad (6.23)$$

where Z is the current and n is the number of points on the image. Mean roughness (R_a) can be determined from the arithmetic average of the absolute values of the surface height deviations measured from the main plane:

$$R_a = \sum_{j=1}^{n} |Z_j| \quad (6.24)$$

The analysis of surfaces can also be carried out in terms of power spectral density (PSD) by considering the surfaces as a superimposition of spatial waves [17]. The variation of height in real space can be considered in terms of the power spectrum in frequency space through the use of the Fourier transform, which is a well-known means of relating real space to frequency space. The surfaces are analyzed by a series of line scans of N_x steps, yielding a profile $h(x)$, for each value from 1 to N_y in the y-direction. The power (P) is determined by taking the Fourier transform of each of these line scans and squaring the results. Once the power (P) is obtained, it may be used to derive one-dimensional power spectral density (1DPSD) values as follows:

$$PSD = \frac{P}{\Delta f} \quad (6.25)$$

Δf is derived by progressively sampling the data from the image's fast Fourier transform center [18]. 1DPSD is expressed in length^3. Two-dimensional PSDs (2DPSD) is determined by taking an annular average of the square of 2D Fourier transform.

6.1.7.4 X-ray diffraction

X-ray diffraction (XRD) is a fast and nondestructive test that is frequently used to characterize thermoset materials and their composites. Crystalline materials are characterized by sharp peaks whereas amorphous materials show broad hump. Thus the degree of crystallinity can be estimated. When a crystalline material like clay is dispersed in a thermoset matrix, one can study the intercalation and exfoliation behavior as will be discussed elaborately in subsequent chapters. If the crystallites of the power are very small the peaks of the pattern will be broadened. From this broadening one can determine an average crystallite size using the Debye–Scherrer equation:

$$d = \frac{0.9\lambda}{B\cos\theta} \quad (6.26)$$

where *B* is the broadening of the diffraction line measured at half of its maximum intensity (radians), θ is the Bragg angle, and λ is the wavelength of the X-ray used.

6.1.8 Thermal analysis

Thermal analysis is an essential tool to study the necessary thermal transitions in polymers without which the processing and the fabrication of polymeric materials are not possible. The thermal behavior of a polymer has significant technological importance. For example, the glass transition temperature of rubber determines the lower limit of the use of rubber and the upper limit of the use of an amorphous thermoplastic. Most important techniques, such as differential scanning calorimetry (DSC), dynamic mechanical analysis (DMA) and thermogravimetric analysis (TGA), will be briefly discussed below.

6.1.8.1 Differential scanning calorimetry

Differential scanning calorimetry (DSC) involves the measurement of relative changes in temperature and energy either under isothermal or adiabatic conditions or in other words the heat capacity of the sample at a constant pressure. It is a technique measuring the energy necessary to establish a nearly zero temperature difference between a substance and an inert reference material as both the specimen and the reference are subjected to identical temperature regimes in the same environment (heated or cooled at a controlled rate).

Two types of DSC systems are commonly used such as power compensation and heat flux DSC. In power-compensation DSC, the temperatures of the sample and the reference are made identical by varying the power input to the two furnaces; the energy required to do this is a measure of enthalpy or heat capacity changes in the sample relative to the reference. In heat-flux DSC, a low resistance heat flow path connects the sample and the reference. The assembly is enclosed in a furnace. The enthalpy or heat capacity change in a sample causes a difference in its temperature relative to the reference. DSC can be used for analyzing the uncured or partially cured thermosets as the curing involves crosslinking which is accompanied by the evolution of heat that can be easily monitored.

DSC is widely used to get information about glass transition temperature (T_g), melting point (T_m), crystallization temperature (T_c), the heat of crystallization (H_c), the heat of melting (H_f), and heat absorbed or evolved during the cure reactions or decomposition reactions. A typical DSC plot showing T_g, T_m, and T_c is shown in Fig. 6.1. During heating, as a solid sample melts and transforms to a liquid state, it absorbs more energy (endothermic) than that of the reference. Similarly, as the crystallization takes place, excess energy is given out (exothermic). As a result, the melting and crystallization

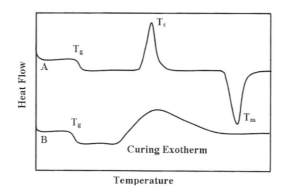

Figure 6.1
A typical DSC plot showing T_g, T_m, and T_c of a polymer sample.

processes appear as peaks (endothermic or exothermic) and the area under the curve gives a direct measure of the heat of fusion or crystallization. Similarly, heat of curing and degradation can be measured.

6.1.8.2 Dynamic mechanical analysis

DMA provides the most significant information on the viscoelastic behavior of a polymer in addition to the thermal transitions. In principle, a sinusoidal strain or stress is applied to a sample and the response is monitored as a function of frequency and temperature. A viscoanalyzer is commonly used to apply a displacement $d(w)$ at the upper end of a sample and measure the force $F(w)$ transmitted to the fixed lower end. $F(w)$ is measured by a dynamic force sensor and $d(w)$ is measured by displacement or acceleration sensor. By measuring the upstream displacement and downstream force, the measurement method has the advantage of being capable of obtaining the stiffness, irrespective of the weight of the sample. There are machines of a very wide range of forces, but generally up to 500 N maximum. Several designs for grip, suspension, sensing, liquid nitrogen flow system, and microprocessor control arrangements are available in commercial machines.

The strain of a viscoelastic body is out of phase with the applied stress, by the phase angle, δ. This phase lag is due to the excess time necessary for molecular motions and relaxations to occur. For a perfectly elastic material, the value of δ is 0 degree and for perfectly viscous material, the value is 90 degrees. Polymers being viscoelastic in nature, the value of δ falls in between 0 and 90 degrees. The phase angle $\delta(w)$, that is, the phase shift between the dynamic force and dynamic displacement can be calculated using the processing of the signals $F(w)$, $d(w)$ according to the first Fourier transform (FFT). Like mechanical testing, DMA can be carried out using various modes like tension, flexure, shear or creep, etc.

The viscoelastic properties may be expressed in terms of a dynamic storage modulus (E'), dynamic loss modulus (E''), and mechanical damping factor ($\tan\delta$). Mathematically they are defined as follows:

$$E' = \frac{\sigma_0}{\varepsilon_0 \cos(\delta)} \tag{6.27}$$

$$E'' = \frac{\sigma_0}{\varepsilon_0 \sin(\delta)} \tag{6.28}$$

$$\tan\delta = \frac{E''}{E'} \tag{6.29}$$

The storage modulus is often associated with the "stiffness" of material and refers to the energy stored in the sample elastically after applying the stress. The dynamic loss modulus is associated with the "internal friction" arising out of the segmental motions and is very sensitive to the different kinds of molecular motions, relaxation processes, transitions, morphology, and other structural heterogeneities. The $\tan\delta$ is the energy of dissipation during the stress cycle relative to the energy stored elastically in the material. A typical DMA plot of a cured thermoset resin is shown in Fig. 6.2. From the figure, it can be seen that as the temperature is increased, the $\tan\delta$ goes through a maximum in the transition region and then decreases in the rubbery region. The damping is low below T_g as the chain segments in that region are frozen. Below T_g, the deformations are thus primarily elastic and the molecular slips resulting in viscous flow is very low. Also above T_g, in the rubbery region, the damping is low because the molecular segments are free to move, and consequently, there is a little resistance to flow. The maximum damping occurs in a region where most of the chain segments take part in this cooperative micro-Brownian motion under harmonic stress. For thermosets, the position and height of the loss tangent peak in the relaxation spectra of a polymer are indicatives of the structure and the extent to which the polymer is crosslinked.

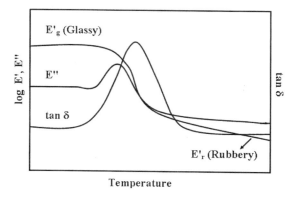

Figure 6.2
A typical DMA plot of a cured thermoset resin.

6.1.8.3 Time−temperature superposition

The principle of time−temperature superposition (TTS) lies in the equivalency of time (frequency) and temperature. Due to the various limitations, one cannot perform the experiments at conditions such as at very low frequencies and very high temperatures or vice versa. For that purpose, TTS is used to get the data at different conditions to save the experimental time. The viscoelastic data of one temperature can be related to the higher or lower temperature using a shift factor (a_T), to the right side or to the left side of the time axis using a reference temperature (T_{ref}). A fully overlapped curve can be obtained for any reference temperature; this curve is called a master curve. It is also widely accepted that a minor vertical shift factor may also be applied to more accurately model master curves.

The related shift factor a_T is given by:

$$a_T = \frac{E'(T)}{E'(T_{\text{ref}})} \tag{6.30}$$

The master curve in form of stiffness versus frequency can be created by fitting the experimentally determined shift factors to a mathematical model. With a multifrequency measurement, frequencies beyond the measurable range of the DMA can be achieved by using the superposition method based on the Williams−Landel−Ferry (WLF) equation [19,20]. For a temperature range above the glass transition temperature, it is generally accepted that the shift factor−temperature relationship is best described by WLF equation:

$$\log a_T = \log\left(\frac{f}{f_0}\right) = \frac{-C_1(T - T_{\text{ref}})}{C_2 + (T - T_{\text{ref}})} \tag{6.31}$$

where C_1 and C_2 are constants. The general values of the constants, $C_1 = 17.4$ and $C_2 = 51.6$, are used. For the temperature range below the glass transition temperature, the Arrhenius equation is generally considered as suitable to describe the relationship between the shift factors of the master curve and the temperature. In the latter case, the activation energy (E_a) for shifting the curves can be obtained by the following equation

$$\ln a_T = \frac{E_a}{R}\left(\frac{1}{T} - \frac{T_{\text{ref}}}{T_{\text{ref}}}\right) \tag{6.32}$$

6.1.8.4 Example of TTS by graphical method

Fig. 6.3 shows an example of typical manual shifting of storage modulus data in temperature scale to frequency scale taking reference temperature as 90°C. In this example, E' was measured for an epoxy thermoset from 47.5°C to 104°C at fixed frequencies such as

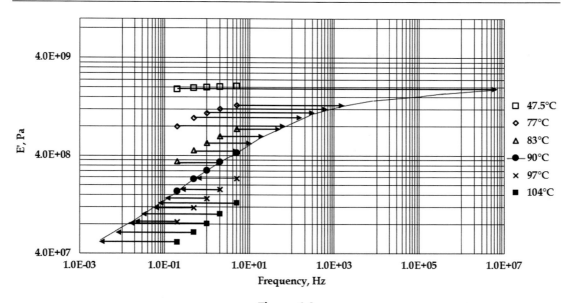

Figure 6.3
Graphical method of time–temperature superposition of an epoxy thermoset at a reference temperature of 90°C.

0.2, 0.5, 1.0, 2.0, and 5.0 Hz. The shift factors are calculated taking the log of the ratio of shifted frequency and test frequency for each temperature. Details of the calculation of shift factors from the graphical shifting lines is given in a book by the present authors [21,22]. Here, the reference temperature taken as 90°C is only to show the process of frequency shifting. The calculated shift factors are plotted against frequency and presented in Fig. 6.4. The plot of shift factor versus temperature shows a linear relationship $r^2 > 0.99$, indicating very precise shifting done manually using graphical shift. Similarly, the loss factor versus frequency plot can be generated. Subsequently, the new (shifted) frequency versus E' is obtained by calculating the new frequency as

$$\text{New frequency} = \text{antilog}[\log(a_T) + \log(\text{test frequency})] \tag{6.33}$$

In this context, it is worth mentioning some of the critical considerations to reduce errors in DMA experiments. The critical considerations are given below:

1. DMA uses small samples and low force. Therefore, to arrive at correct properties, the dimensions should be very uniform and accurate in measurement. Thermosets have an average dynamic modulus of 1–3 GPa at ambient; hence the accuracy and uniformity are more important.
2. DMA calculations are based on the assumption of the linear viscoelastic behavior of thermosets. Therefore, a strain versus modulus sweep is to be performed to ensure a

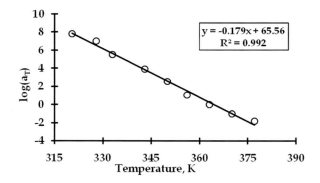

Figure 6.4
Log(a_T) versus temperature plot for graphically shifted isotherms of Fig. 6.3.

limit of the strain of constant modulus. The DMA study should be carried out within this strain limit.

3. For WLF method, the reference temperature for frequency shift should be $T_g + 50°C$. In the case of thermosets, frequency scale data are required at ambient temperature (30°C–40°C), which is far below T_g and does not satisfy the condition of validity of WLF. Hence manual shifting is better.
4. In some machines, the double shear fixture is quite heavy. It is suggested to use lighter fixtures to avoid its effect on modulus and phase angle. The user should check tensile and shear mode data to compare the α-relaxation temperature in both cases. The temperature should be almost the same.
5. The thermal stability of the thermoset should be checked by TGA study prior to deciding the maximum temperature of DMA study so that the material does not thermally degrade in the machine.

6.1.8.5 Thermogravimetric analysis

In this technique, the change in the weight of the sample with respect to the temperature and time is monitored when heated under a controlled program. The program is often of linear increase in temperature or time. This technique is quantitative and is very useful to study the curing reactions/degradations taking place while heating the sample. A thermogravimetric analysis (TGA) instrument consists of a thermobalance, which can records weight with sensitivity around one microgram and a capacity of about a few hundred milligrams and a furnace operated in a temperature range of 50°C–800°C with a heating rate up to 100°C min^{-1}. The thermal stability of a material can be studied in an inert or oxidative atmosphere. The thermogravimetric curves elucidate the decomposition mechanism. There are lots of factors that affect the thermogravimetric curves. The primary factors, which have a major effect, are the heating rate and size of the sample. As the heating rate and sample size increase, the decomposition temperature of the sample also increases. The particle size of the sample, the way in which it is packed, the crucible shape, and the gas flow rate can also affect the kinetics of the reaction. Thus while comparing the thermal stability of two materials, the identical condition with respect to the above-mentioned variables must be maintained.

6.1.9 Thermal stability

The thermal stability of a thermoset is generally analyzed [23–25] by thermogravimetric analysis (TGA). In this technique, the change in the weight of the sample with respect to the temperature and time is monitored when heated under a controlled program. The program is often of linear increase in temperature or time. This technique is quantitative and is very useful to study the curing reactions/degradations taking place while heating the sample. A thermogravimetric analysis (TGA) instrument (used for assessing thermal stability) consists of a thermo balance, which can record weight with sensitivity around one microgram and a capacity of about a few hundred milligrams and a furnace operated in a temperature range of 50°C–800°C with a heating rate up to 100 °C min^{-1}. The thermal stability of a material can be studied in an inert or oxidative atmosphere.

Thermoset resins are generally made from a volatile monomer or curing agent [22,25,26]. Hence there is a possibility of such unreacted liquid remaining absorbed in the network. Moreover, various additives are often added to manipulate the mechanical properties of the final network. It is necessary to know at what temperature it evaporates out and the consequence of this on the functional properties of the network. TGA analysis is very informative in this respect as it provides precise weight loss data. When a polymer is heated in an inert atmosphere, beyond a certain temperature (depending on the chemical structure) it decomposes due to the cleavage of $-C-C-$ bonds and the formation of smaller molecules [23]. These molecules evaporate out and the weight loss is reflected using TGA. The temperature at which 10% weight loss occurs is termed as initial degradation temperature (T_i). Thus from the T_i values determined from TGA, we can assess the thermal stability of a group of materials. If the service temperature is higher than or close to the T_i of the resin, it cannot be used for that particular application. The thermal stability of a thermoset resin depends on the chemical structure. Aliphatic chains degrade more easily compared to aromatic structures [27]. That is why the thermal stability of a phenolic resin is much higher than unsaturated polyester or epoxy resins. Polyimides and cyanate esters are known for their higher thermal stability and are used for high-temperature applications [28,29].

The thermal stability of a thermoset can be enhanced further by using antioxidants [30–34]. Antioxidants interfere with the free radical oxidative cycle to inhibit or retard the oxidation mechanism. UV absorbers such as benzophenone and benzotriazoles compounds absorb ultraviolet (UV) radiation and protect the resin from the harmful derivative effect of UV light. On the basis of the mechanisms by which antioxidants function, they can be classified into two categories Manley primary or chain breaking (CB) antioxidants and secondary or preventive antioxidants. Chain breaking antioxidants are of two types: chain-breaking donor (CB-D) antioxidants and chain-breaking acceptor (CB-A) antioxidants. Details of mechanism on which an antioxidant works is already discussed in Chapter 2. Hence the study of thermal stability is extremely important for the quality control of a thermoset resin for various applications. Assessment of thermal stability is helpful for selecting a thermoset resin considering the service temperature. While comparing the thermal stability of various thermoset by using TGA, the parameters like heating rate, sample size, and gas flow rate should be kept identical because the value of T_i depends on such factors.

6.1.10 Rheological characterization

Rheology is the study of the flow of matter. The name derives from the Greek word "rheo" which means "to flow." A solid will usually respond to a force by deforming and storing energy elastically. A liquid, however, will flow and dissipate energy continuously in viscous losses. A Newtonian liquid has a linear relationship between the shear rate and stress. Complex fluids are interesting because they exhibit both elastic and viscous responses and exhibit a nonlinear relationship between the shear rate and strain. One of the properties often dealt with within rheology is viscosity which measures the resistance to flow. Some materials are intermediate between solids and fluids and the viscosity is not enough to characterize them. A solid material can be described by its elasticity or resilience: when it is deformed, it will store the energy. For example, a spring regains its original shape after being deformed. The other extreme is a fluid that stores no energy while deformed and just flows.

A viscoelastic material is intermediate and stores some energy and flows a little when deformed. A polymer network is generally viscoelastic, with a complex shear modulus having both elastic (solid-like) and viscous (liquid-like) components of similar magnitudes over a large range of frequencies. Rheological testing provides a precise means for measuring intrinsic mechanical material properties that can be related to a material's processing characteristics and performance. The rheological tests for thermosetting resins are carried out in a parallel plate rheometer. The resin system is placed in between the plate and the strain is applied to it. The three rheological test modes, steady, dynamic, and transient, are distinguished by the manner in which the strain is applied to the sample. A steady test uses continuous rotation to apply the strain and provide a constant shear rate. The resultant stress is then measured when the sample reaches a steady state. In a dynamic test, an oscillatory strain is applied to a sample and the resulting stress is measured. Dynamic tests can be made using free oscillations at the resonance frequency of the test material (for example, the torsion pendulum), or with sinusoidal (or other waveform) oscillations at a forced frequency chosen from a wide available range. In a transient test, the response of a material as a function of time is measured after subjecting the material to an instantaneous change in strain, strain rate or stress.

The advantage of the dynamic test is that it allows the measurement of storage modulus (G') and loss modulus (G'') in addition to viscosity. Usually, the rheological properties of a viscoelastic material are independent of strain up to a critical strain level. Beyond this critical strain level, the behavior is nonlinear and the moduli decline. Therefore before carrying out the dynamic rheological experiment it is necessary to define the linear viscoelastic regime. Strain sweep measurements are used to define the linear viscoelastic regime. A generalized plot of the strain sweep experiment is shown in Fig. 6.5, where the storage and loss moduli are plotted as a function of strain. The moduli remain almost unchanged up to a critical strain level, above which the moduli tend to decrease. Hence a strain level less than the critical strain has to be used for determination of storage and loss moduli as a function of frequency.

A generalized plot showing the change in storage and loss moduli as a function of frequency is shown in Fig. 6.6. At low frequencies, G'' is higher than G'. G' increases with increasing frequency because the molecules do not get sufficient time to relax. As a result, the difference between G' and G'' decreases as the frequency increases. G' intersects with G'' at a certain frequency within the transition zone. This crossover frequency between storage and loss moduli indicates the transition from a liquid-like to solid-like behavior. The crossover point is sometimes defined as the gelation of a thermoset network.

The behavior of thermoplastics during the melt processing operations is governed mainly by the polymer's molecular weight, molecular weight distribution, degree of branching and filler content. For thermosetting polymers, the rheological changes occurring during curing reaction can be measured as the resin transforms from a low melting solid to a low

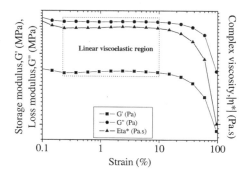

Figure 6.5
A generalized plot of strain sweep experiment where the storage and loss moduli are plotted as a function of strain.

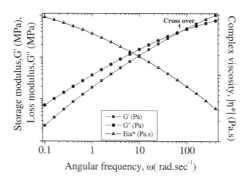

Figure 6.6
A generalized plots showing the change in storage and loss moduli of a polymer as a function of frequency obtained from a parallel plate rheometer.

viscosity liquid, then through the gel point without disrupting the gel structure, and finally to a highly crosslinked stiff solid. In fact, the entire curing process can be simulated by the rheometer to provide guidelines for production as discussed elaborately in Chapter 2.

6.2 Testing of thermoset resins

With the development of polymer science and technology and the proliferation of polymer-based products, testings of materials and components have become an integral part of not only the research and development but also of product design and manufacturing. Testings are necessary for verification of manufacturing process, design concept, quality control, and the reproducibility of a process. It is also used to assess a new product, to prove the design concepts and to ensure the safety and reliability of the products. Testing is carried out to assess the suitability of material followed by testing as a component. However, component testing technology is a vast area and detailed discussion about the same is beyond the scope of this book. Testing of thermoset resins as a material will be discussed briefly here. Detailed procedures for testing can be obtained from the American Society for Testing and Materials (ASTM) handbooks. For the convenience of the readers, the ASTM standards for the relevant testings are presented in Table 6.1.

6.2.1 Mechanical properties

In virtually all applications, polymer materials have to be subjected to a loading force in some way or other. Hence the assessment of the mechanical property is very important for the design of polymer materials for various applications. It also is noted that polymer materials are more sensitive to the service temperature and other environmental effects compared to conventional materials. Hence the data for mechanical properties, measured using conditions similar to the service environment should be used for design rather than

Table 6.1: ASTM standards for various tests used for evaluation of thermoset resins.

Testing	ASTM standard	Testing	ASTM standard
Tensile	D-638	Fatigue	D-671
Flexural	D-790	Heat deflection temperature	D-648
Fracture toughness	E 5045-99	Dielectric constant	D-150
Pendulum impact	D-256	Arc resistance	D-495
Falling weight impact (dart)	D-1709	Dielectric strength	D-149
Falling weight impact (tup)	D-2444	Cone calorimetry	E-1354-92
Creep	D-2990 or D-2991	Limiting oxygen index	D-2863
Shear test (V-notched)	ASTM 5379	Poisson's ratio	ASTM D638, ISO 527

the standard data available in the literature. Of course, the knowledge on standard mechanical property and theoretical understanding of the mechanical behavior of the materials can be important guidelines.

When an elastic body is subjected to a tensile or compressive stress then the strain changes proportionally with stress according to Hooke's law as given below.

$$\sigma = E.\varepsilon \tag{6.34}$$

where σ is the longitudinal stress, ε is the corresponding strain, and E is called Young's modulus or modulus of elasticity. Similarly, in the case of shear deformation, the modulus is called shear modulus or modulus of rigidity (G). When a hydrostatic force is applied, a third elastic modulus is used called modulus of compressibility or bulk modulus (K). It is defined as the ratio of hydrostatic pressure to the volume strain. A deformation (elongation or compression) caused by an axial force is always associated with an opposite deformation (contraction or expansion) in the lateral direction. The ratio of the lateral strain to the longitudinal strain is the fourth elastic constant called Poisson's ratio (v). For a small deformation, the elastic parameters can be correlated in the following way:

$$K = \frac{E}{3(1-2v)} \tag{6.35}$$

$$G = \frac{E}{2(1+v)} \tag{6.36}$$

In the case of the rubbery network, the Poisson's ratio is often close to 0.5 and hence shear modulus can be determined by dividing the elastic modulus by 3.

6.2.1.1 Tensile test

The tensile test is most widely used to assess the performance of a material or identify the material for a particular application. The test indicates the ability of a material to withstand the pull-out forces and used to determine the extent of stretching before the break. The tensile test is carried out using ASTM D-638. The test is carried out using a universal testing machine (UTM). The machine consists of two cross-heads held in a rigid metallic frame and base. One of the cross-heads moves along with the vertical frame and is fitted with a load cell. The specimen is clamped onto the grip of the fixtures attached to the cross-heads. There are different fixtures for different modes of testing. The strain is measured by many optional systems like optical sensor (Laser), contact type extensometer, etc. In the case of measuring Poisson's ratio, two strain measuring devices are attached to the specimen, one axial and one transverse. A uniaxial force is applied at a predetermined speed. The movement of the cross-head is controlled by a motor. The operation is fully computer-controlled. The data acquisition and analysis are also done by computer with suitable software. The data and graphics on load−deflection, stress−strain, creep, and stress relaxation can be obtained. Dumbbell specimens of different sizes as specified by the standard can be used by fixing the specimen in two grips (one stationary and one movable) as illustrated in Fig. 6.7. Proper tightening and

Figure 6.7
A tensile test set-up with a dumb-bell shape specimen.

alignment is very important, otherwise, the test may lead to erroneous results due to the slippage or breakage at the grip. Depending on the nature of the sample different speeds of testing can be used. The most commonly used speed of testing is 5 mm min^{-1}. Tensile strength is determined by dividing the force (measured by the load cell) by the cross-sectional area and strain is calculated from the ratio of elongation to gage length. Calculations of various parameters (like stress, elongation, modulus, etc.) are done automatically by using the computerized instrument. The definition of elongation, elastic limit, and modulus of elasticity as per ASTM standard are quoted below (with permission).

Elongation: The increase in length produced in the gage length of the test specimen by a tensile load. It is expressed in units of length, usually millimeters (inches) (also known as an extension).

Elastic limit: The greatest stress that a material is capable of sustaining without any permanent strain remaining upon complete release of the stress. It is expressed in force per unit area, usually megapascals (pounds-force per square inch).

Modulus of elasticity: The ratio of stress (nominal) to corresponding strain below the proportional limit of a material. It is expressed in force per unit area, usually megapascals (pounds-force per square inch) (also known as elastic modulus or Young's modulus).

Typical stress–strain plots observed in polymer materials are shown in Fig. 6.8. The initial linear portion of the curve represents the elastic behavior where stress is proportional to strain (Hooke's Law). Up to the elastic limit, the strain is recoverable. The deformation is associated with the bending and stretching of interatomic bonds of the polymer chains. Beyond the elastic limit, permanent deformation or yielding takes place as a result of the permanent displacement of molecules relative to each other. This strain is not recoverable completely and instantaneously. Due to permanent deformation, the strain increases keeping the stress almost the same until failure. If the resin is highly crosslinked then failure takes place without any significant permanent deformation and is called brittle failure. This is because the high crosslink density resists the displacement of molecules from their relative positions. On the other hand, a lightly crosslinked network exhibits a significant permanent deformation before failure. This type of failure is called ductile failure. The area under the stress–strain curve is a measure of toughness. Naturally, the area under the stress–strain curve is higher for a ductile (optimally crosslinked) material compared to a brittle (highly crosslinked) material. In some cases, the stress increases before failure due to the development of some order arrangements like strain-induced crystallization. The mechanical behavior of polymer is generally explained by the spring and dashpot model. Spring represents elastic behavior and dashpot represents viscous flow.

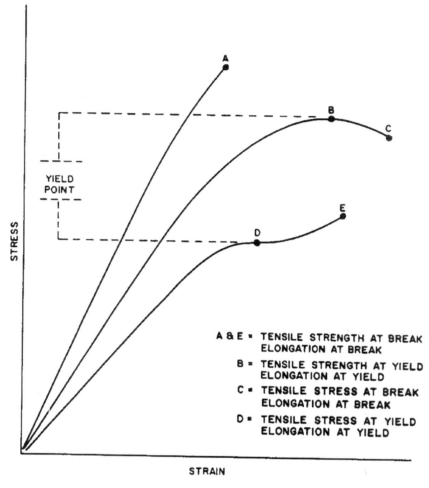

Figure 6.8
Tensile designations for plastic samples. Source: *Reprinted with permission from ASTM D-256-04, ©ASTM International, 100 Barr Harbor Drive, West Conshohocken PA 19428. A copy of the complete standard may be obtained from ASTM (http://www.astm.org).*

6.2.1.2 Flexural test

Flexural properties give an idea of the response of the material under bending load. The test is carried out by a three-point bending method as per ASTM 790. The test setup for this is very simple in which a rectangular sample is placed between two supports and load is applied by means of a loading nose from the midway between the supports as represented in Fig. 6.9. Unlike the tensile test, the flexural test allows easy sample preparation and does not involve clamping of the sample. A very tight clamping of tensile

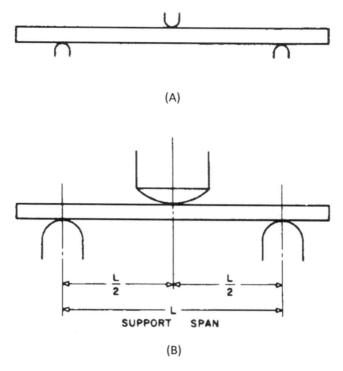

Figure 6.9
A representation of test set-up for a flexural test. Source: *Reprinted with permission from ASTM D-790-07, ©ASTM International, 100 Barr Harbor Drive, West Conshohocken PA 19428. A copy of the complete standard may be obtained from ASTM (http://www.astm.org).*

specimens, which is necessary for the testing of high-strength materials, may generate stress concentration points leading to erroneous results. From a flexural test, generally, three test parameters are reported namely flexural strength, flexural strain, and flexural modulus. The area under the stress–strain curves reflects the toughness of a material. Flexural strength is the ability of a material to withstand the bending force applied perpendicular to its longitudinal axis. The fracture strength ($F.S$) was determined from the peak load (kg) using the following formula:

$$F.S = \frac{\frac{3}{2} \times \text{peakload} \times \text{span} \times 9.8}{\text{width} \times (\text{thickness})_2} \tag{6.37}$$

Flexural strain reflects the flexibility of a material. Highly flexible resins like rubbery epoxy, PU, modified resins do not break even after a large deflection and determination of ultimate flexural strength are impractical for them. The relevance of the mechanical test depends on the form in which the material is going to be used. If a material is to be used as a beam then the flexural test is more relevant than a tensile test.

6.2.1.3 Compression test

Compression is the reverse of tension loading. However, scientifically, compression results in more dense packing of molecules in a viscoelastic body and hence the resistance to deformation increases till the material yields to crush. For polymers, the phenomenon of compression is associated with mode conversion as the material expands in the transverse direction on compression. This mode conversion is more for higher Poisson's ratio. For example, an elastomer will bulge sidewise, which is shear deformation. Most elastomers have Poisson's ratio of 0.45–0.49, indicating a very high side bulge and also exhibit high bulk modulus; they are therefore almost incompressible in bulk.

Universal testing machine (UTM), which is used for tensile testing as described above, is the most appropriate instrument for the study of load–deflection and stress–strain characteristics in compression mode. Unlike tensile mode, the strain level in compression is much lower because the force needed to compress even to 20% is very high as a result of the packing of the polymer chains. Therefore, the measurement of strain must be more precise to record the small change in deformation. The load cell records the resistance of the sample in the reverse direction of the tensile mode and the cross-head movement is also in opposite direction in compression mode. The programming of the run and processing of data are to be done accordingly.

6.2.1.4 Creep and stress relaxation test

Consideration of creep properties is very important to predict the long-term load-bearing capacity and to design the polymer-based products for such application. When a polymer material is subjected to a constant load, it deforms quickly to a strain depending on its physicochemical properties and then continues to deform slowly until failure. This property is called creep. A plastic thread, very tightly tied between the two supports, is found to often become loose after few weeks. This is a manifestation of creep. Creep is an inherent property of all polymers. Because of the presence of chemical crosslink, a cured thermoset network exhibits much lower creep compared to a thermoplastic resin and the creep decreases with an increase in the crosslink density of a thermoset network.

Creep measurements are done by measuring a constant tensile or flexural or compression load to a respective specimen as discussed above and measuring the strain as a function of time. In a typical creep plot, the % creep strain is plotted against time. The apparent creep modulus at a particular time can be calculated by dividing the stress by strain at that particular time. Creep compliance is determined by dividing the strain by stress. For the tensile test, the simple way to measure extension is to make two gage marks on the tensile specimen and note the distance between the marks at different intervals. However, an accurate measurement of extension requires the use of an optical or laser extensometer. In a flexural measurement, the strain is generally calculated with the help of a linear variable differential transformer (LVDT) system.

A stress relaxation experiment can be done using the same type of specimen in a strain control mode. In this case, the cross-head is moved quickly enough to a predetermined deflection value and the cross-head is stopped at that point. The load decreases with time and after a certain period, attains a plateau. The experiment is terminated when the stress is not changing more than 1% of the plateau value per hour. At least three experiments at different constant strain levels should be done to find the average relaxation time of the material.

6.2.1.5 Fatigue test

For many applications (e.g., pressure vessel, hinges, tubing, vibration equipment, and automobiles), polymer materials have to encounter cyclic loading in terms of flexing, stretching, twisting, or compressing, repeatedly. The fatigue test is intended to predict the behavior of the material under such conditions. Like conventional loading, cyclic loading also generates mechanical deterioration, which leads to a complete fracture in the due course. For this test, a material is subjected to a cyclic loading with a particular oscillating condition and the number of cycles required to produce the failure is determined. Thus the stress is plotted as a function of the number of cycles. Fatigue life is defined as the number of cycles of deformation required to bring about the failure of the test specimen under a given set of oscillations. Thus the fatigue life will be longer at a low-stress level and fatigue life will be shorter when the stress level is higher. The limiting stress below which the material will never fail is called fatigue endurance limit. Fatigue tests can be carried out in both tensile and flexural modes (ASTM D671). The flexural fatigue test is carried out in the constant deflection principle and tensile fatigue is carried out in constant load (stress) condition. In flexural fatigue test the specimen is held as a cantilever beam in a vice at the one end and bent by a concentrated load applied through a yoke fastened to the opposite end. In the tensile fatigue test, both the ends of the specimen are mounted in the machine and the same is rotated between two spindles. A typical specimen dimension of the flexural fatigue test is given in Fig. 6.10.

6.2.2 Fracture toughness

Fracture toughness (K_{IC}) is one of the most important properties of a material, which is used to design materials for dynamic applications where the material has to encounter a lot of mechanical shocks and vibrations. The toughness of a material is defined as the energy absorbed by a material before fracture and expressed as impact energy (G_{IC}). Fracture toughness describes the resistance of a material with cracks to fracture. Since it is almost

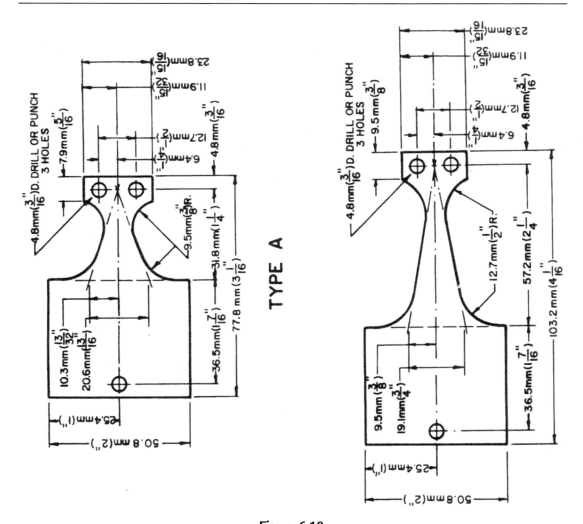

Figure 6.10
Dimensions of constant force flexural fatigue specimen. Source: *Reprinted with permission from ASTM D-671-93, ©ASTM International, 100 Barr Harbor Drive, West Conshohocken PA 19428. A copy of the complete standard may be obtained from ASTM (http://www.astm.org).*

impossible to make a material for practical purposes free from cracks/defects, fracture toughness analysis is extremely important for design applications. The fracture properties of a thermoset material are determined using ASTM 5045-99. The critical stress intensity factor, K_{IC}, and the impact energy, G_{IC}, are determined. The tests were carried out with a universal testing machine using a compact tension or bending mode. An initial crack was machined in a rectangular specimen and a natural crack was generated by tapping on a fresh razor blade placed in the notch. A typical specimen configuration for bending and

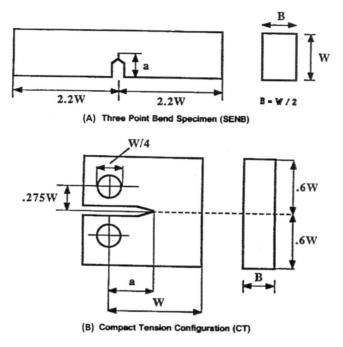

Figure 6.11
Typical specimen configurations for bending and compact tension mode for measurement fracture toughness. Source: *Reprinted with permission from ASTM E 399-06e1, ©ASTM International, 100 Barr Harbor Drive, West Conshohocken PA 19428. A copy of the complete standard may be obtained from ASTM (http://www.astm.org).*

compact tension mode are shown in Fig. 6.11. The parameters can be expressed mathematically as follows:

$$K_{IC} = \frac{P_Q}{BW^{1/2}} f(x) \tag{6.38}$$

$$G_{IC} = \frac{1 - v^2}{E} K_{IC}^2 \tag{6.39}$$

where P_Q is the critical load for crack propagation, B and W are the thickness and width of the specimen; E is elastic modulus; and $f(x)$ is a nondimensional shape factor given by:

$$f(x) = 6\sqrt{x} \frac{\left(1.99 - x(1-x) \times \left(2.15 - 3.93x + 2.7x^2\right)\right)}{(1 + 2x)(1-x)^{3/2}} \tag{6.40}$$

The thickness of the specimen should be higher than the critical value below which the material shows a plane stress behavior. The K_{IC} and G_{IC} values of a given material are functions of testing speed and temperature. Furthermore, the values may be different under cyclic load. Therefore the application of K_{IC} and G_{IC} in the design of service components should be made considering the differences that may exist between the test conditions and the field conditions.

6.2.3 Impact test

The impact tests are used to determine the behavior of a material subjected to shock loading in bending, tension and torsion. The impact tests can be divided into two classes, namely, pendulum impact tests and drop-weight impact tests. Each test can again be subdivided into noninstrumented and instrumented ones. A noninstrumented test displays only one, which is the total impact energy. It cannot provide additional information on fracture behavior, ductility, yielding or dynamic toughness. On the other hand, an instrumented tester measures force continuously while the specimen is penetrated. Thus in addition to the energy required to break the material, it provides a total load history and failure during the impact event can be recorded.

6.2.3.1 Pendulum impact test

The pendulum impact test employs a pendulum type of hammer and is carried out in four different ways such as Izod, Charpy, Chip, and Tensile. The apparatus used for the pendulum test consists of a heavy base with a vice for clamping the specimen and a pendulum hammer swinging at a sample. The machine base has to be kept on a rigid platform to prevent vibrational energy losses.

These Izod/Charpy tests are widely applied in industries due to the ease of sample preparation and it is possible to generate comparative data very quickly. Both the tests are done as per ASTM-256 using a notched specimen. In the Izod test, the specimen is fixed as a cantilever with the help of vice as shown in Fig. 6.12. Charpy test is also carried out in the same apparatus and specimen. The only difference is that in the Charpy test, the specimen is supported as a simple beam. The apparatus consists of a heavy base with a vice for clamping the specimen and a pendulum hammer swinging at a notched sample. The energy transferred to the material can be inferred by comparing the difference in height of the hammer before and after the fracture. Depending on the instrument, the impact energy (J m^{-1}) or the load history during the impact event can be recorded. The test can be carried out using different combinations of impact mass and incident impact velocity to generate the data on damage tolerance as functions of impact parameters, which are helpful for the design of composite materials for a particular application [35].

Chip impact test is carried out in a similar way as Izod test but without any notch. The test specimen is usually 25.4 mm long and 12.7 mm and 1.65 mm thick. The tests are carried out using ASTM 4508. Since the test involved the unnotched sample, it reflects the material toughness rather than the notch sensitivity measured for Izod and Charpy test. Tensile

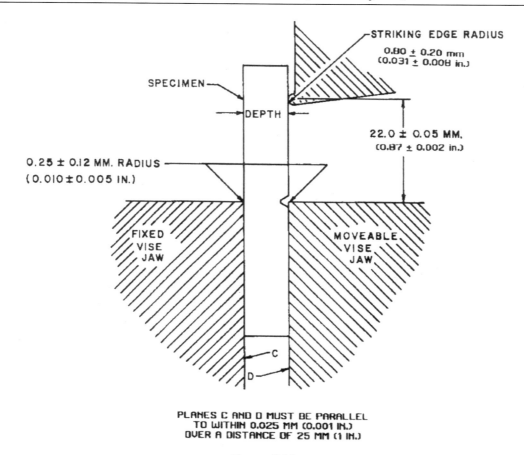

Figure 6.12
Relationship of vice, specimen and striking edge to each other for Izod test methods. Source: Reprinted with permission from ASTM D 256-04, ©ASTM International, 100 Barr Harbor Drive, West Conshohocken PA 19428. A copy of the complete standard may be obtained from ASTM (http://www.astm.org).

impact test is carried for unnotched samples in a test set-up where one end of the specimen is mounted in the pendulum and other end is gripped by a crosshead member, which travels with the pendulum until the instant of impact. The tensile force is exerted by the pendulum. The advantage of the tensile impact test over the Izod or Charpy test is that the thinner and ductile samples can also be tested by tensile impact test whereas Charpy/Izod tests are limited to only thicker and rigid samples.

6.2.3.2 Falling weight impact test

Another test, which is used to evaluate thermoset resins and composites, is the falling-weight or drop-weight impact strength. The falling weight may be a ball, a tup with a chemical nose or a dart. Depending on the nature of weight and the application, two ASTM standards D-5420 (ball, rigid specimen) and D-1709 (dart, film) are used for the testing of thermoset materials. ASTM D 2444 (tup) is used for the evaluation of thermoplastic material. In a dart impact test the dart is allowed to fall on a specimen kept in a fixture with an annular hole typically ranging from 2.5 to 7.5 cm. The output of the load transducer can be directly fed into a signal processor and the impact energy or the entire loading history can be recorded using an instrumented machine. It can also be coupled with ultrasonic c-scan and microscopic techniques to study the postfailure analysis [36,37]. The biggest advantage of the falling-weight impact test over the pendulum impact test is its ability to duplicate the multidirectional impact stresses that a structure or part would be subjected to in the actual service condition. Second, it offers flexibility to use specimens of different sizes and shapes and a component as a whole can also be tested. It may be noted that for critical applications, rigorous testing has to be carried out considering the service conditions, to ensure the specified service life [38].

6.2.4 Hardness test

Hardness is defined as the ability of a material surface to resist a force against indentation or deflection. It is somewhat similar to Young's modulus of a material but has an arbitrary unit according to some well-defined scales. The measure of hardness is actually the depth of indentation by a defined force using a standard presser foot on a uniform, plane surface of the polymer. The hardness of a material can be related to Young's modulus by empirical equations. The device used for soft thermoset is Shore durometer having indenter consists of an indenter, which are cylindrical in shape with tapering at the tip for indentation. Higher numbers in its scale indicate a greater resistance to indentation and thus harder materials, and lower number indicates softer material.

The scale was defined by Albert Ferdinand Shore, who developed a suitable device to measure hardness in the 1920s. The standard used for this purposes is ASTM D2240 type "A" and type "D" scales. There are 12 scales, depending on the intended use: types A, B, C, D, DO, E, M, O, OO, OOO, OOO-S, and R. Each scale results in a value between 0 and 100, with higher values indicating a harder material. The test requires applying the force by an indenter having a spring attached to it. The maximum force of the spring is the force transferred to the surface of the polymer by the indenter. The hardness in Shore "A" is related to the depth of indentation as:

$$d(\text{cm}) = 0.254 - 2.54 \times 10^{-3} S \qquad (6.41)$$

The indentation of 2.54×10^{-3} cm is taken as shore "A" value of 1 degree. Following the above equation, if the indentation (d) on a rubber surface is 0.10 cm, then the Shore "A" Durometer hardness is about 60.6. The material under test should be a minimum of 6.4 mm

(0.25 in.) thick. A minimum time of 15 s should be allowed before any measurement. While testing, the force is to be applied in a consistent manner, without shock.

The Shore "D" durometer is used to measure the thickness of hard thermosets like epoxy, phenolics, etc. The Shore "A" indenter has a flat tip of 0.79 mm diameter (cross section) and the angle of the truncated cone is 35 degrees, while the indenter of Shore "D" has a round tip of 0.10 mm radius of curvature and the angle of the cone is 30 degrees. The force applied with the Shore "A" indenter is 8.064 N and for Shore "D" indenter, it is 44.64 N. The final value of the hardness depends on the depth of the indenter after it has been applied for 15 s on the material. If the indenter penetrates 2.54 mm (0.100 in.) or more into the material, the durometer is 0 for that scale. If it does not penetrate at all, then the durometer is 100 for that scale. The thermoset network having Shore "A" hardness exceeding 90 should be measured in Shore "D" scale. The value in Shore "D" is approximately Shore "A" + 50. Durometer is a dimensionless quantity, and there is no simple relationship between a material's durometer in one scale, and its durometer in any other scale, or by any other hardness test. Typical Shore "D" hardness of rubbery thermoset is 40–60 and glassy thermoset is in the range 75–90.

For hard thermoset, Rockwell hardness tester is also used where indentation is done under a high force for a limited time and the hardness is recorded after a few seconds of recovery. The indenter is either a conical diamond or a hard steel ball. Different indenter ball diameters range from 1/16 to 1/2 in. It determines the hardness by measuring the depth of penetration of an indenter under a large load compared to the penetration made by a preload. There are four categories of Plastic Rockwell hardness tester: HRE, HRK, HRL, and HRM, in which HRE and HRK are most commonly used. HRE uses a load of 100 kgf (980.7 N) with an indenter steel ball of 3.175 mm diameter and HRK uses 150 kg (1471 N) with an indenter steel ball of 3.175 mm diameter. The most common standard for Rockwell hardness test is ISO 2039. Hardness is tested in the macrorange with this method, with a test force between 49 and 980.7 N. It is a differential-depth method. This means that the residual depth of indentation left by the indenter is measured to determine the hardness value of a test specimen.

The size of the specimen can be 50 mm × 50 mm lateral dimension and 6.5 mm thickness. Typically, the tip is rested on the sample surface at a nominal preload force of 49 N for 10 s and subsequently a force of 980.7 N is applied to indent the plastic surface for 15 s and the reading is taken after 15 s of removing the load up to the initial load value (49 N). These time periods in general match with the relaxation or retardation time of most common plastics at ambient temperature. The readings are dynamically taken with time while the indenter tip is moving down and up. In the short time period, data recording by the human eye may give rise to error; hence microprocessor-controlled movement measurement such as LVDT will be more desirable. The typical value of Rockwell hardness of phenolic is 95–105 and glass-filled epoxy 105–110.

6.2.5 Thermal properties

Polymer materials are inferior in thermal properties compared to conventional materials like metals and ceramics. That is why the thermal property plays an important role in deciding the applications (as replacement of conventional materials or the case where the other materials cannot be used) of a polymer in general and thermoset resins in particular [23,24]. The objective of the present chapter is to discuss the thermal properties of thermoset resins. Unlike thermoplastic where thermal property depends upon only molecular weight, thermal properties of a thermoset depend on various factors namely crosslink density, chemical nature of the crosslink (−C−C− or C−O), molecular weight between the crosslinks, uniformity of the network, etc. [1,39]. Hence thermoset resins offer a wide scope for modification of thermal properties and are more challenging to the expertise of polymer scientists. The thermal properties of thermosets will be discussed in different heads such as thermal conductivity, thermal expansion coefficient, thermal stability, heat distortion temperature, flammability, thermal cure characteristics, and transition behavior (especially glass transition temperature).

6.2.5.1 Thermal conductivity

Thermal conductivity (k) is the property of a material that indicates its ability to conduct heat. It can be derived from Fourier's Law for heat conduction [40]

$$k = \frac{\Delta Q}{\Delta t} \cdot \frac{1}{A} \cdot \frac{x}{\Delta T} = \frac{\Delta Q}{A . \Delta t} \cdot \frac{x}{\Delta t} \tag{6.42}$$

where $\Delta Q/\Delta t$ is the rate of heat flow, k is the thermal conductivity, A is the total cross sectional area of conducting surface, ΔT is temperature difference, and x is the thickness of conducting surface separating the two temperatures. In other words, it is defined as the quantity of heat, ΔQ, transmitted during time Δt, through a thickness x, in a direction normal to a surface of area A, due to a temperature difference ΔT, under steady-state conditions and when the heat transfer is dependent only on the temperature gradient. Therefore, thermal conductivity is defined as the rate at which heat is transferred by conduction through a unit cross-section area of a material when a temperature gradient exists perpendicular to the area. The unit of thermal conductivity [40,41] is W m^{-1} K^{-1}. Thermal conductivity may be classified as a transport property since it is the indication of the energy transport in a solid or fluid. In solid, the transport of energy takes place by movement of free electrons and lattice vibration, that is, phonons, electron, or photon transport energy. Phonons, quantized modes of vibration occurring in a rigid crystal lattice, are the primary mechanisms of heat conduction in most polymer since free movement of electron is not possible [42].

Thermosets being mostly amorphous exhibit very low thermal conductivity (0.18−0.24 W m^{-1} K^{-1}) [43,44] due to the phonon scattering from numerous defects. It is necessary to enhance the thermal conductivity of thermoset resins for various applications, namely, circuit board in power electronics, heat exchangers, electronics, and machinery, electric motor and power generators for replacing metal parts. For heat sink application in electric and electronic system [45], the materials need to have thermal conductivity ranging from 1 to 30 W m^{-1} K^{-1}. Improvement in thermal conductivity helps in optimizing/controlling the addition or removal of heat during the processing. The fillers consisting of metal particles and glass fiber are used to improve the thermal conductivity of thermosets [46–49]. The current interest to improve the thermal conductivity of polymer is focused on the selective addition of nanofillers with high thermal conductivity carbon nanotube (CNTs) [50].

There are various methods for determining the thermal conductivity of a polymer. These techniques are described in ASTM standards (ASTM C201, C177). Another method that is mostly used for composite is the 3ω method [51]. In the 3ω method, specimens are prepared by cutting 1–2 mm thick samples. Alternating current at a frequency ω was supplied to the platinum wire using a four-wire probe electroplated to the sample. The power produced at frequency 2ω caused temperature oscillations at $2\omega + \Phi$. The corresponding resistance changes of the platinum wire cause the temperature changes in a linear fashion. The change in the 2ω-resistance component together with the supplied ω current component produces a 3ω-voltage component. This 3ω-voltage component is reflective of the actual material properties. At each setpoint temperature, the technique performs a frequency sweep between 10 and 1000 Hz. Test frequency is inversely proportional to the depth of penetration. The 3ω component changes as a function of frequency is used to obtain the sample thermal conductivity at each set point temperature.

Many theoretical and empirical models have been proposed to predict the effective thermal conductivity of two-phase mixtures. The thermal conductivity models for two-component materials found in the literature are based on one or more of three basic structural models, namely, the series/parallel, Maxwell [52], and Effective Medium Theory (EMT) models [53,54]. For a two-component composite, the simplest alternatives would be with the materials arranged in either parallel or series with respect to heat flow, which gives the upper or lower bounds of effective thermal conductivity. For series conduction model:

$$\frac{1}{k_c} = \frac{1-\phi}{k_m} + \frac{\phi}{k_f} \tag{6.43}$$

For parallel model:

$$k_c = (1-\phi)k_m + \phi k_f \tag{6.44}$$

where k_c, k_m, and k_f are the thermal conductivities of the composite, the matrix, and the filler, respectively, and ϕ is the volume fraction of the filler. According to the Maxwell model for an infinitely dilute composite of spherical particles, the exact expression for the effective thermal conductivity is given as:

$$\frac{k_c}{k_m} = 1 + 3\left(\frac{k_f - k_m}{k_f + 2k_m}\right) \tag{6.45}$$

where k_c, k_m, and k_f are thermal conductivities of composite, continuous-phase (matrix), and dispersed-phase (filler), respectively. Effective Medium Theory is used to get a better idea of the thermal events that have been taking place in the samples, a model incorporating the effective interfacial resistance between the filler and resin was used. It can be stated as [53,54]:

$$\frac{k_c - k_m}{3k_m} = \phi\frac{\gamma - 1}{\gamma - 2} \tag{6.46}$$

where k_c and k_m are the thermal conductivities of the composite and resin, respectively, in W m^{-1} K^{-1}; ϕ is the absolute volume fraction of filler; and γ is a function of the radius of the filler.

6.2.5.2 Thermal expansion coefficient

Thermal expansion is the tendency of a matter to change in volume in response to a change in temperature. When a substance is heated, its particles begin moving more and thus usually maintain a greater average separation. Since thermosets are used in the solid form, one might only be concerned with the change along a length, that is, linear thermal expansion. The degree of expansion divided by the change in temperature is called the material's linear coefficient of thermal expansion (α) and expressed in μm m^{-1} °C^{-1} as per the following equation [55]

$$\alpha_L = \frac{1}{L}\frac{dL}{dT} \tag{6.47}$$

where L is the linear dimension (e.g., length) and dL/dT is the rate of change of that linear dimension per unit change in temperature.

Thermomechanical analysis (TMA) and dilatometer are capable of providing quantitative values of linear coefficient of thermal expansion (LCTE). However, precision and accuracy will depend on factors such as sample dimension in the direction of measurement, amount of expansion ($\alpha\Delta T$), and background noise and vibration. It was reported that the dilatometer was more accurate for measurement of LCTE below 9 μm m^{-1} °C^{-1} while the faster and easier TMA measurements were sufficiently accurate for higher expansion values [56]. Thermoset resins shows LCTE (40–150 μm m^{-1} °C^{-1}) much higher than metals [56,57]. It is very important to consider LCTE while designing a component. For example, thermal expansion can cause significant stress in a component if the design does not allow for expansion and contraction of the component. When a product is designed with close tolerance fit, then thermal expansion has to be taken into consideration seriously as these tolerances will change with increase or decrease in temperature.

6.2.5.3 Heat distortion temperature

Heat distortion temperature or heat deflection temperature (HDT) indicates the temperature range over which the polymer material begins to soften and gets deformed under the application of load. Thus, HDT can identify the materials which lose their rigidity and load bearing capacity over a narrow temperature range. This is commonly used for quality control and assessing materials for short-term heat resistance. The service temperature of a load-bearing component must be lower than the HDT of the composed material. HDT is determined as per ASTM D-648. The temperature, at which a standard test bar deflects 0.01 in. under a load of 66 or 264 psi, is recorded as HDT. HDT is a single point measurement and does not indicate long term heat resistance of the material.

6.2.6 Electrical properties

A material conducts electricity due to the movement of charge carriers like electrons or ions under an electric field. Metallic materials are highly conducting because of the presence of loosely bound electrons and both the valance band and conduction band are overlapped with each other. However, thermoset plastic does not have any free electrons and that is why they exhibit very low conductivity typically less than $10^{-8}\ \Omega^{-1}\ cm^{-1}$ and considered as an insulator. In a thermoset, the outermost electrons of the atoms are covalently bonded. When an electric field is applied, the firmly bound charges in such materials can be displaced by only a limited extent from their equilibrium position depending on the amplitude of the electric field and polarity of the bonds present in the polymer network. The bound charges exert an opposing force toward the displacement of charges and the corresponding energy can be stored in the material. The ability to store energy in this way refers to the dielectric property. The ratio of charge stored in an insulating material placed between the two metallic plates to the charges stored when the insulating material is replaced by air or vacuum is called permittivity or dielectric constant of the material. Dielectric constant of a material can be determined by using ASTM D-150. The value of dielectric constant increases with the increase in polarity of the polymer. For instance, the typical dielectric constant values for more polar thermosets like melamine and phenolic resins are 5.2–7.9 and 3.5–7.0, respectively, whereas the same values for epoxy and polyester resins are 3.5–5.4 and 4.3–5.1, respectively. Because of the higher dielectric constant compared to thermoplastic resins and other materials, they are widely used as insulators/dielectrics for various applications.

6.2.6.1 Electrical conductivity

Thermosetting resins are inherently insulators and widely used in printed circuit boards and microelectronics. However, for many applications of electrostatic dissipation (antistatic materials) or electromagnetic radiation shielding, the electrical conductivity of the cured resin is a crucial factor. The electrical properties of a thermoset network can be altered by blending with conducting fillers like carbon black, metallic powder, or conducting polymers like polyaniline, polythiophene, etc. Recently, carbon nanotube (CNT) has drawn considerable attention for the development of conducting composites. The thermoset/CNT composites will be elaborated in Chapter 7. Hence for both types of applications, the measurement of electrical resistance or conductivity of thermoset materials is very important. Electrical resistance can be measured by a two-probe or a four-probe method. The AC electrical resistivity can be determined by using an AC impedance spectroscopy. Conductivity can be calculated from the resistivity data using the following equation:

$$\sigma = \left(\frac{1}{R}\right)\left(\frac{t}{A}\right) \tag{6.48}$$

where t is the thickness of the thermoset resin film, A is the area of the electrode, and R is the resistance.

The resistance offered by a thermoset material is the combined effect of volume and surface resistances, which always act in parallel. Volume resistivity or specific volume resistance is the resistance to leakage of current through the body of the material and is function of thickness of the material. Volume resistivity is defined as the resistance between two electrodes covering opposite faces of a centimeter cube. On the other hand, surface resistance is the resistance to passage of current through the surface. Thus surface conductivity is influenced by the surface finish and contamination. Samples having rough surface or contaminated with oil, moisture, etc. display lower value of surface resistance compared the samples with smooth and cleaned surfaces. It may be noted that conductivity measurement is affected by the temperature and relative humidity of the atmosphere. Hence, for comparing the conductivity data for different samples, the experiments have to be carried out in similar conditions.

6.2.6.2 Dielectric strength

Ideally, the insulator should not allow any current to pass through it. However, due to the presence of defects, every material is associated with some amount of leakage especially at a higher voltage. The smaller the leakage better is the material. Such leakage generates heat and further increases the extent of leakage. This process slowly accelerates with time leading to the total breakdown of the insulator. The minimum voltage at which dielectric break down occurs is called break down voltage. The dielectric strength is defined as the break down voltage per unit thickness and expressed as $V\ mil^{-1}$ (1 mil = 1/1000 in.). Dielectric strength reflects the electrical strength of an insulator and thus it is an important characteristic of insulating material. The higher is the dielectric strength better is the insulator. The dielectric strength of phenolic resin is in the range of $240-340\ V\ mil^{-1}$.

The dielectric strength is determined using ASTM D-149. The measurement of dielectric strength is usually carried out by short-time method or step by step method. In the first method voltage is increased continuously at a uniform rate until the failure occurs. In step-by-step testing, definite voltages are applied to the sample for a definitive time, starting with a value that is half of that obtained by a short time testing with equal increments of voltage until failure occurs. Since step-by-step is associated with a longer exposure to the electric field, dielectric strength values obtained by this method are always lower that those obtained by short-time test. However, a step-by-step method conforms the actual service condition more, compared to a short-time method and thus the values are useful for designing the thermoset materials for the dielectric application.

6.2.6.3 Arc resistance

Thermoset materials are used for switches, contact bushes and circuit breaker. For these applications, the insulating materials are likely to be subjected to arcing. Arcing tends to produce a conducting carbonized path on the surface and destroys the insulator property. In cases of acrylic resins, instead of carbonization, decomposition to ignitable gases takes place resulting in the failure in a short time. Thus arc resistance is an important criterion for a thermoset material to be used for insulator application. Arc resistance is the ability of a material to resist the action of high voltage electrical arc. The arc resistance of an insulator can be defined as the time in seconds that an arc may play across the surface without forming a conducting path or loss of integrity. The test to measure arc resistance can be conducted in various ways. ASTM D 495 describes the measurement of arc resistance. The resin which easily carbonizes such as phenolics have poor arc resistance. On the other hand, unsaturated polyester and melamine are excellent arc-resistant.

6.2.7 Flammability and smoke tests

In general, polymers are flammable in nature. However, they can be suitably modified to reduce the flammability significantly. Assessing the flammability of a polymer is extremely important from a safety point of view, especially where the material is prone to catch fire. There are many standards and governing regulatory bodies controlling the level of flame retardancy required for various applications as described in earlier issue of Rapra review report [55]. The three most commonly used flame tests, namely, underwriter's laboratory test (UL-94), cone calorimetry, and limiting oxygen index (LOI) test, are very briefly described here.

6.2.7.1 UL-94 flammability test

The UL-94 test is performed on a plastic sample (125 mm by 13 mm with various thicknesses up to 13 mm) suspended vertically above a cotton patch. The plastic is subjected to a flame exposure for 10 s with a calibrated flame in a unit, which is free from the effect of external currents. After the first 10 s exposure, the flame is removed and the time for the sample to self-extinguish is recorded. Cotton ignition is noted if polymer dripping ensues; dripping is permissible if no cotton ignites. Then the second ignition is performed for the same sample and the self-extinguishing time and dripping characteristic recorded. If the plastic self-extinguishes in less than 10 s after each ignition with no dripping, it is classified as V-0. If it self-extinguishes in less than 30 s after each ignition with no dripping, it is classified as V-1 and if the cotton ignites then it is classified as V-2. If the sample does not self-extinguish before burning completely, it is classified as failed (F).

6.2.7.2 Cone calorimetry

Cone calorimeter experiments were performed at an incident heat flask of 50 kW m^{-1}. Combustion behavior was assessed according to ASTM E 1354-92. During the test, the materials were subjected to irradiated heat plus the feedback heat from the flame starting from the ignition of the volatile products. The aim is to simulate the conditions likely to occur in a real fire. The data were collected for first 250 s, this being regarded as representative of the initial stage of fire when it can still be stopped before becoming uncontrollable after flashover. The heat released was calculated from the consumption of oxygen due to combustion. Various parameters like peak heat release rate (HRR), mass loss rate (MLR), specific extinction area (SEA), ignition time (t_{ign}), carbon dioxide yield, carbon monoxide yield, and specific heat of combustion data are calculated. The results obtained from cone calorimeter are considered reproducible to within 10% error when measured at 50 kW m^{-2}. The average heat release rate is correlated to the heat released in a room where the flammable materials are not ignited in the same time. The data should be reported for a minimum of three replicated instruments.

6.2.7.3 LOI test

LOI test is one of the oldest flammability tests and frequently used to compare the flame retardant properties of polymer samples. It was first developed by Fenimore and Martin. This test method has been standardized in North America as ASTM D 2863. The dimensions of the specimen to be tested should have 70–150 mm length, 6.5 mm width, and 3 mm thickness [58]. The method of operation is to select the desired initial concentration of oxygen based on past experience with a similar material. The gas is allowed to flow for 30 s to purge the system. The specimen is ignited so that the entire tip is burning. The relative flammability is determined by adjusting the concentration of oxygen, which will permit the specimen to burn. Limiting oxygen index (LOI) is calculated from the following formula:

$$LOI = \frac{V_o}{V_n + V_o} \times 100 \qquad (6.49)$$

where V_o is the volume of oxygen and V_N is the volume of nitrogen. Thus LOI of a polymer sample 28 indicates that 28% of the oxygen/nitrogen mixture was required to be oxygen in order to support continued combustion of the sample.

6.2.7.4 Smoke test

Smoke density measurements are done by measuring the transmission of light through the smoke produced in the laboratory in controlled and standardized conditions [59]. The smoke density of a thermoset is tested using Smoke Density Tester as per the standard ASTM D 2843. The samples to be tested should have dimensions 25 mm × 25 mm × 4 mm. Before the start of testing, the specimens are conditioned in accordance with procedure A of ASTM D 618-61. According to this procedure the specimens are preconditioned at 23°C + 2°C and 50% relative humidity for a minimum of 40 h. The samples are kept flat and burnt with the help of a burner. There is a light source and a photoreceptor that are present on the either side of the chamber and the percentage of light absorbed is displayed through a linear amplifier digital electronic photometer with 0%–100% absorbance scale. This technique helps in the measurement of relative amount of smoke during the burning or decomposition of the materials. It is determined by plotting graph of the percentage light absorption as a function of time.

6.3 Performance evaluation

6.3.1 Evaluation as a surface coating

When a thermoset resin is to be used for surface coating application, it is necessary to thoroughly characterize it for its various mechanical and thermal properties as discussed above. The reader may refer Table 6.1 for ASTM standards for relevant mechanical tests. In addition, various specific tests are to be conducted to assess their suitability for surface coating applications. Some of the important tests are briefly discussed below.

6.3.1.1 Pull-off adhesion strength

Adhesion strength of the thermoset resin and their pigmented compositions are determined as per the method described in Indian Standard IS 101 (pull-off method) or ASTM D-4545. The method was modified by reducing the area of one dolly to facilitate the debonding of coating from one side. Measurements can be carried out using any Universal Testing Machine. In this method, the coating is applied onto burnished mild steel specimens by a brush or by spraying for a uniform dry film thickness of about 100 μm. Then for each coating, 10 specimens were tested for adhesion strength and average of five higher readings has been reported as the adhesion strength.

6.3.1.2 Cross-hatch adhesion test

A cross-hatch adhesion testis used to establish whether the adhesion of a coating is at an adequate level or not. A cross-hatch cutter is used to cut the film to produce a lattice pattern. After making a cross-cut, the film is brushed with a soft tissue to remove any detached flakes or ribbons of coating. Then an adhesive tape is placed on the cross-cut. The tape is to be removed rapidly back upon itself at an angle of 180 degrees as close as possible. The cross-cut area is then inspected for the removal of coating if any from the substrate and rated for the extent of adhesion according to the scale as per ASTM D 3359.

6.3.1.3 Impact resistance test

Impact test is extremely useful in predicting the performance of an organic coating for its ability to resist cracking caused by impacts during the manufacture of articles and their use in service. A Gardner impact tester is generally used for testing as per ASTM-2794. A standard weight (1 + 0.9 kg) is dropped onto a plate coated with the resin or paint from a specified distance. The value of potential energy at which failure occurs by cracking is reported as the value of impact strength. If a panel coated with the resin/paint sustains the impact from falling height of 1 m, the coating is considered as good in impact property.

6.3.1.4 Bend test

The test is useful in rating the attached coating for its ability to resist cracking when it is subjected to bending. A mandrel bending tester is used for the bending test. The coated metal plates were fastened and bent over the mandrel. The length of the crack formed on the coated panel is reported as per ASTM D-522. If any coating does not show cracking after subjecting the bend test with 0.75 in., then it is considered as passed for the bend test.

6.3.1.5 Scratch resistance test

In order to assess the sensitivity of a surface coating to scratch deformation, a scratch is introduced onto the specimen's surface with a defined tip under a certain normal load and speed. Thus, parameters like applied load and scratch depth can be varied and the resulting deformation of the surface can be monitored. The resulting tangential force, penetration depth, and residual depth after viscoelastic recovery are recorded. Thus, different specimen can be compared in terms of scratch resistance. Any universal scratch hardness tester can be employed for this test using the BS-3900 standard.

6.3.1.6 Gloss measurement test

Gloss is associated with the capacity of a surface to reflect light in some directions than others. The directions associated with specular gloss reflection normally have the highest reflectance. Measurements by this test method correlate with visual observations of surface shine made at roughly the corresponding angles. A RHOPOINT gloss meter can be used to measure the gloss values of coatings on metal plates. This test depends on the measurement of the intensity of light reflected from coated plates in three different directions. For transparent coating it is recommended that haze should be less than 5.

6.3.1.7 Accelerated weathering test

An accelerated weathering test is carried to predict the performance of a coating in long run. The specimens are placed in a QAV weatherometer and tested as per ASTM G-53. UV-radiation (wavelength 285–315 nm) and high humidity condition in a QUV-accelerated weatherometer are simulated accordingly. A weathering cycle, comprising o2 h of condensation (temperature 45°C ± 5°C) and 4 h of UV exposure (temperature 60°C ± 5°C), was maintained during the study. After completion of exposure in the weatherometer physical property, adhesion and gloss were measured and compared with the data taken before exposure. It is very difficult to correlate the weathering study with the life of coating in actual condition. Generally, if a coating passes 500 h exposure, it can be related to 5 years of life in outdoor exposure.

6.3.1.8 Corrosion resistance test

If the coating is meant for corrosion protection, then it is necessary to evaluate by exposing them in a salt spray chamber, humidity cabinet, or immersion in water or seawater depending on the actual application of the coating.

6.3.2 Evaluation of thermoset as adhesive

The main qualifying characteristics [60] of an adhesive is its cohesive strength (defined by the elastic modulus, ultimate mechanical strengths) and its adhesive bond strength (defined by T-peel strength, pull-off strength, lap shear strength, fracture toughness, impact energy, and dynamic viscoelastic properties). In addition, the adhesive-bonded substrate is required to be evaluated for functional performance in vibration and shock. Since the adhesive is always used as a core in sandwich construction, the vibration study and prediction have to be done using constrained layer damping arrangement and shock response is to be studied as a damped structure with the constrained layer damping (CLD) loss factor obtained by experiment or by theoretical prediction. Therefore, there should be two levels of experimental evaluation to qualify an adhesive for high-performance structural application, namely, lab-level coupon testing and functional evaluation as an adhesive bonded replica of the actual item. Functional tests are beyond the scope of this book. The functional evaluation methods are described with examples by the present authors elsewhere [61]. Among the lab-level tests, mechanical, dynamic mechanical, and impact energy are already discussed in previous sections of this chapter. Some specific tests, which are required to done for a thermoset based adhesive, are discussed below. The adhesive tests include lap shear test, T-peel test, thick adherent shear test (TAST), and fracture toughness tests for determination of critical fracture energy (G_{IC}) and critical stress intensity factor (K_{IC}), are discussed here.

6.3.2.1 T-peel test

The most common and simple test of an adhesive joint is the T-peel test. A portion of the two adherend strips are joined by the adhesive and the free ends are pulled apart as a T shape as shown in Fig. 6.13. The dimension and the methodology are given in detail in ASTM D1876. This test is especially suitable for elastomers and thin metal strips. The thickness of metals should be low enough so that the strips must not crack of fail while bending 90 degrees, and, on the other hand, must withstand the adhesive failure force. Typically, the metal strips should be 25 mm in width, minimum 150 mm in bond length with thicknesses of 0.5 mm for steel and 0.7 mm for aluminum. If rubber-to-rubber bond using a thermoset-based adhesive is being tested, then the rubber strips could be 4−5 mm thick, depending on their inherent tensile strength. The stress concentration at the bond line of T-junction might be higher; hence a fillet of the adhesive can be applied to reduce stress concentration. The peel strength is determined by the force at failure divided by the width of the bond (25 mm, normally). Therefore, peel strength can be expressed by $kN\ m^{-1}$. UTM is required for such test, fitted with noncontact extensometer and computer-controlled operation, data acquisition, and analysis. The pulling rate may be slow enough, such as

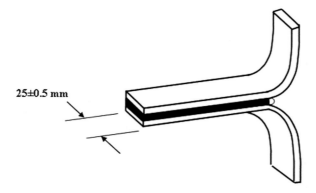

Figure 6.13
T-peel test for adhesive bonded thin metal strips.

10 mm min^{-1} to obtain a plot of force versus peel length, which ultimately shows the maximum force required to peel and consequently the peel strength. For rubbers, this peel rate is fast enough compared to the relaxation time, which is about 10^{-3} s or even less for common elastomers.

6.3.2.2 Lap shear tests

A normal single lap shear test of adhesive bonded metal strips is done for thin adhesive layers only. This method is dependent on the mechanical stiffness (modulus) of the adherend, since this influences the stress concentration at the end of the bond line. Increasing the thickness or modulus results in a wrong value of the adhesive bond strength, as it may fail in cohesion. Minimum possible length of the bond is given by:

$$L > \frac{\sigma_y t}{1.5\tau} \tag{6.50}$$

where σ_y is the yield stress, τ is the expected shear strength of the adhesive, and t is the thickness of the adherend. However, the same method cannot be applied for thick adhesive core sandwiched between two metal strips.

6.3.2.3 Thick adherent shear test

Thick adherent shear test (TAST) is carried out following two commonly used standards, namely, ISO 11003-2 and ASTM D 3165. According to ISO 11003-2, the specimen length should be 110 mm, width 25 mm, and the overlap length 5 mm. The adherend thickness is 6 mm and the bond line thickness is typically 0.5 mm. Slots should be 1.5 mm wide. Load is introduced to the specimen via two 12.7 mm diameter pins inserted in holes 80 mm apart. Also, a new method in compression (also in tension) has been suggested based on ASTM D 695 using 80 mm length, 10 mm width, 5 mm thick adherents, 2 notches 6.5 mm apart, 1.5 mm notch. A typical sample for TAST is shown in Fig. 6.14. The holes should be accurately drilled in the center of each adherend, to ensure the correctness of load data. The slots must be accurately machined through only one adherend.

For this shear test, a UTM is generally used, fitted with appropriate load cells and noncontact extensometers, especially employing either a laser beam or LVDT. Fig. 6.15 shows a sketch of the arrangement. During the test, the entire load is transmitted by the shear force along the bond line between the notches or slots. At the corners of the overlap, the stress is peel stress and has a high concentration. It can be reduced by providing a fillet at the edge of the bond line. The average shear stress mentioned above is given by:

$$\tau = \frac{P}{bL} \tag{6.51}$$

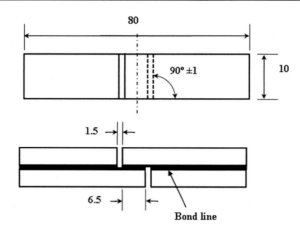

Figure 6.14
Modified design of TAST.

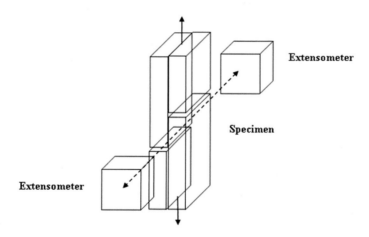

Figure 6.15
A typical measuring arrangement of a bonded joint strength with two extensometers.

where P is the applied load, b is the width of the specimen, and L is the overlap length. The strain is measured by two extensometers fitted opposite sides of the specimen overlap (Fig. 6.15) and averaged for obtaining reasonably accurate strain data. This is very important for thermoset-based adhesives, where the strain is restricted to not more than 10% before failure. In the course of the test, it is possible that the adhesive fails in cohesive mode. The strength of cohesive failure is, then, the strength of the bonded adherend.

6.3.2.4 Fracture toughness of adhesive bonded structures

Brittle failure is very common among highly crosslinked thermosets, such as epoxy, polyester, vinyl ester, phenolics, etc. The failure is catastrophic due to uncontrolled propagation of a small crack, when the local stress at the tip of the crack exceeds the ultimate tensile strength of the thermoset adhesive, although the gross tensile stress may still show below the ultimate limit. Therefore, fracture mechanics has evolved as a necessary study for the design and formulation of adhesives. For a brittle thermoset, if a flexible moiety is introduced in the matrix, the crack tip propagation can be arrested by the flexible particle as it would exhibit large strain due to the local stress and hence, the strain energy is irreversibly absorbed as a viscoelastic loss. The extent of resistance to crack propagation is the fracture toughness of the thermoset adhesive. It is therefore important to evaluate the fracture toughness of a thermoset based adhesive for avoiding catastrophic failure of the bonded structure. Fracture toughness is defined by two important properties: (1) critical strain energy release rate G_{IC} and (2) critical stress intensity factor (K_{IC}).

6.3.2.5 Critical strain energy release rate

When a crack or flaw of length "a" in a substrate propagates due to an applied load as shown in Fig. 6.16 below, then the stress to failure is related to the critical strain energy release rate G_{IC} by Griffith equation [62]:

$$\sigma_f = \sqrt{\frac{EG_C}{\pi a}} \qquad (6.52)$$

where E is the Young's Modulus of the adherend.

To determine G_{IC} of an adhesive, a small linear flaw of length "a" is made into an adherend assembly with a cured adhesive bond line as shown in Fig. 6.16 and the experiment is done in tensile or shear mode. The stress to failure is used to calculate G_{IC} using a compliance calibration method [63]. In this method, the compliance, defined as the ratio of deflection to the force ($C = \delta/P$), is measured at small incremental flaw size

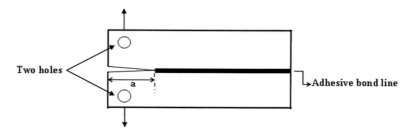

Figure 6.16
A typical sample with a flaw made for fracture toughness experiment.

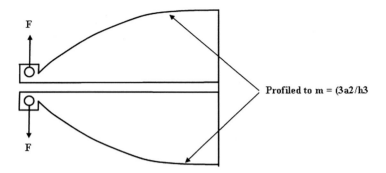

Figure 6.17
A double cantilever beam arrangement.

$(a + \delta a)$ as the force is applied, and plotted against the flaw size "a". The slope at each point of the curve($\delta C/\delta a$) versus a is used to calculate the rate of strain energy release by the following equation:

$$G = \frac{1}{2}P^2 \frac{\partial C}{\partial a} \tag{6.53}$$

At the failure force P_C, the critical crack length is a_C, and the critical strain energy release rate G_{IC} is defined by:

$$G_{IC} = \frac{1}{2}P_C^2 \frac{\partial C}{\partial a} \text{ at } a = a_C \tag{6.54}$$

The unit of G_{IC} is N m. The fracture toughness of a thermoset adhesive can be measured using a double cantilever beam arrangement, as shown in Fig. 6.17. The cantilever arms are profiled to give a constant strain energy release rate when fracture is initiated, regardless of the length of the initial crack. Hence, the specific fracture energy G is calculated directly, using the following equations:

$$G_{IC} = \left(\frac{4F^2}{b^2 E}\right) m \tag{6.55}$$

$$m = \frac{3a^2}{h^3} + \frac{1}{h} \tag{6.56}$$

where F is the applied force at fracture, b is beam width, E is the elastic modulus of the adhesive, a is the crack length, and h is the beam height at the tip of the crack.

6.3.2.6 Critical stress intensity factor (K_{IC})

The stress intensity factor is used in design and analysis assuming that the material can withstand crack tip stresses up to a critical value of stress intensity, termed K_{IC}, beyond which the crack propagates rapidly. This critical stress intensity factor is then a measure of material toughness. The failure stress σ_f is then related to the crack length a and the fracture toughness by:

$$\sigma_f = \frac{K_{IC}}{\alpha\sqrt{\pi a}} \tag{6.57}$$

where α is a geometrical parameter equal to 1 for edge crack and also in many situations. Expressions for α are tabulated for a wide variety of specimen and crack geometries, and specialty finite element methods are available to compute it for new situations.

The stress intensity and energy viewpoints are interrelated, as can be seen by comparing equations of K_{IC} and G_{IC} (with $\alpha = 1$):

$$\sigma_f = \sqrt{\frac{EG_C}{\pi a}} = \frac{K_{IC}}{\alpha\sqrt{\pi a}} \rightarrow K_{IC}^2 = EG_C \tag{6.58}$$

This relation is for plain stress. For plain strain, the expression is different:

$$K_{IC}^2 = EG_C(1 - \nu^2) \tag{6.59}$$

where, ν is Poisson's ratio. For metals with $\nu = 0.3$, $(1 - \nu^2) = 0.91$. This is not a big change; however, the numerical values of G_{IC} or K_{IC} are very different in-plane stress or plane strain situations like thermoplastic elastomers. For thermosets, the both values are almost same. The unit of K_{IC} is N m$^{-1/2}$ or MPa m$^{-1/2}$.

Evaluation of K_{IC} is therefore done in the same experimental procedure as G_{IC} described above. ISO 13586 and ASTM D5045-99 provide two methods for toughness measurement: single edge-notched bending (SENB) and compact tension (CT). CT method is already discussed in Section 6.2.2. SENB is much more popular in the case of testing of adhesive joints because of its simplicity. But a common problem exists as to how a sharp crack can be initiated in a brittle thermoset. A modified sample of SENB is used by Ma et al. [64] eliminating the notch in a brittle and toughened epoxy compositions. Alternately, a double cantilever beam arrangement also can be used. The K_{IC} values were determined using the following relationship:

$$K_{IC} = \left(\frac{P}{Bw^{1/2}}\right) f(x) \tag{6.60}$$

$$f(x) = 6x^{1/2} \frac{\left[1.99 - x(1-x)(2.15 - 3.93x + 2.7x^2)\right]}{(1-2x)(1-x)^{3/2}} \tag{6.61}$$

where P is the force, B is the thickness of the specimen, w is the width of the specimen, a is the crack length, x is the ratio of the crack length to the depth of specimen, a/w, and $f(x)$ is the calibration factor.

6.3.3 Evaluation as a vibration damping material

Thermosets are, in general, characterized by high stiffness and low flexibility compared to thermoplastics as discussed in earlier chapters. The most common thermosets are highly crosslinked, for example, epoxy, phenolic, polyurethane, polyester, and vinyl ester resins, which are widely used for structural elements as such or with fiber reinforcements. The average Young's modulus of thermosets is in the range of 1–3 GPa and their glass transition temperature is generally above 80°C. This means that their relaxation time (or retardation time) is much higher than the experimental time scale. Therefore, such thermosets are not suitable for damping application. However, rubbery epoxies and polyurethanes are widely used for vibration damping applications. For special thermosets, which are modified by mixing a toughening agent such as bitumen, low molecular weight modifiers such as poly(ether) amide, amine-terminated polysulfones, amine-terminated polybutadiene (ATBN), carboxyl-terminated polybutadiene (CTBN) or chemical modifier like carboxyl-terminated poly(ethylene glycol adipate), carboxyl-terminated poly(2-ethyl hexyl acrylate) (CTPEHA), isocyanate capped hydroxyl-terminated polybutadiene (HTPB) [5,65–69], the glass transition, and the Young's modulus would be somewhat lower and some viscoelastic loss at ambient temperature is observed. Details of the dynamic viscoelasticity and vibration damping by toughened epoxy thermoset are described in a book by the present authors [70].

It is therefore important to experimentally evaluate the thermosets intended to use as damped structural elements. The evaluation must be done in two distinct levels such as laboratory-level characterization and functional evaluation in lab or shop floor using a scaled model or actual size for ascertaining the damping as predicted from lab-scale study. Laboratory levels are discussed here.

6.3.3.1 Mechanical property

Methods for evaluating mechanical properties are already discussed. The reader may select the ASTM standard from Table 6.1. The testing and evaluation depend on some critical considerations to avoid errors. These are described below in brief:

1. Sampling procedure: There can be user specified sampling methods, apart from standard sampling procedures. Samples from every batch must be evaluated for reliability in basic properties and functional performance.
2. Sample dimensions: as accurate as possible. Machining, if allowed, should be precise.
3. Conditioning of the sample, as described in the standards to avoid the effect of environmental changes.
4. Instruments: Instruments and the machines must be calibrated regularly. Sometimes dimensional tolerances are very critical such as thin laminates or high-pressure seals.

5. Human error: In case of any reading by the operator himself, care should be taken to follow the exact procedure repeatedly, such as measuring the dimensions of a test sample by a Vernier caliper.
6. In the study of service life estimation, tests are to be carried out periodically after acclimatization at the same condition as the original sample.

6.3.3.2 Dynamic mechanical property

Dynamic mechanical thermal analysis has been elaborated in Section 6.1.8.2. From DMTA, we get the output parameters, namely, (1) storage modulus, loss modulus and loss factor at fixed frequencies against time and temperature scale, (2) static and dynamic strain at all frequencies and temperatures, and (3) force at all frequencies and temperatures. All other parameters such as DMA T_g (α-relaxation temperature), frequency scale master curve, hysteresis, and useful temperature limit for structural application are calculated from the above basic outputs. In evaluation and estimation of structural damping by a thermoset, these basic molecular-level properties are required to be evaluated before vibration tests. The basic viscoelastic properties are required to theoretically estimate damping performance and for the structural design of the system to arrive at construction parameters such as thickness, number of layers, extent of coverage, and performance parameters such as stresses and strains in various portions of the object etc.

6.3.3.3 Vibration tests

Vibration test is the functional test for a vibration-damping thermoset/composite. It is a standard test for a beam held as a single cantilever and is excited by a dynamic force or impact. Two types of vibration tests can be done: (1) modal test using free vibration and (2) frequency sweep using forced vibration for both small beams and large actual objects, up to a certain weight. In the modal test, the sample is subjected to a hammer strike or by a string of a modal shaker and the substrate is allowed to vibrate freely. Generally, the test sample is a small beam of about 300 mm length, 10–12 mm width, and 3–5 mm thickness.

The vibration intensity is picked up as acceleration by an accelerometer. The time-domain response is transformed to frequency domain by Fast Fourier transform (FFT) software and several resonance peaks are obtained. From the peaks, the modal frequency is identified and taking -3dB bandwidth, inverse quality factor is calculated, which is termed as system loss factor. It is mathematically defined as:

$$\eta = \frac{\Delta f_{|-3\text{dB}|}}{f_n} \quad (6.62)$$

Since the hammer strike is manual, the acceleration is normalized by the force of strike and several strikes are done to take the average and to ascertain high coherence of data ($>97\%$). In the hammer strike method, since the beam is small, the added mass of the accelerometer results in an error in resonance frequency and damping, if the sensor adds to more than 10% mass of the beam/plate. Therefore, noncontact electromagnetic sensors are used which works on the current difference. But this method cannot be used for common nonmagnetic substrates like aluminum or austenitic stainless steel or any thermoset/FRP. Second, these sensors are not linear at high intensity of displacement (at resonance) [71]. In the case of the nonmagnetic substrate, a small mild steel plate adheres to the beam for magnetic sensor. Here again, the added mass should be as low as possible and nonlinearity at resonance also is a limitation. It is best to use a noncontact laser vibrometer to detect the dynamic displacement of the beam.

In the frequency sweep method, an electrodynamic shaker is used to vibrate the substrate with a continuous frequency sweep from 10 to 5000 Hz. Major components of an electrodynamic shaker are:

1. Armature: The armature is the heart of the machine. It is a suspended coil winding, which produces a dynamic magnetic field. The force produced is directly proportional to the number of turns of the coil.
2. Suspension system: The armature is secured in a suspension system along the axis of the cylindrical shaker body to achieve maximum displacement under magnetic force.
3. Field structure: Field coils are housed in the stationary shaker body and produce a constant magnetic field. The field coil is fed with direct current to produce a static magnetic field. The field coil produces the majority of heat in the shaker. Therefore, big shakers are provided with air- or water-cooling systems.
4. Vibration controller: The controller is used to produce a test profile on the computer as programmed. A sensor on the shaker provides negative feedback as a closed loop control system.
5. Power amplifier: A power amplifier is required to amplify the output of vibration.
6. Apart from these main components, there are fixtures to fix the test object on the shaker table. Additionally, the user may make suitable fixtures for a nonstandard object.

The forced vibration is more suitable to find the difference in vibration intensity of undamped and damped structures over the total frequency band. Resonances are also identified in this method, as the peak intensities are clearly seen at resonances. This method can be used for large objects also, restricted to a weight limit of the machine. For a large machine, a mass of up to about 1000 kg can also be accommodated. The laboratory method for vibration test is known as "Oberst Beam" having defined dimensions as laid down in ASTM E756-98. The measurement result is a frequency response function (FRF) on both the bare and damped beams. An FRF is a special form of a transfer function. It is the ratio of the Fourier spectrum of a dynamic output (acceleration, velocity, or displacement response) to the Fourier spectrum of the input (the force that caused the response) or a/N.

6.3.3.4 Example of vibration sensitivity

As an example of the damping ability of thermosets compared to metals, an experiment was conducted with a steel beam held in a single cantilever mode and allowed free vibration using an instrumented hammer strike. The coherence of the three strikes was consistent >0.97. The FRF was recorded from 10 to 2000 Hz. Similar experiment was done with the steel beam coated with an epoxy thermoset made from LY556 (epoxy equivalent = 180) and Jeffamine 400 (amine equivalent = 100). Fig. 6.18 shows the frequency-dependent vibration power of steel and epoxy thermoset coated steel. Models are distinctly seen as peaks for both the cases. The dynamic modulus of the epoxy thermoset was 0.8 GPa and the average loss factor was 0.1 at ambient conditions. The frequency response of the acceleration (normalized by striking force) was used to calculate vibration power in dB with reference to 1 g acceleration ($= 9.81$ m s^{-2}). The system loss factors of bare steel and coated steel was calculated from the modal peaks as given in Eq. (6.62). The different modal frequencies and corresponding loss factors (measure of damping) are listed in Table 6.2 below. It is seen from the table that the loss factor has improved by minimum 1.5 times for thermoset coated steel compared to the bare steel at lowest and highest modes, and for 200–650 Hz range, the damping is 2.5–4 times the bare metal.

The damped structures of thermosets not only reduce vibration and radiated noise in the atmosphere, but also improves the fatigue life of a structure under constant vibration. However, the above example is not for a stand-alone thermoset beam. If structures of fully thermoset elements or their composites are evaluated for a comparison with metals, that would provide the extent of improvement by a composite structure compared to the metal. This necessitates an evaluation of all basic properties and the vibration sensitivity

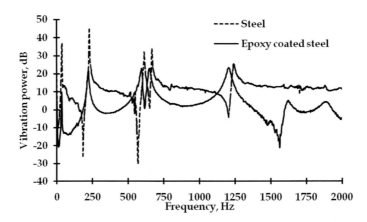

Figure 6.18
Vibration power for the free vibration of a steel beam: bare and coated with an epoxy thermoset held as a cantilever.

Table 6.2: Natural frequencies and loss factors of the bare and composite beam.

Bare steel beam		Epoxy thermoset-coated steel beam	
fn (Hz)	ηS	fn (Hz)	ηS
36	0.0414	34	0.0644
227	0.0079	220	0.0346
612	0.0082	593	0.0203
666	0.0050	652	0.0238
1241	0.0129	1204	0.0242

experiments for a plane beam by either hammer method or modal shaker and the actual object or a replica by an electrodynamic shaker.

6.4 Lifetime prediction

In order to design a polymeric item for a specific application, it is essential to theoretically predict the approximate life of the item under service or storage, since all polymers undergo degradation with time. The rate of degradation depends on the environment of use or storage and stresses on it like thermal, mechanical, diffusion of fluid, radiation, etc. The rate enhances due to combined effects of environment and stress, a typical example being IC engine mount for shock/vibration damping, which is under a static compression and dynamic load in an engine room, where it may come into contact with hydrocarbon oils and moderate temperature above ambient. In many occasions, it is not possible to study the combination of all such external stresses and environment simultaneously to estimate the service life of the item. Additionally, more than one important property of the item is likely to degrade during such complex conditions of use. With above example of a shock/vibration mount, its static stiffness and hysteresis both are important properties, which change due to the heat, compression creep, and absorption of hydrocarbon oil. Therefore, at least these three important influencing factors and the two properties are required to decide its service life. The property that deteriorates at the fastest rate to an acceptable limiting value should decide the service life.

Since thermosets and their composites are very widely used in structural application, high temperature items, electrical insulators, and heavy-duty coatings, it is necessary to study their service life by a suitable method depending on the most important property, which is required to be stringent. The design of the item generally indicates a minimum acceptable value of the property for fail-safe service. A critical example is the composite structure or coating used in air intake of a high-speed aircraft. The temperature of the inside surface can be as high as 300°C, and additionally, abrasive force is also to be considered due to the impingement of fine particles accompanying the high velocity air stream. Similarly, a wing tip of a high-speed aircraft can undergo both high abrasion by fine particles of air, a UV radiation and ozone effect in atmosphere and a thermal shock of −40°C to +150°C. A wind mill blade made of composite is subjected to UV and ozone degradation due to exposure in sunlight continuously together with creep of the blade tip due to self-weight, air thrust and rotational force. A sea water carrying composite pipe can degrade due to water diffusion and vibrational fatigue. In all these situations, no common strategy can be applied to decide the life of these items; rather different methods should be adopted for accelerated ageing study.

In general, following steps may be considered for a comprehensive lifetime prediction of a thermoset:

1. Define the most important functional property, which will be studied for arriving at life of the material under aging.
2. Define at what temperature and environment it has to perform or stored.
3. Define the physical or chemical phenomenon like water ingress, evaporation of small molecules, thermal decomposition, UV, or ozone effects, which cause degradation of the desired property.
4. Mechanism of degradation to be established from pyrolytic GC-Ms, FTIR, and NMR in case of chemical change; mechanism must not be different in the test envelop. For example, if the mechanism of degradation at 100°C and 200°C is different, then same mathematical model cannot be used for these two temperatures.
5. Study the functional property that will be required to predict the life.
6. Define the lowest acceptable numerical value of the desired property upon aging, for example, Young's modulus minimum 80% of initial value.
7. Resort to time−temperature superposition, such as Arrhenius or WLF or graphical method, where changes in mechanical, viscoelastic, or other properties are studied in accelerated mode. For thermal degradation study, kinetics of degradation with Arrhenius function is to be used.
8. Extrapolate for storage/service temperature/condition.
9. Check the theoretical estimation and experimental results.
10. Ageing temperature selection should be as nearer to service temperature as practically possible to obtain a reasonably correct result.

Presently, two useful methods of life estimation of thermosets are being discussed here:

1. Life estimation by thermal degradation using TGA kinetics
2. Study using mechanical stress (tensile) at variable strain rate.
3. There can, however, be many functional properties appropriate to the use of the thermoset, such as diffusion of gas/liquid, determination of burning rate of a solid rocket propellant based on thermoset polyurethane, electrical insulation property of crosslinked polyethylene or FRP circuit boards to name a few.

4. There are many thermal degradation studies with kinetic analysis. An example of a fire-retardant composite based on thermoset of unsaturated polyester and its IPN is by Shih and Jeng [72]. A very comprehensive review of accelerated ageing and lifetime prediction techniques for polymer materials in general and thermoset composites is discussed by Maxwell et al. [73]. The authors discussed aging mechanisms of composites, change of glass transition due to moisture absorption, degradation of fiber due to the harsh environment, stress corrosion, reduction in mechanical strength, and overall thermal degradation.

6.4.1 Thermal degradation kinetics

The process through which the polymer molecular weight reduces is called degradation of the polymer. Thermal degradation takes place when the vibrational energy of the molecule exceeds the primary bond energy among the atoms. The degradation can be by depolymerization or random chain scission. Influencing factors for degradation can be heat, light, oxygen, ozone, mechanical force, chemical reagents, ultrasonic waves, and microorganisms. Since chemical structure of a material decides all the properties, destruction of the structure must be the most important aspect of study for estimation of life of a polymer.

A simple approach is to carry out experiments on the short time heat ageing and simultaneous monitoring the functional property of aged samples. The lifetime is decided when the property reaches the lowest acceptable limit. For example, if for a crosslinked epoxy, 5% weight loss corresponds to maximum allowable 20% reduction in Young's modulus compared to its unaged state, then taking 5% weight loss as the basis of study, a detailed kinetics study can be done using thermogravimetric analysis (TGA). The kinetics of decomposition by TGA for life prediction is discussed in literature, a few are listed in Refs. [74–77]. There are some deviations from the classical decomposition kinetics due to consideration of induction period [78], which is a source of error in the extrapolation of temperature, both high and low, and due to diffusion-limited oxidation phenomenon particularly at low temperature for elastomers [79]. However, these two limitations are not so important for highly crosslinked thermosets and composites at temperatures of common application. A description of the TGA kinetics and lifetime estimation using the kinetic data is given below.

TGA measures the change in weight of the sample with change in temperature in either isothermal or nonisothermal mode under any gaseous atmosphere, for example, nitrogen, argon, oxygen, air, etc. Thermoset polymers are solid and have both dense crosslinking and strong secondary valence bonds. Therefore, they undergo decomposition rather than melting or evaporation. Under a nonoxidizing atmosphere, the decomposition can be fragmentation to smaller molecular weight species, which subsequently evaporate. Under oxidative atmosphere, the decomposition is accelerated by oxidation of all vulnerable groups to gaseous oxides. The decomposition study under inert atmosphere gives the important information on inherent bond energy of the molecule. Therefore, most TGA studies on polymers are carried out in nitrogen or argon atmosphere. TGA balance being very sensitive, the flow of such gas in the TGA sample chamber is quite low and the flow rate should be as accurate and uniform as possible. TGA studies can be done using appropriate temperature and time program to navigate a thermal run. In an isothermal study, for example, the temperature of the sample cell is brought to the desired temperature by a "Jump" command, which increases the temperature at a very high rate to stabilize at the desired temperature and the "Iso" command navigates the run at the isothermal temperature for a predetermined time period. For kinetic analysis, isothermal runs are taken for at least four temperatures. In a nonisothermal study, the temperature is increased at a steady heating rate up to a final temperature. To study at different heating rates, same experiments are repeated with fresh samples. At least four heating rates should be used. It should be noted that the sample size, shape, and condition must be uniform for all runs, to avoid errors in experimental results.

A general decomposition reaction of a solid, such as a cured thermoset, can be written as:

$$-\frac{dC}{dt} = k(T)f(C) \tag{6.63}$$

Introducing Arrhenius expression for rate constant $k(T)$, Eq. (6.63) becomes:

$$-\frac{dC}{dt} = A\exp\left[-\frac{E}{RT}\right]f(C) \tag{6.64}$$

where C is the fractional weight remaining at any instant t during the decomposition process, and hence $(1 - C)$ represents faction decomposed, which is fractional weight loss or fractional conversion α.

$$C = \frac{w}{w_0}$$

Hence, $a = (1 - C) = \dfrac{w_0 - w}{w_0}$ \hfill (6.65)

$$\text{and } -\frac{dC}{dt} = \frac{d\alpha}{dt} \tag{6.66}$$

For convenience, we shall use the form of Eq. (6.64) with fractional conversion α and hence using Eqs. (6.65) and (6.66), we get Eq. (6.67) as our working equation:

$$\frac{d\alpha}{dt} = A\exp\left[-\frac{E}{RT}\right]f(\alpha) \tag{6.67}$$

For a thermal degradation reaction, which may be complex and has multistep mechanism, it is better to study isoconversion kinetics, which does not need a model of the function $f(\alpha)$

to estimate activation energy. The term "Isoconversion" means that kinetic data are extracted from identical conversion for all heating rates. The activation energy can be obtained without any kinetic model $f(\alpha)$ in this method. Therefore, in such model-free kinetics, one gets a conversion-dependent activation energy and hence, a more realistic estimation of lifetime at different extents of conversion. The derivation of a model-free kinetic expression is quite simple and effective. This method is described here briefly.

6.4.2 Differential form of isoconversion kinetics

6.4.2.1 Friedman method

Friedman [80] used this method for the determination of activation energy. Let the extent of degradation be α at a time t and $T_1, T_2, T_3 \ldots$ and $\beta_1, \beta_2, \beta_3 \ldots$ are the temperatures and heating rates used in the TGA experiments. We can then write from Eq. (6.67) after introducing heating rate (β):

$$\left(\frac{d\alpha}{dT}\right)\left(\frac{dT}{dt}\right) = A\exp\left[-\frac{E_\alpha}{RT_\alpha}\right]f(\alpha) \tag{6.68}$$

Taking logarithm and noting that $\beta = \frac{dT}{dt}$ (constant heating rate), we get:

$$\ln\left(\beta\frac{d\alpha}{dT}\right)_{\alpha,i} = \ln[Af(\alpha)] - \frac{E_\alpha}{RT_{\alpha,i}} \tag{6.69}$$

The subscript i denotes a heating rate ($i = 1,2,3\ldots$). The activation energy is calculated from the slope of the plot $\ln(\beta d\alpha/dT)$ versus $1/T$. Since the data of $(\beta d\alpha/dT)$ and $(1/T)$ are taken at the same conversion "α," the activation energy thus obtained is conversion dependent. This method of estimating the activation energy is independent of the reaction model $f(\alpha)$, and also eliminates the term Arrhenius preexponential factor A. Following steps are to be performed to determine the activation energy using Eq. (6.69).

1. The fractional degradation (conversion) α is calculated from weight % data (refer Fig. 6.19).
2. $\beta d\alpha/dT$ is calculated from the α versus T data for all heating rates corresponding to α.
3. Subsequently, $\ln(\beta d\alpha/dT)$ and corresponding $1/T$ are used for all heating rates and hence the set of data can be used to readily obtain a kinetic plot.

The plot is essentially a straight line, with a negative slope, if $(1/T)$ is taken and a positive slope if $(-1/T)$ is taken in X axis. In any case, the activation energy is calculated from the slope as

$$E_\alpha = \text{slope} \times R \tag{6.70}$$

where R is the universal gas constant ($R = 8.314$ J mol^{-1} K^{-1}). The activation energy can be found out at any such conversion and while estimating lifetime, the activation energy

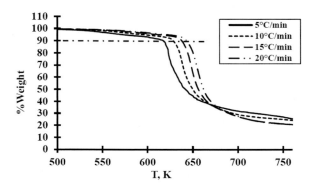

Figure 6.19
TGA thermogram of the epoxy resin-based coating.

corresponding to the selected conversion should be used, unlike other methods, where a constant activation energy is estimated for the entire degradation thermogram.

The advantage of this method is that if a different mechanism of degradation becomes prominent at a temperature, which may be within T_1 to T_i (corresponding to heating rates β_1 to β_i) at the selected conversion α, then it will reflect in the $\ln(\beta d\alpha/dT)$ data and a scatter in the kinetic plot will result. Therefore, activation energy at a series of conversions can be tested by this method to ascertain the validity of the degradation study by kinetics. The differential method by Friedman does not need any assumption; hence the activation energy obtained by this method is quite reliable for particular conversion. However, the most important difficulty in using this method is that any fluctuation of the data in consecutive readings will create large error in $d\alpha/dt$ and consequently in activation energy. Since TGA uses balance of very high sensitivity, the data fluctuations cannot be avoided as there will be flow of a gas media, whatever low in flow rate. Therefore, this method is not used in subsequent example here.

6.4.2.2 Peak rate method (Kissinger)

Considering Eq. (6.67), Kissinger [81] used the condition of the conversion rate maxima by taking derivative of the rate and equating to zero:

$$\frac{d}{dt}\left(\frac{d\alpha}{dt}\right) = 0 = \frac{d}{dt}\left[Ae^{-E/RT}f(\alpha)\right] \qquad (6.71)$$

On differentiation of the right-hand side and equating to zero, we get:

$$\frac{E\beta}{RT_p^2} = A\left[\frac{d}{dt}f(\alpha)\right]_p \exp\left(-\frac{E}{RT_P}\right) \qquad (6.72)$$

The Arrhenius preexponent A and the derivative of $f(\alpha)$ at the peak are constant due to the fixed conversion at the peak. Therefore, taking the logarithm of Eq. (6.72), we get:

$$\ln\left(\frac{\beta}{T_p^2}\right) = \ln\left(A\left[\frac{d}{dt}f(\alpha)\right]_p\right) - \ln\left(\frac{E}{R}\right) - \frac{E}{RT_p} \quad (6.73)$$

A plot of the left-hand side against $1/T$ would result in a straight line and the activation energy can be calculated from the slope of the line. Since the activation energy is calculated without $f(\alpha)$, it is also a model-free method. Generally, if the mechanism of degradation is same for the envelop of the temperature considered, then it has been observed that the peak rate corresponds to a particular conversion irrespective of heating rate (independent of β). Therefore, this method is also a differential method of iso conversion kinetics. The only limitation is that the activation energy thus calculated may not be applicable to conversions other than at peak rate. However, being a differential method, it should yield similar result as Friedman method at the identical conversion (at the peak rate).

6.4.3 Integral form of isoconversion kinetics

The model-free kinetics of isoconversion can be in the integrated form of Eq. (6.67). The integration of Eq. (6.67) gives:

$$g(\alpha) = \int_0^\alpha \frac{d\alpha}{f(\alpha)} = \frac{A}{\beta}\int_{T_0}^T \exp\left[-\frac{E_\alpha}{RT}\right]dT \quad (6.74)$$

$$\text{or, } g(\alpha) = \left(\frac{AE_\alpha}{\beta R}\right)p\left(\frac{E_\alpha}{RT}\right) = \left(\frac{AE_\alpha}{\beta R}\right)p(x) \quad (6.75)$$

where $x = \frac{E}{RT}$ and $p(x)$ are the integral, defined by:

$$p(x) = \left[\frac{e^{-E/RT}}{E/RT} + \int_{-\infty}^{-E/RT}(e^{-x}/x)dx\right] \quad (6.76)$$

The integral in Eq. (6.76) cannot be analytically solved. However, there can be either a numerical integration or an approximate value of the integral may be used. Doyle [82], Murray, and White [83] gave approximate values. A rational approximation of up to four degrees of x is also used which is given in the literature [84,85]. Various solutions to Eq. (6.75) based on approximations give minor variations in the value of activation energy. From Eq. (6.75), taking the logarithm,

$$\log[g(\alpha)] = \log\left(\frac{AE}{R}\right) - \log\beta + \log p[x] \quad (6.77)$$

A table of $x \ (= E/RT)$ and corresponding $-\log p(x)$ is given by Toop [86] for a first approximate value. With Doyle's approximation [82] for the value of $\log p(x)$:

$$\log p\left(\frac{E}{RT_i}\right) \cong -2.315 - 0.457\frac{E}{RT_i} \tag{6.78}$$

and value of $g(\alpha)$:

$$g(\alpha) = 7.03\frac{AE_\alpha}{\beta R}(10^{-3})\exp(-1.052x) \tag{6.79}$$

6.4.3.1 Flynn–Wall–Ozawa method

Flynn–Wall–Ozawa (FWO) method [87,88] is based on Doyle's approximation as shown in Eq. (6.79). Taking the logarithm of Eq. (6.79), we have:

$$\ln(\beta) = \ln\left(\frac{AE_\alpha}{R}\right) - \ln[g(\alpha)] - 4.9575 - 1.052\frac{E_\alpha}{RT} \tag{6.80}$$

At a constant conversion α, the function $g(\alpha)$, A, and E_α are constant. Therefore, Eq. (6.80) can be written as:

$$\ln(\beta) = \text{Cons.} - 1.052\frac{E_\alpha}{RT} \tag{6.81}$$

Therefore, when a particular extent of decomposition is to be taken for lifetime assessment, the TGA runs for several heating rates $\beta_1, \beta_2, \ldots$ should be taken and corresponding temperatures $T_1, T_2, \ldots$ should be noted at that constant extent of conversion (weight loss). Subsequently, a plot of $\log(\beta)$ versus $1/T$ should yield activation energy E_a from the slope of the curve. If more than one weight loss is taken, then we should ideally have parallel lines for each weight loss since the activation energy should ideally be constant and independent of the extent of degradation. FWO method is not so accurate because of the approximation. However, it gives a comparatively acceptable result.

6.4.3.2 Kissinger–Akahira–Sunose method

The Kissinger–Akahira–Sunose (KAS) method [89] is based on the Coats–Redfern approximation [75] according to which $p(x)$ is expressed as:

$$p(x) \cong \frac{e^{-x}}{x^2} \tag{6.82}$$

The KAS equation is supposed to give more accurate estimation of activation energy.

$$\ln\left(\frac{\beta}{T^2}\right) = \ln\left(\frac{AR}{E_\alpha}\right) - \ln[g(\alpha)] - \frac{E_\alpha}{RT} \tag{6.83}$$

$$\ln\left(\frac{\beta}{T^2}\right) = \text{Cons.} - \frac{E_\alpha}{RT} \qquad (6.84)$$

Therefore, a plot of $\ln(\beta/T^2)$ versus $1/T$ should give a straight line with a slope of $-E_a/R$ and subsequently, the activation energy E_a can be calculated.

6.4.3.3 Starink method

Another approximation is given by Starink [90] for $p(x)$, which is supposed to be more accurate estimation of the integral:

$$p(x) \cong \frac{\exp(-1.0008x - 0.312)}{x^{1.92}} \qquad (6.85)$$

and corresponding expression for estimate of activation energy:

$$\ln\left(\frac{\beta}{T^{1.92}}\right) = \text{Cons.} - 1.0008\frac{E_\alpha}{RT} \qquad (6.86)$$

6.4.3.4 MacCallum method for isothermal study

MacCallum [91] method is based on multiple isothermal degradation studies and relates the time to attain a fixed degradation to the inverse of isothermal temperature by the following relationship:

$$\ln(t) = a + \frac{E_\alpha}{RT} \qquad (6.87)$$

where T is the temperature of the isothermal runs, E_α is the activation energy at the conversion α, and the constant a is related to fractional degradation or conversion α and Arrhenius preexponent factor (A) as:

$$a = \ln[F(1-\alpha)] - \ln A \qquad (6.88)$$

where F is a function of conversion. Since the data are taken at a constant conversion from all the isothermal runs, the numerical value of the function $F(1-\alpha)$ is constant. Therefore, a plot of $\ln(t)$ at same conversion versus $1/T$ of a thermogram should be a straight line with a slope E_α/R, and thus the activation energy can be calculated. This method is also based on the integral form of Eq. (6.67) while the study is done at various isothermal temperatures. The time to attain a fixed degree of degradation is taken for subsequent calculations. Since there is no dynamic run, the values of time to attain the degradation cannot be taken from the nonisothermal run, as it is seen that heating rate is a major factor for the time to attain a fixed degradation. In the present case, we only consider nonisothermal kinetics for lifetime prediction.

6.4.4 Advanced isoconversion methods

The activation energy calculated by all above methods for nonisothermal kinetics has a drawback that the integration over 0 to α at a wide range of temperature, resulting in accumulation of errors due to error in temperature and heat flow, thereby resulting in so-called "averaging" of the values of E_α and considerable error. Vyazovkin [92] proposed an advanced isoconversion kinetic equation where the integration is done for a small range of α and temperature:

$$g(\alpha, \alpha - \Delta\alpha) = \int_{\alpha-\Delta\alpha}^{\alpha} \frac{d\alpha}{f(\alpha)} = \frac{A_\alpha}{\beta} \int_{T_{\alpha-\Delta\alpha}}^{T_\alpha} e^{-E/RT} dT \qquad (6.89)$$

and nonlinear process where minimization of the following function leads to a more accurate value of conversion-dependent activation energy:

$$\Phi(E_\alpha) = \sum_{i=1}^{n} \sum_{j \neq i}^{n} \frac{\beta_j \int_{T_{\alpha-\Delta\alpha,i}}^{T_{\alpha,i}} e^{-E_\alpha/RT} dT}{\beta_i \int_{T_{\alpha-\Delta\alpha,j}}^{T_{\alpha,j}} e^{-E_\alpha/RT} dT} \qquad (6.90)$$

However, the minimization of the above function is a time-consuming process of iteration.

A more convenient linear method, still using small range of temperature and the integral as in Eq. (6.89), is proposed by Cai and Chen [93], which reduces computation time significantly and also claims to be more accurate than Vyazovkin method. The final equation derived by Cai and Chen is:

$$\ln\left\{\frac{\beta_i}{T_{\alpha,i}^2\left[h(x_{\alpha,i}) - \frac{x_\alpha^2 e^{x_{\alpha,i}}}{x_{\alpha-\Delta\alpha}^2 e^{x_{\alpha-\alpha,i}}} h(x_{\alpha-\Delta\alpha,i})\right]}\right\} = \ln\left[\frac{A_{\alpha-\Delta\alpha/2} R}{E_{\alpha-\Delta\alpha/2} g(\alpha, \alpha - \Delta\alpha)}\right] - \frac{E_{\alpha-\Delta\alpha/2}}{RT_{\alpha,i}} \qquad (6.91)$$

where $h(x)$ is defined by the following integral:

$$h(x) = x^2 e^x \int_x^\infty \frac{e^{-x}}{x^2} dx \qquad (6.92)$$

An approximate value of the $h(x)$ is given by a 4th-order polynomial rational as:

$$h(x) \cong \frac{x^4 + 18x^3 + 86x^2 + 96x}{x^4 + 20x^3 + 120x^2 + 240x + 120} \qquad (6.93)$$

The Eq. (6.91) is a modification of that by Kissinger–Akahira–Sunose, where the term T^2 is multiplied by a factor, which is again dependent on E_α. Therefore, the iterative process is used till the $E_{\alpha-\Delta\alpha/2}$ value of consecutive iterations are equal or deviate by not more than 0.001 kJ mol^{-1}. The numerical value of $\Delta\alpha$ is important since the authors have shown that a step beyond $\Delta\alpha = 0.2$ increases the relative error beyond 1% at the high extent of conversion ($\alpha > 0.7$). To obtain a constant, consistent conversion-dependent activation energy, $E_{\alpha-\Delta\alpha/2}$, the following steps are to be performed:

Step 1. Initial value of $E_{\alpha-\Delta\alpha/2}$ and T_α, $T_{\alpha-\Delta\alpha}$ are used to calculate $h(x_{\alpha i})$, $h(x_{\alpha-\delta\alpha,i})$, x_α, and $x_{\alpha-\Delta\alpha}$, where

$$x_\alpha = \frac{E_{\alpha-\Delta\alpha/2}}{RT_\alpha} \tag{6.94}$$

$$\text{and } x_{\alpha-\Delta\alpha} = \frac{E_{\alpha-\Delta\alpha/2}}{RT_{\alpha-\Delta\alpha}} \tag{6.95}$$

Step 2. Calculate new $E_{\alpha-\Delta\alpha/2}$ using the slope of the straight line according to Eq. (6.91).

Step 3. Repeat the calculations using the new value of $E_{\alpha-\Delta\alpha/2}$ obtained in Step 2. This iteration will continue till the two last consecutive values of $E_{\alpha-\Delta\alpha/2}$ are equal or deviate not more than 0.001 kJ mol^{-1}. Initial values of activation energy ($E_{\alpha-\Delta\alpha/2}$) can be taken from any method such as Friedman, Flynn−Wall−Ozawa, Kissinger−Akahira−Sunose, or Starink because the final consistent value would be the same. Like the Friedman method, these advanced isokinetics by Vyazovkin or Cai and Chen also have the same limitation of large error in case of data fluctuation or noise as it is common in TGA experiments. Therefore, these methods are not used in the subsequent example in this chapter.

6.4.5 Estimation of a lifetime from TGA kinetics

For ageing of a polymer at a particular temperature, for example the service temperature of a thermoset to be used in a high-temperature coating, $g(\alpha)$ will be given by:

$$g(\alpha) = kt_i \tag{6.96}$$

where the specific rate constant k is given by:

$$k = A_\alpha \exp\left(-\frac{E_\alpha}{RT_i}\right) \tag{6.97}$$

Therefore, at identical conditions, for the same $g(\alpha)$, α will be the same for both isothermal aging and constant heating rate dynamic run in TGA, and at the isothermal time t_i (in the aging study) at the TGA temperature T for the same weight loss, it can be said that:

$$kt_i = \frac{A_\alpha E_\alpha}{\beta R} p(x_i) \tag{6.98}$$

where $x_i = E_\alpha/RT$.

Substituting Eq. (6.97) in Eq. (6.98), we get:

$$t_i = \frac{E_\alpha}{\beta R} e^{E_\alpha/RT_i} p(x_i) \quad (6.99)$$

where T_i is the aging temperature (absolute scale), for example, service temperature for the polymer. Since activation energy E_α is known from any of the methods described above, x_i can be calculated and corresponding log $p(x_i)$ can be found out from the approximations used by different researchers and also from the table given by Toop [86]. A comparison of all approximations of $p(x)$ was done by Starink [90] where it is seen that all the approximations are derived from the series expansion of $p(x)$ by different researchers. However, the following "rational" approximation as a ratio of two polynomials (fourth-degree) by Senum and Yang [94] is the best as shown by Starink. The approximation by Starink (Eq. 6.82) gives the same result for $x > 20$ and up to 100.

$$p(x) \cong \frac{\exp(-x)}{x^2} \frac{x^4 + 18x^3 + 86x^2 + 96x}{x^4 + 20x^3 + 120x^2 + 240x + 120} \quad (6.100)$$

Hence t_i gives the life of the polymer corresponding to a minimum acceptable value of a functional property at any service temperature T. A more convenient form of the Eq. (6.99) is given by Toop [86]:

$$\log t_i = \frac{E_\alpha}{2.303 RT_i} + \log\left[\frac{E_\alpha}{\beta R}\right] + \log[p(x_i)] \quad (6.101)$$

where t_i is isothermal lifetime for conversion a, T_i is isothermal temperature (Kelvin) at which lifetime is to be estimated, E_α is activation energy at the conversion α, β_i is heating rate selected for lifetime calculation, generally the slowest, x_i is E_α/RT, T is TGA temperature corresponding to the conversion (α) at the heating rate β_i. T is taken from the thermogram of the slowest heating rate. $p(x_i)$ is the value of the integral of $\exp(-x_i)\,dT$ corresponding to x_i.

Log$p(x_i)$ is calculated using Eq. (6.100) above. Alternately, if the value of log$p(x_i)$ is taken from the table by Toop in Ref. [86] corresponding to $x_i = E_\alpha/RT$, the numerical value from the table should be used with a minus sign in Eq. (6.101). $E_\alpha/(2.303 RT_i)$ is calculated taking the isothermal aging temperature T_i in Eq. (6.101).

Rearranging Eq. (6.101), the allowable service temperature for a set service life (t_i) is given as:

$$T_i = \frac{E/2.303R}{\log t_i - \log\left[\frac{E}{\beta R}\right] - \log[p(x_i)]} \quad (6.102)$$

6.4.6 Example of lifetime estimation from TGA kinetics

An epoxy-based thermoset is to be used as an FRP automobile part as a replacement for aluminum. The FRP is not subjected to any mechanical stress or any harsh environment other than heat during use. The material is to be qualified only on the basis of life due to possible thermal degradation in atmospheric air. Maximum 10% degradation is allowed. The manufacturer wants to ascertain the service life at various service temperatures from 150°C to 250°C.

6.4.6.1 Solution

A typical TGA kinetics and life estimation is done to ascertain the service life at 150°C and up to 250°C. Fig. 6.19 shows the TGA Thermogram of the epoxy thermoset carried out in an air atmosphere using four heating rates 5, 10, 15, and 20 °C min^{-1}. The dashed horizontal line is drawn for constant ordinate at 10% weight loss and vertical dashed lines are drawn to indicate four temperatures at which 10% weight loss takes place for four selected heating rates. With increase in heating rate, the 10% weight loss progressively occurs at higher temperature. It is assumed that the mechanism of degradation of the epoxy resin is same from 10% to at least 20%. The lowest temperature of 10% weight loss is 617.88°C, corresponding to the thermogram with 5 °C min^{-1} heating rate.

The conversion (fractional weight loss) is plotted against absolute temperature in Fig. 6.20. From Figs. 6.19 and 6.20, the following kinetic data listed in Table 6.3 are obtained for estimation of activation energy by different methods. Conversion (α) is in fraction and heating rate is expressed in K s^{-1}. Subsequently, the activation energies are calculated by the methods of (1) FWO, (2) KAS, and (3) Starink, following Eqs. (6.81), (6.84), and (6.86), respectively, at a conversion of 0.10 taking the calculated data for each method as listed in

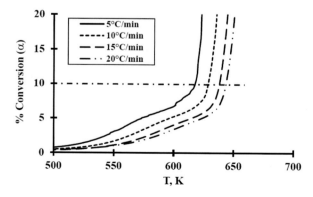

Figure 6.20
Conversion (weight loss) of epoxy thermoset at different heating rate in Example 6.4.

Table 6.3: Parameters for epoxy thermoset at 10% degradation.

β (K s^{-1})	T (K)	$1/T$ (K^{-1})	$\ln(\beta)$	$\ln(\beta/T^2)$	$\ln(\beta/T_{1.92})$
0.083	617.9	0.00168	−2.4849	−15.337	−14.823
0.167	628.8	0.00159	−1.7918	−14.679	−14.164
0.250	638.2	0.00157	−1.3863	−14.304	−13.787
0.333	643.6	0.00155	−1.0986	−14.033	−13.515

Table 6.4: Activation energy of epoxy thermoset determined at constant degradation.

Method	Degradation or conversion (α)	Activation energy (E_α) (kJ mol^{-1})
Flynn−Wall−Ozawa (FWO)	0.10	166.7
Kissinger−Akahira−Sunose (KAS)	0.10	164.9
Starink	0.10	165.2

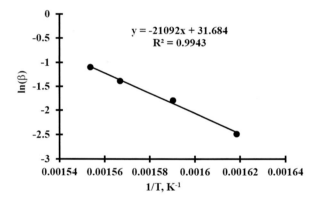

Figure 6.21
Kinetic plot according to Eq. (6.80) by Flynn−Wall−Ozawa method.

Table 6.4. One representative kinetic plot according to Eq. (6.81) by FWO method is shown in Fig. 6.21. The values of E_α calculated from the slopes of the respective plots are listed in Table 6.4. The value of R (universal gas constant) is taken as 8.314 J mol^{-1} K^{-1}.

It is seen from Table 6.4 that the activation energies by different methods are not much different, since the methods employ integrated kinetic equations with almost identical temperature integral approximation. There will be a slight difference in the FWO method, as the approximation is somewhat not so accurate. Further, KAS and Starink showed almost the same value, differing only by 0.18% and FWO method is differing by 0.92% compared to that for the Starink method. Since all these activation energies obtained so far correspond to 10% decomposition, the values are used subsequently to determine lifetime of the epoxy thermoset at temperatures from 150°C to 250°C. Eq. (6.101) was used to calculate the lifetime step by step as described in Section 6.4.5.

Table 6.5: Lifetime estimation of the epoxy thermoset in Example 1.5 by isoconversion kinetics.

Method	T_i (°C)	$E_\alpha/2.303RT_i$	$x_i = E_\alpha/RT$	Log $p(x_i)$	Log($E\alpha/\beta R$)	t_i (s)
FWO	150	20.5810	32.4487	−17.1400	5.381283	6.64E + 08
	175	19.4325	32.4487	−17.1400	5.381283	4.72E + 07
	200	18.4054	32.4487	−17.1400	5.381283	4.43E + 06
	225	17.4815	32.4487	−17.1400	5.381283	5.28E + 05
	250	16.6458	32.4487	−17.1400	5.381283	7.71E + 04
KAS	150	20.3568	32.0952	−16.9772	5.376526	5.704E + 08
	175	19.2209	32.0952	−16.9772	5.376526	4.170E + 07
	200	18.2050	32.0952	−16.9772	5.376526	4.020E + 06
	225	17.2910	32.0952	−16.9772	5.376526	4.902E + 05
	250	16.4645	32.0952	−16.9772	5.376526	7.309E + 04
Starink	150	20.3929	32.1520	−17.0034	5.377294	5.85E + 08
	175	19.2549	32.1520	−17.0034	5.377294	4.25E + 07
	200	18.2372	32.1520	−17.0034	5.377294	4.08E + 06
	225	17.3216	32.1520	−17.0034	5.377294	4.96E + 05
	250	16.4937	32.1520	−17.0034	5.377294	7.37E + 04

In order to calculate the lifetime, the value of log $p(x_i)$ is first calculated using the table by Toop [86] and using Eq. (6.99) by Senum and Yang [94]. The values are almost the same, as −21.8565 and −21.8567, respectively, for $E/RT = 42.77$. Therefore, the lifetime was calculated taking the log $p(x_i)$ values using Eq. (6.99). Table 6.5 shows the calculated parameters of Eq. (6.101) and estimated lifetime at various isothermal aging temperatures 150°C−250°C at an interval of 25°C for all cases. The lifetimes estimated by the various methods are listed in Table 6.5. The lifetime versus service temperature (isothermal aging temperature) for FWO and Starink are only shown in Fig. 6.22, since the values obtained by both the methods of KAS and Starink are almost coinciding as a result of their almost identical activation energy. However, the line for FWO is somewhat different. The logarithm of time linearly varies with temperature, which is typical of $t-T$ superposition principle of any thermal process.

It is seen from Fig. 6.22 that if activation energies are different, the difference in predicted life is higher at lower aging temperature, that is, when the aging temperature is far below the TGA temperature of degradation. Hence, at 150°C, the lifetime estimated by FWO (E_α = 166.69 kJ mol^{-1}) is 6.64 × 10^8 s (29 years), which is higher compared to the same for Starink (E_α = 165 kJ mol^{-1}) at about 5.85 × 10^8 s (18 years), while at 250°C, the both the lifetimes are much closer, about 7.71 × 10^4 s (29 h) and 7.37 × 10^4 s (20 h), respectively. From TGA thermogram of 5 °C min^{-1}, the degradation temperature at 10% conversion (weight loss) is 617.8K.

For reliable results by interpolation to low temperature, the activation energy must be accurately determined. There will be large variation in the calculated lifetime determined by different integral methods, even if there is 1% variation in activation energy. Generally,

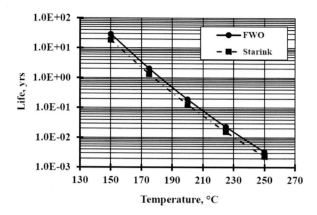

Figure 6.22
Estimated life of the epoxy thermoset of Example 6.4 at various ageing temperatures by different methods for 10% degradation.

experimental data from TGA are the source of error, as there can be difference in the sample and reference temperature, heat loss, stability of the balance, accuracy of heating rate, variation in the gas flow in the sample chamber, base line stability at the full range of temperature even beyond the experimental range, etc. For thermosets, one common error can be incomplete curing. This can be ensured by repeated DSC runs and design a proper curing cycle according to DSC plots. The other type of error can be data sampling. It is desired that very small and constant interval of time should be used for data sampling, since the first-derivative $d\alpha/dt$ is calculated by numerical method. In this case, the data sampling interval was 0.50 s, corresponding to the temperature interval of 0.165°C at the highest temperature ramp (20 °C min^{-1}), about 1200 data for the temperature span of 200°C (from 450°C to 650°C), which can be considered sufficient for fairly accurate analysis.

6.5 Conclusion

The thermal degradation kinetics by dynamic TGA at various heating rate is a standard method, with a caution of accuracy of the instrument and data. Integral methods are preferred over differential methods because of possible data fluctuation. Generally, the lowest lifetime predicted among all methods is taken as safe approach, provided the methods result in quite close values of activation energy (<1% variation). One most important point here is that the TGA should be done in air environment, since the practical thermal aging of any item takes place in atmosphere and not under an inert gas envelop, as normally done in all TGA experiments. The present example was thus done in air atmosphere. Moreover, TGA study of a thermoset is to be combined with the study of its functional property, such as toughness or coating integrity, with ageing time at various isothermal aging temperatures.

6.5.1 Ageing study under stress

The degradation of polymers is a very complicated process. Further complicacy arises when a polymer is subjected to mechanical loading during the service. The tensile load can pull the radicals apart, produced during chain scission, to hinder recombination and hence facilitate initiation of new reaction chains, while a compressive load may have a packing effect, bringing the radicals closer and facilitate recombination and discourage new chain reactions. Stress may have an effect on the diffusion of reactants and hence the reaction that depends on the diffusion process. Finally, stress can change the activation energy of durability. It is expressed by a simple modification of the Arrhenius equation, which includes a term related to stress. This modified Arrhenius expression is known as Euring–Zhurkov expression [95]:

$$\tau(\sigma(t)) = \tau_0 \exp\left(\frac{U - \gamma\sigma}{RT}\right) \quad (6.103)$$

This equation is widely used for the prediction of lifetime of polymers under isothermal conditions (T) and under a constant stress σ. The coefficient γ of the stress in the exponential term has a dimension of volume and is often called *activation volume*. The Arrhenius preexponent τ_0 is the period of submolecular oscillation at $T \to \infty$, typically about 10^{-12} s for α-relaxation of segments in polymers. In absence of any stress, the durability $\tau(\sigma(t))$ is related to the temperature by the original Arrhenius expression. Durability is the result of thermofluctuations, which degrades a material even without applying stress. The stress only decreases the potential barrier U in the stress direction and therefore facilitates the thermofluctuations.

6.5.1.1 Bailey criteria

When the durability (or lifetime) is to be determined for a polymer under time-dependent simple loading conditions and under isothermal aging, it is possible to use the Bailey criterion of durability [96,97]:

$$\int_0^{t^*} \frac{dt}{\tau(\sigma(t))} = 1 \quad (6.104)$$

Here $\tau(\sigma(t))$ is the durability function determined experimentally under constant stress σ and t^* is the durability under a stress $\sigma(t)$. The above equation is subject to the following conditions:

1. The damage process (to failure of the polymer) is irreversible.
2. The criteria are independent of the various damage accumulations.
3. There are no memory effects in every damage accumulation, which means that Boltzmann superposition of stress is not significant during the process of degradation.

The above criteria can be applied to engineering plastics, thermoset resins, and highly extended polymeric fibers. These hard materials have such common features as incremental deformations and almost absence of any memory effect. With the above three conditions, Eq. (6.104) might be applied to various different kinetic processes, whose behaviors are possible to describe using discrete processes. Indeed, Eq. (6.104) has been used by Leonov et al. [98] to describe the nonisothermal cure kinetics for rubbers and thermosets.

Bailey criteria are applicable to Eq. (6.103) not only when $\sigma = \sigma(t)$, but also when temperature variations are applied with or without the time-dependent stress [99,100]. However, in this section, a brief discussion is being done to observe the effect of constant stress on the aging of the thermoset polymer due to thermofluctuation. In the discussion, the time to break a thermoset sample under constant stress is determined and the life is also decided by the time taken by the thermoset to break at a particular service temperature.

In order to apply Baily Criteria on breaking of a sample, Eq. (6.103) is considered here for the case of any thermoset, assuming that the stresses in the polymer are not superimposed during the experiment according to Boltzmann Superposition Principle, since the relaxation time of the thermoset polymer is almost infinity at ambient temperature and the glass transition temperature is very high compared to ambient. Eq. (6.103) is rewritten as:

$$\tau = \tau_0 \exp\left[\frac{U}{RT}\right] \exp\left[-\frac{\gamma\sigma}{RT}\right] \tag{6.105}$$

$$\text{or, } \tau = \tau_0 \exp\left[\frac{U}{RT}\right] \exp[-a\sigma], \quad \text{where } a = \frac{\gamma}{RT} \tag{6.106}$$

$$\text{or, } \tau = B\exp[-a\sigma] \tag{6.107}$$

$$\text{where } B = \tau_0 \exp\left[\frac{U}{RT}\right] \tag{6.108}$$

Now, we can assume, with slight approximation,

$$e^{-a\sigma} = b\sigma^{-\kappa} \tag{6.109}$$

taking logarithm (ln), we get,

$$a\,\sigma = \ln(b') + \kappa \ln(\sigma) \tag{6.110}$$

The above is a linear equation relating σ and $\ln(\sigma)$ with a slope of κ/a. Since the temperature is constant, the term a is also a constant. Now, the assumption for Eq. (6.109) should be valid, if

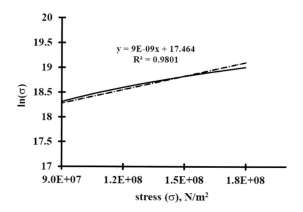

Figure 6.23
Plot of UTS versus ln(UTS) for a thermoset having Young's modulus 2–4 GPa.

the slope κ/a is constant for the ultimate tensile stresses in minimum 3 decades of stretching rate (mm min^{-1}). For thermosets of high T_g, the difference in ultimate tensile strength (UTS) is not more than 200% for even 3 decades of stretching rate. However, we check for a thermoset having Young's modulus of 2–6 GPa and maximum elongation at break of 3.3%. Let us say, that the UTS varies from 60 to 150 MPa due to increase in stretching rate. The $\ln(\sigma)$ is plotted against σ in Fig. 6.23 below. It is seen that the relationship is linear ($r^2 = 0.9801$) in this range. Hence, the assumption is approximately correct.

Therefore, we rewrite Eq. (6.108), after introducing Eq. (6.109) in it,

$$\tau(\sigma(t)) = (Bb)\,\sigma^{-\kappa} \qquad (6.111)$$

$$\text{or,}\ \tau(\sigma(t)) = C\sigma^{-\kappa} \qquad (6.112)$$

where $C = Bb$, a constant at a constant temperature. In addition, the quantity γ in Eq. (6.104) is also constant since there is no significant volume change for the segments of the polymer at a constant temperature.

Now, introducing Eq. (6.112) in Baily Criteria described by Eq. (6.104), we get,

$$\int_0^{t^*} \frac{dt}{C\sigma^{-\kappa}} = 1 \qquad (6.113)$$

Let t_r be the time to failure at a stress σ_r. By denoting $\sigma_r = w_r t_r$, where w_r is the stretching rate (MPa min^{-1}) for σ_r. On substituting it in Eq. (6.113), we obtain, after integration:

$$\int_0^{\tau} \frac{dt}{C\sigma^{-\kappa}} = \int_0^{t_r} \frac{dt}{C(wt)^{-\kappa}} = \frac{(w_r t_r)^{\kappa+1}}{Cw_r(\kappa+1)} = 1 \qquad (6.114)$$

Rearranging and taking the logarithm, we get the workable expression as

$$\log \sigma_r = \frac{1}{\kappa + 1}(\log C + \log(\kappa + 1)) + \frac{1}{\kappa + 1}\log w_r \qquad (6.115)$$

Eq. (6.115) represents the linear relationship between the $\log(\sigma_r)$ and $\log(w_r)$, whose values can be obtained experimentally.

6.5.1.2 Lifetime estimation

In order to determine the lifetime, a series of experiments are done as below:

1. Determination of ultimate tensile strength for a thermoset strip of ASTM dimension, at multiple rates of elongation (mm min^{-1}).
2. Time t_r to break is recorded for each loading rate experiment. If there is a yield point in the stress—strain curve, the % elongation, time to reach yield point, and the yield stress should be taken since, in practice, the useable limit of an engineering thermoset is up to the yield point.
3. The ultimate tensile stress or yield stress value divided by t_r gives w_r (stretching rate) for Eq. (6.115).
4. These experiments can also be carried out after systematic aging.

Experimental data is plotted according to Eq. (6.115) and the slope and intercept, C and κ are calculated. However, to find out a lifetime, Eq. (6.115) as such is used with the numerical values of the slope and the intercept. For example, if the thermoset, in any application, is used under high loading, then that stress (σ_r) is used to find out w_r from the equation. Subsequently, using the relation $\sigma_r = w_r t_r$, the time to fail (t_r) is calculated.

References

[1] F.W. Billmeyer, Textbook of Polymer Science, third ed., Wiley, New York, 1984.
[2] D.W. Van Krevelen, Properties of Polymers, Elsevier, Amsterdam, 1976.
[3] D. Ratna, Handbook of Thermoset Resins, Smithers Rapra Technol, United Kingdom, 2009, pp. 1–427.
[4] C. Hansen, Hansen Solubility Parameters: A User's Handbook, second ed., CRC Press, Boca Raton, FL, 2007.
[5] D. Ratna, M. Patri, B.C. Chakrabarty, P.C. Deb, J. Appl. Polym. Sci. 65 (1997) 901–907.
[6] D. Ratna, A.K. Banthia, N.C. Maity, Ind. J. Chem. Tech. 2 (1995) 253.
[7] G. Casiraghi, G. Casnati, M. Cornia, G. Sartori, F. Bigi, Makromol. Chem. 182 (1981) 2973.
[8] D. Ratna, G.P. Simon, J. Appl. Polym. Sci. 117 (2010) 557–564.
[9] S.K. Sharma, P.K. Pujari, Prog. Polym. Sci. 75 (2017) 31–47.
[10] P.J. Flory, Principle of Polymer Chemistry, Cornell University Press, Ithaka, 1953.
[11] A.V. Tobolsky, D.W. Carlson, N. Indictor, J. Polym. Sci. 54 (1961) 175.
[12] O. Gryshchuk, J. Karger-Kocsis, R. Thomann, Z. Konya, I. Kiricsi, Compos. A 37 (9) (2006) 1251.
[13] M. Maiti, A.K. Bhowmick, Polymer 47 (2006) 6156.
[14] E.A. Stefanescu, A. Dundigalla, V. Ferreiro, E. Loizou, L. Porcar, I. Negulescu, et al., Phys. Chem. Chem. Phys. 8 (2006) 1739.

[15] E. Loizou, P. Butler, L. Porcar, G. Schmidt, Macromolecules 38 (2005) 2047.
[16] D. Ratna, O. Becker, R. Krishnamurty, G.P. Simon, R. Varley, Polymer 44 (24) (2003) 7449.
[17] G. Binnig, C.F. Quate, C.H. Gerber, Phys. Rev. Lett. 56 (1986) 930.
[18] C.J. Buchko, K.M. Kozlo, D.C. Martin, Biomaterials 22 (2001) 1289.
[19] T. Murayama, Dynamic Mechanical Analysis of Polymeric Materials, Materials Science Monograph 1, Elsevier Science, New York, 1978.
[20] J.D. Ferry, Viscoelastic Properties of Polymers, John Wiley, New York, 1980.
[21] B.C. Chakraborty, D. Ratna, ViscoelasticityChapter 3 Polymers for Vibration Damping Applications, Elsevier, United Kingdom, 2020pp. 122–125.
[22] R. Bruce Prime, Thermosets, in: second ed., E.A. Turi (Ed.), vol. 2, Academic Press, New York, 2017.
[23] E.A. Turi (Ed.), Thermal Characterization of Polymeric Materials, Academic Press, 1997.
[24] R.F. Speyer, Thermal Analysis of Materials, Marcel Dekker, New York, 1992.
[25] H. Saleem, A. Edathil, T. Ncube, J. Pokhrel, S. Khoori, A. Abraham, et al., Macromol. Mater. Eng. 301 (2016) 231–259.
[26] I. Havlicek, K. Dusek, in: B. Sedlecek, J. Kahobek (Eds.), Crosslinked Epoxies, Walter de Gruyter, Berlin, 1987, p. 359.
[27] P.P. Klemchuk, Antioxidants, Ulman's Encyclopedia of Industrial Chemistry, vol. A3, Wiley-VCH Verlag Gmbh & Co KgaA, Wienheim, 2002.
[28] I.K. Varma, in: P. Ramakrishna (Ed.), Advances in Composite Materials, IBH, Oxford, 1991, pp. 85–93.
[29] S. Robitaille, Cynate ester resin, ASTM Handbook, Composites, ASM, 2002, pp. 126–131.
[30] H. Zweifel, Macromol. Symp. 115 (1997) 181.
[31] P.P. Nicholas, A.M. Luxader, L.A. Brooks, P.A. Hammes, Antioxidants and antiozonants, in: M. Grayson (Ed.), Kirk-Othmer Encyclopedia of Chemical Technology, Wiley, New York, 1978, p. 128.
[32] S. Al-Malaika, Chapter 2 in: G. Scott (Ed.), Atmospheric Oxidation and Antioxidants, vol. 1, Elsevier Applied Science Publisher, London, 1993.
[33] J. Pospisil, Ch. 1 in: G. Scott (Ed.), Developments in Polymer Stabilization, vol. 1, Applied Science Publishers, London, 1979.
[34] R. Bagheri, K.B. Chakraborty, G. Scott, Polym. Deg. Stab. 4 (1) (1982) 1.
[35] N.K. Naik, S.V. Borade, H. Arya, M. Sailendra, S.V. Prabhu, J. Reinf. Past. Compos. 21 (2002) 1347.
[36] M.L. Karasek, L.H. Strait, M.F. Amateau, J.P. Runt, J. Compos. Technol. Res. 17 (1995) 11.
[37] D. Hui, Conference Proceedings ICCM/9 on Composite Behaviour, 1993, vol. 5, pp. 321–323.
[38] D. Ratna, Rapra Rev. Rep. 16 (5) (2005) 1.
[39] R.P. Chartoff, in: second ed., E.A. Turi (Ed.), Thermoplastic Polymers, vol. 1, Academic Press, New York, 1997.
[40] T.M. Tritt (Ed.), Thermal Conductivity Theory, Properties and Applications, Plenum Publisher, New York, 2004.
[41] D.E. Thermal, J. Polym. Sci. 50 (1961) 441–450.
[42] A. Majumder, Microscale transport phenomena, in: W.M. Rohsenow, J.R. Hartnet (Eds.), Handbook of Heat Transfer, 3rd ed., MacGraw Hill, New York, 1998.
[43] D. Hansen, C. Ho, J. Polym. Sci. Part A 3 (2) (1965) 659–670.
[44] H. Zhidong, F. Alberto, Prog. Polym. Sci. 36 (2011) 914–944.
[45] J.A. King, K.W. Tucker, B.D. Vogt, E.H. Weber, C. Quan, Polym. Compos. 20 (1999) 643–654.
[46] H.O. Pierson, Handbook of Carbon, Graphite, Diamond and Fullerenes: Properties, Processing and Applications, Noyes Publications, New Jersey, 1993.
[47] G. Wypych, Handbook of Fillers: Physical Properties of Fillers and Filled Materials, Chem Tech Publishing, Tpronto, 2000.
[48] J.E. Fischer, Carbon nanotubes: structure and properties, in: Y. Gogotsi (Ed.), Carbon Nanomaterials, Taylor and Francis Group, New York, 2006, pp. 51–58.
[49] P.J.F. Harris, Int. Mater. Rev. 49 (2004) 31–43.
[50] M. Monoruzzanan M., K.I. Winey, Macromolecules 39 (2006) 5194–5205.

[51] R. Nayak, T. Dora, A. Satapathy, Comp. Mater. Sci. 48 (2010) 576–581.
[52] J.C. Maxwell, Reprinted A Satapathy on Electricity and Magnetism, third ed., Dover Publications Inc, New York, 1954 (Chapter 9).
[53] C.J.F. Bottcher, Theoryof Electric Polarization, Elsevier, Houston, TX, 1952, pp. 415–420.
[54] R. Landauer, J. Appl. Phys. 23 (1952) 779–784.
[55] E. James, Mark (Eds.), Polymer Data Handbook, University Press, 1999.
[56] I. Rhoney, R. Pethrick, J. Mater. Des. Appl. 226 (2012) 76–87.
[57] G. Schwarz, Cryogenics 28 (1988) 248.
[58] J. Innes, A. Innes, Rapra Rev. Rep. (No. 168) 14 (12) (2003).
[59] D. Ratna, Rapra Rev. Rep. (No. 168) 14 (12) (2003).
[60] D. Ratna, H. Dutta, A.K. Banthia, J. Adhes. Sci. Technol. 20 (2006) 555–561.
[61] B.C. Chakraborty, D. Ratna, Impact, shock and vibration characteristics of epoxy based composites for structural adhesive joints. Chapter 9 in: K.L. Mittal, S.K. Panigrahi (Eds.), Structural Adhesive Joint—Design, Analysis and Testing, John Wiley & Scrivener, Beverly, MA, 2020, pp. 265–287.
[62] A.A. Griffith, Philos. Trans. Ser. A 221 (1920) 163–198.
[63] D. Roylance, Introduction to Fracture Mechanics, Massachusetts Institute of Technology, Cambridge, MA, 2001, p. 02139.
[64] J. Maa, Q. Qi, J. Bayley, Xu-S. Du, M.-S. Mo, L.-Q. Zhang, Development of SENB toughness measurement for thermoset resins, Polym. Test. 26 (2007) 445–450.
[65] S. Zavareh, G. Vahdat, J. Reinf. Plast. Compos. 31 (2012) 247–258.
[66] K.P. Unnikrishnan, E.T. Thachil, Toughening of epoxy resins, Designed Monomers Polym. 9 (2) (2006) 129–152.
[67] D. Ratna, A.K. Banthia, P.C. Deb, J. Appl. Polym. Sci. 80 (2001) 1792–1801.
[68] M. Murali, D. Ratna, A.B. Samui, B.C. Chakraborty, Synth. J. Appl. Polym. Sci. 103 (2007) 1723–1730.
[69] R.D. Brooker, A.J. Kinloch, A.C. Taylor, The morphology and fracture properties of thermoplastic toughened epoxy polymers, J. Adhes. 86 (9) (2010) 726–741.
[70] B.C. Chakraborty, D. Ratna, Polymers for Vibration Damping Applications, Elsevier, United Kingdom, 2020.
[71] J.L. Wojtowicki, L. Jaouen, R. Panneton, Rev. Sci. Instrum. (AIP J.) 75 (8) (2004) 2569–2574.
[72] Y.F. Shih, R.J. Jeng, Polym. Degrad. Stab. 91 (2006) 823–831.
[73] A.S. Maxwell, W.R. Broughton, G. Dean, G.D. Sims, Review of accelerated ageingmethods and lifetimeprediction techniques forpolymeric materials, NPL Report DEPC MPR 016 (2005) 1–84, National Physical Laboratory, Middlesex, United Kingdom.
[74] C.D. Doyle, Kinetic analysis of thermogravimetric data, J. Appl. Polym. Sci. V. 15 (1961) 285–292.
[75] A.W. Coats, J.P. Redfern, Nature 201 (1964) 68–69.
[76] S. Vyazovkin, Alternative description of process kinetics, Thermochim. Acta 211 (1992) 181–187.
[77] S. Vyazovkin, W. Linert, Evaluation and application of isokinetic relationships: the thermal decomposition of solids under nonisothermal conditions, J. Chem. Inf. Comput. Sci. 34 (1994) 1273–1278.
[78] I. Blanco, Materials 11 (2018) 1383.
[79] K.T. Gillen, R. Bernstein, M. Celina, Rubber Chem. Technol. 88 (1) (2015) 1–27.
[80] H.L. Friedman, J. Polym. Sci. Part B Polym. Lett. 7 (1969) 41–46.
[81] H.E. Kissinger, Anal. Chem. 29 (11) (1957) 1702–1706.
[82] C.D. Doyle, J. Appl. Polym. Sci. 6 (24) (1962) 639–642.
[83] P. Murray, J. White, Trans. Brit. Ceram. Soc. 54 (1955) 204.
[84] G.I. Senum, R.T.J. Yang, Therm. Anal. 11 (1977) 445–449.
[85] L.A. Perez-Maqueda, J.M. Criado, J. Therm. Anal. Calorim. 60 (2000) 909–915.
[86] D.J. Toop, Theory of life testing and use of thermogravimetric analysis to predict the thermal life of wire enamels, IEEE Trans. Electr. Insul. EI-6 (1) (1971) 2–14.
[87] J.H. Flynn, L.A. Wall, Polym. Lett. 4 (1966) 323–328.

[88] T. Ozawa, Bull. Chem. Soc. Jpn. 1965 (1881) 38.
[89] T. Akahira, T.T. Sunuse, Joint Convention of Four Electrical Institutes, 16, Research Report, Chiba Institute of Technology, Chiba, 1971, pp. 22–31.
[90] M.J. Starink, Thermochim. Acta 404 (2003) 163–176.
[91] J.R. MacCallum, Thermogravimetric analysis. Ch. 37 in: G. Allen, J.C. Bevington (Eds.), Comprehensive Polymer Science, 1, Pergamon Press, Oxford, 1996, pp. 903–909.
[92] S. Vyazovkin, Comput. Chem. 22 (2001) 178.
[93] J. Caiand, S. Chen, J. Comput. Chem. 30 (2009) 1986–1991.
[94] G.I. Senum, R.T. Yang, J. Therm. Anal. 11 (3) (1977) 445–447.
[95] R. White, in: K. Mallinson (Ed.), The Use of the Zhurkov Equation to Describe Polymer Ageing Processes in Ageing Studies and Lifetime Extension of Materials, Academic Plenum Publishers, 2001, pp. 475–481.
[96] J. Bailey, An attempt to correlate some tensile strength measurements on glass: III, Glas. Ind. 20 (1939) 95–99.
[97] A.D. Freed, A.I. Leonov, The Bailey criterion: Statistical derivation and applications to interpretations of durability tests and chemical kinetics, Z. Angew. Math. Phys. 53 (2002) 160–166.
[98] A.I. Leonov, N.I. Basov, Y.V. Kazankov, Fundamentals of Injection Molding of Thermosets and Rubbers. (In Russian), Chimiya, Moscow, 1977, pp. 56–59.
[99] G.M. Bartenev, Yu. S. Zuyev, Strength and Failure of Visco-Elastic Materials, PergamonPress, New York, 1968.
[100] S. Glasston, K.G. Laidler, H. Eyring, The Theory of Rate Processes, McGraw Hill, NewYork, 1941.

Index

Note: Page numbers followed by "*f*" and "*t*" refer to figures and tables, respectively.

A

Accelerated weathering test, 549
Accelerators, 14
Acetic acid, 505
Acid-catalyzed reaction, 8–9
Acid value, 504–505
Acrylamide azobenzene (AAAB), 232–233
Acrylate-based toughening agents, 332–342
　acrylate-modified epoxy, 334–342
　synthesis of functionalized acrylate rubbers, 333–334
Acrylate-modified epoxy, 334–342
　comparison of toughening effects of various modifiers for epoxy matrix, 339*t*
　effect of CTPEHA modification, 336*t*
Acrylic block copolymers, 454
Acrylonitrile, 341–342
Acrylonitrile–butadiene–styrene copolymer (ABS copolymer), 280, 296
Acrylopimaric acid (APA), 26
Activation volume, 576
Addition polyimide, 116–117
　precursor, end groups, and thermal properties of, 119*t*
　scheme for synthesis, 118*f*
Additives
　phenolic resin as, 91–96
　for rubber, 91–92
　for thermoset resin, 269–273
　　antioxidants, 269–271
　　blowing agent, 271–272
　　colorant, 273
　　coupling agent, 272
　　filler, 271
　　surfactants, 273
Adhesive
　evaluation of thermoset as, 549–555
　　critical strain energy release rate, 553–554
　　critical stress intensity factor, 554–555
　　fracture toughness of adhesive bonded structures, 553
　　lap shear tests, 551
　　T-peel test, 550–551
　　TAST, 551–552
　fracture toughness of adhesive bonded structures, 553
　polyurethane, 81
Agar-overlay methods, 473–475
Ageing study under stress, 576–579
　bailey criteria, 576–579
　lifetime estimation, 579
Aliphatic amine, 41–43, 42*t*
Aliphatic urethane diacrylate (AUD), 282
Alkyd resin, 19–23
　synthesis, 20*f*
1-alkyl-3-methyl imidazolium (BMIM), 49
Allyl-terminated hyperbranched polyimide (AT-PAEKI), 323–324
α-relaxation, 195–196
Aluminum trihydrate (ATH), 488
American Society for Testing and Materials (ASTM), 390, 526
Amido amine-functionalized single-wall carbon nanotubes (Amido amine FCNTs), 457
Aminated grapeseed oil (AGSO), 57
Amine-terminated poly(2-ethylhexyl acrylate) (ATPEHA), 333
Amine-terminated polybutadiene (ATBN), 319, 556
Amine-terminated polysulfone (ATPS), 326
Amine(s), 41
　amine-based extender, 67–68
　amine-functionalized grapheme, 471
Amino resins, 98–100
12-aminolauric acid, 427–428
3-aminopropyltriethoxysilane (APES/APTS), 72, 459
3-aminopropyltrimethoxysilane (APTMS), 445–449
Ammonium bentonite (E + T + TMA), 485
Amorphous thermoplastics, 352–353
Aniline (AN), 103, 106–107
Antifouling release coating, 79–81
Antioxidants, 269–271
Arc resistance, 545
Aromatic amine, 41–43, 42*t*
Aromatic dicarboxylic acid, 352
Aromatic isocyanates, 64–65
Aromatic polyesters, 352
　polyols, 64–65
Aromatic proton, 507–508
Arrhenius equation, 181–182, 576
ASTM D 2444, 537–538

A

Asymmetric BMI systems, 133–134
Atomic force microscopy (AFM), 431, 515–516
Atomic network structure, 175
Attenuated total reflectance (ATR), 506–507
Autocatalytic reaction, 41
Autonomic self-healing material, 226
"Averaging" of values, 569
Azeotrope vapor, 5
4,4'-Azobis (4-cyanovaleric acid) (ABCVA), 333

B

Bailey criteria, 576–579
Basalt fiber–reinforced thermoset composite (BFRT), 398
Bend test, 548
3,3',4,4'-benzophenone tetracarboxylic acid anhydride (BTDA), 117–120
Benzoxazine resin (Bz resin), 101–115, 108t, 324–325
Benzoyl peroxide (BPO), 14, 29
Berger model, 215–216
β-cyclodextrine (β-CD), 232–233
Bicarbonates, 272
Binder jetting technology (BJ technology), 280
Bio-based anhydride hardner, 57–58, 58f
Bio-based epoxy, 52–58
Bio-based polyurethanes, 72–76
Bio-based unsaturated polyester, 19–23
  alkyd resin, 19–23
Biomineralization, 421
Bis (2-hydroxy-3-methacryloxypropyl) ether (BisGMA), 223–224
2,2-bis (4-hydroxyphenyl) propane, 313
2,2-*bis* (methylol) propionic acid (BMPA), 22–23
4,4'-bis methylenebis (2-chloroaniline) (MOCA), 444
2-bis-[4-(4-aminophenoxy) phenyl] propane (BAPP), 445–449
4,4'-bis(2-hydroxyethoxy)biphenyl (BEBP), 237–238
Bis(2-hydroxyethyl) methyl, 475–476
Bismaleimide (BMI), 239–240
  resin, 126–139
    applications, 137–139
    bio-based BMIs with selected properties blended with BDM, 136t
    curing of, 129
    properties, 129
    structure–property relationship, 129–136
    toughening of, 323–325
4,4-bismaleimidodiphenylmethane (BMI-MDA), 127
Bisphenol-A (BPA), 10–11, 37, 103, 181
Bisphenol-A bismaleimide resin (MPA-BMI), 323–324
Bisphenol-A dicyanate ester (BADCy), 223–224
BJ technology. *See* Binder jetting technology (BJ technology)
Block copolymers (BCP), 296–297, 449–451
  block copolymer-based nanocomposite, 449–455
    BCP worm-like micelle-modified epoxy, 453f
    block copolymer-modified epoxy systems, 456t
    bulk fracture and tensile properties of modified epoxies, 454t
    formation of sphere-like nanodomains, 451f
    microphase separation processes, 450f
Blocking agent, 67
Blowing agent, 271–272
"Blowing-out effect", 263–264, 267
Boltzmann constant, 304
Bragg's relation, 428–429
Break down voltage, 544–545
Brittle failure, 529, 553
Brittle polymer, toughening mechanism of, 297–299
  crazing, 299
  rubber cavitation, 298
  shear yielding, 297–298
Brönsted-acid catalytic effect, 431–432
BTDA. *See* 3,3',4,4'-benzophenone tetracarboxylic acid anhydride (BTDA)
1,4 butanediol (BD), 71–72
Butyl acrylate (BA), 341–342

C

CAD. *See* Computer-aided design (CAD)
Camphoric anhydride (CA), 57–58, 58f
CANs. *See* Covalent adaptable networks (CANs)
Carbohydrate-derived isosorbide, 26
Carbon-epoxy prepreg, 192
Carbon fiber (CF), 374–375, 378–379, 415
  various surface treatment techniques for modification of fibers, 378f
Carbon fiber-reinforced composite (CFRC), 402–403
Carbon fiber-reinforced plastic composites (CFRP composites), 399
Carbon fiber-reinforced thermoset (CFRT), 398
Carbon nanotube (CNT), 247–248, 420–422, 455–456, 466–468, 544
  CNT-based nanocomposites, 455–468
    π–π interaction, 461f
    cation–π interactions, 463f
    CNT/EP composite film, 467f
    epoxy/MWCNT composites, 464f
    TEM images, 462f
    tensile properties of epoxy/MWCNT/SAHA, 465t
Carbonization, 378
Carbonyl iron (CI), 466–468
Carboxyl groups, 7

Carboxyl-randomized poly(2-ethylhexyl acrylate) (CRPEHA), 333
Carboxyl-terminated copolymer of acrylonitrile and butadiene (CTBN), 313, 319, 347–348, 556
Carboxyl-terminated poly(2-ethylhexyl acrylate) (CTPEHA), 333, 556
Carboxyl-terminated poly(ethylene glycol adipate), 556
Carboxyl-terminated poly(ethylene glycol) (CTPEGA), 409–410
Carboxyl-terminated poly (propylene glycol) adipate (CTPPGA), 331–332
Cation exchange capacity (CEC), 424
Cation–π interactions, 462–466
Cavitations, 307
CE resin. See Cyanate ester resin (CE resin)
Ceramic–matrix composite (CMC), 372
Chain breaking (CB), 523–524
　antioxidants, 269–270, 523–524
Chain breaking donor antioxidants (CB-D antioxidants), 269–270, 523–524
Chain extender, 67–72
Chain-breaking acceptor antioxidants (CB-A antioxidants), 269–270, 523–524
Chain-extended bismaleimide (CBMI), 239–240
Charpy tests, 536
Chemical modification, 326–327
　effect on thermomechanical properties of epoxy, 328t
Chemical vapor deposition (CVD), 402–403
Chemorheology, 204, 475–476
Chip impact test, 536–537
Chiral epoxy, 52
Chlorendic resin, 10–11
Chlorinated paraffins (CP), 257

Chlorofluro carbon compounds (CFCs), 271–272
Cis-9,10-epoxy-18-hydroxy octadecanoic acid, 55
Closite-hexadecyl amine (HDA), 440–441
Closite-octadecyl amine (ODA), 440–441
Closite-tetradecyl amine (TDA), 440–441
Cloud point (ηcp), 302, 304
Coating, 78–79
　antimicrobial, 473
　conventional alkyd, 20
　epoxy-based microwave-absorbing coating, 466–468
　HBUA/SNP nanocomposite-based antimicrobial coating, 474f
　polyurethane-based coating, 78–82
　antifouling and foul release coating, 79–81
　polyurethane adhesive, 81
Cocontinuous structure, 350–351
Colbert reaction mechanism, 101
Combination organoclay, 486–488
Commercial toughening agents, 328–329
Common solvent technique, 427
Compact tension (CT), 555
Composite, 372
　interface, 375–379
　　representation of interface in fiber-reinforced thermoset composites, 376f
　　surface tension and contact angle, 376–377
　　surface treatment of fibers, 377–379
　mechanical testing of, 391
　　ASTM standards for tests of thermoset composites, 391t
　processing of, 380–390, 380t
Compression
　molding, 274–275
　test, 532
Computer-aided design (CAD), 275–276
Condensation techniques, 511

Cone calorimetry, 546
Constrained layer damping (CLD), 217–219, 549–550
Contact angle, 376–377
Contact molding, 381
Continuous fiber–reinforced composites, 374–375
Conversion, 572–573
"Convex meniscus", 376
Core–shell particles (CSPs), 300
　toughening, 346–347
Core–shell rubber (CSR), 477–478
Core–shell rubber particle (CSRP), 415
Corresponding strain, 527
Corrosion resistance test, 549
Cotton ignition, 546
Couchman's analysis, 197–198
Coupling agent, 272
Covalent adaptable networks (CANs), 247
Crack propagation process, 358
Craze, 299
Creep relaxation test, 532–533
Critical fracture energy, 549–550
Critical strain energy release rate, 553–554
Critical stress intensity factor (K_{IC}), 549–550, 554–555
Cross polarized optical microscopy (CPOM), 94
Cross-hatch adhesion test, 548
Cross-linking
　of phenolic resin, 86–89
　of polyimide, 122
Crosslink points, 508–509
Crosslinked density, 508–510
Crosslinked networks, 175
Crosslinker, 67–72
Cryogenic microscopy (cryo-TEM), 514–515
Crystalline materials, 516–517
Crystalline thermoplastics, 353–354
Crystallization temperature (T_c), 517
Cure/Curing, 2–3
　agents, 39–46, 40t
　epoxy, 59t

Cure/Curing (Continued)
    Lewis acid-catalyzed curing of epoxy by alcohol, 41f
    mechanism of curing of epoxy resins, 40f
    of bismaleimides, 129
    characteristics, 179–193
        dynamic cure kinetics, 184–189
        isothermal cure kinetics, 180–184
        modeling from pressure volume temperature approach, 191–193
        vitrification effect on polymerization rate, 189–191
    cycle, 193–195
    of cyanate ester resin, 140–141
    of polyimide resin, 122–123
    of unsaturated polyester resin, 14–17
    of vinyl ester resins, 29–35
Cyanate ester resin (CE resin), 139–154, 223–224
    application, 152–154
    chemical structures of commercially available, 140f
    curing of, 140–141
    modified difunctional CE resins, 143–148
    modified multifunctional CE resins, 148–152
    properties, 142–143
    toughening of, 323–325
Cycloaliphatic amine, 41–43, 42t

D

Debulking process, 384
Debye–Scherrer equation, 110, 516–517
Decabromobiphenyl oxide (DB), 257
Degradation of polymer, 562
Degree of polymerization (DP_n), 177
Dendrimer, 20–21
Dendritic hyperbranched polymer-based toughening agents (HBP-based toughening agents), 342
Deoxyribonuclic acid (DNA), 510
"Derakane", 23–25
Diamino diphenyl methane (DDM), 43–44, 48–49, 211, 265–266
Diamino diphenyl sulfone (DDS), 43–44, 239–240, 341–342, 349–350
3,3′-diamino diphenyl sulfone (33DDS), 55–57
4,4′-diaminodiphenylmethane (DAM), 107–109
Diamond, 175
Dibenedetto equation, 197–198
3,3′-dichloro-4,4′-diamino diphenylmethane, 469–471
Dichloromethane (DCM), 49
Dicyanamide (Dicy), 47, 349–350
Dicyclopentadiene (DCPD), 226–227
Die casting, 273–274
Dielectric constant, 543
Dielectric strength, 544–545
Diels–Alder reaction (DA reaction), 122, 230, 449
Diethyl toluene diamine (DETDA), 205–206
    DETDA-based nanocomposite, 429–431
Diethylene glycol, 9–10
Differential scanning calorimetry (DSC), 179, 517–518
Difunctional benzoxazines, 104
Difunctional epoxies, 38, 200, 399
    chemical structures of, 38f
    physicochemical properties of, 39t
Digital light processing (DLP), 278
Diglycidyl ether of bisphenol-A (DGEBA), 37, 200, 205–206, 211, 257, 449–450
9,10 dihydro-9-oxa-10-phosphaphenanthrene-10-oxide and 1-(3-aminopropyl)-imidazole (DOPO-AI), 258–259
9,10-dihydro-9,10-oxa-10-phosphaphenanthrene-10-oxide (DOPO), 48
Dilatometer, 542–543
Dimethacrylated-epoxidized-sucrose-soyate (DMESS), 27–28
Dimethyl acetamide (DMAc), 427
Dimethylaniline system, 29
Dimethylol propionic acid (DMPA), 237–238
Dioctyltin catalyst, 247–248
Diol extender, 67–68
Dipentaerythritol (DPE), 22–23
Diphenic anhydride (DPA), 57–58, 58f
Diphenylmethane diisocyanate (MDI), 48–49, 66–67, 202–204
Dipodal silane (DSi), 459
Direct-ink-write approach (DIW approach), 282
Disk diffusion methods, 473–475
Dithiodiglycolic acid (DTDGA), 333
Dodecaphenyl-POSS (DP-POSS), 449
Double cantilever beam (DCB), 402–403
Ductile failure, 529
Durometer, 539
Dyestuffs, 273
Dynamic cure kinetics, 184–189
Dynamic loss modulus (E''), 519
Dynamic mechanical analysis (DMA), 396–397, 517–519
    plot of cured thermoset resin, 519f
Dynamic mechanical thermal analysis, 557
Dynamic storage modulus (E'), 519
Dynamic viscoelasticity, 215–216

E

Effective medium theory, 541–542
Effective Medium Theory models (EMT models), 541
Elastic layered MOFs (ELFs), 252
Elastic modulus, 527, 529
Elastomeric epoxy network, 224
Elastomeric thermoset, 213–219

Index

rubbery thermoset application, 214–219
Elastomers, 321–322
Electrical conductivity, 544
Electrochemical technique, 378–379
Electromagnetic interference (EMI), 494
Electron beam melting technology (EBM technology), 279–280
Elongation, 527–529
End group analysis, 512
Energy dissipating mechanism, 358
Energy-dispersive X-ray spectroscopy (EDX spectroscopy), 514
Engineering thermoplastics, 348–351
Epichlorohydrin (ECD), 37, 39
Epocryl, 23–25
Epoxidized plant oils, 54
Epoxidized poly(styrene-bisoprene-b-styrene) (eSIS), 449–450
Epoxidized soybean oil (ESO), 331–332
Epoxy
  determination of epoxy equivalent, 505–506
  epoxy-based shape-memory thermoset, 239–240
  epoxy-based systems, 316–318
  epoxy/CTBN systems, 304
  epoxy/graphene nanocomposite, 468
Epoxy aliphatic acrylate (EAA), 282
Epoxy resins (EP), 37–63, 222–223, 239–240, 485, 505
  application, 58–63
  curing agents, 39–46, 40t
  epoxy formulation, 39
  modified epoxy systems, 46–58
Escherichia coli, 473
eSIS-amine-reactive functionalities (eSIS-AEP), 449–450
1,2-ethane diol, 352
Etherification, 41

1-ethyl-3-methylimidazolium dicyanamide ([EMIM]N(CN)$_2$), 49–50, 51f
Ethylene propylene diene rubber (EPDM), 92
Eugenol-based epoxy resin, 55–57
Euring–Zhurkov expression, 576
Ex situ approach, 473
Extragalery reactions, 433–435
Extrinsic self-healing process, 226–229
Extrusion process, 385–387

F

Fabrication process, 380
Failure stress, 554–555
Falling weight impact test, 537–538
Fast Fourier transform (FFT), 518, 557
Fatigue endurance limit, 533
Fatigue test, 533
Fatty acids, 54
Fiber-reinforced composite (FRC), 383–384
Fiber-reinforced plastics (FRP), 217–219, 295, 358, 372
  constituents of FRP composites, 374–375
    physical properties of man-made fibers and natural fibers, 375t
Fibers
  hybridization, 374–375, 401–402
  reinforcement, 374
  surface treatment of, 377–379
    carbon fiber, 378–379
    glass fiber, 377–378
    polymeric fiber, 379
Filament winding technique, 387
Filler, 271
Finite element analysis (FEA), 458
First-generation hyperbranched alkyd resins (HBA1 resins), 23
Flame retardancy, 481–482
Flame retardants (FR), 486–488
  thermoset, 255–269
    halogen-containing flame retardants, 257–258

inorganic flame retardants, 256
phosphorous-containing flame retardant, 258–269
Flammability test, 545–547
  cone calorimetry, 546
  LOI test, 546–547
  UL-94 flammability test, 546
Flexibilization, 295
Flexural strain reflects flexibility, 531
Flexural test, 530–531, 531f
Flory–Huggins equation, 302
Flory–Rehner theory, 509
Fluorinated ethynyl-terminated imide oligomers (FETI oligomers), 224
Flynn–Wall–Ozawa method (FWO method), 567
Forced vibration, 558
Foul release coating, 79–81
Fourier transform infrared (FTIR), 179
Fourier transformed infrared spectroscopy (FTIR spectroscopy), 93–94, 506–507
Fox equation, 322
Fracture energy (K_{IC}), 356
Fracture strength of thermoset resins, 530–531
Fracture toughness of thermoset resins, 533–536, 553
Free phenol, 507–508
Free-layer damping (FLD), 217–219
Freeze-drying process, 424–425
Frequency response function (FRF), 558
Friedman method, 564–565
Fumarate double bonds, 3–5
Fumarate resin, 10
Fumeric acid, 3
Functional groups, 378–379
Functionalized acrylate rubbers, synthesis of, 333–334, 336t
Furan resins, 100–101
Furfuryl alcohol (FFA), 100, 100f
Fused deposition model (FDM), 278

G

Gas chromatography (GC), 122
Gas chromatography-mass spectrometry (GC-Ms), 486–488
Gate pressure, 252
Gel permeation chromatography (GPC), 513–514
Gel point (GP), 178, 206–207
Gelation, 175–178
Glass fiber, 374–375, 377–378
Glass fiber-reinforced composite (GFRC), 413–415
Glass fiber-reinforced plastic composite (GFRP composite), 397
Glass fiber–reinforced thermoset (GFRT), 398
Glass transition temperature (T_g), 195–196, 345–346, 517
  effect of cure conversion on, 197–199
Gloss measurement test, 549
Glutaric anhydride (GA), 57–58, 58f, 239–240
Glycidyl methacrylate (GMA), 331, 346–347, 449–450
Glycidyloxypropyl-heptaisobutyl POSS (iBu-GlyPOSS), 445
Glycidyloxypropyl-heptaphenyl POSS (iPh-GlyPOSS), 445
Glycol, 5
Graft copolymerization, 296
"Graft from" approach, 458–459
"Graft to" approach, 458–459
Graphene, 491
  graphene-based nanocomposite, 468–473
    PU/GO/EP nanocomposites, 470f
    WVP vs. exposure period plots, 469f
Graphene oxide (GO), 402, 469–472
Graphene oxide nanosheets (GONS), 469–471
Graphitization, 378
Green flame retardants, 257–258
Grubbs' catalyst, 226–227

H

Half-neutralized salt of adipic acid (HNAA), 460–462
Halogen-containing flame retardants, 257–258
Halpin–Tsai equations, 396
Hansen's iteration method, 506
Hard segment (HS), 202–204
Hardness test, 538–539
Heat deflection temperature (HDT), 543
"Heat flux cell", 182–184
Heat release capacity (HRC), 488
Heat release rate (HRR), 483–484, 546
4,4′-(hexafluoroisopropylidene)-dipthalic acid (HFDE), 120
Hexamethoxymethyl melamine (HMMM), 23
1,6 hexamethylene diisocyanate (HDI), 66–67, 202–204
Hexamethylene tetramine (HMTA), 83, 86–87, 87f
High processing temperatures, 432–433
High-impact polystyrene (HIPS), 296, 305–306
Hildebrand equation, 302
Hooke's law, 527
Host–guest interactions, 232–233
Hybrid nanocomposites, 421–422
Hydrogen silsesquioxane (HSQ), 132
2-hydroxyethyl methacrylate (HEMA), 240
Hydroxyl group (OH), 7, 505
  value, 505
Hydroxyl-functionalized hyperbranched polymers, 64–65
Hydroxyl-terminated natural rubber (HTNR), 319
Hydroxyl-terminated polybutadiene (HTBN/HTPB), 313, 319, 556
Hydroxyl-terminated polyether (THE), 314
Hydroxylated MWCNT (OH-MWCNTs), 459
Hyperbranched alkyd urethane (HBAU), 73–74, 74f
  mechanical and weathering properties, 76t
Hyperbranched poly(butylene adipate) (HPBA), 345–346
Hyperbranched polyether epoxy (HBPEE), 324–325, 345–346
Hyperbranched polymer (HBP), 20–21, 314, 410–411, 476–477
  hyperbranched polymer-based toughening agents, 342–346
Hyperbranched polyols, 20–21
  synthesis of, 22f
Hyperbranched polyurethane (HBPU), 244
  synthesis of one-pack–NCO-terminated hyperbranched polyurethane, 75f
Hyperbranched urethane alkyd (HBUA), 473–475

I

Ignition time, 546
Imidazoles, 48
Imidazolium-type ILs, 49–50
Impact energy, 401, 533–535
Impact resistance test, 548
Impact test, 536–538
  falling weight impact test, 537–538
  pendulum impact test, 536–537
In situ polymerization of monomeric reactants, 117–122
Infrared spectroscopy (IR spectroscopy), 506–507
Inhibitor, 13–14
Injection technology, 413–415
Inorganic flame retardants, 256
Insoluble colorant. See Pigment
Integrated circuit board (ICB), 113
Interconnected polymer networks (ICN), 219–220
Interlaminar shear stress (ILSS), 391–392
Interlayer expansion, 435
Internal cavitation, 298
"Internal friction", 519
Interpenetrating polymer networks (IPNs), 219–225, 319, 480–481
Interphase, 375–376

Intragallery reactions, 433–435
Intramolecular cyclization, 316–318
Intrinsic self-healing, 229–233
Ionic liquids (ILs), 49
　chemical structure of, 50t
Isobutanol (IBM), 23
Isoconversion kinetics, 564–566
　Friedman method, 564–565
　integral form of, 566–568
　　FWO method, 567
　　KAS method, 567–568
　　MacCallum method for isothermal study, 568
　　Starink method, 568
　peak rate method, 565–566
Isoconversion methods, 179–180, 569–570
Isocyanates, 66–67
　chemical structures used for PU synthesis, 66f
　chemistry of blocking of, 68f
　reaction with water, 70f
Isophthalic resin, 10–11
Isoreticular frameworks, 253
Isothermal cure kinetics, 180–184
Isothermal kinetic analysis of thermoset, 179–180
Itaconic anhydride (IA), 57–58, 58f
Izod/Charpy tests, 536

K

Kamal's model, 181
Kelvin–Voigt model (KV model), 215–216
Kevlar fiber-reinforced thermoset (KFRT), 398
Kevlar fibers, 374–375
Kinetic model, 563–564
Kissinger method, 565–566
Kissinger–Akahira–Sunose method (KAS method), 567–570

L

Lap shear tests, 551
Laser scanning dilatometer, 191–192
Latent cure epoxy, 47–52
Lattice, 175
Lewis acid, 431
Lifetime estimation, 579
　solution, 572–575
　　conversion, 572f
　　estimated life of epoxy thermoset, 573f
　　lifetime estimation of epoxy thermoset, 574t
　from TGA kinetics, 572–575
Light-scattering calibration constants, 513
Lignin, 27–28, 28f, 55
Limiting oxygen index test (LOI test), 263, 545–547
Linear coefficient of thermal expansion (LCTE), 542–543
Linear variable differential transformer system (LVDT system), 532–533
Liquid composites molding process (LCM process), 402
Liquid crystalline epoxy networks (LC epoxy networks), 210
Liquid crystalline epoxy resin (LCEP), 211
Liquid crystalline thermoset, 210–213
Liquid natural rubber (LNR), 480
Liquid oxidizing agents, 378–379
Liquid rubber toughening, 300–313
　list of liquid rubbers used for toughening of thermoset, 301t
　mechanism of toughening of thermoset, 304–309
　　cavitation and shear yielding, 307–309
　　crazing, 305–306
　　rubber bridging and tearing, 305
　　shear yielding and crazing, 306–307
　microstructural features, 309–313
　　interfacial adhesion, 312–313
　　matrix ligament thickness, 311–312
　　parameters influencing rubber toughening, 309t
　　particle size, 310–311
　　volume fraction, 310
　　reaction-induced phase separation, 301–304
Liquid rubber-modified epoxy systems, 354
Log–log function, 198–199
Long continuous fibers composites, 374–375
Longitudinal stress, 527
Loss modulus (G''), 525
Low-profile additives (LPAs), 316, 480
　LPA-rich phase, 480

M

m-phenylene diamine (MPDA), 262–263
MacCallum method for isothermal study, 568
Magic angle spinning (MAS), 103
Maleic anhydride, 3
Maleopimaric acid (MPA), 26
　synthesis of maleopimaric acid-based vinyl ester resin, 27f
Mass loss rate (MLR), 546
Material's linear coefficient, 542
Matrix crosslink density effect on rubber toughening, 356
Matrix ligament thickness (MLT), 300, 311
Matrix-assisted laser desorption-ionization mass spectrometric analysis (MALDI mass spectrometric analysis), 201–202
Matrix–fiber delamination, 401
Matrix–rubber particle adhesion, 312
Maxwell's model, 215–216, 541
"Maya blue" painting, 421
Mean roughness, 515–516
Mechanical damping factor, 519
Melamine-formaldehyde resins (MF resins), 98, 99f
Melt elasticity, 511
Melting point (T_m), 195–196, 517
Membrane osmometry, 512
Metal nanoparticles decorated graphene hybrids, 490–491

Metal oxide, 489–490
Metal plastics, 217–219
Metallic materials, 543
Metal–matrix composite (MMC), 372
Metal–organic frameworks (MOFs), 252–255
  applications, 255
  synthesis, 253–254
Methacryl-POSS (M-POSS), 449
Methacrylate-styrene-butadiene copolymers, 296
Methyl acrylate (MA), 449–450
Methyl ethyl ketone peroxide (MEKP), 14, 29
Methyl methacrylate (MMA), 223–224, 240
4,4′-methylene bis [3-chloro 2,6-diethylaniline] (MCDEA), 43–44
Methylene diphenyl diisocyanate (MDI), 247–248, 469–471
4,4′-methylenebis[2,6 diethylaniline] (MDEA), 43–44
Methylethylketone (MEK), 23
MF resins. *See* Melamine-formaldehyde resins (MF resins)
Microcapsule-type latent curing agent, 49
Microcellular injection molding method (MIM method), 488
Microcomposites, 420–421
Microdeformation mechanisms, 307
Microencapsulation technology, 227
Microfibrillated cellulose (MFC), 481
"Microgel particles", 15–16
Microwave-absorbing coating (MAC), 466–468
Model-free kinetics of isoconversion, 566
Modified difunctional cyanate ester resins, 143–148
Modified epoxy systems, 46–58
  bio-based epoxy, 52–58
  chiral epoxy, 52
  latent cure epoxy, 47–52

Modified multifunctional cyanate ester resins, 148–152
Modulus of elasticity, 529
Molar volume of solvent, 509
Molecular network structure, 175
Molecular weight (M_w), 353
  distribution, 510–514
    end group analysis, 512
    GPC, 513–514
    light scattering, 513
    membrane osmometry, 512
    viscometry, 511–512
    VPO, 512
Monofunctional benzoxazine, 104
Monomer, 11–13
"Monomer flooded process", 221–222
Monte Carlo simulation, 201–202
Montmorillonite, 424
Multidirectional tape-prepreg, 389
Multifunctional epoxy, 38
  network, 207
  resins, 359–360
Multiphase thermoset network, 309–310
Multiwall carbon nanotubes (MWCNT), 455–456, 458

N

N-phenylmaleimide–styrene copolymer (PMS), 323
Nanocarbon-based flame retardants, 488–491
Nanoclays, 440–441
  nanoclay-based flame retardant, 483–485
    cone calorimeter profiles, 484f
    epoxy samples, 487f
    ignited polymer, 486f
Nanocomposite, 420–421, 423, 427–428
  application, 491–494
  polymer, 492t
  synthesis methods, 427–428
    thermoset/clay nanocomposite, 428f
  technology, 421
Nanofillers, 489
Nanomaterials, 420–421

Nanoreinforcement and toughening, 475–481
  fracture surfaces, 478f
  neat epoxy and epoxy composites, 482f
  room temperature mechanical tests, 479f
  ternary blends of epoxy, epoxy-functionalized HBP, and nanoclay, 477t
  thermoset resins, 476f
  transmission electron micrographs, 477f
Nanoscale mesoporous silica (NMS), 421–422, 445–449
Nanosilica (NS), 413–415
Nanostructured inorganic organic hybrid materials, 421
Nanotechnology and flammability, 481–482
Natural epoxidized plant oil, 55
Natural fibers, 400
  natural fiber-based composites, 400–401
Natural rubber (NR), 92, 321–322
Nitric acid, 459–460
Nitrile rubber, 321–322
Nonautonomic self-healing materials, 226
Noncatalytic mechanism, 41
"Nonidet P-40", 468–469
Nonisothermal cure kinetic method, 186–187
Nonlinear optical polymers (NLO polymers), 125, 494
Nonoxidative methods, 378–379
Nonreactive plasticizers, 88–89
Novolac phenolic resins, 83–85, 320–321
Nuclear magnetic resonance spectroscopy (NMR spectroscopy), 103, 179, 431, 507–508
Number-average molecular weight (ΣMn), 510
Nylon 6, 353–354

O

"Oberst Beam" test, 558
Octavinyl-POSS (OV-POSS), 449

Old manmade hybrid material, 421
Olefin metathesis reaction, 248
One-dimensional power spectral density values (1DPSD values), 516
One-stage method, acrylate-modified epoxy, 334
Onium ions, 425–426
Opposing forces, 435–436
Organic hydrazines, 47
Organic ligands, 252
Organoclay, 433–435
Organofunctional silanes, 377–378
Organophosphorous compounds, 258
Orthopositronium(o-Ps), 200
Osmotic pressure, 512
Oxidative methods, 378–379
2,3,3′4′-oxydiphthalic dianhydride (a-ODPA), 134–135
4,4′-oxydipthalic anhydride (ODPA), 445–449
Ozawa analysis, 187–189

P

P-cresol formaldehyde (PCF), 321–322
Paal–Knorr reaction, 230
Particle–matrix morphology, 310
PBA. *See* Terephthaloyl bis (4-oxybenzoic) acid (PBA)
Peak rate method, 565–566
Pendulum impact test, 536–537
Permeability of nanocomposite, 442
Permittivity constant, 543
Phase morphology, 299–300
Phase separation time (t_{ps}), 304
Phenol, 320–321
Phenol formaldehyde resin (PF resin). *See* Phenolic resins
Phenolic hydroxyl groups, 111
Phenolic resins, 82–98
　addition cure phenolic resin, 97–98
　as additives, 91–96
　　additives for rubber, 91–92
　　modifier for poly(ethylene oxide), 92–96
　applications, 89–91
　cross-linking of, 86–89
　novolac, 83–85
　resole, 85
　toughening of, 320–322
　　flexural properties of PF resole/PEO blends, 322*t*
Phenolic-nitrile rubber (NBR), 321–322
Phenyl-ethynyl-terminated imide (PETI), 223–224
Phlorotannins, 28–29
Phosphonium bentonite (E + T + TPP), 485
Phosphonium derivative, 49–50
Phosphorous-containing flame retardant, 258–269
Photocurable thermosets, 282
Phthalic anhydride (PA), 57–58, 58*f*
Phthalonitrile resin (PN resin), 154–157
　properties, 156–157, 157*t*
　synthesis, 155–156
Phthalonitrile-etherified resole resin (PNR), 157
Pigment, 273
Pipyridine (PIP), 340–341
Plasma polymerization process, 379
"Plastic tooling", 58–62
Plasticizers, 88–89, 294–295
　for phenolic resin, 89*f*
Plunger-type instrument, 191–192
Poisson's ratio, 527, 532
Poly ((tetramethylene fumarate)-*co*-(tetramethylene phthalate)] (BFPA), 16–17
Poly (ether ketones) (PEK), 351
Poly (methyl methacrylate) (PMMA), 48–49
Poly (methyl methacylate-b-butylacrylate-b-methyl methacrylate), 454
Poly (pethylethylene)-b-poly (ethylene oxide) (OP), 449–450, 454
Poly (propylene oxide-block-ethylene oxide), 475–476
Poly [(tetramethylene maleate)-*co*-(tetramethylene phthalate)] (BMPA), 16–17
Poly dimethyl siloxane-blockpoly (ε-caprolactone) (PDMS-PCL), 455
Poly(2-ethylhexyl acrylate), 333
Poly(2-hydroxyethyl methacrylate) (PHEMA), 21–22
Poly(acrylonitrile–butadiene–styrene) (ABS), 353
Poly(butyl acrylate), 454–455
Poly(butylene terephthalate) (PBT), 298, 353–354
Poly(dimethylsiloxane) oligomer (PDMS), 330–331
Poly(ether imide) (PEI), 349–350
Poly(ether sulfone) (PES), 349–350
Poly(ether-block-amide) (PEBA), 406–407
Poly(ethylene glycol) (PEG)
Poly(ethylene glycol) adipate (CTPEGA), 326
Poly(ethylene oxide) (PEO), 219–220, 321–322, 353–354, 449–450
　modifier for, 92–96
　PEO-based block copolymers, 452
Poly(ethylene oxide)-b-poly (butylene oxide) (PEO-PBO), 451–452, 454–455
Poly(ethylene-alt-propylene) (PEP), 323, 449–450
Poly(ethylene-alt-propylene)-b-poly(ethylene oxide) (PEP-PEO), 452
Poly(methyl methacrylate) (PMMA), 346–347
Poly(methyl methacrylate)-b-poly (butyl acrylate)-b-poly(methyl methacrylate) (MAM), 454
Poly(methyl) methacrylate, 454–455
Poly(phenylene oxide) (PPO), 296
Poly(styrene-b-butadiene-b-methyl methacylate), 454
Poly(tetramethylene glycol) (PTMG), 469–471

Poly(vinyl acetate) (PVAc), 316
Poly(vinyl chloride) (PVC), 296
Poly(vinylidene fluoride) (PVDF), 353–354
Polyamide-6 (PA6), 402
Polyamides, 297, 321–322
Polybenzoxazines (PBZs), 101
　application, 113–115
　main-chain polybenzoxazines, 104–110
　polybenzoxazine-based blends, 111–113
　precursor used for various types of, 106t
Polybutadiene-based polyols, 64–65
Polycaprolactone (PCL), 54, 229, 236–237, 331, 444
Polycarbonate (PC), 296
Polydimethyl siloxane (PDMS), 62, 80
Polydispersity index, 511
Polydopamine (PDA), 49
Polyepichlorohydrin (PECH), 331
Polyester structure, 6–7
Polyesterification kinetics, 8–9
Polyesters, 3
　types of, 9–11
　　general-purpose resin, 9–10
　　specialty polyester resin, 10–11
Polyether ether ketone (PEEK), 351
Polyether polyols, 63–65
Polyethylene glycol (PEG), 223, 326
Polyhedral oligomeric silsesquioxanes (POSS), 262–263, 421–422, 443–449
　epoxy network modified with iBu-GlyPOSS, 446f
　NMS/polyimide nanocomposite, 447f
　PI/NMS nanocomposite, 448f
　POSS-based nanostructured epoxy thermoset, 444f
Polyimide (PI), 115–126, 445–449
　addition, 116–117
　application, 124–126
　cross-linking of, 122
　curing of, 122–123
　in situ polymerization of monomeric reactants, 117–122
　polyimide-based nanocomposites, 427
　properties of commercial thermoplastic polyimides, 116t
　scheme for synthesis, 115f
　toughening of, 323–325
Polyketone (PK), 230
Polymer, 217–219, 493
　fiber, 379
　　interactions, 374–375
　materials, 540
　matrix, 473
　nanocomposites, 493
　network, 175, 524–525
　　polymer networks-based shape-memory thermoset, 240–245
　polymer-based reinforcing fibers, 379
　transition, 195–196
Polymer/clay nanocomposites (PCN), 421–443, 456–457, 483–484, 493–494
　characterization of, 428–431
　　TEM micrographs, 430f
　　XRD plots, 429f
　clay with alkyl ammonium cation by ion-exchange method, 426f
　controlling factors for nanocomposite formation, 431–437
　　curing reaction of epoxy/clay nanocomposites, 437f
　　forces acting on pair of clay layers, 436f
　intercalated and exfoliated nanocomposites, 426f
　methods of nanocomposite synthesis, 427–428
　model structure for montmorillonite clay, 424f
　principle of, 424–427
　properties of, 437–443
　　clay loading effect on tensile properties, 440f
　　dynamic modulus vs. temperature plots, 439f
　　stress–strain diagram, 441f
　　TEM microphotographs, 438f
　　tortuous model, 442f
　thermoset resin with pristine clay, 425f
Polymerization, 29, 296
　vitrification effect on polymerization rate, 189–191
Polymer–matrix composite (PMC), 372
Polymethyl methacrylate (PMMA), 224, 316
Polyol, 63–66
　chemical structures of, 64f
　synthesis of star-shaped polyester polyols, 65f
Polyphosphoric acid (PPA), 491
Polypropylene (PP), 39, 432
Polypyrrole (Ppy), 493–494
Polystyrene (PS), 296
Polystyrene-b-poly(ethylene oxide) (PS-PEO), 451–452
Polysulfone, 321–322
polytetramethylene ether glycol (PTMEG), 244
Polyurethane (PU), 63–82, 275, 319, 441–442, 469–471
　applications, 76–82
　　polyurethane foam, 76–78
　　polyurethane-based coating, 78–82
　bio-based polyurethanes, 72–76
　chain extender and crosslinker, 67–72
　isocyanates, 66–67
　oligomer, 332
　polyol, 63–66
　polyurethane-based shape-memory polymer, 235–239
　prepolymers, 67
　PU-based microcellular nanocomposite, 488
Positron annihilation lifetime spectroscopy (PALS), 200, 224–225, 508
Postcuring heat treatment, 302

Index

Powdered rubber (PR), 347
Power spectral density (PSD), 516
Prepolymers, 67
Prepreg molding, 389–390
Prepreg-molded composites, 390
Prepregs-based composites, 389
Prereact methods, 445
Pressure volume temperature (PVT) approach, modeling from, 191–193
Pull-off adhesion strength, 547–548
Pultrusion process, 385–387
    typical pultruded FRP profiles, 386f

Q

Quinones, 269–270

R

Radio frequency (R/F), 507–508
Raman spectroscopic method, 458
Rapid additive manufacturing. *See* 3D printing
Rault's law, 512
Reaction injection molding (RIM), 82, 275, 385
Reactive diluents, 11–13
Reactive liquid polymers, 321–322
Reactive plasticizers, 88–89
Resin
    impregnation, 389
    medium, 304
    toughening composites, 409–415
        impact and ILSS data for various toughened DGEBA epoxy composites, 410t
Resin film infusion (RFI), 387–389
    RFI element with thermocouple and fiber optic, 388f
Resin infusion flexible tooling method (RIFT method), 415
Resin transfer molding (RTM), 106–107, 382–383, 382f
Resole phenolic resins, 85, 86f
Resultant stress, 524–525
Retro-DA reaction (r-DA reaction), 230

Rheological characterization of thermoset, 524–526, 525f
Rheological properties of thermoset resin, 204–210
Rheological testing, 205
Rigid fillers, 295
Rigid particle toughening of epoxy, 356–358
Rigid-particle-toughened thermoset system, 358
Rockwell hardness tester, 539
Root mean square (RMS), 515–516
Rosin-based anhydrides, 57
Rotational casting, 274
Roughness calculation (R_{rms}), 515–516
Rubber
    additives for, 91–92
    cavitations, 297–298, 312
        micromechanism, 299–300
    rubber-based IPNs, 221–222
    rubber-based nanocomposites, 427
    rubber-toughened epoxy systems, 304–305
    rubber–bridging mechanism, 309
    toughening, 328–346
        acrylate-based toughening agents, 332–342
        commercial toughening agents, 328–329
        hyperbranched polymer-based toughening agents, 342–346
        rubber-based toughening agents, 330–332
Rubbery epoxy network, 213–214
Rubbery thermoset application, 214–219
    dynamic viscoelasticity, 215–216
    vibration-damping behavior, 216–219

S

S-2-chloro,4-methyl pentanoic acid (CMPA), 52
    chiral properties of CMPA-modified epoxy, 53t

Scanning electron microscope (SEM), 300–301, 431, 471, 514
Scratch resistance test, 548
Second-generation hyperbranched alkyd resins (HBA2 resins), 23
Secondary alcohols, 41
Secondary building units (SBU), 252
Selective laser sintering technology (SLS technology), 278–279
Self-catalyzed reaction, 8
Self-healing thermoset, 226–246
    epoxy-based shape-memory thermoset, 239–240
    extrinsic self-healing, 226–229
    intrinsic self-healing, 229–233
    polyurethane-based shape-memory polymer, 235–239
    SMT, 233–235
        application, 245–246
        interpenetrating polymer networks-based SMT, 240–245
Serratia marcescens, 473–475
Service temperature (T_s), 196
Sestak–Berggren model, 181
Shape memory polymers (SMP), 92–93, 493
Shape memory polyurethanes (SMPU), 235–236
Shape-memory alloys (SMA), 235
Shape-memory polymers (SMPs), 245
Shape-memory thermosets (SMT), 233–235
    application, 245–246
Shear localization process, 308
Shift factor, 520–521
Shore durometer, 538
Short fiber composites, 374–375
Short-beam shear test (SBS test), 391
Short-fiber reinforcement technique, 406–407
Silica-based nanocomposites, 443–449
Silicate layers, 437–439
Silicon carbide (SiC), 466–468
Silicone resin, 3D printing of, 283–284

Siloxane rubbers, 330–331
Silver nanoparticle-based nanocomposites (SNP-based nanocomposites), 473–475
  HBUA/SNP nanocomposite-based antimicrobial coating, 474f
Single edge-notched bending (SENB), 555
Single-wall carbon nanotubes (SWCNT), 455–457
Small-angle X-ray scattering (SAX), 429–431
Smart thermoset, 226–246
  self-healing thermoset, 226–246
Smoke
  density measurements, 547
  test, 547
Sodium chloride cell, 506–507
Sodium salts of 6-aminohexanoic acid (SAHA), 463–464
Soft porous crystals (SPCs), 252
Solubility parameter, 506
Soluble colorant. See Dyestuffs
Solution polymerization, 334
Solvent impregnation process, 389
Solvent-assisted self-healing technology, 227
Specialty polyester resin, 10–11
Specific extinction area (SEA), 546
Staphylococcus aureus, 473
Starink method, 568
Stereolithography (SLA), 275–277
Stimuli-responsive self-healing materials. See Nonautonomic self-healing materials
Stokes–Einstein equation, 304
Storage modulus (G'), 524–525
Strain sweep measurements, 525
Stress
  ageing study under, 576–579
  intensity factor, 554–555
  relaxation
    experiment, 533
    test, 532–533
Stress–strain curve, 529
Styrene (St), 11, 25–26, 341–342
  emission, 13
styrene-acrylonitrile copolymer, 296
Styrene-butadiene rubber (SBR), 92, 224–225, 296
Succinic anhydride (SA), 57–58, 58f
Sulfonic graphene oxide (SGO), 471
Sulfonylhydazides, 272
Supramolecular polymers (SPs), 403–406
Surface coating, thermoset resin, 547–549
  accelerated weathering test, 549
  bend test, 548
  corrosion resistance test, 549
  cross-hatch adhesion test, 548
  gloss measurement test, 549
  impact resistance test, 548
  pull-off adhesion strength, 547–548
  scratch resistance test, 548
Surface modification techniques, 378–379
Surface tension, 376–377
Surfactants, 273
Synthetic polymers, 2–3
Synthetic rubber compounds, 92

T

t-butyl peroxide (TBO), 14
T-peel test, 550–551, 550f
Tack property, 92
Tactoids, 424
Tannic acid, 57
TDA. See Closite-tetradecyl amine (TDA)
Tensile force, 536–537
Tensile test, 527–529
  dumb-bell shape specimen, 528f
  tensile designations for plastic samples, 530f
Terephthaloyl bis (4-oxybenzoic) acid (PBA), 211
Terephthaloyl chloride, 211
Testing of thermoset resins, 526–547
  electrical properties, 543–545
    arc resistance, 545
    dielectric strength, 544–545
    electrical conductivity, 544
  flammability and smoke tests, 545–547
  fracture toughness, 533–536
  hardness test, 538–539
  impact test, 536–538
  mechanical properties, 526–533
    ASTM standards, 527t
    compression test, 532
    creep and stress relaxation test, 532–533
    fatigue test, 533
    flexural test, 530–531
    tensile test, 527–529
  thermal properties, 540–543
    HDT, 543
    thermal conductivity, 540–542
    thermal expansion coefficient, 542–543
TETA. See Triethylenetetramine (TETA)
Tetra bromobisphenol-A (TBBPA), 181, 257
Tetra-functional epoxy resins, 38
Tetrachorobisphenol-A (TCMPA), 181
Tetraethoxysilane (TEOS), 421–422
Tetraglycidyl methylene dianiline (TGMDA), 340–341
Tetraglycidylether of 4,4′ diaminodiphenyl methane (TGDDM), 200, 205–206, 356
  TGDDM-based composites, 399
Textile-reinforced epoxy composites, 402
Theoretical models, 541
Thermal analysis of thermoset, 517–523
  DMA, 518–519
  DSC, 517–518
  example of TTS by graphical method, 520–522
  graphical method, 521f
  TGA, 522–523
  TTS, 520
Thermal conductivity of thermoset resins, 540–542

Thermal degradation kinetics, 562–564
Thermal degradation reaction, 563–564
Thermal expansion coefficient, 542–543
Thermal property of thermoset resin, 195–199
  effect of cure conversion on T_g, 197–199
  T_g vs. service temperature, 196
Thermal stability of thermoset, 523–524
Thermodynamic miscibility, 304
Thermogravimetric analysis (TGA), 122, 517, 522–523, 562
  lifetime from TGA kinetics, 570–571
Thermomechanical analysis (TMA), 542–543
Thermomechanical properties
  mechanical properties, 397–401
    comparison of fiber-reinforced thermoset composites and steel, 399t
    mechanical properties of FRP composites, 399t
    properties of glass fiber-reinforced UPE resin, 398t
  thermal properties, 396–397
  of thermoset composites, 396–401
  of thermoset resin, 199–204
    UPE, vinyl ester, and epoxy, 200t
Thermoplastics, 321–322
  3D printing of, 280–281
  polymers, 2–3
  toughening, 296–297, 347–356
    amorphous thermoplastics, 352–353
    crystalline thermoplastics, 353–354
    engineering thermoplastics, 348–351
    effect of matrix crosslink density, 356
    mechanism of toughening, 355–356
  morphology and microstructural aspects, 354–355
Thermoset composites
  analysis and testing of composites, 390–392
    determination of glass content, 390
    ILSS, 391–392
    mechanical testing of composites, 391
  composite interface, 375–379
  constituents of FRP composites, 374–375
  and manufacturing processes, 373t
  prediction of composite strength and rigidity, 393–396
  processing of, 380–390
    contact molding, 381
    filament winding, 387
    prepreg molding, 389–390
    pultrusion, 385–387
    RFI, 387–389
    RIM, 385
    RTM, 382–383
    VARTM, 383–384
    wet lay-up molding, 381–382
  thermomechanical properties of, 396–401
  toughened composites, 401–415
Thermoset nanocomposites, 421–496
  block copolymer-based nanocomposite, 449–455
  CNT-based nanocomposites, 455–468
  combination organoclay and flame retardants, 486–488
  graphene-based nanocomposite, 468–473
  nanocarbon-based flame retardants, 488–491
  nanoclay-based flame retardant, 483–485
  nanocomposites, application of, 491–494
  nanoreinforcement and toughening, 475–481
  nanotechnology and flammability, 481–482
  PCN, 423–443
  POSS and silica-based nanocomposites, 443–449
  scheme of main chemical routes, 422f
  silver nanoparticle-based nanocomposites, 473–475
Thermoset resin, 3, 374–375, 381, 475, 504
  3D printing, 275–284
  ageing study under stress, 576–579
  amino resins, 98–100
  analysis and characterization, 504–526
    crosslinked density, 508–510
    infrared spectroscopy, 506–507
    molecular weight distribution, 510–514
    NMR spectroscopy, 507–508
    PALS, 508
    rheological characterization, 524–526
    thermal analysis, 517–523
    thermal stability, 523–524
    titration, 504–506
  benzoxazine resins, 101–115
  bio-based unsaturated polyester, 19–23
  bismaleimide resin, 126–139
  cyanate ester resin, 139–154
  epoxy resins, 37–63
  and field of application, 4t
  furan resins, 100–101
  lifetime prediction, 560–575
    advanced isoconversion methods, 569–570
    differential form of isoconversion kinetics, 564–566
    integral form of isoconversion kinetics, 566–568
    lifetime estimation from TGA kinetics, 572–575
    from TGA kinetics, estimation of, 570–571

Thermoset resin (*Continued*)
  thermal degradation kinetics, 562–564
  morphological characterization, 514–517
    AFM, 515–516
    SEM, 514
    TEM, 514–515
    XRD, 516–517
  performance evaluation, 547–560
    evaluation as surface coating, 547–549
    evaluation as vibration damping material, 556–560
    evaluation of thermoset as adhesive, 549–555
  phenolic resins, 82–98
  phthalonitrile resin, 154–157
  polyimides, 115–126
  processing of, 273–275
    compression molding, 274–275
    die casting, 273–274
    RIM process, 275
    rotational casting, 274
  properties
    additives, 269–273
    cure characteristics, 179–193
    cure cycle, 193–195
    elastomeric thermoset, 213–219
    FR thermoset, 255–269
    gelation and vitrification, 175–178
    IPNs, 219–225
    liquid crystalline thermoset, 210–213
    MOFs, 252–255
    rheological properties, 204–210
    smart thermoset, 226–246
    thermal property, 195–199
    thermomechanical properties, 199–204
    vitrimers, 246–251
  PUs, 63–82
  testing of, 526–547
  UPE resin, 3–19
  VE resin, 23–37

Thermosets, 294–295
  materials, 545
  matrix, 305–306
  networks, 295
  polymers, 2–3
  thermoset-based IPNs, 222–223
  thermoset-based liquid crystalline materials, 494
  toughening of, 300
Thermosetting resin, 3, 544
  composition, 481–482
  matrix, 374
Thick adherent shear test (TAST), 549–552
Thio-disulfide reaction, 230–231
Three dimension (3D)
  network, 2–3, 372
  powdered rubber, 481
  printing, 275–284
    of thermoplastic, 280–281
    of thermoset, 282–284
    types, 277–280
      BJ technology, 280
      DLP, 278
      EBM technology, 279–280
      FDM, 278
      SLA, 277
      SLM technology, 279
      SLS technology, 278–279
Time–temperature superposition (TTS), 520
  by graphical method, 520–522
Time–temperature–transformation diagram (TTT diagram), 207–208
Titration, 504–506
  acid value, 504–505
  determination of epoxy equivalent, 505–506
  hydroxyl value, 505
  solubility parameter, 506
2,4-tolylene diisocyanate (TDI), 244
Toughened composites, 401–415
  improvements in mode I interlaminar fracture energy, 405t
  mechanical properties of control composite laminate, 403t
  resin toughening, 409–415

Toughened epoxy resins, 325–326
Toughened thermoset resins
  chemical modification, 326–327
  core–shell particle toughening, 346–347
  liquid rubber toughening, 300–313
  mechanism of toughening of brittle polymer, 297–299
  modification of UPE resin, 316–319
  morphological aspects, 299–300
  rigid particle toughening of epoxy, 356–358
  rubber toughening, 328–346
  thermoplastic toughening, 347–356
  toughened epoxy resins, 325–326
  toughening
    of phenolic resin, 320–322
    of PI, BMI, and CE resins, 323–325
    of thermoplastics, 296–297
    of thermosets, 300
    of vinyl ester resin, 313–316
Trade Hycar CTBN company, 302
Traditional polymer processing techniques, 427
Transesterification, 5, 6f
Transmission electron microscope (TEM), 299, 428–431, 514–515
Tri-functional epoxy resins, 38
Triethylene glycol dimethacrylate (TEDGMA/TEGDM), 223–224, 240
Triethylenetetramine (TETA), 41–43
4,4′-(2,2,2-trifluoro-1-phenylethyle dine) dipthalic anhydride (3 FDA), 120
Trifunctional epoxy (TGAP-based composites), 399–400
Trifunctional eugenol-based epoxy monomer (TEU-EP), 55–57
Triglyceride oils, 331–332
Triglycidyl amino phenol (TGAP), 205–206, 356

Trihexyltetra-decylphosphonium dicyanamide ([THTDP]N(CN)$_2$), 49–50
Trimethylolpropane (TMP), 64–65
Triol-based extender, 67–68
Triphenylphosphine (TPP), 326
"Trommsdorff effect", 29–35
Two-dimensional PSDs (2DPSD), 516
Two-stage method, acrylate-modified epoxy, 334
2D chemically reduced graphene oxide (2D CRGO), 481

U

UL-94 flammability test, 546
Ultimate tensile strength (UTS), 577–578
Ultraviolet (UV), 277, 523–524
  radiation, 271
  UV-assisted DIW fabrication approach, 282
Uniaxial force, 527–529
Unidirectional prepreg, 389
Universal testing machine (UTM), 527–529, 532
Unsaturated polyester (UPE), 397–398, 480
  modification of, 316–319
  resin, 3–19
    application, 17–19, 18t
    bio-based unsaturated polyester, 19–23
    curing of, 14–17
    formation of long and short branches in, 6f
    inhibitor, 13–14
    polyester structure, 6–7
    polyesterification kinetics, 8–9
    reactive diluents or monomer, 11–13
    synthesis, 5f
    types of polyester, 9–11
    unsaturated polyester, 3–5
  UPE–PVAc systems, 316
Urea formaldehyde resins (UF resins), 98, 99f
US Environmental Protection Agency, 25–26

V

Vacuum bagging technique, 387–389
Vacuum-assisted resin transfer molding, 383–384
Vacuum-assisted resin transfer molding process (VARTM process), 383–384, 488
Vanillin, 27–28, 28f, 55
Vapor pressure osmometry (VPO), 512
Variable frequency microwave (VFM), 122
Vernoria oil, 331–332
Vibration, 214
  intensity, 557
  sensitivity, 559–560, 560t
  tests, 557–558
  vibration-damping behavior, 216–219
Vibration damping material, evaluation as, 556–560
  dynamic mechanical property, 557
  example of vibration sensitivity, 559–560
  mechanical property, 556–557
  vibration tests, 557–558
Vinyl benzyl glycidyl ether (VBGE), 341–342
Vinyl ester (VE), 23–37, 222–223, 229, 480–481
  applications, 36–37
  curing of, 29–35, 30t
  properties, 36
  resin
    liquid rubber toughening, 313–316
    toughening, 313–316
  synthesis of, 25f
Vinyl ester-urethane hybrid (VEUH), 314
Vinyl ester/epoxy interpenetrating networks (VE/EP IPN), 480–481
Vinyl triethoxy silane (VTES), 263
Vinyl-ended prepolymer, 25–26
Vinyl-terminated poly (butadiene–acrylonitrile rubber) (VTBN), 313
Vinyl-trimethoxysilane (VTS), 265
Viscoelastic material, 524–525
Viscometry, 511–512
Viscosity-average molecular weight (ΣMv), 510–512
Vitrification, 175–178
  effect on polymerization rate, 189–191
Vitrimerization, 249, 251f
Vitrimers, 246–251
  application, 249–251
  materials and healing conditions, 250t
  synthesis of, 247–249
Volatile organic compound (VOC), 13, 20, 25–26
Volume fraction of polymer, 509
Volume resistivity, 544

W

Water vapor permeability (WVP), 468–469
Weight fraction of PEO (WPEO), 452
Weight-average molecular weight (ΣMw), 510
Wet lay-up molding method, 381–382, 384
Wetting process, 377
Wide-angle X-ray diffraction (WAX), 428–429
Williams–Landel–Ferry equation (WLF equation), 216, 520
Work of adhesion (W_A), 376–377
Woven fabric prepreg, 389

X

X-ray diffraction (XRD), 428–429, 516–517

Y

Young's modulus, 527, 529

Z

Z-average molecular weight (ΣMz), 510
z-direction fibers, 407
Zenner model, 215–216

Printed in the United States
by Baker & Taylor Publisher Services